NONLINEAR OPTICAL PARAMETRIC PROCESSES IN LIQUIDS AND GASES

QUANTUM ELECTRONICS — PRINCIPLES AND APPLICATIONS

A Series of Monographs

EDITED BY

PAUL F. LIAO
Bell Telephone Laboratories
Murray Hill, New Jersey

PAUL KELLEY
Lincoln Laboratory
Massachusetts Institute of Technology
Lexington, Massachusetts

N. S. Kapany and J. J. Burke. OPTICAL WAVEGUIDES, 1972

Dietrich Marcuse. THEORY OF DIELECTRIC OPTICAL WAVEGUIDES, 1974

Benjamin Chu. LASER LIGHT SCATTERING, 1974

Bruno Crosignani, Paolo Di Porto, and Mario Bertolotti. STATISTICAL PROPERTIES OF SCATTERED LIGHT, 1975

John D. Anderson, Jr. GASDYNAMIC LASERS: AN INTRODUCTION, 1976

W. W. Duley. CO_2 LASERS: EFFECTS AND APPLICATIONS, 1976

Henry Kressel and J. K. Butler. SEMICONDUCTOR LASERS AND HETEROJUNCTION LEDs, 1977

H C. Casey and M. B. Panish. HETEROSTRUCTURE LASERS: PART A, FUNDAMENTAL PRINCIPLES; PART B, MATERIALS AND OPERATING CHARACTERISTICS, 1978

Robert K. Erf (Ed.). SPECKLE METROLOGY, 1979

Marc D. Levenson. INTRODUCTION TO NONLINEAR LASER SPECTROSCOPY, 1982

David S. Kliger (Ed.). ULTRASENSITIVE LASER SPECTROSCOPY, 1983

Robert A. Fisher (Ed.). OPTICAL PHASE CONJUGATION, 1983

John F. Reintjes. NONLINEAR OPTICAL PARAMETRIC PROCESSES IN LIQUIDS AND GASES, 1984

In Preparation

S. H. Lin, Y. Fujimura, H. J. Neusser, and E. W. Schlag. MULTIPHOTON SPECTROSCOPY OF MOLECULES, 1984

YOH-HAN PAO
Case Western Reserve University
Cleveland, Ohio

Founding Editor
1972–1979

NONLINEAR OPTICAL PARAMETRIC PROCESSES IN LIQUIDS AND GASES

JOHN F. REINTJES

Naval Research Laboratory
Washington, D.C.

1984

ACADEMIC PRESS, INC.

(Harcourt Brace Jovanovich, Publishers)

Orlando San Diego San Francisco New York London
Toronto Montreal Sydney Tokyo São Paulo

ACADEMIC PRESS, INC.
Orlando, Florida 32887

United Kingdom Edition published by
ACADEMIC PRESS, INC. (LONDON) LTD.
24/28 Oval Road, London NW1 7DX

Library of Congress Cataloging in Publication Data

Reintjes, John F.
Nonlinear optical parametric processes in liquids and gases.

(Quantum electronics--principles and applications)
Includes bibliographical references and index.
1. Liquids--Optical properties. 2. Gases--Optical properties. 3. Nonlinear optics. I. Title. II. Series.
QC145.4.O6R44 1984 530.4'2 84-11603
ISBN 0-12-585980-5

PRINTED IN THE UNITED STATES OF AMERICA

84 85 86 87 9 8 7 6 5 4 3 2 1

To Maura and Chris

CONTENTS

PREFACE

Extensive advances have been made in the understanding and application of nonlinear optical interactions since the invention of the laser. This book covers an important subset of these interactions—parametric processes that occur in liquids and gases. The book is written at a level that assumes familiarity with university-level mathematics, atomic physics, and electromagnetic theory. While many of the sections draw on mathematical results, they are in many cases accompanied by physical interpretations of various phenomena. The mathematical results are intended primarily for their usefulness in quantifying the physical interpretations and in aiding in quantitative application of the various concepts to actual systems.

In Chapter 1 a general discussion of nonlinear optical processes is given. Although brief, it should give readers who are familiar with optical interactions in linear dielectrics an understanding of the basis for nonlinear optical interactions. References are made at the end of the chapter to more extensive developments of nonlinear optical theory for those who wish a more extensive elementary treatment.

Specific types of nonlinear interactions are presented in subsequent chapters. In Chapter 2 various third-order frequency mixing processes are described. Basic properties of nonlinear interactions such as phase matching and resonant enhancement are described here in detail and are used throughout the book. In Chapter 3 the discussion is extended to higher-order interactions. In both chapters the processes of frequency mixing and harmonic generation are used as illustrations of the basic principles. Emphasis in these chapters is on the frequency conversion processes as radiation sources. Relevant experimental results are also described in an attempt to illustrate the application of the principles to real systems.

Important aspects of nonlinear optical frequency conversion in liquids and gases are the limitations that are imposed by competing linear and nonlinear interactions. The various competing processes and the effects which they can have on the frequency conversion interactions are discussed in Chapter 4. Emphasis here is on the limitations which the various competing mechanisms impose on the efficiency of the generation of new radiation. Again, experimental results are described to illustrate the performance of real systems.

In Chapter 5 the class of nonlinear interactions that occur at the same frequency as the incident radiation is discussed. Self-action effects, including self-focusing, self-defocusing, and self-phase modulation, are discussed briefly since they are covered extensively in the references given at the end of the chapter. The more recently developed interaction of phase conjugation by degenerate frequency mixing is discussed in more detail.

In Chapter 6 applications of various nonlinear optical interactions are treated. Emphasis here is on the use of nonlinear optical processes to obtain information about a material system or to control the propagation of optical waves. These discussions are more qualitative in nature and are again accompanied by reports of experimental results. The list of applications is by necessity limited in scope and reflects some of the interests of the author. They are intended to give indications of the uses to which nonlinear optics can be put.

Equations throughout the book are given consistently in Gaussian units. They may be evaluated for specific systems by using values for susceptibilities in electrostatic units, wavelengths and other length dimensions in centimeters, laser powers in ergs (watts times 10^7), and laser intensities in ergs per centimeter squared-second (watts per centimeter squared times 10^7).

ACKNOWLEDGMENTS

It is a pleasure to acknowledge the assistance of many colleagues who have helped in the preparation of this book. I acknowledge in particular helpful discussions with R. H. Lehmberg, L. L. Tankersley, J.-C. Diels, J. J. Wynne, R. C. Eckardt, C. Y. She, M. D. Duncan, and M. L. Dlabal. I also acknowledge the assistance of L. J. Palumbo who performed numerical calculations for some of the data presented. I would especially like to thank Professor N. Bloembergen who provided me with excellent training in the field of nonlinear optics. Finally I would like to thank all those who freely gave their permission to reproduce their results.

I would also like to express my appreciation to my wife and family for their patience and understanding while this book was in preparation.

CHAPTER 1

INTRODUCTION

1. INTRODUCTION

The field of nonlinear optics has expanded tremendously in more than 20 years since the development of the laser, growing from a subject that was primarily studied in the laboratory to one that is making contributions to such diverse fields of study as high-resolution atomic, molecular, and solid-state spectroscopy, kinetics of energy transfer in various material systems, optical microscopy, remote pollution monitoring, signal processing, image transmission, and combustion analysis. The concept of nonlinear optics is actually considerably older than the laser. Optical interactions involving the simultaneous absorption of two quanta of radiation (two-photon absorption) were described theoretically as early as 1931 by Goeppert-Mayer,[1] and other nonlinear optical effects such as the saturation of an atomic transition were observed experimentally well before the laser was invented. Nonlinear optical interactions, however, are generally weak when compared to linear optical effects, and experimental observation and verification of many nonlinear optical effects, which so often provide an impetus for extensive development of theory and understanding, had to wait for the intense coherent fields that are generated by the laser.

Nonlinear optics involves a class of interactions between optical radiation and matter in which the response of the medium, as manifested by the polarization induced by the incident fields, depends on the second- and higher-order powers of the optical fields. More familiar interactions, such as linear absorption, refraction, and scattering, have in common the property that they arise from oscillating polarizations in the medium that are proportional to the first power of the optical fields.

Some of the nonlinear optical processes are counterparts to linear optical processes, e.g., nonlinear absorption and nonlinear refraction. However, the more general involvement of the optical fields in the nonlinear optical interactions can result in effects that are qualitatively different from those evident in linear optics. A useful classification scheme for the various nonlinear effects divides the processes into two broad categories involving inelastic and elastic interactions with the medium. In the first type of process, the nonlinear interaction involves a net transfer of energy between the optical waves and the material system. Examples of such effects are various stimulated scattering processes such as stimulated Raman, Brillouin, Rayleigh or polariton scattering, multiphoton absorption or emission, and dielectric breakdown.

In the second type of process, energy is exchanged among the various optical waves, but no energy is given to or gained from the medium. It merely serves to determine the direction and amount of energy flow among the various optical waves. Examples of this type of interaction are optical harmonic generation, frequency mixing, and self-action effects such as self-focusing, self-defocusing, and self-phase-modulation.

Many of the nonlinear optical processes are actually optical frequency analogs of effects that are well established in other regions of the electromagnetic spectrum. As such they will be familiar in principle to readers who are familiar with other electronic fields. For example, optical rectification, which occurs in certain types of anisotropic crystals, finds an exact analogy in the electronic rectification of lower-frequency electronic signals with diodes. Optical harmonic generation similarly finds a counterpart in the harmonic distortions that can be present in audio power amplifiers.

The study of nonlinear optical effects has not only provided us with a wealth of information about the interaction of intense radiation with matter but also with new technologies that can be used to accomplish tasks that are more difficult or impossible with linear optics alone. Examples of such developing technologies can be found in areas such as optical-frequency conversion in which coherent radiation that is generated by lasers in one part of the spectrum is converted to coherent radiation in an entirely different spectral region. In this way coherent radiation becomes available in spectral ranges that may otherwise be inaccessible by primary lasers. A related area involves the application of nonlinear optics to spectroscopy, for which a multitude of techniques have been developed for increasing signal levels and spectroscopic resolution. An example of such a development is the use of nonlinear optical interactions to obtain spectra of atomic gases with resolutions below the Doppler limit.

More recently, nonlinear techniques for wave front correction have been developed in which distortions that are introduced by aberrations or dispersion in linear optical systems can be removed. In a sense nonlinear optical systems can be used to make the performance of a real optical system approach that of an ideal distortion-free system without the need of holding impossibly precise tolerances in manufacture. Techniques are also on the horizon for control and processing of optical signals in direct analogy to the electrical operation of transistors at lower frequencies. Thus active optical circuitry in integrated circuit geometries may well be possible after the manner of present-day electrical circuits.

In this book we shall consider the details of a number of elastic, or parametric, nonlinear optical interactions. We shall further restrict our considerations to interactions that occur in liquids and gases. This is a rather arbitrary restriction but it is useful in a certain sense. The subject of nonlinear optics has been included in many books on quantum electronics in recent years and these treatments usually include the second-order effects that can occur in anisotropic solids.

There has been considerable theoretical and experimental study of the nonlinear properties of liquids and gases in recent years and, while they have been included in some books and book chapters, not nearly as much has been written about liquids and gases as has been written about the anisotropic solids.

Liquids and gases have a number of properties that set them apart from other materials in both a practical and a fundamental sense. Because they are spatially isotropic, their dominant nonlinear response usually comes from third-order terms in the nonlinear susceptibility. In this sense they are unlike the classic prototype nonlinear materials that exhibit second-harmonic generation, a subject that is usually covered in quantum electronics textbooks. Rather, they are similar to isotropic solids (calcite, diamond, NaCl), but unlike the crystalline solids, they are available in virtually unlimited spatial dimensions. Thus, as a practical matter, greater interaction lengths can be used to avoid the higher intensities that are generally anticipated to be needed in working with third-order responses. In addition, liquids and gases are self-healing and thus are not subject to the limitations on intensities that occur with solids because of optical damage.

The characteristics of resonant enhancement, by which the strength of a nonlinear interaction can increase by many orders of magnitude when a laser frequency or one of its multiples is tuned near an appropriate energy level in the medium, can be manifested very strongly in liquids and gases, especially atomic gases whose absorption lines are very sharp. As a result a number of nonlinear processes can be studied using relatively low-power dye lasers, rather than high-power fixed-frequency pulsed lasers. It is this property of gaseous media that has led to widespread use of nonlinear optical techniques in such areas as spectroscopic source production and to an overall improvement in the convenience of nonlinear optical experiments.

In the remainder of this chapter we shall outline the mathematical foundations of the nonlinear optical response of the materials that will enable us to describe the various nonlinear processes. We shall also examine briefly the physical origins of the optical nonlinearities of the material systems, discuss some of their general properties, and describe several specific nonlinear effects to which they give rise.

In subsequent chapters we shall examine specific nonlinear interactions. In Chapter 2 we shall consider third-order frequency mixing, of which third-harmonic generation is a special case. Emphasis in this chapter will be on the characteristics of the radiation that is generated as a result of the nonlinear interactions. In Chapter 3 we shall discuss similar processes that occur as a result of higher-order interactions. In Chapter 4 we shall consider the efficiency with which the processes discussed in Chapters 2 and 3 can proceed. Attention will be given to competing linear and nonlinear optical processes that serve to limit the efficiency of the frequency conversion process.

In Chapter 5 we shall consider nonlinearities that produce effects on waves at the same frequencies as those in the incident radiation. Self-action effects such as self-focusing and self-phase-modulation as well as degenerate four-wave mixing will be discussed. Finally, in Chapter 6 we shall discuss some additional applications of the nonlinear processes that are described in the previous chapters. For example, we return to four-wave frequency mixing as a spectroscopic tool. Here emphasis is placed on the information that can be obtained about the medium, rather than on the mechanisms of the optical-wave generation. We also discuss the use of optical nonlinearities in bistable devices.

For discussions of effects that are not covered in this book, as well as for a more extensive treatment of the basic theory of the nonlinear polarizations, the reader is referred to several fine books and book chapters listed at the end of the first chapter.[2–7]

2. OPTICAL NONLINEARITIES—PHYSICAL ORIGINS

The optical properties of materials are determined by the response of electrons and ions to incident light. The interaction between radiation and matter can be described conceptually as follows. Light that is incident on a medium induces in the electron or ion distribution an oscillating dipole moment. The induced dipole moment, or polarization, in turn radiates a second optical field that can interfere with the incident field. In many materials the behavior of the electrons or ions can be described as if they were bound about equilibrium positions in a potential well that is harmonic for small displacements. In all real materials, however, the potential well becomes anharmonic for sufficiently large displacements from equilibrium, as

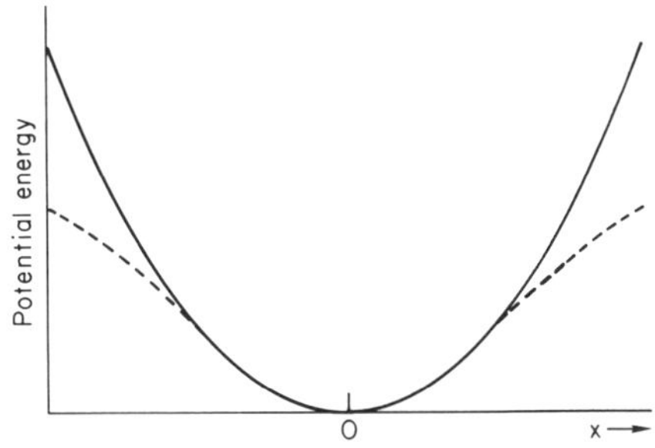

Fig. 1. Schematic representation of a potential well that can be responsible for optical nonlinearities. The solid line represents a harmonic well ($V = ax^2$), which allows linear optical effects. The dashed line shows an anharmonic well ($V = ax^2 - bx^4$), which can result in third-order optical nonlinearities.

illustrated in Fig. 1. This model has been used by Bloembergen[8] and Akhmanov[9] to estimate the second- and third-order nonlinear coefficients.

When the incident fields are weak the electrons move in the harmonic part of the well. Under these conditions the induced polarization depends linearly on the incident field strength and radiates secondary fields only at the same frequencies as those that are contained in the incident waves. The interference between the incident and secondary waves results in the familiar linear optical properties of scattering, refraction, and linear absorption. When the incident fields are sufficiently strong, however, displacements of the electrons are larger, carrying them into the anharmonic region of the potential well. The polarization developed in this regime contains components that depend on the second and higher powers of the incident fields. The secondary fields radiated by the induced polarization can now contain frequencies that are different from those in the incident fields and give rise to the nonlinear effects discussed in this book. In the next section we shall put these qualitative concepts on a more quantitative basis by giving a mathematical description of the optical nonlinearities.

3. MATHEMATICAL FOUNDATIONS

The propagation of an optical wave through a nonconducting medium is described by Maxwell's equations

$$\nabla \times \mathscr{E} = -\frac{1}{c}\frac{\partial \mathscr{B}}{\partial t}, \tag{1a}$$

$$\nabla \times \mathscr{H} = \frac{1}{c}\frac{\partial \mathscr{D}}{\partial t}, \tag{1b}$$

where $\mathscr{E}$, $\mathscr{D}$, $\mathscr{H}$, and $\mathscr{B}$ are the usual electric and magnetic vectors associated with the light wave. The response of the medium can be accounted for by defining electric and magnetic polarization vectors $\mathscr{P}$ and $\mathscr{M}$ as

$$\mathscr{D} = \mathscr{E} + 4\pi\mathscr{P}, \tag{2a}$$

$$\mathscr{B} = \mathscr{H} + 4\pi\mathscr{M}. \tag{2b}$$

The majority of the nonlinear interactions that have been studied have resulted from the electric polarization. Therefore we shall consider only these effects from hereon and set $\mathscr{M}$ equal to zero. A description of the magnetic effects can be developed in a manner similar to the one to be given for the electric effects by retaining the terms containing $\mathscr{M}$ in the appropriate equations.

Substitution of Eqs. (2) (with $\mathscr{M}$ set equal to zero) into Eqs. (1) gives an equation for the electric field of the light wave in terms of the electric polarization:

$$\nabla^2\mathscr{E} - \frac{1}{c^2}\frac{\partial^2\mathscr{E}}{\partial t^2} = \frac{4\pi}{c^2}\frac{\partial^2\mathscr{P}}{\partial t^2}. \tag{3}$$

We can use the slowly varying envelope approximation for the radiation from most laser sources and represent the electric field as the product of two components, one of which oscillates at the optical frequency of the laser, and a second whose time variations occur on a much slower scale. For a wave propagating in the $+\hat{z}$ direction, we have

$$\mathscr{E}(r, z, t) = \tfrac{1}{2}[A(r, z, t)e^{-i(\omega t - kz)} + \text{c.c.}], \tag{4}$$

where A is the peak amplitude of the real electric field, k the wave vector of the radiation, related to the wavelength by $k = 2\pi/\lambda$, and r the transverse spatial coordinate. The amplitude A in Eq. (4) is allowed to be complex to account for time- or space-dependent phase changes that can either be present in the incident radiation or be introduced by interactions in the medium.

The slowly varying envelope approximation is generally very well satisfied by laser radiation available today. For example, the fastest amplitude structure that has been produced on visible laser pulses has a characteristic time of the order of 5×10^{-14} sec (this is actually the duration of the shortest pulses that have been produced), while the period of oscillation of light at visible wavelengths ($\lambda = 500$ nm) is $T = \lambda/c = 2.6 \times 10^{-16}$ sec, about 200 times shorter.

In problems involving nonlinear optical processes we anticipate the presence of radiation at more than one optical frequency, either because it is present in the incident radiation or because it is generated in the course of the nonlinear interactions. Optical fields whose components are not too closely spaced in wavelength can be expressed simply as the sum of terms similar to the expression in Eq. (4), with the total field given by

$$\mathscr{E}_{\mathrm{T}} = \frac{1}{2}\left[\sum_i A_i(r, z, t)e^{-i(\omega_i t - k_i z)} + \text{c.c.}\right]. \tag{5}$$

Roughly speaking, this representation is valid when the frequency spectra of the various field components do not overlap. A similar expression can be written down for the polarization that is induced by the fields:

$$\mathscr{P} = \frac{1}{2}\left[\sum_j \mathbf{P}_j(r, z, t)e^{-i(\omega_j t - k_j^{\mathrm{p}} z)} + \text{c.c.}\right]. \tag{6}$$

The wave vector of the polarization in Eq. (6) has been specifically designated as belonging to the polarization to emphasize that it can in general differ from the wave vector of an optical field at the same frequency. This difference in wave vectors, which arises from a difference in phase velocities between the polarization and the field that drives it, can affect the direction of energy flow among the various waves.

We shall now proceed to derive a set of equations for the field amplitudes in terms of the polarization amplitudes based on a set of specific assumptions. In the next section we shall discuss the limits of the validity of those assumptions. A list of these assumptions, along with the simplification they provide and the situation under which they can fail is given in Table 1.

The first assumption involves the separation of the polarization into the sum of linear and nonlinear parts:

$$\mathscr{P} = \mathscr{P}^{\mathrm{L}} + \mathscr{P}^{\mathrm{NL}}, \tag{7}$$

and the second involves specifying the linear polarization to be proportional to the electric field:

$$\mathscr{P}^{\mathrm{L}} = \chi^{(1)}\mathscr{E}, \tag{8}$$

where $\chi^{(1)}$ is the linear susceptibility. These two assumptions are valid for a wide range of interactions in both liquids and gases, although there are some situations in which one or both of them are not valid. Such situations will be discussed in the next section.

If we substitute Eqs. (5)–(8) into Eq. (3), we can obtain a set of coupled equations for the field amplitudes in terms of the nonlinear polarizations:

$$\begin{aligned}
\nabla_\perp^2 A_1 + 2ik_1\frac{\partial A_1}{\partial z} + \frac{2in_1^2\omega_1}{c^2}\frac{\partial A_1}{\partial t} &= -\frac{4\pi\omega_1^2}{c^2}\mathbf{P}_1^{\mathrm{NL}}e^{-i\Delta k_1 z},\\
&\vdots\\
\nabla_\perp^2 A_j + 2ik_j\frac{\partial A_j}{\partial z} + \frac{2in_j^2\omega_j}{c^2}\frac{\partial A_j}{\partial t} &= -\frac{4\pi\omega_j^2}{c^2}\mathbf{P}_j^{\mathrm{NL}}e^{-i\Delta k_j z},\\
&\vdots
\end{aligned} \tag{9}$$

where $\Delta k_j = k_j - k_j^{\mathrm{p}}$ is the wave vector mismatch for the jth wave.

TABLE 1

Approximations Used in Derivation of Equations for Field Amplitudes

Approximation	Simplification	Condition for validity	Situations where approximation can fail
Polarization represented as sum of linear and nonlinear parts	Separation of linear and nonlinear effects	Weak pump fields	Saturation of optical transitions
Linear polarization proportional to electric field	Removes temporal nonlocality for linear response	Pulse duration long compared with response time of linear polarization	Short pulses combined with single-photon resonance
Slowly varying envelope approximation	Differential equations reduce to first order	Variation of field amplitudes small in one wavelength or optical period	Generally valid for most laser radiation
Equal group velocities at all frequencies	Equations for pulsed fields reduce to time-independent equations	Frequency spread of pulses contained within phase-matching width of medium	Pulses with short duration or wide bandwidth
Multipole expansion–dipole approximation	Removes spatial nonlocality	Wavelengths large compared to size of atomic scatterers	Very short wavelength radiation, dominant resonance with higher multipole transition
Expansion of nonlinear polarization in powers of optical fields	Nonlinear response separates into terms of different order	Weak optical fields	Saturation of nonlinear transitions, dielectric breakdown
Time-independent nonlinear susceptibilities	Removes temporal nonlocality from nonlinear response	Pulse duration long compared to response time of nonlinear polarization	Resonant nonlinear interactions

In deriving Eqs. (9) we have used the approximations

$$\frac{\partial^2 A_j}{\partial z^2} \ll k_j \frac{\partial A_j}{\partial z}, \qquad \frac{\partial^2 A_j}{\partial t^2} \ll \omega_j \frac{\partial A_j}{\partial t},$$

which are a direct result of the slowly varying envelope approximation. We have also used the relation

$$k_j^2 = \frac{\omega_j^2}{c^2}(1 + 4\pi\chi_j^{(1)}) = \frac{\omega_j^2}{c^2} n_j^2, \tag{10}$$

where n_j is the usual linear refractive index at the frequency of the jth wave.

The index j covers all frequencies, including those that are in the incident radiation fields and those that are generated in the medium in the course of the nonlinear interaction. Equations (9) thus represent an infinite set of equations that describe all the possible fields that can arise from arbitrary combinations of the incident laser frequencies. In most situations, however, only a finite subset of the equations in the system (9) need be considered. One commonly encountered situation, for example, is when the incident fields are sufficiently weak that their amplitudes are not significantly changed during the interactions. In many situations of this type the various components of the nonlinear polarization will depend only on the amplitudes of the incident fields and not on those of the generated fields. As a result the equations for the incident field amplitudes in (9) may be dropped, while the equations for the various generated field amplitudes decouple from one another.

A second commonly encountered situation is one in which a particular nonlinear process, such as third-harmonic generation, dominates the interaction. In this case the amplitude of the generated wave may grow to be comparable to that of the incident wave while the incident wave can be depleted. In such a situation only those equations in (9) that apply to the two dominant waves need be retained. Other waves can of course be generated in weaker nonlinear interactions, but they will generally not affect the dominant one. If needed, the amplitudes of the weaker generated waves can be solved for in the usual way after suitable expressions for the amplitudes of the strong waves are obtained.

Equations (9) were derived as if the linear susceptibilities were dispersionless. Actually most materials will have some degree of dispersion in their susceptibilities. Under these more realistic conditions, Eqs. (9) will still be valid if the field amplitudes are time independent, that is, if the waves are truly monochromatic. If the field amplitudes are time dependent, however, then the waves will have a finite frequency spread. If the frequency spread is small, then the dispersion in the linear susceptibilities may be accounted for by introducing the group velocity:

$$v_{g_i} = \frac{c/n_i}{1 + (\omega_i/n_i)\, d\omega_i/dn_i}. \tag{11}$$

The equations for the time-dependent field amplitudes will then have a form similar to (9), but with the phase velocities replaced with the group velocities:

$$\begin{aligned}
\nabla_\perp^2 A_1 + 2ik_1\left(\frac{\partial A_1}{\partial z} + \frac{1}{v_{g_1}}\frac{\partial A_1}{\partial t}\right) &= -\frac{4\pi\omega_1^2}{c^2}\mathbf{P}_1^{\mathrm{NL}} e^{-i\Delta k_1 z},\\
&\vdots\\
\nabla_\perp^2 A_j + 2ik_j\left(\frac{\partial A_j}{\partial z} + \frac{1}{v_{g_j}}\frac{\partial A_j}{\partial t}\right) &= -\frac{4\pi\omega_j^2}{c^2}\mathbf{P}_j^{\mathrm{NL}} e^{-i\Delta k_j z},\\
&\vdots
\end{aligned} \tag{12}$$

In general we can expect that the group velocities of the various spectral components will be different. These differences become important when the bandwidth of the incident or generated radiation becomes too large to satisfy appropriate phase matching conditions (see Section 2-3). In most situations, however, it is normally assumed that the group velocities of the various waves are equal. Under this assumption we can transform to a coordinate system that propagates at the common group velocity:

$$z' = z, \qquad t' = t - z/v_g.$$

The explicit time derivative on the left-hand side in Eqs. (12) can be eliminated, giving

$$\begin{aligned}
\nabla_\perp^2 A_1(r, z', t') + 2ik_1 \frac{\partial}{\partial z'} A_1(r, z', t') &= -\frac{4\pi\omega_1^2}{c^2}\mathbf{P}_1^{\mathrm{NL}} e^{-i\Delta k_1 z'},\\
&\vdots\\
\nabla_\perp^2 A_j(r, z', t') + 2ik_j \frac{\partial}{\partial z'} A_j(r, z', t') &= -\frac{4\pi\omega_j^2}{c^2}\mathbf{P}_j^{\mathrm{NL}} e^{-i\Delta k_j z'},\\
&\vdots
\end{aligned} \tag{13}$$

These equations are of the same form as those that describe interaction with steady-state fields.

3.1. Constitutive Relations

In order to complete our description of the nonlinear optical interaction we must give constitutive relations that specify the dependence of the polarization on the optical fields. In its most general form the polarization is a non-local functional of the optical fields:

$$\mathscr{P}(r, z, t) = f(\mathscr{E}(r, z, t), \mathscr{E}(r'', z'', t'')). \tag{14}$$

Determination of an exact expression for the polarization in a general situation usually requires solution of a complex set of differential equations that describe the interaction of the fields with all the individual charges in the medium. For most of the problems of interest in this book we can make several assumptions that simplify the form of the polarization considerably.

First, it is convenient to make a multipole expansion in which the polarization is expressed in terms of moments of the charge distribution in the medium.[10] In this approximation we have

$$\mathscr{P}(r, z, t) = \mathscr{P}(r, z, t) + \nabla\mathscr{Q}(r, z, t) + \nabla : \nabla\mathscr{O} + \cdots. \tag{15}$$

The first term in Eq. (15) corresponds to the dipole polarization, the second to the quadrupole polarization, the next to the octupole polarization, and so on. Because typical wavelengths of laser radiation (0.5×10^{-4} cm for visible light, for example) are much larger than the characteristic dimensions of the atomic oscillators in the medium (typically of the order of 10^{-8} cm) this series converges very rapidly in the wavelength range of interest in this book, with the dipole polarization usually dominating the response of the medium. There are situations, however, in which the higher-order multipole terms can be significant or even make the dominant contribution to particular interactions.

It remains for us now to specify the relation between the polarization components in Eq. (15) and the optical fields. In the general case the time-dependent polarizations that are needed in Eqs. (12) or (13) can be quite complicated. It is useful therefore to look first at the Fourier components of the polarization. In most situations that we shall encounter the optical fields are weak compared to the fields that bind the electrons in the medium. Under these conditions the Fourier components of each term in Eq. (15) can be expanded in the Fourier components of the powers of the optical fields. Thus we have for the Fourier transform of the total polarization

$$\begin{aligned}\mathscr{P}_{\mathrm{T}}(r, z, \omega) = {} & \chi^{(1)}(\omega)\mathscr{E}_{\mathrm{T}}(r, z, \omega) + \{\chi^{(2)}\mathscr{E}_{\mathrm{T}}^2(r, z)\}_\omega + \{\chi^{(3)}\mathscr{E}_{\mathrm{T}}^3(r, z)\}_\omega + \cdots \\ & + Q^{(1)}(\omega)\nabla\mathscr{E}_{\mathrm{T}}(r, z, \omega) + \{Q^{(2)}\nabla_{\mathrm{T}}\mathscr{E}_{\mathrm{T}}^2(r, z)\}_\omega \\ & + \{Q^{(3)}\nabla_{\mathrm{T}}\mathscr{E}^3(r, z)\}_\omega + \cdots. \end{aligned} \tag{16}$$

Here $\mathscr{E}_{\mathrm{T}}(\omega)$ is the Fourier transform of the total optical field of Eq. (5) and the symbol $\{\ \}_\omega$ indicates that only the Fourier component at frequency ω of the quantity inside the braces is to be considered. The coefficients of the various terms in Eq. (16) are known as the *n*th-order *m*-pole susceptibilities. Because the dipole terms are usually much stronger than the corresponding higher multipole terms, their coefficients are usually termed simply nth-order susceptibilities without special designation.

The linear susceptibility has only one frequency argument and describes linear polarizations at frequency ω, which arise from fields at the same frequency. The nonlinear susceptibilities connect polarizations at a frequency ω to a number of fields that can be at different frequencies. For example, a polarization at frequency $\omega_3 = \omega_1 + \omega_2$ can be generated through a second-order term by fields at frequency ω_1 and ω_2. The nonlinear susceptibility for this processes is specified as $\chi^{(2)}(-\omega_3, \omega_1, \omega_2)$. The usual convention for the frequency arguments of the nonlinear susceptibilities is to write the frequency of the resultant polarization first with a minus sign. The frequencies of all the fields that combine to form the resultant polarization are written next, with the algebraic sum of all the frequencies in the argument equaling zero.

If we designate the linear polarization as

$$\mathscr{P}_{\mathrm{T}}^{\mathrm{L}}(\omega) = \chi^{(1)}(\omega)\mathscr{E}_1(\omega) + Q^{(1)}(\omega)\,\nabla\mathscr{E}_{\mathrm{T}}(\omega) + \cdots \tag{17}$$

and the rest of the terms in Eq. (16) as the nonlinear polarization

$$\begin{aligned}\mathscr{P}_{\mathrm{T}}^{\mathrm{NL}}(\omega) = {} & \{\chi^{(2)}\mathscr{E}_{\mathrm{T}}^2\}_\omega + \{\chi^{(3)}\mathscr{E}_{\mathrm{T}}^3\}_\omega + \cdots \\ & + \{Q^{(2)}\,\nabla\mathscr{E}_{\mathrm{T}}^2\}_\omega + \{Q^{(3)}\,\nabla\mathscr{E}_{\mathrm{T}}^3\}_\omega + \cdots,\end{aligned} \tag{18}$$

then Eq. (16) can be written as

$$\mathscr{P}_{\mathrm{T}}(\omega) = \mathscr{P}_{\mathrm{T}}^{\mathrm{L}}(\omega) + \mathscr{P}_{\mathrm{T}}^{\mathrm{NL}}(\omega).$$

This is exactly the form for the polarization, although expressed in the frequency domain, that was used in deriving Eqs. (9)–(13) for the field amplitudes. Thus we see that one of the conditions for the set of amplitude equations in (13) to be valid is that the optical fields be weak enough for the perturbation expansion in (16) to converge.

From this point on we shall retain only the dipole polarization terms as these usually form the dominant interaction in liquids and gases.

We can obtain a perturbation expansion for the nonlinear polarization in the time domain by taking the inverse transform of each term in Eq. (16). The resultant equation will involve convolutions of powers of the time-dependent optical fields with time-dependent response functions of the medium which correspond to inverse transforms of the frequency-dependent susceptibilities in Eq. (16). When the characteristic response time of the polarization components is much faster than the time scale on which the field amplitudes vary, the convolution integrals can be replaced with simple products. The time-dependent dipole polarization can then be written as

$$\mathscr{P}_{\mathrm{T}}(t) = \boldsymbol{\chi}^{(1)}\mathscr{E}_{\mathrm{T}}(t) + \boldsymbol{\chi}^{(2)}\mathscr{E}_{\mathrm{T}}^2(t) + \boldsymbol{\chi}^{(3)}\mathscr{E}_{\mathrm{T}}^3(t) + \cdots. \tag{19}$$

The coefficients of the powers of the fields in Eq. (19) play the same role in the time-dependent equations as do the frequency-dependent susceptibilities in the frequency domain. Within the approximations that led to Eq. (19) they are simple constants that are determined by the properties of the medium. They are written in boldface type to indicate that they represent the time-dependent response of the medium rather than the response in the frequency domain, as in Eq. (16).

3.2. Degeneracy Factors

When the polarization is written in terms of the powers of the total optical field it is given properly by Eq. (19). It is normal, however, to express the various terms as products or powers of the individual field components that are contained in Eq. (5). When this is done it is necessary to take into account the number of distinct frequencies that are present. If we consider, for example, the second-order process of second-harmonic generation, then the appropriate component of the nonlinear polarization can be written as

$$\mathscr{P}^{(2)}(t) = \boldsymbol{\chi}^{(2)}\mathscr{E}_1^2. \tag{20}$$

If, on the other hand, we consider the sum-frequency process $\omega_3 = \omega_1 + \omega_2$ due to the combination of incident fields at ω_1 and ω_2, then the total field is

$$\mathscr{E}_\mathrm{T}(t) = \mathscr{E}_1(t) + \mathscr{E}_2(t). \tag{21}$$

The polarization component at the sum frequency ω_3 is given by

$$\mathscr{P}^{(2)}(t) = 2\boldsymbol{\chi}^{(2)}\mathscr{E}_1(t)\mathscr{E}_2(t), \tag{22}$$

where the extra factor 2 has come from the two possible orders in which the pump waves may be taken. In general the nonlinear polarization for a particular process may be written as

$$\mathscr{P}^{(q)}(t) = g\boldsymbol{\chi}^{(q)}\mathscr{E}_1(t)\cdots\mathscr{E}_q(t), \tag{23}$$

where the degeneracy factor g corresponds to the number of distinct permutations of the incident fields. For third-order polarizations these factors are 1, 3, and 6 for one, two, or three distinct pump fields, respectively.

A similar relation applies between the Fourier components of the nonlinear polarization and the Fourier components of the incident fields. Thus for a particular nonlinear process the nonlinear polarization amplitude at $\omega_q = \omega_1 + \omega_2 + \cdots + \omega_n$ can be written as

$$\mathscr{P}^{(q)}(\omega_q) = g\chi^{(q)}(-\omega_q, \omega_1, \ldots, \omega_n)\mathscr{E}(\omega_1)\mathscr{E}(\omega_2)\cdots\mathscr{E}(\omega_n), \tag{24}$$

where $\mathscr{E}(\omega_i)$ is the Fourier transform of the optical field at ω_i. When written in this manner it is easy to see that the time-dependent response in Eq. (19)

for a given nonlinear process is the same as the frequency-dependent susceptibility in Eq. (16). Thus, for example, for the sum-frequency process $\omega_q = \omega_1 + \omega_2 + \cdots + \omega_h$ we have

$$\chi^{(q)}(-\omega_q, \omega_1, \ldots, \omega_n) = \boldsymbol{\chi}^{(q)}, \tag{25}$$

where $\boldsymbol{\chi}^{(q)}$ is the coefficient for this process in Eq. (19). From here on we shall not distinguish between the frequency-dependent susceptibilities and the coefficients in the time-dependent equation (19) except when required to do so by the failure of the approximations that led to Eq. (19).

We can proceed one step further and express the amplitude of the nonlinear polarization in terms of the amplitudes of the various field components. For the sum-frequency process described in Eq. (24) we have

$$P^{\mathrm{NL}}(\omega_q) = \frac{g}{2^{n-1}}\, \chi^{(q)}(-\omega_q, \omega_1, \omega_2, \ldots, \omega_n) A_1 A_2 \cdots A_n. \tag{26}$$

The factors of 2 in the denominator have come from the definition of the field amplitudes in Eq. (5). This is the standard form for the nonlinear polarization used in most of the literature.

Equations (13) and (19) will form the basis for most of our discussions of specific nonlinear optical effects. There are some situations, however, in which alternate equations must be used. One commonly encountered example when these approximations can fail is when one of the incident frequencies, or a sum- or difference-frequency combination or a frequency multiple, falls sufficiently close to a resonance in the medium. When the resonance occurs at a multiple or combination frequency, as can happen, for example, in resonantly enhanced frequency mixing, harmonic generation, or Raman scattering, and when the duration of the incident radiation is sufficiently short (picosecond pulses or, in some cases, even nanosecond pulses), the nonlinear polarization can fail to respond rapidly enough to follow the applied fields. Under these conditions Eqs. (9) and the form of the expansion of Eq. (16) are still valid. However, because the nonlinear polarization cannot follow the fields, the time-dependent expression in Eq. (19) is not valid. The time-dependent components of the nonlinear polarization must be solved for as dynamical variables along with the field amplitudes.

When the resonance is between an allowed transition and an incident frequency, linear absorption must also be accounted for, and, in some cases, the linear polarization will not be simply proportional to the instantaneous value of the optical fields. Here again the linear polarization must be treated as a dynamic variable to be solved for self-consistently with the field amplitudes.

In other situations, for example, with problems involving dielectric breakdown or saturation of an excited level, the electric fields are too intense for the perturbation expansion of Eq. (16) to converge rapidly and, although the form of Eqs. (9) remain valid, the nonlinear polarization no longer has a simple power dependence on the optical fields.

4. PROPERTIES OF THE NONLINEAR SUSCEPTIBILITIES

Before proceeding to the solutions of Eqs. (13) and (19) for specific systems that will be discussed in subsequent chapters let us consider some of the properties of the nonlinear susceptibilities in Eq. (16).

4.1. Tensor Properties—Spatial Symmetries

The equations in the preceding sections were written as if all the quantities were scalars. Actually the electric fields are vectors and the nth-order susceptibility is an $(n + 1)$-rank tensor with in general $3^{(n+1)}$ components. For the most part in this book we shall ignore the tensor aspect of the nonlinear equations, dropping the tensor subscripts and treating all quantities as if they were truly scalars. However, the tensor components that are used in a particular equation must be the ones appropriate to the situation. In certain cases we shall make use of some of the tensor properties of the equations, but then all quantities will have their tensor or vector components expressly identified.

The symmetry properties of the nonlinear susceptibilities have been discussed by a number of authors. (See, for example, Butcher,[3] Kleinman,[11] or Bloembergen.[2] Flytzanis[12] has given a general discussion of the spatial symmetry properties of the susceptibilities of different order.) Here we shall discuss only a few consequences of the symmetries of the nonlinear susceptibilities.

4.1.1. Symmetry Properties in Isotropic Media

4.1.1a. EVEN-ORDER PROCESSES. We shall first show that the second-order dipole susceptibility is zero in media with inversion symmetry. To see this let us write the second-order polarization

$$\mathscr{P}_{\mathrm{T}}^{(2)} = \chi^{(2)}\mathscr{E}_{\mathrm{T}}^{2}. \tag{27}$$

If we invert the coordinate system and write the inverted variables with primes, then Eq. (27) becomes

$$\mathscr{P}_{\mathrm{T}}^{\prime(2)} = \chi^{(2)\prime}(\mathscr{E}_{\mathrm{T}}^{\prime})^{2}. \tag{28}$$

When the coordinates are inverted both $\mathscr{P}$ and $\mathscr{E}$ must change sign while $\chi^{(2)}$, which carries the symmetry of the medium, cannot. Thus we have

$\mathscr{P}_{\rm T} = -\mathscr{P}'_{\rm T}$, $\mathscr{E}_{\rm T} = -\mathscr{E}'_{\rm T}$, $\chi^{(2)} = \chi^{(2)\prime}$, and Eq. (28) becomes

$$\mathscr{P}_{\rm T} = -\chi^{(2)}\mathscr{E}_{\rm T}^2. \tag{29}$$

Equations (27) and (29) are compatible only if $\chi^{(2)} = 0$. Similar reasoning may be used to show that all even-order dipole susceptibilities are zero in media with inversion symmetry. These restrictions do not apply to quadrupole susceptibilities, and second-order effects from these terms can be observed in isotropic media. For the rest of this book, however, we shall restrict our attention to the more commonly encountered dipole susceptibility.

4.1.1b. Third-Order Susceptibilities. Since the third-order susceptibility is the lowest-order nonzero component in isotropic media, we shall consider some of its properties. The third-order susceptibility $\chi^{(3)}_{ijkl}$ has in general 81 separate tensor components, but inversion symmetry reduces these to 21 nonzero elements, of which four can be different and only three are independent.[12] They are

$$\begin{aligned}
&\chi_{xxxx} = \chi_{yyyy} = \chi_{zzzz},\\
&\chi_{xxyy} = \chi_{xxzz} = \chi_{yyxx} = \chi_{yyzz} = \chi_{zzxx} = \chi_{zzyy},\\
&\chi_{xyxy} = \chi_{xzxz} = \chi_{yxyx} = \chi_{yzyz} = \chi_{zyzy} = \chi_{zxzx},\\
&\chi_{xyyx} = \chi_{xzzx} = \chi_{yxxy} = \chi_{yzzy} = \chi_{zxxz} = \chi_{zyyz},\\
&\chi_{xxxx} = \chi_{xxyy} + \chi_{xyxy} + \chi_{xyyx}.
\end{aligned} \tag{30}$$

These relations show that a medium that is isotropic for first-order interactions need not be for the higher-order interactions.

4.1.1c. Circular Polarization. We can also show that there is no third harmonic generated from a circularly polarized beam in an isotropic medium. This property is useful in suppressing unwanted nonlinear effects when two or more processes are possible simultaneously. We represent the total incident wave as

$$\mathscr{E}_{\rm T} = \hat{x}E_x + \hat{y}E_y, \tag{31}$$

where E_x and E_y are the x and y components of the incident field. For circular polarization

$$E_x = E_y. \tag{32}$$

We calculate the x and y components of the nonlinear polarization at 3ω from the third-order term in Eq. (19):

$$\begin{aligned}
\mathscr{P}_x(3\omega) &= (\chi \,\vdots\, \mathscr{E}_{\rm T}^3)_x\\
&= \chi_{xxxx}E_x^3 - \chi_{xxyy}E_xE_y^2\\
&\quad -\chi_{xyxy}E_yE_xE_y - \chi_{xyyx}E_y^2E_x\\
&\quad - i(\chi_{yyyy}E_y^3 - \chi_{yyxx}E_yE_x^2\\
&\quad - \chi_{yxyx}E_xE_yE_x - \chi_{yxxy}E_x^2E_y).
\end{aligned} \tag{33}$$

Here we have not used the degeneracy factors but have written out each term in the cubic product explicitly. If we use Eq. (32) and the last relation in Eq. (30), then we obtain

$$\mathscr{P}_x = 0.$$

A similar calculation for $\mathscr{P}_y$ shows that $\mathscr{P}_y = 0$ also. A similar cancellation does not occur for third-order sum- or difference-frequency mixing if the two input waves have opposite senses of circular polarization.

4.2. Permutation Symmetries

No physical significance can be attributed to the order in which the pump fields at different frequencies are written in Eq. (26). As a result the susceptibilities must remain unchanged when the indices and frequencies of the pump fields are simultaneously interchanged. For example, the third-order susceptibility obeys the relation

$$\begin{aligned}\chi^{(3)}_{ijkl}(-\omega_4, \omega_1, \omega_2, \omega_3) &= \chi^{(3)}_{iklj}(-\omega_4, \omega_2, \omega_3, \omega_1) = \chi^{(3)}_{iljk}(-\omega_4, \omega_3, \omega_1, \omega_2)\\ &= \chi^{(3)}_{ijlk}(-\omega_4, \omega_1, \omega_3, \omega_2) = \chi^{(3)}_{ilkj}(-\omega_4, \omega_3, \omega_2, \omega_1)\\ &= \chi^{(3)}_{ikjl}(-\omega_4, \omega_2, \omega_1, \omega_3). \qquad (34)\end{aligned}$$

This property is generally referred to as intrinsic permutation symmetry. When more than one frequency is present in the incident fields the susceptibility is usually written in the form

$$\chi^{(q)}_{ijkl}(-\omega_q, \omega_1, \ldots, \omega_n) = \frac{1}{n!}\,\mathbf{S}\{X^{(q)}_{ijkl}(-\omega_q, \omega_1, \ldots, \omega_n)\}, \qquad (35)$$

where $X^{(q)}_{ijkl}$ is the susceptibility calculated for a particular order of fields, frequencies, and indices and the operator $\mathbf{S}$ represents a sum over all n distinguishable simultaneous permutations of the fields and frequencies and all indices except the first ones. When written in this manner the susceptibility automatically contains the proper intrinsic permutation symmetry.

Under certain conditions the nonlinear susceptibilities have another kind of permutation symmetry, termed overall permutation symmetry. Under this type of symmetry the susceptibility remains unchanged under simultaneous interchange of all the frequencies and indices. For the third-harmonic susceptibility this property can be expressed as

$$\begin{aligned}\chi^{(3)}_{ijkl}(-\omega_4, \omega_1, \omega_2, \omega_3) &= \chi^{(3)}_{jkli}(-\omega_1, \omega_2, \omega_3, \omega_4)\\ &= \chi_{klij}(-\omega_2, \omega_3, \omega_4, \omega_1)\\ &= \chi_{lijk}(-\omega_3, \omega_4, \omega_1, \omega_2). \qquad (36)\end{aligned}$$

The susceptibilities will generally have overall permutation symmetry unless there are resonances between material transitions and the incident or generated frequencies or their multiples or sum or difference combinations.

4.3. Local Field Corrections

We have discussed the susceptibilities in this section as if the fields to which they respond are the same as those that are incident on the medium. A microscopic analysis shows that the field that any particular molecule or atom sees consists of a depolarization field due to the polarization of all the other atoms or molecules in addition to the incident fields.[13] These depolarizing fields lead to local field corrections for the various susceptibilities in a manner analogous to the local field corrections that are used for linear susceptibilities.

For isotropic media such as liquids or gases the effects of the depolarizing fields can be accounted for by multiplying each field component and the generated polarization by the factor[14]

$$L = \frac{\epsilon(\omega_i) + 2}{3}, \tag{37}$$

where $\epsilon(\omega_i)$ is the linear dielectric constant at frequency ω_i. A macroscopic susceptibility can be constructed by transferring these factors from the fields to the microscopic atomic or molecular susceptibilities. Thus we have

$$\begin{aligned}
&\text{Linear:} && \chi^{(1)}_{\text{mac}} = \left(\frac{\epsilon(\omega) + 2}{3}\right)^2 \chi^{(1)}_{\text{mic}}(\omega),\\
&\text{Third order:} && \chi^{(3)}_{\text{mac}} = \left[\frac{\epsilon(\omega_4) + 2}{3}\right]\left[\frac{\epsilon(\omega_1) + 2}{3}\right]\left[\frac{\epsilon(\omega_2) + 2}{3}\right]\\
&&& \qquad \times \left[\frac{\epsilon(\omega_3) + 2}{3}\right]\chi^{(3)}_{\text{mic}}(-\omega_4, \omega_1, \omega_2, \omega_3),\\
&\text{Fifth order:} && \chi^{(5)}_{\text{mac}} = \prod_{i=1,6}\left[\frac{\epsilon(\omega_i) + 2}{3}\right]\chi^{(5)}_{\text{mic}}(-\omega_6, \omega_1, \ldots, \omega_5).
\end{aligned} \tag{38}$$

It is these macroscopic susceptibilities that should be used in Eqs. (4)–(19). In rarefied media such as gases where the dielectric constant is never much different from unity these corrections are small and can usually be neglected. They are much larger in dense media such as liquids and can have a significant effect on the relation of experimentally determined susceptibilities to ones calculated from microscopic theories. They do not, however, affect the form of the polarization in Eq. (19), and so they will be dropped in our subsequent discussions. The reader should remain aware, however, that they should be used in appropriate circumstances.

5. SPECIFIC NONLINEAR EFFECTS

Very often, specific optical processes can be associated with individual components of the linear and nonlinear susceptibilities. For example, the first term in Eq. (16) describes the linear dipole response of the medium and contains only those frequencies that are present in the incident radiation. Its real part gives rise to the linear refractive index, whereas its imaginary part is responsible for linear absorption (or gain in media with population inversions). Linear elastic scattering, that is, scattering without a frequency shift, is also described by this term. Various forms of inelastic linear scattering, such as spontaneous Raman, Brillouin, or Rayleigh wing scattering, that produce scattered light at a shifted wavelength involve excitation of internal modes of the medium and are not included in this term.

The second-order term carries frequencies that are different from those in the incident fields. If the incident light has radiation at the frequencies ω_m and ω_n we can identify the following amplitudes of the second-order polarization using the notation of Eqs. (5) and (6):

$$\begin{aligned} P_i(\omega_3 = 2\omega_m) &= \tfrac{1}{2}\chi^{(2)}_{ijk}(-\omega_3, \omega_m, \omega_m)A_j^{\omega_m}A_k^{\omega_m}, \\ P_i(\omega_3 = \omega_m + \omega_n) &= \chi^{(2)}_{ijk}(-\omega_3, \omega_m, \omega_n)A_j^{\omega_m}A_k^{\omega_n}, \\ P_i(\omega_3 = \omega_m - \omega_n) &= \chi^{(2)}_{ijk}(-\omega_3, \omega_m, -\omega_n)A_j^{\omega_m}(A_k^{\omega_n})^*, \\ P_i(\omega_3 = 0) &= \chi^{(2)}_{ijk}(0, \omega_m, -\omega_m)A_j^{\omega_m}(A_k^{\omega_m})^*. \end{aligned} \tag{39}$$

Here $A_j^{\omega_m}$ is the amplitude of the jth component of the field at frequency ω_m and P_i is the amplitude of the ith component of the nonlinear polarization.

The first of these terms represents second-harmonic generation, the next two represent second-order sum- and difference-frequency mixing. The last term represents optical rectification.

Similar identification can be made for third-order interactions. If the incident radiation contains the frequencies ω_m, ω_n, and ω_p, we can identify the following components of the polarization:

third-harmonic generation,

$$P_i(\omega_4 = 3\omega_m) = (1/2^2)\chi_{ijkl}(-\omega_4, \omega_m, \omega_m, \omega_m)A_j^{\omega_m}A_k^{\omega_m}A_l^{\omega_m}; \tag{40}$$

sum- and difference-frequency mixing,

$$\begin{aligned} P_i(\omega_4 = 2\omega_m + \omega_n) &= (3/2^2)\chi_{ijkl}(-\omega_4, \omega_m, \omega_m, \omega_n)A_j^{\omega_m}A_k^{\omega_m}A_p^{\omega_m}, \\ P_i(\omega_4 = 2\omega_m - \omega_n) &= (3/2^2)\chi_{ijkl}(-\omega_4, \omega_m, \omega_m, -\omega_n)A_j^{\omega_m}A_k^{\omega_m}(A_l^{\omega_n})^*, \\ P_i(\omega_4 = \omega_m + \omega_n + \omega_p) &= (6/2^2)\chi_{ijkl}(-\omega_4, \omega_m, \omega_n, \omega_p)A_j^{\omega_m}A_k^{\omega_n}(A_l^{\omega_p})^*, \\ P_i(\omega_4 = \omega_m + \omega_n - \omega_p) &= (6/2^2)\chi_{ijkl}(-\omega_4, \omega_m, \omega_n, -\omega_p)A_j^{\omega_m}A_k^{\omega_n}(A_l^{\omega_p})^*, \\ P_i(\omega_4 = \omega_m - \omega_n - \omega_p) &= (6/2^2)\chi_{ijkl}(-\omega_4, \omega_m, -\omega_n, -\omega_p) \\ &\quad \times A_j^{\omega_m}(A_k^{\omega_n})^*(A_l^{\omega_p})^*; \end{aligned} \tag{41}$$

self-action effects:

self-focusing or defocusing (nonlinear refractive index),

$$P_i(\omega_m) = (3/2^2)\,\mathrm{Re}\{\chi_{iikk}(-\omega_m, \omega_m, \omega_m, -\omega_m)\}A_i^{\omega_m}|A_k^{\omega_m}|^2, \qquad (42a)$$

two-photon absorption,

$$P_i(\omega_m) = (3/4)\,\mathrm{Im}\{\chi_{iikk}^{(3)}(-\omega_m, \omega_m, \omega_m, -\omega_m)\}A_i^{\omega_m}|A_k^{\omega_m}|^2; \qquad (42b)$$

degenerate four-wave mixing,

$$P_i(\omega_m) = (3/4)\chi_{ijkl}^{(3)}(-\omega_m, \omega_m, \omega_m, -\omega_m)A_j^{\omega_m}A_k^{\omega_m}(A_l^{\omega_m})^*; \qquad (43)$$

Raman effects:

Raman-induced Kerr effect,

$$P_i(\omega_m) = (3/4)\,\mathrm{Re}\{\chi_{iikk}^{(3)}(-\omega_m, \omega_m, \omega_n, -\omega_n)\}A_i^{\omega_m}|A_k^{\omega_n}|^2; \qquad (44)$$

Stokes–Raman scattering,

$$P_i(\omega_S) = (3/2)\,\mathrm{Im}\{\chi_{iikk}^{(3)}(-\omega_S, \omega_S, \omega_L, -\omega_L)\}A_i^{\omega_S}|A_k^{\omega_L}|^2; \qquad (45)$$

anti-Stokes–Raman scattering,

$$P_i(\omega_{AS}) = (3/2)\,\mathrm{Im}\{\chi_{ijkl}^{(3)}(-\omega_{AS}, \omega_L, \omega_L, -\omega_S)\}A_j^{\omega_L}A_k^{\omega_L}(A_l^{\omega_S})^*. \qquad (46)$$

The last two Raman processes occur primarily when the difference of the two incident frequencies matches that of an appropriate material excitation:

$$\omega_{mat} = \omega_L - \omega_S.$$

In the last two expressions the designations of the optical frequencies were changed to conform to the usual usage for Raman interactions: L = laser, S = Stokes, and AS = anti-Stokes.

The interactions represented in Eqs. (40)–(46) can be represented diagramatically as shown in Fig. 2. In these diagrams the upward arrows indicate the energy of the photons (frequencies) that are added in the nonlinear polarizations, while the downward pointing arrows indicate frequencies that are subtracted. The direction of power flow among the various waves cannot be determined from these diagrams but must be determined from solutions of the appropriate equations with the proper boundary conditions. The interactions in Eqs. (40), (41), (42a), (43), and (44) are parametric interactions that do not necessarily involve energy transfer to the medium. They can take place in a region of the spectrum in which the medium is transparent for both linear and nonlinear absorption. Two-photon absorption [Eq. (42b)] and stimulated Raman scattering [Eq. (45)] involve energy transfer to the medium and the frequencies involved must be matched to appropriate levels in the medium. The last process, anti-Stokes–Raman scattering, is a form of four-wave parametric mixing that is resonantly enhanced by a real intermediate Raman transition.

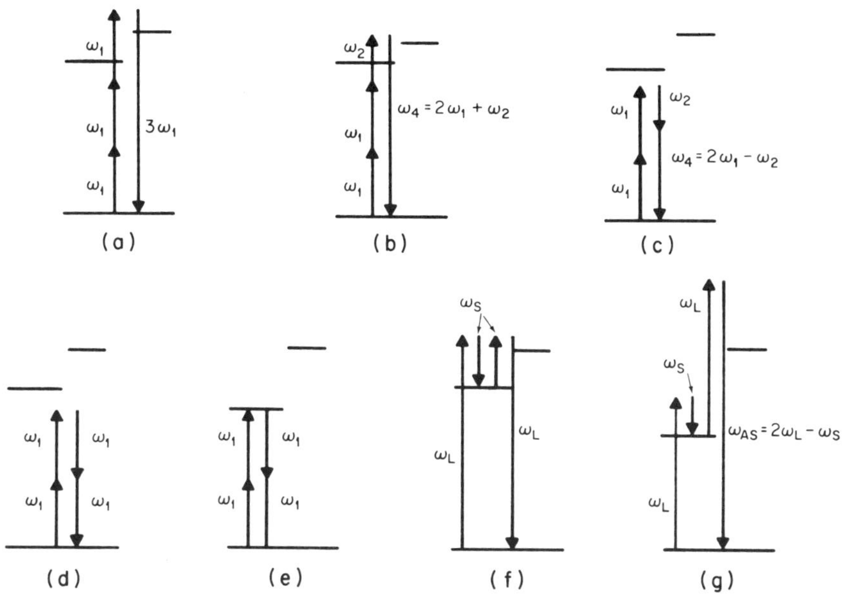

Fig. 2. Diagrammatic representation of various nonlinear optical interactions: (a) third-harmonic generation, (b) four-wave sum-frequency mixing, (c) four-wave difference-frequency mixing, (d) nonlinear refractive index, (e) two-photon absorption, (f) stimulated Stokes–Raman scattering, and (g) anti-Stokes–Raman scattering. The arrows represent photon energies and the horizontal lines represent energy states of the medium.

6. CALCULATION OF NONLINEAR SUSCEPTIBILITIES

We shall now consider techniques for calculating nonlinear susceptibilities for various types of materials. The determination of the magnitude of the nonlinear susceptibilities represents a significant area of both experimental and theoretical research in nonlinear optics. For some materials, such as atomic gases, calculation of the various components of the susceptibility tensor can be accomplished by a straightforward, if sometimes tedious, application of perturbation theory. In other cases, such as for some molecular gases and liquids, calculation of the nonlinear response from first principles is not yet possible, and here we usually have to rely on experimentally determined values of the nonlinear susceptibilities.

Formal equations for the nonlinear susceptibilities in a medium with isolated quantum levels can be obtained from time-dependent perturbation theory. The development of such equations have been discussed by Bloembergen,[2] Armstrong *et al.*,[15] Bloembergen and Shen,[16] Orr and Ward,[17] Flytzanis,[12] Yuratich and Hanna,[18] Hanna *et al.*,[7] and Puell and Vidal.[19] Calculations can be performed using either the perturbed wave functions or the density matrix approach. It is usually easier to calculate the higher-order susceptibilities using the perturbed wave functions. However, introduction of damping due to dephasing and population decay is more straightforward in the density matrix approach. In this section we shall outline the essentials of such a calculation for the dipole susceptibilities using the density matrix formalism.

The polarization is given by the expectation value of the dipole moment operator:

$$\mathscr{P} = \operatorname{tr} \mu\rho = \sum_{i,j} \mu_{ij}\rho_{ji}. \tag{47}$$

The components of the density operator are determined in turn from the equations of motion:

$$i\hbar \frac{\partial}{\partial t} \rho + \frac{i\hbar}{T} \rho = [H, \rho], \tag{48}$$

where the damping constant T represents either a population decay time for diagonal elements or a dephasing time for off-diagonal elements. In the dipole approximation the Hamiltonian is given by the expression

$$H = H^0 - \mu\mathscr{E}_{\mathrm{T}}, \tag{49}$$

where H^0 is the unperturbed Hamiltonian that described the material system in the absence of the optical fields, $\mathscr{E}_{\mathrm{T}}$ is the total electric field of Eq. (5), and μ is the dipole moment operator whose matrix elements μ_{ab} are the transition dipole matrix elements between states a and b.

The elements of the density matrix can be determined by assuming a perturbation expansion of the form

$$\rho = \rho^{(0)} + \rho^{(1)} + \rho^{(2)} + \rho^{(3)} + \cdots. \tag{50}$$

By substituting Eqs. (50) and (49) into (48) we can obtain a hierarchy of equations for terms of increasing order:

$$\begin{aligned}
i\hbar \frac{\partial}{\partial t} \rho^{(0)} + \frac{i\hbar}{T} \rho^{(0)} &= [H^{(0)}, \rho^{(0)}], \\
i\hbar \frac{\partial}{\partial t} \rho^{(1)} + \frac{i\hbar}{T} \rho^{(1)} &= [H^0, \rho^{(1)}] - [\mu\mathscr{E}_{\mathrm{T}}, \rho^{(0)}], \\
i\hbar \frac{\partial}{\partial t} \rho^{(2)} + \frac{i\hbar}{T} \rho^{(2)} &= [H^0, \rho^{(2)}] - [\mu\mathscr{E}_{\mathrm{T}}, \rho^{(1)}], \\
i\hbar \frac{\partial}{\partial t} \rho^{(3)} + \frac{i\hbar}{T} \rho^{(3)} &= [H^0, \rho^{(3)}] - [\mu\mathscr{E}_{\mathrm{T}}, \rho^{(2)}], \\
&\vdots \\
i\hbar \frac{\partial}{\partial t} \rho^{(n)} + \frac{i\hbar}{T} \rho^{(n)} &= [H^0, \rho^{(n)}] - [\mu\mathscr{E}_{\mathrm{T}}, \rho^{(n-1)}], \\
&\vdots
\end{aligned} \tag{51}$$

A formal solution for the component $\rho_{mn}^{(i)}$ at frequency $\pm\omega$ has been given by Puell and Vidal[19] in terms of an iterated integral:

$$\rho_{mn}^{(i)}(t) = \frac{i}{2\hbar} e^{-i\omega_{mn}t} \int_{-\infty}^{t} A(t')\{e^{-i(\omega-\omega_{mn})t'} + e^{i(\omega+\omega_{mn})t'}\} \times [\mu, \rho^{(i-1)}]\, dt', \tag{52}$$

where $\omega_{mn} = (E_m - E_n)/\hbar - i/T_{mn}$ and $E_m - E_n$ is the energy difference between levels m and n. The formal solution of Eq. (52) can be overly complicated for calculating the susceptibilities in the quasistatic approximation, for which explicit solution of the set of equations in (51) is often easier. The solution in Eq. (52) is useful, however, for handling resonant interactions in which the quasistatic approximation cannot be used.

Explicit formulas can be derived for the susceptibilities that describe specific nonlinear processes in media with discrete quantum levels. Formulas have been given in Armstrong *et al.*[15] for the third-order susceptibility for the four-wave mixing processes $\omega_4 = \omega_1 + \omega_2 + \omega_3$ and $\omega_4 = 2\omega_1 + \omega_2$ under conditions in which all frequency combinations were far enough away from resonances that the damping could be ignored. These were later extended to situations when damping constants had to be included by Puell and Vidal[19] for the third-harmonic susceptibility and by Bloembergen *et al.*[20] for the four-wave mixing interactions. Bloembergen and Shen[16] have given expressions for the susceptibilities that describe Raman processes. Flytzanis[12] has given extensive formulas for second- and third-order susceptibilities in solids and gases.

We shall finish this chapter with an explicit calculation of the susceptibility for third-harmonic generation in the steady-state approximation for a single atom with the energy levels shown in Fig. 3. We shall assume that the population starts in the ground state ($\rho_{00}^{(0)} = 1$; $\rho_{nn}^{(0)} = 0$, $n \neq 0$). Solutions to Eqs. (51) are usually obtained by calculating all the density matrix

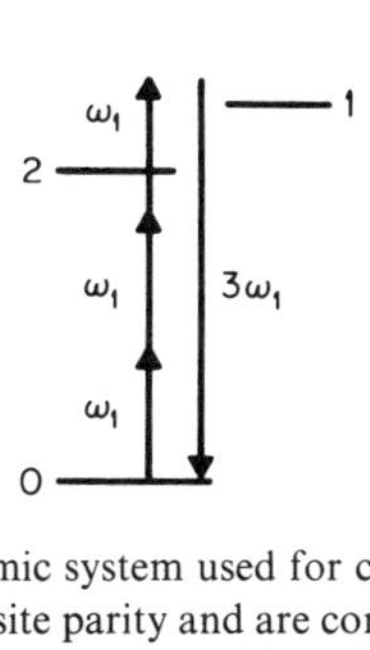

Fig. 3. Simplified diagram of atomic system used for calculating nonlinear susceptibility in Section 6. Levels 0 and 1 are of opposite parity and are connected by a dipole transition. Levels 0 and 2 are of the same parity and are not connected by a dipole transition.

elements in increasing order (see, for example, Armstrong *et al.*,[15] Bloembergen and Shen,[16] and Puell and Vidal[19]). Here we shall give a slightly different approach in which we start with the term we wish to evaluate and calculate it in terms of lower-order quantities, continuing the procedure until we reach the zero-order starting conditions. Such a procedure enables us to identify exactly which lower-order terms are necessary in order to calculate the desired term and can reduce the computational effort by eliminating the calculation of unnecessary terms. In this example we shall also write out each term explicitly in an attempt to indicate how diagonal and off-diagonal terms are combined in the expressions that are found in the literature.

The polarization at the third harmonic is given by

$$\begin{aligned}\mathscr{P}(3\omega) &= \operatorname{tr} \mu\rho(3\omega) = \sum_{m,n} \mu_{mn}\rho_{nm}(3\omega) \\ &= \mu_{10}\rho_{01}^{(3)}(3\omega) + \mu_{01}\rho_{10}^{(3)}(3\omega) \\ &\quad + \mu_{12}\rho_{21}^{(3)}(3\omega) + \mu_{21}\rho_{12}^{(3)}(3\omega). \end{aligned} \tag{53}$$

We calculate first the third-order components of the density matrix in terms of the lower-order elements. We assume for the third-order elements a variation of the form

$$\rho_{mn}^{(3)}(3\omega) = \rho_{mn}^{(3)-} e^{-i3\omega t} + \rho_{mn}^{(3)+} e^{i3\omega t}. \tag{54}$$

Using the appropriate equation from the set (51) we find

$$\begin{aligned}\rho_{10}^{(3)-}(3\omega) = {}& \frac{1}{2\hbar(\omega_{10} - 3\omega - i/T_{01})} [\mu_{12}\rho_{20}^{(2)-}(2\omega)A_1 \\ & + \mu_{10}(\rho_{00}^{(2)-}(2\omega) - \rho_{11}^{(2)-}(2\omega))A_1], \end{aligned} \tag{55a}$$

$$\begin{aligned}\rho_{01}^{(3)-}(3\omega) = {}& \frac{1}{2\hbar(\omega_{10} + 3\omega + i/T_{01})} [\mu_{21}\rho_{02}^{(2)-}(2\omega)A_1 \\ & + \mu_{01}(\rho_{00}^{(2)-}(2\omega) - \rho_{11}^{(2)-}(2\omega))A_1], \end{aligned} \tag{55b}$$

$$\begin{aligned}\rho_{21}^{(3)-}(3\omega) = {}& \frac{1}{2\hbar(\omega_{21} - 3\omega - i/T_{21})} [\mu_{21}(\rho_{11}^{(2)-}(2\omega) - \rho_{22}^{(2)-}(2\omega))A_1 \\ & - \mu_{01}\rho_{20}^{(2)-}(2\omega)A_1], \end{aligned} \tag{55c}$$

$$\begin{aligned}\rho_{12}^{(3)-}(3\omega) = {}& \frac{1}{2\hbar(\omega_{21} + 3\omega + i/T_{21})} [\mu_{12}(\rho_{11}^{(2)-}(2\omega) - \rho_{22}^{(2)-}(2\omega))A_1 \\ & - \mu_{10}\rho_{02}^{(2)-}(2\omega)A_1]. \end{aligned} \tag{55d}$$

For the second-order components we assume a variation of the form

$$\rho_{mn}^{(2)}(2\omega) = \rho_{mn}^{(2)-}(2\omega)e^{-i2\omega t} + \rho_{mn}^{(2)+}e^{i2\omega t}. \tag{56}$$

Using the second-order equations from the set (51) we find the following components:

$$\rho_{20}^{(2)-}(2\omega) = \frac{1}{2\hbar(\omega_{20} - 2\omega - i/T_{20})}[\mu_{21}\rho_{10}^{(1)-}(\omega)A_1 - \rho_{21}^{(1)-}(\omega)\mu_{10}A_1], \tag{57a}$$

$$\rho_{02}^{(2)-}(2\omega) = \frac{1}{2\hbar(\omega_{20} + 2\omega + i/T_{20})}[\mu_{12}\rho_{01}^{(1)-}(\omega)A_1 - \rho_{12}^{(1)-}(\omega)\mu_{01}A_1], \tag{57b}$$

$$\rho_{00}^{(2)-}(2\omega) = \frac{1}{2\hbar(2\omega + i/T_0)}[\rho_{01}^{(1)-}(\omega)\mu_{10}A_1 - \mu_{01}\rho_{10}^{(1)-}(\omega)A_1], \tag{57c}$$

$$\rho_{11}^{(2)-}(2\omega) = \frac{1}{2\hbar(2\omega + i/T_1)}\begin{bmatrix}\mu_{01}\rho_{10}^{(1)-}(\omega)A_1 & -\mu_{10}\rho_{01}^{(1)-}(\omega)A_1 \\ +\mu_{21}\rho_{12}^{(1)}(\omega)A_1 & -\mu_{12}\rho_{21}^{(1)}(\omega)A_1\end{bmatrix}, \tag{57d}$$

$$\rho_{22}^{(2)-}(2\omega) = 0. \tag{57e}$$

Finally, in first order we assume the form

$$\rho_{mn}^{(1)}(\omega) = \rho_{mn}^{(1)-}(\omega)e^{-i\omega t} + \rho_{mn}^{(1)+}(\omega)e^{i\omega t} \tag{58}$$

and find

$$\rho_{10}^{(1)-}(\omega) = \frac{\mu_{10}A_1}{2\hbar(\omega_{10} - \omega - i/T_{10})}, \tag{59a}$$

$$\rho_{01}^{(1)-}(\omega) = \frac{\mu_{01}A}{2\hbar(\omega_{10} + \omega + i/T_{10})}, \tag{59b}$$

$$\rho_{21}^{(1)-}(\omega) = \rho_{12}^{(1)-}(\omega) = 0. \tag{59c}$$

In deriving the first-order expressions we have used the assumption that the only zero-order term that is nonzero is $\rho_{00}^{(0)}$. Also, in Eqs. (55)–(59) damping terms with one subscript refer to lifetimes of populations in the corresponding level, whereas damping terms with two subscripts refer to dephasing times for transitions between the two levels indicated by the subscripts.

In order to calculate the polarization for the third-harmonic process we need to combine the expressions in Eqs. (55), (57), and (59). From hereon we shall drop the damping terms, assuming that all frequency combinations are far enough from resonance that they are negligible. Corresponding expressions including the damping terms can be found in Puell and Vidal[19] and in Bloembergen *et al.*[20]

Using Eqs. (59a) and (59b) in Eqs. (57) we obtain

$$\rho_{20}^{(2)-}(2\omega) = \frac{\mu_{21}\mu_{10}A_1^2}{4\hbar^2(\omega_{20} - 2\omega)(\omega_{10} - \omega)}, \tag{60a}$$

$$\rho_{02}^{(2)-}(2\omega) = \frac{\mu_{01}\mu_{12}A_1^2}{4\hbar^2(\omega_{20} + 2\omega)(\omega_{10} + \omega)}, \tag{60b}$$

$$\rho_{11}^{(2)-}(2\omega) = \frac{|\mu_{01}|^2}{4\hbar^2}\frac{1}{(\omega_{10} - \omega)(\omega_{10} + \omega)}, \tag{60c}$$

$$\rho_{00}^{(2)-}(2\omega) - \rho_{11}^{(2)-}(2\omega) = -\frac{|\mu_{01}|^2 A_1^2}{2\hbar^2(\omega_{10} - \omega)(\omega_{10} + \omega)}. \tag{60d}$$

Using Eqs. (60) in Eqs. (55) we obtain

$$\rho_{12}^{(3)-}(3\omega) = \frac{\mu_{12}|\mu_{10}|^2 A_1^3}{8\hbar^3(\omega_{10} + \omega)(\omega_{10} - \omega)(\omega_{20} + 2\omega)}, \tag{61a}$$

$$\rho_{21}^{(3)-}(3\omega) = \frac{\mu_{21}|\mu_{10}|^2 A_1^3}{8\hbar^3(\omega_{10} + \omega)(\omega_{10} - \omega)(\omega_{20} - 2\omega)}, \tag{61b}$$

$$\rho_{10}^{(3)-}(3\omega) = \frac{\mu_{10}A_1^3}{8\hbar^3}\left\{\frac{|\mu_{21}|^2}{(\omega_{10} - 3\omega)(\omega_{20} - 2\omega)(\omega_{10} - \omega)} - \frac{|\mu_{10}|^2}{(\omega_{10} - 3\omega)(2\omega)(\omega_{10} - \omega)} + \frac{|\mu_{10}|^2}{(\omega_{10} + \omega)(2\omega)(\omega_{10} - \omega)}\right\}, \tag{61c}$$

$$\rho_{01}^{(3)-}(3\omega) = \frac{\mu_{01}A_1^3}{8\hbar^3}\left\{\frac{|\mu_{12}|^2}{(\omega_{10} + 3\omega)(\omega_{20} + 2\omega)(\omega_{10} + \omega)} + \frac{|\mu_{01}|^2}{(\omega_{10} + 3\omega)(2\omega)(\omega_{10} + \omega)} - \frac{|\mu_{01}|^2}{(\omega_{10} - \omega)(2\omega)(\omega_{10} + \omega)}\right\}. \tag{61d}$$

The susceptibility can now be obtained from the definition in Eq. (26) (with $g = 1$) along with Eqs. (53) and (61). In making the final combinations that are indicated in Eq. (56) we can represent the various terms in the form of a summation over intermediate states:

$$\chi^{(3)}(-3\omega, \omega, \omega, \omega) = \frac{A_1^3}{8\hbar^3} \sum_{i=0,2} \mu_{01}\mu_{1i}\mu_{i1}\mu_{10} \times \left[\frac{1}{(\omega_{10} - 3\omega)(\omega_{i0} - 2\omega)(\omega_{10} - \omega)} + \frac{1}{(\omega_{10} + \omega)(\omega_{i0} - 2\omega)(\omega_{10} - \omega)} + \frac{1}{(\omega_{10} + \omega)(\omega_{i0} + 2\omega)(\omega_{10} - \omega)} + \frac{1}{(\omega_{10} + \omega)(\omega_{10} + 2\omega)(\omega_{10} + 3\omega)}\right]. \tag{62}$$

When written in this form each term involves a combination of diagonal and off-diagonal matrix elements.

For a medium with more than one level connected to the ground state by single-photon transitions the susceptibility will involve two more summations over these levels. In order to perform the intermediate sums correctly the signs of the matrix elements as well as their magnitudes must be known. This requirement makes the calculation of the nonlinear susceptibilities considerably more complicated than the calculation of linear susceptibilities unless a particular path through near-resonant levels dominates the interaction, eliminating the need for the summation.

Calculation of the susceptibilities for other nonlinear interactions can be accomplished by the same procedure outlined here. One additional complication can arise in the calculation of susceptibilities for interactions that involve field combinations of the form $(A^{\omega}A^{\omega *})$. When the ground state is used as the second intermediate state for terms with this combination, the formula corresponding to Eq. (62) would give a zero in the denominator. The treatment of such secular terms, as they are called, is discussed by Orr and Ward,[17] Puell and Vidal,[19] and Hanna *et al.*[7]

The total polarization in the medium takes into account the total number of scattering centers. For gases we can ignore the local field corrections as discussed earlier and the total susceptibility is simply

$$\chi_{\mathrm{T}} = N\chi_{\mathrm{a}},$$

where N is the number density of the scattering centers. In subsequent chapters we shall use equations that involve both the total susceptibility and the single-atomic or -molecular susceptibility. If an equation does not contain the number density explicitly, the total susceptibility will be understood unless the susceptibility is specifically designated as belonging to a single atom or molecule by the subscript a, i.e., unless it is written as χ_a. In equations which explicitly contain the number density, on the other hand, the susceptibility will be understood to be the atomic or molecular susceptibility. Explicit expressions that are given for the nonlinear susceptibilities of various systems will be for the individual atoms or molecules.

Expressions such as Eq. (62) are most useful for calculating the susceptibilities of atomic or molecular gases for which a sufficient number of dipole matrix elements are known. Examples of such materials are some of the rare gases (e.g., He, Ne, Ar), the alkali metals (Li, Na, K, Rb, and Cs), and some of the alkaline earths. For materials such as liquids or some gases whose appropriate dipole matrix elements are either not known or cannot be calculated, equations like (62) cannot be used to give quantitative values for the susceptibilities. In such situations we must rely on experimentally determined susceptibilities. Often relative measurements of a particular element of the susceptibility can be made very accurately for a wide range of materials using a single, well-known susceptibility of a particular material as an absolute reference. Several techniques have been developed for doing these measurements, some of which will be discussed in Chapter 6.

Considerable progress has been made in calculations of the susceptibilities of crystalline solids using various models (see, for example, Flytzanis[12]). Extension of these techniques to liquids and other amorphous materials has not yet been fruitful due mainly to a lack of detailed knowledge of the disordered state. As microscopic theories of liquids are developed, however, similar techniques may make calculations of nonlinear susceptibilities of liquids from first principles possible in the future.

REFERENCES

1. M. Goeppert-Mayer, *Ann. Phys. (Leipzig)* **9**, 273 (1931).
2. N. Bloembergen, "Nonlinear Optics." Benjamin, New York, 1965.
3. P. N. Butcher, "Nonlinear Optical Phenomena," Bull. No. 200. Eng. Exp. St., Ohio State Univ., Columbus, 1965.
4. A. Yariv, "Quantum Electronics," 2nd ed. Wiley, New York, 1975.
5. F. T. Arecchi and E. O. Schultz-Dubois, eds., "Laser Handbook," Vols. 1 and 2. North-Holland Publ., Amsterdam, 1972 and 1972.
6. H. Rabin and C. L. Tang, eds., "Quantum Electronics," Vol. 1, Parts A and B. Academic Press, New York, 1975.
7. D. C. Hanna, M. A. Yuratich, and D. Cotter, "Nonlinear Optics of Free Atoms and Molecules." Springer-Verlag, Berlin and New York, 1979.
8. N. Bloembergen, "Nonlinear Optics," p. 5. Benjamin, New York, 1965.
9. S. A. Akhmanov, *Nonlinear Spectrosc., Corso Soc. Ital. Fis., 64th, Bologna* pp. 239–254 (1977).
10. J. D. Jackson, "Classical Electrodynamics," p. 101. Wiley, New York, 1963.

11. D. A. Kleinman, *Phys. Rev.* **126**, 1977 (1962).
12. C. Flytzanis, *in* "Quantum Electronics" (H. Rabin and C. L. Tang, eds.), Vol. 1, Part A, p. 30. Academic Press, New York, 1975.
13. J. D. Jackson, "Classical Electrodynamics," p. 119. Wiley, New York, 1963.
14. N. Bloembergen, "Nonlinear Optics," p. 69. Benjamin, New York, 1965.
15. J. A. Armstrong, N. Bloembergen, J. Ducuing, and P. S. Pershan, *Phys. Rev.* **127**, 1918 (1962).
16. N. Bloembergen and Y. R. Shen, *Phys. Rev.* **133**, A37, (1964).
17. B. J. Orr and J. F. Ward, *Mol. Phys.* **20**, 513 (1971).
18. M. A. Yuratich and D. C. Hanna, *J. Phys. B* **9**, 729 (1976).
19. H. Puell and C. R. Vidal, *Phys. Rev. A* **14**, 2225 (1976).
20. N. Bloembergen, H. Lotem, and R. T. Lynch, Jr., *Indian J. Pure Appl. Phys.* **16**, 151 (1978).

CHAPTER 2

THIRD-HARMONIC GENERATION AND THIRD-ORDER FREQUENCY MIXING

1. INTRODUCTION

In this chapter we shall discuss frequency mixing and harmonic conversion interactions that arise from the third-order term in the dipole susceptibility. Harmonic generation is one of the oldest and most widely studied of the nonlinear optical effects. Second-harmonic generation in solids was first reported in 1961 with the frequency doubling of ruby laser radiation.[1] Since then it has been widely used as a source of visible and near-ultraviolet radiation using Nd:YAG ($\lambda = 1.06\ \mu m$), ruby ($\lambda = 0.6943\ \mu m$), and dye lasers as pump sources.

Third-harmonic generation and its more general form of four-wave frequency mixing have also been studied and used rather extensively. In the ultraviolet and infrared regions of the spectrum, the selection of transparent nonlinear crystals that are suitable for second-harmonic generation is not as widespread as in the visible. Gases that are transparent in the ultraviolet and gases and liquids that are transparent in the infrared have made it possible to extend the study of optical nonlinearities to these spectral ranges through third-order processes. In addition, the very large nonlinear susceptibilities provided by the sharp resonances present in media such as atomic gases make it possible to observe these nonlinear effects with relatively low power tunable dye lasers. As a result, spectroscopic applications and extensive studies of the dispersion of the nonlinear susceptibilities have become possible.

Harmonic generation and four-wave mixing have been used both as a source of radiation and as a means to study the properties of the nonlinear susceptibilities and the materials themselves. In this chapter we shall describe these effects from the point of view of a radiation source, giving details of the process of generation of new waves. In Chapter 6 we shall return to these same processes and indicate how they can be used to study the properties of the materials in which they take place.

We shall assume for most of our discussion that the quasi-steady-state approximation is valid and that the nonlinear response of the material is described by Eqs. (1-13) and (1-19) with constant susceptibilities. This approximation is valid for many interactions that are studied in the laboratory. In Sections 5 and 7 we shall consider specific systems in which this approximation is invalid and indicate methods of accounting for its failure. In later sections we shall summarize experimental results that have been obtained for specific types of lasers.

One important aspect of these nonlinear interactions is the conversion efficiency, a measure of the effectiveness with which the power in the pump waves is transferred to the generated wave. A complete discussion of conversion efficiency must include other processes, both linear and nonlinear, that can affect the nonlinear interaction. These competing effects are much more important for gases and some liquids than for solids, where an accurate description of conversion efficiency can generally be obtained by considering only a single nonlinear interaction. We shall defer a complete discussion of conversion efficiency until Chapter 4, where we shall treat limitations on third- and higher-order effects simultaneously. In this chapter we shall discuss conversion efficiency for somewhat idealized systems in which only the frequency-mixing process is considered. Our discussion here will serve as a basis for comparison with the more complete description of Chapter 4.

2. FOUR-WAVE FREQUENCY MIXING

Four-wave mixing includes several effects, depending on the exact situation that one is interested in. For an incident wave that includes three separate frequencies, ω_1, ω_2, and ω_3, with an electric field given by

$$\mathscr{E}_{\text{inc}}(r, z, t) = \tfrac{1}{2}\{A_1(r, z)e^{-i(\omega_1 t - k_1 z)} + A_2(r, z)e^{-i(\omega_2 t - k_2 z)} + A_3(r, z)e^{-i(\omega_3 t - k_3 z)} + \text{c.c.}\}, \tag{1}$$

the following nonlinear polarization amplitudes, each of which drives a new wave at ω_4, can be created:

$$P^{\text{NL}}(\omega_4 = 3\omega_i) = \tfrac{1}{4}\chi^{(3)}(-3\omega_i, \omega_i, \omega_i, \omega_i)A_i^3, \tag{2a}$$

$$P^{\mathrm{NL}}(\omega_4 = 2\omega_i + \omega_j) = \tfrac{3}{4}\chi^{(3)}(-\omega_4, \omega_i, \omega_i, \omega_j)A_i^2 A_j, \tag{2b}$$

$$P^{\mathrm{NL}}(\omega_4 = 2\omega_i - \omega_j) = \tfrac{3}{4}\chi^{(3)}(-\omega_4, \omega_i, \omega_i, -\omega_j)A_i^2 A_j^*, \tag{2c}$$

$$P^{\mathrm{NL}}(\omega_4 = \omega_i + \omega_j + \omega_k) = \tfrac{3}{2}\chi^{(3)}(-\omega_4, \omega_i, \omega_j, \omega_k)A_i A_j A_k, \tag{2d}$$

$$P^{\mathrm{NL}}(\omega_4 = \omega_i + \omega_j - \omega_k) = \tfrac{3}{2}\chi^{(3)}(-\omega_4, \omega_i, \omega_j, -\omega_k)A_i A_j A_k^*, \tag{2e}$$

$$P^{\mathrm{NL}}(\omega_4 = \omega_i - \omega_j - \omega_k) = \tfrac{3}{2}\chi^{(3)}(-\omega_4, \omega_i, -\omega_j, -\omega_k)A_i A_j^* A_k^*. \tag{2f}$$

The processes in Eqs. (2a)–(2c) are special cases of the last three interactions and can occur when only one or two frequencies are present in the incident field.

If we consider each of these processes independently, then the total field can be expressed as

$$\begin{aligned}\mathscr{E}_{\mathrm{T}} &= \mathscr{E}_{\mathrm{inc}} + \mathscr{E}_4\\ &= \tfrac{1}{2}\{A_1(r, z)e^{-i(\omega_1 t - k_1 z)} + A_2(r, z)e^{-i(\omega_2 t - k_2 z)}\\ &\quad + A_3(r, z)e^{-i(\omega_3 t - k_3 z)} + A_4(r, z)e^{-i(\omega_4 t - k_4 z)} + \text{c.c.}\},\end{aligned}$$

where ϵ_4 is the field generated by one of the processes in Eqs. (2a)–(2f).

There will also be nonlinear polarizations produced at the incident frequencies which represent the action of the generated fields back on the incident fields. For example, for the process $\omega_4 = \omega_1 + \omega_2 + \omega_3$, the nonlinear polarization amplitudes at each of the incident frequencies are given by

$$\begin{aligned}P^{\mathrm{NL}}(\omega_1) &= \tfrac{3}{2}\chi^{(3)}(-\omega_1, \omega_4, -\omega_2, -\omega_3)A_4 A_2^* A_3^*,\\ P^{\mathrm{NL}}(\omega_2) &= \tfrac{3}{2}\chi(-\omega_2, \omega_4, -\omega_1, -\omega_3)A_4 A_1^* A_3^*,\\ P^{\mathrm{NL}}(\omega_3) &= \tfrac{3}{2}\chi(-\omega_3, \omega_4, -\omega_1, -\omega_2)A_4 A_1^* A_2^*.\end{aligned}$$

In the quasistatic approximation the field amplitudes are given by Eqs. (1-13), repeated here for convenience:

$$\nabla_\perp^2 A_1 + 2ik_1\frac{\partial A_1}{\partial z} = -\frac{4\pi\omega_1^2}{c^2} P^{\mathrm{NL}}(\omega_1)e^{-i\Delta k_1 z}, \tag{3a}$$

$$\nabla_\perp^2 A_2 + 2ik_2\frac{\partial A_2}{\partial z} = -\frac{4\pi\omega_2^2}{c^2} P^{\mathrm{NL}}(\omega_2)e^{-i\Delta k_2 z}, \tag{3b}$$

$$\nabla_\perp^2 A_3 + 2ik_3\frac{\partial A_3}{\partial z} = -\frac{4\pi\omega_3^2}{c^2} P^{\mathrm{NL}}(\omega_3)e^{-i\Delta k_3 z}, \tag{3c}$$

$$\nabla_\perp^2 A_4 + 2ik_4\frac{\partial A_4}{\partial z} = -\frac{4\pi\omega_4^2}{c^2} P^{\mathrm{NL}}(\omega_4)e^{-i\Delta k_4 z}, \tag{3d}$$

where $P^{\mathrm{NL}}(\omega_4)$ is one of the polarization amplitudes from Eqs. (2a)–(2f).

2.1. Third-Harmonic Generation

We shall consider first the special case of third-harmonic generation. We assume that the total field has the form

$$\mathscr{E}_{\mathrm{T}}(r, z, t) = \tfrac{1}{2}\{A_1(r, z)e^{-i(\omega_1 t - k_1 z)} + A_3(r, z)e^{-i(\omega_3 t - k_3 z)} + \text{c.c.}\}, \tag{4}$$

where A_1 is the amplitude of the incident, or fundamental, field and A_3 is the amplitude of the harmonic field. With the use of Eq. (4), Eqs. (3) become

$$\nabla_\perp^2 A_1 + 2ik_1 \frac{\partial A_1}{\partial z} = -\frac{3\pi\omega_1^2}{c^2}\chi^{(3)}(-\omega_1, 3\omega_1, -\omega_1, -\omega_1)A_3 A_1^{*2} e^{i\Delta kz}, \tag{5a}$$

$$\nabla_\perp^2 A_3 + 2ik_3 \frac{\partial A_3}{\partial z} = -\frac{\pi\omega_3^2}{c^2}\chi^{(3)}(-3\omega_1, \omega_1, \omega_1, \omega_1)A_1^3 e^{-i\Delta kz}, \tag{5b}$$

where $\Delta k = k_3 - 3k_1$.

2.1.1. *Plane-Wave Solutions*

Let us first consider solutions of Eqs. (5) when the incident field consists of a plane wave:

$$A_1(r, z) = A_1(z). \tag{6}$$

Equations (5a) and (5b) can now be written

$$\frac{dA_1}{dz} = i\frac{3\pi^2}{n_1\lambda_1}\chi_{\mathrm{h}}^{(3)} A_3(A_1^*)^2 e^{i\Delta kz}, \tag{7a}$$

$$\frac{dA_3}{dz} = i\frac{3\pi^2}{n_3\lambda_1}\chi_{\mathrm{h}}^{(3)} A_1^3 e^{-i\Delta kz}, \tag{7b}$$

where $\chi_{\mathrm{h}}^{(3)} = \chi^{(3)}(-3\omega_1, \omega_1, \omega_1, \omega_1)$, n_i is the refractive index at ω_i, and the wave vector mismatch for the third-harmonic process is

$$\Delta k = 6\pi/\lambda_1\,(n_3 - n_1). \tag{8}$$

2.1.1a. Weak Pump Fields. When the incident field is relatively weak, it will remain undepleted in the interaction. It can then be treated as a constant in Eq. (7b), whereas Eq. (7a) can be neglected. The solution of Eq. (7b) for a nonlinear medium that extends from $z = 0$ to $z = L$ is

$$A_3(z) = i\frac{3\pi^2}{n_3\lambda_1}\chi_{\mathrm{h}}^{(3)} A_1^3 \frac{(1 - e^{-i\Delta kL})}{i\Delta k}. \tag{9}$$

The harmonic field consists of two components, one a free wave that propagates with the usual phase velocity at the third-harmonic frequency, and the other a driven wave that propagates with the phase velocity of the fundamental wave. The form of Eq. (9) emphasizes the presence of these two waves at the third harmonic. Because of dispersion in the refractive indexes, these two waves will in general have different phase velocities and will interfere constructively or destructively in different parts of the nonlinear medium. The resulting third-harmonic wave has a phase velocity that is in general different from that of the free wave at the harmonic frequency, with a wave vector that is given by

$$k_h = k_3 - \Delta k/2. \tag{10}$$

Equation (10) shows that the phase velocity of the harmonic wave has its usual value only when the phase velocities of the harmonic and fundamental waves are equal (i.e., $\Delta k = 0$). The process is then said to be phase matched.

The harmonic intensity, related to the field amplitude by

$$I_3 = \frac{cn_3}{8\pi}|A_3|^2, \tag{11}$$

is

$$I_3 = \frac{576\pi^6}{n_3 n_1^3 \lambda_1^2 c^2}|\chi_h^{(3)}|^2 I_1^3 L^2 \frac{\sin^2(\Delta kL/2)}{(\Delta kL/2)^2}. \tag{12}$$

The dependence of the harmonic intensity on the wave vector mismatch for a medium of constant length is illustrated in Fig. 1. The harmonic intensity is a symmetric function about $\Delta k = 0$ with a maximum at $\Delta k = 0$. Equation (8) shows that perfect phase matching ($\Delta k = 0$) requires that the refractive indices at the two frequencies be equal:

$$n_3 = n_1. \tag{13}$$

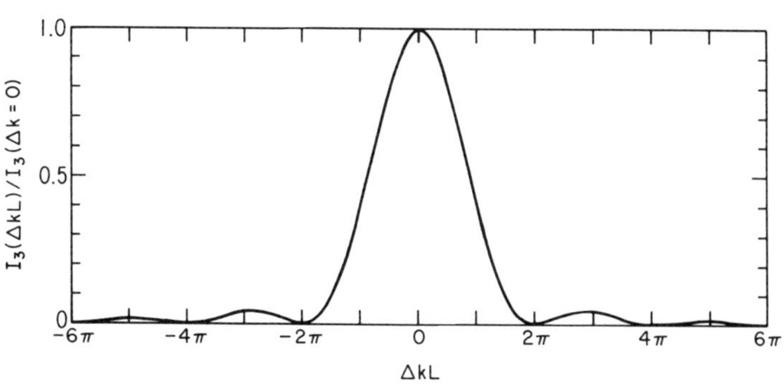

Fig. 1. Normalized third-harmonic intensity as a function of wave vector mismatch.

For finite values of phase mismatch the harmonic intensity decreases from its value at $\Delta k = 0$ and undergoes damped oscillations. This behavior is a result of the fact that when the process is phase mismatched the harmonic field gets out of phase with the polarization that drives it as the waves propagate through the medium. Alternatively it can be viewed as the result of alternate constructive and destructive interference between the free and forced waves, as discussed earlier.

The full width of the phase-matching peak at the point at which the harmonic intensity has dropped to one-half of its maximum value is given by

$$\Delta k_{1/2} = 5.6/L. \tag{14}$$

This value serves as a useful guide for the amount of wave vector mismatch that can be tolerated in a specific material and still allow effective harmonic generation. In practice the wave vector mismatch can be a function of various experimental parameters, such as the pressure of a gas, the wavelength of a pump beam, or the angle between two pump beams (see Section 3.3). Equation (14) can be used as a measure of how accurately these quantities must be chosen.

For values of Δk given by

$$\Delta k = 2n\pi/L, \qquad n = 1, 2, \ldots, \tag{15}$$

the harmonic intensity drops to zero. At these values the shift in phase between the harmonic field and the polarization that drives it has resulted in a complete return of the harmonic energy to the fundamental wave.

When viewed as a function of the length of the medium at a fixed value of Δk, the harmonic intensity behaves as illustrated in Fig. 2. At perfect phase matching the harmonic field maintains a constant phase relation with its

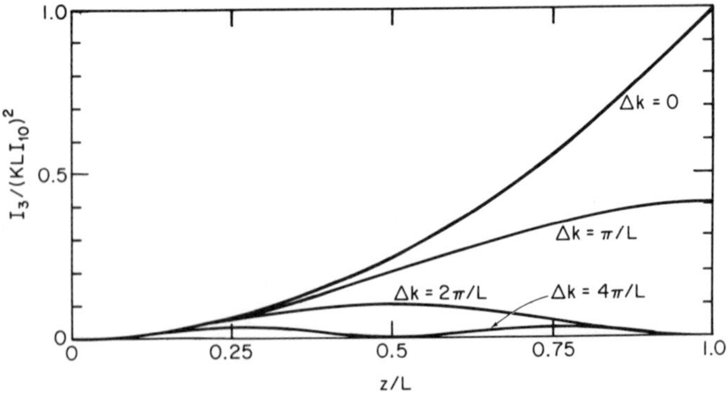

Fig. 2. Normalized third-harmonic intensity as a function of distance for various values of Δk.

driving polarization and the harmonic intensity grows as the square of the interaction length. At nonzero values of Δk, however, the harmonic intensity oscillates as a function of interaction length. Maxima in the harmonic intensity occur at odd multiples of a coherence length L_c, which is given by

$$L_c = \pi/\Delta k. \tag{16}$$

At this distance the harmonic field has slipped in phase by $\pi/2$ rad relative to its driving polarization. If further propagation occurs, then the nonlinear polarization will take power from the harmonic wave and return it to the fundamental until another phase change of $\pi/2$ occurs. For a general value of Δk net harmonic radiation can be generated only from the last pair of coherence lengths, or fraction thereof. The coherence length decreases with increasing Δk, leading to the decreasing amplitude of the oscillations in Fig. 1.

2.1.1b. PUMP DEPLETION. When the interaction is strong enough to convert a significant fraction of the incident radiation to the harmonic, depletion of the pump wave must be accounted for. At perfect phase matching the intensities of the two waves are given by

$$I_3(z) = \frac{I_{10}^3 K^2 z^2}{1 + K^2 I_{10}^2 z^2}, \tag{17a}$$

$$I_1(z) = \frac{I_{10}}{1 + K^2 I_{10}^2 z^2}, \tag{17b}$$

where

$$K = \frac{24\pi^3 |\chi_{\mathrm{h}}^{(3)}|}{n^2 c \lambda_1}, \tag{17c}$$

n is the common refractive index for the two waves, and I_{10} the incident fundamental intensity.

The dependence of the intensities of the two waves on the interaction length is illustrated in Fig. 3. Also shown is the extrapolated harmonic intensity that would be calculated in the absence of pump depletion. From these results it can be seen that it is possible in principle to obtain 100% conversion of the pump wave to its harmonic.

We can define a depletion length for the third harmonic:

$$L_d = \frac{1}{K I_{10}} = \frac{c n^2 \lambda_1}{24\pi^3 |\chi_{\mathrm{h}}^{(3)}| I_{10}}. \tag{18}$$

At this distance one-half of the energy in the incident wave is converted to the harmonic. In a real system efficient conversion is possible as long as the

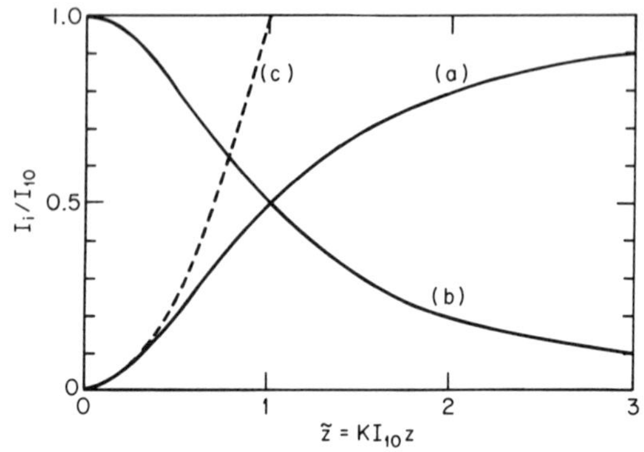

Fig. 3. Ratio of the third-harmonic intensity (a) and fundamental intensity (b) to the incident fundamental intensity as a function of the interaction length for $\Delta k = 0$. (c) Calculated harmonic intensity in the absence of pump depletion.

coherence length of Eq. (16) is longer than the depletion length. For interaction lengths that are less than about one-fifth the depletion length, corresponding to a conversion of about 0.04, the harmonic intensity can be approximated to within about 10% by the weak field expression of Eq. (12).

The form of more general solutions of Eqs. (7a) and (7b) for nonperfect phase matching have been discussed by Armstrong *et al.*[2] From their considerations we can conclude that the conversion efficiency will remain small unless the system is phase matched very well. This aspect is very important determining limitations on conversion efficiency and we shall return to it in later sections.

2.1.2. *Focused Beams*

For many applications the use of focused beams is preferred over the use of collimated beams. In some situations the harmonic power that can be generated in a medium of a given length with a focused beam is significantly greater than that which can be generated with a collimated beam. In addition, in certain materials studies such as measurements of nonlinear susceptibilities, focusing at the center of the cell can discriminate against harmonic generation in the cell windows.

The results of the preceding section were derived for the rather idealized case of an infinite plane wave. For beams of finite dimensions they can be used with only minor modifications if the beam remains effectively collimated throughout the nonlinear interaction region. For beams that focus or diverge within the nonlinear medium, however, the results must be modified significantly. In particular, the optimum phase-matching conditions are changed when the focal depth of the beam becomes comparable to or smaller than the extent of the nonlinear medium.

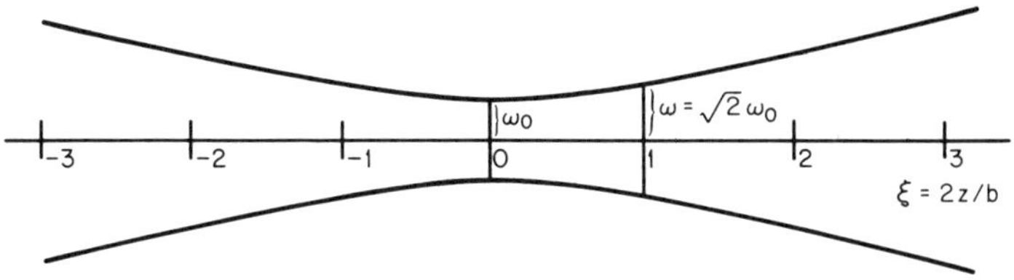

Fig. 4. Parameters of a Gaussian beam near its waist. $A = A_0/\sqrt{1+\xi^2}\,e^{-i\tan^{-1}\xi}$; $\omega(z) = \omega_0\sqrt{1+\xi^2}$.

The description of harmonic generation for a focused beam of arbitrary cross section can be extremely complex. As a result most of the discussions in the literature have been for beams with Gaussian profiles. Third-harmonic conversion with focused Gaussian beams have been discussed by Ward and New[3] and by Puell *et al.*[4] Bjorklund[5] has given an extensive discussion of four-wave sum- and difference-frequency mixing with focused beams. Third-harmonic generation with focused beams that contain higher-order transverse spatial modes has been treated by Yiu *et al.*[6]

In this section we shall discuss third-harmonic conversion of a focused Gaussian beam whose parameters are defined in Fig. 4. Solutions of the coupled equations including pump depletion requires numerical computation. This problem has been discussed by Kildal and Brueck.[7] Here we shall assume that pump depletion can be neglected and that the harmonic conversion process is described entirely by Eq. (5b).

The amplitude of the pump beam has the form

$$A_1(r, z) = \frac{A_{10}}{(1 + i\xi)}\, e^{-k_1 r^2/b(1 + i\xi)}, \tag{19}$$

where $b = 2\pi\omega_0^2 n_1/\lambda_1$ is the confocal parameter of the focused beam, ω_0 the spot size of the beam waist ($1/e$ field radius), and $\xi = 2(z - z_0)/b$, with z_0 being the position of the beam waist. Substitution of Eq. (19) into Eq. (5b) gives

$$\nabla_\perp^2 A_3 + 2ik_3\frac{\partial A_3}{\partial z} = -\frac{\pi\omega_3^2}{c^2}\chi_{\mathrm{h}}^{(3)}\frac{A_{10}^3}{(1 + i\xi)^3}\, e^{-3k_1 r^2/b(1 + i\xi)} e^{-i\Delta k z}. \tag{20}$$

Solutions to equations of the form in (20) have been given by Ward and New,[3] Puell *et al.*[4] and Bjorklund.[5] Kleinman *et al.*[8] have discussed solution of similar equations for second-harmonic generation. Here we shall summarize the results.

If we assume that the wave vector mismatch is much smaller than the wave vector of either the harmonic or fundamental waves,

$$\Delta k \ll k_3, k_1$$

the harmonic wave will have a Gaussian profile with the same confocal parameter and beam waist location as the fundamental. Its amplitude is

$$A_3(r) = i\,\frac{3\pi^2 b}{2n_3\lambda_1}\,\chi_{\rm h}^{(3)}\,\frac{A_{10}^3}{(1+i\xi)}\,e^{-3k_1r^2/b(1+i\xi)}e^{-i\Delta k z_0} \times \int_{-\xi'}^{\xi}\frac{e^{-ib\Delta k\xi''/2}}{(1+i\xi'')^2}\,d\,\xi'', \tag{21}$$

where the medium extends from $-\xi'$ to ξ. The total harmonic power is obtained by integrating the power density over the beam profile:

$$\mathscr{P}_3 = \frac{cn_3}{8\pi}\int 2\pi r|A_3(r)|^2\,dr. \tag{22}$$

For the field amplitude of the form in Eq. (21) the total power is

$$\mathscr{P}_3 = \frac{768\pi^6|\chi_{\rm h}^{(3)}|^2}{n_1n_3\lambda_1^4c^2}\,\mathscr{P}_{10}^3|F_1(\xi',\xi,b\Delta k,b/L)|^2, \tag{23}$$

where

$$|F_1(\xi',\xi,b\Delta k,b/L)|^2 = \left|\int_{-\xi'}^{\xi}\frac{e^{-ib\Delta k\xi''/2}}{(1+i\xi'')^2}\,d\xi''\right|^2. \tag{24}$$

We can understand the terms in Eq. (23) as follows. The harmonic power varies as the cube of the fundamental power just as for plane waves. The phase-matching integral is, however, replaced by the function in Eq. (24). Its form is the same as that of the corresponding expression for plane waves except for the denominator in the integrand, which accounts for the effects of focusing. Detailed behavior of the phase-matching integral has been given by Bjorklund.[5] Here we shall describe its behavior in certain limits.

2.1.2a. COLLIMATED BEAMS OF FINITE EXTENT. When the length of the nonlinear medium is much shorter than the confocal parameter of the pump beam, the beams remain effectively collimated throughout the nonlinear medium. This situation is encountered in a large number of experimental configurations. For this case the magnitude of ξ'' always remains small compared to unity and can be neglected in the denominator of the integrand in Eq. (24). For a beam whose waist is centered in a cell of length

$$L(\xi' = \xi = L/2)$$

the phase-matching integral $|F_1|^2$ is given by

$$|F_1|^2 = \frac{4L^2}{b^2}\,\frac{\sin^2[(\Delta k + 4/b)L/2]}{[(\Delta k + 4/b)L/2]^2}. \tag{25}$$

This result is similar to the phase-matching expression for plane waves in Eq. (12). However, the optimum phase mismatch is no longer at $\Delta k = 0$ but at

$$\Delta k = -4/b, \tag{26}$$

as pointed out by Bjorklund.[5]

This offset of the optimum phase mismatch from zero is characteristic of focused beams except for certain circumstances that will be described in the next section. It is a result of the phase shift that occurs in a focused beam relative to an unfocused plane wave. The on-axis phase increment of a focused Gaussian beam relative to a plane wave is given by

$$\Delta\phi = \phi_{\text{Gauss}} - kz = -\tan^{-1}\xi. \tag{27}$$

The dependence of this phase increment on distance is illustrated in Fig. 5. For large values of ξ the phase increment is almost constant, while near the waist it has the value

$$\Delta\phi = -2(z - z_0)/b.$$

Near the waist, then, the beam behaves as if it has an effective wave vector given by

$$k_{\text{eff}} = k_0 - (2/b),$$

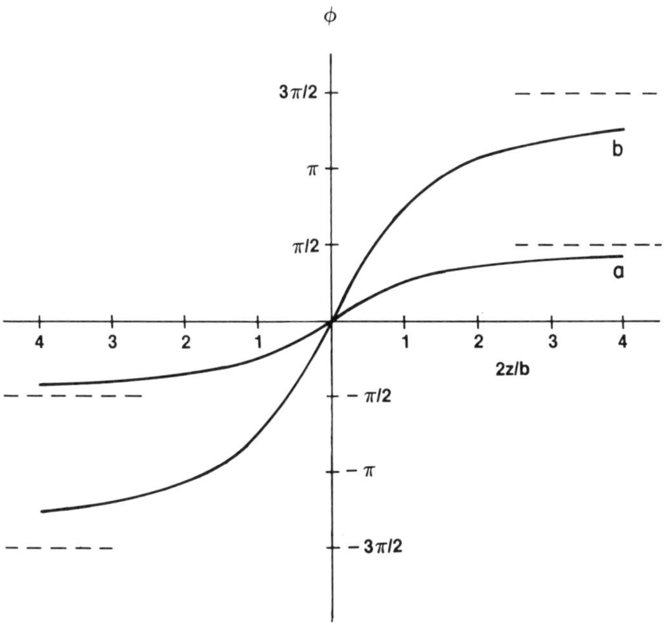

Fig. 5. On-axis phase shift of a focused Gaussian beam (a) and the third-harmonic polarization (b) as a function of distance near the waist.

where k_0 is the wave vector of a plane wave at the same frequency. The nonlinear polarization at the third harmonic depends on the cube of the fundamental field and therefore has a phase increment that is three times as large as that of the field amplitude (see Fig. 5). Near the beam waist the effective phase mismatch is given by

$$\begin{aligned}\Delta k_{\text{eff}} &= k_{\text{field, eff}} - k_{\text{pol, eff}} \\ &= (k_{\text{field, 0}} - 2/b) - (k_{\text{pol, 0}} - 6/b) = \Delta k_0 + 4/b, \end{aligned} \tag{28}$$

which, when the optimum phase-matching condition $\Delta k_{\text{eff}} = 0$ is used, is identical to the result in Eq. (26). The optimum wave vector offset can be expressed in terms of the width of the phase-matching peak [see Eq. (14)] as

$$\Delta k = 0.71\,(L/b)\,\Delta k_{1/2}.$$

When the condition for collimated beams is well satisfied, $b \gg L$ and the offset is much smaller than the width of the phase-matching peak. As a practical consideration it should not disturb the plane wave phase-matching conditions too severely. Nevertheless, observation of the offset has been reported in some experiments.[4]

2.1.2b. Tight Focusing. In the tight-focusing limit the confocal parameter is much smaller than the length of the nonlinear medium. Under this condition the limits on the phase-matching integral in Eq. (24) can be replaced by $\xi' = \xi = \infty$ and the integration can be performed exactly with the result

$$|F_1|^2 = \begin{cases} \pi^2 (b\Delta k)^2 \, e^{b\Delta k}, & \Delta k < 0, \\ 0, & \Delta k > 0. \end{cases} \tag{29}$$

This function is shown in Fig. 6. Harmonic power is generated only for negative values of Δk with the peak occurring at $b\Delta k = -2$. No harmonic power is generated for positive dispersion or for perfect phase matching ($\Delta k = 0$).

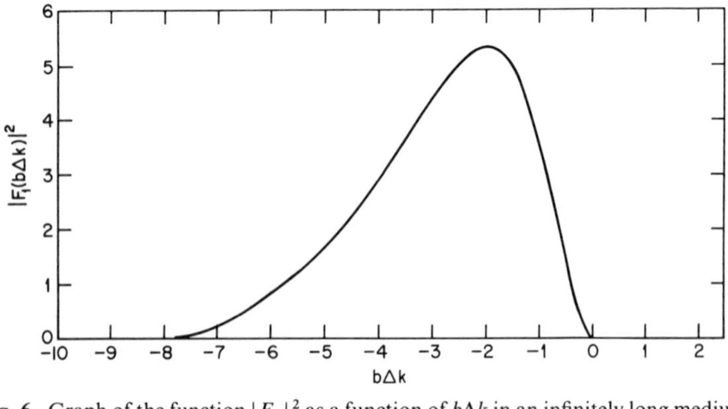

Fig. 6. Graph of the function $|F_1|^2$ as a function of $b\Delta k$ in an infinitely long medium.

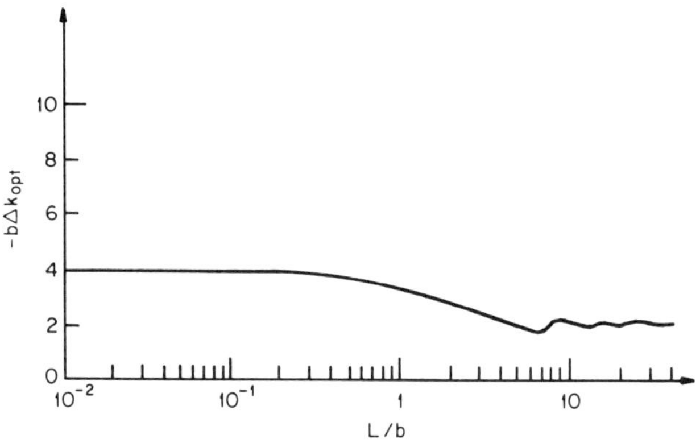

Fig. 7. Optimal value of the phase-mismatch parameter $b\Delta k$ as a function of the focusing parameter L/b for third-harmonic generation. Tight-focused regime is at right, plane-wave regime at left. (Adapted from Ref. 9.)

This behavior also results from the phase shift that the beams undergo as a result of the focus. We consider a dispersionless medium in which the only phase mismatch arises from the effects of focusing. From Fig. 5 we see that the phase shift between the harmonic field and its driving polarization has a value π at the waist relative to its value at $\xi = -\infty$ and a value 2π at $\xi = \infty$. This is the same situation that was encountered for conversion of plane waves when the medium was exactly two coherence lengths long. The harmonic power that is generated in the second half of the medium is out of phase with, and exactly cancels, the power that is generated in the first half. In order to counteract this cancellation, dispersion of the proper sign must be introduced into the medium to offset the phase slippage in the focus. Because the phase slippage is not uniform with distance from the waist, this compensation can only be done in an average sense, and the optimum value of $b\Delta k$ will depend on the extent of the nonlinear medium relative to the confocal parameter. Fischer and Wieczorek[9] have calculated the optimum values of $b\Delta k$ for various harmonic processes for beams that are focused at the center of the nonlinear medium. Their results for third-harmonic generation are shown in Fig. 7.

2.1.2c. Intermediate Focusing. Asymptotic solutions to Eq. (24) for a beam focused at the center of a medium of arbitrary length with boundaries defined by $\xi' = m$ and $\xi = m$ have been given by Akhmanov *et al.*[10] and Fischer and Wieczorek.[9]

The variation of $|F_1(b\Delta k, b/L, f/L)|^2$ for various values of b/L are shown in Fig. 8. As can be expected the behavior of $|F_1|^2$ for finite values of b/L is intermediate between the tight focus regime and the plane-wave regime.

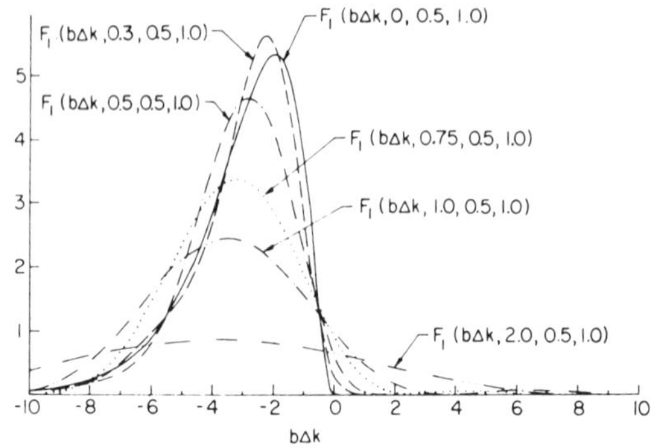

Fig. 8. Variation of the function $|F_1|^2$ as a function of the phase-mismatch parameter $b\Delta k$ for various values of b/L. (From Ref. 5, © 1975 IEEE.)

For nonzero values of b/L finite conversion can be obtained for positive values of the dispersion. The damped oscillations that are evident in the curves for nonzero b/L result from the abrupt termination of the nonlinear medium at $\xi = \pm m$. More extensive discussion of the behavior of the phase-matching integral for focused beams in other limits has been given by Bjorklund.[5]

2.2. Four-Wave Mixing

Let us now turn to the more general processes of four-wave mixing involving two or more different laser frequencies. In this section we shall consider the following processes:

$$\text{I} \qquad \omega_1 + \omega_2 + \omega_3 \rightarrow \omega_4, \tag{30a}$$

$$\text{II} \qquad \omega_1 + \omega_2 - \omega_3 \rightarrow \omega_4, \tag{30b}$$

$$\text{III} \qquad \omega_1 - \omega_2 - \omega_3 \rightarrow \omega_4, \tag{30c}$$

Special cases of these processes can be observed when only two input frequencies are present:

$$\text{IV} \qquad 2\omega_1 + \omega_2 \rightarrow \omega_4, \tag{31a}$$

$$\text{V} \qquad 2\omega_1 - \omega_2 \rightarrow \omega_4. \tag{31b}$$

The concepts of phase matching and focusing that were discussed in the previous section on harmonic generation also apply to the four-wave mixing interactions and the results will be summarized in this section without detailed derivation.

Nonlinear Polarizations. The appropriate nonlinear polarization amplitudes for the processes in Eqs. (30) and (31) are

Process	Polarization amplitude	
I	$P_4 = \frac{3}{2}\chi^{(3)}(-\omega_4, \omega_1, \omega_2, \omega_3)A_1 A_2 A_3,$	(32a)
II	$P_4 = \frac{3}{2}\chi^{(3)}(-\omega_4, \omega_1, \omega_2, -\omega_3)A_1 A_2 A_3^*,$	(32b)
III	$P_4 = \frac{3}{2}\chi^{(3)}(-\omega_4, \omega_1, -\omega_2, -\omega_3)A_1 A_2^* A_3^*,$	(32c)
IV	$P_4 = \frac{3}{4}\chi^{(3)}(-\omega_4, \omega_1, \omega_1, \omega_2)A_1^2 A_2,$	(33a)
V	$P_4 = \frac{3}{4}\chi^{(3)}(-\omega_4, \omega_1, \omega_1, -\omega_2)A_1^2 A_2^*.$	(33b)

2.2.1. *Plane-Wave Solutions*

For incident plane waves the solutions for the intensities of the generated waves for all of these processes are similar to those for third-harmonic generation. Using Eqs. (32) of (33) we obtain from Eq. (3d) in the plane-wave approximation

Process	Intensity	
I, II, III	$I_4 = \dfrac{2304\pi^6 \lvert\chi^{(3)}\rvert^2}{n_1 n_2 n_3 n_4 \lambda_4^2 c^2} I_1 I_2 I_3 \dfrac{L^2 \sin^2(\Delta kL/2)}{(\Delta kL/2)^2},$	(34a)
IV, V	$I_4 = \dfrac{576\pi^6 \lvert\chi^{(3)}\rvert^2}{n_1^2 n_2 n_4 \lambda_4^2 c^2} I_1^2 I_2 \dfrac{L^2 \sin^2(\Delta kL/2)}{(\Delta kL/2)^2}.$	(34b)

The form of these solutions is similar to that for the third-harmonic conversion of plane waves and the discussion of phase-matching properties in Section 2.1.1 applies for these interactions as well. The phase-matching condition $\Delta k = 0$ for each of these processes is given by

Process	Condition	
I	$\dfrac{n_4}{\lambda_4} = \dfrac{n_1}{\lambda_1} + \dfrac{n_2}{\lambda_2} + \dfrac{n_3}{\lambda_3},$	(35a)
II	$\dfrac{n_4}{\lambda_4} = \dfrac{n_1}{\lambda_1} + \dfrac{n_2}{\lambda_2} - \dfrac{n_3}{\lambda_3},$	(35b)
III	$\dfrac{n_4}{\lambda_4} = \dfrac{n_1}{\lambda_1} - \dfrac{n_2}{\lambda_2} - \dfrac{n_3}{\lambda_3},$	(35c)
IV	$\dfrac{n_4}{\lambda_4} = \dfrac{2n_1}{\lambda_1} + \dfrac{n_2}{\lambda_2},$	(35d)
V	$\dfrac{n_4}{\lambda_4} = \dfrac{2n_1}{\lambda_1} - \dfrac{n_2}{\lambda_2}.$	(35e)

2.2.2. *Focused Beams*

A detailed discussion of the behavior of four-wave sum- and difference-frequency mixing with focused beams has been given by Bjorklund.[5] Sum-frequency conversion with focused beams is similar to third-harmonic generation. The sum-frequency power generated by incident beams with identical confocal parameters and waist locations is given by

Process Generated power

I $$\mathscr{P}_4 = \frac{9216\pi^6|\chi^{(3)}|^2}{n_4 c^2 \lambda_1 \lambda_2 \lambda_3 \lambda_4} \frac{k_0}{k'} \mathscr{P}_{10} \mathscr{P}_{20} \mathscr{P}_{30} |F_1(b\Delta k)|^2, \tag{36a}$$

IV $$\mathscr{P}_4 = \frac{2304\pi^6|\chi^{(3)}|^2}{n_4 c^2 \lambda_1^2 \lambda_2 \lambda_4} \frac{k_0}{k'} \mathscr{P}_{10}^2 \mathscr{P}_{20} |F_1(b\Delta k)|^2, \tag{36b}$$

where $k_0 = \omega_4/c = 2\pi/\lambda_4$, $k' = 2k_1 + k_2$ for process IV, $k' = k_1 + k_2 + k_3$ for process I, $\mathscr{P}_{i0}$ is the incident power in the ith wave, $F_1(b\Delta k)$ is the integral given in Eq. (24), and Δk is given for each process by

Process Wave vector mismatch

I $$\Delta k = k_4 - (k_1 + k_2 + k_3), \tag{37a}$$

IV $$\Delta k = k_4 - (2k_1 + k_2). \tag{37b}$$

For these interactions the generated beam has a Gaussian cross section with the same confocal parameter as the incident beams and the discussion of phase optimization in Section 2.1.2b applies.

The difference-frequency mixing processes II, III, and V exhibit somewhat different behavior for focused beams. For process III the generated power is given by[5]

$$\mathscr{P}_4 = \frac{576\pi^2}{c^2} \frac{k_0^4 k_1 k_2 k_3}{k_4^2 k' n_1 n_2 n_3} |\chi^{(3)}|^2 \mathscr{P}_{10} \mathscr{P}_{20} \mathscr{P}_{30} |F_3(b\Delta k, k', k'')|^2, \tag{38}$$

where

$$\left| F_3\left(b\Delta k, \frac{b}{L}, \frac{f}{L}, \frac{k''}{k'}\right)\right|^2 = \frac{2k'}{\pi b} \int_0^\infty 2\pi r\, dr \left| \int_{-\xi'}^{\xi} \frac{e^{-i(b\Delta k \xi''/2)} e^{-r^2/bH(\xi,\xi'')}}{(1 - i\xi'')(k'' - ik'\xi'')H(\xi,\xi'')} d\xi'' \right|^2, \tag{39a}$$

$$H(\xi, \xi'') = \frac{1 + \xi''^2}{(k'' - ik'\xi'')} - \frac{i(\xi'' - \xi)}{k'}, \tag{39b}$$

and $k' = k_1 - k_2 - k_3$, $k'' = k_1 + k_2 + k_3$, $k_0 = \omega_4/c$.

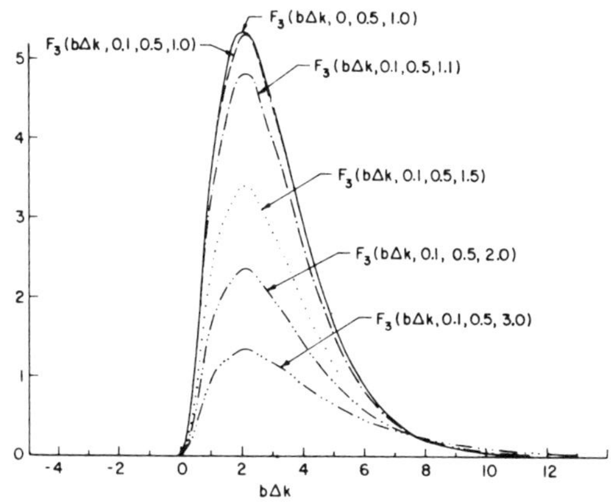

Fig. 9. Variation of the function $|F_3|^2$ as a function of $b\Delta k$ for various values of k''/k' and small values of b/L. (From Ref. 5, © 1975 IEEE.)

The phase-matching integral $|F_3|^2$ is illustrated in Fig. 9 for a tightly focused beam. Its behavior is similar to that of $|F_1|^2$ for the sum-frequency processes (see Fig. 6), except that the sign of the dispersion that is needed for optimum conversion has changed. In this case optimum conversion occurs at a value of $b\Delta k = +2$ and no conversion is obtained for negative dispersion.

The difference-frequency process II and V are special cases in that their phase optimization condition is always $\Delta k = 0$. The power generated at ω_4 for these processes is given by

Process Generated power

II
$$\mathscr{P}_4 = \frac{576\pi^2 n_4}{c^2} \frac{k_0^4 k_1 k_2 k_3}{k_4^2 k' n_1 n_2 n_3} |\chi^{(3)}|^2 \mathscr{P}_{10} \mathscr{P}_{20} \mathscr{P}_{30} |F_2|^2, \tag{40}$$

where $k' = k_1 + k_2 - k_3$ and

V
$$\mathscr{P}_4 = \frac{144\pi^2 n_4}{c^2} \frac{k_0^4 k_1^2 k_2}{k_4^2 k' n_1^2 n_2} |\chi^{(3)}|^2 \mathscr{P}_{10}^2 \mathscr{P}_2 |F_2|^2, \tag{41}$$

where $k' = 2k_1 - k_2$. For these expressions the phase-matching integral is given by

$$\left|F_2\left(b\Delta k, \frac{b}{L}, \frac{f}{L}, \frac{k''}{k'}\right)\right|^2 = \frac{2k'}{\pi b} \int_0^{2\pi} 2\pi r\, dr \left| \int_{-\xi}^{\xi} \frac{e^{-i(b\Delta k \xi''/2)} e^{-r^2/bH}}{(1 + i\xi'')(k'' - ik'\xi'')H} d\xi'' \right|^2, \tag{42}$$

where H is defined in Eq. (39b) and $k'' = k_1 + k_2 + k_3$ for process II and $k'' = 2k_1 + k_2$ for process V.

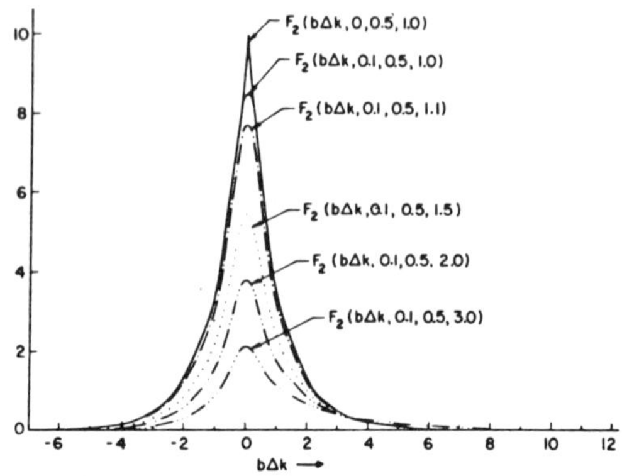

Fig. 10. Variation of the function $|F_2|^2$ as a function of $b\Delta k$ for various values of k''/k' and small values of b/L. (From Ref. 5, © 1975 IEEE.)

The behavior of $|F_2|^2$ for a tightly focused beam is illustrated in Fig. 10. The optimum phase mismatch for these processes is at $\Delta k = 0$ just as for plane waves.

The profile of the beams generated in the difference-frequency processes is in general not Gaussian, but varies from a near-Gaussian profile for small negative values of Δk to an annular ring structure for large positive values of Δk. This behavior is illustrated in Fig. 11 for process V. The development of annular rings is related to the mismatch of the beam waist dimensions and the confocal parameter of the generated beam when the phase mismatch becomes large.

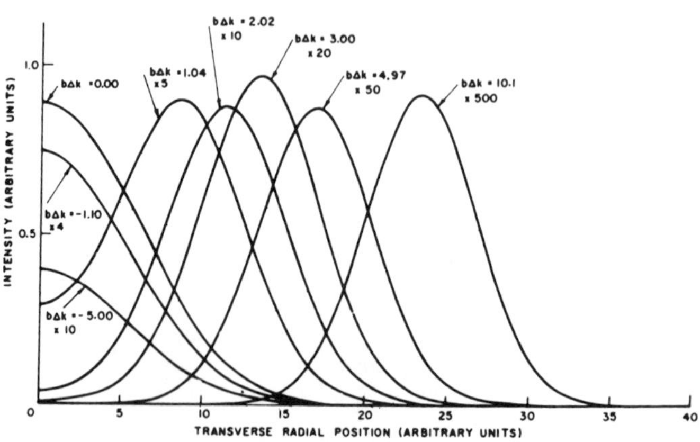

Fig. 11. Far-field profile of a focused beam generated through the process $\omega_4 = \omega_1 + \omega_2 - \omega_3$ with $k''/k' = 3$ and various values of $b\Delta k$. The pump beam was initially Gaussian and was focused at the center of the nonlinear medium with $b/L = 0.1$. (From Ref. 5, © 1975 IEEE.)

3. TECHNIQUES FOR PHASE OPTIMIZATION

In the previous sections we have seen that the effectiveness of the frequency conversion processes depends on the magnitude and sign of the wave vector mismatch. In this section we shall discuss techniques for adjusting the phase mismatch in liquids and gases.

Phase mismatches arise from dispersion in the refractive indexes between the incident and the generated frequencies. In the transparent region of most liquids and in the spectral range below the first excited levels in gases the dispersion is positive, or normal, and the refractive index increases with increasing frequency. This dispersion is exactly the kind that is needed for phase matching the difference frequency process of the type in Eq. (30c). It can also be used under certain circumstances with the difference-frequency processes in Eqs. (30b) and (31b). It cannot be used for the sum-frequency processes or harmonic conversion in an infinite medium, however. Because gases and liquids are isotropic, phase matching cannot be accomplished by using refractive index differences associated with birefringence, as is commonly done with some solids. There are other techniques, however, which can be used for phase matching in gases and liquids in restricted regions of the spectrum.

Dispersion in gases can be controlled by the use of mixtures of components which have different dispersive properties and also by adjusting the pressure of a single-component medium. Both of these techniques affect the wave vector mismatch through the macroscopic refractive indexes. A third technique for phase matching in gases is to select a combination of incident and generated wavelengths so that the dispersion in each atom or molecule is zero. This technique is useful for phase matching those processes in which the optimum value of Δk is zero.

These techniques are not as readily used in liquids where the densities are more or less fixed, although phase matching by using mixtures of dye molecules or dissolved gases in liquids has been reported.[11,12] A more common technique in liquids is to use noncollinear pump beams to control the wave vector mismatch.

3.1. Phase Matching with Gas Mixtures

In order to be able to adjust the wave vector mismatch by using mixtures of two or more gases it is necessary that at least one of them have negative dispersion for the process in question while another have positive dispersion. Negative dispersion can be obtained by taking advantage of the anomalous dispersion that accompanies allowed dipole transitions. This dispersion is illustrated schematically in Fig. 12 for the third-harmonic conversion process.

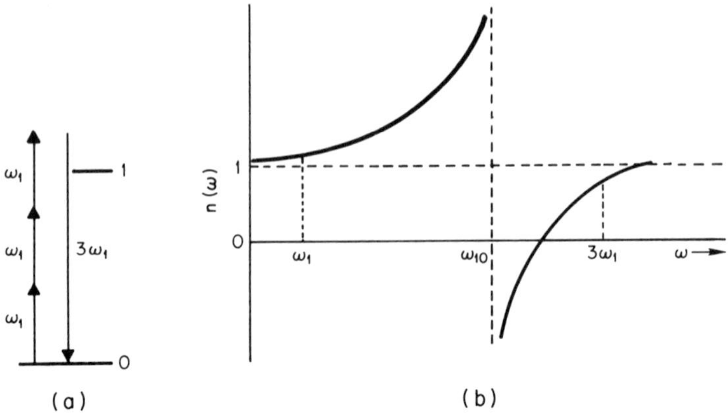

Fig. 12. (a) Energy level diagram showing suitable relation between incident and generated frequencies and energy levels for negative dispersion in third-harmonic generation. (b) Variation of refractive index for level diagram in (a). For the frequencies shown the refractive index at the harmonic is less than that at the fundamental.

If the frequency of the incident wave lies below the upper dispersive level, while that of the generated wave is close to, but above, the upper level, the refractive index at the generated frequency will be less than that at the incident frequency and the wave vector mismatch will be negative.

For third-harmonic generation the wave vector mismatch is given by

$$\Delta k = (6\pi/\lambda_1)(n_3 - n_1),$$

where n_i is the refractive index at ω_i. The refractive indexes for many atomic and molecular gases at frequencies that are not too close to allowed transitions can be represented by a Sellmeier equation

$$n_i - 1 = 2\pi N_i \chi^{(1)}(\omega_i), \tag{43}$$

where $\chi^{(1)}(\omega_i)$ is the linear susceptibility per atom at ω_i given by

$$\chi^{(1)}(\lambda_i) = \frac{r_e}{4\pi^2} \sum_n \frac{f_n}{1/\lambda_n^2 - 1/\lambda_i^2}. \tag{44}$$

Here $r_e = 2.818 \times 10^{-13}$ cm, λ_n is the wavelength of the nth transition in centimetres, f_n is the oscillator strength of the nth transition, and λ_i is the wavelength of the radiation at ω_i in centimeters. The requirement that the dispersion be negative can be expressed as

$$\chi^{(1)}(3\omega) < \chi^{(1)}(\omega).$$

Regions of negative dispersion can usually, although not always, be found for some frequency range above all levels that have allowed transitions to the ground state. In some materials, such as the alkali metals, the oscillator strength of the resonance transitions is very large and extensive regions of negative dispersion exist for generated wavelengths above the first excited state. For other materials, such as the rare gases, the oscillator strengths of the discrete transitions are not as large and negative dispersion will exist only in restricted frequency ranges above the transition frequencies with the largest oscillator strengths.

If a second medium with positive dispersion is introduced, then the value of the total wave vector mismatch can be adjusted by varying the relative concentration of the two components. Positive dispersion will be present in a medium if all the wavelengths lie below the lowest excited level or if the generated wavelength lies close to but below a dispersive level. These situations are shown in Figs. 13a and 13b, respectively. The net wave vector mismatch is given by

$$\Delta k = \frac{12\pi^2}{\lambda_1} N_1 \left[\left(\chi_1^{(1)}(3\omega) + \frac{N_2}{N_1} \chi_2^{(1)}(3\omega) \right) - \left(\chi_1^{(1)}(\omega) + \frac{N_2}{N_1} \chi_2^{(1)}(\omega) \right) \right], \tag{45}$$

where N_i is the concentration of the ith component and the linear susceptibilities may be calculated from the appropriate Sellmeier equation.

Phase matching by gas mixtures is a very versatile technique. It can be used to obtain arbitrary values of positive or negative wave vector mismatches as well as zero wave vector mismatch over a wide range of densities of the nonlinear medium. As a result it can be used with collimated or focused beams for any of the harmonic or sum- or difference-frequency mixing processes that were discussed earlier. Because of the need to use media that have opposite signs of dispersion for the same process, however, the choice of particular materials or nonlinear processes that can be used at particular

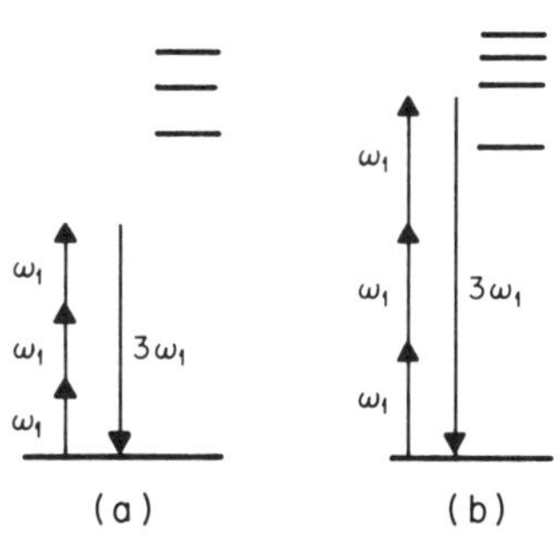

Fig. 13. Energy level diagram showing relation between optical frequencies and energy levels for positive dispersion. (a) All frequencies below all the upper levels. (b) Fundamental frequency below all upper levels; harmonic frequency close to, but below, higher-lying dispersive level.

wavelengths can be restricted. Restrictions can also arise due to limitations on the maximum pressure that can be used because, for example, of the capacity of the experimental system or, in the case of metal vapors, of condensation of the nonlinear medium at high buffer gas pressures. In addition, limitations on conversion efficiency due to incomplete mixing of the gases has also been reported.

Optimization of conversion in gas mixtures is usually done by adjusting the ratio of concentrations of the two components to obtain the desired value of Δk. In many situations only one of the components contributes to the nonlinear interaction, while both components contribute to the dispersion. This is the situation that we shall discuss here. Phase-matching behavior when both components contribute to the nonlinear interaction as well as to the dispersion has been discussed by Hilbig and Wallenstein[13] and Batishche *et al.*[14]

Phase matching of third-harmonic conversion of plane waves requires that the ratio of the concentrations of the components of a two-component mixture be given by

$$\frac{N_2}{N_1} = -\frac{[\chi_a^{(1)}(\omega_1) - \chi_a^{(1)}(3\omega_1)]_1}{[\chi_a^{(1)}(\omega_1) - \chi_a^{(1)}(3\omega_1)]_2}, \tag{46}$$

where $\chi_a^{(1)}$ is the linear susceptibility per atom and the subscripts outside the square brackets refer to the different components. This same relation is approximately valid for collimated beams of finite dimension when the depth of focus is much longer than the length of the nonlinear medium as discussed in Section 2.1.2.

3.1.1. Focused Beams

Optimization of the frequency conversion processes in focused beams requires adjustment of the phase mismatch to maximize the expressions in Eqs. (23), (36a), (36b), (38), (40), or (41), depending on the process involved. When optimization is done in gas mixtures and only one of the components contributes to the nonlinear interaction, the frequency conversion processes are optimized by maximizing the integrals $|F_1|^2$, $|F_2|^2$, or $|F_3|^2$ that are given in Eqs. (24), (42), and (39) according to the discussions in the last section. It should again be emphasized that the optimum value of $b\Delta k$ for each of the interactions considered here depends on the focusing geometry, as was illustrated in Fig. 7 for third-harmonic generation.

The relative concentrations of the two components depends on the magnitude of the linear susceptibility of each component at the wavelengths involved as well as on the geometry. Usually the component that contributes to the nonlinear interaction has a much larger susceptibility per atom than

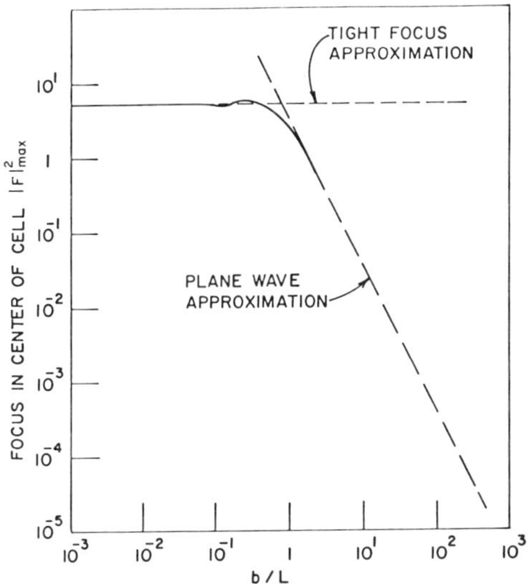

Fig. 14. Variation of the maximum value of $|F_1|^2$ (at the optimal value of $b\Delta k$) with b/L. (From Ref. 30.)

the linear component. As a result the concentration of the linear component is usually much higher. Because of the dependence of the optimum value of Δk on the focusing geometry, the concentration of the linear medium that is required to phase match a collimated beam interaction will usually be larger than that needed to phase match the same interaction with tightly focused beams.

The variation of the peak value of $|F_1|^2$ as a function of b/L for a beam focused at the center of the cell is shown in Fig. 14. In the plane-wave limit, $b/L \gg 1$, the maximum value of $|F_1|^2$ decreases as $1/b^2$ and the harmonic intensity conversion efficiency is independent of b, as expected from Eqs. (25) and (23). In the tight focus limit the peak value of $|F_1|^2$ is independent of b/L. As a result, when the beam is focused with a confocal parameter that is much shorter than the cell length the power in the generated wave is independent of how tightly the beam is focused. An absolute maximum value of $|F_1|^2$ occurs at approximately $b/L = 0.3$ but its peak value here is only about 5% larger than its peak value in the tight focus limit. Thus, for a given total power in the pump waves, optimum conversion can be effectively obtained for any focusing geometry between $b/L = 0$ and $b/L = 0.3$. A choice of a particular geometry involves comparing the effects associated with the higher intensities that occur in the tight-focus limit (see Chapter 4, for example) with any problems that might accompany the higher pressures that are usually needed to phase match collimated interactions.

3.2. Phase Matching by Pressure Adjustment

The magnitude of the wave vector mismatch in a single-component medium can be adjusted by varying its pressure. This technique is experimentally easier than the use of gas mixtures. However, the dispersion of a given medium will usually be either positive or negative for a given nonlinear interaction and a given set of incident and generated frequencies. As a result not all nonlinear interactions can be optimized in a given medium. In addition, zero phase mismatch cannot be achieved except under certain circumstances which will be discussed in Section 3.4.

This technique can be used to optimize harmonic generation and sum- and difference-frequency conversion corresponding to processes I, III, and IV in Eqs. (30) and (31). Since it cannot give zero wave vector mismatch in a general situation, it cannot be used to obtain perfect phase matching in collimated beams or in focused beams for processes II and IV in Eqs. (30) and (31). However, because both the dispersion and the conversion efficiency depend on the density of the medium, a form of optimization can be obtained even for those processes that require zero phase mismatch.

We shall again illustrate the principle by considering third-harmonic generation. The wave vector mismatch is given by

$$\Delta k = \frac{12\pi^2 N}{\lambda_1}[\chi_a^{(1)}(3\omega) - \chi_a^{(1)}(\omega)].$$

The optimization of conversion in a single-component medium is qualitatively similar to optimization in gas mixtures. However, because both the harmonic power and the wave vector mismatch depend on the density, the conditions for optimization are different. Equation (23) can be rewritten to reflect this dependence as

$$\mathscr{P}_3 = \frac{4\pi^4}{3n_1 n_3 c^2}\left|\frac{N\chi_a^{(3)}}{n_1 - n_3}\right|^2 \left(\frac{4\mathscr{P}_{10}}{b\lambda_1}\right)^2 \mathscr{P}_{10}|b\Delta k F_1(b\Delta k)|^2, \tag{47}$$

where $(n_1 - n_3)/N$ is the dispersion per atom, independent of the density N, $\chi_a^{(3)}$ is the nonlinear susceptibility per atom, and the quantity $(4\mathscr{P}_{10}n_1/b\lambda_1)$ is the peak on-axis intensity of the pump beam at the focus. All of the dependence of the harmonic power on the density and the wave vector mismatch is now contained in the quantity $|b\Delta k F_1|^2$.

Bjorklund[5] has given extensive calculations of this quantity for various focusing geometries. Curves for tight-focusing geometries are reproduced for processes I, II, and III in Fig. 15. For sum-frequency processes the phase optimization still occurs for negative dispersion with the peak value occurring at a pressure corresponding to the condition $b\Delta k = -4$. Similarly the difference-frequency process III has an optimum for positive Δk at the condition $b\Delta k = 4$ for tight focusing and values of k''/k' near unity [see Eq. (39)].

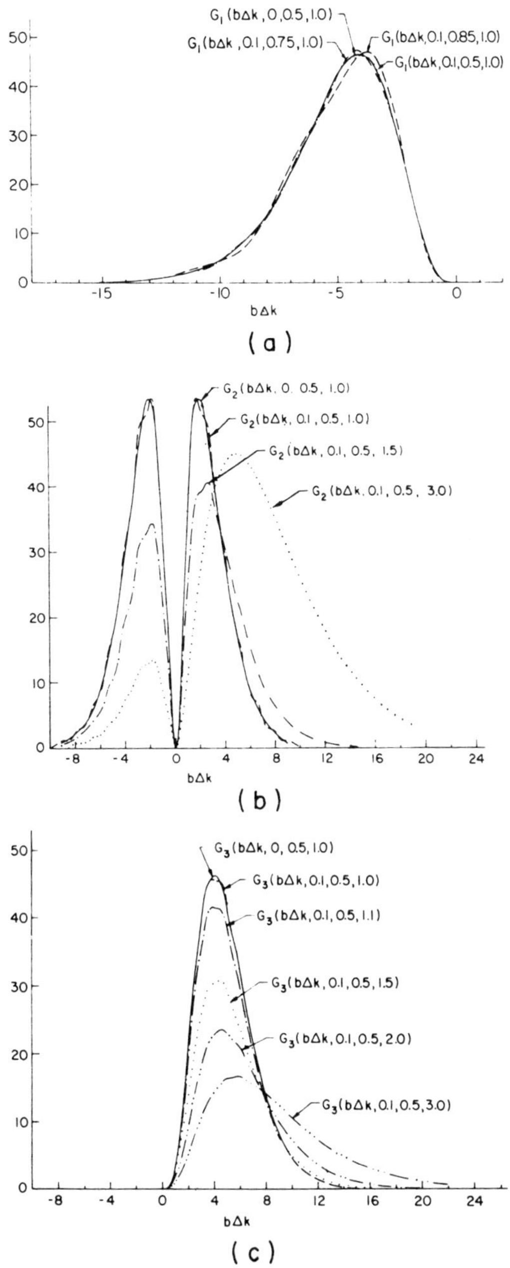

Fig. 15. Variation of $|b\Delta k F_1|^2$ (a), $|b\Delta k F_2|^2$ (b), and $|b\Delta k F_3|^2$ (c), with phase-mismatch parameter $b\Delta k$ for small values of b/L and various values of k''/k'. (From Ref. 5, © 1975 IEEE.)

For process II or IV the maximum occurs for either positive or negative dispersion at a value of $b\Delta k$ near ± 2.5, with the exact value depending on k''/k' [see Eqs. (40)–(42)]. For each of these interactions optimal conversion occurs at only one pressure for a fixed value of b. For optimal conversion at a given pressure the value of b must be adjusted to satisfy the given conditions.

The maximum conversion in a single-component medium in the tight focus limit increases monotonically as the beam is focused more tightly, in contrast to the behavior encountered in phase-matched mixtures. This is a result of the fact that the optimization of the conversion occurs at progressively higher pressures as the beam is focused more tightly.

3.3. Noncollinear Phase Matching

Noncollinear phase matching can be used with both liquids and gases. It has been used extensively for phase-matching difference-frequency mixing in liquids. Noncollinear phase matching makes use of the fact that the wave vector mismatch is really a vector relation:

$$\Delta\vec{k} = \vec{k}_{\text{pol}} - \vec{k}_{\text{field}}.$$

This relation is illustrated in Fig. 16 for the difference-frequency process $\omega_4 = 2\omega_1 - \omega_2$, for which the wave vector mismatch is given by

$$\Delta\vec{k} = \vec{k}_4 - (2\vec{k}_1 - \vec{k}_2). \tag{48}$$

Satisfaction of the plane-wave phase-matching condition $\Delta\vec{k} = 0$ requires that the wave vector diagram close, as illustrated in Fig. 16b. The difference-frequency process illustrated in these figures can be phase matched in media with positive dispersion, making noncollinear phase matching appropriate for use in transparent liquids. In the small angle limit the angles that are

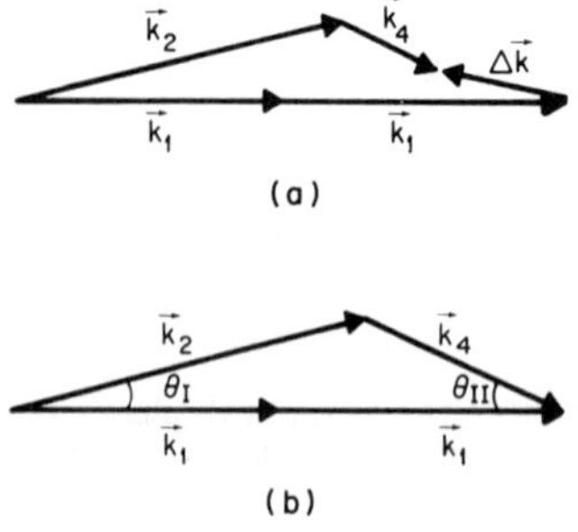

Fig. 16. K-vector diagrams for noncollinear phase matching of the process $\omega_4 = 2\omega_1 - \omega_2$. (a) General case with a nonzero phase mismatch $\Delta\vec{k}$. (b) Perfect phase matching in medium with positive dispersion.

defined in Fig. 16b are given in terms of the refractive indexes by the relations

$$\theta_{\text{I}} = \left\{\frac{(n_4/\lambda_4)^2 - [(2n_1/\lambda_1) - (n_2/\lambda_2)]^2}{2n_1 n_2/(\lambda_1 \lambda_2)}\right\}^{1/2}, \tag{49a}$$

$$\theta_{\text{II}} = \left\{\frac{(n_2/\lambda_2)^2 - [(2n_1/\lambda_1) - (n_4/\lambda_4]^2}{2n_1 n_4/(\lambda_1 \lambda_4)}\right\}^{1/2}. \tag{49b}$$

This noncollinear phase matching is responsible for the rings that are observed in anti-Stokes–Raman scattering and plays an important role in various types of nonlinear spectroscopy that will be discussed in Chapter 6.

Noncollinear phase matching can also be used to optimize the efficiency of sum-frequency processes in media with negative dispersion. This type of geometry is illustrated in Fig. 17 for the process $\omega_4 = 2\omega_1 + \omega_2$. The appropriate angles are now given in the small angle limit by

$$\theta_{\text{I}} = \left\{\frac{[(2n_1/\lambda_1) + (n_2/\lambda_2)]^2 - (n_4/\lambda_4)^2}{2n_1 n_2/(\lambda_1 \lambda_2)}\right\}^{1/2}, \tag{50a}$$

$$\theta_{\text{II}} = \left\{\frac{(n_2/\lambda_2)^2 - [(2n_1/\lambda_1) - (n_4/\lambda_4)]^2}{2n_1 n_4/(\lambda_1 \lambda_4)}\right\}^{1/2}. \tag{50b}$$

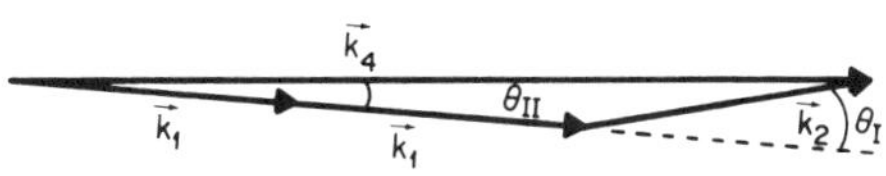

Fig. 17. *K*-vector diagram for noncollinear phase matching of sum-frequency process $\omega_4 = 2\omega_1 + \omega_2$ in a negatively dispersive medium.

3.4. Phase Matching by Frequency Selection

Another technique that can be used to obtain perfect phase matching in atomic and molecular gases for sum-frequency processes and harmonic generation and in liquids for difference-frequency interactions is to choose a combination of incident and generated frequencies such that there is zero phase mismatch in each atom or molecule in a collinear geometry. This technique was first suggested by Armstrong *et al.*[2] and was later used experimentally by Bjorklund *et al.*[15,16]

A suitable combination of wavelengths arranged about a dispersive resonance for phase-matching third-harmonic conversion by this technique is illustrated in Fig. 18a. In order for the third-harmonic process to be phase matchable with this technique it is necessary for the generated frequency to lie between two dispersive resonances. It is also evident that for a given medium phase matching of third-harmonic conversion can be accomplished only for a single input wavelength for each pair of suitable resonances.

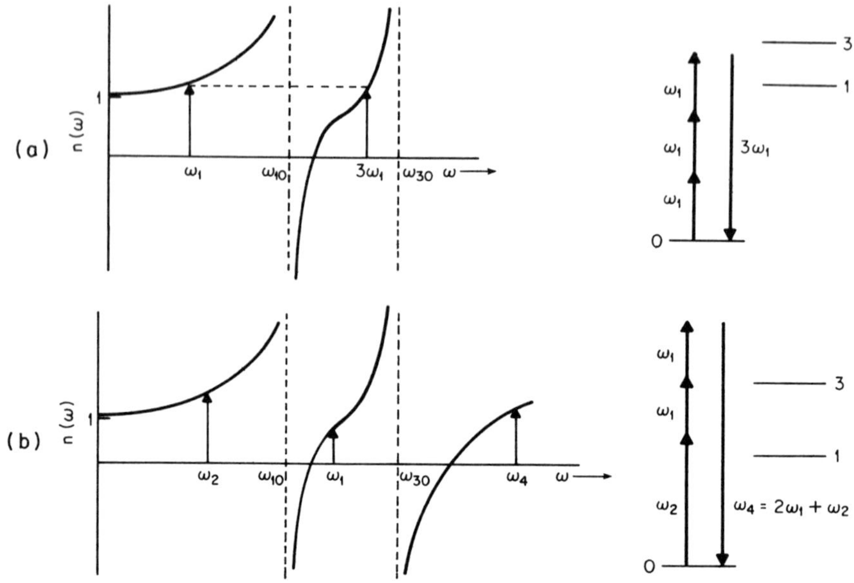

Fig. 18. Phase matching by frequency selection. (a) Third-harmonic generation. (b) Sum-frequency process $\omega_4 = 2\omega_1 + \omega_2$. In each case the pump and generated frequencies are chosen to satisfy the plane-wave phase-matching condition ($\Delta k = 0$) of Eq. (13) or (35d).

More flexibility can be obtained if two input frequencies are used for the process $\omega_4 = 2\omega_1 + \omega_2$. This situation is shown in Fig. 18b. If the two input frequencies are chosen to straddle a strong dispersive resonance, then sufficient flexibility exists to allow phase matching at several output frequencies providing both ω_1 and ω_2 are tuned. Alternatively a fixed output frequency can be produced with the frequencies ω_1 and ω_2 chosen to satisfy a two-photon resonance condition simultaneously with the phase-matching condition. If three incident frequencies are used, then conditions can be chosen to provide simultaneous two-photon resonance and phase matching while allowing the output frequency to be tuned.

For a general medium the combination of incident frequencies that are required to generate a specific sum frequency can be in widely different parts of the spectrum, e.g., in the visible and mid-infrared. In some cases this can make the technique difficult to work with. Nevertheless it has certain advantages when it can be used. For example, it overcomes one of the major limitations that is encountered in phase matching by mixtures: that of incomplete mixing of the components, which prevents the phase-matching conditions from being maintained throughout the nonlinear medium. With the technique of frequency selection phase matching can be maintained throughout the medium regardless of density variations, thereby increasing the potential conversion efficiency.

4. NONLINEAR SUSCEPTIBILITIES

Expressions for the nonlinear susceptibilities that describe the processes discussed in this chapter within the quasistatic approximation can be derived by the techniques described at the end of Chapter 1. Expressions for third-order susceptibilities for four-wave mixing interactions with pump waves at ω_1, ω_2, and ω_3 have been given by Armstrong *et al.*[2] and Butcher[17] under conditions[5] in which the damping terms associated with resonances can be ignored. Puell and Vidal[18] have given derivations for steady-state susceptibilities for various nonlinear interactions up to order 4 for both resonant and nonresonant conditions under the assumption that all the diagonal elements of the density matrix had the same relaxation rate. They also treated transient interactions. Bloembergen *et al.*[19] have given more complete expressions for the steady-state third-order susceptibility for four-wave mixing, which include correction factors due to damping terms. These correction terms are important when collisional damping is significant and a resonance exists between two excited states. Flytzanis[20] has given general expressions for third-order susceptibilities that include the contribution of nuclear motions and Raman modes. Third-order nonlinear susceptibilities have also been discussed by Yuratich and Hanna[21] and by Hanna *et al.*[22]

Three of the more widely studied interactions are third-harmonic generation and four-wave mixing of the form $\omega_4 = 2\omega_1 \pm \omega_2$. We give here explicit expressions for the steady-state susceptibilities for these processes for a system with isolated quantum levels. In these expressions we shall neglect the correction terms of Bloembergen *et al.*[19] and shall assume, following Puell and Vidal,[18] that there is only one damping constant for the diagonal matrix elements. We shall also write the susceptibilities with complex frequencies of the form $\omega_{ab} = \omega_{ab}^0 - i\Gamma_{ab}$ where ω_{ab}^0 is the frequency of the transition between levels a and b and Γ_{ab} is the damping rate for the transition. Under this assumption $\omega_{ba} = -(\omega_{ab})^*$.

With these assumptions and approximations the steady-state susceptibility per atom for third-harmonic generation is

$$\chi_a^{(3)}(-3\omega, \omega, \omega, \omega) = \frac{(ea_0)^4}{(\hbar c)^3} \sum_{i,j,k} r_{0i} r_{ij} r_{jk} r_{k0} \times \left\{ \frac{1}{(\omega_{i0} - \omega)(\omega_{j0} - 2\omega)} \left[\frac{1}{\omega_{k0} - 3\omega} + \frac{1}{\omega_{k0}^* + \omega} \right] + \frac{1}{(\omega_{i0}^* + \omega)(\omega_{j0}^* + 2\omega)} \left[\frac{1}{\omega_{k0}^* + 3\omega} + \frac{1}{\omega_{i0} - \omega} \right] \right\}. \tag{51}$$

For the four-wave mixing interactions $\omega_4 = 2\omega_1 \pm \omega_2$ the corresponding expressions are

$$\chi^{(3)}(-\omega_4, \omega_1, \omega_1 \pm \omega_2) = \frac{1}{3}\frac{(ea_0)^4}{(\hbar c)^3} \sum r_{0i} r_{ij} r_{jk} r_{k0}$$

$$\times \Bigg[\Bigg(\frac{1}{[\omega_{i0} \mp \omega_2][\omega_{j0} - (\omega_1 \pm \omega_2)]}$$

$$+ \frac{1}{[\omega_{i0} - \omega_1][\omega_{j0} - (\omega_1 \pm \omega_2)]}\Bigg)$$

$$\times \left(\frac{1}{\omega_{k0} - \omega_4} + \frac{1}{\omega_{k0}^* + \omega_1}\right)$$

$$+ \frac{1}{[\omega_{i0} - \omega_1][\omega_{j0} - 2\omega_1]}\Bigg(\frac{1}{\omega_{k0} - \omega_4}$$

$$+ \frac{1}{\omega_{k0}^* \pm \omega_2}\Bigg) + \Bigg(\frac{1}{[\omega_{i0}^* \pm \omega_2][\omega_{j0}^* + (\omega_1 \pm \omega_2)]}$$

$$+ \frac{1}{[\omega_{i0}^* + \omega_1][\omega_{j0}^* + (\omega_1 \pm \omega_2)]}\Bigg)$$

$$\times \left(\frac{1}{\omega_{k0}^* + \omega_4} + \frac{1}{\omega_{k0} - \omega_1}\right)$$

$$+ \frac{1}{(\omega_{i0}^* + \omega_1)(\omega_{j0}^* + 2\omega_1)}$$

$$\times \left(\frac{1}{\omega_{k0}^* + \omega_4} + \frac{1}{\omega_{k0}^* \pm \omega_2}\right)\Bigg]. \qquad (52)$$

In these expressions the denominators contain angular frequencies in units of cm^{-1} and the matrix elements of the distance operator r_{ij} are in units of the Bohr radius. The intermediate sums are understood to be over all states of appropriate parity and should be replaced with an integral when intermediate transitions to a continuum are important. The expression for the third-harmonic susceptibility can be seen to be a generalization of the expression given in Eq. (62). Similar expressions for third-order susceptibilities that describe the nonlinear refractive index will be given in Chapter 4.

These expressions can be used to calculate the susceptibilities for materials for which the appropriate matrix elements are known. Such materials include the alkali and many of the alkaline earth metal vapors, some of the rare gases, as well as other selected metal vapors and some molecular gases. First principle calculations of this type generally cannot be made for liquids. The magnitude of these susceptibilities are usually obtained from experimental measurements. Numerical values of third-order susceptibilities for electronic and Raman interactions in some liquids have been given by Levenson and Bloembergen.[23]

In order to calculate the susceptibility from the expressions in Eqs. (51) or (52) it is necessary to know both the magnitude and the sign of the transition dipole matrix elements of all the transitions in the medium. These can in principle be calculated from the overlap integral of the radial wave functions of the states that are involved if the wave functions are known. Alternatively, the magnitude of the integral can be obtained from tables of oscillator strengths such as those given for certain materials by Stone,[24] Heavens,[25] Anderson and Zilitis[26] or Wiese *et al.*[27] The sign of the matrix elements can be determined from the tables of Bates and Daamgard[28] and the transition frequencies from tables such as those of Moore.[29] The techniques used by Bates and Daamgard to determine the sign of the matrix elements are approximations that were derived for light atoms. Although no serious discrepancies have been reported when they are used for heavy atoms, they should be used with caution for calculation of the matrix elments of materials such as the heavier rare gases.

The magnitude of the radial matrix element r_{ab} can be related to the oscillator strength of the appropriate transition. For linearly polarized light only the z component of the radial matrix element is needed. For transitions between states with the designations a, J, M_J and $b, J+1, M_{J+1}$ the z component of r_{ab} in atomic units is given by

$$|\langle b, J+1, M_J|z|a, J, M_J\rangle| = \left[\frac{3[(J+1)^2 - M_J^2]gf_{ab}}{(J+1)(2J+1)(2J+3)\,\Delta E}\right]^{1/2}, \quad (53a)$$

where f_{ab} is the absorption oscillator strength between states a and b, ΔE is the energy between the states in rydbergs, and g is the degeneracy of the lower state. For transitions between states designated a, J, M_J and b, J, M_J the z component of r_{ab} is

$$|\langle b, J, M_J|z|a, J, M_J\rangle| = \left[\frac{3M_J^2 gf_{ab}}{J(J+1)(2J+1)\,\Delta E}\right]^{1/2}. \quad (53b)$$

When circularly polarized light is used the x and y matrix elements are needed. These are given by

$$|\langle b, J+1, M_J \pm 1|x \pm iy|a, J, M_J\rangle| = \left[\frac{3(J \pm M_J + 1)(J \pm M_J + 2)g f_{ab}}{2(J+1)(2J+1)(2J+3)\,\Delta E}\right]^{1/2}, \tag{53c}$$

$$|\langle b, J, M_J \pm 1|x \pm iy|a, J, M_J\rangle| = \left[\frac{3(J \mp M_J)(J \pm M_J + 1)g f_{ab}}{2J(J+1)(2J+1)\,\Delta E}\right]^{1/2} \tag{53d}$$

for transitions between states

$$(b, J+1, M_J \pm 1) \leftrightarrow (a, J, M_J)$$

and

$$(b, J, M_J \pm 1) \leftrightarrow (a, J, M_J),$$

respectively. In these formulas g is again the degeneracy of the lower state.

Miles and Harris[30] have given similar expressions for the z component of the radial matrix element for transitions in alkali metal vapors when spin–orbit splitting can be neglected. Eicher[31] has given expressions for the matrix elements of the individual levels of the spin–orbit manifolds in the alkali metal vapors.

Miles and Harris[30] have tabulated the matrix elements for many of the transitions of the alkali metal vapors without spin–orbit coupling using the matrix elements of Anderson and Zilitis.[26] They have also calculated the third-harmonic susceptibilities for the alkali metal vapors for pump wavelengths between 2 and 0.6 μm. Yuratich and Hanna[21] and Eicher[31] have given calculations that included the effects of spin–orbit splitting. Eicher[31] has calculated values for the matrix elements of the alkali metal vapors independently and has calculated third-harmonic susceptibilities for the fundamental and harmonic wavelengths of the Nd and I_2 lasers for these materials. His matrix elements are between 1.5 and 2 times larger than those of Miles and Harris[30] and his susceptibilities are larger by about a factor 1.75. Eicher's values appear to be in better agreement with the experimental results on third-harmonic generation reported by Puell *et al.*[4]

Matrix elements between bound and continuum states are not as fully tabulated as are those for transitions between bound states. These transitions are important when the generated wavelength falls in the ionization continuum of the nonlinear medium and the bound–free transition provides the coupling to the ground state in the final step. They can also be expected to be important as intermediate states in materials such as the rare gases for which the continuum states carry a significant fraction of the oscillator strength. In other materials, such as the alkali metal vapors, the continuum contribution should be negligible unless the generated wavelength falls in the ionization continuum.

Appropriate oscillator strengths for transitions from the ground state to the continuum can be determined from photoionization cross sections. These are available, for example, for the alkali metal vapors and the rare gases.[32,33] Direct calculation of the appropriate matrix elements to continuum states of Mg has been reported.[34,35] Calculation of coupling to autoionizing levels, which is important in the generation of radiation that lies in the ionization continuum of many electron atoms, can be done using the theory of Fano or multichannel quantum defect theory. These types of calculations have been used to analyze the line shapes that are associated with the generation of radiation near autoionizing resonances.[36–39]

In some situations the nonlinear susceptibility is dominated by the contribution of only one or a few levels and the calculation is simplified considerably. Examples of such situations are if the oscillator strength is concentrated in one or a few levels, such as in alkali metal vapors, or if a particular level is resonant with an incident frequency or a multiple or sum or difference combination of incident frequencies. In other situations the entire intermediate summation must be retained. In such cases strong cancellations can

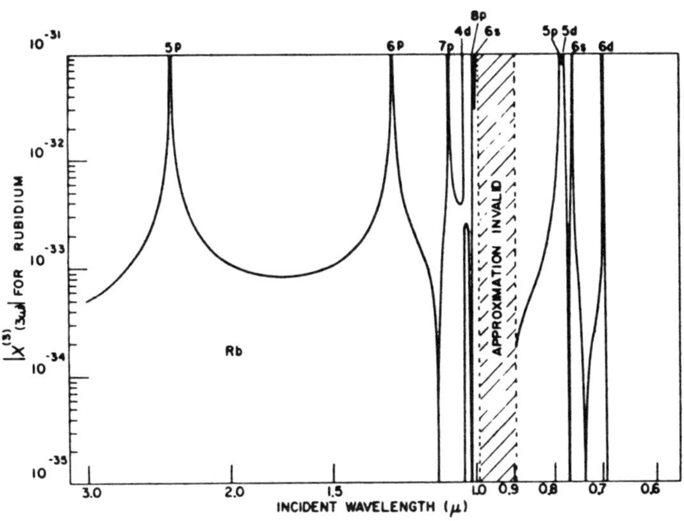

Fig. 19. Calculated value of the magnitude of the third-harmonic susceptibility of Rb as a function of the fundamental wavelength. The numbers above the state designations indicate the number of photons that are involved in the resonance. (From Ref. 30, © 1973 IEEE.)

occur between various terms because of the variation in sign of the appropriate matrix elements. These cancellations can lead to very small values of the susceptibilities even if the frequencies lie near resonant levels. Such cancellations are evident in the third-harmonic susceptibility of Rb calculated by Miles and Harris,[30] which is shown in Fig. 19.

5. RESONANT ENHANCEMENT

The magnitude of the susceptibility can be increased in value, sometimes by as much as eight orders of magnitude, if the incident frequencies or their sum or difference combinations fall near appropriate resonances in the medium. Suitable resonances can arise from electronic transitions in atomic or molecular vapors or from molecular vibrations in molecular gases or liquids. Resonant enhancement is very important in optical-frequency conversion because the large nonlinear susceptibilities that it provides allows many interactions to be observed with relatively low power tunable lasers. In addition, as will be discussed in Chapter 4, the presence of resonance enhancement determines to a large extent the ultimate efficiency that can be obtained for a particular interaction in a given nonlinear medium. In this section we shall examine qualitatively the properties of resonances that can occur in the nonlinear interactions.

We can illustrate the effects of resonant enhancement by considering four-wave sum-frequency mixing in a four-level system with incident frequencies ω_1, ω_2, and ω_3, as shown in Fig. 20. Here frequency ω_1 is assumed to be resonant with level (1), the combination $\omega_1 + \omega_2$ is assumed to fall near level (2), and the generated frequency $\omega_4 = \omega_1 + \omega_2 + \omega_3$ is assumed to fall near, but above, level (3) to provide the negative dispersion needed for phase matching. The ground level (0) is assumed to be coupled to levels (1) and (3) by dipole transition moments μ_{10} and μ_{30}, respectively. Level (2) is coupled to levels (1) and (3) by the dipole moments μ_{21} and μ_{23} but the level pairs (0), (2) and (1), (3) are not connected by dipole transitions.

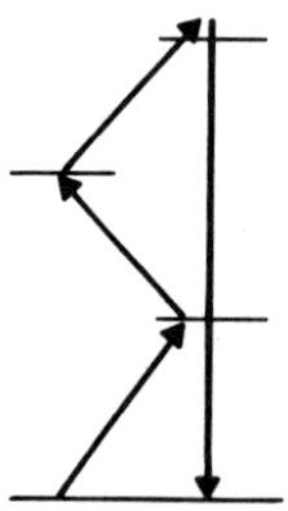

Fig. 20. Level diagram showing states involved in resonant enhancement of sum-frequency process $\omega_4 = \omega_1 + \omega_2 + \omega_3$.

The nonlinear susceptibility for this process can be written as the sum of a resonant and nonresonant term:

$$\chi^{(3)}(-\omega_4, \omega_1, \omega_2, \omega_3) = \chi_{\mathrm{R}}^{(3)} + \chi_{\mathrm{NR}}^{(3)}. \tag{54}$$

Here $\chi_{\mathrm{R}}^{(3)}$ is the part of the expression similar to those in Eq. (51) or (52), which carry frequency denominators of the form $(\omega_{10} - \omega_1)$, $(\omega_{20} - \omega_1 - \omega_2)$, or $(\omega_{30} - \omega_4)$. The nonresonant part makes up the rest of the susceptibility. Enhancement of the susceptibility due to one-, two-, and three-photon resonances is evident in the calculations shown in Fig. 19.

5.1. Single-Photon Resonances

Single-photon resonances occur when the incident frequency matches that of an allowed dipole transition from the ground state. In the example of Fig. 20 a single-photon resonance occurs with ω_1 but in a general situation it could involve ω_2 or ω_3 as well. The selection rules for single-photon resonances are the same as those for single-photon dipole transitions. These are usually given as follows. The upper state must have opposite parity to that of the ground state. In addition, for media in which the total angular momentum J is a good quantum number, single-photon resonances involve transitions with $\Delta J = 0$ or ± 1, with the exclusion of transitions in which J goes from 0 to 0. The magnetic quantum number obeys the rules $\Delta M_J = 0$, ± 1, with the restriction that M_J cannot go from 0 to 0 when $\Delta J = 0$. In media in which L–S coupling is obeyed and single-photon resonances involve transitions with $\Delta S = 0$, $\Delta L = 0$, ± 1, again with the restriction that transitions with $\Delta L = 0 \leftrightarrow 0$ are not allowed.

There are certain limitations on the use of single-photon resonances for enhancing the nonlinear susceptibility, and as a result they are not used as commonly as are two- and three-photon resonances. First, because they are accompanied by strong absorption of the incident light the resonant denominator $(\omega_{10} - \omega_1)$ cannot be made too small or the pump radiation will be lost. Indeed, our treatment of frequency conversion has neglected pump absorption and will be invalid if the absorption is significant (see Chapter 4).

A second property associated with single-photon resonances is their effect on the phase-matching conditions. The variation of the dispersion near a single-photon resonance has proved advantageous in phase matching by frequency selection, as discussed in Section 3.4. However, in a general situation the decrease in coherence length that accompanies a reduction in the single-photon detuning can make maintenance of phase-matching conditions difficult and can prevent the detuning from becoming too small.

Sorokin *et al.*[40] have pointed out that in some situations the smaller detunings that can be used with single-photon transitions that violate one or more of the selection rules given above can make such "forbidden" transitions as effective as the allowed transitions, which have a minimum useful detuning, in enhancing the nonlinear susceptibility.

5.2. Three-Photon Resonances

Three-photon resonances involve a combination of three incident frequencies. In the example in Fig. 20 all three frequencies were different, but in general two or three of them can be equal. Difference-frequency combinations are also allowed. Three-photon resonances are similar in many respects to single-photon resonances. Like the one-photon resonances they involve states of opposite parity. For media in which the total angular momentum is a good quantum number, they can occur for transitions that involve angular momentum changes of $\Delta J = 0, \pm 1, \pm 2$ or ± 3. For media described by L–S coupling the restrictions are $\Delta L = 0, \pm 1, \pm 2$ or ± 3 and $\Delta S = 0$. For the third-order processes being discussed here, however, the resonant state must also be connected to the ground state by a single-photon transition and so the selection rules for the one- and three-photon resonances are the same. Higher-order processes that will be discussed in the next chapter can involve the more general selection rules for the three-photon resonances.

Three-photon resonances are very important for sum-frequency generation and harmonic conversion. Because of the requirement of negative dispersion for phase matching for these processes there will usually be some degree of three-photon resonant enhancement for these interactions. In addition, the existence of a three-photon resonance can determine to a large extent the materials in which these interactions can be observed. Negative dispersion is not a requirement for difference-frequency mixing and these interactions do not usually involve three-photon resonances unless special efforts are made.

Just as with single-photon resonances, the degree of resonant enhancement that can be obtained with three-photon resonances is limited by the absorption of the generated radiation or the decrease in coherence length that accompanies a reduction of the three-photon detuning. These effects will be treated in Chapter 4.

5.3. Two-Photon Resonances

Two-photon resonances involve sum or difference combinations of two input frequencies. Again, in addition to the resonance shown in Fig. 20 resonances with two equal frequencies are possible. Two-photon resonances occur between states of the same parity and when J is a good quantum number, involve transitions with $\Delta J = 0, \pm 1$ or ± 2. Additionally, when L–S coupling is appropriate, the selection rules $\Delta L = 0, \pm 1, \pm 2$ and $\Delta S = 0$ apply. These resonances have a very important place in the study of frequency conversion interactions. The two-photon absorption that accompanies them is generally much weaker than the absorption that accompanies one- or three-photon absorption. As a result much larger enhancements of the susceptibility can be obtained without loss of optical power. However, at sufficiently high powers two-photon absorption can become significant and limit the conversion efficiency as will be discussed in Chapter 4. Secondly, two-photon resonances do not involve levels that have allowed transitions with the ground state, and as a result they do not affect the phase-matching conditions. Thus all phase optimization adjustments can be made independently of the detuning from the two-photon resonance.

Two-photon sum-frequency resonances of the type illustrated in Fig. 20, and also those that involve the frequency combination $2\omega_1$, have been used primarily in four-wave mixing experiments in atomic and molecular vapors. The large enhancements that are attainable in the susceptibility have made possible effective conversion of tunable radiation from low-power dye lasers to provide tunable radiation ranging from the infrared to the vacuum ultraviolet (VUV). Two-photon resonances involving difference-frequency combinations, e.g., $\omega_1 - \omega_2$, have been used to enhance the generation of radiation in the visible and infrared. The difference-frequency resonance also corresponds to a Raman resonance in molecules and is used extensively in areas such as coherent anti-Stokes–Raman scattering (see Chapter 6).

5.4. Resonances between Excited Levels

The enhancements discussed in the preceding sections have all involved resonances between the ground state and an excited state. Bloembergen *et al.*[19] have shown that it is possible for resonant enhancement to occur if the incident frequencies or their sum or difference combinations match the frequency difference between two appropriate excited levels. Such resonances can occur even if all the population remains in the ground state. They occur because of the change in the relation of the damping constants for the various transitions involved in the nonlinear susceptibility when collisions become important, as discussed by Prior *et al.*[41] They have thus been referred to as pressure induced resonances. Observation of such resonances has been reported in measurements of four-wave difference-frequency mixing in Na vapor by Prior *et al.*[41]

6. EFFECTS OF FINITE LASER BANDWIDTH

The results in the previous sections were derived under the assumption that the pump radiation was in the form of a monochromatic wave. Real laser radiation, however, has a finite spectral width that can arise from one or more of a number of sources. Spectral width can result from multimode operation of a laser oscillator, from phase modulation that is developed as a result of propagation through media with nonlinear refractive indexes (see Chapter 5), or simply from a requirement that the laser radiation have sufficient bandwidth to support its pulse duration. This last effect is most prominent with ultrashort pulses from mode-locked lasers.

The effects of the finite spectral width of the pump radiation on the frequency conversion process depends in general on the exact nature of the source of the bandwidth, i.e., on whether it arises from time variations in the amplitude or phase, on whether the time variations are coherent or stochastic and if they are stochastic, on their statistical properties. The effects of the laser bandwidth also depend on the relation of its magnitude to the widths of any phase-matching peaks or material resonances that may be involved.

Various aspects of frequency conversion of radiation with a finite bandwidth have been treated by several authors. Ducuing and Bloembergen[42] derived results for phase-matched nonresonant conversion of randomly modulated radiation. Akhmanov *et al.*[10,43] have given a general discussion of several aspects of conversion of broad-band radiation. Third-order resonant conversion of randomly modulated radiation has been discussed by Anikin *et al*,[44] Stappaerts,[45] Dutta,[46] and Leubner *et al.*[47] Transient resonant processes in which the laser bandwidth arises from a short pulse duration have been treated by Puell and Vidal,[18] Elgin and New,[48] Diels and Georges,[49] Poluektov and Nazarkin,[50] Grischowsky *et al.*,[51] and Georges *et al.*[52]

6.1. Nonresonant Interactions

6.1.1. *Phase-Matched Interactions*

We consider first a situation in which there are no resonant enhancements involved. In this situation the nonlinear polarization is still given by the quasistatic approximation of Eq. (1-19), but now the field amplitudes have a time dependence that reflects the effects of the finite bandwidth. We shall further assume that the entire bandwidth of the pump radiation is much smaller than the spectral width of the phase-matching peak.

Under these assumptions the instantaneous intensity of the generated wave for third-harmonic generation or four-wave mixing is still given by Eqs. (11) or (34). In dealing with multimode radiation we are often interested in the average intensity of the generated wave. For third-harmonic conversion the average harmonic intensity can be expressed as

$$\langle I_3 \rangle = \frac{9\pi^2 c}{8\lambda_1^2 n_3} |\chi^{(3)}|^2 \frac{L^2 \sin^2(\Delta k L/2)}{(\Delta k L/2)^2} \langle (A_1 A_1^*)^3 \rangle, \tag{55}$$

where $\langle\ \rangle$ indicates a time average of the form

$$\langle f(t)\rangle = \frac{1}{T}\int_{t-T/2}^{t+T/2} f(t)\,dt. \tag{56}$$

For steady-state fields the average in Eq. (56) is taken in the limit as $T \to \infty$. For pulsed fields with a large time–bandwidth product the time T can be taken to be large compared to the correlation time, but small compared to the time over which the average intensity changes.

Expressions similar to Eq. (55) for the average intensity of the generated wave in mixing processes of the type $\omega_4 = 2\omega_1 \pm \omega_2$ involve time averages of the form $\langle(A_1 A_1^*)^2 A_2 A_2^*\rangle$ and in processes of the type $\omega_4 = \omega_1 + \omega_2 \pm \omega_3$ involve averages of the form $\langle A_1 A_1^* A_2 A_2^* A_3 A_3^*\rangle$.

Further quantitative analysis requires specific assumptions about the characteristics of the pump radiation. The qualitative nature of the effects of the finite bandwidth depend to a certain extent on whether it results from coherent or incoherent modulation. If the laser radiation carries coherent modulation, then the appropriate averages are readily calculated. We consider the effects of random modulation in the next section.

6.1.1a. Random Modulation. Random modulation of the pump radiation is commonly encountered in the use of multimode lasers. These lasers produce radiation that is distributed over many equally spaced frequencies with phases that vary randomly from one mode to the next. The spectrum of radiation from such a laser is illustrated schematically in Fig. 21. The total field can be represented as a sum of complex field amplitudes over the modes that are present:

$$\mathscr{E}_\mathrm{T} = \tfrac{1}{2}\left\{\sum_i A_i e^{-i(\omega_i t - k_i z + \phi_i)} + \text{c.c.}\right\}, \tag{57}$$

where ϕ_i is the phase of the ith mode. As a function of time the radiation consists of a series of spikes or random amplitude, width, and spacing which arise from interference among the various modes. The minimum width of the spikes is determined by the inverse of the spectral width of the radiation.

When a large number of modes is present the radiation can often be described as a stationary Gaussian random process. The properties of the radiation field are conveniently described in terms of the correlation function

$$R(\tau) = \frac{4}{\langle |A_1|^2\rangle}\int_{-\infty}^{\infty} \mathscr{E}_\mathrm{T}(t)\mathscr{E}_\mathrm{T}^*(t-\tau)\,dt, \tag{58a}$$

which is the Fourier transform of the normalized intensity spectrum. For example, for a Gaussian spectrum the correlation function is given by

$$R(\tau) = e^{-\tau^2/\tau_0^2}, \tag{58b}$$

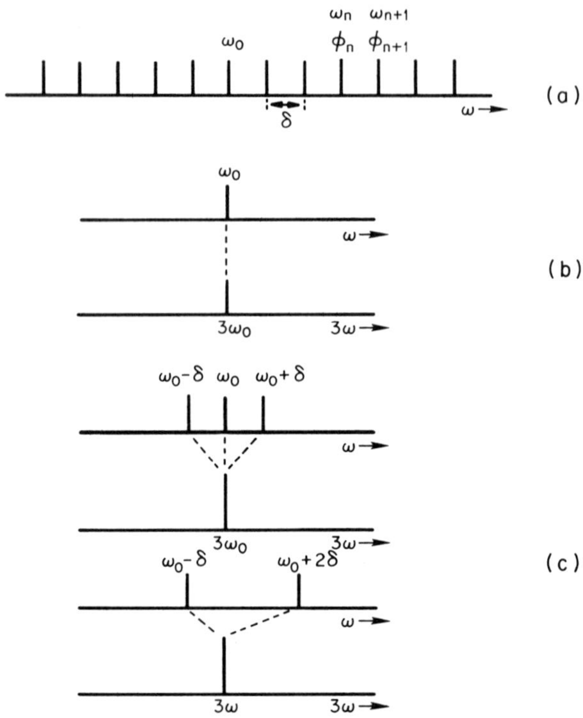

Fig. 21. (a) Distribution of modes in a typical broadband laser. Each mode is characterized by a frequency ω_n and a phase ϕ_n, with the phases of different modes statistically uncorrelated. (b) Third-harmonic generation from the central mode at ω_0. (c) Sum-frequency generation of radiation at frequency $3\omega_0$ from modes that are symmetrically adjacent to the central mode (i) and modes that are not symmetrically distributed about the central mode (ii).

where τ_0 is the correlation time, related to the laser bandwidth (full width of the power spectrum at the $1/e$ point) by

$$\tau_0 = 4/\Delta\omega_L \tag{59}$$

and $\langle |A|^2 \rangle$ is the square of the magnitude of the total field averaged over a time long compared to the correlation time.

The statistical averages that appear in the expressions for the average intensity of the generated wave result in a different average conversion efficiency for multimode radiation than would be obtained from single-mode radiation of the same average intensity. For example, if we consider the average intensity of the generated wave over a time that is long compared to the correlation time but short compared to the time over which the average intensity of the pump radiation changes, then the appropriate averages for the various third-order interactions are

Third-harmonic generation:

$$\omega_3 = 3\omega_1,$$
$$\langle (A_1(t)A_1^*(t))^3 \rangle = 3!\langle |A_1(t)|^2 \rangle^3; \tag{60a}$$

Four-wave mixing:

$$\omega_4 = 2\omega_1 \pm \omega_2,$$
$$\langle (A_1(t)A_1^*(t))^2 A_2(t)A_2^*(t) \rangle = 2!\langle |A_1(t)|^2 \rangle^2 \langle |A_2(t)|^2 \rangle, \tag{60b}$$
$$\omega_4 = \omega_1 \pm \omega_2 \pm \omega_3,$$
$$\langle A_1(t)A_1^*(t)A_2(t)A_2^*(t)A_3(t)A_3^*(t) \rangle = \langle |A_1|^2 \rangle \langle |A_2|^2 \rangle \langle |A_3|^2 \rangle. \tag{60c}$$

Comparison of these expressions with the corresponding ones for single-mode radiation shows that the average intensity of the generated wave is increased with multimode radiation by the factor $q!$, where q is the number of correlated waves in the pump radiation. Ducuing and Bloembergen[42] have shown that this result holds for higher-order harmonic generation as well. The increase in conversion that accompanies multimode radiation occurs because the frequency conversion process is nonlinear and the increased conversion that occurs at the interference maxima more than offsets the decrease in conversion that occurs at the interference minima, leading to an increase in the average conversion efficiency. This property also causes the harmonic radiation to have more pronounced fluctuations than the pump radiation.

It should be emphasized that the factorial factors in Eqs. (60) are a result of the assumed Gaussian nature of the statistical distribution of the pump radiation. When only a few modes are present in the laser radiation this assumption is not well satisfied and the numerical results given above will not be accurate.

6.1.1b. FREQUENCY DOMAIN. Additional insight into these effects can be gained from viewing the process in the frequency domain. The laser radiation can be considered as a collection of equally spaced modes with random phases, as pointed out earlier and as pictured in Fig. 21. Let us consider the generation of radiation at a frequency that is three times the center frequency $\omega_4 = 3\omega_0$. Such radiation can be produced by third-harmonic conversion of the pump mode at ω_0, as illustrated in Fig. 21b, and also by various sum-frequency conversion processes of the type

$$3\omega_0 = \omega_0 + (\omega_0 + \delta) + (\omega_0 - \delta),$$
$$3\omega_0 = \omega_0 + (\omega_0 + 2\delta) + (\omega_0 - 2\delta),$$
$$3\omega_0 = \omega_0 + 2\delta + (\omega_0 - \delta) + (\omega_0 - \delta),$$

etc., involving modes separated from the center one by an amount $n\delta$. These processes are illustrated in Fig. 21c. When the number of modes present in the pump radiation is large the sum-frequency processes will dominate the harmonic process, leading to the increases in conversion efficiency that were discussed in the previous section.

6.1.2. *Laser Spectral Width Wider than Phase-Matching Peak*

When the spectral width of the laser radiation is greater than that of the phase-matching peak the conversion efficiency can be reduced by the presence of multimode structure. The effects of incomplete phase matching of multimode radiation are related to the mismatch of the group velocities at the two frequencies. To illustrate these effects we consider third-harmonic generation. The spectral width of the phase-matching peak is related to the difference of the group velocities at the two frequencies by

$$\Delta\omega_{\mathrm{pm}} = \frac{5.6}{3L}\left|\frac{1}{v_{\mathrm{g},3}} - \frac{1}{v_{\mathrm{g},1}}\right|^{-1}, \tag{61}$$

where $v_{\mathrm{g},i}$ is the group velocity of the wave at ω_i, L the length of the nonlinear medium, and $\Delta\omega_{\mathrm{pm}}$ the full width at half-maximum of the phase-matching peak. We can define a walk-off time due to the difference in group velocities as

$$\Delta t_{\mathrm{w}} = L\Delta\left(\frac{1}{v_{\mathrm{g}}}\right) = L\left(\frac{1}{v_{\mathrm{g},3}} - \frac{1}{v_{\mathrm{g},1}}\right). \tag{62}$$

The walk-off time is a measure of how far the envelopes of the harmonic and fundamental radiation get out of step in a medium of length L. It is related to the width of the phase-matching peak by

$$\Delta t_{\mathrm{w}} = 5.6/3\Delta\omega_{\mathrm{pm}}. \tag{63}$$

When the spectral width of the laser is much wider than the phase-matching peak of the medium the walk-off time is much longer than the correlation time [see Eq. (59)]:

$$\Delta t_{\mathrm{w}} \gg \frac{5.6}{12}\tau_0, \qquad \Delta\omega_{\mathrm{L}} \gg \Delta\omega_{\mathrm{pm}}. \tag{64}$$

As a result when $\Delta\omega_{\mathrm{L}} \gg \Delta\omega_{\mathrm{pm}}$ the envelopes of the fundamental and harmonic waves get out of step by more than the correlation time of the pump radiation.

When the spectral width of the laser is comparable to or greater than that of the phase-matching peak the individual group velocities must be kept in the equations for the various field amplitudes. Equations (1-12) must then be used in place of Eqs. (1-13).

The qualitative effects of the spectral bandwidth of the pump radiation depend on whether it arises from phase or amplitude modulation. When the bandwidth arises from amplitude modulation and the spectral width of the laser is much larger than the width of the phase-matching peak, the local intensity peaks in the harmonic and fundamental radiation slip out of step in the nonlinear medium. As a result, for a given interaction length the fundamental is less effective in generating the harmonic and the conversion efficiency decreases. However, because the assumption of amplitude modulation implies that the phase of the fundamental is constant in time, the fundamental is always phased to give energy to the harmonic. Consequently complete conversion of the fundamental is still possible in principle for a sufficiently long medium. The spectrum of the harmonic will be narrowed relative to the width it would have if phase matching at all frequency components of the fundamental were perfect and the temporal structure in the fundamental is smoothed out. These effects have been analyzed by Karamzin and Sukhorukov,[53] Akhmanov *et al.*,[10,43] and Eckardt and Reintjes[54] for second-harmonic conversion. Results for third-harmonic conversion should be qualitatively similar. Dutta,[46] for example, has shown that for two-photon resonant third-harmonic generation the spectral width of the harmonic radiation is inversely proportional to the length of the nonlinear medium under these conditions.

6.1.2a. Phase Modulation. If the spectral width of the fundamental radiation arises from phase modulation. Then the conversion efficiency can be severely limited, even in an infinitely long medium. For this situation the appropriate time for comparison with the group velocity walk-off time is not the duration of the pump pulse or the local intensity spikes in the pump radiation but rather the time it takes for the phase of the pump radiation to change by π rad. When significant walk-off occurs between the harmonic and fundamental the harmonic will get out of phase with the fundamental and the direction of power flow will reverse, converting radiation back into the fundamental wave. As a result the effective conversion in a medium of arbitrary length can be limited to a maximum value that can be significantly less than unity. This effect has proved to be of great practical importance in limiting the second-harmonic conversion of mode-locked laser pulses in solids,[55] and it would be expected to be of similar importance in limiting the conversion of such pulses in higher-order processes in media in which the group velocity dispersion is significant.

An important class of interactions involves the frequency conversion of mode-locked laser pulses. Such pulses are often used in nonlinear interactions because they tend to have very high optical intensities. Currently mode-locked laser pulses range in duration from about 30 psec for passively mode-locked Nd : YAG lasers to about 0.03 psec for mode-locked dye lasers.

The minimum spectral width of these pulses is given by

$$\Delta\omega_L = 4/t_p, \tag{65}$$

where t_p is the pulse duration, but it can be wider if additional amplitude or phase modulation is present.

These considerations indicate that complete conversion of a minimum width mode-locked laser pulse is possible even when it is so short that its spectrum is wider than the phase-matching peak. In such situations, however, the harmonic pulse will generally be longer than the fundamental pulse. However, any additional phase modulation that is present on the fundamental pulse, such as might be accumulated by propagation through media that have nonlinear refractive indexes, for example, can, because of group velocity dispersion in the nonlinear mixing medium, reduce the conversion efficiency in relatively short distances.

6.2. Resonant Interactions

The effects of a finite laser bandwidth in resonant third-order processes have been discussed by several authors. Anikin *et al.*[44] have treated two-photon resonant third-harmonic generation and four-wave sum-frequency mixing for homogeneously and inhomogeneously broadened media with a time-domain description using autocorrelation functions for Gaussian random noise. Dutta[46] has given a similar treatment for resonant interactions including coherent anti-Stokes–Raman scattering for different types of statistical properties of the pump radiation. Leubner *et al.*[47] and Stappaerts *et al.*[45] have treated two-photon resonant sum-frequency generation in the frequency domain, starting from a sum over the laser modes. Leubner *et al.* have given analytic expansions for the general integrals.

Two-photon resonant problems involving pulse durations that are short compared to the response time of the nonlinear polarization have been treated by Puell and Vidal,[18] Elgin and New,[48] Diels and Georges,[49] Poluektov and Nazarkin,[50] Grischowsky *et al.*,[51] and Georges *et al.*[52]

In order to illustrate the effects of two-photon resonances we shall consider the situation shown in Fig. 22. The medium is assumed to have a near-two-

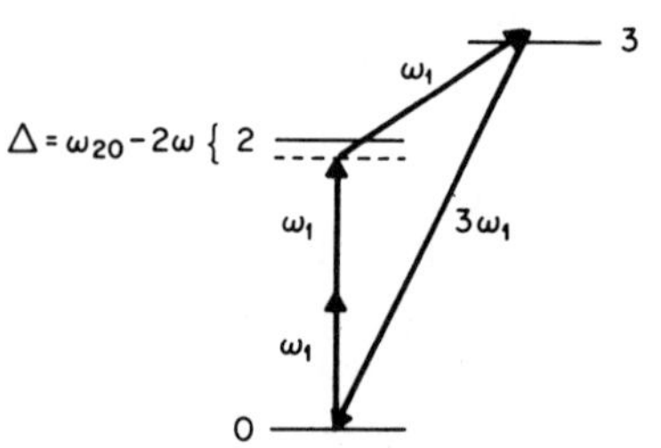

Fig. 22. Level diagram showing a two-photon resonance with $2\omega_1$.

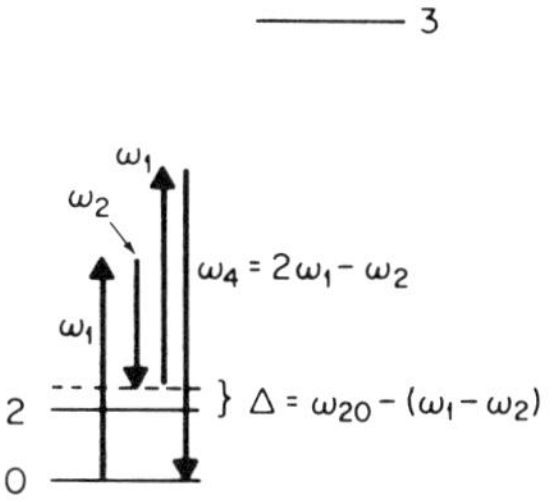

Fig. 23. Level diagram showing two-photon resonance with the frequency combination $\omega_1 - \omega_2$.

photon resonance with the pump radiation at ω_1. Either third-harmonic generation or four-wave sum- or difference-frequency mixing can occur. A similar situation exists for coherent anti-Stokes–Raman scattering, illustrated in Fig. 23, in which the two-photon resonance occurs for the frequency combination $\omega_1 - \omega_2$. We shall consider here third-harmonic generation.

The nonlinear polarization at the harmonic frequency consists of a resonant and nonresonant contribution:

$$P^{\mathrm{NL}}(3\omega) = P_{\mathrm{R}}^{\mathrm{NL}}(3\omega) + P_{\mathrm{NR}}^{\mathrm{NL}}(3\omega). \tag{66}$$

The nonresonant contribution is determined by a nonresonant susceptibility

$$P_{\mathrm{NR}}^{\mathrm{NL}} = \chi_{\mathrm{NR}}^{(3)}(-\omega_3, \omega_1, \omega_1, \omega_1)A_1^3, \tag{67}$$

where $\chi_{\mathrm{NR}}^{(3)}$ is given by expressions of the form in Eq. (51) that do not contain the two-photon resonant term. The resonant contribution to the nonlinear polarization is given in terms of components of the density matrix by

$$P_{\mathrm{R}}^{\mathrm{NL}}(3\omega) = Nq_2'\rho_{20}(t)A_1(t), \tag{68}$$

where

$$q_2' = \frac{1}{\hbar}\sum_i \mu_{0i}\mu_{i2}\left(\frac{1}{\omega_{i0} - 3\omega} + \frac{1}{\omega_{i0} + \omega}\right) \tag{69}$$

and the density matrix component ρ_{20} is described by the equations

$$\frac{\partial \rho_{20}}{\partial t} + i[\omega_{20} - 2\omega + \delta|A_1|^2 - i\Gamma]\rho_{20} = \frac{i}{4\hbar} q_2(\rho_{00} - \rho_{22})A_1^2(t), \tag{70a}$$

$$\frac{\partial}{\partial t}[\rho_{00} - \rho_{22}] + \frac{1}{T_1}[(\rho_{00} - \rho_{22}) - 1] = -\operatorname{Im}\left[\frac{q_2}{\hbar}\rho_{20}(A_1^*(t))^2\right], \tag{70b}$$

where

$$q_2 = \frac{1}{\hbar}\sum \frac{\mu_{0i}\mu_{i2}}{\omega_{i0} - \omega_1} \tag{70c}$$

and δ is a Stark shift parameter given by

$$\delta = \frac{1}{2\hbar}\left(\sum \frac{|\mu_{2i}|^2 \omega_{2i}}{\omega_{2i}^2 - \omega_1^2} + \sum \frac{|\mu_{io}|^2 \omega_{io}}{\omega_{io}^2 - \omega_1^2}\right). \tag{70d}$$

Here we have assumed that all atoms start in the ground state and that the conversion efficiency is small enough that the elements of the density matrix and the Stark shifts are determined only by the pump intensity.

From these equations we can identify conditions under which the quasi-static approximation that was used in the preceding section is valid. Generally speaking the results of the previous section will be valid as long as the time derivatives in Eq. (70a) or (70b) can be neglected. The time derivative of ρ_{20} in Eq. (70a) can be neglected under two conditions: (1) the laser bandwidth is much less than the detuning from resonance; and (2) the laser bandwidth is less than the linewidth Γ of the two-photon transition (equivalent here to the inverse of the dephasing time T_2). The second condition will be valid even if the central frequency of the laser is not detuned by an amount large compared to the bandwidth. The time derivative of $(\rho_{00} - \rho_{22})$ can be neglected if the spectral width of the laser radiation is small compared to the inverse of the lifetime T_1 of if the total population changes are small.

Steady-state solutions to these equations allowing for saturation of the two-photon population will be discussed in Chapter 4. Here we shall consider solutions obtained when the time derivatives cannot be ignored. We consider first a situation in which the pump radiation consists of broadband radiation with an overall duration long compared to T_1 and T_2. We shall also assume that the population changes are small so the equation for $(\rho_{00} - \rho_{22})$ can be dropped. Our solutions follow the treatment of Anikin *et al.*[44]

The harmonic field amplitude is obtained by integrating Eq. (70a) and using that solution along with Eq. (68) in Eq. (3d):

$$A_3(t) = i\frac{3\pi^2}{n_3\lambda_1} NL \frac{q_2 q_2'}{\hbar} \frac{\sin(\Delta kL/2)}{(\Delta kL/2)} A_1(t) \int_{-\infty}^{t} e^{(\Gamma + i\Delta)(t' - t)} A_1^2(t')\, dt'\, e^{-i\Delta kL/2}, \tag{71a}$$

where

$$\Delta = \omega_{20} - 2\omega. \tag{71b}$$

Here we have assumed that the bandwidth of the laser radiation is smaller than the width of the phase-matching peak. The average harmonic intensity is given by

$$\langle I_3 \rangle = \frac{cn_3}{8\pi}\left(\frac{3\pi^2 q_2 q_2' NL}{n_3\lambda_1\hbar}\right)^2 \frac{\sin^2(\Delta kL/2)}{(\Delta kL/2)^2}$$
$$\times \int_0^\infty \int_0^\infty e^{-[(\Gamma + i\Delta)\tau_1 + (\Gamma - i\Delta)\tau_2]} \langle A_1(t) A_1^*(t) A_1^2(t - \tau_1)(A_1^*(t - \tau_2))^2 \rangle$$
$$\times\, d\tau_1\, d\tau_2, \tag{72}$$

where $\langle\ \rangle$ represents a time average, as indicated in Eq. (56).

If inhomogeneous broadening (as, for example, from the Doppler effect) is important, then the nonlinear polarization must be averaged over the inhomogeneous distribution of resonant frequencies. For a Doppler-broadened distribution given by

$$g(\omega_{20} - \omega_{20}^0) = \frac{1}{\sqrt{\pi}\gamma_0} e^{-(\omega_{20}-\omega_{20}^0)^2/\gamma_0^2}, \tag{73}$$

where ω_{20}^0 is the center of the Doppler-broadened line, Anikin *et al.*[44] have shown that the corresponding expression for the harmonic intensity is

$$\begin{aligned}\langle I_3 \rangle = {} & \frac{cn_3}{8\pi}\left(\frac{3\pi^2 q_2 q_2' NL}{n_3 \lambda_1 \hbar}\right)^2 \frac{\sin^2 \Delta kL/2}{(\Delta kL/2)^2} \\ & \times \int_0^\infty \int_0^\infty \exp\left(-\frac{\gamma_0^2}{4}(\tau_1^2 + \tau_2^2) - (\Gamma + i\Delta_0)\tau_1 - (\Gamma - i\Delta_0)\tau_2\right) \\ & \times \langle A_1(t)A_1^*(t)A_1^2(t - \tau_1)(A_1^*(t - \tau_2))^2\rangle \, d\tau_1 \, d\tau_2.\end{aligned} \tag{74}$$

where Δ_0 is the detuning from the center of the Doppler line.

Equation (74) contains the general solution for the third-harmonic intensity generated by a pump of finite bandwidth. Further quantitative analysis requires specific assumptions about the statistical nature of the pump radiation. We shall give examples here for broadband Gaussian noise. Corresponding expressions for other types of distributions have been given by Dutta.[46] He has also given results for Gaussian noise when the laser bandwidth is larger than the phase-matching peak.

For an incident pump wave with a Gaussian noise distribution Eq. (74) takes the form

$$\langle I_3 \rangle = \frac{n_3}{n_1}\left(\frac{24\pi^3 q_2 q_2' NL}{n_1 n_3 \lambda_1 c\hbar}\right)^2 \frac{\sin^2(\Delta kL/2)}{(\Delta kL/2)^2} \langle I_1 \rangle^3 [2J_1 + 4J_2], \tag{75}$$

where

$$J_1 = R^2(\tau_2 - \tau_1),$$
$$J_2 = R(\tau_1)R(\tau_2)R(\tau_2 - \tau_1).$$

For a correlation function of the form in Eq. (58b) these expressions become

$$\begin{aligned}J_1 = {} & \int_0^\infty \int_0^\infty \exp\left\{-\left[\frac{\gamma_0^2}{4}(\tau_1^2 + \tau_2^2) + 2(\tau_1 - \tau_2)^2/\tau_0^2 + (\Gamma + i\Delta_0)\tau_1\right.\right. \\ & \left.\left. + (\Gamma - i\Delta_0)\tau_2\right]\right\} d\tau_1 \, d\tau_2,\end{aligned} \tag{76a}$$

$$\begin{aligned}J_2 = {} & \int_0^\infty \int_0^\infty \exp\left\{-\left[\left(\frac{\gamma_0^2}{4} + \frac{1}{\tau_0^2}\right)(\tau_1^2 + \tau_2^2) + (\tau_1 - \tau_2)^2/\tau_0^2 + (\Gamma - i\Delta_0)\tau_1\right.\right. \\ & \left.\left. + (\Gamma - i\Delta_0)\tau_2\right]\right\} d\tau_1 \, d\tau_2,\end{aligned} \tag{76b}$$

as shown by Anikin *et al.*[44]

When the detuning is larger than either the laser or the material bandwidth, both integrals reduce to $1/\Delta_0^2$ and the statistical factor $3!$ in Eq. (60a) is retrieved. At exact resonance in the limit in which the laser bandwidth is much greater than the material linewidth the integrals become

$$J_1 \approx \frac{2\tau_0}{\sqrt{2}\gamma_0} e^{4\Gamma^2/\gamma_0^2} \int_{-2\Gamma/\gamma_0}^{\infty} e^{-\tau_1'^2}\, d\tau_1' \int_{(\Gamma\tau_0/2\sqrt{2})-2\sqrt{2}\tau_1'/\gamma_0\tau_0}^{\infty} e^{-\tau_2'^2}\, d\tau_2', \tag{77a}$$

$$J_2 \simeq \frac{\tau_0}{\sqrt{2}} \int_0^{\infty} e^{-3\tau_1^2/2\tau_0^2}\, d\tau_1 \int_{(\Gamma\tau_0/2\sqrt{2})-(\tau_1\sqrt{2}\tau_0)}^{\infty} e^{-\tau_2'^2}\, d\tau_2'. \tag{77b}$$

At exact resonance the dominant contribution comes from the integral J_1 and the statistical factor is reduced to 2. When homogeneous broadening is dominant the average harmonic intensity is given by

$$\langle I_3 \rangle = C\frac{\sqrt{\pi}\,\tau_0}{2\sqrt{2}\Gamma} = C\frac{\sqrt{2\pi}}{\Gamma\Delta\omega_{\mathrm{L}}}, \tag{78}$$

while for dominant Doppler broadening the average intensity is

$$\langle I_3 \rangle = C\frac{\pi\tau_0}{2\sqrt{2}\gamma_0} = C\frac{\sqrt{2}\,\pi}{\gamma_0\Delta\omega_{\mathrm{L}}}, \tag{79}$$

where

$$C = \frac{n_3}{n_1}\left(\frac{24\pi^3 q_2 q_2' NL}{n_1 n_3 \lambda_1 c\hbar}\right)^2 \frac{\sin^2(\Delta kL/2)}{(\Delta kL/2)^2}\langle I_1 \rangle^3.$$

These results show that in the limit of large laser linewidths at exact resonance the average harmonic intensity varies inversely with the product of the laser linewidth and the material linewidth. Similar dependences have been obtained by Stappaerts *et al.*[45] and Leubner *et al.*[47] from analyses in the frequency domain. Leubner *et al.* have also derived more complete formulas that can be used when the laser linewidth is comparable to the material linewidth and the detuning is arbitrary. Dutta[46] has examined the behavior for different kinds of statistics and has derived dependences for the bandwidth of the generated radiation and has considered the effects of group velocity mismatches.

As a result of the variation of the harmonic intensity in Eqs. (78) and (79) the conversion efficiency can be expected to increase as the laser linewidth decreases until it becomes narrower than the material linewidth.

However, as pointed out by Stappaerts the expected increase in harmonic intensity may not be fully realized if the reduction in linewidth is accompanied by a reduction of the pump intensity.

6.2.1. *Coherent Transient Effects*

When the laser radiation is in the form of a single pulse with a duration that is shorter than the lifetime T_1 and the dephasing time T_2 of the resonant transition, additional effects can occur because of the coherent transfer of population. We shall treat these briefly here and shall discuss them in more detail in Chapter 4 in relation to the limitations they place on the conversion efficiency.

We shall consider here transient effects related to two-photon resonant third-harmonic generation. We shall assume that the conversion efficiency is small so that the elements of the density matrix are determined only by the pump intensity. Equations (70) are more easily handled in this situation by defining three new variables:

$$r_1 = \rho_{20}(-2\omega) + \rho_{02}(2\omega), \tag{80a}$$

$$r_2 = i(\rho_{20}(-2\omega) - \rho_{02}(2\omega)), \tag{80b}$$

$$r_3 = \rho_{00} - \rho_{22}. \tag{80c}$$

In terms of these variables Eqs. (70) beome

$$\frac{\partial r_1}{\partial t} + \Gamma r_1 = -\Delta r_2, \tag{81a}$$

$$\frac{\partial r_2}{\partial t} + \Gamma r_2 = \Delta r_1 - \frac{q_2}{2\hbar} A_1^2 r_3, \tag{81b}$$

$$\frac{\partial r_3}{\partial t} + \frac{1}{\Gamma_1}(r_3 - r_3^0) = \frac{q_2}{2\hbar} A_1^2 r_2, \tag{81c}$$

where q_2 and Δ are defined in Eqs. (70c) and (71b) and we have chosen the pump wave to be real so that $A_1 = (A_1)^*$.

6.2.1a. NEAR-RESONANT INTERACTIONS. Our initial assumption that the pulse duration is short compared to T_1 and T_2 allows us to neglect the linewidth Γ and the lifetime T_1 in Eqs. (81). We consider first the situation in which the laser is tuned near but not exactly on the two-photon resonance. In particular, we assume that the detuning Δ is larger than both the two-photon linewidth and the laser linewidth as shown in Fig. 24. The condition on the laser linewidth relative to the detuning from the resonance is equivalent to the condition

$$\frac{1}{A_1}\frac{\partial A_1}{\partial t} \ll \Delta. \tag{82}$$

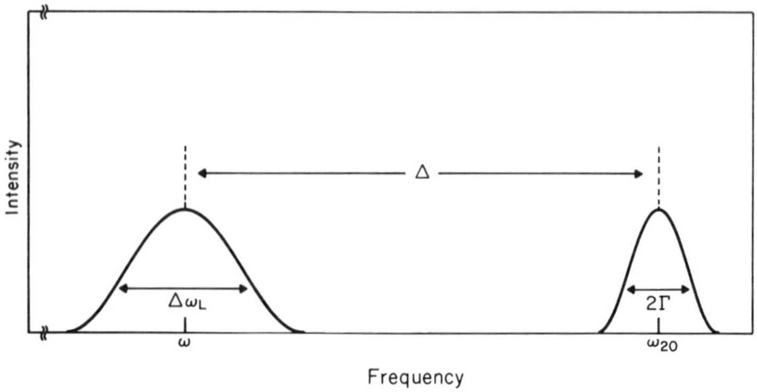

Fig. 24. Schematic illustration for the relative linewidths of the laser and the material system for the near-resonant coherent interactions.

With this approximation the solutions to Eqs. (81) are

$$r_1 = \frac{q_2 A_1^2}{2\hbar[\Delta^2 + (q_2 A_1^2/2\hbar)^2]^{1/2}}, \tag{83a}$$

$$r_2 = 0, \tag{83b}$$

$$r_3 = \frac{\Delta}{[\Delta^2 + (q_2 A_1^2/2\hbar)^2]^{1/2}}, \tag{83c}$$

and the resulting nonlinear polarization from Eqs. (80) and (68) is

$$P^{\mathrm{NL}}(3\omega) = \frac{N q_2 q_2' A_1^3(t)}{4\hbar[\Delta^2 + (q_2 A_1^2/2\hbar)^2]^{1/2}}. \tag{84}$$

At low pump intensities, such that

$$A_1^2 \ll 2\hbar\Delta/q_2, \tag{85}$$

this expression reduces to the same one that is obtained in the quasistatic approximation. At higher pump intensities saturation of the nonlinear polarization occurs because of the coherent transfer of population. This aspect of the solution will be discussed further in Chapter 4. Solutions of the form in Eq. (83) have been discussed by Grischowsky *et al.*[51] and by Anikin *et al.*[44]

6.2.1b. Interactions at Exact Resonance. When the pump wave at ω_1 is tuned to exact resonance and the pulse duration is shorter than the dephasing time of the resonant transition, a nonlinear susceptibility for the time-domain equation (19) can no longer be defined, although the frequency-dependent susceptibility of Eq. (1-15) still exists. Calculation of the nonlinear

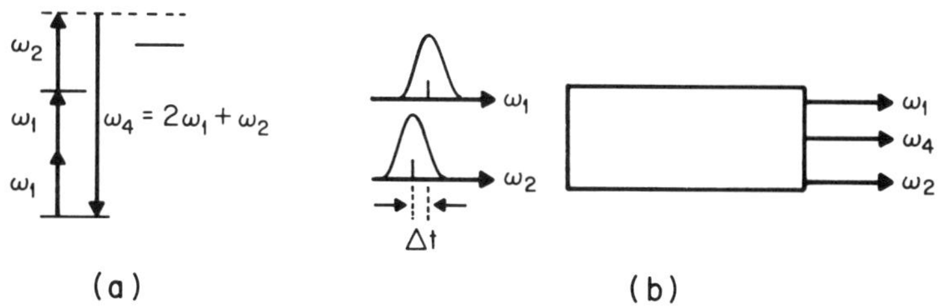

Fig. 25. Schematic diagram showing arrangement for observing retarded response of two-photon resonant third-order sum-frequency generation. (a) Two-photon resonant process. (b) Experimental arrangement. Arrival time of pulse at ω_2 can be varied relative to arrival time of pulse at ω_1.

polarization in the frequency domain now involves convolution integrals of the susceptibility and the pump field components. In the time domain the nonlinear polarization is related to the fields by the integral equation

$$P^{\mathrm{NL}}(t) = i\,\frac{Nq_2q_2'A_1(t)}{4\hbar}\int_{-\infty}^{t} e^{-\Gamma(t-t')}\,[\rho_{00}(t') - \rho_{22}(t')]A_1^2(t')\,dt'.$$

The polarization will in general lag the field, leading to reduced efficiency for third-harmonic conversion.

Solutions for the components of the density matrix have been given by Elgin and New[48] for four-wave sum-frequency mixing of the form $\omega_4 = 2\omega_1 + \omega_2$. They assumed that the two-photon population was determined only by the wave at ω_1, with the second wave completing the frequency conversion process as illustrated in Fig. 25. They also allowed for the pulses at ω_1 and ω_2 to arrive at different times. The components of the density matrix for this process are still described by Eq. (81) and the solutions (with $\Delta = 0$) are, in the limit that the lifetimes can be neglected,

$$r_1 = 0, \tag{86a}$$

$$r_2 = -\sin\theta(t), \tag{86b}$$

$$r_3 = -\cos\theta(t), \tag{86c}$$

where

$$\theta(t) = \frac{q_2}{2\hbar}\int_{-\infty}^{t} A_1^2(t')\,dt'. \tag{87}$$

The corresponding nonlinear polarization is given by

$$P^{\mathrm{NL}}(\omega_4 = 2\omega_1 + \omega_2) = i\,\frac{q_2q_2''}{4\hbar}\,A_2(t)\sin\theta(t), \tag{88}$$

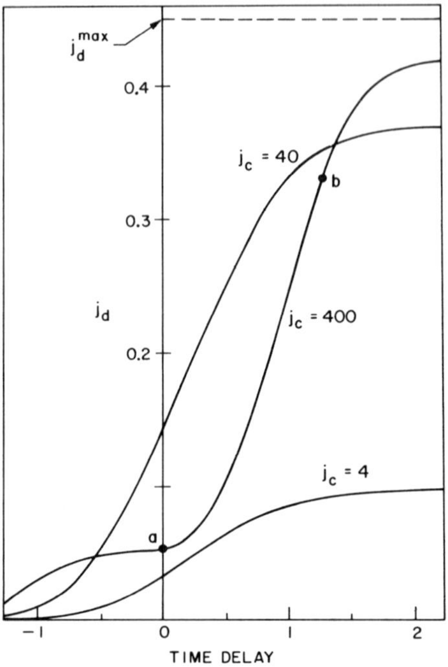

Fig. 26. Calculated values of energy in the pulse at ω_4 generated through the process $\omega_4 = 2\omega_1 + \omega_2$ as a function of the time delay of the pulse at ω_2 for various values of the energy of the pump pulse at ω_2. (From Ref. 56.)

where

$$q_2'' = \frac{1}{\hbar} \sum \mu_{0i} \mu_{i2} \left(\frac{1}{\omega_{i0} - \omega_4} + \frac{1}{\omega_{i0} + \omega_2} \right). \tag{89}$$

These results again show that the development of the nonlinear polarization lags behind the pump pulse because of the integrating effects of the resonance. As a result optimum conversion of the pulse at ω_2 should occur when it arrives near the end of the pulse at ω_1. Numerical calculations done by New[56] illustrate this effect (see Fig. 26). The optimum conversion of the delayed pulse can exceed that of a simultaneous pulse ($\Delta t = 0$) by up to a factor 8 for sufficiently strong pulses for the conditions of these calculations. Experimental studies of two-photon resonantly enhanced conversion in Sr vapor by Royt and Lee[57] have confirmed some of these predictions.

6.2.2. *Two-Photon Vector Model*

The results of the previous two sections can also be obtained within the framework of the vector model of the atom. The application of this model to two-photon problems has been discussed by Hartmann,[58] Belenov and

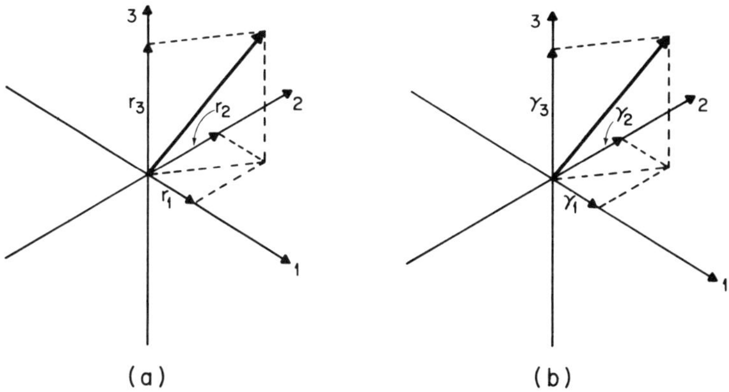

Fig. 27. Diagram showing vector model of the atom. (a) Pseudomoment $\vec{r}$ representing the response of the atom. $r_1 = \rho_{20}(-2\omega) + \rho_{02}(2\omega)$; $r_2 = i[\rho_{20}(-2\omega) - \rho_{02}(2\omega)]$; $r_3 = \rho_{00} - \rho_{22}$. (b) Effective driving field $\vec{\gamma}$. $\gamma_1 = \hbar^{-1}(H'_{20} + H'_{02})$; $\gamma_2 = i\hbar^{-1}(H'_{20} - H'_{02})$; $\gamma_3 = \hbar^{-1}(H'_{22} - H'_{00})$.

Poluektov,[59] Takatsuji,[60] Grischowsky *et al.*,[51] and Courtens.[61] It is especially convenient to use in analyzing resonant and near-resonant interactions for which the equations of the density matrix take on the simple forms given in the previous sections.

In the vector model the response of the atom is again described in a coordinate system in which the rapidly varying parts of the appropriate parts of the driving polarization are transformed away. In this coordinate system the response of the atom is given by a pseudomoment vector $\vec{r}$, whose components (r_1, r_2, r_3) are given in Eq. (80). These vector components are illustrated in Fig. 27a. The first two components describe the polarization components of the atom that are, respectively, in phase with and 90° out of phase with the effective driving field. The third component is proportional to the population difference between the two states involved in the two-photon resonance. Apart from the difference in sign of r_3, this is the same system used by Grischowsky *et al.*[51]

The pseudomoment is driven by an effective field $\vec{\gamma}$, which has the components

$$\gamma_1 = \hbar^{-1}(H'_{20} + H'_{02}), \tag{90a}$$

$$\gamma_2 = i\hbar^{-1}(H'_{20} - H'_{02}), \tag{90b}$$

$$\gamma_3 = \hbar^{-1}(H'_{22} - H'_{00}). \tag{90c}$$

The quantities H'_{ij} are elements of the Hamiltonian in the transformed coordinate system. The components of $\vec{\gamma}$ are illustrated in Fig. 27b.

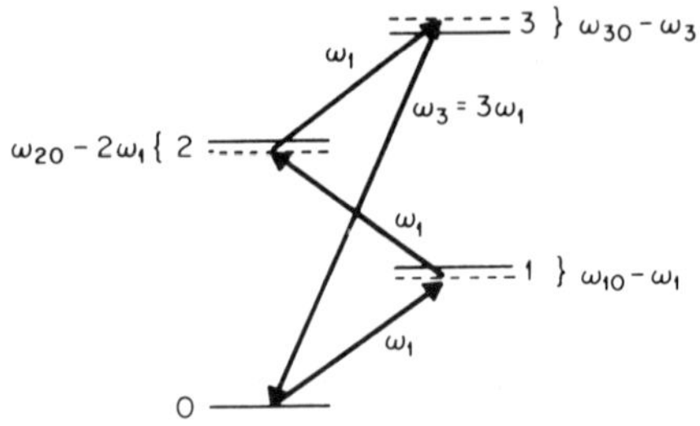

Fig. 28. Level diagram showing two-photon resonance for adiabatic following solution using vector model.

For systems in which there is a two-photon resonance with an incident field at ω_1 the transformed coordinate system is one that oscillates (or rotates for circular polarization) at a frequency $2\omega_1$. The components of $\vec{\gamma}$ are

$$\gamma_1 = (q_2/2\hbar)A_1^2, \tag{91}$$

$$\gamma_2 = 0, \tag{92}$$

$$\gamma_3 = \Delta + \delta|A_1|^2, \tag{93}$$

where δ is a Stark shift parameter given in Eq. (70d). Grischowsky *et al.*[51] have given corresponding expressions for $\vec{\gamma}$ for systems with two-photon resonances at a sum frequency $\omega_1 + \omega_2$.

The equation of motion for $\vec{r}$ is

$$\frac{d\vec{r}}{dt} = \vec{\gamma} \times \vec{r}, \tag{94}$$

where it is understood that the transverse relaxation time T_2 [Γ^{-1} in (81a), (81b)] is to be used with the components r_1 and r_2, while the lifetime T_1 is to be used in the equation for r_3.

Equation (94) describes the precession of the pseudomoment $\vec{r}$ about the effective driving field $\vec{\gamma}$ in the transformed coordinate system. The precession frequency is given by $\Omega = |\gamma|$.

To illustrate the use of this model we can consider the same two interactions that were discussed in the previous sections. We consider the system shown in Fig. 28b with the two-photon detuning $\Delta = \omega_{20} - 2\omega_1$. For near-resonant interactions the effective driving field has the form, neglecting Stark shifts,

$$\gamma = \left(\frac{q_2}{2\hbar} A_1^2, 0, \Delta\right). \tag{95}$$

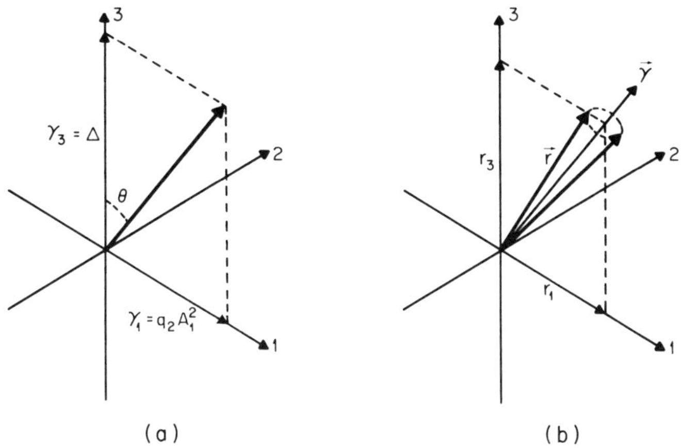

Fig. 29. Response of driving field $\vec{\gamma}$ (a) and pseudomoment $\vec{r}$ (b) for adiabatic following behavior.

The motion of the pseudomoment is illustrated in Fig. 29. Its initial position is straight up, corresponding to the initial condition $\rho_{00} = 1, \rho_{22} = 0$. As the applied field turns on, the effective driving vector rotates in the 1–3 plane. The angle it makes with the 3-axis is given by

$$\tan \theta = \frac{\gamma_1}{\gamma_3} = \frac{q_2 A_1^2}{2\hbar\Delta}. \tag{96}$$

The condition that was assumed in Section 6.2.1

$$\frac{1}{A_1}\frac{dA_1}{dt} \ll \Delta \tag{97}$$

implies that the pseudomoment precesses around the driving field much more rapidly than the driving field changes its direction. As a result the pseudomoment remains approximately aligned along the driving field as it varies in time. It therefore makes effectively the same angle with the 3-axis as the driving field, as shown in Fig. 29b. The r_2 component of the pseudomoment remains small throughout the interaction.

The components of the pseudomoment may be determined from the geometry shown in Fig. 29b. They are

$$r_1 = \sin \theta = \frac{q_2 A_1 A_2}{[\Delta^2 + (q_2 A_1^2/2\hbar)^2]^{1/2}}, \tag{98a}$$

$$r_2 = 0, \tag{98b}$$

$$r_3 = \cos \theta = \frac{\Delta}{[\Delta^2 + (q_2 A_1^2/2\hbar)^2]^{1/2}}. \tag{98c}$$

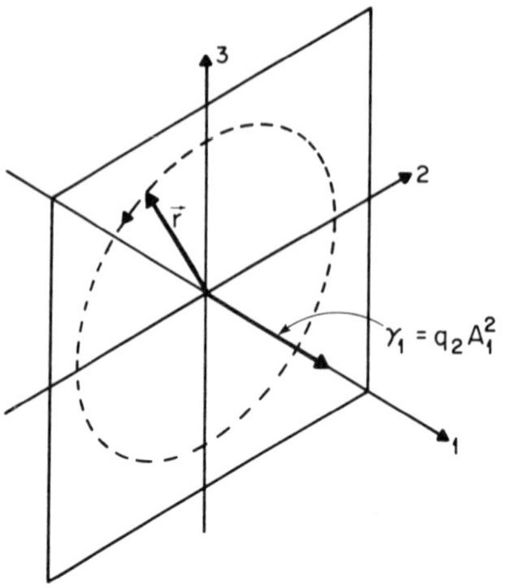

Fig. 30. Motion of pseudomoment $\vec{r}$ for transient behavior at exact resonance.

The resulting nonlinear polarization can be obtained from Eqs. (80) and (68). Note that with the sign convention used here if Δ is negative, the effective driving field will lie in the lower half of the 1–3 plane and the pseudomoment will be approximately aligned antiparallel to γ. These solutions are of the same form as those derived by solution of the differential equations in Section 6.2.1. The response of the atom described here is termed adiabatic following because the pseudomoment follows the effective driving field. It has been analyzed in depth by Grischowsky *et al.*,[51] who have given a careful consideration of the conditions that are necessary for its applicability.

Interactions at Exact Resonance. When the applied field is at exact resonance the effective driving field has the form

$$\vec{\gamma} = \left(\frac{q_2}{2\hbar} A_1^2, 0, 0\right). \tag{99}$$

The motion of the pseudomoment under these conditions is illustrated in Fig. 30 for a constant applied field that is turned on at $t = 0$. It moves in a circle in the 2–3 plane with a precession frequency given by

$$\Omega = (q_2/2\hbar) A_1^2. \tag{100}$$

Its components are

$$r_1 = 0, \tag{101a}$$

$$r_2 = -\sin \Omega t, \tag{101b}$$

$$r_3 = -\cos \Omega t, \tag{101c}$$

which again can be seen to be equivalent to the ones given in Eq. (86) for an incident field of constant amplitude. The nonlinear polarization can again be obtained from the components of the pseudomoment using Eqs. (80) and (68).

7. CONVERSION EFFICIENCY

Although we shall defer a complete discussion of the factors that affect the conversion efficiency until Chapter 4, we shall make a few comments here regarding the conversion efficiencies that can be achieved in principle when only the frequency conversion process is considered. We have seen that virtually 100% conversion is possible for plane waves (see Fig. 3). The same conclusion can be expected to apply to the conversion of the peak on-axis intensity of collimated beams of finite dimension. Usually, however, the total energy conversion of a laser beam is of more interest than simply the peak intensity conversion. When the profile of the beam in space and time is taken into account the total energy conversion can be less than the peak intensity conversion. An example of this effect is illustrated in Fig. 31, where total energy conversion is compared at optimal phase matching for a beam in the plane-wave limit, one with a Gaussian spatial profile but constant in time, and one that has Gaussian profiles in both space and time. The total energy conversion efficiency of the profiled beams is always smaller than that of the plane wave because of the low-intensity wings. This effect can place severe demands on the peak intensity conversion efficiency that must be achieved if even only modest energy conversion is desired. For example, in order to have an energy conversion efficiency of 50% for a beam with a spatial and temporal profile, peak on-axis intensity conversion efficiency

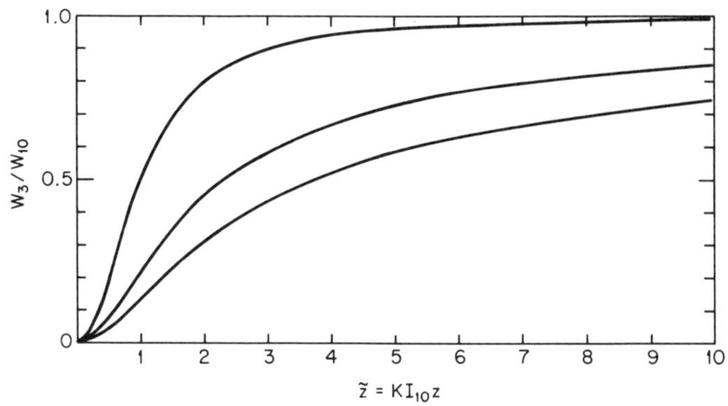

Fig. 31. Third-harmonic energy conversion efficiency for plane waves (a), waves with a Gaussian spatial profile (b), and waves with a Gaussian profile in space and time.

must be in excess of 90%. As a result stringent requirements on phase-matching accuracy and peak intensities must be met.

The energy conversion efficiency of beams with spatial and temporal profiles that are tightly focused have not been examined extensively in the high-depletion regime. Here we could expect additional problems to arise because changes in the profile of the pump beam as it is depleted may affect the optimum phase-matching conditions. White *et al.*[62] have shown that for second-harmonic generation the total power conversion efficiency integrated over a Gaussian profile can be as high as 95%. The optimum focusing conditions for these results were similar to those for the undepleted pump situation, with the confocal parameter comparable to the length of the nonlinear medium. They also discussed mode distortion due to pump depletion for weakly focused Gaussian beams. We may therefore expect similar high power conversion efficiencies for third-order processes for confocal focusing geometries ($b = L$). For more tightly focused geometries, however, the conversion efficiency may be reduced even further.

In practice competing effects have proved to be more effective in limiting conversion efficiency than problems due to beam profiles. These will be discussed in Chapter 4.

8. EXPERIMENTAL RESULTS

In this section we shall summarize some of the experimental results that have been reported on frequency conversion in gases and liquids. We shall examine these results from the point of view of source generation—that is, the production of radiation at new wavelengths for the purpose of creating a new radiation source for additional experiments. As such, the main emphasis on the experiments described in this chapter will be on wavelength ranges, pump sources, and measured conversion efficiencies. In Chapter 4 we shall discuss some of the reasons for observed limitations on efficiency, and in Chapter 6 we shall discuss the frequency conversion processes from the point of view of obtaining information about the nonlinear medium.

Frequency conversion has been accomplished in a wide range of gases, vapors, and liquids with many different kinds of laser sources. I have chosen to classify the experimental results discussed in this section by wavelength range of the generated radiation. Within a wavelength range the conversion processes are further divided according to pump source and nonlinear medium. This choice is somewhat arbitrary and can lead to some redundancy. For example, visible dye lasers can be used to produce both infrared and vacuum ultraviolet radiation by different processes. It is hoped, however, that the organization of results that is provided here can serve as a guide to a researcher who seeks to determine how a particular conversion process can be accomplished using a given pump source, or alternatively, how radiation can be produced in a given wavelength range.

8.1. Near Ultraviolet

Third-harmonic conversion of both ruby and Nd lasers has been used to generate fixed frequency radiation in the near UV.

8.1.1. Conversion of Ruby Laser Radiation

The materials that have been used for third-harmonic conversion of radiation from ruby lasers are listed in Table 1, along with the magnitude of the nonlinear susceptibility, the sign of the dispersion, and any resonances that are involved. Conversion in rare gases was first reported by Ward and New.[3] The rare gases have relatively low susceptibilities in this spectral range and also have positive dispersion. As a result the conversion efficiency was rather low (of the order of 10^{-12}), but these first experiments served to

TABLE 1

Third-Harmonic Conversion of Ruby Laser Radiation

Material	$\chi_a^{(3)}$ (esu)	Dispersion	Resonance	Ref.
Helium	4×10^{-39}	Positive	None	3
Neon	7.6×10^{-39}	Positive	None	3
Argon	124×10^{-39}	Positive	None	3
Krypton	316×10^{-39}	Positive	None	3
Xenon	987×10^{-39}	Positive	None	3
H_2	80×10^{-39}	Positive	None	3
CO_2	156×10^{-39}	Positive	None	3
N_2	107×10^{-39}	Positive	None	3
Cs	10^{-30}		Two-photon $6s^2$–$6s9d\ ^2D_{3/2}$	64

confirm the order of magnitude of the nonlinear susceptibilities that could be calculated from existing data. The experimental values for the susceptibilities that are listed in Table 1 are all in reasonable agreement with the theoretical ones given in Ref. 63.

Ruby laser radiation has also been converted to its third harmonic in Cs vapor by Leung *et al.*[64] There is a near-two-photon resonance with the $9d\ ^2D_{3/2}$ state for this process and as a result the nonlinear susceptibility is several orders of magnitude higher than that of the rare gases.

8.1.2. Conversion of Radiation from Nd: YAG and Nd: Glass Lasers

The materials that have been used to convert radiation from Nd:YAG and Nd: glass lasers to its third harmonic at 354.7 and 353 nm, respectively, are listed in Table 2. The first experiments were done in liquid hexafluoroacetone by Bey *et al.*[11] Here the dispersion is positive and attempts were made to phase match the process by mixing with the dye fuchsin red, whose molecules have negative dispersion for this process. An increase in the conversion efficiency was observed, but perfect phase matching could not be achieved because absorption of the harmonic radiation by the dye prevented the use of sufficiently high dye concentrations.

TABLE 2

Third Harmonic of Nd: YAG and Nd: Glass Laser Radiation

Laser	Material	Phase matching	max η	Ref.
Nd:glass	Hexafluoracetone	Fuchsin red		11
Nd:YAG	Rb	Xe	0.1	67
Nd:YAG	Na	Mg	0.0375	66
Nd:YAG	Na	Xe	0.027	67
Nd:glass	Rb	Xe	0.024	4

Extensive studies of the conversion of Nd laser radiation in alkali metal vapors has been made by Harris and co-workers[5,30,65–68] and by Puell *et al.*[4] The alkali metals are particularly well suited for frequency conversion. Most of their oscillator strength is concentrated in their resonance transitions, leading to relatively high susceptibilities for generating radiation that lies near the resonance levels. In addition, there is a relatively wide range of wavelengths above the resonance levels for which the conversion process has negative dispersion, allowing optimization by phase matching or tight focusing.

Phase matching can be done with any compatible gas or vapor that has positive dispersion for the process under study. The most common materials that have been used are the rare gases. These gases all have positive dispersion in the near IR, visible, and near UV and are transparent throughout this range. The most commonly used rare gas for phase matching the third-harmonic conversion of Nd laser radiation is Xe because it has the largest dispersion per atom of all the rare gases in the near UV and therefore requires the lowest pressure to provide a given amount of total dispersion.

Phase matching has also been done with mixtures of metal vapors.[66] Although this technique is somewhat more complicated than using a mixture of the metal vapor with a rare gas, it does offer the possibility of providing phase matching at a lower total pressure if the generated wavelength lies near, but below, an allowed transition in the second metal vapor.

8.1.3. Heat Pipe Ovens

Before describing the details of the experiments in third-harmonic conversion of radiation from Nd lasers it is appropriate to discuss the use of heat pipe ovens to contain the metal vapors. The use of heat pipe ovens for spectroscopy was first described by Vidal and Cooper.[69] Their operation is illustrated in Fig. 32a. Solid metal is heated in the center of a cell to a temperature above its melting point. Metal vapor flows by convection to a point outside of the heated region, where it condenses onto a wick that lines the inner wall of the cell. The liquid metal then flows back along the wick to the center of the cell by capillary action to be recycled into vapor. The wick usually consists of a fine mesh of any material that is chemically and thermally compatible with the metal vapor being used and which can be wetted by the liquid metal. Stainless steel is a common material that is compatible with many metal vapors. Tantalum and copper are also used.

The metal vapor is confined to the central region of the cell by a buffer gas at the ends of the cell. As long as sufficient heat is supplied to maintain the

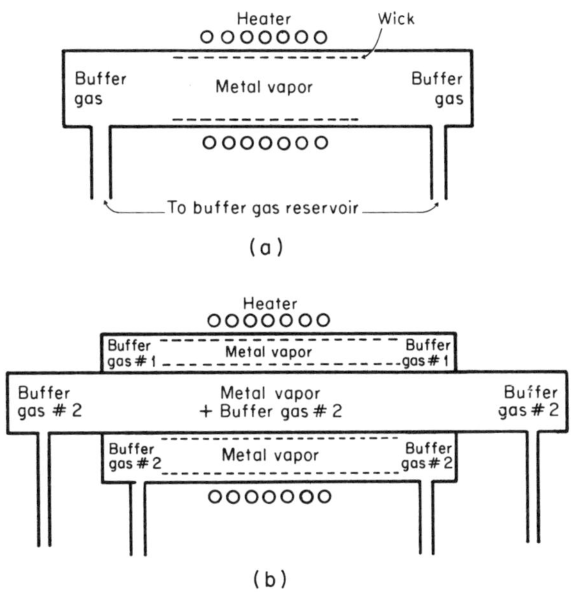

Fig. 32. Diagrams of heat pipe ovens: (a) single metal heat pipe; (b) concentric heat pipe for gas mixtures.

liquid metal at a temperature such that its vapor pressure matches that of the buffer gas, the circulating flow of the metal in its liquid and vapor forms serves to exclude the buffer gas from the center of the cell. The temperature of the central zone is determined by the condition that the vapor pressure of the metal be equal to that of the buffer gas. The density of the metal vapor is controlled by adjusting the pressure of the buffer gas, rather than by changing the heat input to the oven. The extent of the central zone is determined by the amount of heat that is supplied to the oven and the pressure of the buffer gas.

Near the end of the central zone is a transition region over which the metal vapor density drops to zero. The extent of this transition region is of the order of the smaller of the mean free path of the metal atoms or the diameter of the cell. Usually this transition zone is much smaller than the length of the heated central region and so does not have a major effect on the frequency conversion process. It can produce noticeable effects, especially in the plane-wave phase-matching properties,[4] as will be discussed later.

A double heat pipe of the form shown in Fig. 32b can be used for mixtures of metal vapors with other gases.[70] In this configuration the outer section operates as a heat pipe, as described before. The inner section has two regions. Its central zone, which corresponds to the hot region of the outer section, is maintained at the same uniform temperature as the hot central zone of the outer section. The vapor pressure of the metal in this region of the inner section is determined by the temperature of the outer section, which in turn is determined by the pressure of the buffer gas in the outer section. The total pressure in the inner section is determined by the larger of either the metal vapor pressure or the inner buffer gas pressure. If the pressure of the inner buffer gas is larger than that of the inner metal vapor, then the metal vapor will be confined to the central hot region of the inner section. This region will also have sufficient partial pressure of the buffer gas to equilibrate the pressure throughout the inner section of the cell. The ratio of the densities of the two gases in the central portion of the inner section is then determined by the ratio of the pressures of the two buffer gases. A heat pipe configuration that can be used for mixtures of two metal vapors has been described by Vidal and Hessel.[71]

The heat pipe offers advantages over other methods of containing metal vapors. First, because the pressure of the metal vapor is determined by the pressure of the buffer gas, and not by the amount of heat supplied to the oven, the density of the metal vapor can be measured much more accurately than it could be from measurements of the cell temperature. This is especially important because of the difficulties that are usually encountered in finding the coldest part of the cell, which determines the metal vapor density in a conventional arrangement. Second, because the metal vapor is excluded

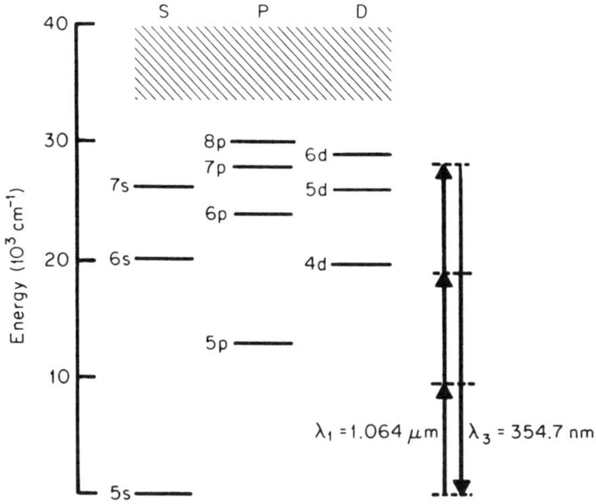

Fig. 33. Level diagram for third-harmonic conversion of Nd:YAG laser radiation in Rb.

from the end region of the cell near the windows, room temperature seals containing rubber O-rings can be used. In addition, highly corrosive metals such as cesium can be used conveniently without degradation of the windows.

8.1.4. Experiments with Metal Vapors

Most of the experiments in third-harmonic conversion of Nd laser radiation have been done with pure Rb and with Rb:Xe mixtures. A partial energy level diagram for the third-harmonic process in Rb is shown in Fig. 33. The third-harmonic frequency lies between the 7p and 8p levels. The oscillator strength of the resonance transition is large enough to make the dispersion for the third-harmonic process negative.

Third-harmonic conversion of radiation from an Nd:YAG laser in pure Rb has been reported by Young *et al.*[65] An example of the variation of the third-harmonic signal with the temperature of a cell containing Rb is shown in Fig. 34. The pump beam was gently focused for these measurements ($b = 40$ cm compared with a cell length of 20 cm) so that the plane wave approximation was almost satisfied. A pump power of 50 kW was used and a third-harmonic power of 3 μW was generated at the optimum temperature, corresponding to a conversion efficiency of 6×10^{-11}.

The alternating maxima and minima in the third-harmonic signal correspond to temperatures at which the cell contains an odd or even number of coherence lengths. The first peak is slightly larger than the rest because of the small but finite amount of focusing of the pump beam. Measurements of this type can be used to determine the sign of the dispersion in materials for which sufficient information is unavailable for calculations. Situations

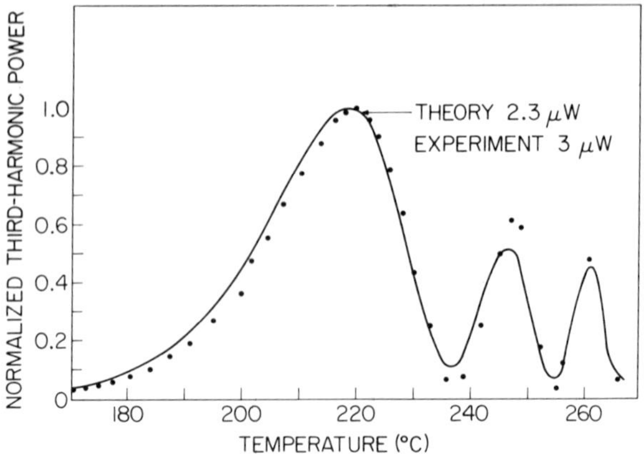

Fig. 34. Third-harmonic conversion of Nd:YAG laser radiation in pure Rb as a function of cell temperature. Pump power was 50 kW, confocal parameter was 40 cm, and the cell length was 20 cm. (From Ref. 65.)

of this type can be encountered, for example, for some materials in the vacuum ultraviolet region of the spectrum. Figure 35 shows curves calculated by Miles and Harris[30] for third-harmonic generation in two materials that have opposite signs of dispersion but otherwise identical properties. The calculations were done for a confocal-focusing geometry ($b = L$). The negatively dispersive medium very obviously shows a dominant peak in the conversion,

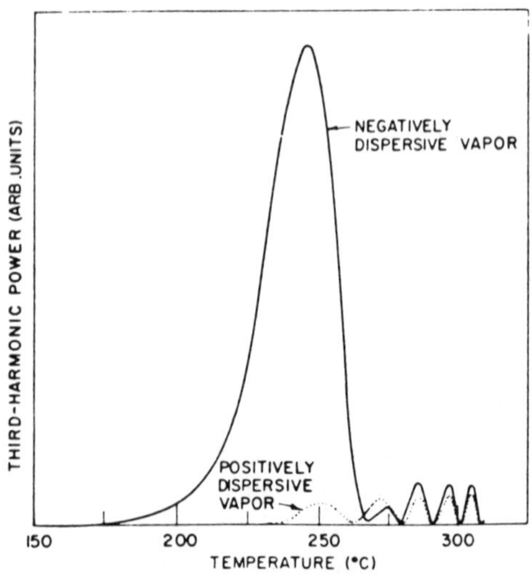

Fig. 35. Calculated third-harmonic conversion for negatively and positively dispersive media with otherwise identical properties. (From Ref. 30, © 1973 IEEE.)

TABLE 3

Calculated Plane-Wave Phase-Matching Ratios for Third-Harmonic Conversion of Nd : YAG Laser Radiation in Mixtures of Xe and Alkali Metal Vapors[a]

Element	Ratio $N_{Xe}:N_{metal}$
Li	154
Na	152
K	326
Rb	414
Cs	658

[a] From Ref. 30.

while all the peaks from the positively dispersive medium are of comparable magnitude.

Phase-matched conversion of radiation from an Nd:YAG laser in Rb:Xe mixtures has been reported by Young *et al.*,[65] Miles and Harris,[30] and Bloom *et al.*[67] Puell *et al.*[4] have reported similar measurements using radiation from an Nd:glass laser. Miles and Harris[30] have given plane-wave phase-matching ratios for mixtures of all the alkali metals with Xe. Their calculated values for the Nd laser wavelength ($\lambda = 1.064\ \mu$m) are listed in Table 3. These values were calculated from the following Sellmeier formulas for the metal vapors and Xe:

$$(n-1)_{\text{metal}} = 4.48 \times 10^{-14}\, N \sum \frac{f_n}{(\lambda_n^{-2} - \lambda^{-2})}, \tag{102}$$

$$(n-1)_{\text{Xe}} = N\left(\frac{1.46 \times 10^{-14}}{4.63012 \times 10^{9} - \lambda^{-2}} + \frac{1.46 \times 10^{-14}}{5.9779 \times 10^{9} - \lambda^{-2}} + \frac{2.75 \times 10^{-13}}{1.39831 \times 10^{10} - \lambda^{-2}}\right). \tag{103}$$

Here λ and λ_n are in cm and N is in cm^{-3}.

An example of phase-matched conversion in the Rb:Xe system is shown in Fig. 36. In these measurements the Xe pressure was fixed at 81 Torr and the Rb density was varied by adjusting the cell temperature. Phase matching was observed at a temperature of 267°C, corresponding to a phase-matching ratio of Xe to Rb of 380:1. The peak conversion in the phase-matched system was 33 times greater than that for the pure Rb. At a pump power of 50 kW from a Q-switched Nd:YAG laser about 100 μW of power was generated at 354.7 nm, corresponding to a conversion efficiency of 2×10^{-9}.

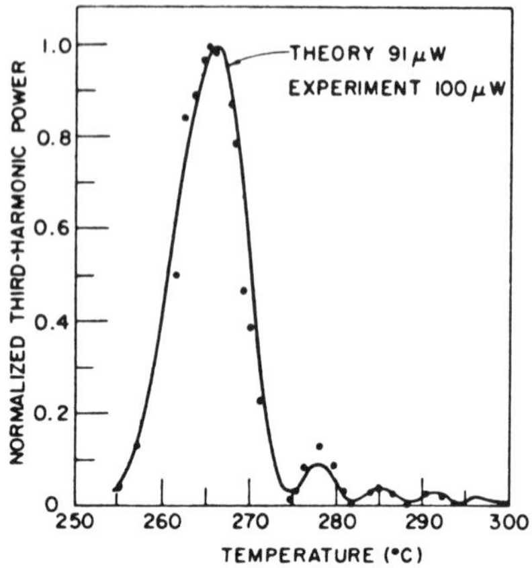

Fig. 36. Phase-matched third-harmonic conversion of Nd:YAG laser radiation in Rb:Xe mixture as a function of cell temperature. (From Ref. 30, © 1973 IEEE.)

Similar measurements reported by Bloom *et al.*[67] have resulted in a conversion efficiency of 0.1, the highest efficiency that has yet been observed in a vapor system. They used a concentric heat pipe to contain the vapor mixture with an active column length of 50 cm. The pump beam was obtained from a mode-locked Nd:YAG laser and had a peak power of 3×10^8 W and a pulse duration of 30 psec. It was focused to an area of about 1.1×10^{-2} cm^2, giving a pump intensity of 10^{10} W/cm^2. The confocal parameter of the pump beam was of the order of 400 cm and so the measurements were done effectively in the plane-wave regime.

The conversion efficiency is shown as a function of pump intensity in Fig. 37. The Rb pressure for these measurements was 3 Torr and the phase-matching ratio of Xe to Rb was 400 : 1. The maximum conversion efficiency of 0.1 was observed at a pump intensity of 10^{10} W/cm^2. The conversion efficiency in these measurements appeared to be limited by an inability to increase the pressure of the Rb because of condensation of the Rb at the Rb–Xe interface. Similar measurements in Na that were reported by the same authors gave a conversion efficiency of 0.027. Here the limitations on efficiency were attributed to multiphoton absorption in the Na.

Measurements by Puell *et al.*[4] of third-harmonic conversion of radiation from an Nd:glass laser have yielded conversion efficiencies of 0.024 with 7-psec-long pump pulses. They used slightly lower Rb pressures (1 Torr) and higher pump intensities (2×10^{11} W/cm^2) and focused their pump beam so that its confocal parameter was comparable to the length of the active medium. Conversion efficiency in these experiments appeared to be limited by breaking of phase matching due to a nonlinear refractive index. These limiting effects will be treated in more detail in Chapter 4.

Comparison of the results reported in these two sets of experiments indicates a trend that seems to be characteristic of conversion in vapor systems.

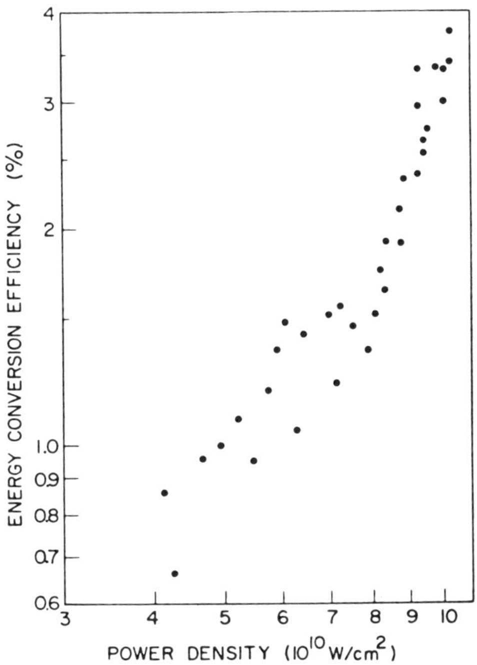

Fig. 37. Variation of third-harmonic energy conversion efficiency in phase-matched Rb:Xe mixture as a function of pump intensity. The cell length was 50 cm, the confocal parameter was about 400 cm, and the Rb pressure was 3 Torr. (From Ref. 67.)

Higher maximum conversion efficiencies appear to be obtainable when phase-matched plane-wave geometries and relatively low peak intensities are used rather than tightly focused geometries and relatively high peak intensities. This trend indicates the importance of considering other non-linear processes in determining the maximum conversion efficiency that can be obtained from a given system.

8.1.5. Comparison of Calculated and Measured Conversion Efficiencies

It is of interest to compare the conversion efficiencies that are calculated from the formulas given in earlier sections with the values that are measured in the laboratory. This comparison can serve as a useful guide to the reliability of the theoretical formulas in predicting actual performance. To this end we can calculate the expected conversion efficiency for the experiments which were reported in Bloom *et al.*,[67] the results of which were given in Fig. 37.

We can calculate the parameter K in Eq. (17c) from the experimental parameters given in Bloom *et al.*[67] and the nonlinear susceptibility calculated by Eicher.[31] In the absence of pump depletion the intensity conversion efficiency is given by $(KI_{10}L)^2$. The numerical values of the quantities needed to evaluate K are $\chi_a^{(3)} = 1.07 \times 10^{-32}$ esu; $N = 1.05 \times 10^{17}\ \text{cm}^{-3}$; and $\lambda_1 = 1.064 \times 10^{-4}$ cm. With these values we find a value for $K = 2.59 \times 10^{-19}\ \text{cm}^{-3}/\text{W}$. At a peak on-axis pump intensity of 10^{10} W/cm^2 $= 10^{17}$ erg/cm^2 sec^{-1} and an interaction length of 50 cm we calculate $KI_{10}L = 1.259$. For a $(KI_{10}L)$ product this large we need to include pump depletion and we find from Eq. (17a) the on-axis intensity conversion efficiency to be

$$\eta_{\text{I, on axis}} = \frac{I_{3,\max}}{I_{10}} = 0.626.$$

The total energy conversion efficiency, which is the quantity that was measured in Ref. 67, can be obtained from Fig. 31 for a beam that is Gaussian in space and time. With a value of $KI_{10}L = 1.29$ the total predicted energy conversion efficiency is

$$\eta_{\text{energy, total}} = 0.18.$$

This value is somewhat higher than the maximum measured value of 0.1 that was reported in Bloom *et al.*[67] Better agreement between theory and experiment can be obtained by using the value for $\chi_a^{(3)}$ of 6.15×10^{-33} esu that is tabulated by Miles and Harris.[30] When this value is used a theoretical efficiency of 0.09 is obtained. The magnitude of the susceptibility given by Eicher,[31] however, is in better agreement with the experiments that are reported in Ref. 4.

The degree of agreement between theory and experiment illustrated in the preceding paragraph is indicative of the accuracy with which conversion efficiencies can be predicted, provided that the necessary parameters of the nonlinear medium, the pump laser, and the experimental geometry are accurately known.

The accuracy of measured energy conversion efficiencies depends on the precision with which the relative sensitivity of a suitable energy meter is known in different spectral regions. Typically the calibration is obtained from manufacturers' specifications and can be accurate to better than 5%.

Calculation of expected conversion efficiency in a given experimental configuration is subject to more sources of error. These arise from uncertainties in the value of the nonlinear susceptibility, the absolute value of the pump intensity, and the profile of the pump beam in space and time. We have seen in the example above the effects of small uncertainties in the magnitude of the nonlinear susceptibility. Measurements of the pump beam parameters are of equal importance. Absolute measurements of the pulse duration and pulse energy can usually be made to an accuracy of 5% and perhaps better. Since the pump intensity enters the conversion efficiency formula squared, uncertainties in the predicted intensity conversion efficiency of up to 10% can be expected from this source. Somewhat lower uncertainties will exist if pump depletion is present.

In dealing with theoretical predictions of expected energy conversion efficiency that can be obtained with a given pump beam, it is necessary to know the profile of the pump beam in space and time accurately, especially in the low-intensity wings. Unfortunately it is just this part of the beam that is hardest to measure accurately. As a result the energy conversion efficiency can easily be overestimated unless special precautions are taken to use a beam with an accurately known profile. The magnitude of the inaccuracies that can be introduced by this source can be appreciated by considering the curves in Fig. 31. At low conversion levels the conversion efficiency of a square beam (no low-intensity wings) is higher than that of a beam with Gaussian wings in space and time by the factor 5.2. Beams with more pronounced wings, such as Lorentzians, will have even less energy conversion efficiency for a given on-axis pump intensity, while beams with less-pronounced wings, such as truncated Gaussians, will have a relatively higher energy conversion efficiency. Further examples of the variation of conversion efficiency of beams with different spatial profiles have been given by Yiu *et al.*[6] The problems mentioned here are most likely to be encountered in predictions of conversion efficiency for relatively high power beams for which it is desired to produce as much harmonic energy as possible. In experiments in which pump beam power can be sacrificed in favor of spatial filtering to produce an accurately known profile, more accurate predictions of conversion efficiency may be possible.

A more serious discrepancy between theory and experiment can result from additional effects which compete with the frequency conversion process and limit its efficiency. When these effects are prominent the formulas in this chapter do not accurately model the frequency conversion process and will not give a reliable estimate of the conversion efficiency that can be obtained in the laboratory.

For example, in the experiments of Puell *et al.*[4] in phase-matched mixtures of Rb:Xe at pump intensities of 10^{11} W/cm^2, the calculated conversion efficiency using formulas similar to those given in this chapter was an order of magnitude larger than that which was measured. The efficiency that was measured at lower pump intensity was well accounted for by these formulas, however. The competing effects that are responsible for these discrepancies generally become more pronounced as the pump intensity is increased. For experiments done at even higher intensities than the 10^{11} W/cm^2 that was used in the experiments of Ref. 4 the disagreement between the simple theory given here and the measured values can be much greater.

8.2. Vacuum Ultraviolet Radiation

Radiation in the vacuum ultraviolet can be generated from pump sources that lie in the visible or near ultraviolet. In this section we shall consider conversion to wavelengths between about 200 nm, where atmospheric absorption sets in, to about 110 nm, the approximate cutoff of the transmission of LiF windows. Fixed-frequency radiation in this wavelength range has been generated using light from pulsed lasers, including harmonics of Nd:YAG lases and pulsed nitrogen lasers. Tunable radiation has been generated using light from dye lasers and parametric generators. For the most part the light that is generated in this spectral range is pulsed, but reports of cw generation have been given. Conversion has been done in rare gases, molecular vapors, and metal vapors, and both resonant and non-resonant interactions have been used.

8.2.1. Fixed-Frequency Radiation

Fixed-frequency radiation has been generated primarily in nonresonant interactions with pump radiation that is obtained from pulsed lasers. The reported interactions are summarized in Table 4.

The harmonics of the Nd:YAG laser have been used to generate radiation at several fixed frequencies in the vacuum ultraviolet. Both metal vapors and rare gases have been used in this frequency range. Kung *et al.*[72] used Cd vapor to generate radiation at 177 nm by third-harmonic conversion of the second harmonic of the YAG laser, at 118.2 nm by third-harmonic conversion of the third harmonic of the YAG laser, and at 152 nm by four-wave sum-frequency mixing of the third harmonic and the fundamental. These processes are all negatively dispersive in Cd and were phase matched with Ar. Conversion efficiencies of 10^{-4} and 10^{-7} were measured for the third-

TABLE 4

Generation of Fixed-Frequency VUV Radiation

Laser	Nonlinear process	Material	Phase matching	λ_g (nm)	max η	Ref.
Nd:YAG	3×532	Cd	Ar	117	10^{-4}	72
	$2 \times 532 + 1060$	Cd	Ar	152	10^{-6}	72
	3×354.7	Cd	Ar	118.2	10^{-7}	72
	3×354.7	Ar	Xe	118.2	0.002[75]	74–76
N_2	3×337.1	Kr	(Tight focus)	112.4	10^{-6}	77
Rare-gas	$2 \times 431 + 308$	Mg[a]		127	10^{-3}–10^{-4}	78
halide + dye	$2 \times 431 + 249$	Mg[a]	Xe	115	10^{-3}–10^{-4}	78

[a] Two-photon resonance with the $3s^2$–$3s3d\ {}^1D_2$ transition.

TABLE 5

Wavelength at which Each of the Rare Gases First Exhibits Negative Dispersion

Gas	λ (nm)
Helium	58.4
Neon	74.4
Argon	106.6
Krypton	123.5
Xenon	147

harmonic processes and an efficiency of 10^{-6} was reported for the sum-frequency process. The pump power at 532 nm was about 16 MW and the power at 354.7 nm was about 2 MW. The values of the nonlinear susceptibilities that were estimated from these measurements were in reasonable agreement with the corresponding theoretical values.

The rare gases themselves can be used for phase-matched frequency conversion in the vacuum ultraviolet for generation of radiation at frequencies above their resonance transitions. The wavelength at which each of the rare gases first becomes negatively dispersive for third-harmonic generation is listed in Table 5. Mahon *et al.*[73] have calculated the wave vector mismatch for third-harmonic generation in some of the rare gases as a function of generated wavelength. These results are reproduced in Fig. 38. The ranges of negative dispersion for each gas are indicated by the shaded regions.

Frequency tripling of the third harmonic of the Nd:YAG laser at 354.7 nm to produce radiation at 118.2 can be accomplished in Xe and can be phase matched with Ar. Reports of this process have been given by Kung *et al.*,[74] Zych and Young,[75] and Berkov *et al.*[76] The experimental arrangement used by Kung *et al.* and by Zych and Young is shown in Fig. 39. Zych and Young[75] reported a maximum conversion efficiency of 0.002 in a phase-matched mixture of Xe and Ar with a pump power of about 45 MW, corresponding to a generated power of about 90 kW. They observed that the conversion efficiency reached a limit when the focused pump intensity reached a level of about 5×10^{12} W/cm^2.

Theoretical calculations of the nonlinear susceptibility of Xe in this wavelength range is complicated by the need to include the contribution of the continuum in the sum over intermediate states. Zych and Young[75] estimated a value of $\chi_a^{(3)}(-3\omega, \omega, \omega, \omega) = 2.3 \times 10^{-35}$ esu from their conversion measurements at low pump intensities. Based on their measured values for the various susceptibilities they estimate that a conversion efficiency of

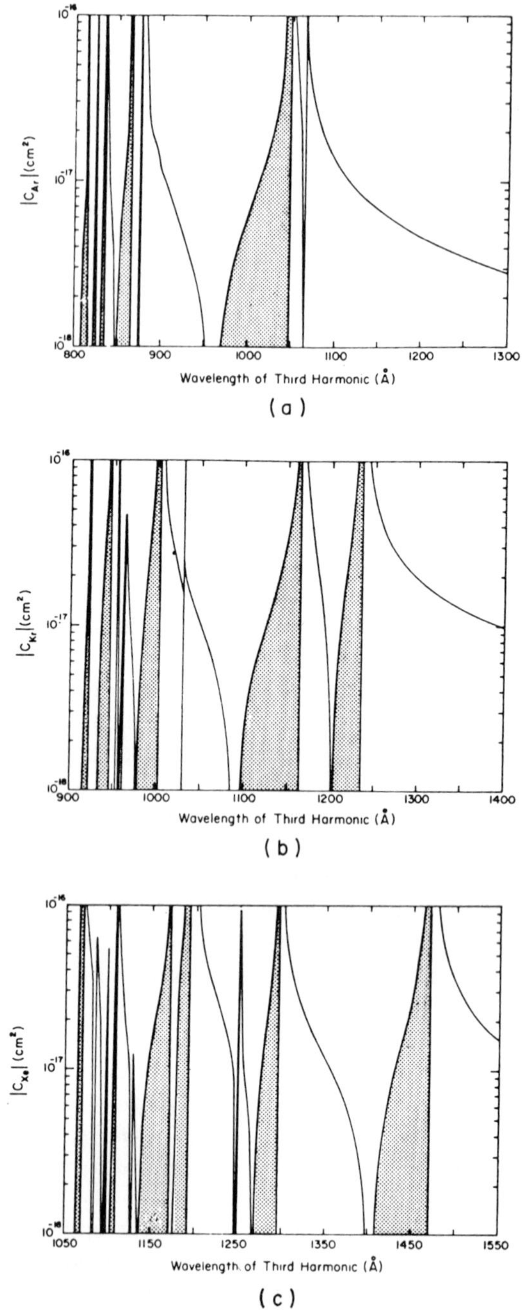

Fig. 38. Wave vector mismatch per atom for third-harmonic generation in the rare gases as a function of the generated wavelength: (a) argon, (b) krypton, (c) xenon. (From Ref. 73, © 1979 IEEE.)

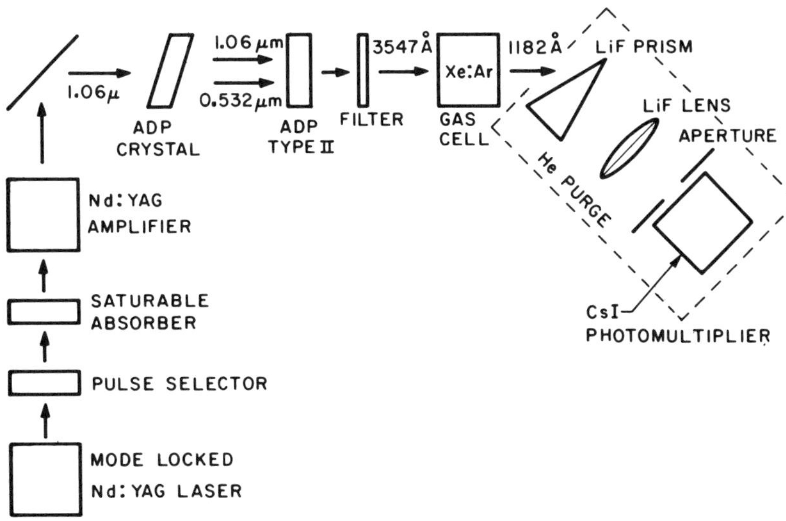

Fig. 39. Experimental arrangement used to study generation of vacuum ultraviolet radiation in rare gases. (From Ref. 74.)

0.009 should be possible for this process in phase-matched Xe. The limitations on efficiency that were observed in these experiments will be discussed in more detail in Chapter 4.

Berkov *et al.*[76] reported a conversion efficiency for this process in a phase-matched mixture of Xe and Ar of 0.001 with a pump power of about 400 MW in a pulse with a duration between 150 and 200 psec.

Cotter[77] has reported third-harmonic conversion of radiation from an N_2 laser, producing radiation at 112.4 nm. Conversion was done in Kr in in a tight-focused geometry. Maximum conversion was observed at a gas pressure of 5 atm. An efficiency of 10^{-6} was reported for a pump power of about 1 MW. Caro *et al.*[78] have used two-photon resonant sum-frequency mixing in Mg vapor to produce radiation at 127 and 115 nm by mixing radiation from a dye laser with radiation from an XeCl laser and a KrF laser.

8.2.2. *Tunable VUV Radiation*

Tunable radiation in the vacuum ultraviolet has been generated through the processes $\omega_4 = 3\omega_1$, $\omega_4 = 2\omega_1 \pm \omega_2$, and $\omega_4 = \omega_1 + \omega_2 \pm \omega_3$ with one or more tunable lasers used for the pump radiation. For third-harmonic conversion tunability of the generated radiation requires that the pump radiation be tuned. For the sum- and difference-frequency mixing processes tunable radiation can be generated if one of the pump frequencies is tunable. The remaining pump frequency or frequencies can be fixed. It can be adjusted to match an appropriate two-photon resonance in the medium to enhance the conversion efficiency or it can be derived from a fixed-frequency laser.

When multiple pump frequencies are used, several third-order processes are present simultaneously and can generate radiation at several wavelengths. For example, when the pump radiation contains the frequencies ω_1 and ω_2, radiation at the frequencies $\omega_4 = 3\omega_1, 3\omega_2, 2\omega_1 \pm \omega_2$, and $2\omega_2 \pm \omega_1$ will in general be produced. In many situations it is desirable, however, that the generated radiation be produced through only one of these processes. Several techniques can be used to suppress the unwanted radiation. For example, the unwanted processes can be discriminated against by appropriate phase-matching adjustments. Additional discrimination can be obtained through the dependence of the conversion efficiency on resonances and polarization of the pump radiation. Two techniques have been used to suppress the third harmonic of either pump beam in favor of one of the mixing processes. In the first, illustrated in Fig. 40a, the pump beams are circularly polarized with opposite senses. The third harmonic of either pump beam cannot be generated, as discussed in Section 1-4.11c. For similar reasons the difference frequency processes will also not occur. The sum-frequency processes will be allowed with the generated wave having circular polarization in the same sense as the pump wave that is used twice in the sum process. If the frequency of one of the pump waves, e.g., ω_1, is chosen to

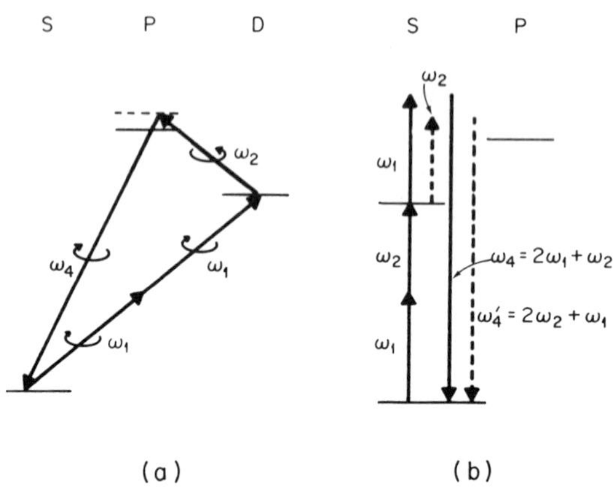

Fig. 40. Methods for suppressing third-harmonic generation in favor of sum-frequency generation. (a) Pump beams at ω_1 and ω_2 are oppositely circularly polarized. Output beam is polarized in same sense as beam at ω_1. Only states with D symmetry can be used for two-photon resonant enhancement. (b) Pump beams linearly polarized. Discrimination is obtained through two-photon resonance at $\omega_1 + \omega_2$. Frequency of the output wave can be chosen to be either at $\omega_4 = 2\omega_1 + \omega_2$ or $\omega_4' = 2\omega_2 + \omega_1$ by phase-matching adjustments. States with either S or D symmetry can be used for the two-photon resonant enhancement.

lie near a two-photon resonance while the other pump frequency is not two-photon resonant, the sum-frequency process $\omega_4 = 2\omega_2 + \omega_1$ will also be suppressed, leaving only the radiation generated at $\omega_4 = 2\omega_1 + \omega_2$. The frequency of the generated wave can be tuned by tuning ω_2 while the resonant enhancement is maintained by ω_1. For media with L–S coupling and a ground state with S symmetry only D states can be used for the two-photon resonance in interactions of this type.

A second technique for suppressing unwanted processes involves the use of two or three tunable pump lasers. It is illustrated in Fig. 40b. Here the two-photon resonance is obtained through the combination $\omega_1 + \omega_2$, while both $2\omega_1$ and $2\omega_2$ are nonresonant. Either of the sum-frequency processes $\omega_4 = 2\omega_1 + \omega_2$ or $\omega_4' = 2\omega_2 + \omega_1$ can be selected by a further resonance at ω_4 or ω_4' or by appropriate phase-matching conditions.

In order to maintain the two-photon resonance with this technique as the output wavelength is varied it is necessary to tune both the pump waves at ω_1 and ω_2. Alternatively a third wave at ω_3 can be used to generate radiation at $\omega_4 = \omega_1 + \omega_2 + \omega_3$. The output wavelength can now be varied by tuning only the wave at ω_3 and leaving the frequencies of the other two waves fixed to match the two-photon resonance condition. This technique allows the use of S-symmetry states for the two-photon resonance, a property that has proved useful in the generation of cw VUV generation.

Tunable radiation in the VUV has been produced through interactions involving two or three tunable dye lasers, a single tunable dye laser, and a combination of radiation from a dye laser with light from fixed-frequency pulsed lasers. Resonant interactions have been used extensively in metal vapors to enhance the conversion of radiation from relatively low power dye lasers. Both resonant and nonresonant interactions have been reported in rare gases using light from dye lasers and parametric generators.

Metal Vapors. Radiation has been generated in metal vapors that can be tuned in steps from about 111 to 200 nm. The wavelength ranges, nonlinear processes, and particular media are summarized in Table 6. In the wavelength range between about 158 and 195 nm most of the work has been done in Sr vapor. Studies of nonlinear conversion in Sr have been given by Hodgson *et al.*,[80] Armstrong and Wynne,[36] Bjorklund *et al.*,[16] Freeman and Bjorklund,[81] Economou *et al.*,[82] Scheingraber *et al.*,[83] and Royt and Lee.[57]

Hodgson *et al.*[80] have reported a series of experiments using two N_2-laser pumped dye lasers as shown in Fig. 41a. The two-photon resonances for the processes listed in Table 6 that involved D states could be accessed by circularly polarized pump radiation at ω_1. Third-harmonic generation was suppressed by the use of two circularly polarized pump waves, as illustrated in Fig. 40a.

TABLE 6

Generation of Tunable VUV Radiation by Four-Wave Mixing in Metal Vapors

Wavelength range (nm)	Interaction (nm)	Material	Resonance	max η	VUV power (W)	Ref.
200	3×600	Na	Two-photon 3s–5s	0.08		79
195.7–190.7[a]	$2 \times 575.7 + \lambda_2$	Sr[d]	Two-photon $5s^2$–$5s5d\ ^1D_2$			80, 83
191.4–187[a]	$2 \times 575.7 + \lambda_2$	Sr[d]	Two-photon $5s^2$–$5s5d\ ^1D_2$			79
187.5–183.3[a]	$2 \times 540.9 + \lambda_2$	Sr[d]	Two-photon $5s^2$–$5s\ ^1D_2$			79
181.7–177.8[a]	$575.7 + 474.2 + \lambda_3$	Sr[d]	Two-photon $5s^2$–$5s7s\ ^1S_0$	3×10^{-5}	5×10^{-5}	79
171.0–168.0[a]	$2 \times 520.8 + \lambda_2$	Sr[d]	Two-photon $5s^2$–$5s7s\ ^1S_0$ autoionizing			16
	$440 + 636.2 + \lambda_3$	Sr[e]	Two-photon $5s^2$–$5s7s\ ^1S_0$ autoionizing			16
169[b]	$2 \times 520.2 + \lambda_2$	Sr[d]	Two-photon $5s^2$–$5s7s\ ^1S_0$ autoionizing			81
167.4–167[b]	$2 \times 520.2 + \lambda_2$	Sr[d]	Two-photon $5s^2$–$5s7s\ ^1S_0$ autoionizing			81

165.6–164.8[c]	$594.5 + 462.5 + \lambda_3$	Sr[d,e]	Two-photon $5s^2$–$5s7s\ ^1S_0$ autoionizing			82
160–135	$2 \times 430.88 + \lambda_2$	Mg	Two-photon $3s^2$–$3s3d\ ^1D_2$	0.003[f]	100[f]	34,34a
160–143	$2 \times 459.7 + \lambda_2$	Mg	Two-photon $3s^2$–$3s4s\ ^1S_0$			34a
129–121	$2 \times 380.3 + \lambda_2$	Mg	Two-photon $3s^2$–$3s5s\ ^1S_0$		10^8 photons/pulse	84
143	3×430.88	Mg[g]	Two-photon $3s^2$–$3s3d\ ^1D_2$	1.2×10^{-4}		35
125.1	$2 \times 312.8 + \lambda_2$	Hg	Two-photon $6s^2$–$6s7s\ ^1S_0$	0.003	5000	86
124.7–125.3, 122.8–123.5, 121.8–122.1	$2\omega_1 + \omega_2$	Hg	Two-photon $6s^2\ ^1S_0$–$6s7s\ ^1S_0$	0.01		87
121.8–119.7	$2 \times 312.5 + \lambda_2$	Hg	Two-photon $6s^2$–$6s7s\ ^1S_0$		2×10^9 photons/pulse	88
188–177	$2 \times \lambda_1 - \lambda_{dye}$	Hg	Two-photon			89
127–119	$2 \times \lambda_1 + \lambda_{dye}$	Hg	Two-photon			89
116–111.7	$2 \times \lambda_1 + \lambda_{dye}$	Hg	Two-photon			89

[a] λ_{VUV} peaks near autoionizing resonances.
[b] Autoionizing resonances tuned with electric field.
[c] Autoionizing resonances tuned with magnetic field.
[d] Can be phase matched with Xe.
[e] Phase matched in tight focus by frequency selection.
[f] Maximum efficiency and power obtained with third-harmonic generation of 143 nm.
[g] Phase matched with Kr.

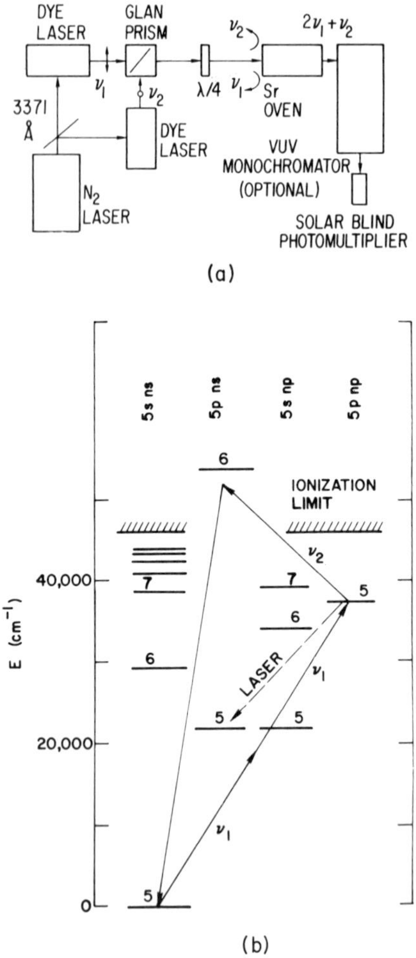

Fig. 41. (a) Schematic illustration of apparatus used to generate tunable VUV radiation in Sr vapor. (b) Energy level diagram showing four-wave frequency conversion processes in Sr. (From Ref. 80.)

Resonances with S states can also be used if the wavelength of both pump waves is adjusted so that the resonance condition is matched by the combination $\omega_1 + \omega_2$ and if the pump radiation is linearly polarized. Hodgson *et al.* reported that their strongest VUV conversion in Sr was obtained in this manner with the resonance occurring the 5s7s $^1S_0$ state. They measured a conversion efficiency of 3×10^{-5} for the process $\omega_4 = 2\omega_1 + \omega_2$, where $\lambda_1 = 575.7$ nm, $\lambda_2 = 474.2$ nm, and $\lambda_4 = 179.1$ nm and the efficiency was measured with respect to the power in the wave at ω_2. Somewhat higher

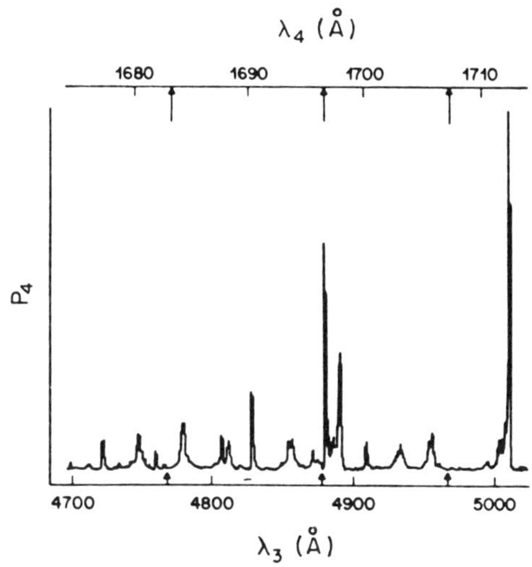

Fig. 42. Spectrum of radiation generated in Sr through the process $\omega_4 = 2\omega_1 + \omega_2$ in the wavelength range near 170 nm. The peaks in the spectrum are due to autoionizing levels with $J = 1$. (From Ref. 94.)

conversion efficiency for conversion in Sr vapor has been reported by Wallace and Zdasiuk,[34] who used a pump beam with a smaller bandwidth, and by Scheingraber *et al.*[83]

The radiation that is generated in Sr lies in the photoionization continuum, as shown in the level diagram in Fig. 41b. The sum-frequency process shows evidence of additional resonant enhancement when the generated frequency is tuned near an autoionizing resonance. A typical spectrum showing the effects of these resonances is given in Fig. 42. Wynne and Armstrong[36] have reported use of the four-wave mixing process for spectroscopic studies of the autoionizing transitions.

It is evident from the spectrum shown in Fig. 42 that the presence of these resonances can restrict the effective tuning range. Freeman and Bjorklund[81] and Economou *et al.*[82] have shown that the position of the resonances can be tuned with an electric or magnetic field to follow the wavelength of the generated radiation, thereby extending the tuning range. A typical demonstration of the range over which the autoionizing levels can be shifted is shown in Figs. 43a and 43b.

Attempts to phase match these interactions near the autoionizing resonances have been reported by Hodgson *et al.*,[80] Scheingraber *et al.*,[83] and Bjorklund *et al.*[16] Hodgson *et al.*[80] used mixtures of Sr and Xe and found an increase in VUV output of only a factor 5 compared to the conversion in pure Sr. Scheingraber *et al.*[83] also used Xe to phase match the interaction

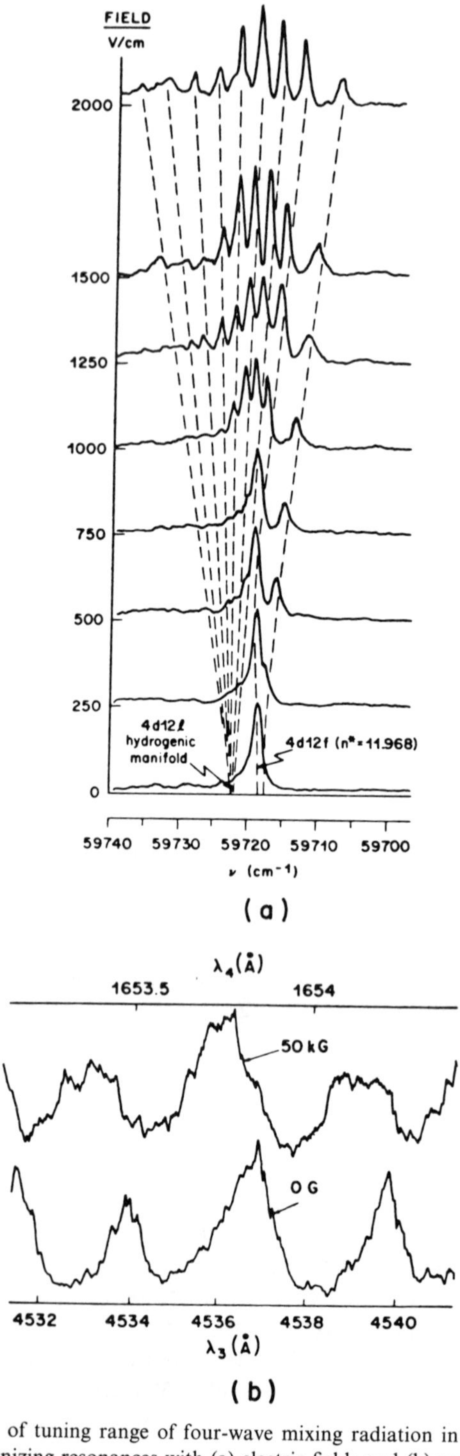

Fig. 43. Extension of tuning range of four-wave mixing radiation in Sr by tuning of the position of the autoionizing resonances with (a) electric fields and (b) magnetic fields. [From Ref. 81 (a) and Ref. 82 (b).]

and found an increase in VUV output of a factor 10^3 over that which they observed in pure Sr. They also demonstrated that increased conversion by phase matching is limited by absorption of the generated radiation by the Sr vapor.

Bjorklund *et al.*[16] have reported conversion in phase-matched interactions in which three pump frequencies were chosen to satisfy the appropriate phase-matching conditions (see Section 3.4). They chose the sum-frequency combination $\omega_1 + \omega_2$ to match the $5s^2 - 5s7s\ ^1S_0$ two-photon resonance ($\lambda_1 = 440$ nm, $\lambda_2 = 636.2$ nm) and swept ω_4 through the autoionizing levels by varying ω_3. Again increases of the order of 10^3 were observed when the frequency combination was correct for phase matching.

Radiation in the range from 120 to 160 nm has been generated in Mg vapor by Wallace and Zdasiuk,[34, 34a] McKee *et al.*,[84] and Junginger *et al.*[35] A partial level diagram for the processes involved is shown in Fig. 44. For generation of radiation in the range 135–160 nm a two-photon resonance with the $3s3d\ ^1D_2$ state was used with radiation from a dye laser tuned to 430.88 nm. The second laser was tuned over the range 430–600 nm using several dyes. Radiation near the long wave end of this region was also generated using pump radiation tuned to the two-photon resonance with the $3s4d\ ^1S_0$ state.[34a] Both third-harmonic generation and four-wave sum-frequency mixing were used.

The generated radiation falls in the photoionization continuum for Mg just as it did for Sr in the experiments at longer wavelengths, but, unlike Sr,

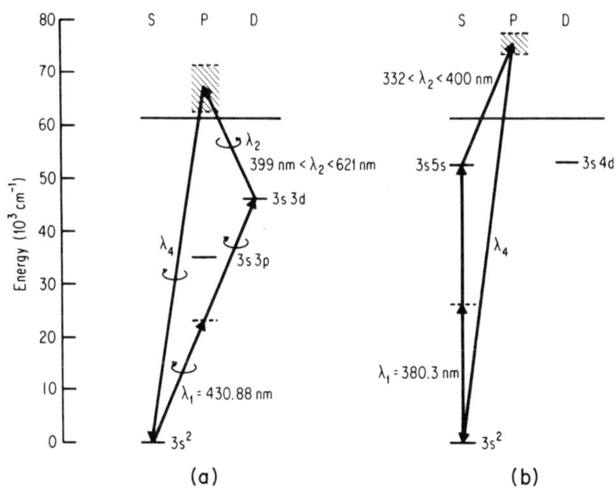

Fig. 44. Level diagram of Mg showing levels used for generation of radiation in the range 140–160 nm (a) and in the range 121–129 nm (b).

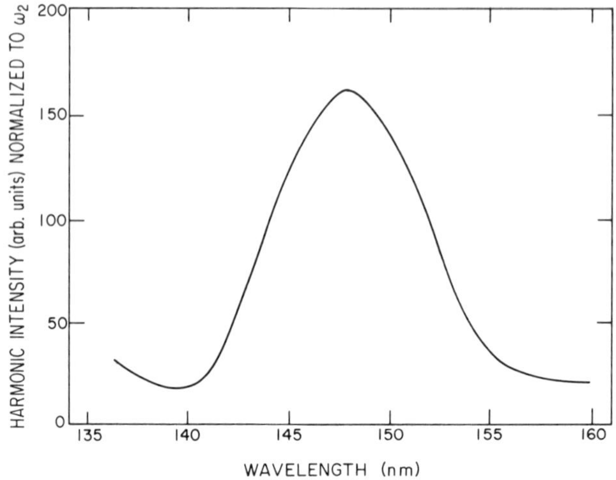

Fig. 45. VUV output spectrum from Mg in the range between 135 and 160 nm using the resonant channel shown in Fig. 44a. (From Ref. 34a, © 1981 by John Wiley & Sons, Inc.)

Mg has no sharp features in this spectral range due to autoionizing levels. As a result the tuning range in Mg can be much broader, as is shown in the spectrum in Fig. 45. Wallace and Zdasiuk[34] report that the conversion efficiency in Mg is about 10^3 higher than that in Sr under similar conditions.

They reported a conversion efficiency of 0.003 for third-harmonic conversion at a pump wavelength of 430.88 nm and a pump power of 30 kW, giving a VUV power of about 100 W. Junginger *et al.*[35] reported a lower conversion efficiency, 1.4×10^{-4}, for the same process in a phase-matched mixture of Mg and Kr.

Radiation in the range 120–130 nm has also been generated in Mg vapor by McKee *et al.*[84] This time the 3s5s $^1S_0$ state was used for the two-photon resonance with λ_1 tuned to 380.3 nm. About 2×10^9 photons could be generated in a single pulse at 127 nm with a pump power of 10 kW.

VUV radiation can also be generated by four-wave frequency mixing in Hg vapor. Several mixing processes in Hg have been proposed by Popov and Timofeev.[85] Those that have been reported in the VUV region of the spectrum are indicated in Table 6. Tomkins and Mahon[86–88] have reported generation of radiation in several wavelength bands ranging from about 119.7 to 125.14 nm and also at 208.57 nm. Conversion efficiencies up to about 0.01 were reported with maximum VUV powers of the order of 5 kW. Hilbig and Wallenstein[89] have also reported generation of radiation in several wavelength bands in the VUV by frequency mixing in Hg vapor. These are also indicated in Table 6.

Frequency Conversion in Molecular Vapors. VUV radiation has also been generated by two-photon resonant conversion of dye laser radiation in the molecular gases NO, Br_2, benzene, and I_2. The use of molecular gases offers the advantage of room temperature operation and a relatively large number of closely spaced even-parity levels that can be used for two-photon resonant enhancement. As a result third-harmonic conversion can be used for generating tunable radiation over a relatively wide spectral range as compared with metal vapors.

Innes *et al.*[90] have reported generation of radiation in the range 130–150 nm in NO, near 163 nm in benzene, and near 177 nm in Br_2. Typical spectra of the third-harmonic intensity in the vicinity of 151 nm generated in NO are shown in Fig. 46 for pressures of 50 Torr and 10 atm. The harmonic conversion was resonantly enhanced by two-photon transitions to the (0, 0) band of the $A^2\Sigma^+$–$X^2\pi_{3/2}$ electronic band system. They estimate the magnitude of the nonlinear susceptibility to be of the order of 10^{-33}–10^{-32} esu. The structure in the low-pressure curve of Fig. 46 is due to structure in the two-photon resonance spectrum. It is washed out at higher pressure.

A conversion efficiency of 10^{-7} was reported, in reasonable agreement with expectations for a pump power of 20 kW and an NO pressure of 90 Torr. This level of conversion corresponds to a generated power of 2 mW, or about 10^7 photons per pulse at a wavelength of 150.9 nm.

The structure in the spectrum of Fig. 46a indicates that third-harmonic conversion can be used for spectroscopic studies of the molecules of the nonlinear medium. Such studies have been reported for NO by Wallace and Innes[91] and for I_2 by Tai *et al.*[92]

Conversion in Rare Gases. Radiation over much of the range from about 118 to about 200 nm has been produced by conversion in rare gases. Third-harmonic generation and four-wave sum- and difference-frequency mixing have been used in both resonant and nonresonant interactions. Conversion of pump radiation from dye lasers and of radiation obtained from tunable parametric generators combined with radiation from fixed-frequency pulsed lasers has been reported. Sum-frequency mixing ($\omega_4 = 2\omega_1 + \omega_2$) can be used in this frequency range for generation of radiation that lies in a region of the spectrum for which one of the rare gases is negatively dispersive (see Section 3 and Fig. 38). Such interactions can be optimized in a tight focus or phase matched with a gas mixture. The difference-frequency process $\omega_4 = 2\omega_1 - \omega_2$, which can be optimized by pressure adjustment for conversion of focused beams in media with dispersion of either sign, has proved to be very useful for generation of radiation in media with positive dispersion. This process has been utilized to obtain radiation over an

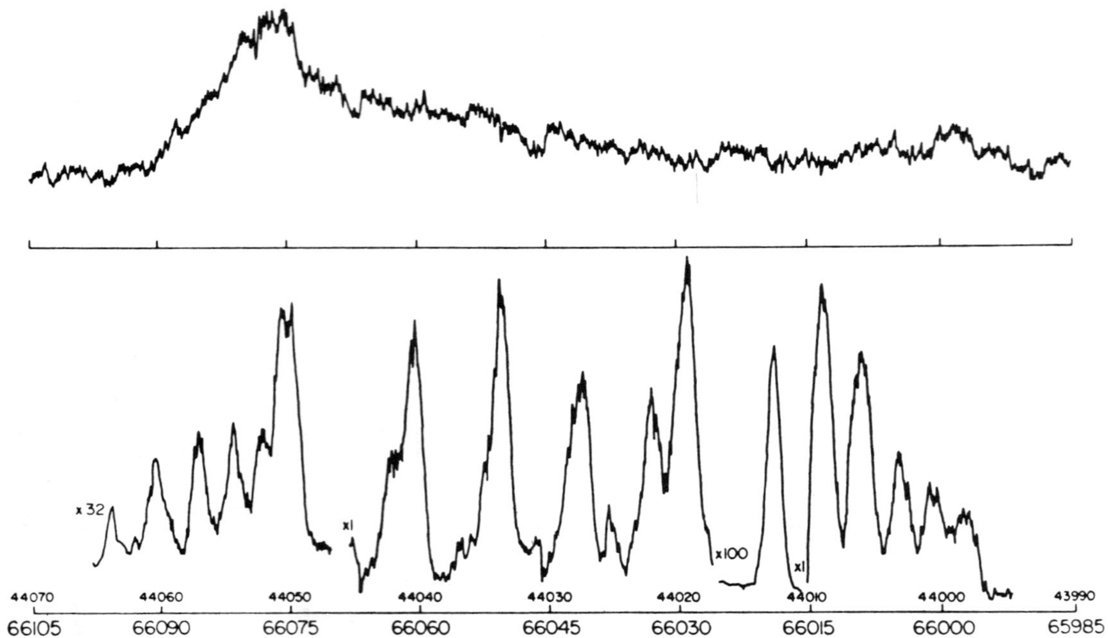

Fig. 46. Typical spectrum of VUV radiation near 150 nm obtained by third-harmonic conversion in NO at a pressure of 50 Torr (bottom) and 10 atm. (From Ref. 90.)

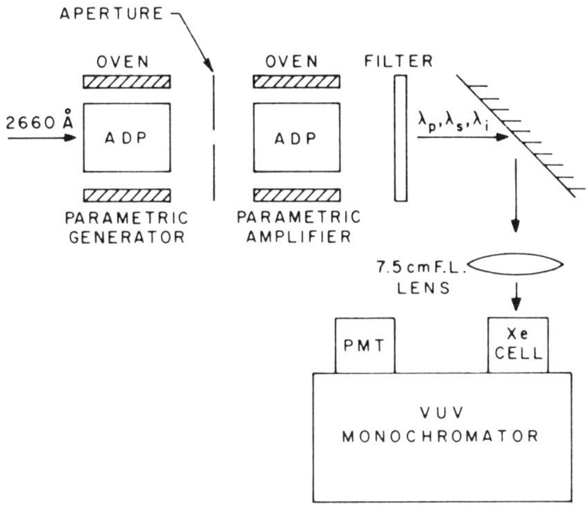

Fig. 47. Experimental arrangement used to generate tunable VUV radiation using light from a parametric generator. (From Ref. 93.)

extensive range of wavelengths that lie below the resonance lines of Xe. A summary of the nonlinear interactions, pump sources, rare gases and wavelength ranges is given in Table 7.

Kung[93] reported the generation of VUV picosecond pulses in Xe that are tunable in steps from about 117 to 195 nm. His tunable pump radiation was obtained from a parametric generator that was pumped by the fourth harmonic of a mode-locked Nd:YAG laser, as shown in Fig. 47. The parametric generator was capable of producing radiation that could be tuned from about 400 to 700 nm with a peak power of 10 MW.

The spectral range over which the VUV radiation was generated with this apparatus is illustrated in Fig. 48. Both sum- and difference-frequency processes of the type $\omega_4 = 2\omega_1 + \omega_2$ and $\omega_4 = 2\omega_1 - \omega_2$ were used, and the radiation from the parametric generator was mixed both with itself and with the fourth harmonic of the Nd:YAG laser as indicated in the figure.

The restrictions on the production of radiation by frequency conversion that are imposed by the sign of the dispersion of the medium as discussed in Section 3.1.1 are clearly illustrated in the results of Kung. The spectral range of the radiation that is produced by the sum-frequency process is in good agreement with the extent of the regions that are predicted to have negative dispersion by Mahon *et al.*[73] The results shown in Fig. 48 also indicate that no radiation was produced in the spectral regions for which the sum-frequency process had positive dispersion. Radiation was produced in regions of positive dispersion through the difference-frequency process, which can be optimized in a tight focus in a medium that has dispersion of

TABLE 7

Generation of Tunable VUV Radiation in Rare Gases

Wavelength range (nm)	Laser	Process (nm)	Material	Phase matching	Resonance	max η	Pulse power (W)	Ref.
194.6–163.1	Nd:YAG + parametric generator	$2 \times 266.1 + \lambda_{par.\,gen.}$	Xe			10^{-7}	1	93
147.0–140	Nd:YAG + parametric generator	$266.1 + 2 \times \lambda_{par.\,gen.}$	Xe			10^{-7}	1	93
129.6–126.5	Nd:YAG + parametric generator	$266.1 + 2 \times \lambda_{par.\,gen.}$	Xe			10^{-7}	1	93
119.2–118	Nd:YAG + parametric generator	$266.1 + 2 \times \lambda_{par.\,gen.}$				10^{-7}	1	93
207–197	Dye (590–620 nm) + second harmonic of dye	$2 \times \lambda_{SH} - \lambda_{dye}$	Xe			3.4×10^{-6}	15	89a
198.3–185	Dye (555–595 nm) second harmonic of dye	$2 \times \lambda_{SH} - \lambda_{dye}$	Xe			1.2×10^{-5}	60	89a
186.6–177.1	Second harmonic of dye (607.5–635 nm) + Nd:YAG	$2 \times \lambda_{SH} - \lambda_{1.06\,\mu m}$	Xe			4.6×10^{-6}	20	89a
178–170.7	Second harmonic of dye (588–610 nm) + Nd:YAG	$2 \times \lambda_{SH} - \lambda_{1.06\,\mu m}$	Xe			6.7×10^{-6}	30	89a
171.2–159.5	Second harmonic of dye (555–590 nm) + Nd:YAG	$2 \times \lambda_{SH} - \lambda_{1.06\,\mu m}$	Xe			3.3×10^{-6}	15	89a
129.4–126	Dye (633–647 nm) + second harmonic of dye	$2 \times \lambda_{SH} + \lambda_{dye}$	Xe			6.6×10^{-6}	22	89a

123.5–121.2	Dye (606–617.5 nm) + second harmonic of dye	$2 \times \lambda_{SH} + \lambda_{dye}$	Kr			2×10^{-6}	10	89a
122.1–120	Dye (600–610.5 nm) + second harmonic of dye	$2 \times \lambda_{SH} + \lambda_{dye}$	Kr			1.7×10^{-6}	8	89a
119.2–117.2	Dye (586–596 nm) + second harmonic of dye	$2 \times \lambda_{SH} + \lambda_{dye}$	Xe			5.3×10^{-6}	20	89a
117–113.4	Dye (567–585 nm) + second harmonic of dye	$2 \times \lambda_{SH} + \lambda_{dye}$	Xe			3.7×10^{-6}	20	89a
116.4–110	Dye (550–582 nm) + second harmonic of dye	$2 \times \lambda_{SH} + \lambda_{dye}$	Kr			0.8×10^{-6}	4	89a
160–122.6	Nd:YAG + dye + second harmonic of dye	$2 \times \lambda_{UV} - \lambda_{dye}$[a] $2 \times \lambda_{UV} - 1.06\ \mu m$[a]	Xe, Kr					89
146.9–140.3	Nd:YAG + dye + second harmonic of dye	$3 \times \lambda_{dye}$	Xe	Kr			30	13
123.6–120.3	Nd:YAG + dye + second harmonic of dye	$3 \times \lambda_{dye}$	Kr	Xe, Ar			16	13
220–155	Nd:YAG + dye + second harmonic of dye	$2 \times 255.94 - \lambda_{dye}$[b]	Xe		Two-photon 5p $^1S_0$–6p[5/2] $J = 2$		500	89
182–135	Nd:YAG + dye + second harmonic of dye	$2 \times \lambda_2 - \lambda_{dye}$[c]	Xe		Two-photon 5p–7p			89
174–130	Nd:YAG + dye + second harmonic of dye	$2 \times \lambda_3 - \lambda_{dye}$[c]	Xe		Two-photon 5p–4f			89
162.6	Dye + Raman shifted second harmonic of dye	$2 \times 252.2 - 564.1$	Xe		Two-photon 5p $^1S_0$–6p$[\frac{3}{2}]$ $J = 2$	0.004	360	89b

[a] λ_{UV} obtained from the process $\lambda_{UV} = \lambda_{SH} + \lambda_{1.06}\ \mu m$.
[b] $300\ nm < \lambda_{dye} < 760\ nm$.
[c] λ_2, λ_3 obtained as second harmonic of dye laser tuned to two-photon resonance in Xe.

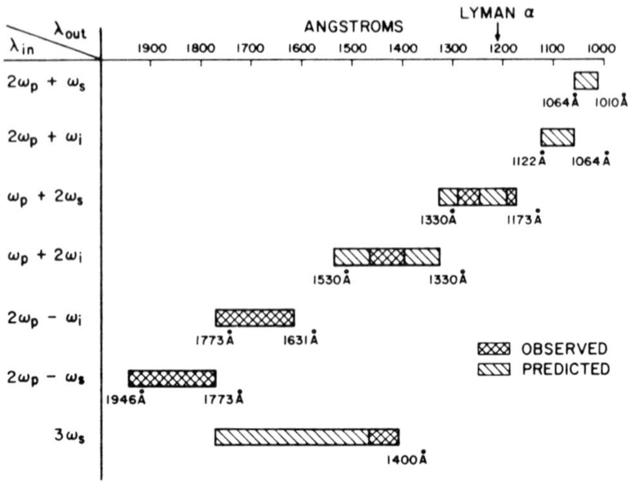

Fig. 48. Wavelength coverage in the VUV from conversion in Xe with the apparatus shown in Fig. 47. $\lambda_p = 266.1$ nm, λ_S and λ_I are obtained from the parametric generator. (From Ref. 93.)

either sign. Conversion efficiency of the order of 10^{-7} was reported, corresponding to a VUV power of about 1 W.

Nonresonant third-harmonic conversion of dye laser radiation and sum- and difference-frequency mixing of the second harmonic of the dye laser radiation with the fundamental of the dye laser and the fundamental of an Nd:YAG laser have been used by Hilbig and Wallenstein[13,89,89a] to generate tunable radiation in several wavelength bands over the range 110–207 nm. The pump radiation for these processes was obtained from a dye laser pumped by the second harmonic of a Nd:YAG laser. The frequency ω_1 was obtained by frequency doubling the dye laser radiation, whereas the frequency ω_2 was obtained directly from the dye laser or from the fundamental of the Nd:YAG laser. For the sum-frequency processes the dye laser radiation was tuned over the range 550–647 nm using Rhodamine 6G and DCM. The pump power in the dye fundamental was typically of the order of 4.5–5 MW and about 1–1.7 MW was obtained in the dye second harmonic. For the difference-frequency processes the dye laser radiation was tuned over the range 555–635 nm using the dyes Rh6G, Rh620, and Rh640. Similar powers were obtained for the pump fundamental and second harmonic.

The sum-frequency process $2\omega_1 + \omega_2$, where ω_1 is the second harmonic of the dye laser and ω_1 is its fundamental, was used to generate radiation between 110 and 130 nm in the negatively dispersive regions of Xe and Kr. The best performance was obtained in Xe, for which a maximum conversion efficiency of 10^{-6}–10^{-5}, corresponding to a VUV power of 20 W, was observed. Somewhat smaller conversion was obtained in Kr.

The difference-frequency process $2\omega_1 - \omega_2$, where ω_1 is the second harmonic of the dye laser and ω_2 is the fundamental of either the dye laser or the Nd:YAG laser, was used to generate radiation in the positively dispersive region of Xe between about 160 and 207 nm. A maximum power output in the VUV of about 60 W was obtained near 190 nm for these experiments, and somewhat lower powers, about 20 W, were obtained in the other wavelength ranges. Conversion in Kr in this wavelength range was found to be less efficient by a factor of about 4.

Hilbig and Wallenstein[89] have also used two-photon resonant conversion in Xe to produce radiation in the range 130–220 nm, as indicated in Table 7. They used an interaction of the form $\omega_{VUV} = 2\omega_{UV} - \omega_D$, where ω_{UV} was the second harmonic of a dye laser pumped by the third harmonic of a Nd:YAG laser and was adjusted to coincide with an appropriate two-photon resonance in Xe. The second frequency, ω_D, was obtained from a second dye laser. Because the difference-frequency process is maximized at zero wave vector mismatch they could optimize the VUV power in this wavelength range by adjusting the Xe pressure, regardless of the dispersion of the medium. They reported VUV powers up to 500 W for $P_{VUV} = 150$ kW, $P_D = 0.9$ MW, and a Xe pressure of 14 Torr, corresponding to a conversion efficiency of 0.0005.

They could also obtain fixed-frequency radiation by mixing the fundamental of the dye laser radiation or the Nd:YAG laser radiation with the two-photon resonant radiation. In mixing processes using the dye laser radiation they obtained a power of 1 kW in the VUV at 170.6 nm with a total input power of 1.3 MW. A somewhat higher conversion efficiency, 0.005, was observed with tighter focusing, but for this arrangement the VUV power was observed to saturate near 200 W when the total pump power was of the order of 100 kW.

Hager and Wallace[89b] have used the difference-frequency process to generate radiation at 162.6 nm with a two-photon resonant interaction in Xe using the $5p^6$–$5p^5\,6p[\frac{3}{2}]$ $J = 2$ transition. The resonant pump radiation at 252.2 nm was obtained by frequency doubling and anti-Stokes–Raman shifting the radiation from a dye laser which was pumped by a harmonic of an Nd:YAG laser. The second frequency was the fundamental of the dye laser radiation. The pump power was of the order of 80 kW for λ_1 and 100 kW for λ_2, and the conversion efficiency was of the order of 0.004, corresponding to a VUV power of about 360 W.

8.2.3. *Continuous Wave Vacuum Ultraviolet Radiation*

Freeman *et al.*[94] have reported the generation of cw radiation in the VUV by mixing radiation from a cw dye laser with radiation from an argon–ion laser in Sr vapor. The level diagram for their process is illustrated in Fig. 49.

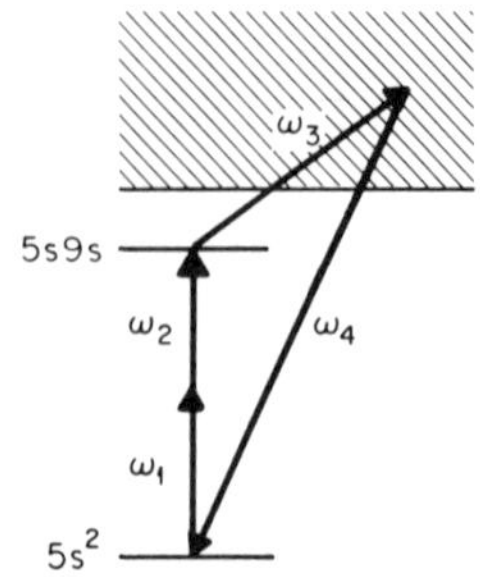

Fig. 49. Level diagram showing processes used for cw VUV generation of strontium. $\lambda_1 = 476.5$ nm; $\lambda_2 = 572.4$ nm; $\lambda_3 = 476.5$, 480, and 496.5 nm.

They used the sum-frequency process $\omega_4 = \omega_1 + \omega_2 + \omega_3$, where ω_1 and ω_3 were obtained from the argon laser and ω_2 was produced by the dye laser. The 5s7s $^1S_0$ state of Sr was used for two-photon resonant enhancement with the sum frequency $\omega_1 + \omega_2$, where λ_1 was the 476.5 nm line of the argon laser and λ_2 was tuned to 572.4 nm. Continuous wave radiation between 168 and 172 nm was generated by using for the pump wave at ω_3 the 476.4-, 488-, and 496.4-nm lines of the argon laser. The radiation that was generated near the autoionizing peak of Sr near 170nm using λ_3=488 nm was enhanced by the factor 20 by tuning the position of the autoionizing resonance with a 55-kG magnetic field, as indicated in Fig. 50. A peak VUV power of about

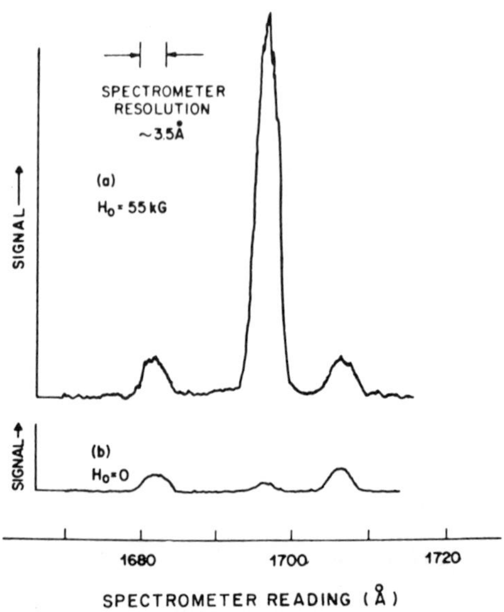

Fig. 50. Spectrum of cw VUV radiation showing the enhancement that is obtained by tuning the position of the autoionizing level of Sr with a 55-kG magnetic field. (From Ref. 94.)

6×10^{-11} W was observed when the magnetic field was used and the pump powers were 0.85 W at ω_3, 0.425 W at ω_1, and 0.66 W at ω_2. Comparison of the efficiency of these experiments with that obtained in the experiments of Bjorklund *et al.*[16] using pulsed lasers indicated that there was a slow process occurring in the cw experiments that competed with the conversion process and limited the efficiency.

8.2.4. Four-Photon Parametric Oscillation

Four-photon parametric oscillation has also been used for generation of radiation in both the near infrared and in the vacuum ultraviolet. This process involves the interaction $\omega_1 + \omega_2 \to \omega_3 + \omega_4$, where the pump waves are at ω_1 and ω_2 and the waves at ω_3 and ω_4 are generated in the interaction. In practice most experiments have been done with $\omega_1 = \omega_2$.

A typical level diagram for this process is shown in Fig. 51. The pump frequency is tuned to a two-photon resonance with a suitable level. Radiation is generated at frequency ω_3 near the 2–1 transition frequency and at ω_4 near the 1–0 transition frequency. The equations that describe the interaction are

$$\frac{1}{2ik_3}\nabla_\perp^2 A_3 + \frac{\partial A_3}{\partial z} = -\frac{\omega_3}{n_3}\kappa_p A_4^* e^{-i\Delta kz}, \tag{104a}$$

$$\frac{1}{2ik_4}\nabla_\perp^2 A_4 + \frac{\partial A_4}{\partial z} = -\frac{\omega_4}{n_4}\kappa_p A_3^* e^{-i\Delta kz}, \tag{104b}$$

where

$$\kappa_p = \frac{\pi|\mu_{01}|^2|\mu_{12}|^2 T_2 A_1^2}{2\hbar^3 c(\omega_{10} - \omega_1)}\left(\frac{1}{\omega_{10} - \omega_3} + \frac{1}{\omega_{10} - \omega_4}\right). \tag{104c}$$

In deriving Eqs. (104a) and (104b) we have assumed that any pump depletion can be neglected and that the pump wave is tuned to the exact two-photon resonance, while the generated waves are detuned from the intermediate resonance by several linewidths. We have also neglected any population changes.

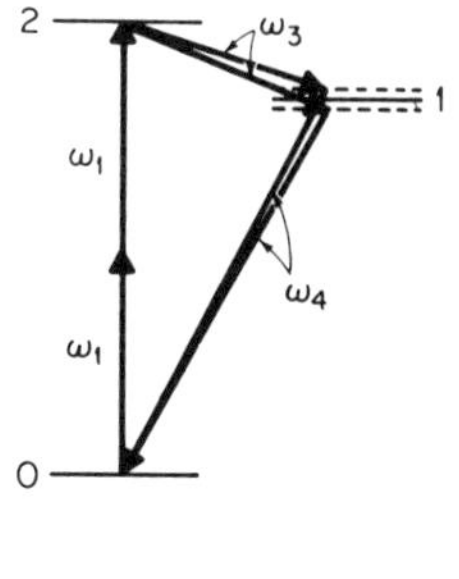

Fig. 51. Level diagram showing four-photon parametric oscillation. Pump wave is at ω_1, while waves at ω_2 and ω_3 are generated in the interaction.

In the plane-wave approximation at exact phase matching Eqs. (104) have the solution

$$A_3 = A_{30}\cosh(\kappa_{\mathrm{p}}\sqrt{\omega_3\omega_4/n_3n_4}\,z) - A_{40}^*\sqrt{\omega_3/\omega_4}\,\sin(\kappa_{\mathrm{p}}\sqrt{\omega_3\omega_4/n_3n_4}\,z), \tag{105a}$$

$$A_4 = A_{40}\cosh(\kappa_{\mathrm{p}}\sqrt{\omega_3\omega_4/n_3n_4}\,z) - A_{30}^*\sqrt{\omega_4/\omega_3}\,\sinh(\kappa_{\mathrm{p}}\sqrt{\omega_3\omega_4/n_3n_4}\,z), \tag{105b}$$

where A_{30} and A_{40} are the values of the generated fields at $z = 0$. The initial values for these fields may be either the zero-point quantum fluctuations, as in the case of three-wave parametric oscillators, or the tail of the emission from the 2–1 or 1–0 transitions.

The four-wave parametric interaction shown in Fig. 51 is doubly resonantly enhanced. Its efficiency depends strongly on the detuning of the pump wave from the two-photon resonance. Generally a weaker dependence is observed on the detuning of the generated waves from the intermediate resonance and radiation can be generated over a band of wavelengths extending to either side of the single-photon resonance.

The values of the frequencies ω_3 and ω_4 are determined by the phase-matching conditions just as in the case of three-wave parametric oscillators. When gases are used for the nonlinear medium, phase matching can be done for a particular pair of frequencies by varying the relative concentration of the components of a gas mixture or, if focused beams are used, by adjusting the pressure of the gas. The output wavelengths can be tuned to a certain extent by varying the phase-matching conditions.

In normal operation a population inversion is created on the 2–1 transition by the two-photon pumping of the upper level. As a result laser emission at ω_{21} usually accompanies the four-wave parametric interaction. The laser emission usually can be observed in both directions, while the radiation generated in the parametric interaction is observed only in the forward direction.

Four-wave parametric generation has been reported by several authors for the generation of radiation in the IR and in the UV and VUV. The IR results will be discussed in Section 8.3. In the VUV Tomkins and Mahon[86] and Bokor *et al.*[95] have reported four-wave parametric oscillation in Hg vapor. The pump and generated wavelengths as well as the resonant transitions for these interactions are summarized in Table 8. Both groups used dye lasers for pumping most of the transitions, although Bokor *et al.* were able to used a KrF laser for pumping the $6s^2\ {}^1S_0$–$6s10s\ {}^1S_0$ transition. Bokor *et al.*[95] reported a conversion efficiency of 10^{-4} for the generation of radiation at 143.5 nm with pump pulses of 2 MW and an efficiency of 10^{-5} for generation of radiation at 125.9 nm with pump pulses of 20 MW.

TABLE 8

Generation of VUV Radiation by Four-Wave Parametric Oscillation

VUV wavelength (nm)	Laser	Material	Two-photon resonance	Single-photon resonance	VUV power	Ref.
184.9	Dye	Hg	$6s^2$–$6s7s\ ^1S_0$	$6s6p\ ^1P_1$	10^8 photons/pulse	86
184.9	Dye	Hg	$6s^2$–$6s6d\ ^1D_2$	$6s6p\ ^1P_1$		95
143.5	Dye	Hg	$6s^2$–$6s6d\ ^1D_2$	$6s7p\ ^3P_1$	1 μJ/200 W	95
140.1	KrF	Hg	$6s^2$–$6s10s\ ^1S_0$	$6s7p\ ^1P_1$		95
130.7	KrF	Hg	$6s^2$–$6s10s\ ^1S_0$	$6s8p\ ^3P_1$		95
130.1	KrF	Hg	$6s^2$–$6s10s\ ^1S_0$	$6s8p\ ^1P_1$		95
125.9	KrF	Hg	$6s^2$–$6s10s\ ^1S_0$	$6s9p\ ^3P_1$	1 μJ/200 W	95
125.0	KrF	Hg	$6s^2$–$6s10s\ ^1S_0$	$6s9p\ ^1P_1$		95

8.2.5. *Anti-Stokes–Raman Scattering*

Anti-Stokes–Raman scattering is a type of two-photon resonant four-wave parametric mixing of the form $\omega_{AS} = 2\omega_L - \omega_S$ in which the intermediate state is a Raman active transition that is resonant with the frequency difference $\omega_L - \omega_S$. In a typical experiment the laser radiation at ω_L is supplied externally and the Stokes radiation at ω_S is generated by stimulated Raman scattering. The growth of the anti-Stokes–Raman beam in this type of interaction has been described by Bloembergen.[96]

Anti-Stokes–Raman scattering has been used to produce coherent radiation at several wavelengths in the ultraviolet and vacuum ultraviolet using radiation from dye lasers and various rare-gas halide lasers in combination with various molecular gases.[97,98] A summary of the combinations of laser wavelengths, generated wavelengths, and gases is given in Table 9. Usually several orders of both Stokes and anti-Stokes radiation are generated in a single interaction. Interactions of this type have been used to generate radiation down to 138 nm through a fifth-order anti-Stokes shift in H_2 of radiation from an ArF laser.[98] Similar interactions have been used to generate radiation at longer wavelengths in the near ultraviolet and visible.

8.2.6. *Ly–α Radiation*

Considerable effort has been spent by several authors in producing narrow-band tunable radiation at the wavelength of the Ly–α transition of atomic hydrogen ($\lambda = 121.6$ nm). This effort is motivated in part by a need to measure the density of neutral H atoms in Tokamak fusion plasmas by resonance fluorescence. Some of the reported results are given in Table 10.

TABLE 9

Generation of VUV Radiation by Anti-Stokes–Raman Scattering

Laser	λ_{pump} (nm)	Nonlinear medium	Anti-Stokes order	λ_{gen} (nm)	Ref.
Dye	300[a] (Second harmonic of Rh 101)	H_2	5	185[a]	97
Dye	600[a] (Rh 101)	H_2	9	185[a]	97
Dye	275[a] (second harmonic of Rh 6G)	H_2	4	189[a]	97
Dye	600[a]	H_2	5	175[a]	97
KrF	249[b]	H_2	3	190[b]	98
ArF	193[b]	H_2	1	179[b]	98
ArF	193[b]	D_2	1	182[b]	98
ArF	193[b]	H_2	2	166[b]	98
ArF	193[b]	H_2	3	155[b]	98
ArF	193[b]	H_2	4	146[b]	98
ArF	193[b]	H_2	5	138[b]	98

[a] Tunable over dye laser band.
[b] Tunable over excimer laser band.

Two-photon resonant conversion has been reported in Mg vapor (McKee *et al.*)[84], Hg vapor (Tomkins and Mahon)[88], and Be vapor (Mahon *et al.*)[99] using radiation from dye lasers. Ly–α generation by third harmonic conversion of dye laser radiation has been reported in rare gases[100–104] and in phase-matched mixtures of rare gases.[13, 14, 73, 103, 105] At a wavelength of 121.6 nm, Kr is the only rare gas that has negative dispersion. The third-harmonic process for generating Ly–α radiation can be phase matched with a mixture of Kr and either Ar or Xe.

The limitations on the bandwidth and tuning range of the converted radiation due to dispersion in the nonlinear medium have been analyzed by Mahon *et al.*[73] Generally it is found that wider bandwidths and larger tuning ranges can be obtained with focused beams than with collimated beams. The phase-matching properties of various rare-gas mixtures have been investigated by Hilbig and Wallenstein,[13] Batishche *et al.*,[14] Mahon *et al.*,[73] Langer *et al.*,[103] and Mahon and Yiu.[105] The usefulness of Xe as a phase-matching gas at this wavelength is limited by its absorption in the VUV.[13]

Hilbig and Wallenstein[13] reported the generation of pulses with a power of 16 W in a phase-matched mixture of Kr and Ar using pump pulses with a power of 1.5 MW. The line width of the pump radiation was of the order of 0.87×10^{-3}–$3.4 \times 10^{-2}\ \mathrm{cm}^{-1}$. Langer *et al.*[103] reported a conversion efficiency of 1.4×10^{-4} in a phase-matched mixture of Kr and Ar, which would correspond to a power level for their parameters of 420 W.

TABLE 10

Nonlinear Generation of Ly–α Radiation ($\lambda = 121.6$ nm)

Interaction (nm)	Nonlinear medium	Resonance	Phase matching	max η	Pulse energy/power (photons/pulse)	Ref.
$2 \times 380.3 + 336.7$	Mg	Two-photon $3s^2$–$3s5s\ ^1S_0$			10^8	84
$2 \times 310.4 + 560.6$	Be	Two-photon $2s^2$–$2s3d\ ^1D_2$		3×10^{-7}		99
$2 \times 312.5 + 520$	Hg	Two-photon $6s^2$–$6s7s\ ^1S_0$			2×10^9	88
3×364.8	Kr			10^{-5}	7×10^{11} (60 W)[a]	100–104
3×364.8	Kr		Ar	1.4×10^{-4}	420 W[b]	13, 14, 73, 103, 105
3×364.8	Kr		Ar	10^{-5}	16 W[c]	13, 14, 73, 103, 105
3×364.8	Kr		Xe[d]			

[a] Ref. 101.
[b] Ref. 103.
[c] Ref. 13.
[d] Usefulness for phase matching at Ly–α limited by absorption of the generated light.

8.2.7. Extreme Ultraviolet Radiation

Third-order frequency conversion has also been used to generate radiation at wavelengths near and below 100 nm, in the extreme ultraviolet (XUV) region of the spectrum. A summary of these interactions, along with the pump lasers that were used and the wavelengths that were generated is given in Table 11. Higher-order processes have also been used to generate radiation in this wavelength range. These interactions will be discussed in Chapter 3.

The first demonstration of conversion to wavelengths in the XUV was done by Harris *et al.*,[68] who frequency tripled the fourth harmonic of an Nd:YAG laser in Ar, producing radiation at 88.7 nm. Since then several other laser sources have been used to produce radiation at wavelengths below 100 nm. Reintjes *et al.*[106] also generated the third harmonic of the fourth harmonic of the Nd:YAG laser, using He and Ne as nonlinear materials as well as Ar. They also mixed the fourth harmonic of the YAG laser with lower-order harmonics to generate radiation at somewhat longer wavelengths. Hutchinson *et al.*[107] used radiation from an Xe_2 excimer laser at about 171 nm to generate radiation at 57 nm by third-harmonic conversion. In these experiments the wavelength of the pump laser could be tuned over a sufficient range to take advantage of a two-photon resonance with the 3p ^{5}5p$[\frac{5}{2}]J = 2$ state of Ar to increase the intensity of the generated radiation. The two-photon resonance is illustrated in Fig. 52.

Rare-gas halide (RGH) lasers are an efficient source of medium- to high-power radiation in the near ultraviolet and as such are excellent pump lasers for producing radiation in the XUV. Reintjes[108–110] has reported the generation of radiation near 102.6 nm by third-harmonic conversion of radiation from an XeCl laser in Ar and Kr. Egger *et al.*[111] and Rothschild *et al.*[112] have reported third-harmonic conversion of radiation from a KrF laser in Xe, producing radiation near 83 nm. Pummer *et al.*[113] and Rhodes[114] have reported frequency tripling of radiation from an ArF laser in Ar, Kr, Xe, H_2 and CO, producing radiation at 64.3 nm. In these last experiments[114] pump powers of 30 MW in 10-nsec duration pulses were used, producing XUV powers of the order of 30 W. When the pump pulse duration was reduced to about 10 psec and the pump power was raised to about 1 GW, the XUV power was reported to increase to about 20 kW.

The radiation from the RGH lasers is tunable over a limited range in the near ultraviolet, typically of the order of 0.1–1% of the output wavelength. As a result the XUV radiation that is generated by third-harmonic conversion of the radiation from these lasers will also be tunable over a restricted range. Such tunability has been demonstrated for XeCl by Reintjes,[109] for KrF by Egger *et al.*[111] and Rothschild *et al.*,[112] and for ArF by Pummer *et al.*[113] A wider tuning range can be achieved by mixing the output from the RGH

TABLE 11

Generation of XUV Radiation by Third-Order Processes

λ_{gen} (nm)	Laser	Interaction (nm)	Material	Resonance	max η	XUV power	Ref.
106.4	Nd:YAG	$2 \times 266.1 + 532$	He, Ne				106
102.6[a]	XeCl	3×308	Ar, Kr		10^{-8}	5 mW	108–110
102.6[b]	KrF + dye	$2 \times 249 - 583$	Xe	Two-photon			115
101–73 nm[b]	Dye + second harmonic	$2 \times \lambda_{SH} + \lambda_{dye}$	Xe	Two-photon			89
97.2	Dye	$2 \times 243 + 486$	H	Two-photon	10^{-9}		119
93	Dye	3×279	Hg	Two-photon			118
89.6	Nd:glass	3×268.8	Hg	Two-photon	10^{-6}–10^{-9}		117
88.7	Nd:YAG	3×266.1	Ar				68
			He, Ne				106
83[a]	KrF	3×249	Xe	Two-photon	10^{-8}	40 mW	111, 112
			Hg	Two-photon			118
79[b]	ArF + dye	$2 \times 193 + \lambda_{dye}$	H_2			200 mW	116
64.3[a]	ArF	3×193	H_2, D_2, Ar, Kr	Two-photon	10^{-6}	30 W ($t_p = 7$ nsec) 20 kW ($t_p = 10$ psec)	113
			CO	Nonresonant			114
57	Xe_2	3×171	Ar	Two-photon			107

[a] Tunable over excimer laser bandwidth.
[b] Tunable over dye laser bandwidth.

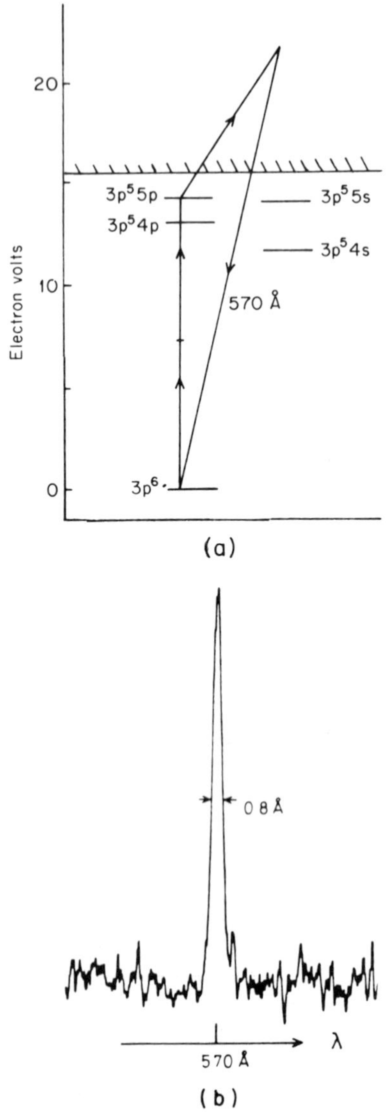

Fig. 52. Generation of extreme ultraviolet radiation by third-harmonic conversion of light from an Xe_2 laser in Ar. (a) Level diagram showing two-photon resonant enhancement. (b) Variation of power generated at 57 nm as the laser is tuned through the two-photon resonance. (From Ref. 107.)

lasers with radiation from dye lasers. Such interactions have been demonstrated by Pummer *et al.*[115] in mixing radiation from a KrF laser with light from a Rhodamine 6G laser to produce tunable radiation near 102.6 nm. A similar interaction using light from an ArF laser and radiation from a stilbene laser between 446 and 458 nm was used to produce tunable radiation near 79 nm.[116]

Tunable XUV radiation with a very narrow bandwidth (about 0.01 cm^{-1}) has been produced in these interactions by using radiation from a narrow-band cw dye laser to control the output of the RGH laser. A schematic of such a system is shown in Fig. 53a. High-resolution spectra of H_2, HD, and D_2 molecules in the XUV have been obtained with this system with spectral resolution well below the Doppler limit.[111,112] An example of such a spectrum is shown in Fig. 53b.

Several four-wave mixing processes for generating XUV radiation using resonant conversion in Hg vapor have been proposed by Popov and Timofeev.[85] Slabko *et al.*[117] have used one that involves a two-photon resonance with the 6s8s $^1S_0$ state for radiation from the fourth harmonic of an Nd : glass laser to generate tunable radiation near 89.6 nm. Freeman and Bokor[118] have generated radiation near 93 nm using similar interactions to convert radiation from dye lasers and near 83 nm for conversion of radiation from a KrF laser.

Tunable radiation over the range 101–73 nm using two-photon resonant conversion in Xe has been reported by Hilbig and Wallenstein.[89]

Troshin *et al.*[119] have reported the generation of radiation at 97.2 nm in atomic hydrogen. This interaction involved radiation at 486.2 nm, and its second harmonic and was enhanced with both a two-photon resonance with the 2s level and a single-photon resonance between the 2s and 4p levels. The dye laser power was about 50 kW, the second harmonic power about 1 kW, and the conversion efficiency was reported to be about 10^{-9}.

8.3. Infrared Radiation

Frequency conversion in the infrared region of the spectrum has been used both to change the wavelength of infrared lasers and to generate infrared radiation from visible lasers. We shall discuss these types of interactions in this section. Related experiments involving the up-conversion of infrared radiation to the visible will be discussed in Chapter 6.

Frequency conversion experiments in the infrared have been done in atomic metal vapors, molecular gases, and molecular liquids. The interactions in the metal vapors use electronic energy levels and are similar to the ones discussed in earlier sections for UV and VUV generation. Interactions in molecular liquids or gases can make use of vibrational–rotational transitions as well as electronic ones for resonant enhancement of frequency conversion processes. Homonuclear diatomic molecules do not have electric dipole transitions between vibrational levels in the ground electronic manifold, and intermediate one- and three-photon coupling must take place through states in an excited electronic manifold as shown in Fig. 54a. Two-photon resonances between vibrational levels in the ground electronic

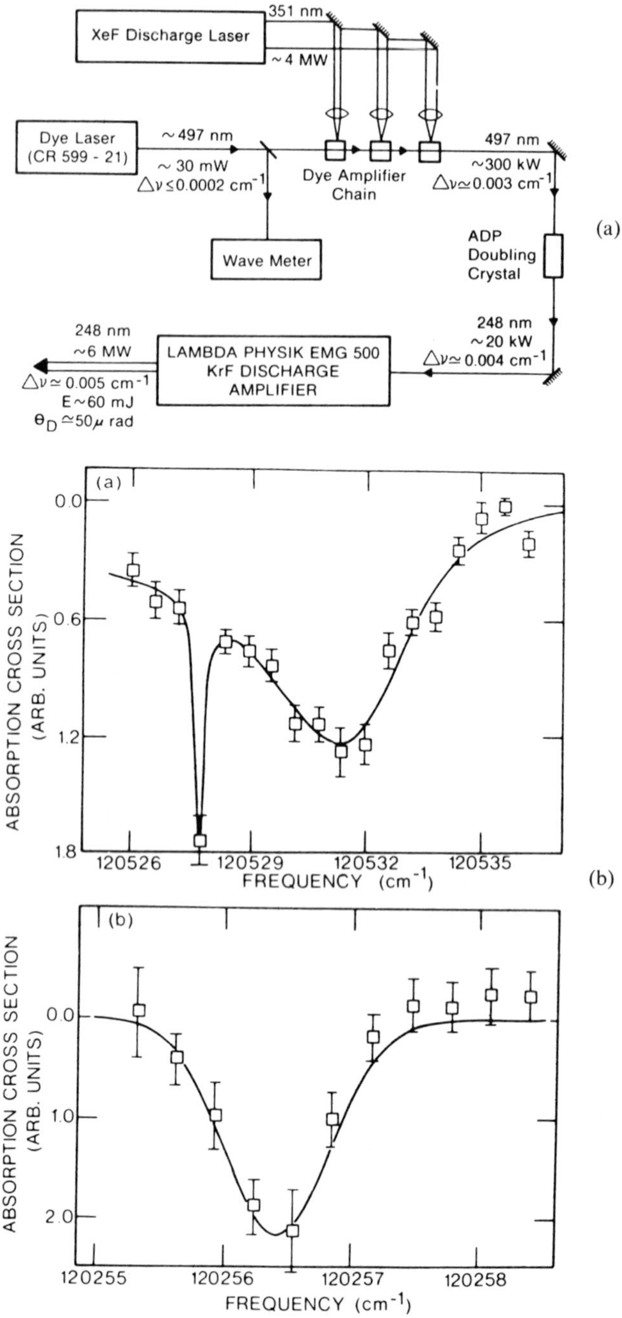

Fig. 53. (a) Schematic illustration of method used to obtain narrow-bandwidth high-power radiation from rare-gas halide lasers for generation of narrow-bandwidth tunable VUV radiation. (b) Examples of high resolution spectra in the XUV obtained with the apparatus shown in (a). Top: Absorption of the D → X (5, 0) R(2) transition in H_2 (broad feature), along with the D → X (5, 0) Q(1) transition (narrow feature). Bottom: Absorption of the D″ → X (0, 0) R(1) transition. The solid curves were calculated from the theory described in Ref. 112. (From Ref. 112.)

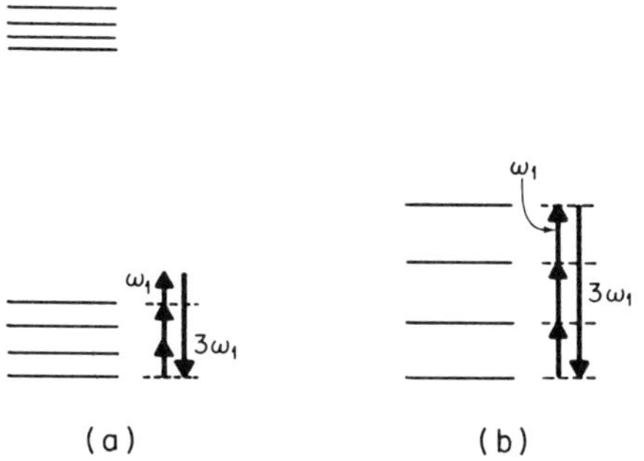

Fig. 54. Level diagrams showing third-harmonic conversion using molecular vibrational levels. (a) Homonuclear molecules. Intermediate one- and three-photon coupling occurs through excited electronic manifold. (b) Heteronuclear molecules. Intermediate one- and three-photon coupling can occur between vibrational levels of ground-state electronic manifold.

manifold may occur. Dispersion in such molecules is usually positive for infrared radiation. Harmonic conversion and four-wave sum-frequency mixing must therefore be done in one coherence length. Four-wave difference-frequency mixing processes can be optimized in a tight focus. Mixtures with media that have negative dispersion or smaller positive dispersion can be used for phase matching or to extend the coherence length.

In heteronuclear molecules one- and three-photon coupling can take place between vibrational levels of the ground electronic manifold as well as with levels in the excited electronic manifold, as illustrated in Fig. 54b. The oscillator strengths for these transitions are generally smaller than those for the electronic transitions but the nonlinear susceptibility can be increased through the presence of one- and three-photon resonances that generally do not occur with the upper electronic manifolds for infrared radiation. For a typical molecule two-photon resonances are provided by Q-branch transitions while P and R branch transitions provide both single- and three-photon resonances and determine the dispersion of the medium. The dispersion can be either positive or negative depending on the relation of the pump and generated frequencies to the vibrational levels that are involved.

Ueda and Shimoda[120] have given theoretical expressions for the third-order susceptibility of diatomic molecules due to vibrational structure. For diatomic and linear polyatomic molecules the expressions for the susceptibilities are similar to those in Eqs. (94)–(96) given earlier for atomic systems except that the transition dipole moments to be used are those that are appropriate for the vibrational transitions. For symmetric and asymmetric top molecules terms that describe the contributions of the permanent dipole moment to the nonlinear susceptibility must also be included. Ueda and Shimoda point out that these terms can often be of comparable importance to the terms that are due solely to the induced dipoles.

The susceptibility for third-harmonic generation of a single molecule using linearly polarized light is

$$\chi^{(3)}_{zzzz}(-3\omega, \omega, \omega, \omega) = \frac{e^4}{\hbar^3} \sum_J \sum_{l,m,n} N_J z_{Jn} z_{nm} z_{ml} z_{lJ}$$
$$\times \left\{ \frac{1}{(\omega_{lJ} - \omega)(\omega_{mJ} - 2\omega)(\omega_{nJ} - 3\omega)} + \frac{1}{(\omega_{lJ} - \omega)(\omega_{mJ} - 2\omega)(\omega_{nJ} + \omega)} + \frac{1}{(\omega_{lJ} - \omega)(\omega_{mJ} + 2\omega)(\omega_{nJ} + \omega)} + \frac{1}{(\omega_{lJ} + \omega)(\omega_{mJ} + 2\omega)(\omega_{nJ} + 3\omega)} \right\}. \quad (106)$$

Here J is a rotational quantum number of the lowest vibrational rotational level; l, m, and n, intermediate rotational–vibrational levels; and N_J, the population in rotational state J.

A typical matrix element for the z component of the dipole moment is

$$\mu_z(e, v; e', v') = \mu_0 \, \delta_{ev, e'v'} + \frac{\partial \mu}{\partial x} \langle e, v | z | e'v' \rangle \langle R | \cos \theta | R' \rangle. \quad (107)$$

Here $\langle e, v | z | e', v' \rangle$ is the z component of the matrix element of the radial wave function between a vibrational level v in an electronic manifold e and a vibrational level v' in an electronic manifold e'. The quantity $\langle R | \cos \theta | R' \rangle$ is an average of the z component of the dipole moment over rotational states. The quantity μ_0 is the permanent dipole moment of the molecule in state e, v.

Ueda and Shimoda[120] have given expressions for the matrix elements for transitions between vibrational levels of the same electronic manifold. They have shown that the contribution of the rotational matrix elements to the susceptibility when averaged over all possible orientations, or, equivalently, over all quantum numbers M (M is the projection of J onto an axis of quantization) for linear molecules is

$$\frac{1}{2J + 1} \sum_{J', J'', J'''} \sum_{M=-J}^{J} \langle J, M | \cos \theta | J', M \rangle \langle J', M | \cos \theta | J'', M \rangle$$
$$\times \langle J'', M | \cos \theta | J''', M \rangle \langle J''', M | \cos \theta | J, M \rangle = \tfrac{1}{5}. \quad (108)$$

Here we have used the property that the quantum number M does not change in the intermediate transitions for linearly polarized light.

The z components of the radial matrix elements for a molecule that can be described as an anharmonic oscillator as given by Ueda and Shimoda[120] are

$$\langle v|z|v\rangle = -[\hbar f/(2m^2\omega_e)](2v + 1), \tag{109a}$$

$$\langle v|z|v+1\rangle = \{[\hbar/(2m\omega_e)](v + 1)\}^{1/2}, \tag{109b}$$

$$\langle v|z|v+2\rangle = [\hbar f/(6m^2\omega_e^3)][(v + 1)(v + 2)]^{1/2}, \tag{109c}$$

$$\langle v|z|v+3\rangle = (f^2/24)\{[\hbar^3/(2m^2\omega_e'')](v + 1)(v + 2)(v + 3)\}^{1/2}, \tag{109d}$$

where

$$f = (9/2\hbar)(2\pi c)^5 m^3 \omega_e^4(\omega_e x_e),$$

ω_e is the frequency of the vibrational mode and $\omega_e x_e$ is the anharmonicity.

Vibrational transitions among states in different electronic manifolds involve electronic oscillator strengths. These are at present available only for a few molecules. She and Billman[121] have given estimates of the efficiency of third-harmonic generation and four-wave sum-frequency mixing in the infrared in H_2. The appropriate level diagrams are shown in Fig. 55. Using oscillator strengths for the electronic transitions from Allison and Dalgarno[122] they calculate a susceptibility for third-harmonic conversion of radiation at $\nu = 2077.6\ \text{cm}^{-1}$ ($\lambda = 4.8\ \mu$m), the frequency for the two-photon resonance, to be $\chi^{(3)} = 3.89 \times 10^{-58}\ \text{m}^5/\text{V}^2$ (4.8×10^{-28} esu). They estimate that conversion efficiencies of 0.1 should be possible for a laser power of 0.88 MW and a gas pressure of 4 atm.

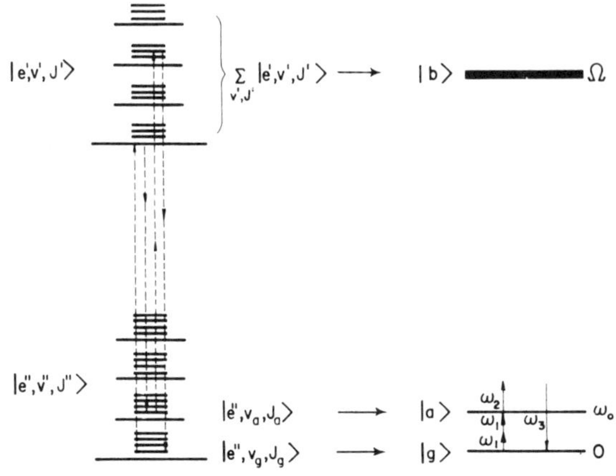

Fig. 55. Level diagram for two-photon resonant third-harmonic conversion in H_2. [From Ref. 121.]

TABLE 12

Third-Harmonic Conversion of CO_2 Laser Radiation

Band	Line	Material	Resonance	Phase matching	max η	Ref.
10.4 μm	P(20)–P(30)	SF_6	One-photon, three-photon		5×10^{-12} P(20)	123
	P(14), P(16)–P(26)	SF_6	One-photon, three-photon		4×10^{-11} P(20)	124
	P(14)–P(34), R(14)–R(32)	BCl_3	One-photon, three-photon		4×10^{-12} P(26)	123
	P(24)–P(28)	NO	Two-photon		3.8×10^{-9} P(28)	125
9.4 μm	R(8)–R(14)	CO	Two-photon		3.8×10^{-9} R(10)	123
	R(8)–R(14)	CO	Two-photon	SF_6	1.9×10^{-8} R(10)	125
	P(20)–P(26)	DCl	Two-photon	CF_4		126
	P(22)	DCl	Two-photon		2.6×10^{-9} P(22)	127
	R(6)	CO (liquid) mixed with O_2	Two-photon		0.003	128
	R(6)	CO (liquid)–O_2	Two-photon	SF_6	0.02	12
	R(6)	CO (liquid)–O_2	Two-photon	SF_6 (double-pass geometry)	0.08	129

8.3.1. *Third-Harmonic Conversion of CO_2 Laser Radiation*

Radiation from CO_2 lasers has been converted in third-harmonic processes in several molecular gases and cryogenic liquids. Conversion efficiencies up to 0.08 have been reported at pump power levels of the order of 600 kW. The various materials that have been used along with the wavelengths involved, phase-matching conditions, and conversion efficiencies are summarized in Table 12.

Conversion in Molecular Gases. Third-harmonic conversion of various lines of the P and R branches of the 10.4-μm band of the CO_2 laser has been observed in gaseous SF_6,[123,124] BCl_3,[123] NO,[125] CO,[123,125] and DCl.[126,127] Third-harmonic conversion in SF_6 involves near-one-, two-, and three-photon resonances between several lines of the P branch of the CO_2 laser and the vibrational levels of the ν_3 mode (Fig. 56). The P(20) line ($\lambda = 10.59$ μm) is very near a two-photon resonance with the $v = 2$ vibrational level ($\Delta\nu = 1.2$ cm^{-1}), while negative dispersion is provided by the $v = 3$ level for laser lines with wavelengths shorter than about that of the P(24) line ($\lambda = 10.63$ μm).

Measurements of conversion in SF_6 have been reported by Kildal and Deutsch[123] and by Kang *et al.*[124] Both groups reported the greatest conversion on the P(20) line which had the strongest two-photon resonant enhancement with conversion dropping off rapidly for lines on either side of P(20), as shown in Fig. 57. Kildal and Deutsch[123] reported a conversion efficiency of 5×10^{-12} with a pump power of 4 MW at a SF_6 pressure of

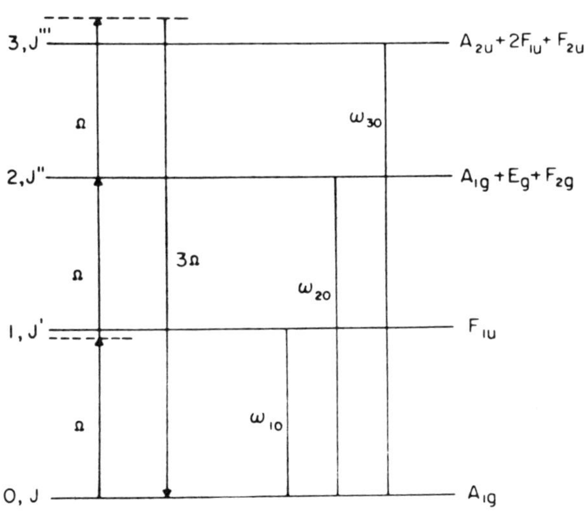

Fig. 56. Energy level diagram for third-harmonic conversion of 10.6-μm radiation in SF_6. (From Ref. 124.)

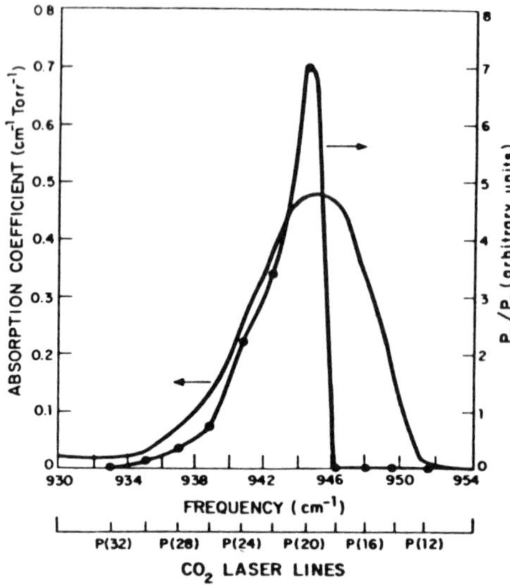

Fig. 57. Variation of third-harmonic power generated in SF_6 with wavelength of the CO_2 laser pump radiation. (From Ref. 123, © 1976 IEEE.)

about 4.5 Torr. They estimated the nonlinear susceptibility for this process to be about $\chi^{(3)} = 3.2 \times 10^{-35}$ neglecting rotational structure. This value gave good agreement with their experimental observations.

Kang *et al.*[124] used lower laser power ($P = 0.25$ MW) and higher gas pressure (up to 1250 Torr) and a shorter cell length ($L = 1.9$ cm). They measured a slightly higher conversion efficiency, 4×10^{-11}, at a higher pressure (500 Torr).

Kildal and Deutsch[123] have reported third-harmonic conversion of several P- and R-branch lines in BCl_3 with an efficiency up to 4×10^{-12} for the P(28) line. Kildal[125] has reported conversion of the P-branch lines P(24) to P(28) in NO with an efficiency up to 4×10^{-9}.

Kildal and Deutsch[123] and Kildal[125] have reported third-harmonic conversion of several R-branch lines in gaseous CO. As with SF_6 and BCl_3,

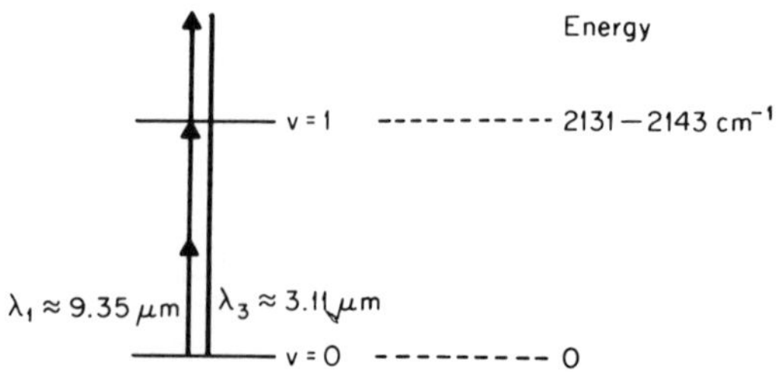

Fig. 58. Energy level diagram for third-harmonic conversion of 9.4-μm laser radiation in CO.

CO has a two-photon resonance to enhance the third-harmonic conversion, but it does not have one- or three-photon resonances to absorb the incident or generated radiation, as shown in Fig. 58. Near-two-photon resonances exist between several levels of the Q branch of the 0–1 transition of the CO molecule and various R branch lines of the CO_2 laser, as indicated in Table 13. Negative dispersion for the third-harmonic process is provided by the P and R branches of the 0–1 vibrational transitions in the CO molecule. Conversion efficiencies up to 1.9×10^{-8} were observed on the R(10) line for a phase-matched mixture of CO and SF_6 at a pump power of 1.2 MW.

TABLE 13

Two-Photon Resonances between R-Branch Lines of the 9.4-μm Band of the CO_2 Laser and Transitions of the Q Branch in CO[a]

CO (0–1) Q branch		CO_2 (00°1–02°0)	
J	$\nu_{Q(J)}$ (cm^{-1})	Line	$2\nu_{CO_2}$ (cm^{-1})
		R(10)	2143.768
0	2143.272		
1	2143.237		
2	2143.166		
3	2143.061		
4	2142.921		
5	2142.746		
6	2142.536		
7	2142.291		
8	2142.011		
9	2141.696		
10	2141.346		
11	2140.961	R(8)	2140.925
12	2140.541		
13	2140.086		
14	2139.596		
15	2139.071		
16	2138.511		
17	2137.916	R(6)	2138.028
18	2137.285		
19	2136.620		
20	2135.920		
21	2135.185	R(4)	2135.078
22	2134.415		
23	2133.610		
24	2132.770		
25	2131.895		

[a] From Ref. 123, © 1976 IEEE.

Near-two-photon resonances exist in DCl for the P(22) and P(24) lines of the 9.4-μm branch of the CO_2 laser. A conversion efficiency of 2.6×10^{-9} was reported for the P(22) line in pure DCl.[127] Comparison of conversion in phase-matched DCl–CF_4 mixtures and CO–SF_6 mixtures indicated that significantly higher conversion could be obtained in the DCl system.[126]

Conversion in Liquid CO. Kildal and Brueck have also studied third-harmonic conversion of CO_2 laser radiation in cryogenic mixtures of CO and O_2,[128] and of CO, O_2, and SF_6.[7,12,129] In the liquid phase CO continues to have the negative dispersion and sharp resonances that are present in its gas phase. In their early experiments Kildal and Brueck[128] attempted to reduce the magnitude of the dispersion of CO by diluting it with O_2. The O_2 molecules also contributed to the nonlinear susceptibility, but more important, the dilution of the liquid CO changed the linewidth and position of the two-photon resonance and its integrated strength. Conversion efficiency with this system was limited to 0.003 on the R(6) line of the CO_2 laser.

Better performance was obtained with the ternary mixture of liquid CO with O_2 and SF_6.[12] As before the O_2 could be used to control the position of the two-photon resonance, but now the wave vector mismatch could be controlled by the SF_6, which has positive dispersion for third-harmonic conversion of the 9.35-μm line of the CO_2 laser. A typical phase-matching curve for this mixture is shown in Fig. 59. Sufficient SF_6 could be added to obtain perfect phase matching. A conversion efficiency of 0.02 was observed in a mixture containing a SF_6 density of $1.4 \times 10^{19}\ cm^{-3}$ with a pump power of 600 kW in a tight-focus geometry.

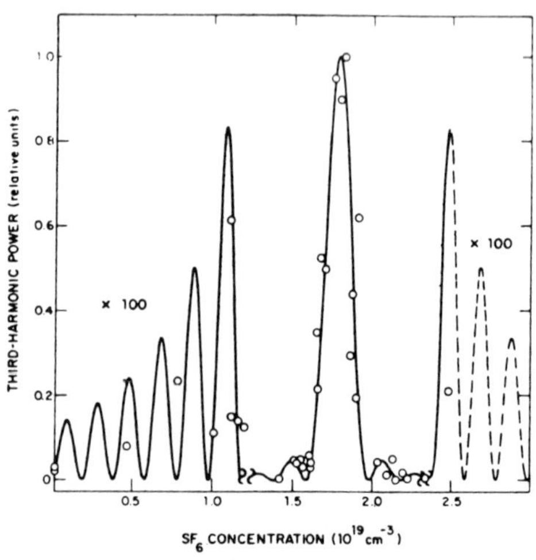

Fig. 59. Third-harmonic conversion of CO_2 laser radiation in liquid CO–O_2–SF_6 mixture as a function of the SF_6 concentration. (From Ref. 12.)

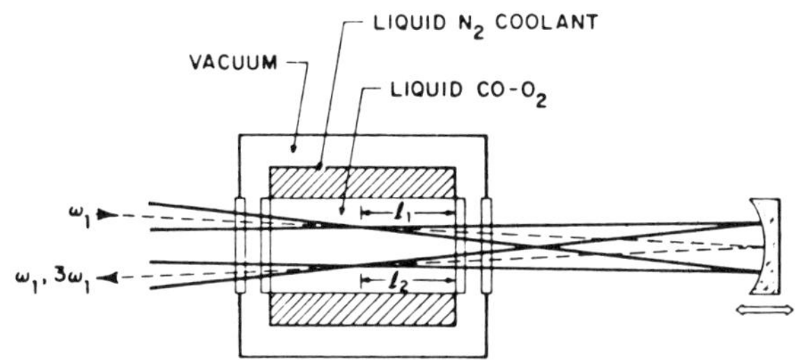

Fig. 60. Schematic diagram showing double-pass focusing geometry used to increase efficiency of third-harmonic conversion. (From Ref. 129.)

Further increases in the conversion efficiency were obtained by using a double-pass configuration, as shown in Fig. 60.[129] In this arrangement advantage is taken of the fact the most of the pump radiation is transmitted through the cell. The transmitted pump and the harmonic radiation that is generated on the first pass are returned to the cell for further conversion. By choosing the distance between the cell and the mirror properly the relative phase of the generated wave and the pump wave can be adjusted to ensure that further conversion occurs on the second pass. Brueck and Kildal[129] report an increase of a factor of 4 in conversion efficiency with this technique, bringing the net efficiency for the process to about 0.08.

8.3.2. Infrared Generation

Various four-wave difference-frequency mixing processes have been used to generate radiation over an extensive range in the infrared. Some of the processes that have been used are illustrated in Fig. 61. The process in Fig. 61a

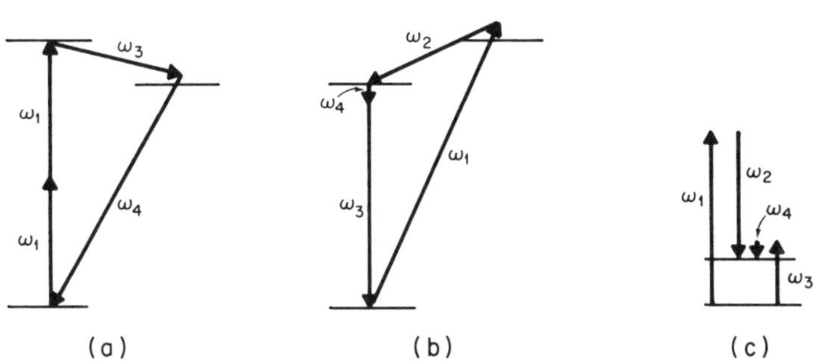

Fig. 61. Energy level diagrams for various processes used to generate infrared radiation by difference-frequency interactions. (a) Four-wave parametric oscillation. (b) Four-wave difference-frequency mixing $\omega_4 = \omega_1 - \omega_2 - \omega_3$. Radiation at ω_2 is supplied by stimulated Raman scattering. (c) Four-wave difference-frequency mixing $\omega_4 = \omega_3 - \omega_1 + \omega_2$. Radiation at frequency ω_2 is obtained from stimulated Raman scattering using ω_1 as a pump in a separate interaction.

is the same four-wave parametric oscillation interaction that was described in Fig. 51. In infrared generation experiments, however, it is the wave at ω_3 that is considered the signal. The difference-frequency interaction in Fig. 61b generates radiation at the frequency combination $\omega_4 = \omega_1 - \omega_2 - \omega_3$, while that in Fig. 61c produces radiation at $\omega_4 = \omega_3 - (\omega_1 - \omega_2)$. In each of these last two interactions the pump waves at ω_1 and ω_3 are supplied by pump lasers, while the wave at ω_2 is obtained by stimulated Raman scattering in the nonlinear medium. The output wavelength at ω_4 can be tuned by tuning the pump wave at ω_3. The pump wave at ω_1 can be a fixed frequency or it can be tuned to obtain resonant enhancement of the stimulated Raman process or to satisfy phase-matching requirements. Because the pump wave at ω_2 is obtained from the stimulated Raman process the output wavelength is independent of ω_1 and the conditions for two-photon resonant enhancement of the four-wave mixing process are automatically satisfied. Some of the reported wavelength ranges over which these interactions have been used and the appropriate nonlinear materials are given in Table 14.

Hartig[130] has reported generation of radiation near 2.34 and 3.415 μm through four-wave parametric oscillation. For these interactions the pump radiation was obtained from a dye laser tuned to two-photon resonances in Na vapor.

Sorokin *et al.*[131] have reported experiments with the interaction in Fig. 61b in metal vapors generating radiation over the range from 2 to 30 μm. Specific combinations of the dye lasers needed to cover this range have been given by Stokes *et al.*[132] Brosnan *et al.*[133] have used the interaction in Fig. 61c to generate tunable radiation in the wavelength from 3.5 to 13 μm in H_2. Sorokin *et al.*[134] and Frey *et al.*[135] have also reported use of processes similar to that shown in Fig. 61c to generate radiation at 16 μm using H_2 as the nonlinear medium. Conditions that affect the efficiency and saturation of the interaction in Fig. 58b have been investigated by Tyler *et al.*[136]

Sorokin *et al.*[131] have reported measurements using the interaction in Fig. 61b in K and Rb vapors. They chose the frequency ω_1 to lie on the high-frequency side of an allowed resonance, as illustrated in Fig. 62. In their initial experiments they reported a tuning range from 2 to 4 μm in K and from 2.9 to 5.4 μm in Rb. In order to maintain their best conversion they had to tune the wave at ω_1 farther from the (n + 1)p resonance to maintain phase matching as the output wavelength increased. Eventually the decrease of the intensity in the Stokes wave at ω_2 caused by the progressive detuning of the wave at ω_1 from the (n + 1)p resonance provided a long wave limit for the process. They subsequently extended the tuning range to 30 μm (the long wave limit of their detector) by using phase-matched mixtures of Na and K. Now the frequency of the pump wave at ω_1 could be chosen to give strong stimulated Raman scattering and the concentration of the Na

TABLE 14

Generation of IR Radiation by Four-Wave Mixing

Wavelength range (μm)	Laser	Interaction	Material	Resonance	Phase matching	IR power	Ref.
2.34	Dye	$2\omega_1 \rightarrow \omega_{IR} + \omega_2$[a]	Na	Two-photon 3s–4d One-photon 4d–4p			130
3.415	Dye	$2\omega_1 \rightarrow \omega_{IR} + \omega_2$[b]	Na	Two-photon 3s–5s One-photon 5s–4p			130
1.181	Dye	$\omega_{IR} = \omega_1 - \omega_2 - \omega_3$[c]	Na	Two-photon 3s–4s One-photon 3s–4p			130
2.9–5.4	Dye	$\omega_{IR} = \omega_1 - \omega_2 - \omega_3$[d]	Rb	Two-photon 5s–6s One-photon 5s–5p[d]			131
2–4	Dye	$\omega_{IR} = \omega_1 - \omega_2 - \omega_3$[d]	K	Two-photon 4s–5s One-photon 4s–5p[d]			131
2.2–25	Dye	$\omega_{IR} = \omega_1 - \omega_2 - \omega_3$[d]	K	4s–5p[d]	Na	100 mW	131
2–4	Dye	$\omega_{IR} = \omega_1 - \omega_2 - \omega_3$[d]	K	4s–5p[d]		9.5 W	136
3.5–13	Nd:YAG + OPO	$\omega_{IR} = \omega_{OPO} - (\omega_1 - \omega_2)$[e]	H_2	Two-photon		6 kW (5 nsec)	133

[a] $\lambda_1 = 578.74$ nm; λ_2 generated in four-wave parametric interaction.
[b] $\lambda_1 = 602.4$ nm; λ_2 generated in four-wave parametric interaction.
[c] $\lambda_1 = 330.2$ nm, generated in four-wave parametric interaction; λ_2 generated by stimulated Raman scattering from λ_1; λ_3 supplied by dye laser.
[d] λ_2 supplied by stimulated Raman scattering from λ_1; λ_1 nearly resonant with indicated one-photon transition.
[e] λ_{OPO} supplied by parametric oscillator; $\lambda_1 = 1.06$ μm; λ_2 supplied by stimulated Raman scattering from λ_1 in H_2.

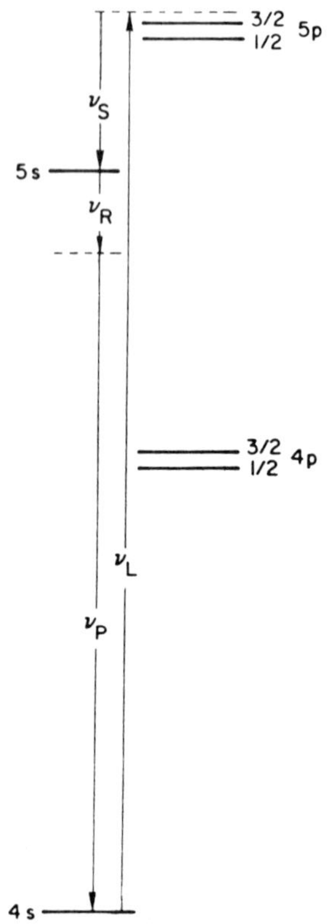

Fig. 62. Level diagram for generation of infrared radiation in potassium. (From Ref. 131.)

could be varied to maintain phase matching as the output wavelength was varied. Details of the phase-matching conditions are discussed in Sorokin *et al.*[131] Sorokin *et al.*[134] reported IR power levels of about 0.1 W at 2.3 μm in the K–Na system using a pump power of 1 kW at ω_1. Their power level at 25 μm was estimated to be about 0.1 mW.

Higher conversion efficiencies have been reported by Tyler *et al.*[136] Theoretical analyses have shown that at low-power levels the conversion efficiency of the wave at ω_3 to the infrared wave at ω_4 is given by

$$\eta = \frac{I_4}{I_3} = \left(\frac{\omega_4}{\omega_2}\right)^2 \left|\frac{\chi^{(3)}(-\omega_4, \omega_1, -\omega_2, -\omega_3)}{\chi^{(3)}(-\omega_2, \omega_1, -\omega_1, \omega_2)}\right|^2 \frac{I_2}{I_1}. \tag{110}$$

Tyler observed, however, that for conversion in K the power in the infrared wave becomes independent of the power in the pump wave at ω_1 for power levels 2.6 kW, but continues to grow with the pump power at ω_3. This saturation in the four-wave mixing efficiency has been attributed to a saturation in the stimulated Raman process that generates the pump wave at ω_2.

By appropriately apportioning their available pump power between ω_1 and ω_3 they were able to obtain a 95-fold increase in IR power, producing up to 9.5 W at a wavelength of 2.2 μm, and an increase in conversion efficiency of a factor of 47.

Brosnan *et al.*[133] have used the process illustrated in Fig. 61c in H_2 to generate tunable IR radiation after an initial suggestion by Harris and Byer.[137] In their experiments the wave at ω_1 was obtained from an Nd:YAG laser at 1.06 μm and the wave at ω_3 was obtained from a $LiNbO_3$ parametric oscillator which was pumped by the YAG laser. The output wavelength theoretically could be tuned from 3.5 μm to infinity as the parametric oscillator is tuned over its available range from 1.4 to 2.4 μm. Experiments reported by Brosnan *et al.* showed a tuning range from 3.5 to 13 μm as the parametric oscillator was tuned from 14 to 2.12 μm. A power level of 6 kW (3 μJ in 5 nsec) was measured in the generated beam for a pump power of 2 MW (10 mJ in 5 nsec) at ω_3. The pump wave at ω_1 had a power of about 10 MW (100 mJ in 10 nsec). The maximum conversion efficiency to the wave at ω_2 was about 0.22.

Fleming[138] has shown for this process that if the the Stokes gain at ω_2 is sufficiently large compared to the wave vector mismatch

$$\Delta k = -(k_1 - k_2) + (k_3 - k_4), \tag{111}$$

then phase matching does not play a dominant role in the four-wave mixing process. As a result the infrared wavelength can be tuned over a broad range without optimizing the phase mismatch at each step. The use of the relatively high powers of the Nd:YAG laser combined with the nonresonant Raman process in H_2 offers the further advantage that the pump wavelengths are not restricted to values that are near resonances.

REFERENCES

1. P. Franken, A. Hill, C. Peters, and G. Weinreich, *Phys. Rev. Lett.* **7**, 118 (1961).
2. J. A. Armstrong, N. Bloembergen, J. Ducuing, and P. S. Pershan, *Phys. Rev.* **127**, 1918 (1962).
3. J. F. Ward and G. H. C. New, *Phys. Rev.* **185**, 57 (1969).
4. H. Puell, K. Spanner, W. Falkenstein, W. Kaiser, and C. R. Vidal, *Phys. Rev. A* **14**, 2240 (1976).
5. G. C. Bjorklund, *IEEE J. Quantum Electron.* **QE-11**, 287 (1975).
6. Y. M. Yiu, T. J. McIlrath, and R. Mahon, *Phys. Rev. A* **20**, 2470 (1979).
7. H. Kildal and S. R. J. Brueck, *IEEE J. Quantum Electron.* **QE-16**, 566 (1980).
8. D. A. Kleinman, A. Ashkin, and G. D. Boyd, *Phys. Rev.* **145**, 338 (1966).
9. R. Fischer and L.-W. Wieczorek, *Ann. Phys.* (*Leipzig*) **35**, 389 (1978).
10. S. A. Akhmanov, A. I. Kovrygin, and A. P. Sukhorukov, *in* "Quantum Electronics" (H. Rabin and C. L. Tang, eds.), Vol. 1, Part B, p. 522. Academic Press, New York, 1975.
11. P. P. Bey, J. F. Giuliani, and H. Rabin, *Phys. Rev. Lett.* **11**, 14 (1963).

12. S. R. J. Brueck and H. Kildal, *Opt. Lett.* **2**, 33 (1978).
13. R. Hilbig and R. Wallenstein, *IEEE J. Quantum Electron.* **QE-17**, 1566 (1981).
14. S. A. Batishche, V. S. Burakov, Y. V. Kostenich, V. A. Mostovnikov, P. A. Naumenkov, N. V. Tarasenko, V. I. Gladushchak, S. A. Moshkalev, G. T. Razdobarin, V. V. Semenov, and E. J. Shreider, *Opt. Commun.* **38**, 71 (1981).
15. G. C. Bjorklund, J. E. Bjorkholm, P. F. Liao, and R. H. Storz, *Appl. Phys. Lett.* **29**, 729 (1976).
16. G. C. Bjorklund, J. E. Bjorkholm, R. R. Freeman, and P. F. Liao, *Appl. Phys. Lett.* **31**, 330 (1977).
17. P. N. Butcher, "Nonlinear Optical Phenomena," Bull. No. 200. Eng. Exp. St., Ohio State Univ., Columbus, 1965.
18. H. Puell and C. R. Vidal, *Phys. Rev. A* **14**, 2225 (1976).
19. N. Bloembergen, H. Lotem, and R. T. Lynch, *Indian J. Pure Appl. Phys.* **16**, 151 (1978).
20. C. Flytzanis, *in* "Quantum Electronics" (H. Rabin and C. L. Tang, eds.), Vol. 1, Part A, p. 37. Academic Press, New York, 1975.
21. M. A. Yuratich and D. C. Hanna, *J. Phys. B* **9**, 729 (1976).
22. D. C. Hanna, M. A. Yuratich, and D. Cotter, "Nonlinear Optics of Free Atoms and Molecules." Springer-Verlag, Berlin and New York, 1979.
23. M. D. Levenson and N. Bloembergen, *Phys. Rev. B* **10**, 4447 (1974).
24. P. M. Stone, *Phys. Rev.* **127**, 1151 (1962).
25. O. S. Heavens, *J. Opt. Soc. Am.* **51**, 1058 (1961).
26. E. M. Anderson and V. A. Zilitis, *Opt. Spectrosc.* **16**, 211 (1964).
27. W. L. Wiese, M. W. Smith, and B. M. Miles, "Atomic Transition Probabilities." U.S. Natl. Bur. Stand., U.S. Gov. Print. Off., Washington, D.C., 1969.
28. D. R. Bates and A. Daamgaard, *Philos. Trans. R. Soc. London, Ser. A* **242**, 101 (1949).
29. C. E. Moore, "Atomic Energy Levels." U.S. Natl. Bur. Stand., U.S. Gov. Print. Off., Washington, D.C., 1971.
30. R. B. Miles and S. E. Harris, *IEEE J. Quantum Electron.* **QE-9**, 470 (1973); also R. B. Miles, Ph.D. Thesis, Stanford Univ. (1972).
31. H. Eicher, *IEEE J. Quantum Electron.* **QE-11**, 121 (1975).
32. G. L. Weissler, "Encyclopedia of Physics" (S. Flügge, ed.), Vol. 21, pp. 304–341. Springer-Verlag, Berlin and New York, 1956.
33. J. A. R. Samson, *Adv. At. Mol. Phys.* **2**, 178–263 (1966).
34. S. C. Wallace and G. Zdasiuk, *Appl. Phys. Lett.* **28**, 449 (1976).
34a. S. C. Wallace, Photoselective chemistry, part 2, *Adv. Chem. Phys.* **47**, 153 (1981).
35. H. Junginger, H. B. Puell, H. Scheingraber, and C. R. Vidal, *IEEE J. Quantum Electron.* **QE-16**, 1132 (1980).
36. J. A. Armstrong and J. J. Wynne, *Phys. Rev. Lett.* **33**, 1183 (1974).
37. P. Esherick, *Phys. Rev. A* **15**, 1920 (1977).
38. J. J. Wynne and J. P. Hermann, *Opt. Lett.* **4**, 106 (1979).
39. J. Sugar, T. B. Lucatorto, T. J. McIlrath, and A. W. Weiss, *Opt. Lett.* **4**, 109 (1979).
40. P. P. Sorokin, J. A. Armstrong, R. W. Dreyfus, R. T. Hodgson, J. R. Lankard, L. H. Manganaro, and J. J. Wynne, *in* "Laser Spectroscopy" (S. Haroche, J. C. Pebay-Peyroula, T. W. Hansch, and S. E. Harris, eds.), p. 46. Springer-Verlag, Berlin and New York, 1975.
41. Y. Prior, A. R. Bogdan, M. Dagenais, and N. Bloembergen, *Phys. Rev. Lett.* **46**, 111 (1981).
42. J. Ducuing and N. Bloembergen, *Phys. Rev.* 133, A1493 (1964).
43. S. A. Akhmanov, A. S. Chirkin, K. N. Drabovich, A. I. Kovrygin, R. V. Khokhlov, and A. P. Sukhorukov, *IEEE J. Quantum Electron.* **QE-4**, 598 (1964).
44. V. I. Anikin, S. A. Akhmanov, K. N. Drabovich, and A. N. Dubovik, *Kvantovaya Elektron. (Moscow)* **3**, 2014 (1976) [*Sov. J. Quantum Electron.* (*Engl. Transl.*) **6**, 1096 (1976)].

45. E. A. Stappaerts, G. W. Bekkers, J. F. Young, and S. E. Harris, *IEEE J. Quantum Electron.* **QE-12**, 330 (1976).
46. N. Dutta, *J. Phys. B* **13**, 411 (1980).
47. C. Leubner, H. Scheingraber, and C. R. Vidal, *Opt. Commun.* **36**, 205 (1981).
48. J. N. Elgin and G. H. C. New, *Opt. Commun.* **16**, 242 (1976).
49. J.-C. Diels and A. T. Georges, *Phys. Rev. A* **19**, 1589 (1979).
50. I. A. Poluektov and Nazarkin, *Kvantovaya Elektron.* (*Moscow*) **6**, 2525 (1979) [*Sov. J. Quantum Electron* (*Engl. Transl.* **9**, 1495 (1979)].
51. D. Grischowsky, M. M. T. Loy, and P. F. Liao, *Phys. Rev. A* **12**, 2514 (1975).
52. A. Georges, P. Lambropoulos, and J. H. Marburger, *Opt. Commun.* **18**, 509 (1976).
53. Y. N. Karamzin and A. P. Sukhorukov, *Kvantovaya Elektron.* (*Moscow*) **2**, 912 (1975) [*Sov. J. Quantum Electron.* (*Engl. Transl.*) **5**, 496 (1975)].
54. R. C. Eckardt and J. Reintjes, *NRL Memo. Rep.* **NRL-MR-3482**, pp. 7–29 (1977).
55. J. Reintjes and R. C. Eckardt, *Appl. Phys. Lett.* **30**, 91 (1977).
56. G. H. C. New, *Opt. Commun.* **19**, 177 (1976).
57. T. R. Royt and C. H. Lee, *Appl. Phys. Lett.* **30**, 332 (1977).
58. S. R. Hartmann, *IEEE J. Quantum Electron.* **QE-4**, 802 (1968).
59. E. M. Belenov and I. A. Poluektov, *Zh. Eksp. Teor. Fiz.* **56**, 1407 (1969) [*Sov. Phys.—JETP* (*Engl. Transl.*) **29**, 754 (1969)].
60. M. Takatsuji, *Phys. Rev. A* **4**, 808 (1971).
61. E. Courtens, *in* "Laser Handbook" (F. T. Arecchi and E. O. Schultz-Dubois, eds.), Vol. 2, p. 1259. North-Holland Publ., Amsterdam, 1972.
62. D. R. White, E. L. Dawes, and J. H. Marburger, *IEEE J. Quantum Electron.* **QE-6**, 793 (1970).
63. E. L. Dawes, *Phys. Rev.* **169**, 47 (1968); P. W. Langhoff, J. D. Lyons, and R. P. Hurst, *Phys. Rev.* **148**, 18 (1966); M. N. Grasso, K. T. Chung, and R. P. Hurst, *Phys. Rev.* **167**, 1 (1968).
64. K. M. Leung, J. F. Ward, and B. J. Orr, *Phys. Rev. A* **9**, 2440 (1974).
65. J. F. Young, G. C. Bjorklund, A. H. Kung, R. B. Miles, and S. E. Harris, *Phys. Rev. Lett.* **27**, 1551 (1971).
66. D. M. Bloom, J. F. Young, and S. E. Harris, *Appl. Phys. Lett.* **27**, 390 (1975).
67. D. M. Bloom, G. W. Bekkers, J. F. Young, and S. E. Harris, *Appl. Phys. Lett.* **26**, 687 (1975).
68. S. E. Harris, J. F. Young, A. H. Kung, D. M. Bloom, and G. C. Bjorklund, *in* "Laser Applications to Optics and Spectroscopy" (S. F. Jacobs, M. Sargent, III, J. F. Scott, and M. O. Scully, eds.), Physics of Quantum Electronics, Vol. 2, p. 181. Addison-Wesley, Reading, Massachusetts, 1975.
69. C. R. Vidal and J. Cooper, *J. Appl. Phys.* **40**, 3370 (1969).
70. C. R. Vidal and F. B. Haller, *Rev. Sci. Instrum.* **42**, 1779 (1971).
71. C. R. Vidal and M. M. Hessel, *J. Appl. Phys.* **43**, 2776 (1972).
72. A. H. Kung, J. F. Young, G. C. Bjorklund, and S. E. Harris, *Phys. Rev. Lett.* **29**, 985 (1972).
73. R. Mahon, T. J. McIlrath, V. P. Meyerscough, and D. W. Koopman, *IEEE J. Quantum Electron.* **QE-15**, 444 (1979).
74. A. H. Kung, J. F. Young, and S. E. Harris, *Appl. Phys. Lett.* **22**, 301 (1973).
75. L. J. Zych and J. F. Young, *IEEE J. Quantum Electron.* **QE-14**, 147 (1978).
76. V. I. Berkov, V. V. Korobkin, Y. V. Korobkin, A. S. Markin, A. V. Prokhindeev, and V. B. Studenov, *Prib. Tekh. Eksp.* **23**, 200 (1980) [*Instrum. Exp. Tech.* (*Engl. Transl.*) **23**, 1019 (1980)].
77. D. Cotter, *Opt. Lett.* **4**, 134 (1979).
78. R. G. Caro, A. Costela, and C. E. Webb, *Opt. Lett.* **6**, 464 (1981).

79. K. N. Drabovich, D. I. Metchkov, V. M. Mitev, L. I. Pavlov, and K. V. Stamenov, *Opt. Commun.* **20**, 350 (1977).
80. R. T. Hodgson, P. P. Sorokin, and J. J. Wynne, *Phys. Rev. Lett.* **32**, 343 (1974).
81. R. R. Freeman and Bjorklund, *Phys. Rev. Lett.* **40**, 118 (1978).
82. N. P. Economou, R. R. Freeman, and G. C. Bjorklund, *Opt. Lett.* **3**, 209 (1978).
83. H. Scheingraber, H. Puell, and C. R. Vidal, *Phys. Rev. A* **18**, 2585 (1978).
84. T. J. McKee, B. P. Stoicheff, and S. C. Wallace, *Opt. Lett.* **3**, 207 (1978).
85. A. K. Popov and V. P. Timofeev, *Opt. Commun.* **20**, 94 (1977).
86. F. S. Tomkins and R. Mahon, *Opt. Lett.* **6**, 179 (1981).
87. R. Mahon and F. S. Tomkins, *IEEE J. Quantum Electron.* **QE-18,** 913 (1982).
88. F. S. Tomkins and R. Mahon, *Top. Meet. Laser Tech. Extreme Ultraviolet Spectrosc., Boulder, Colo.* Pap. TuC4 (1982).
89. R. Hilbig and R. Wallenstein, *Top. Meet. Laser Tech. Extreme Ultraviolet Spectrosc., Boulder, Colo.* Pap. WB1 (1982).
89a. R. Hilbig and R. Wallenstein, *Appl. Opt.* **21**, 913 (1982).
89b. J. Hager and S. C. Wallace, *Chem. Phys. Lett.* **90**, 472 (1982).
90. K. K. Innes, B. P. Stoicheff, and S. C. Wallace, *Appl. Phys. Lett.* **29**, 715 (1976).
91. S. C. Wallace and K. K. Innes, *J. Chem. Phys.* **72**, 4805 (1980).
92. C. Tai, F. W. Talby, and G. L. Giles, *Phys. Rev. A* **20**, 233 (1979).
93. A. H. Kung, *Appl. Phys. Lett.* **25**, 653 (1974).
94. R. R. Freeman, G. C. Bjorklund, N. P. Economou, P. F. Liao, and J. E. Bjorkholm, *Appl. Phys. Lett.* **33**, 739 (1978).
95. J. Bokor, R. R. Freeman, R. L. Panock, and J. C. White, *Opt. Lett.* **6**, 182 (1981).
96. N. Bloembergen, "Nonlinear Optics." Benjamin, New York, 1965.
97. V. Wilke and W. Schmidt, *Appl. Phys.* **18**, 177 (1979).
98. R. S. Hargrove and J. A. Paisner, *Proc. Top. Meet. Excimer Lasers, Charlston, S.C.* Pap. ThA6–1 (1979).
99. R. Mahon, T. J. McIlrath, F. S. Tomkins, and D. E. Kelleher, *Opt. Lett.* **4**, 360 (1979).
100. S. A. Batishche, V. S. Burakov, V. G. Voronin, V. I. Galdushchak, V. A. Mostovnikov, P. A. Naumenkov, G. T. Rasdobarin, A. N. Rubinov, V. V. Semenov, N. V. Tarasenko, and E. J. Shreider, *Sov. Tech. Phys. Lett.* (*Engl. Transl.*) **3**, 473 (1977).
101. R. Mahon, T. J. McIlrath, and D. W. Koopman, *Appl. Phys. Lett.* **33**, 305 (1978).
102. R. Wallenstein, *Opt. Commun.* **33**, 119 (1980).
103. H. Langer, H. Puell, and H. Rohr, *Opt. Commun.* **34**, 137 (1980).
104. D. Cotter, *Opt. Lett.* **4**, 134 (1979).
105. R. Mahon and Y. M. Yiu, *Opt. Lett.* **5**, 279 (1980).
106. J. Reintjes, C. Y. She, and R. C. Eckardt, *IEEE J. Quantum Electron.* **QE-14**, 581 (1978).
107. M. H. R. Hutchinson, C. C. Ling, and D. J. Bradley, *Opt. Commun.* **18**, 203 (1976).
108. J. Reintjes, *Opt. Lett.* **4**, 242 (1979).
109. J. Reintjes, *Opt. Lett.* **5**, 342 (1980).
110. J. Reintjes, *Appl. Opt.* **19**, 3889 (1980).
111. H. Egger, R. T. Hawkins, J. Bokor, H. Pummer, M. Rothschild, and C. K. Rhodes, *Opt. Lett.* **5**, 282 (1980).
112. M. Rothschild, H. Egger, R. T. Hawkins, J. Bokor, H. Pummer, and C. K. Rhodes, *Phys. Rev. A* **23**, 206 (1981).
113. H. Pummer, T. Srinivasan, H. Egger, K. Boyer, T. S. Luk, and C. K. Rhodes, *Opt. Lett.* **7**, 93 (1982).
114. C. K. Rhodes, *Top. Meet. Laser Tech. Extreme Ultraviolet Spectrosc., Boulder, Colo.* Pap. MC1 (1982).
115. H. Pummer, H. Egger, M. Rothschild, and C. K. Rhodes, *J. Opt. Soc. Am.* **70**, 1587 (1980).

116. H. Egger, T. Srinivasan, K. Boyer, H. Pummer, and C. K. Rhodes, *Top. Meet. Laser Tech. Extreme Ultraviolet Spectrosc., Boulder, Colo.* Pap. WB2 (1982).
117. V. V. Slabko, A. K. Popov, and V. F. Lukinykh, *Appl. Phys.* **15**, 239 (1977).
118. R. R. Freeman and J. Bokor, *Top. Meet. Laser Tech. Extreme Ultraviolet Spectrosc., Boulder, Colo.* Pap. WA2 (1982).
119. B. I. Troshin, V. P. Chebotaev, and A. A. Chernenko, *Pis'ma Zh. Eksp. Teor. Fiz.* **27**, 293 (1978) [*JETP Lett.* (*Engl. Transl.*) **27**, 273 (1978)].
120. Y. Ueda and K. Shimoda, *J. Phys. Soc. Jpn.* **28**, 196 (1970).
121. C. Y. She and K. W. Billman, *Appl. Phys. Lett.* **27**, 76 (1975).
122. A. C. Allison and A. Dalgarno, *J. Quant. Spectrosc. Radiat. Transfer* **9**, 1543 (1969).
123. H. Kildal and T. F. Deutsch, *IEEE J. Quantum Electron.* **QE-12**, 429 (1976).
124. M. H. Kang, K. M. Chung, and M. F. Becker, *J. Appl. Phys.* **47**, 4944 (1976).
125. H. Kildal, *IEEE J. Quantum Electron.* **QE-13**, 109 (1977).
126. M. H. Kang, V. T. Nguyen, T. Y. Chang, T. C. Damen, and E. G. Burkhardt, *Appl. Phys. Lett.* **33**, 303 (1978).
127. R. T. V. Kung and S. A. Mani, *J. Appl. Phys.* **50**, 2546 (1979).
128. H. Kildal and S. R. J. Brueck, *Phys. Rev. Lett.* **38**, 347 (1977).
129. S. R. J. Brueck and H. Kildal, *Appl. Phys. Lett.* **33**, 928 (1978).
130. W. Hartig, *Appl. Phys.* **15**, 427 (1978).
131. P. P. Sorokin, J. J. Wynne, and J. R. Lankard, *Appl. Phys. Lett.* **22**, 342 (1973); J. J. Wynne, P. P. Sorokin, and J. R. Lankard, *in* "Laser Spectroscopy" (R. G. Brewer and A. Mooradian, eds.), pp. 103–111. Plenum, New York, 1974; J. J. Wynne and P. P. Sorokin, *in* "Nonlinear Infrared Generation" (Y. R. Shen, ed.), Topics in Applied Physics, Vol. 16, p. 159. Springer-Verlag, Berlin and New York, 1977.
132. E. D. Stokes, F. B. Dunning, R. F. Stebbings, G. K. Walters, and R. D. Dundel, *Opt. Commun.* **5**, 267 (1972).
133. S. J. Brosnan, R. N. Fleming, R. L. Herbst, and R. L. Byer, *Phys. Rev. Lett.* **30**, 330 (1977).
134. P. P. Sorokin, M. M. T. Loy, and J. R. Lankard, *IEEE J. Quantum Electron.* **QE-13**, 871 (1977).
135. R. Frey, F. Pradere, and J. Ducuing, *Opt. Commun.* **18**, 204 (1976).
136. I. L. Tyler, R. W. Alexander, and R. J. Bell, *Appl. Phys. Lett.* **27**, 346 (1975).
137. S. E. Harris and R. L. Byer, *Stanford Univ. Microwave Lab. Rep.* No. 1918 (1970) (unpublished).
138. R. N. Fleming, *Stanford Univ. Microwave Lab. Rep.* No. 2521 (1976) (unpublished).

CHAPTER 3

HIGHER-ORDER PROCESSES

1. INTRODUCTION

In this chapter we shall consider frequency conversion through processes that are of order higher than three. The use of higher-order interactions can be advantageous in frequency conversion when a large frequency shift from the fundamental is required. The higher-order processes allow a larger step along the frequency scale to be taken in a single step, thereby reducing the number of cascade frequency conversion stages that are required, or eliminating their need altogether. In this regard the higher-order processes have been particularly useful in the generation of radiation in the extreme ultraviolet region of the spectrum.

The higher-order interactions are also important because they contribute to the saturation of the conversion efficiency of third-order processes. Thus an understanding of the nature and strength of the higher-order processes can contribute to identification of the conditions that must be met for optimization of the conversion efficiency in the lower-order interactions.

The higher-order processes have not been studied in as great detail as have the third-order ones. They usually require higher intensities to be observed than do the lower-order effects, and many times they can be masked by stronger third-order interactions. In addition, many effects that are due to the higher-order nonlinearities can also be created by multiple interactions involving lower-order processes. Under these conditions it can be difficult to identify unambiguously a given effect as arising from a higher-order nonlinearity.

Higher-order processes have been discussed experimentally and theoretically by a number of authors. General properties of the higher-order interactions have been treated by Akhmanov.[1] Tomov and Richardson[2] have given calculations specifically for fifth-harmonic conversion of Nd: glass laser radiation in metal vapors. Harris[3] has suggested the use of the higher-order interactions for the generation of radiation in the extreme ultraviolet and soft x-ray special ranges. Observation of fifth-harmonic conversion of radiation at 1.06 μm from an Nd:glass laser have been reported by Metchov *et al.*[4] Reintjes *et al.*[5–14] have generated coherent radiation in the extreme ultraviolet at several wavelengths extending to 38 nm by fifth- and seventh-order harmonic generation and frequency-mixing processes. Ninth-harmonic conversion of Nd:glass radiation has been observed by Grozeva *et al.*[15] Puell *et al.*[16] have discussed the role of fifth-order processes in limiting third-harmonic conversion. Higher-order interactions have also been used in some coherent Raman spectroscopy studies.[17,18]

In this chapter we shall discuss frequency conversion through the higher-order processes. First we shall give general theoretical results for harmonic generation of arbitrary order and compare their properties to third-harmonic generation. Then we shall summarize specific theoretical and experimental results that have been reported in the literature.

2. HIGHER-ORDER HARMONIC GENERATION

The generation of radiation at the higher-order harmonic frequencies can be accomplished through a direct process involving the higher-order nonlinear susceptibility and also through indirect, or multiple-step, processes in which radiation at one or more lower-order harmonic frequencies is generated through lower-order processes and is subsequently combined with the fundamental radiation in another low-order process to yield radiation at the same frequency as is generated in the direct higher-order process.

In many situations the power of the radiation generated in the indirect process can equal or exceed that generated in the direct process. This generally presents no problem in frequency conversion experiments in which the goal is to generate as much radiation as possible through any interaction that is effective. It does mean, however, that proper care must be exercised in inferring values of the higher-order susceptibilities from measurements of the generated harmonic power. It can also have bearing on attempts to compare experimental results with theory, to predict optimum conversion conditions, or to assess the role of the higher-order interactions in limiting the efficiency of the lower-order ones.

2.1. Direct Higher-Order Generation

We shall first consider the process of direct higher-order harmonic generation. The nonlinear polarization at the qth harmonic arises from the qth-order susceptibility, $\chi^{(q)}(-q\omega_1, \omega_1, \omega_1, \ldots, \omega_1)$. Its amplitude is given by

$$P^{\mathrm{NL}}(q\omega_1) = \frac{1}{2^{q-1}} \chi^{(q)}(-q\omega_1, \omega_1, \ldots, \omega_1) A_1^q. \tag{1}$$

Calculation of the higher-order nonlinear susceptibilities can be done in a manner similar to that used for the lower-order ones. For example, the fifth-harmonic susceptibility of a quantum system with discrete levels is given in the quasistatic limit in atomic units by

$$\begin{aligned}
\chi_{\mathrm{a}}^{(5)}(-5\omega, \omega, \omega, \omega, \omega, \omega) &= \frac{(ea_0)^6}{(\hbar c)^5} \sum_{g,a,b,c,d,e} \mu_{ga}\,\mu_{ab}\,\mu_{bc}\,\mu_{cd}\,\mu_{de}\,\mu_{eg} \\
&\times \left[\frac{1}{(\omega_{ag} - \omega)(\omega_{bg} - 2\omega)(\omega_{cg} - 3\omega)(\omega_{dg} - 4\omega)(\omega_{eg} - 5\omega)} \right. \\
&+ \frac{1}{(\omega_{ag} - \omega)(\omega_{bg} - 2\omega)(\omega_{cg} - 3\omega)(\omega_{dg} - 4\omega)(\omega_{eg} + \omega)} \\
&+ \frac{1}{(\omega_{ag} - \omega)(\omega_{bg} - 2\omega)(\omega_{cg} - 3\omega)(\omega_{dg} + 2\omega)(\omega_{eg} + \omega)} \\
&+ \frac{1}{(\omega_{ag} - \omega)(\omega_{bg} - 2\omega)(\omega_{cg} + 3\omega)(\omega_{dg} + 2\omega)(\omega_{eg} + \omega)} \\
&+ \frac{1}{(\omega_{ag} - \omega)(\omega_{bg} + 4\omega)(\omega_{cg} + 3\omega)(\omega_{dg} + 2\omega)(\omega_{eg} + \omega)} \\
&\left. + \frac{1}{(\omega_{ag} + 5\omega)(\omega_{bg} + 4\omega)(\omega_{cg} + 3\omega)(\omega_{dg} + 2\omega)(\omega_{eg} + \omega)} \right].
\end{aligned} \tag{2}$$

2.1.1. *Plane-Wave Solutions*

The amplitude of the higher-order harmonics can be calculated from Eqs. (1) and (1-13). In the plane-wave limit the equations for the amplitudes at the fundamental and harmonic frequencies become

$$\frac{dA_q}{dz} = i \frac{q\pi^2 N}{2^{q-3} n_q \lambda_1} \chi_{\mathrm{a}}^{(q)}(-q\omega_1, \omega_1, \omega_1, \ldots, \omega_1) A_1^q e^{-i\Delta kz}, \tag{3a}$$

$$\frac{dA_1}{dz} = i \frac{q\pi^2 N}{2^{q-3} n_1 \lambda_1} \chi_{\mathrm{a}}^{(q)}(-\omega_1, q\omega_1, -\omega_1, \ldots, -\omega_1) A_q (A_1^*)^{q-1} e^{i\Delta kz}, \tag{3b}$$

where $\Delta k = k_q - qk_1$.

2.1.1a. Weak Field Limit. When depletion of the pump can be neglected the amplitude of the qth-harmonic field in a medium of length L is given by

$$A_q = i\frac{q\pi^2 NL}{2^{q-3} n_q \lambda_1}\chi_a^{(q)}(-q\omega_1, \omega_1, \ldots, \omega_1)A_1^q \frac{\sin(\Delta kL/2)}{(\Delta kL/2)} e^{-i\Delta kL/2} \tag{4}$$

and the intensity of the qth harmonic is

$$I_q = \frac{(2\pi)^{q+3} q^2 N^2 L^2 |\chi_a^{(q)}|^2 I_1^q}{n_q n_1^q \lambda_1^2 c^{q-1}} \frac{\sin^2(\Delta kL/2)}{(\Delta kL/2)^2}. \tag{5}$$

These expressions are similar to those derived in the last chapter for third-harmonic generation (see Section 2-2.1.1). The qth-harmonic intensity varies as the qth power of the fundamental intensity. The phase-matching properties of the higher-order harmonics in the plane-wave limit are described by the same $(\sin x/x)^2$ function as were those of the third-order interactions, and the discussion of plane-wave phase matching of Section 2-2.1.1 applies. The wave vector mismatch for the qth-harmonic process is given by

$$\Delta k_q = k_q - qk_1 = (2\pi q/\lambda_1)(n_q - n_1) \tag{6}$$

and the condition for optimum phase matching, $\Delta k_q = 0$, is equivalent to the condition

$$n_q = n_1. \tag{7}$$

2.1.1b. Pump Depletion. When the fundamental intensity is sufficiently great the depletion of the pump wave must be accounted for and Eqs. (3a) and (3b) must be solved simultaneously. Closed form analytic solutions for the higher-order harmonics are difficult to obtain. However, the amplitude of the qth harmonic field at perfect phase matching can be found from the following transcendental equation:

$$\tan\theta\left\{\frac{1}{(q-2)\cos^{q-3}\theta} + \frac{q-3}{(q-2)(q-4)\cos^{q-5}\theta} + \frac{(q-3)(q-5)}{(q-2)(q-4)(q-6)\cos^{q-7}\theta} + \cdots + \frac{(q-3)(q-5)\ldots(2)}{(q-2)(q-4)(q-6)\ldots(1)}\right\} = \kappa_q A_1(0)^{q-1} z. \tag{8}$$

Here $A_q(z)/A_1(0) = i\sin\theta$, $A_1(z)/A_1(0) = \cos\theta$, and

$$\kappa_q = \frac{Nq\pi^2\chi_a^{(q)}}{2^{q-3} n_q \lambda_1}. \tag{9}$$

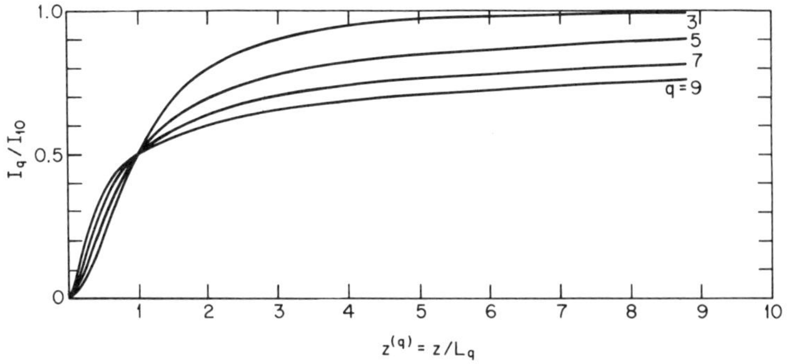

Fig. 1. Conversion efficiency as a function of normalized propagation distance for harmonics three–nine. Propagation constant L_q is the distance required for each harmonic to reach 50% conversion [see Eq. (10)].

Solutions for the intensity conversion efficiency $\eta = I_q(z)/I_1(0)$ are shown in Fig. 1 for the third–ninth harmonics. The horizontal scale is normalized for each harmonic to its own depletion length, defined as the distance required for 50% conversion: $z^{(q)} = z/L_q$. The depletion lengths for the various harmonics are given by

$$L_3 = [K_3 I_1(0)]^{-1}, \tag{10a}$$

$$L_5 = 1.333[K_5 I_1(0)^2]^{-1}, \tag{10b}$$

$$L_7 = 1.866[K_7 I_1(0)^3]^{-1}, \tag{10c}$$

$$L_9 = 2.743[K_9 I_1(0)^4]^{-1}, \tag{10d}$$

where $K_q = (8\pi/cn_1)^{(q-1)/2}\kappa_q$.

Generally one can expect the intensity that is required for significant conversion to take place to increase as the order of the harmonic increases. It is also evident from Fig. 1 that while complete conversion of the pump to any one of the harmonics is theoretically possible it becomes increasingly difficult to obtain complete conversion as the harmonic order increases. This trend should be even more pronounced for the energy conversion of beams with temporal and spatial profiles.

2.1.2. *Focused Beams*

The generation of radiation in the higher-order harmonics through the direct processes is usually done in focused beams to take advantage of the stronger interactions that accompany the higher intensities that are present in the focus. A general treatment of the generation of harmonics of arbitrary order in focused beams in the absence of pump depletion has been given by Ward and New.[19] The solutions for the higher harmonics bear many similarities to those which were discussed in the last chapter for the third-order interactions. The equation for the amplitude of the qth harmonic can be obtained from Eqs. (1-13) and (1):

$$\nabla_{\perp}^2 A_q(r, z) + 2ik_q \frac{\partial A_q}{\partial z} = -\frac{4\pi\omega_q^2 N}{2^{q-1}c^2} \chi_{\mathrm{a}}^{(q)} A_1^q(r, z) e^{-i\Delta kz}. \quad (11)$$

If we assume that the fundamental has a Gaussian profile of the form

$$A_1(r, z) = A_{10}(1 + i\xi)^{-1} \exp\{-[kr^2/b(1 + i\xi)]\}, \quad (12)$$

where $\xi = 2z/b$ and b is the confocal parameter of the focused beam, then the amplitude of the qth harmonic that is generated in a medium extending from $-\xi'$ to $+\xi$ is given by

$$A_q(r, z) = i \frac{q\pi^2}{2^{q-2} n_q \lambda_1} \frac{N\chi_{\mathrm{a}}^{(q)} A_{10}^q}{(1 + i\xi)} b \exp\left[-\left(\frac{qk_1 r^2}{b(1 + i\xi)}\right)\right] F_q(b\Delta k, \xi', \xi), \quad (13)$$

where the phase optimization integral F_q is given by

$$F_q(b\Delta k, \xi', \xi) = \int_{-\xi'}^{\xi} (1 + i\xi'')^{1-q} e^{-i(b\Delta k\xi''/2)}\, d\xi''. \quad (14)$$

These expressions are similar to the corresponding ones given in the last chapter for third-harmonic generation (see Section 2-2.1.1). Like the third harmonic, the qth-harmonic amplitude has a Gaussian profile with the same confocal parameter and waist location as the fundamental. This is a consequence of the approximation $\Delta k_q \ll k_1, k_q$, which was used in deriving Eq. (13).

The total power in the qth harmonic, obtained by integrating the square of the magnitude of the field amplitude over the Gaussian profile, is given by

$$\mathscr{P}_q = \frac{2^{q+5}\pi^{q+3}q}{n_1 n_q \lambda_1^4 c^{q-1}} |N\chi_{\mathrm{a}}^{(q)}|^2 \left(\frac{4\mathscr{P}_1}{b\lambda_1}\right)^{q-3} \mathscr{P}_1^3 |F_q(b\Delta k)|^2. \quad (15)$$

Here $\mathscr{P}_1$ is the total fundamental power and the quantity $(4\mathscr{P}_1/b\lambda_1)$ is the peak focused intensity. Again this expression is similar to the corresponding one for the third-harmonic conversion given in Section 2-2.1.1.

2.1.2a. PHASE MATCHING. The higher-order interactions can be phase matched by the same techniques that were discussed in Section 2-3 for third-order processes. Plane waves or collimated beams of finite diameter can be phase matched for higher-order harmonic conversion, either in gas mixtures or by frequency selection about a dispersive resonance. The plane-wave phase-matching condition

$$n_q = n_1$$

is the same for all harmonics. The conditions for optimizing the phase mismatch in focused beams or in collimated beams of finite diameter depend on the order of the harmonic. These conditions will be discussed in the next two sections.

2.1.2b. PHASE OPTIMIZATION MIXTURES. Optimization of the phase-matching integral can be accomplished in mixtures of two or more media with dispersions of different sign. When only one of the two media contributes to the nonlinear susceptibility and phase optimization is done by adjusting the density of the second (linear) medium, the conversion is optimized by maximizing the magnitude of the integral F_q. This integral can be seen to be a generalization of the corresponding integral for third-harmonic conversion and many of the properties of the general integral are qualitatively similar to those of the third-order integral as discussed in Section 2-2.1.2. We shall discuss its behavior here in various limits.

Collimated Beams. In the limit of weak focusing ($b/L \gg 1$) the beam remains effectively collimated throughout the nonlinear medium and $\xi'' \ll 1$ everywhere. The phase optimization integral F_q takes the form

$$F_q(b\Delta k, \xi', \xi) = \int_{-\xi'}^{\xi} \exp(-i\{[b\Delta k/2 + (q-1)]\xi''\})\, d\xi''. \quad (16)$$

For a beam focused to the center of a medium of length L the solution to Eq. (16) is

$$F_q(b\Delta k, \xi', \xi) = \frac{2L}{b} \frac{\sin[\{\Delta k + 2(q-1)/b\}L/2]}{[\Delta k + 2(q-1)/b]L/2}. \quad (17)$$

This function has the same $\sin x/x$ form as was obtained for the plane wave approximation [Eq. (4)] except for an offset to the wave vector mismatch:

$$\Delta k_{\text{eff}} = \Delta k + 2(q-1)/b.$$

This is the same kind of offset that was discussed for third-harmonic conversion of collimated beams of finite diameter. The effective wave vector mismatch increases with increasing order of the harmonic, reflecting the greater mismatch near the beam waist in the higher orders between the harmonic field and the polarization that drives it. Just as with the third harmonic, however, the phase mismatch introduced by the wave vector offset in the higher orders is small compared to the width of the phase-matching peak when the approximation for collinated beams, $b/L \gg 1$, is well satisfied.

Tightly Focused Beams. In the limit of tight focusing, $b/L \ll 1$, the integral has the solution[19]

$$F_q(b\Delta k) = \begin{cases} \dfrac{2\pi}{(q-2)!}(b\Delta k/2)^{q-2}e^{b\Delta k/2}, & \Delta k < 0, \\ 0, & \Delta k > 0. \end{cases} \tag{18}$$

The variation of the square of the magnitude of F_q with the phase-matching parameter $b\Delta k_q$ is shown in Fig. 2 for the third–ninth harmonics. For all the harmonics conversion is possible in the tight-focus limit only in media with negative dispersion. The source of this behavior is identical to that which was discussed in Section 2-2.2.2 in connection with the third-order processes, namely, the phase shift associated with the Gaussian beam as it goes through the focus.

The optimum value of $b\Delta k_q$ is a function of the harmonic order and also of the limits of the integral in Eq. (15). In the tight-focus limit the optimum value of $b\Delta k_q$ is given by

$$b\Delta k_{\text{opt, tight focus}} = -2(q-2). \tag{19}$$

This value increases with increasing harmonic order, again reflecting the greater mismatch between the harmonic and its driving polarization in the higher orders.

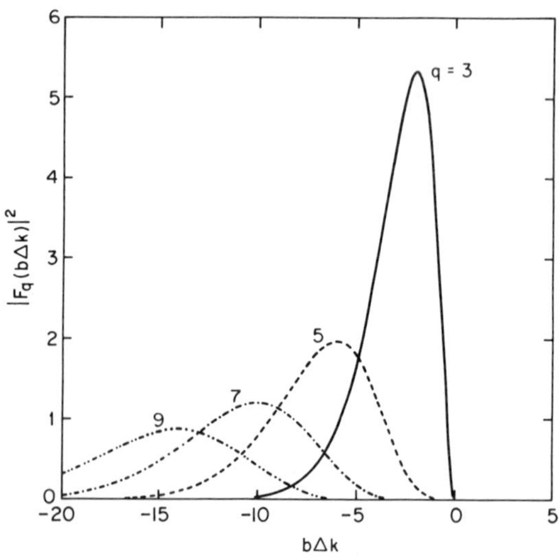

Fig. 2. Variation of the quantity $|F_q|^2$ with the phase-mismatch parameter $b\Delta k$ in an infinite medium for harmonics three–nine.

Fischer and Wieczorek[20] have calculated the optimum value of $b\Delta k_q$ for harmonics 3–12 as a function of the ratio L/b for beams that are focused at the center of a cell of length L. Their results are shown in Fig. 3a for harmonics 3–7. At small values of L/b the optimum value of $b\Delta k_q$ approaches its collimated beam limit of $-2(q-1)$. For large values of L/b the optimum value of $b\Delta k_q$ approaches its tight-focus limit of $-2(q-2)$. The transition from collimated beam behavior to tight-focused behavior occurs generally for values of L/b falling between 1 and 5, with the exact value depending on the order of the harmonic. The transition moves to smaller values of L/b and becomes narrower as the order of the harmonic increases, reflecting the fact that more of the conversion occurs in the vicinity of the beam waist for the higher harmonics.

The value of the phase optimization integral at the optimum value of $b\Delta k_q$ has also been calculated for the 3rd–12th harmonics by Fischer and Wiczorek.[20] Curves are shown in Fig. 3b for harmonics 3–7. In the plane-wave limit the integral has the same value, $4L/b$, for all of the harmonics. In the tight-focus limit the integral for each harmonic becomes independent of L/b, approaching the value

$$\lim_{L/b\to\infty} F_q(b\Delta k, \xi', \xi) = \frac{2\pi(q-2)^{q-2}e^{(2-q)}}{(q-2)!}. \tag{20}$$

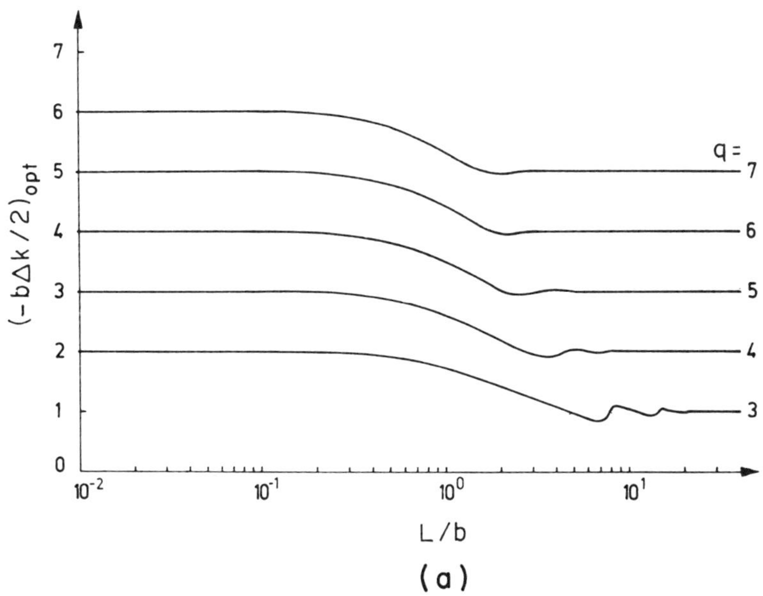

Fig. 3a. Optimum value of $(b\Delta k/2)$ for harmonics three–seven as a function of the focusing parameter $m = L/b$ for a beam focused at the center of the nonlinear medium. (From Ref. 20.)

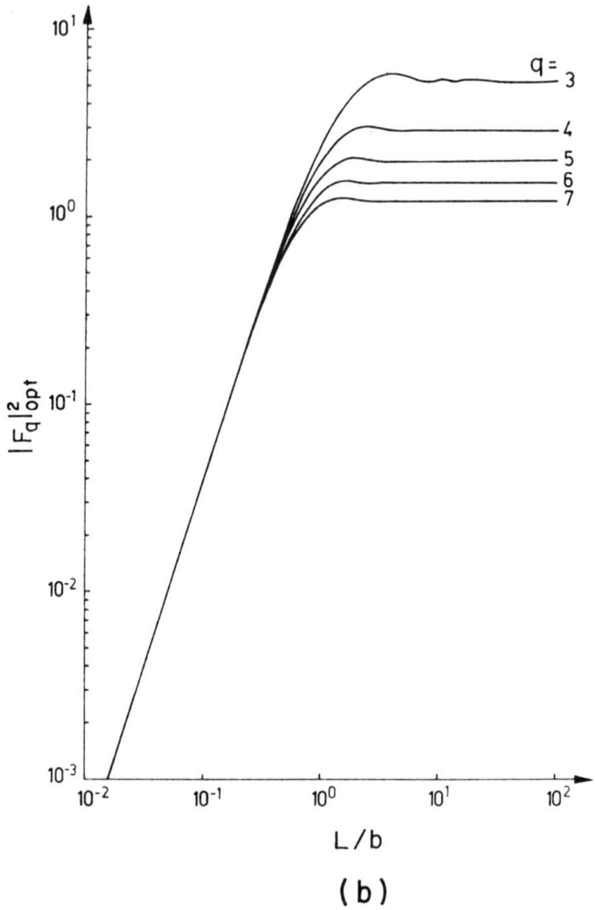

Fig. 3b. Maximum value of $|F_q|^2$ as a function of $m = L/b$ for a beam focused at the center of the nonlinear medium. (From Ref. 20.)

This value decreases as the order of the processes increases, reflecting the fact that the phase mismatch near the focus can be less well compensated in the higher-order interactions.

Although the phase optimization integral is independent of the ratio L/b for all of the harmonics in the tight-focus limit, the conversion efficiency is not. At constant pump power and optimum value of $b\Delta k_q$ the power in the qth harmonic varies as $(1/b)^{q-3}$ [see Eq. (15)]. This factor, which is unity for the third harmonic, increases with tighter focusing for the fifth and higher-order harmonics. Thus, conversion to the higher-order harmonics, unlike the third harmonic, is favored by the use of as tight a focus as possible, even when the phase optimization is done by varying the relative concentration of components in a mixture.

In some situations conversion is done in a medium that is effectively semi-infinite, extending to one side of the beam waist. This configuration is encountered, for example, in the generation of radiation in the XUV part of the spectrum. In this spectral range there are no convenient windows that will simultaneously contain the nonlinear medium and transmit the generated radiation. A common solution to this problem is to use differential pumping with a small pinhole, which allows a significant gas pressure to be maintained on one side and a relative vacuum on the other. The pump beam is typically focused at the pinhole. In this situation the integral in Eq. (14) must be evaluated between the limits $\xi' = -\infty$ and $\xi = 0$. The square of the magnitude of F_q under such conditions is shown in Fig. 4 for the third–seventh harmonics. Again optimum conversion is obtained only in media with negative dispersion, with the maximum conversion occurring at $b\Delta k = -0.8, -4.8,$ and -8.8 for the third, fifth and seventh harmonics, respectively. For the semi-infinite medium, however, some conversion is also allowed for media with negative dispersion. This property can be important in regions of the spectrum in which there are no transparent, negatively despersive media. The effective window that is created in the differential-pumping process can be advantageous in some circumstances, e.g., conversion in a medium that is absorbing at the harmonic, since it

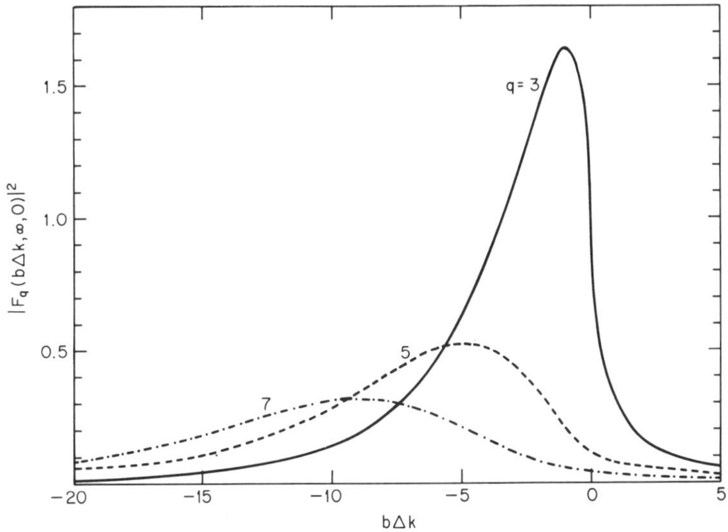

Fig. 4. Variation of the quantity $|F_q|^2$ with the phase-mismatch parameter $b\Delta k$ in a semi-infinite medium for harmonics three–seven.

allows maximum intensity to be focused at the effective window without causing damage, or conversion in a positively dispersive medium.

2.1.2c. PHASE OPTIMIZATION IN SINGLE-COMPONENT MEDIA. Phase optimization for higher-order harmonic conversion of focused beams can also be accomplished by adjusting the density of a single-component medium (see Section 2-3.2). In this method the phase mismatch is proportional to the density N of the nonlinear medium:

$$\Delta k = N \Delta k_{\text{atom}}.$$

The power in the qth harmonic can be rewritten from Eq. (15) to reflect this dependence as

$$\mathscr{P}q = \frac{2^{q-1}\pi^{q+1}}{q n_1 n_q c^{q-1}} \left| \frac{N}{n_1 - n_q} \chi_{\mathrm{a}}^{(q)} \right|^2 |b\Delta k F_q(b\Delta k)|^2 \left(\frac{4\mathscr{P}_1}{b\lambda_1} \right)^{q-1} \mathscr{P}_1. \qquad (21)$$

The quantity $(n_1 - n_q)/N$ is the dispersion per atom and is independent of the number density. The quantity $(4\mathscr{P}_1/b\lambda_1)$ is the peak focused pump intensity. The entire dependence of the harmonic power on the density, phase mismatch and focusing is now contained in the quantity $|b\Delta k_q F_q(b\Delta k_q)|^2$.

For a tightly focused beam in an infinite medium this dependence is given by

$$|G_q|^2 = |b\Delta k F_q(b\Delta k, \infty, \infty)|^2 = \begin{cases} \left[\dfrac{4\pi}{(q-2)!} (b\Delta k/2)^{q-1} e^{b\Delta k/2} \right]^2, & \Delta k < 0 \\ 0, & \Delta k > 0. \end{cases} \qquad (22)$$

This function is shown in Fig. 5a for harmonics three–nine. Conversion to the qth harmonic is optimized for a value of $b\Delta k_q$ given by

$$b\Delta k_q = -2(q-1). \qquad (23)$$

For a constant value of the confocal parameter the pressure at which optimum conversion occurs is higher for the higher harmonics. Optimum conversion in a single-component medium for all the harmonics is favored by use of as small a confocal parameter as possible, accompanied by a corresponding increase in pressure to maintain the optimum value of the product $b\Delta k_q$.

Again, generation of radiation in certain regions of the spectrum such as the XUV can involve conversion in a semi-infinite medium, as discussed in Section 2.1.2b. The variation of the appropriate phase optimization integral $|G_q(b\Delta k, -\infty, 0)|^2$ is shown in Fig. 5b for the third–seventh harmonics. Again, optimum conversion is obtained in media with negative dispersion, with the maximum conversion occurring at $b\Delta k = -6.2$, -10.4, and -14.6

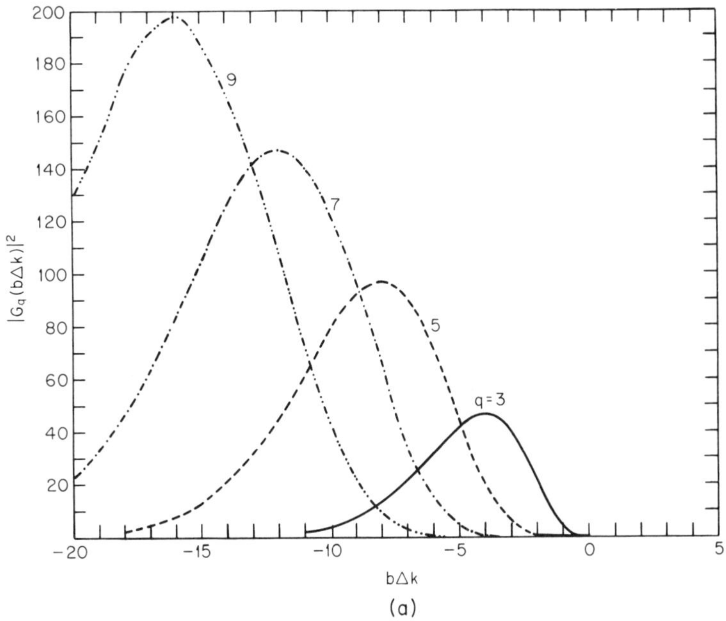

Fig. 5a. Variation of the quantity $|G_q(b\Delta k)|^2 = |b\Delta k F_q(b\Delta k)|^2$ as a function of $b\Delta k$ in an infinite medium for harmonics three–nine.

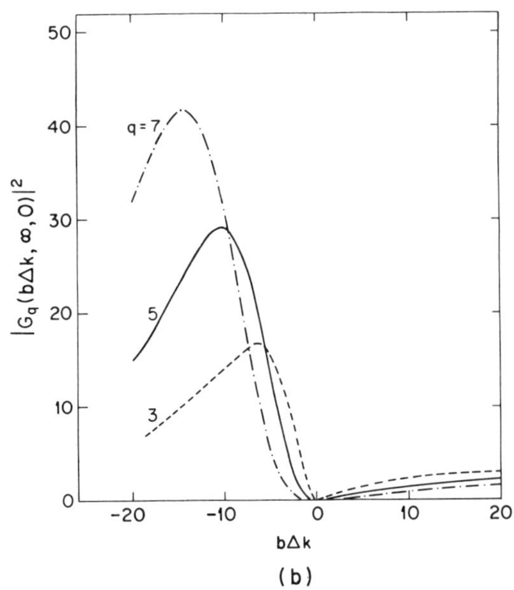

Fig. 5b. Variation of the quantity $|G_q|^2$ with the phase-mismatch parameter $b\Delta k$ in a semi-infinite medium for harmonics three–seven.

for the third, fifth and seventh harmonics, respectively. Again, a somewhat smaller, but nonzero, amount of radiation can be generated in media with positive dispersion.

2.2. Higher-Order Sum- and Difference-Frequency Mixing Processes

The phase-matching requirements for the various sum- and difference-frequency mixing processes in the higher orders can be obtained from analyses similar to those given in Sections 2-2.2.1 and 2-2.2.2 for the four-wave processes. The plane-wave phase-matching conditions are a straightforward extension of the conditions for the four-wave frequency-mixing interactions. The appropriate condition for interactions of the form $\omega_m = \omega_1 \pm \omega_2 \pm \ldots \pm \omega_n$ is

$$\frac{n(\omega_m)}{\lambda_m} = \frac{n(\omega_1)}{\lambda_1} \pm \frac{n(\omega_2)}{\lambda_2} \pm \cdots \pm \frac{n(\omega_n)}{\lambda_n}.$$

The dispersion requirements for the various higher-order frequency-mixing processes with focused beams can be derived by analyses similar to that given by Bjorklund.[21] The phase-matching integrals, and therefore the requirements on the wave vector mismatch, for the sum-frequency processes are the same as those for the corresponding harmonic processes. The phase-matching integrals for the higher-order difference-frequency interactions depend on the relative values of k' and k'', where k' is the wave number of the generated radiation and k'' is the sum of the wave numbers of the pump fields, just as do the integrals for the corresponding third-order processes (see Section 2.2.2). Dispersion requirements for interactions of the form $\omega_6 = n\omega_1 \pm m\omega_2$ are summarized in the following table for tightly focused beams in infinite and semi-infinite media in the limit that $k'' \approx k'$, as can occur, for example, when the nonlinear interaction involves pump fields at greatly differing wavelengths.

Process	Requirement on Δk	
	Infinite medium	Semi-infinite medium
$n\omega_1 + m\omega_2$	<0	<0 Optimum >0 Allowed
$n\omega_1 - \omega_2$	<0 Optimum >0 Allowed	<0 Optimum >0 Allowed
$n\omega_1 - m\omega_2,\ n, m \neq 1,\ m - n \neq 1$	<0 Optimum >0 Optimum	<0 Optimum >0 Optimum
$n\omega_1 - m\omega_2,\ m - n = 1$	>0 Optimum <0 Allowed	>0 Optimum <0 Allowed
$\omega_1 - m\omega_2$	>0	>0 Optimum <0 Allowed

The phase optimization integral F for the process $n\omega_1 - \omega_2$ in an infinite medium has a maximum for $\Delta k < 0$ and its magnitude decreases exponentially for $\Delta k > 0$. This process can be optimized in gas mixtures with $\Delta k < 0$ and in media with either sign of dispersion by pressure adjustment. The phase optimization integrals for processes of the type $n\omega_1 - m\omega_2$ with $n, m \neq 1$ and $m - n \neq 1$ have maxima for both positive and negative values of Δk. These interactions can be optimized in media with either sign of dispersion using either gas mixtures or pressure adjustment. The process $n\omega_1 - m\omega_2$ with $n - m = 1$ has a symmetric phase optimization integral in the limit that $k' \approx k''$. Otherwise the phase optimization integrals for these processes are in general asymmetric about $\Delta k = 0$. The process $n\omega_1 - m\omega_2$ with $m - n = 1$ has a phase optimization integral with a maximum for positive values of Δk. Its magnitude decreases exponentially with increasing negative values for Δk. The phase optimization integral for the process $\omega_1 - m\omega_2$ has an optimum value for positive Δk and is zero in the tight-focus limit in infinite media with negative values of Δk.

2.3. Indirect Processes

When multistep processes are important the nonlinear polarization at the frequency of the higher-order harmonic contains additional terms arising from multiple lower-order interactions. The net harmonic field is determined by the interference of the fields that are generated by the two types of interaction. We can illustrate the interplay of the direct and indirect processes by considering fifth-harmonic generation.

The direct and indirect processes are illustrated in Fig. 6. In the direct process the fifth harmonic is generated in the interaction $\omega_5 = \omega_1 + \omega_1 + \omega_1 + \omega_1 + \omega_1$ through the fifth-order nonlinear susceptibility $\chi^{(5)}(-5\omega_1, \omega_1, \omega_1, \omega_1, \omega_1, \omega_1)$, as was discussed in the last section. In the indirect process the third harmonic is generated through the third-order interaction

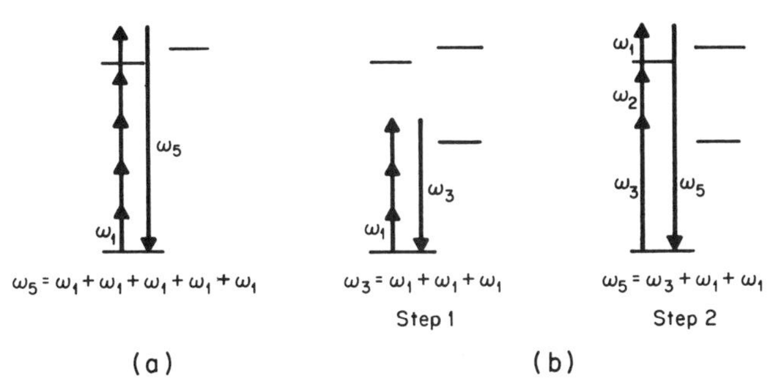

Fig. 6. Diagrammatic representation of direct (a) and indirect (b) fifth-harmonic processes. Four-photon resonant levels need not be present. Three- and five-photon resonant levels are necessary to ensure proper phase matching.

$\omega_3 = \omega_1 + \omega_1 + \omega_1$, and radiation at the fifth-harmonic frequency is generated through another third-order interaction $\omega_5 = \omega_3 + \omega_1 + \omega_1$. These steps proceed through the third-order susceptibilities $\chi^{(3)}(-3\omega_1, \omega_1, \omega_1, \omega_1)$ and $\chi^{(3)}(-5\omega_1, \omega_3, \omega_1, \omega_1)$.

The two-step process illustrated in Fig. 6 is to be distinguished from the cascade harmonic system illustrated in Fig. 7, in which the lower-order harmonic is generated in one medium and the higher-order harmonic is generated in a lower-order mixing process in a second medium. In the cascade process illustrated in Fig. 7 the fields that are generated in each of the lower-order interactions do not interfere with each other and can be identified and optimized with respect to the overall interaction in a straightforward manner by the techniques that were discussed in the last chapter. Cascade generation of the fifth harmonic of radiation from an Nd:glass laser has been described by Tunkin *et al.*[22] Optimization procedures for cascade fifth-harmonic generation have been discussed by Tomov and Richardson.[2] By contrast, the two types of processes occur simultaneously in the same medium in the two-step interaction illustrated in Fig. 6 and the relative contribution of the direct and indirect processes is much harder to determine.

One of the difficulties associated with separating the contributions of the direct process from those of the indirect process, as we shall discuss in the next section, is that both processes make a contribution to the phase-matched fifth harmonic. It is a general property of the higher-order interactions that whenever the direct higher-order process is phase matched, so are all the lower-order indirect processes, even if the individual steps of the indirect processes are not phase matched. Separation of the contributions of the direct and indirect processes usually requires knowledge of the susceptibilities for all of the interactions involved. These can be obtained by calculation or by auxiliary measurements. For example, phase-matching conditions usually exist for each of the indirect processes separately and measurement of the relative harmonic intensity that is generated under each of the phase-matching conditions can provide sufficient information to determine the contribution of the direct interaction to the harmonic generation process.

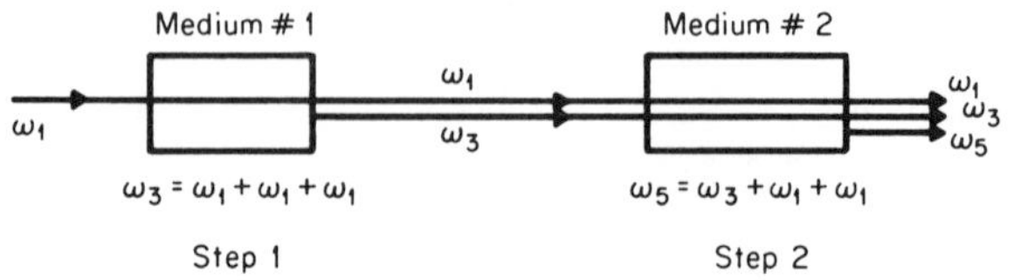

Fig. 7. Schematic diagram of higher-order harmonic generation done in a cascade geometry, involving two separate steps.

For the interaction illustrated in Fig. 6 the polarization amplitude at the fifth-harmonic frequency is

$$P_5 = (N/16)\chi_a^{(5)}(-5\omega, \omega, \omega, \omega, \omega, \omega)A_1^5 + (3N/4)\chi_a^{(3)}(-5\omega, \omega, \omega, 3\omega)A_1^2 A_3, \tag{24}$$

the polarization at the third-harmonic frequency is

$$P_3 = (N/4)\chi_a^{(3)}(-3\omega, \omega, \omega, \omega)A_1^3 + (3N/4)\chi_a^{(3)}(-3\omega, 5\omega, -\omega, -\omega)A_5(A_1^*)^2, \tag{25}$$

and the polarization at the fundamental is

$$P_1 = (3N/4)\chi_a^{(3)}(-\omega_1, 3\omega_1, -\omega_1, -\omega_1)A_3(A_1^*)^2 + (5N/16)\chi_a^{(5)}(-\omega, 5\omega, -\omega, -\omega, -\omega, -\omega)A_5(A_1^*)^4 + (6N/4)\chi_a^{(3)}(-\omega, 5\omega, -\omega, -3\omega)A_5 A_3^* A_1^*. \tag{26}$$

The nonlinear susceptibility for the sum-frequency part of the indirect process $\chi_a^{(3)}(-5\omega, 3\omega, \omega, \omega)$ is in general different from the third-harmonic susceptibility $\chi_a^{(3)}(-3\omega, \omega, \omega, \omega)$. The sum-frequency susceptibility carries all the resonant enhancements of the third-harmonic susceptibility. It can also carry any additional resonances with higher-lying states that are present in the fifth-harmonic susceptibility. As a result the sum-frequency susceptibility can be larger than the third-harmonic susceptibility, especially if the fifth-order susceptibility is resonantly enhanced. In other materials where such additional resonances are not present the sum-frequency susceptibility may be smaller than the third-harmonic susceptibility.

The equations for the field amplitudes at the three frequencies in the plane-wave limit are

$$\frac{dA_5}{dz} = i\frac{5\pi^2 N}{4n_5\lambda_1}\chi_a^{(5)}A_1^5 e^{-i\Delta k_{15}z} + i\frac{15\pi^2 N}{n_5\lambda_1}\chi_a^{(3)\prime}A_3 A_1^2 e^{-i\Delta k_{135}z}, \tag{27a}$$

$$\frac{dA_3}{dz} = i\frac{3\pi^2 N}{n_3\lambda_1}\chi_a^{(3)}A_1^3 e^{-i\Delta k_{13}z} + i\frac{9\pi^2 N}{n_3\lambda_1}\chi_a^{(3)\prime}A_5(A_1^*)^2 e^{i\Delta k_{135}z}, \tag{27b}$$

$$\frac{dA_1}{dz} = i\frac{5\pi^2 N}{4n_1\lambda_1}\chi_a^{(5)}A_5(A_1^*)^4 e^{i\Delta k_{15}z} + i\frac{3\pi^2 N}{n_1\lambda_1}\chi_a^{(3)}A_3(A_1^*)^2 e^{i\Delta k_{13}z} + i\frac{6\pi^2 N}{n_1\lambda_1}\chi_a^{(3)\prime}A_5A_3A_1^* e^{i\Delta k_{135}z}, \tag{27c}$$

where

$$\chi_a^{(5)} = \chi_a^{(5)}(-5\omega, \omega, \omega, \omega, \omega, \omega), \tag{28a}$$

$$\chi_a^{(3)} = \chi_a^{(3)}(-3\omega, \omega, \omega, \omega), \tag{28b}$$

$$\chi_a^{(3)\prime} = \chi_a^{(3)}(-5\omega, \omega, \omega, 3\omega), \tag{28c}$$

and

$$\Delta k_{15} = k_5 - 5k_1, \tag{29a}$$

$$\Delta k_{13} = k_3 - 3k_1, \tag{29b}$$

$$\Delta k_{135} = k_5 - k_3 - 2k_1 = \Delta k_{15} - \Delta k_{13}. \tag{29c}$$

The nature of the solutions to these equations depends on the phase mismatches for all the steps involved in the direct and indirect processes. In single-component media or in binary mixtures one generally has control over only one of the wave vector mismatches. For these conditions it is appropriate to examine the equations under the assumption that one of the phase-mismatch parameters is optimized and the others are large. In ternary mixtures one can have the ability to adjust all of the phase-mismatch parameters independently, subject to the interrelation expressed in Eq. (29c). Under these conditions it is appropriate to examine the equations when all of the steps are phase matched. The behavior of the solutions also depends on the distance required for energy to be transferred from the third- to the fifth-harmonic field in the second step of the indirect process relative to the distance required for depletion of the fundamental through direct third- and fifth-harmonic generation.

We consider first a situation in which the third-harmonic process is not phase matched, i.e., $|\Delta k_{13}| \gg 0$. If we assume that depletion of the fundamental and of the third harmonic in the second step of the indirect process can be neglected, then the fifth-harmonic field amplitude is given by

$$A_5 = i\frac{5\pi^2 N}{4n_5\lambda_1}\left(\chi_a^{(5)} - \frac{36\pi^2 N\chi_a^{(3)}\chi_a^{(3)\prime}}{n_3\lambda_1\Delta k_{13}}\right)LA_1^5\frac{\sin(\Delta k_{15}L/2)}{(\Delta k_{15}L/2)}e^{-i(\Delta k_{15}L/2)} + i\frac{45\pi^4 N^2\chi_a^{(3)}\chi_a^{(3)\prime}}{n_3n_5\lambda_1^2\Delta k_{13}}LA_1^5\frac{\sin(\Delta k_{135}L/2)}{(\Delta k_{135}L/2)}e^{-i(\Delta k_{135}L/2)}. \tag{30}$$

If the direct fifth-harmonic process is phase matched, i.e., $\Delta k_{15} \approx 0$, the second term will, by asumption, be phase mismatched:

$$\Delta k_{135} = \Delta k_{15} - \Delta k_{13} \neq 0, \tag{31}$$

and the first term will make the dominant contribution to the fifth-harmonic process. This term can be expressed as

$$A_5 = i\frac{5\pi^2}{4n_5\lambda_1} NL\chi_{\text{eff}}^{(5)} A_1^5 \frac{\sin(\Delta k_{15}L/2)}{(\Delta k_{15}L/2)} e^{-i(\Delta k_{15}L/2)}, \tag{32}$$

where

$$\chi_{\text{eff}} = \chi_{\text{a}}^{(5)} - \frac{36\pi^2 N\chi_{\text{a}}^{(3)}\chi_{\text{a}}^{(3)\prime}}{n_3\lambda_1\Delta k_{13}}. \tag{33}$$

The corresponding fifth-harmonic intensity is

$$I_5 = \frac{6400\pi^8 N^2 L^2 |\chi_{\text{eff}}|^2 I_{10}^5}{n_5 n_1^5 \lambda_1^2 c^4} \frac{\sin^2(\Delta kL/2)}{(\Delta kL/2)^2}. \tag{34}$$

The phase-matched fifth harmonic is thus generated through an effective nonlinear susceptibility which includes contributions from both the direct and two-step process, even though each of the individual steps in the indirect process is not phase matched. This behavior results from the fact that the third-harmonic field consists of a free wave, which propagates with the phase velocity that is normally associated with the third-harmonic frequency, and a driven wave, which propagates with a phase velocity that is determined by the fundamental. When the direct fifth harmonic is phase matched the part of the indirect process that involves the driven third-harmonic wave is also phase matched:

$$\begin{aligned}\Delta k_{\text{indirect}} &= k_5 - (2k_1 + k_{3,\text{driven}})\\ &= k_5 - (2k_1 + 3k_1) = \Delta k_{15}.\end{aligned}$$

The fifth-harmonic fields generated in the two types of interactions are completely indistinguishable. The relative importance of the two contributions can only be determined from a knowledge of the relative magnitudes of the susceptibilities obtained from calculations or independent measurements.

The ratio of the contribution of the indirect process to that of the direct process is given by

$$\frac{A_{5,\text{indirect}}}{A_{5,\text{direct}}} = \frac{36\pi^2 N\chi_{\text{a}}^{(3)}\chi_{\text{a}}^{(3)\prime}}{n_3\lambda_1\Delta k_{13}\chi_{\text{a}}^{(5)}}. \tag{35}$$

The indirect process is favored by smaller values of Δk_{13}, while the direct process is favored by larger values of Δk_{13}. When Δk_{13} is independent of the density of the nonlinear medium, such as in gas mixtures, the indirect process will become more important relative to the direct process as the density increases. If, on the other hand, Δk_{13} is proportional to N, as, for example, in a single-component medium, then the relative importance of the two types of process is independent of N. In some instances, such as when four- or five-photon resonances are involved or when two different materials with the same type of resonant structure are compared, all the susceptibilities will scale together, with the ratio of the fifth- to the third-order susceptibilities remaining constant. Under such conditions the indirect process is favored by larger absolute values of the susceptibilities, while the direct process is favored by smaller values of the susceptibilities.

Wang and Baardsen[23] have shown that for nonresonant third-harmonic conversion in solids the direct and indirect processes make comparable contributions. Similar results have been obtained by Yablonovitch *et al.*[24] for four-wave mixing in semiconductors. Consideration of fifth-harmonic conversion of focused beams in low-pressure gases has shown that either the direct[8] or indirect process[2] can dominate depending on the exact situation that is involved.

The indirect process can also be phase matched independently of the direct process. In general this can be done in two ways. If the condition $\Delta k_{135} \approx 0$ is met then the second term in Eq. (30) will be phase matched. Because of our original assumption that $|\Delta k_{13}| \gg 0$ the direct process will not be phase matched and the dominant contribution to the fifth harmonic will arise from the indirect process, as described by the second term in Eq. (30). In this case the fifth-harmonic field is given by

$$A_5 = i\,\frac{45\pi^4 N^2 \chi_{\mathrm{a}}^{(3)} \chi_{\mathrm{a}}^{(3)\prime} L}{n_5 n_3 \lambda_1^2 \Delta k_{13}} \frac{A_1^5 \sin(\Delta k_{135} L/2)}{(\Delta k_{135} L/2)} e^{-i\Delta k_{135} L/2} \tag{36}$$

and the fifth-harmonic intensity is

$$I_5 = \frac{8.2944 \times 10^6 \pi^{12} N^4 |\chi_{\mathrm{a}}^{(3)} \chi_{\mathrm{a}}^{(3)\prime}|^2 L^2 I_1^5}{n_5 n_3^2 n_1^5 c^4 \lambda_1^4 \Delta k_{13}^2} \frac{\sin^2(\Delta k_{135} L/2)}{(\Delta k_{135} L/2)^2}. \tag{37}$$

In this situation the fifth harmonic is generated through a phase-matched interaction involving the free third-harmonic wave and the fundamental wave.

The indirect process can also be phase matched for fifth-harmonic generation by phase matching the third-harmonic process, i.e., by making $\Delta k_{13} \approx 0$. If we assume that the fifth harmonic is not simultaneously phase matched, i.e., if $\Delta k_{15} \neq 0$, then the dominant term in the fifth harmonic will again arise from the indirect interaction. This time the fifth-harmonic field receives contributions from both indirect terms in Eq. (30) and is given by

$$A_5 = -i\frac{45\pi^4 N^2 \chi_a^{(3)} \chi_a^{(3)'} L\, A_1^5}{n_3 n_5 \lambda_1^2 \Delta k_{135}} \frac{\sin(\Delta k_{13} L/2)}{(\Delta k_{13} L/2)^2} e^{-i(\Delta k_{135} + \Delta k_{15})L/2}. \tag{38}$$

The corresponding fifth-harmonic intensity is

$$I_5 = \frac{8.2944 \times 10^6 \pi^{12} N^4 |\chi_a^{(3)} \chi_a^{(3)'}| L^2 I_1^5}{n_5 n_3^2 n_1^5 c^4 \lambda_1^4 \Delta k_{135}^2} \frac{\sin^2(\Delta k_{13} L/2)}{(\Delta k_{13} L/2)}. \tag{39}$$

Measurements of the relative fifth-harmonic intensity that is generated for each of the three phase-matching conditions can be used to determine the magnitude of the true fifth-order susceptibility.

When the intermediate third-harmonic step is simultaneously phase matched with the direct fifth-harmonic process ($\Delta k_{13} = \Delta k_{15} = \Delta k_{135} = 0$) the fifth- and third-harmonic amplitudes are given by

$$A_5 = i\frac{5\chi_a^{(5)}}{12\sqrt{15}\chi_a^{(3)'}} A_1^3 \sin(3\sqrt{15}\pi^2 N \chi_a^{(3)'} A_1^2 L/n\lambda_1) - \frac{2}{3}\frac{\chi_a^{(3)}}{\chi_a^{(3)'}} A_1^3 \sin^2(3\sqrt{15}\pi^2 N \chi_a^{(3)'} A_1^2 L/2n\lambda_1), \tag{40a}$$

$$A_3 = i\frac{1}{\sqrt{15}}\frac{\chi_a^{(3)}}{\chi_a^{(3)'}} A_1 \sin(3\sqrt{15}\pi^2 N \chi_a^{(3)'} A_1^2 L/n\lambda_1) - \frac{1}{6}\frac{\chi_a^{(5)}}{\chi^{(3)'}} A_1^3 \sin^2(3\sqrt{15}\pi^2 N \chi_a^{(3)'} A_1^2 L/2n\lambda_1), \tag{40b}$$

where we have used $n_1 = n_3 = n_5$ and we have assumed that A_1 is real. In addition, we have again neglected the depletion of the fundamental wave but have allowed the third harmonic to be depleted in the indirect process.

For sufficiently small values of L, or equivalently for sufficiently small pump fields, the amplitudes of the harmonic fields are given by

$$A_5 = i\frac{5\pi^2 N \chi_a^{(5)} A_1^5 L}{4n\lambda_1} - \frac{45\pi^4 N^2 \chi_a^{(3)} \chi_a^{(3)'}}{2n^2 \lambda_1^2} A_1^5 L^2, \tag{41a}$$

$$A_3 = i\frac{3\pi^2 N \chi_a^{(3)} A_1^3 L}{n\lambda_1} - \frac{45\pi^4 N^2 \chi_a^{(5)} \chi_a^{(3)'}}{8n^2 \lambda_1^2} A_1^7 L^2. \tag{41b}$$

The fifth-harmonic fields generated in the direct and indirect processes now have different dependences on interaction length. The field generated in the direct process grows linearly with distance and is favored by relatively short interaction lengths, while the field generated in the indirect process grows as L^2 and is favored by relatively long interaction lengths. The relative contributions of the two types of processes can now be separated in principle at least by their different dependences on the interaction length.

For larger values of L, or equivalently for sufficiently large values of the pump field, the third-harmonic field can be depleted in the second step of the indirect process. This leads ultimately to the oscillatory behavior that is contained in the expressions in Eq. (40). It is convenient to define a characteristic length for the transfer of energy from the third to the fifth harmonic:

$$\begin{aligned} L_{35} &= (6\sqrt{15}\pi N \chi_a^{(3)\prime} A_1^2 / n\lambda_1)^{-1} \\ &= (48\sqrt{15}\pi^2 N \chi_a^{(3)\prime} I_1 / n^2 c \lambda_1)^{-1}. \end{aligned} \tag{42}$$

In terms of this length the expressions in Eqs. (40a) and (40b) become

$$\begin{aligned} A_5 = {} & i \frac{5\pi N \chi_a^{(5)} A_1^5 L_{35}}{2n\lambda_1} \sin(\pi L/2L_{35}) \\ & - \frac{2}{3} \frac{\chi_a^{(3)}}{\chi_a^{(3)\prime}} A_1 \sin^2(\pi L/4L_{35}), \end{aligned} \tag{43a}$$

$$\begin{aligned} A_3 = {} & i \frac{6\pi N \chi_a^{(3)}}{n\lambda_1} A_1^3 L_{35} \sin(\pi L/2L_{35}) \\ & - \frac{1}{6} \frac{\chi_a^{(3)}}{\chi_a^{(3)\prime}} A_1^3 \sin^2(\pi L/4L_{35}). \end{aligned} \tag{43b}$$

The quantity L_{35} plays a role that is similar to that of the coherence length which arose in our discussions of the individual frequency mixing processes. The oscillations in Eq. (43) do not result from phase differences caused by a mismatch in the refractive indexes, however, but rather from the difference in phase of the field components that are generated through the direct and indirect processes.

The energy in the optical fields oscillates among the fundamental and the harmonics as a function of distance. When the direct harmonic processes are dominant the maximum energy in both the third and fifth harmonics occurs at $L = (2m + 1)L_{35}$, and the energy is returned to the fundamental when $L = mL_{35}$. When both the direct and indirect processes are important the fifth and third harmonics will be maximized at different distances. Nevertheless all the energy can again be returned to the fundamental whenever $L = 4mL_{35}$.

The interference between the third- and fifth-harmonic fields can result in a fundamental limitation on the fifth-harmonic conversion efficiency at a value which under certain conditions is considerably less than unity. To see this we examine the behavior of Eq. (43) in the limits where either the indirect or direct process makes the dominant contribution to the fifth harmonic.

If the direct fifth-harmonic process can be neglected we can set $\chi_a^{(5)} = 0$ and the fifth- and third-harmonic intensities are given by

$$I_5 = \frac{4}{9}\left|\frac{\chi_a^{(3)}}{\chi_a^{(3)'}}\right|^2 I_1 \sin^4(\pi L/4L_{35}), \tag{44a}$$

$$I_3 = \left(\frac{8\pi}{cn}\right)^2 I_1^3 \left|\frac{36\pi^2 N^2(\chi_a^{(3)})^2 L_{35}^2}{n^2\lambda_1^2}\sin^2(\pi L/2L_{35}) + \frac{1}{36}\left(\frac{\chi_a^{(5)}}{\chi_a^{(3)'}}\right)^2 \sin^4(\pi L/4L_{35})\right|^2. \tag{44b}$$

The maximum value of the fifth harmonic occurs when $L = 2mL_{35}$. If this distance is less than the distance required for depletion of the fundamental wave through the direct third harmonic process L_3 [see Eq. (10a)], then our original assumption that pump depletion can be neglected will be valid and the maximum value of the fifth-harmonic intensity will be, from Eq. (44a),

$$I_{5,\max} = \frac{4}{9}\left|\frac{\chi_a^{(3)}}{\chi_a^{(3)'}}\right|^2 I_1. \tag{45}$$

The condition $L_{35} \ll L_3$ is equivalent to the condition that the sum-frequency susceptibility be greater than the third-harmonic susceptibility:

$$\begin{gathered} L_{35} \ll L_3, \\ \frac{n\lambda_1}{6\sqrt{15}\,\pi N\chi_a^{(3)'}A_1^2} \ll \frac{n\lambda_1}{3\pi^2 N\chi_a^{(3)}A_1^2}, \\ \chi_a^{(3)}/\chi_a^{(3)'} \ll 2\sqrt{15}/\pi \approx 2.5. \end{gathered} \tag{46}$$

This ratio is a property of the nonlinear medium and cannot be affected by choice of pump intensity, interaction length, or medium density. In materials in which Eq. (46) is satisfied pump depletion will be negligible and Eq. (45) shows that the efficiency of conversion to the fifth harmonic will be limited to a value considerably less than unity when both parts of the indirect process are phase matched. In materials in which Eq. (46) is not satisfied the pump will be depleted before the energy is transferred to the fifth harmonic through the two-step process and larger conversion is possible, as will be discussed in the next section.

When the direct fifth-harmonic process is dominant over the indirect process we can neglect the second term in Eq. (43a). The fifth-harmonic intensity is now given by

$$I_5 = \frac{2.56 \times 10^4 \pi^6 N^2 |\chi_a^{(5)}|^2 L_{35}^2 I_1^5}{n^6 \lambda_1^2 c^4} \sin^2(\pi L/2L_{35}). \tag{47}$$

It has a maximum value of

$$I_{5,\max} = \frac{2.56 \times 10^4 \pi^6 N^2 |\chi_a^{(5)}|^2 L_{35}^2 I_1^5}{n^6 \lambda_1^2 c^4}, \tag{48}$$

which occurs at a value of L given by

$$L = (2m + 1)L_{35}, \qquad m = 0, 1, \ldots . \tag{49}$$

The maximum value of the fifth-harmonic intensity that can be obtained in a given medium is determined by the relative values of L_{35} and L_5, the distance required for depletion of the fundamental through the direct fifth-harmonic process [see Eq. (10b)]. When

$$L_{35} \ll L_5 \tag{50}$$

the expression in Eq. (48) will be valid and condition (50) can place a limit on the maximum fifth-harmonic intensity. Equation (48) can be expressed as

$$\frac{I_{5,\max}}{I_1} = \frac{64\pi^2}{9} \left(\frac{L_{35}}{L_5}\right)^2. \tag{51}$$

Inspection of Eq. (51) shows that if inequality (50) is very well satisfied, then the fifth-harmonic conversion efficiency can be limited to a value much less than unity by the transfer of energy to the third harmonic even though the fifth harmonic is generated through the direct process.

If inequality (50) is not satisfied, then the fundamental can be depleted by the direct process first and virtually complete conversion can be obtained. This situation can be ensured by using a sufficiently large pump intensity to satisfy the condition

$$I_1 \gg \frac{4\sqrt{15}\, n_5 c \chi_a^{(3)'}}{5\pi^2 \chi_a^{(5)}}. \tag{52}$$

When the direct and indirect processes are of comparable importance the fifth-harmonic intensity has a maximum when

$$\cos(\pi L/2L_{35}) = B^2/(B^2 - 4C^2), \tag{53a}$$

where $C = 5\pi N \chi_a^{(5)} A_1^5 L_{35}/2n\lambda_1$ and $B = 2\chi_a^{(3)} A_1/3\chi_a^{(3)}$, provided $4C^2 > B^2$. The maximum fifth-harmonic intensity is

$$I_5 = \frac{cn_5}{2\pi} \frac{c^4}{4C^2 - B^2}. \tag{53b}$$

The net result of these considerations is that simultaneous phase matching of the third- and fifth-harmonic processes to increase the ultimate fifth-harmonic intensity in the plane-wave geometry may not have the desired effect. In materials in which the direct fifth harmonic is dominant, complete conversion to the fifth harmonic can be ensured only by adjusting the pump intensity to obtain pump depletion through the direct harmonic process. In this situation the indirect process is not needed and in fact it can place a limit on conversion efficiency by causing power to flow into the third harmonic from the fifth. When the direct process is negligible, on the other hand, complete conversion to the fifth harmonic is possible only for materials for which $\chi_a^{(3)} \gg \chi_a^{(3)\prime}$. In materials for which $\chi_a^{(3)} \ll \chi_a^{(3)\prime}$ the conversion to the fifth harmonic will always be considerably less than unity.

2.3.1. Pump Depletion

Solutions of Eqs. (27a)–(27c) when pump depletion is significant requires numerical computation. Tomov and Richardson[2] have given typical solutions for values of $\chi_a^{(3)}$, $\chi_a^{(3)\prime}$, $\chi_a^{(5)}$ and the various wave vector mismatches that are appropriate for conversion of radiation from an Nd : glass laser in mixtures of Na, Cd, and Xe. The ternary mixture allows arbitrary adjustment of the three-wave vector mismatches that appear in the equations. The amplitudes of the fundamental, third harmonic and fifth harmonic are shown in Fig. 8 for the situation in which the third- and fifth-harmonic processes are simultaneously phase matched, and in Fig. 9 for the situation in which only the direct fifth-harmonic process is phase matched. Calculations are shown in

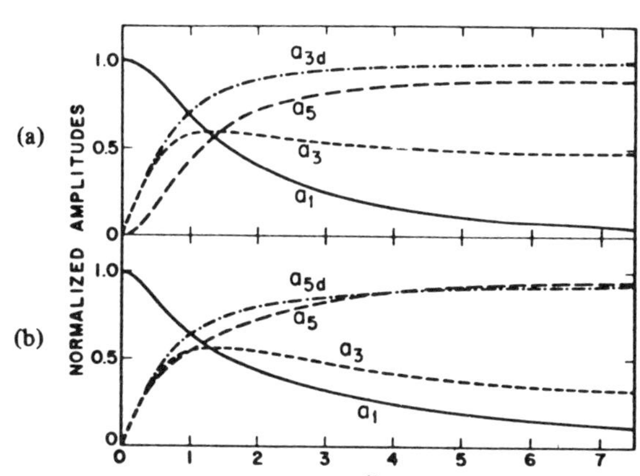

Fig. 8. Normalized field amplitudes for the third and fifth harmonics a function of the normalized propagation distance for the third harmonic for direct and indirect fifth-harmonic generation with plane waves. Curves labeled *d* are for the direct process alone. Both steps in the indirect process are assumed to be phase matched. The curves in (a) are calculated for a fifth-harmonic depletion length $L_5 = (K_5|A_1|^4)^{-1}$ [see Eq. (10)] that is 45 times longer than the third-harmonic depletion length. The curves in (b) were calculated for equal depletion lengths for the two harmonics: $L_5 = (K_5|A_1|^4)^{-1} = L_3 = (K_3|A_1|^2)^{-1}$. (From Ref. 2, © 1976 IEEE.)

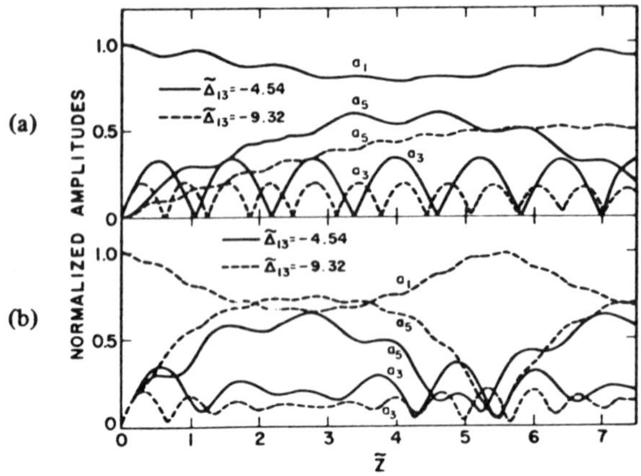

Fig. 9. Variation of the field amplitudes of the third and fifth harmonics with the normalized distance for plane waves assuming that the intermediate step in the indirect process is not phase matched. In (a) $L_3 = 0.0221 L_5$. In (b) $L_3 = L_5$. (From Ref. 2, © 1976 IEEE.)

each figure for two levels of pump intensity. The horizontal scale for each figure is the distance in units of the depletion length for the third-harmonic process $\tilde{z} = L_3^{-1} z$. In Fig. 8a the fifth-harmonic depletion length [defined[2] as $L_5^{-1} = K_5 |A_1|^4$] is approximately 50 times longer than that of the third harmonic and almost the entire fifth harmonic is due to the indirect process. In Fig. 8b a higher pump intensity was used such that the depletion lengths for the third and fifth harmonics were equal. The growth of the third harmonic in Fig. 9a and the fifth harmonic in Fig. 9b by direct processes are also shown. In either situation almost complete conversion to the fifth harmonic appears to be possible. The oscillations and limitations that were discussed in the previous section are not evident because of the strong pump depletion that occurs on the same scale of interaction length as the transfer of energy between the third and fifth harmonics.

When the third harmonic is not phase matched (Fig. 9b), oscillation of the third- and fifth-harmonic amplitudes is more pronounced. Significant conversion to the fifth harmonic is still possible. When the indirect process is prominent the fifth-harmonic conversion increases with a decrease in the third-harmonic phase mismatch Δk_{13}. When the direct fifth-harmonic process is significant the net conversion can decrease as the third-harmonic process becomes better phase matched.

2.3.2. *Focused Beams*

Fifth-harmonic generation in focused beams through direct and indirect processes has been treated theoretically by Tomov and Richardson,[2] who specifically discussed the conversion of Nd:glass laser radiation in metal vapors. When depletion of both the fundamental and third-harmonic waves can be neglected Tomov and Richardson have shown that the fifth-harmonic amplitude from a focused Gaussian beam is given by

$$A_5 = i\frac{5\pi^2 b A_{10}^5}{8n_5\lambda_1(1+i\xi)} e^{-5k_1r^2/b(1+i\xi)} N\chi_a^{(5)} \times \left[F_{5,\text{direct}} + i\frac{18\pi^2 N\chi_a^{(3)}\chi_a^{(3)\prime}}{n_3\lambda_1\chi_a^{(5)}} bF_{5,\text{indirect}}\right], \quad (54)$$

where

$$F_{5,\text{direct}} = \int_{-\xi'}^{\xi} \frac{e^{-ib\Delta k_{15}\xi''/2}}{(1+i\xi'')^4} d\xi'' \quad (55)$$

and

$$F_{5,\text{indirect}} = \int_{-\xi'}^{\xi} \frac{e^{-ib\Delta k_{135}\xi''/2}}{(1+i\xi'')^2} d\xi'' \int_{-\xi'}^{\xi''} \frac{e^{-ib\Delta k_{13}\eta/2}}{(1+i\eta)^2} d\eta. \quad (56)$$

Here, as before, b is the confocal parameter of the focused beam, $\xi, \eta = 2z/b$, the medium extends from $-\xi'$ to ξ, and the wave vector mismatches are as defined in Section 2.3.

The first term in Eq. (54) is due to the direct process and its properties were discussed in section 2.12. The second term is due to the indirect process. The behavior of this term will depend in general on the values of $b\Delta k_{13}$, $b\Delta k_{135}$, and on the ratio of the length of the nonlinear medium to the confocal parameter $m = L/b$. Just as for plane waves the characteristics of this term will depend on whether the phase-mismatch parameters can be optimized independently or only one at a time.

2.3.2a. Intermediate Step Badly Mismatched. When either $b|\Delta k_{13}| \gg 1$ or $b|\Delta k_{135}| \gg 1$ the integrals in Eq. (56) can be expressed as

$$\begin{aligned} F_{5,\text{indirect}} &= \frac{2i}{b\Delta k_{13}} \int_{-\xi'}^{\xi} \frac{e^{-ib\Delta k_{15}\xi''/2}}{(1+i\xi'')^4} d\xi'' - \frac{2ie^{ib\Delta k_{13}\xi'/2}}{b\Delta k_{13}(1-i\xi')^2} \int_{-\xi'}^{\xi} \frac{e^{-ib\Delta k_{135}\xi''/2}}{(1+i\xi'')^2} d\xi'' \\ &= \frac{2i}{b\Delta k_{13}} F_{5,\text{direct}} - \frac{2ie^{ib\Delta k_{13}\xi'/2}}{b\Delta k_{13}(1-i\xi')^2} F_3(b\Delta k_{135}) \end{aligned} \quad (57)$$

when $b|\Delta k_{13}| \gg 1$ or

$$F_{5,\text{indirect}} = \frac{2ie^{-i(b\Delta k_{135}\xi/2)}}{b\Delta k_{135}(1+i\xi)^2} F_3(b\Delta k_{13}) \quad (58)$$

when $b|\Delta k_{135}| \gg 1$. The integrals $F_3(b\Delta k)$ are the same ones that were discussed in Section 2-2.1.2.

When $b|\Delta k_{13}| \gg 1$ the fifth-harmonic amplitude is given by

$$A_5(r, \xi) = i\,\frac{5\pi^2}{8n_5\lambda_1}\,bN\chi_{\mathrm{a}}^{(5)}\,\frac{A_{10}^5}{(1+i\xi)}\,e^{-5k_1r^2/b(1+i\xi)}\left[1 - \frac{36\pi^2N\chi_{\mathrm{a}}^{(3)}\chi_{\mathrm{a}}^{(3)\prime}}{n_3\lambda_1\chi_{\mathrm{a}}^{(5)}\Delta k_{13}}\right]$$
$$\times F_{5,\mathrm{direct}}(b\Delta k, \xi', \xi)$$
$$+ i\,\frac{45\pi^4bN^2\chi_{\mathrm{a}}^{(3)}\chi_{\mathrm{a}}^{(3)\prime}A_{10}^5e^{ib\Delta k_{13}\xi'/2}e^{-5k_1r^2/b(1+i\xi)}}{2n_5n_3\lambda_1^2\Delta k_{13}(1-i\xi')^2(1+i\xi)}\,F_3(b\Delta k_{135}, \xi', \xi). \tag{59}$$

The first term carries contributions from both the direct and indirect processes, the latter arising from the interaction involving the driven third-harmonic wave in a manner similar to that which was encountered in the plane-wave solutions [see Eq. (30)]. When phase matching is done independently of the density of the nonlinear medium, e.g., by component mixtures, the contribution of this term is optimized for both the direct and indirect processes by maximizing the quantity $|F_{5,\mathrm{direct}}(b\Delta k_{15}, -\xi', \xi)|$, as was discussed in section 2.12b. Again the relative contributions of the direct and indirect processes cannot be separated by measuring only the phase-matched fifth-harmonic conversion efficiency.

The second term in Eq. (59) arises from the indirect process alone. Its phase-matching properties are determined by the function

$$F_3(b\Delta k_{135}, \xi', \xi)$$

resulting from phase matching of the sum-frequency process $\omega_5 = \omega_3 + \omega_1 + \omega_1$ using the free third-harmonic wave at $\omega_3 = 3\omega_1$. Another contribution from the indirect process arises from phase matching the third-harmonic step $\omega_3 = \omega_1 + \omega_1 + \omega_1$ and allowing a phase mismatch in both the second step of the indirect process and in the direct process. This contribution is determined by the integral given in Eq. (58). If $b\Delta k_{15} \gg 1$ and $b\Delta k_{135} \gg 1$ the fifth-harmonic field amplitude is given by

$$A_5 = -i\,\frac{45\pi^4bN^2\chi_{\mathrm{a}}^{(3)}\chi_{\mathrm{a}}^{(3)\prime}e^{-ib\Delta k_{135}\xi/2}}{2n_5n_3\lambda_1^2\Delta k_{135}(1+i\xi)^3}\,A_{10}^5e^{-5k_1r^2/b(1+i\xi)}$$
$$\times F_3(b\Delta k_{13}, \xi', \xi). \tag{60}$$

In an infinite medium the contribution from the indirect process through either the second term in Eq. (59) or the expression in Eq. (60) is zero because of the factors involving $(1 + i\xi)$ and $(1 - i\xi')$ in the denominators. In this situation the only nonzero contribution from the indirect process is contained in the first term in Eq. (59). A nonzero contribution from the indirect process

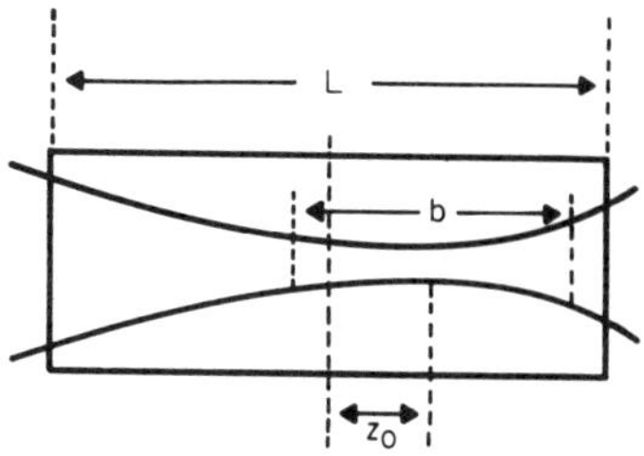

Fig. 10. Diagram illustrating the focusing geometry for analyzing the indirect contributions of Eqs. (61a) and (61b).

through the second term in Eq. (59) or the term in Eq. (60) does exist, however, in media of finite length. The magnitude of this contribution and the conditions under which it is optimized depend on the position of the focus relative to the edges of the nonlinear medium. The optimization conditions are in general different from those that optimize the direct fifth-harmonic process.

To examine this contribution of the indirect process we consider a beam focused into a nonlinear medium, as shown in Fig. 10. The beam is assumed to be focused at a position z_0 relative to the center of the medium and the ratio of the length of the medium to the confocal parameter is defined as $m = L/b$. In terms of these quantities the fifth-harmonic field amplitudes due to the indirect processes from Eqs. (59) and (60) can be written as

$$A_{5,\,\mathrm{indirect}}(r,\xi) = i\,\frac{45\pi^4 N^2 \chi_{\mathrm{a}}^{(3)}\chi_{\mathrm{a}}^{(3)'} b A_{10}^5 e^{ib\Delta k_{13}m(1+2l)/2} e^{-5k_1 r^2/b[1+im(1-2l)]}}{2n_3 n_5 \lambda_1^2 \Delta k_{13}[1-im(1+2l)]^2[1+im(1-2l)]} \times F_3(b\Delta k_{135}, m(1+2l), m(1-2l)), \qquad |b\Delta k_{13}| \gg 1, \tag{61a}$$

$$A_{5,\,\mathrm{indirect}} = -i\,\frac{45\pi^4 N^2 \chi_{\mathrm{a}}^{(3)}\chi_{\mathrm{a}}^{(3)'} b A_{10}^5 e^{-ib\Delta k_{135}m(1-2l)/2}}{2n_3 n_5 \lambda_1^2 \Delta k_{135}[1+im(1-2l)]^3} \times e^{-5k_1 r^2/b[1+im(1-2l)]} \times F_3(b\Delta k_{13}, m(1+2l), m(1-2l)), \qquad |b\Delta k_{135}| \gg 1, \tag{61b}$$

where $l = z_0/L$, $\xi = m(1-2l)$, and $\xi' = m(1+2l)$. The corresponding fifth-harmonic powers are

$$\mathscr{P}_5 = \frac{6.63552 \times 10^6 \pi^{12} N^4 |\chi_{\mathrm{a}}^{(3)}\chi_{\mathrm{a}}^{(3)'}|^2}{n_5 n_3^2 n_1^5 \lambda_1^6 c^4 \Delta k_{13}^2 [1+m^2(1+2l)^2]^2} (4\mathscr{P}_1/b\lambda_1)^2 \mathscr{P}_1^3 \times |F_3(b\Delta k_{135}, m(1+2l), m(1-2l))|^2, \quad |b\Delta k_{13}| \gg 1, \tag{62a}$$

and

$$\mathscr{P}_5 = \frac{6.63552 \times 10^6 \pi^{12} N^4 |\chi_a^{(3)} \chi_a^{(3)'}|^2}{n_5 n_3^2 n_1^5 \lambda_1^6 c^4 \Delta k_{135}[1 + m^2(1 - 2l)^2]^2} (4\mathscr{P}_1/b\lambda_1)^2 \mathscr{P}_1^3$$

$$\times |F_3(b\Delta k_{13}, m(1 + 2l), m(1 - 2l)|^2, \qquad |b\Delta k_{135}| \gg 1. \quad (62b)$$

When the beam is focused at the center of the nonlinear medium the optimum value of $b\Delta k_{13}$ for Eq. (62a) or $b\Delta k_{135}$ for Eq. (62b) can be obtained from Fig. 3. For large values of m the magnitude of the contribution of both of these terms decreases as $1/m^2$, approaching zero in the tight-focus limit.

Larger contributions from the indirect processes can be obtained by focusing the pump beam at one end of the nonlinear medium. When the process $\omega_5 = \omega_3 + \omega_1 + \omega_1$ is phase matched and $b\Delta k_{13} \gg 1$, the indirect process is described by Eq. (61a). The behavior of the optimum power as given in Eq. (62a) is shown in Fig. 11 as a function of m for a pump beam focused at the center of the medium and at either edge. The contribution of this term is maximum when the pump beam is focused at the entrance face of the nonlinear medium ($l = -\frac{1}{2}$). In this case the fifth-harmonic power is given by

$$\mathscr{P}_5 = \frac{6.63552 \times 10^6 \pi^{12} N^4 |\chi_a^{(3)} \chi_a^{(3)'}|^2}{n_5 n_3^2 n_1^5 \lambda_1^6 c^4 \Delta k_{13}^2} \left(\frac{4\mathscr{P}_1}{b\lambda_1}\right)^2 \mathscr{P}_1^3 |F_3(b\Delta k_{135}, 0, 2m)|^2. \tag{63}$$

When the third-harmonic step $\omega_3 = \omega_1 + \omega_1 + \omega_1$ is phase matched and $b\Delta k_{135} \gg 1$, the indirect contribution is described by Eq. (62b). The behavior of this integral for pump beams focused at the center or either edge of the nonlinear medium can be obtained from Fig. 11 if the roles of $b\Delta k_{13}$ and

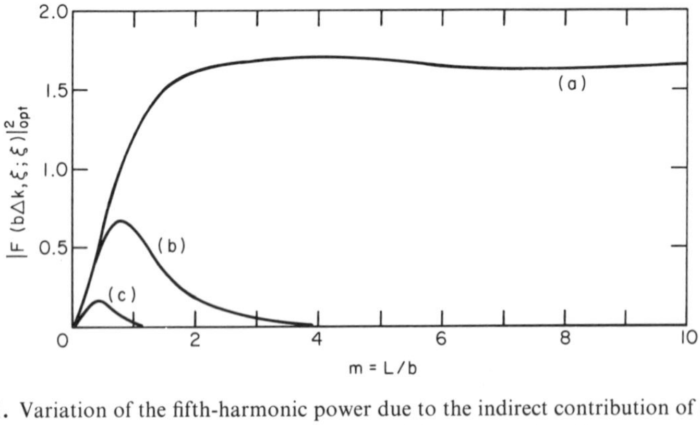

Fig. 11. Variation of the fifth-harmonic power due to the indirect contribution of Eq. (62a) when pump beam is focused at the front (a), the center (b), or the end (c) of the nonlinear medium.

$b\Delta k_{135}$ are interchanged and the curves for focusing at the front and back edges are also interchanged. The contribution of the term in Eq. (62b) is a maximum when the pump beam is focused at the exit face of the nonlinear medium. In this situation the fifth-harmonic amplitude is given by

$$\mathscr{P}_5 = \frac{6.63552 \times 10^6 \pi^{12} N^4 |\chi_a^{(3)} \chi_a^{(3)\prime}|^2}{n_5 n_3^2 n_1^5 \lambda_1^6 c^4 (\Delta k_{135})^2} \left(\frac{4\mathscr{P}_1}{b\lambda_1}\right)^2 \mathscr{P}_1^3 |F_3(b\Delta k_{13}, 2m, 0)|^2. \tag{64}$$

If the pump beam is focused at the "correct" edge of the nonlinear medium, then the fifth-harmonic field amplitude from the indirect processes does not approach zero in the tight-focus limit. For this situation the fifth-harmonic power in a semi-infinite medium extending to one side of the beam waist is given for the process in (62b) by

$$\mathscr{P}_5 = \frac{6.63552 \times 10^6 \pi^{12} N^4 |\chi_a^{(3)} \chi_a^{(3)\prime}|^2}{n_5 n_3^2 n_1^5 \lambda_1^6 c^4 (\Delta k_{135})^2} \left(\frac{4\mathscr{P}_1}{b\lambda_1}\right)^2 \mathscr{P}_1^3$$
$$\times \begin{cases} \left| \pi \dfrac{b\Delta k_{13}}{2} e^{b\Delta k_{13}/2} + i\left[1 + \dfrac{b\Delta k_{13}}{2} e^{b\Delta k_{13}/2} \operatorname{Ei}(-b\Delta k_{13}/2)\right.\right|^2, & \Delta k < 0 \\ \left| 1 - \dfrac{b\Delta k_{13}}{2} e^{b\Delta k_{13}/2} E_1(b\Delta k_{13}/2)\right|^2, & \Delta k > 0, \end{cases} \tag{65}$$

where $\operatorname{Ei}(x)$ and $E_1(x)$ are exponential integrals. The optimum value of $b\Delta k_{13}$ is given by

$$b\Delta k_{13} \approx -6.2. \tag{66}$$

For the process in Eq. (62a) the fifth-harmonic power in the tight-focus limit in a semi-infinite medium is identical to that in Eq. (65) if $b\Delta k_{13}$ and $b\Delta k_{135}$ are interchanged and the beam is focused at the other end of the medium.

A similar treatment of third-harmonic generation by second-order indirect processes has been given by Rostovtseva *et al.*[25] and similar behavior was also predicted for those processes.

2.3.2b. Intermediate Step Phase Matched. The relative behavior of the direct and indirect processes for focused beams that was discussed in the previous section is similar to that which was discussed in Section 2.3 for plane waves. In ternary mixtures or in other systems in which the values of $b\Delta k_{13}$ and $b\Delta k_{135}$ can be optimized simultaneously the relative behavior of the direct and indirect processes is somewhat different. This situation has been discussed by Tomov and Richardson.[2]

Returning to the expression in Eq. (54) for the fifth-harmonic field amplitude, we see that if the values of $b\Delta k_{13}$ and $b\Delta k_{135}$ can be optimized simultaneously, then the net fifth-harmonic field will not depend on Δk_{13} or Δk_{135} independently. In this situation the relative magnitude of the fields contributed by the direct and indirect processes is given by the quantity

$$R = \frac{18\pi^2 N\chi_a^{(3)}\chi_a^{(3)'}b}{n_3\lambda_1\chi_a^{(5)}}\left|\frac{F_{5,\text{indirect}}}{F_{5,\text{direct}}}\right|. \tag{67}$$

This expression is similar to that given in Eq. (35) for the relative importance of the two types of processes in plane-wave geometries, except that the factor $1/\Delta k_{13}$ has been replaced by the factor b. Just as for plane waves, then, the indirect process in focused beams is favored by larger values of the susceptibilities. The indirect process is also favored by higher densities. In addition, when both wave vector mismatches can be optimized simultaneously the indirect process is favored by weaker focusing, while the direct process is favored by tighter focusing.

Tomov and Richardson[2] have given calculations of fifth-harmonic conversion of Nd:glass laser radiation in metal vapors for focused beams when both $b\Delta k_{13}$ and $b\Delta k_{135}$ can take on arbitrary values. We discuss here some of the optimization characteristics of the integrals under these conditions.

We recall from Section 2.1.2b that in an infinite medium the integral for the direct fifth-harmonic process is real and has the value

$$F_{5,\text{direct}} = \begin{cases} (\pi/24)(b\Delta k_{15})^3 e^{b\Delta k_{15}/2}, & \Delta k_{15} < 0, \\ 0, & \Delta k_{15} > 0. \end{cases} \tag{68}$$

It is optimized when $b\Delta k_{15} = -6$. The real part of the integral for the indirect process in an infinite medium from Eq. (56) can be expressed as[2]

$$\text{Re}(F_{5,\text{indirect}}) = \frac{1}{2}\int_{-\infty}^{\infty} \frac{e^{-ib\Delta k_{135}\xi''/2}}{(1+i\xi'')^2}\,d\xi'' \int_{-\infty}^{\infty} \frac{e^{-ib\Delta k_{13}\eta/2}}{(1+i\eta)^2}\,d\eta.$$

It has the value

$$\text{Re}[F_{5,\text{indirect}}(b\Delta k_{13}, b\Delta k_{135})] = \begin{cases} \frac{1}{2}\pi^2 b^2 \Delta k_{13}\Delta k_{135} e^{b(\Delta k_{13}+\Delta k_{135})/2}, & \Delta k_{13},\ \Delta k_{135} < 0 \\ 0, & \Delta k_{13} > 0 \ \text{or}\ \Delta k_{135} > 0. \end{cases} \tag{69}$$

Note that our sign convention for the wave vector mismatches is the opposite of that used by Tomov and Richardson.[2]

The maximum value of $\text{Re}[F_{5,\text{indirect}}]$ occurs when $b\Delta k_{13} = b\Delta k_{135} = -2$, corresponding to a value for $b\Delta k_{15}$ of -4. This value can be contrasted with the optima value of $b\Delta k_{15} = -6$ for the direct process.

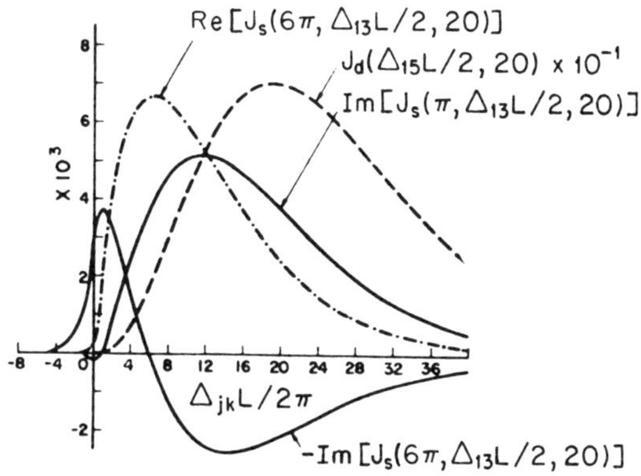

Fig. 12. Variation of the quantity $\mathrm{Im}[F_{5,\mathrm{indirect}}\,(\Delta k_{135}L/2,\ \Delta k_{13}L/2,\ m)]$ with phase mismatch $\Delta k_{13}L/2$ for different values of $\Delta k_{135}L/2$. Also shown are $\mathrm{Re}[F_{5,\mathrm{indirect}}(\Delta k_{135}L/2,\ \Delta k_{13}L/2,\ m)]$ for $\Delta k_{135}L/2 = 6\pi$, and $F_{5,\mathrm{direct}}\,(\Delta k_{15}L/2, m)$ as a function of $k_{15}L/2$. The focusing parameter for all the curves was $m = L/b = 20$. $J_s = F_{5,\mathrm{indirect}}$; $J_d = F_{5,\mathrm{direct}}$. (From Ref. 2, © 1976 IEEE.)

The imaginary part of $F_{5,\mathrm{indirect}}$ is to be combined with the real part of $F_{5,\mathrm{direct}}$. Because the indirect process makes use of the third-harmonic field that is generated throughout the nonlinear medium, the imaginary part of the two-step integral is not necessarily zero in an infinite medium, as is the imaginary part of the direct integral. Analytic expressions for the imaginary part of $F_{5,\mathrm{indirect}}$ have not been given. Tomov and Richardson[2] have examined the properties of $\mathrm{Im}[F_{5,\mathrm{indirect}}]$ numerically for a medium in which $m = L/b = 20$. Some of their results are shown in Fig. 12. Their analysis shows that $\mathrm{Im}[F_{5,\mathrm{indirect}}]$ has a significant magnitude in only two regions of the Δk_{13}–Δk_{135} space. In one region, $\Delta k_{135} < 0$ and $|\Delta k_{13}| < |\Delta k_{135}|$, the maximum value of $\mathrm{Im}[F_{5,\mathrm{indirect}}]$ occurs at $\Delta k_{135}L/2 = -12\pi$, $\Delta k_{13}L/2 = -\pi$ and is negative. In the other region $\Delta k_{13} < 0$ and $|\Delta k_{135}| < |\Delta k_{13}|$. Here the maximum value of $\mathrm{Im}[F_{5,\mathrm{indirect}}]$ occurs when $\Delta k_{135}L/2 = -\pi$, $\Delta k_{13} = -12\pi$ and is positive. For $\Delta k_{13} = \Delta k_{135}$ the imaginary part of $F_{5,\mathrm{indirect}}$ is zero. Again the difference in the sign convention for the wave vector mismatches between that used here and that used in Ref. 2 should be pointed out. Through further analysis they have also determined that the imaginary part of $F_{5,\mathrm{indirect}}$ has an absolute maximum at the points $(\Delta k_{135}L/2 = \pi, \Delta k_{13} = 4/b)$ and $(\Delta k_{135} = 4/b, \Delta k_{13}L/2 = \pi)$ when the focusing parameter $m = 20$.

They have also given calculations for the net fifth harmonic generated from focused beams when both the direct and indirect processes are important. Their curves, shown in Fig. 13, show an increase in total fifth harmonic and movement of the peak of the curve toward smaller values of $b\Delta k_{15}$ as the indirect process becomes important.

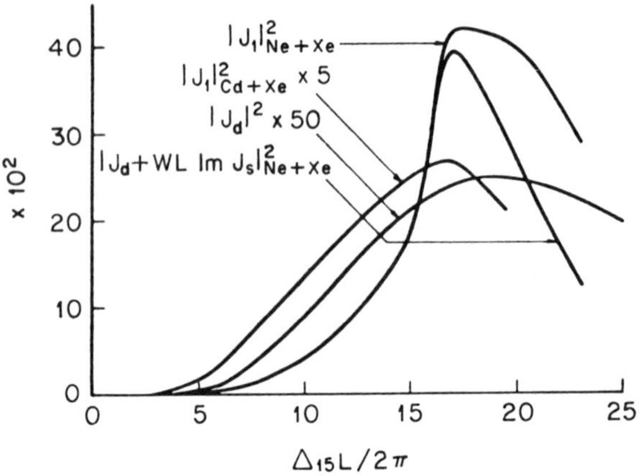

Fig. 13. Calculated fifth-harmonic power in focused beams due to direct and indirect processes for conversion in mixtures of Na:Xe and Cd:Xe. $J_1 = F_{5,\text{direct}} + iWLF_{5,\text{indirect}}$; $J_d = F_{5,\text{direct}}$, $J_s = F_{5,\text{indirect}}$; $W = \dfrac{18\pi^2 N\chi_a^{(3)}\chi_a^{(3)\prime}b}{n_3\lambda_1\chi_a^5 L}$. (From Ref. 2, © 1976 IEEE.)

2.3.3. *Higher-Order Harmonics*

As the order of the harmonic increases the number of indirect processes that can contribute to the generated field increases rapidly. For example, the seventh harmonic can be generated through the direct process

$$\omega_7 = \omega_1 + \omega_1 + \omega_1 + \omega_1 + \omega_1 + \omega_1 + \omega_1 \tag{70}$$

and through the following indirect processes:

Two-step:

$$\begin{aligned} \omega_3 &= \omega_1 + \omega_1 + \omega_1, \\ \omega_7 &= \omega_3 + \omega_1 + \omega_1 + \omega_1 + \omega_1 + \omega_1, \\ \omega_3 &= \omega_1 + \omega_1 + \omega_1, \\ \omega_7 &= \omega_3 + \omega_3 + \omega_1, \\ \omega_5 &= \omega_1 + \omega_1 + \omega_1 + \omega_1 + \omega_1, \\ \omega_7 &= \omega_5 + \omega_1 + \omega_1; \end{aligned} \tag{71}$$

Three-step:

$$\begin{aligned} \omega_3 &= \omega_1 + \omega_1 + \omega_1, \\ \omega_5 &= \omega_3 + \omega_1 + \omega_1, \\ \omega_7 &= \omega_5 + \omega_1 + \omega_1. \end{aligned} \tag{72}$$

As a result unambiguous identification of the higher-order susceptibilities becomes increasingly difficult as the order of the harmonic increases.

3. RELATIVE IMPORTANCE OF HARMONIC PROCESSES OF DIFFERENT ORDER

The higher-order processes have been proposed as a useful method for generating radiation at wavelengths that are considerably shorter than the pump radiation.[3,8,11,13] It is of interest, then, to compare the effectiveness of processes of different order in generating harmonic radiation. From the perturbation expansion of Eq. (1-19) it can be seen that the higher-order interactions will be more effective at higher pump intensities. Generally when conversion is done in solids the pump intensity is limited to a maximum value by the damage threshold of the medium, typically of the order of 10^9–10^{10} W/cm^2, depending on the wavelength and pulse duration of the pump radiation. In these media comparisons[1,26] have always shown that conversion efficiency decreases as the order of the interaction increases. In low-pressure gases, however, it is possible for higher pump intensities to be used and conversion in the higher-order processes can be more significant.

The relative power that is generated in harmonics of different order is determined by a combination of pump intensity, phase matching, and the nonlinear susceptibilities for the various processes. The appropriate phase-matching conditions can generally be chosen to be satisfied for one of the processes under comparison while the others remain phase mismatched. The higher-order processes can therefore be favored by choice of phase-matching conditions. Usually, however, it is the relative magnitudes of the susceptibilities that determine whether significant power can be generated in the higher-order processes.

Let us consider phase-matched conversion for the qth harmonic through a direct process. In the absence of pump depletion the intensity of the qth harmonic is, from Eq. (5),

$$I_q = \frac{(2\pi)^{q+3} q^2 N^2}{n_q n_1^q \lambda_1^2 c^{q-1}} |\chi_{\rm a}^{(q)}|^2 L^2 I_1^q. \tag{73}$$

The ratio of the intensity of the $(q + 2)$th harmonic to that of the qth harmonic is given approximately by

$$\frac{I_{q+2}}{I_q} \approx \left(\frac{2\pi}{c}\right)^2 \left(\frac{q+2}{q}\right)^2 \left|\frac{\chi_{\rm a}^{(q+2)}}{\chi_{\rm a}^{(q)}}\right|^2 I_1^2. \tag{74}$$

The intensity of the harmonics of successive odd order will be comparable when

$$I_1^2 \approx \left(\frac{c}{2\pi}\right)^2 \left|\frac{\chi_{\rm a}^{(q)}}{\chi_{\rm a}^{(q+2)}}\right|^2, \tag{75}$$

assuming that each harmonic is phase matched.

We can estimate the order of magnitude of the pump field that is required for the two harmonic intensities to be comparable as follows. Using the anharmonic oscillator model Bloembergen[27] has shown that the ratio of the nonresonant harmonic susceptibilities of successive order is approximately

$$\chi_{\mathrm{a}}^{(q+1)}/\chi_{\mathrm{a}}^{(q)} \approx 1/E_{\mathrm{atomic}}, \tag{76}$$

where E_{atomic} is the electric field that binds the electrons to the nuclei, typically of the order of 3×10^8 V/cm ($\sim 10^6$ esu). Akhmanov[1] has given experimental data that suggests that Eq. (76) should be replaced with the following expression:

$$\chi_{\mathrm{a}}^{(q+2)}/\chi_{\mathrm{a}}^{(q)} \approx F(q)/E_{\mathrm{atomic}}, \tag{77}$$

where $F(q)$ is a decreasing function of q.

If we apply (76) twice to obtain the ratio $\chi_{\mathrm{a}}^{(q+2)}/\chi_{a}^{(q)}$ we find that the $(q+2)$th-harmonic intensity is comparable to the qth-harmonic intensity when the electric field in the pump wave is of the order of E_{atomic}, corresponding to an intensity of 5×10^{14} W/cm^2. Intensities of this order of magnitude are presently available in the focal regions of intense picosecond pulses from mode-locked lasers.

It can be seen from these considerations that the electric fields in the optical wave required to provide comparable intensity in successive harmonics are comparable to those that bind the electrons in the nonlinear medium. As a result one can expect that, when nonresonant interactions are involved, the use of such field strengths would violate the conditions required for the perturbation expansion for the nonlinear polarization in Eq. (1-19). Thus we should expect that as long as the perturbation expansion is valid the intensity generated in the harmonic wave will decrease as the order of the harmonic increases.

If resonances are present in the higher-order susceptibilities but not in the lower-order ones, then it is possible that the ratio $\chi_{\mathrm{a}}^{(q+2)}/\chi_{\mathrm{a}}^{(q)}$ will be larger than that given in Eq. (76). Under these conditions it is possible that a higher intensity can be generated in a higher-order harmonic than in a lower-order one without violating the conditions required for the validity of the perturbation expansion in Eq. (1-19). For example, in experiments that will be described in the next section involving fifth-harmonic conversion in helium gas, there was a moderate five-photon resonance in the fifth-order susceptibility but none in the third, and the ratio of the fifth- to the third-order susceptibilities was about 10^{-10} esu. For this situation the fifth-order effect would be expected to become comparable to the third-order effect at a field strength of 3×10^7 V/cm, corresponding to a pump intensity of 5×10^{12} W/cm^2. Intensities of this magnitude are readily achieved and can be used in some low-pressure gases without destroying the medium through dielectric breakdown. Eventually, however, when no new resonances appear in the susceptibility as the order increases, the intensity in the succeeding higher orders will again decrease. It appears then that it should be possible to choose a combination of pump wavelengths and energy levels to optimize conversion to a harmonic of arbitrary order.

These considerations must be regarded as only very approximate since they fail to include the effects of many other nonlinear interactions that can affect the harmonic conversion process. Such competing interactions will be treated in some detail in Chapter 4. These competing mechanisms can prevent effective conversion to higher-order harmonics even though sufficient pump intensity is available and can be used without destroying the medium. Nevertheless, this discussion can be used as a general guide for the minimum conditions that must be satisfied if conversion to the higher-order harmonics is to equal or to exceed conversion to the lower-order ones.

4. EXPERIMENTAL AND THEORETICAL RESULTS

Fifth- and higher-order harmonic conversion and sum- and difference-frequency generation have been studied most extensively in gases, although fourth-harmonic conversion of Nd : glass laser radiation in lithium formate[26] and fifth-harmonic conversion in calcite[28] have been reported by Akhmanov *et al.* The high intensities that are needed to obtain significant conversion in the higher-order processes are more suited to gases rather than to liquids or solids, which generally have lower breakdown thresholds. Cascade generation of the form shown in Fig. 7 is usually used to produce higher-order harmonics in solids. Higher-order Raman processes have also been observed in gases and liquids.

4.1. Fifth Harmonic of Nd:Glass Laser Radiation

Fifth-harmonic conversion of radiation from an Nd : glass laser at 1.06 μm in mixtures of Na:Cd:Xe has been considered theoretically by Tomov and Richardson[2] and experimentally in mixtures of Na:Xe by Metchkov *et al.*[4] and Doitcheva *et al.*[29] Level diagrams for fifth-harmonic conversion of radiation at 1.06 μm by both direct and indirect processes in these materials are shown in Fig. 14. Both the third- and fifth-harmonic processes are negatively dispersive in Na and either can be phase matched by mixing with Xe in which both processes are positively dispersive. In Cd the third-harmonic process is positively dispersive while the fifth-harmonic process is negatively dispersive. Mixtures of Cd:Xe can be used to phase match the fifth-harmonic and ternary mixtures of Na:Cd:Xe can be used to provide simultaneous phase matching for the third- and fifth-harmonic processes. In the plane-wave approximation the appropriate Na:Cd:Xe mixture for simultaneous

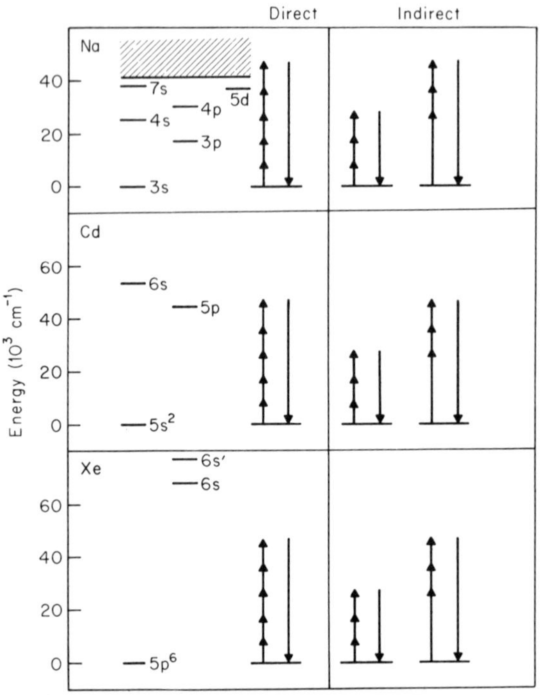

Fig. 14. Level diagrams for direct and indirect fifth-harmonic conversion or radiation from an Nd:glass laser ($\lambda = 1.06\ \mu$m) in Na, Cd, and Xe.

phase matching of third- and fifth-harmonic conversion ($\Delta k_{13} = \Delta k_{15} = \Delta k_{135} = 0$) is 1:2.86:117.5, as given in Ref. 2. The refractive index variation with wavelength for this mixture is shown in Fig. 15.

The third-harmonic susceptibility in Na as calculated by Eicher[30] is $\chi_a^{(3)}(-3\omega_1, \omega_1, \omega_1, \omega) = \chi_a^{(3)} = 8.5 \times 10^{-34}$ esu. Tomov and Richardson[2] have given a value for the fifth-harmonic susceptibility of $\chi_a^{(5)} = 6.7 \times 10^{-45}$ esu. In their calculations of the indirect process they assumed that the sum-frequency susceptibility $\chi_a^{(3)}(-5\omega, 3\omega, \omega, \omega) = \chi_a^{(3)\prime}$ in Na was the same as the third-harmonic susceptibility. They also assumed that the values for all the susceptibilities in Cd were the same as the corresponding ones in Na.

Their calculations for fifth-harmonic conversion have been presented in Figs. 8 and 9 for collimated beams and in Fig. 13 for focused beams. The calculations reported in Fig. 8 were done for the doubly phase-matched mixture. The intensity used in the calculations shown in Fig. 8a was 2×10^{11} W/cm^2, giving a third-harmonic depletion length of 6.3 cm and a fifth-harmonic depletion length of 283.6 cm. In the calculation presented in Fig. 8b the pump intensity was 9.1×10^{12} W/cm^2, giving a depletion length

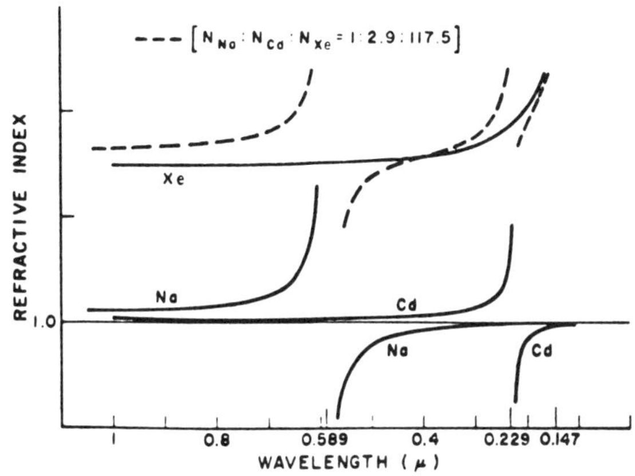

Fig. 15. Variation of the refractive indexes with wavelength for a mixture of Na:Cd:Xe which phase matches the intermediate step in the indirect fifth-harmonic conversion of radiation from an Nd:glass laser. (From Ref. 2.)

of 0.14 cm for both the third and fifth harmonics. As was discussed earlier virtually complete conversion is predicted for each situation. It must be remembered, however, that at the higher intensities used in these calculations, competing processes that were not included in the theory in this chapter may prevent complete conversion in either process. Such competing processes are discussed in the next chapter.

The calculations presented in Fig. 9 allow a phase mismatch in the third-harmonic step. For one of the curves the phase mismatch ($\Delta k_{13} = -4.54\ \text{cm}^{-1}$) is appropriate for an Na:Cd:Xe mixture at a temperature of 810 K in which both the Na and Cd vapors are saturated with a relative density of 1:3.5. The Na density at this temperature is $10^{17}\ \text{cm}^{-3}$. The second curve in these figures is for a mixture of Cd:Xe with a Cd density of $10^{17}\ \text{cm}^{-3}$. Again substantial fifth-harmonic conversion is predicted. When the direct fifth-harmonic process is important (the dotted curve in Fig. 9b) conversion in the Cd:Xe mixture exceeds the conversion in the ternary mixture.

Numerical calculations for focused beams for the direct and indirect processes were presented in Fig. 13. The parameters used were appropriate for mixtures of Na:Xe or Cd:Xe. Conversion in the Na:Xe mixture was calculated to be higher than in the Cd:Xe mixture. For both mixtures the contribution of the indirect process was greater than that of the direct process.

Fifth-harmonic conversion of Nd:glass laser radiation in Na:Xe mixtures was studied experimentally by Metchkov *et al.*[4] Their experiments were done in the plane-wave regime for which the appropriate phase-matching mixtures for the direct and indirect processes are shown in Table 1. They observed phase-matching peaks for the direct and both indirect phase-matching

TABLE 1

Plane-Wave Phase-Matching Ratios For Third- and Fifth-Harmonic Conversion of Nd:Glass Laser Radiation in Mixtures of Xe and Various Metal Vapors[a]

	N_{Xe}/N_{metal}			
	Interaction			
Metal	$\omega + \omega + \omega = 3\omega$	$\omega + \omega + 3\omega = 5\omega$	$2\omega + 2\omega + \omega = 5\omega$	$\omega + \omega + \omega + \omega + \omega = 5\omega$
Li	152.7	12.6	—	33.3
Na	150.4	9.6	—	30.2
K	325.3	31.5	—	75.6
Rb	412.5	39.8	—	95.8
Cs	660.7	70.9	11.8	160.1
Ca	242.4	—	58.3	31.9
Cd	—	34.4	33.1	31.1
Mg	—	28.7	19.1	15.6
Zn	—	255.6	243.7	215.2

[a] From Ref. 2, © 1976 IEEE.

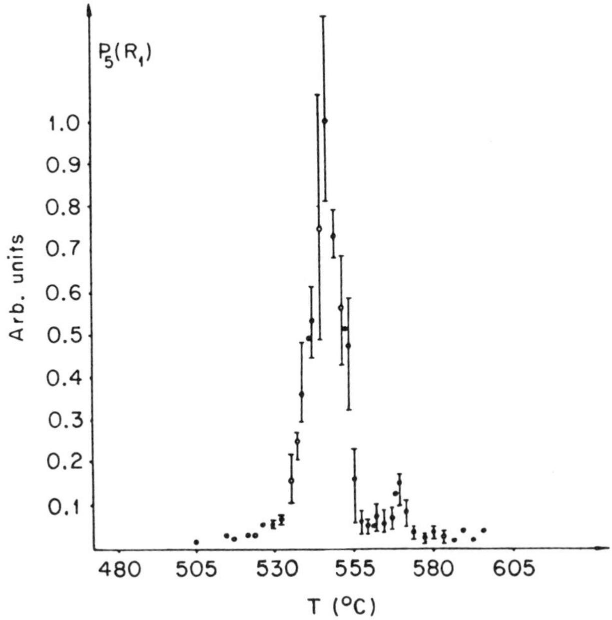

Fig. 16. Phase-matching peak for the direct fifth-harmonic process for conversion of Nd : glass laser radiation in a mixture of Na and Xe. (From Ref. 4.)

conditions ($\Delta k_{15} = 0$, $\Delta k_{13} = 0$, and $\Delta k_{135} = 0$) by varying the cell temperature with a fixed amount of Xe. The Xe pressure determined which phase-matching condition was satisfied at a given pressure of Na. The phase-matching peak for the direct process is shown in Fig. 16.

They observed a conversion efficiency of 6×10^{-7} at the phase-matching peak for the direct process, giving a fifth-harmonic power of 150 W for a pump power of 2.5×10^{8} W. The pulse duration was about 10 psec. The conversion at the phase-matching peaks for the indirect processes was smaller. They reported a fifth-harmonic power of 35 W, corresponding to a conversion efficiency of 1.4×10^{-7} when the condition Δk_{135} was satisfied, and a fifth-harmonic power of 15 W, corresponding to a conversion efficiency of 6×10^{-8}, when the condition $\Delta k_{13} = 0$ was satisfied. The pressure of the Xe was adjusted so that the phase matching for each process occurred at an Na pressure of 10 Torr.

They reported considerably higher conversion efficiency for a two-stage arrangement in which the third harmonic was first generated in nonlinear crystals and then mixed with the fundamental in the Na vapor in the process $\omega_5 = 3\omega + \omega + \omega$. For this arrangement a conversion efficiency of 2.8×10^{-5} was observed, giving a fifth-harmonic power of 7 kW.

From a comparison of the conversion efficiencies that they obtained when the direct and indirect processes were phase matched, they estimated the value of the fifth-harmonic susceptibility to be $3 \pm 2 \times 10^{-45}$ esu, in reasonable agreement with the value calculated by Tomov and Richardson.[2]

Doitcheva *et al.*[29] reported a conversion efficiency of 5×10^{-5} for fifth-harmonic conversion of pulses from a mode-locked Nd:glass laser in phase-matched mixtures of Na:Xe at a pump power of 2.5×10^8 W, corresponding to a fifth-harmonic power of 12.5 kW. Saturation of the conversion efficiency at high pump powers was observed.

4.2. Resonant Fifth-Harmonic Conversion

Two-photon resonant fifth-harmonic conversion of dye laser radiation has been reported by Dinev *et al.*[31] Pump radiation was obtained from a mode-locked dye laser at a wavelength near 602 nm. An enhancement of the fifth-harmonic signal of about 10^4 was observed as the laser wavelength was tuned through the $3s^2$–3s5s two-photon resonance in Na. The process was phase matched with Ar. Saturation of the conversion was observed for pump intensities above 10^8 W/cm^2.

4.3. Extreme Ultraviolet Generation

One of the most extensive experimental studies of the higher-order interactions has been in the generation of coherent radiation in the extreme ultraviolet region of the spectrum at wavelengths below 100 nm. Second-harmonic generation and frequency mixing in nonlinear crystals has been very successful in generating radiation at wavelengths down to about 185 nm. Below this wavelength suitable nonlinear crystals become opaque and further frequency conversion must be done in gases and vapors. Third-harmonic generation and third-order sum- and difference-frequency mixing have been used in metal vapors to generate coherent radiation virtually throughout the entire range between 120 and 200 nm. Third-order processes have also been used to generate coherent radiation at specific wavelengths and in restricted wavelength ranges down to about 57 nm (see Section 2-8.3). However, continued use of the third-order interactions for short-wave generation becomes increasingly difficult because sufficiently powerful lasers at the correspondingly shorter pumping wavelengths become scarce.

The higher-order processes offer an attractive alternative to the third-order interactions because they allow a larger step along the frequency scale to be taken in a single interaction. As a result a shorter wavelength can be reached from a given pumping wavelength, extending the wavelength range that is accessible with a given pump laser.

Harris[3] suggested the use of several higher-order processes in rare gases

TABLE 2

Calculated Conversion Efficiencies for UV and XUV Generation through Higher-Order Processes[a]

Process	Species and path	Limiting P/A (W/cm^2)	Conversion efficiency (%)
3 × 5320 Å → 1773 Å	Xe,[b] 5p-6s-5p-6s-6p-8d-c	1.94×10^{12}	0.084
5 × 5320 Å → 1064 Å	As above	As above	0.051
5 1182 Å → 236 Å	Li^+,[c] 1s-2p-1s-2p-3s-2p-3s-4p-c	1.68×10^{15}	0.002
7 × 1182 Å → 169 Å	As above	As above	0.004
15 × 2660 Å → 177 Å	Li^+,[d] $(1s\text{-}2p)^7\text{-}(2p\text{-}3s)^7$-3p-4d-c	3.47×10^{15}	4×10^{-7}

[a] From Ref. 3.
[b] 5p = $5p^6[^1S]0$; 6s = $6s[1\frac{1}{2}]^0 1$; 6p = $6p[2\frac{1}{2}]2$; 8d = $8d[2\frac{1}{2}]^0 3$; c = continuum.
[c] 1s = $1s^2[1S]0$; 2p = $2p[^1P^0]1$; 3s = $3s[^1S]0$; 4p = $4p[^1P^0]1$; c = continuum.
3p = $3p[^1P^0]1$, others as in *c*.

and ionized metal vapors for generating radiation at specific wavelengths ranging from 106.4 to 17.7 nm. The processes which he considered, all of which involved various harmonics of radiation from an Nd:YAG laser, along with the estimated conversion efficiencies, are given in Table 2. The conversion efficiency shown for each process was calculated with the assumption that the efficiency would be limited by multiphoton absorption or multiphoton ionization at the pump power indicated (see Section 4-3). It was also assumed that conversion took place in one coherence length. The processes in Xe did not involve resonant enhancement, while the 15th-order process in Li^+ involved a fortuitous resonance between the generated wavelength and the $1s^2[^1S]_0$–$1s3p[^1P^0]_1$ transition. The calculated conversion efficiencies at the longer wavelengths are only a little smaller than the efficiencies that have been observed in the vacuum ultraviolet for third-order interactions,[32,33] and the estimated efficiency for the 15th-harmonic conversion of 266 nm radiation in Li^+ is comparable to the conversion observed in the early experiments involving sum-frequency generation of radiation near 190 nm in Sr.[34]

Radiation at several wavelengths ranging from 38 to 118.2 nm has been generated through higher-order processes involving harmonics of Nd:YAG,[5–13,32] Nd:glass,[15] and XeCl lasers.[14] The wavelengths of the generated radiation along with the pumping wavelengths, the pumping lasers, and the nonlinear processes are listed in Table 3.

4.3.1. *Harmonics of Nd:YAG Laser Radiation*

Radiation at specific wavelengths between 38 and 76 nm have been generated from harmonics of an Nd:YAG laser through fifth- and seventh-order interactions.[5–13] The experimental arrangement used for these

TABLE 3

Generation of XUV Radiation by Higher-Order Processes

Generated wavelength (nm)	Laser	Interaction	Material	Maximum efficiency	Ref.
118.2	Nd:glass	9 × 1.06 μm	Na:Ar	10^{-16}	15
106	Nd:glass	5 × 532 nm	Na:Ar	10^{-7}	31
76	Nd:YAG	4 × 266.1 nm −532 nm	He, Ne		7, 8
71	Nd:YAG	4 × 266.1 nm −1.06 μm	He, Ne		7, 8
62.6	Nd:YAG	4 × 266.1 nm +1.06 μm	He, Ne		7, 8
61.6	XeCl	5 × 308 nm	Ar, Kr		14
59.1	Nd:YAG	4 × 266.1 nm +532 nm	He, Ne		7, 8
53.2	Nd:YAG	5 × 266.1 nm	He, Ne, Ar, Kr	10^{-5}	5, 8
38.0	Nd:YAG	7 × 266.1 nm	He	10^{-6}	6, 8

experiments is shown in Fig. 17. In this arrangement the pump radiation was focused with a CaF_2 lens at the front of a VUV spectrometer. A 250-μm-diameter pinhole replaced the entrance slit of the spectrometer and allowed the creation of an effective window through differential pumping that would pass the XUV radiation since most window materials are opaque to radiation at wavelengths below 105 nm.

The radiation was dispersed by the grating and was detected at the exit of the spectrometer with either photographic film or a solar blind photomultiplier. Aluminum films of 80–160 nm thickness were used in front of the photographic film to block the long-wave pump radiation.

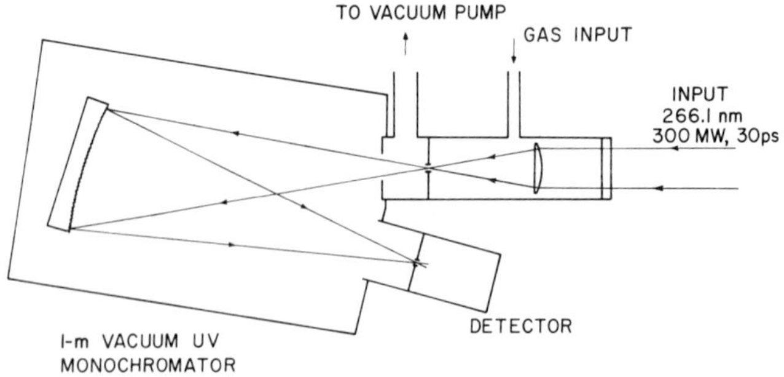

Fig. 17. Experimental diagram for generating coherent XUV radiation from harmonics of an Nd:YAG laser. (From Ref. 8.)

The pump radiation for these experiments was obtained from a mode-locked Nd : YAG laser. Fourth-harmonic radiation at 266.1 nm was produced by two successive stages of second-harmonic generation in KDP and KD*P. The second-harmonic radiation at 532 nm and the fundamental radiation at 1.06 μm were used with the fourth-harmonic radiation for fifth-order frequency mixing. The mode-locked laser produced pulses of about 30 psec in duration with a power level of 2 GW. Internal conversion efficiencies of about 75 % in each of the second-harmonic stages were obtained and, after surface losses, about 300 MW of power was available at 266.1 nm.

4.3.1a. HIGHER-ORDER HARMONIC GENERATION. Fifth-harmonic conversion of the YAG fourth harmonic, producing radiation at 53.2 nm, was observed in four rare gases: He, Ne, Ar, and Kr. Seventh-harmonic conversion, producing radiation at 38 nm, was also observed in He. Level diagrams for these processes are shown in Fig. 18. In He the fifth-harmonic

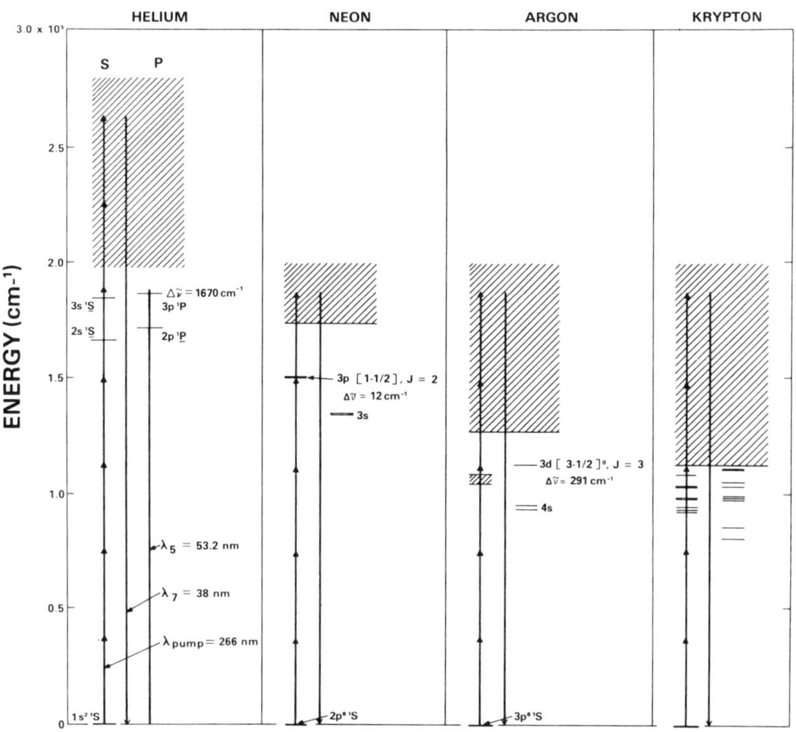

Fig. 18. Level diagrams for generating the fifth harmonic of the fourth harmonic of an Nd:YAG laser (λ_1 = 266.1 nm) in four of the rare gases. Seventh-harmonic conversion in helium is also shown. (From Ref. 8.)

wavelength lies just above the 1s3p $^1P^0$ level, while the fundamental lies well below all the excited levels, so the process can be expected to be negatively dispersive. A modest resonance exists between the generated wavelength and the 1s3p $^1P^0$ level ($\Delta\tilde{\nu} = 1760\ \mathrm{cm}^{-1}$) but there are no other resonances to enhance the fifth-order susceptibility. For the remaining rare gases the fifth-harmonic wavelength lies in the photoionization continuum, making absorption at the harmonic a potential problem at pressures that are sufficiently high that the absorption length is comparable to or shorter than the confocal parameter (see Section 4-4.1.1a). In Ne there is a very close accidental four-photon resonance with the $3p[1\frac{1}{2}]J = 2$ level ($\Delta\tilde{\nu} = 12\ \mathrm{cm}^{-1}$). In the absence of strong Stark shifts (see Section 4-3.4) considerable enhancement of the susceptibility can be expected from this resonance.

The seventh harmonic lies in the continuum for all the rare gases. Helium has the smallest photoabsorption cross section of all the rare gases at 38 nm ($\sigma \approx 4.5$ Mb at $N = 2.68 \times 10^{19}\ \mathrm{cm}^{-3}$) and He was the only medium in which the seventh harmonic was observed.

Calculated refractive indexes for He and Ne for wavelengths up to and just beyond the photoabsorption limit are shown in Fig. 19.

4.3.1b. Indirect versus Direct Processes in He. The relative contribution of the direct and indirect processes that can be expected in these experiments can be estimated from Eq. (67). Using Eqs. (28), (51), (52), and (2) the nonlinear susceptibilities per atom for the direct fifth-harmonic and the indirect process are calculated in the quasistatic approximation to be $\chi_a^{(5)} = -3.39 \times 10^{-48}$ esu, $\chi_a^{(3)} = 3.86 \times 10^{-38}$ esu, and $\chi_a^{(3)\prime} = -3.4 \times 10^{-37}$ esu (for the definitions of these quantities see Section 2.3).

The calculated fifth-harmonic field amplitude, which is proportional to the quantity $|b\Delta k F|$ for conversion in a single-component medium, is shown in Fig. 20 as a function of the parameter $b\Delta k$ for the direct process and for the combined direct and indirect processes. Here $F = F_{5,\mathrm{direct}}$ for the direct process [see Eq. (55)] and $F = F_{5,\mathrm{direct}} + iF_{5,\mathrm{indirect}}$ for the combined processes. These calculations were done for a beam that is focused at the end of the nonlinear medium (see the discussion in Section 2.3.2). With fixed values of b and Δk per atom the horizontal scale is proportional to the helium pressure. The pressure scale is appropriate for a confocal parameter of 0.5 mm, corresponding to the use of a 5-cm focal length lens in the experiments described in Ref. 8. Values of Δk per atom were calculated from tabulated oscillator strengths and included the contributions of the continuum.

These calculations show that for the helium pressures at which the experiments of Reintjes *et al.*[8] were performed ($P \leq 200$ Torr), the indirect process makes a negligible contribution to the fifth-harmonic generation. It should

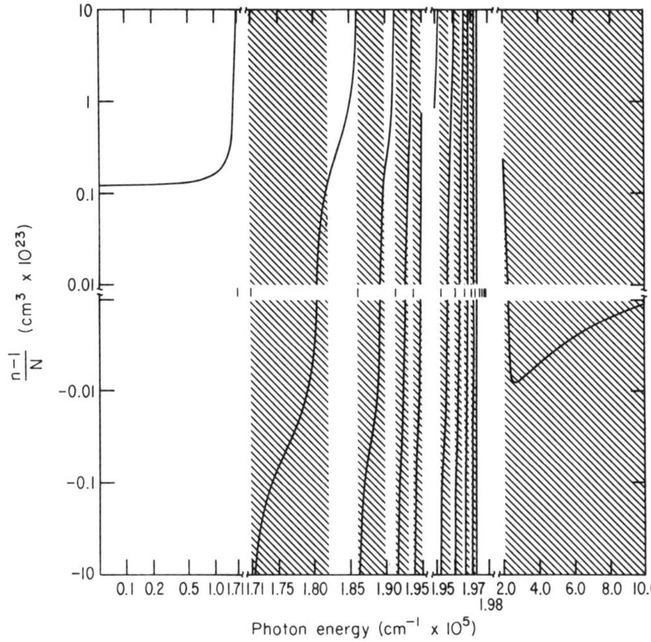

(a)

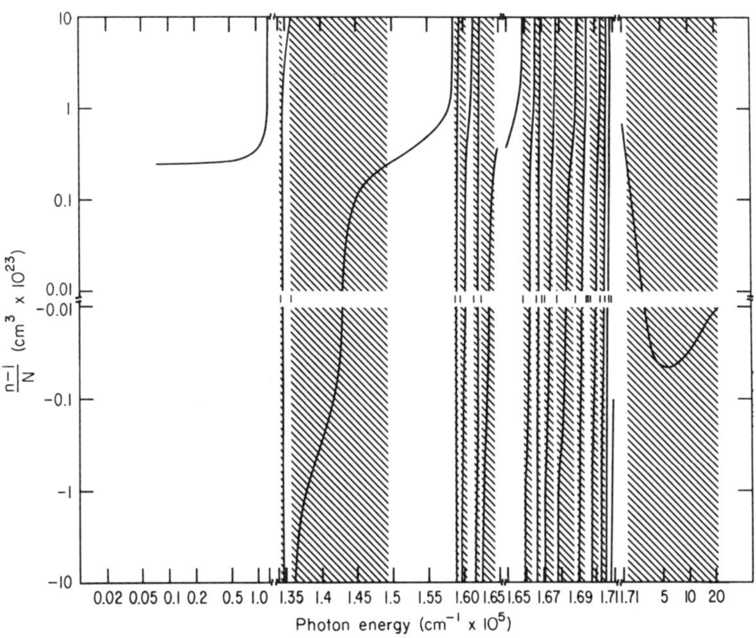

(b)

Fig. 19. Variation of the refractive indexes of He (a) and Ne (b) with photon energy. Shaded region has negative dispersion for fifth-harmonic conversion. Vertical bars in center indicate atomic transitions.

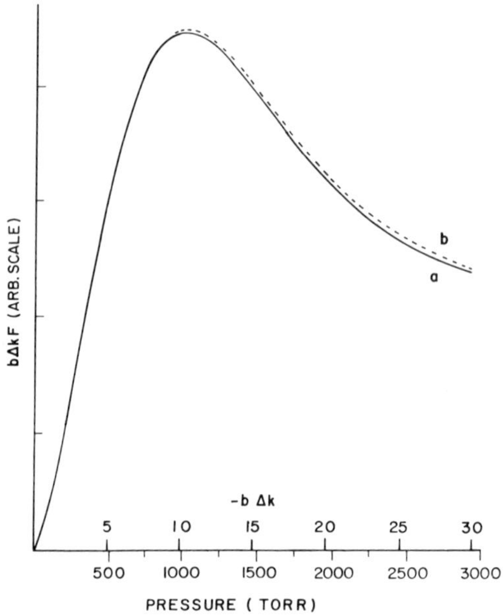

Fig. 20. Variation of the fifth harmonic of pump radiation at 266.1 nm in He due to the direct process (a) and the direct and indirect processes combined (b). The pressure scale is appropriate for the conditions of the experiments described in Refs. 5 and 8. (From Ref. 8.)

be remembered, however, that the relative contributions of the direct and indirect processes depend on the value of the confocal parameter and the absolute magnitudes of the susceptibilities as well as their ratios. Thus it is possible for the indirect process to make a more prominent contribution if different experimental conditions or different materials are used.

4.3.1c. Experimental Results. Typical experimental results for fifth-harmonic generation in neon are shown in Figs. 21a and 21b.[5] The spectrum of the harmonic radiation as shown in Fig. 21a has a single sharp line at the spectral position of the fifth-harmonic wavelength of 53.2 nm. A spectrum from a helium discharge lamp in the same spectral region is shown for comparison. Three helium emission lines are evident.

Measurements of the fifth-harmonic energy as a function of pump energy are shown in Fig. 21b. A least-squares fit to the data shows a power law dependence of 4.7, in good agreement with the expected fifth-power law behavior.

Similar measurements of the seventh-harmonic radiation generated in helium are shown in Figs. 22a and 22b.[6] The spectrum in Fig. 22a shows the fifth- and seventh-harmonic lines at 53.2 and 38 nm, respectively. Measurements of the seventh-harmonic pulse energy as a function of pump pulse

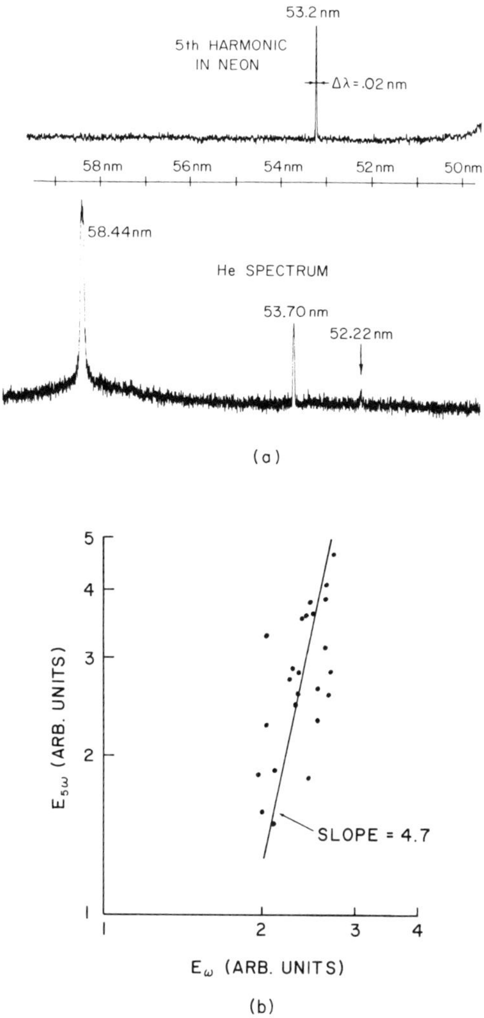

Fig. 21. (a) Spectrum of the fifth-harmonic radiation generated from pump radiation at 266.1 nm (top). Helium emission spectrum near 53.2 nm (bottom). (b) Dependence of the fifth-harmonic power on the pump power. Least-squares fit gives a power law dependence with an exponent of 4.7. $P = 40$ Torr. (From Ref. 5.)

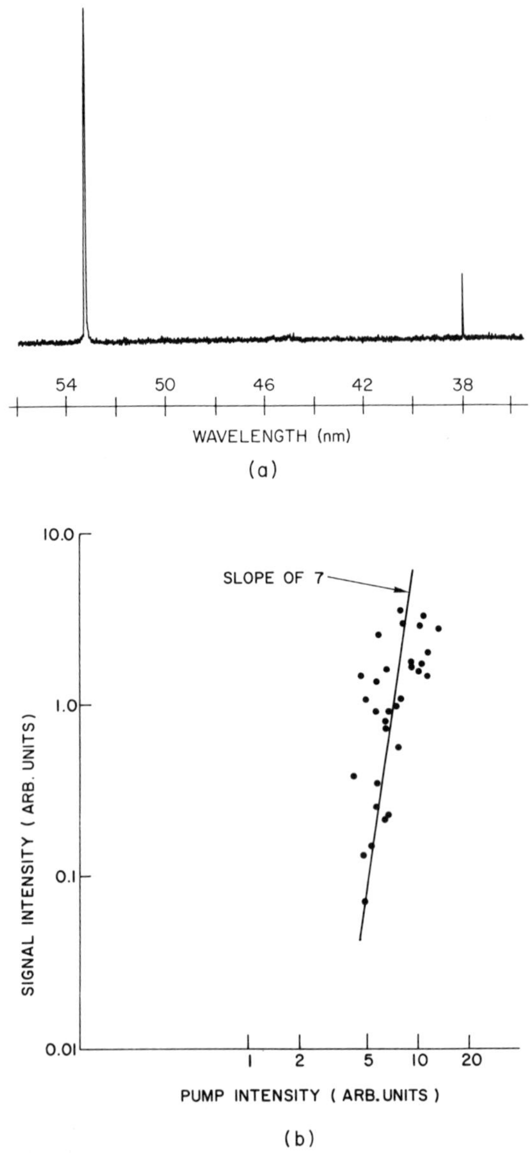

Fig. 22. (a) Spectrum of fifth-harmonic and seventh-harmonic radiation in helium. (b) Dependence of the seventh-harmonic power on pump power. Straight line has slope of seven. (From Ref. 6.)

energy show reasonable agreement with the expected seventh power law behavior.

4.3.2. *Six-Wave Mixing*

Radiation at additional wavelengths in the XUV has been generated by six-wave mixing, utilizing combinations of the fundamental, second, and fourth harmonics from the Nd:YAG laser.[7] Because of strong depletion of the fundamental and second harmonics in the generation of the fourth harmonic, the class of interactions that is expected to be the strongest involves frequency combinations of the type

$$\omega_5 = 4\omega_{UV} \pm \omega_{vis,IR},$$

where ω_{UV} signifies a photon at the fourth-harmonic wavelength and $\omega_{vis,IR}$ signifies a photon at either the fundamental or second-harmonic wavelengths.

Level diagrams for these processes along with the fifth- and third-harmonic processes in helium and neon are shown in Fig. 23. Helium is transparent at all of the generated wavelengths, but all except the fifth harmonic lie below the lowest excited state. As a result the mixing processes will be positively dispersive in helium. Neon is transparent to the third harmonic and to all of the fifth-order lines except the fifth harmonic. Various detunings from odd-parity levels, indicating possible resonant enhancements, as well as both

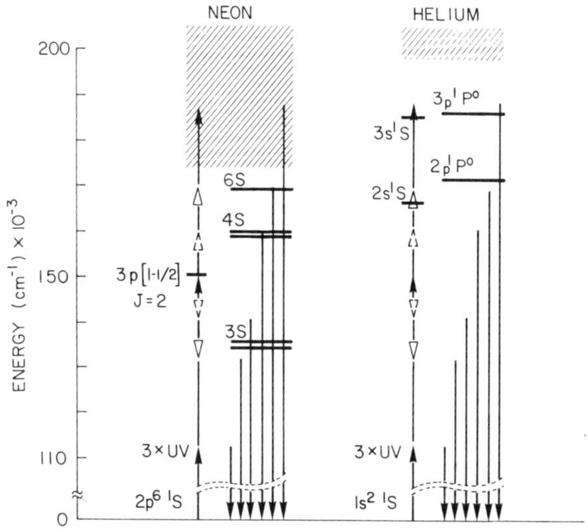

Fig. 23. Level diagrams for frequency mixing the harmonics of an Nd:YAG laser in He and Ne according to the rule $\omega_5 = 4\omega_{UV} \pm \omega_{vis,IR}$, where ω_{UV} corresponds to a photon at 266.1 nm and $\omega_{vis,IR}$ corresponds to a photon at either 532.2 nm or 1.06 μm. (From Ref. 7.)

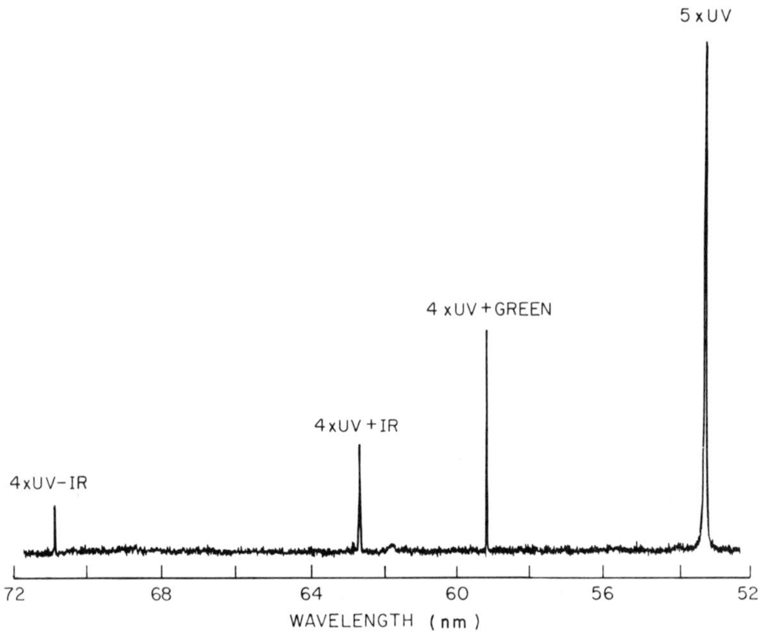

Fig. 24. Spectrum of the radiation obtained from the frequency-mixing processes of Fig. 23. (From Ref. 7.)

positive and negative dispersion, are encountered for the various wavelength combinations. In addition, there is near-four-photon resonance with the $3p[1\frac{1}{2}]J = 2$ level, indicating possible resonant enhancement. However, data will be given in this section that indicate the absence of significant enhancement from this level, possibly as the result of strong Stark shifts.

A densitometer tracing of a photograph of a spectrum taken in neon at a pressure of 30 Torr using a 5-cm focal length lens is shown in Fig. 24. Three of the fifth-order mixing lines, along with the fifth-harmonic line, are evident. The longer-wavelength mixing line and the third-harmonic line could not be photographed because of the increasing absorption of the Al filter.

The variation of the intensities of the various lines generated in helium and in neon with various focal length lenses are shown in Fig. 25. The signals were recorded with a windowless solar blind photomultiplier and for purposes of comparison were corrected for the spectral response of the phototube and the spectrometer.

Data taken with only the UV pump present show only harmonics of the UV signal, indicating that the other lines did indeed require the presence of the other pump wavelengths. A detailed theoretical understanding of the relative intensities that are present in the various lines in these measurements

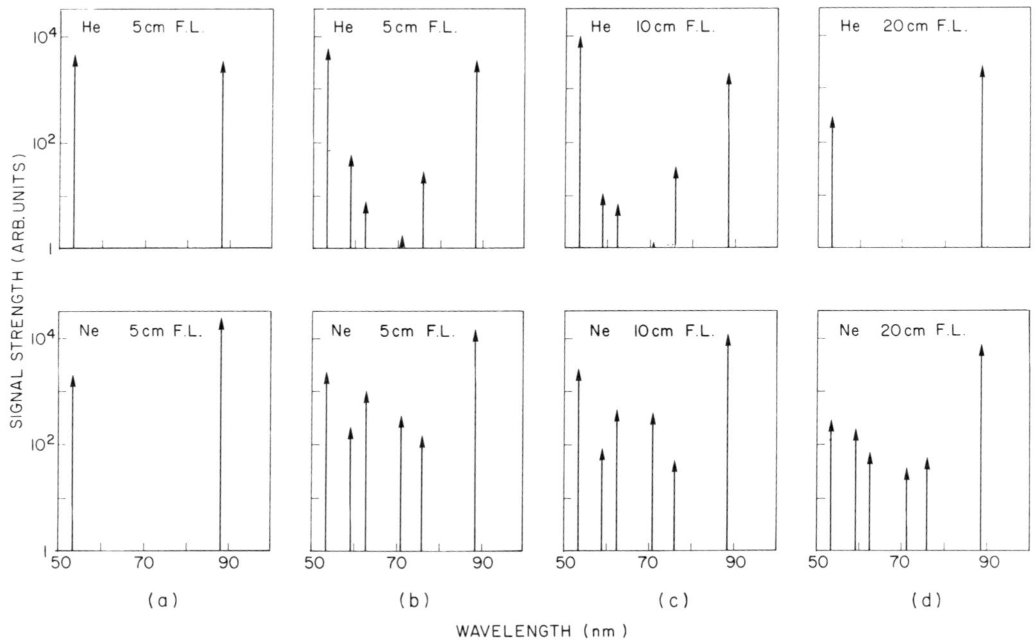

Fig. 25. Comparison of the strength of the fifth-order mixing lines from an Nd : YAG laser in helium and neon at a pressure of 30 Torr for various focusing lenses. Data in (a) were taken with only the 266.1-nm radiation present. (From Ref. 7.)

requires exact knowledge of the energy and beam waist dimensions of each of the pump beams, the degree of spatial and temporal overlap of the various harmonics at the focus of the lens, and the gas density variation near the focus. Comparison of the signal strengths of the various lines can, however, give qualitative indications of some of the major features of the interactions.

For example, the mixing lines are stronger in neon than in helium for each of the lenses that was used. This is especially evident in the data taken with the 20-cm lens. This observed difference is likely to be due to a larger non-linear susceptibility in neon arising from a larger nonresonant susceptibility combined with possible resonant enhancements for the mixing lines. In helium, on the other hand, the mixing lines may be discriminated against by the positive dispersion which exists for the combination processes. The importance of dispersion may also be reflected in the dominance of the fifth harmonic over the mixing lines in helium. A similar dominance is not evident in neon. This characteristic may be due to unfavorable dispersion for the harmonic process or to loss due to photoabsorption of the harmonic in the continuum. Finally, comparison of the data taken with the 5- and 10-cm focal length lenses in helium shows that the fifth harmonic is stronger than the third. This observation will be discussed in more detail in a later section.

4.3.3. Comparison of Fifth-Harmonic Conversion in Rare Gases

The relative strength of the fifth harmonic that was generated in the different rare gases is shown as a function of gas pressure in Fig. 26.[8] The data were taken with a 10-cm focal length lens ($b = 0.2$ cm) at full laser power (peak on-axis pump intensity of about 2.5×10^{14} W/cm^2). All of the gases except He absorb the harmonic radiation by photoionization. The vertical line near the peak of the curves for these gases shows the pressure at which the absorption length at 53.2 nm is equal to one-quarter of the confocal parameter. The good correlation between this pressure and the pressure at which the conversion in each of these gases peaks indicates that the limitation on the conversion efficiency in these gases is related to the absorption of the harmonic.

At pressures well below those for which photoabsorption is important the fifth-harmonic signal increases with increasing atomic wight of the media (except for the order of the conversion in He and Ne). This order of increase is the same as that of the nonresonant third-order susceptibility in these gases and the trend observed here likely reflects the order of increasing nonresonant fifth-order susceptibility. The relatively low conversion in neon in this pressure range is surprising in light of the enhancement expected from the near-four-photon resonance. Strong Stark shifts of the near-resonant level in the focus of the pump beam may be responsible for this observation.

At higher pressures where the conversion in the heavier gases is limited by

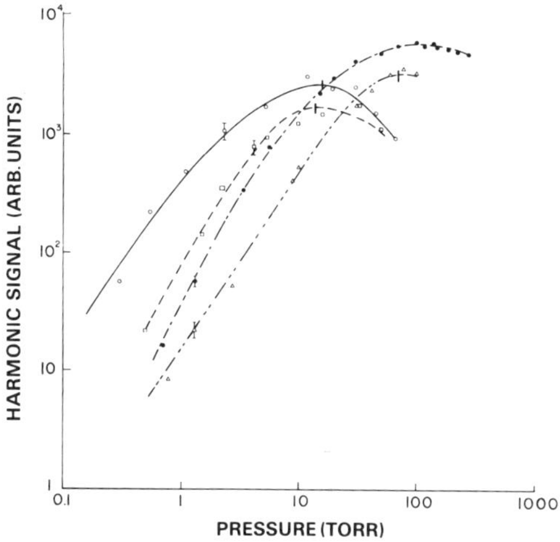

Fig. 26. Comparison of the fifth-harmonic power in four of the rare gases as a function of pressure (●, He; △, Ne; □, Ar; ○, Kr). Vertical line near the peak of each curve indicates the pressure at which the absorption length at 53.2 nm equals one-quarter of the confocal parameter. (From Ref. 8.)

absorption the highest conversion is obtained in helium. Conversion in helium shows a peak as a function of pressure that occurs near 100 Torr. Since helium is transparent at 53.2 nm it is not likely that this peak is due to absorption. It has been attributed to a disturbance of the phase optimization conditions by the nonlinear refractive index. These effects will be discussed in more detail in Section 4-4.2.

4.3.4. *Comparison of Conversion to Harmonics of Different Order*

A comparison of the efficiency of conversion to harmonics of three different orders can be made in helium. Similar comparisons have been made in crystals for lower orders and the results have always shown that the efficiency decreased as the order of the interaction increased.[1,28] Such measurements, however, were limited to intensities below the damage threshold of the nonlinear material. Considerably higher pump intensities can be used for conversion in helium. In addition, the presence of a moderate resonance in the fifth-order susceptibility that is not present in the third-order one can lead to the possibility of higher conversion in the fifth harmonic than in the third.

4.3.4a. Comparison of Fifth- and Third-Harmonic Conversion. Fifth- and third-harmonic conversion of the Nd:YAG fourth-harmonic radiation in helium have been compared by Reintjes and She.[9] In the experiments reported in Refs. 5–14 the aperture at the front of the spectrometer formed an effective window so that the nonlinear medium appeared to be semi-infinite, extending to one side of the beam waist (see Fig. 17). Finite conversion to the third harmonic could therefore be obtained, even though the third-harmonic process is positively dispersive in He.

The power in the third and fifth harmonics can be expressed as

$$P_q = G_q |F_q(b\Delta k)|^2, \tag{78}$$

where F_q is given in Eq. (14) and

$$G_q = \frac{2^{q+5}\pi^{q+3}}{n_1 n_q \lambda_1^4 c^{q-1}} q |N\chi_a^{(q)}|^3 \left(\frac{4\mathscr{P}_1}{b\lambda_1}\right)^{q-3} \mathscr{P}_1^3. \tag{79}$$

The phase optimization integrals F_3 and F_5 for a semi-infinite medium are given by

$$F_3(b\Delta k_{13}, \infty, 0) = -\pi(b\Delta k_{13}/2)e^{b\Delta k_{13}/2} + i[1 + (b\Delta k_{13}/2)e^{b\Delta k_{13}/2}\,\mathrm{Ei}(-b\Delta k_{13}/2), \qquad \Delta k_{13} < 0, \tag{80a}$$

$$F_3(b\Delta k_{13}, \infty, 0) = i[1 - (b\Delta k_{13}/2)e^{b\Delta k_{13}/2}E_1(b\Delta k_{13}/2)], \qquad \Delta k_{13} > 0, \tag{80b}$$

$$F_5(b\Delta k_{15}, \infty, 0) = (-\pi/6)(b\Delta k_{15}/2)^3 e^{b\Delta k_{15}/2} + (i/3)[1 - b\Delta k_{15}/4 + (1/2)(b\Delta k_{15}/2)^2 + (1/2)(b\Delta k_{15}/2)^3 e^{b\Delta k_{15}/2}\,\mathrm{Ei}(-b\Delta k_{15}/2)], \qquad \Delta k_{15} < 0, \tag{81a}$$

$$F_5(b\Delta k_{15}, \infty, 0) = (i/3)[1 - b\Delta k_{15}/4 + (1/2)(b\Delta k_{15}/2)^2 - (1/2)(b\Delta k_{15}/2)^3 e^{b\Delta k_{15}/2} \times E_1(b\Delta k_{15}/2)], \qquad \Delta k_{15} > 0, \tag{81b}$$

where $\mathrm{Ei}(x)$ and $E_1(x)$ are exponential integrals.

The phase optimization integrals for the two harmonics are shown in Fig. 27. The arrows indicate the values of $b\Delta k$ which are appropriate for the experimental conditions reported in Reintjes and She[9] ($P = 18$ Torr, $b = 0.5$ mm). These curves show that the phase-matching conditions actually favor the third-order process over the fifth even though it has the "wrong" dispersion for phase optimization.

The ratio of the squared magnitudes of the phase optimization integrals for the two harmonics evaluated at the appropriate values of $b\Delta k$ is

$$\left|\frac{F_5(b\Delta k_{15})}{F_3(b\Delta k_{13})}\right|^2 = 0.127. \tag{82}$$

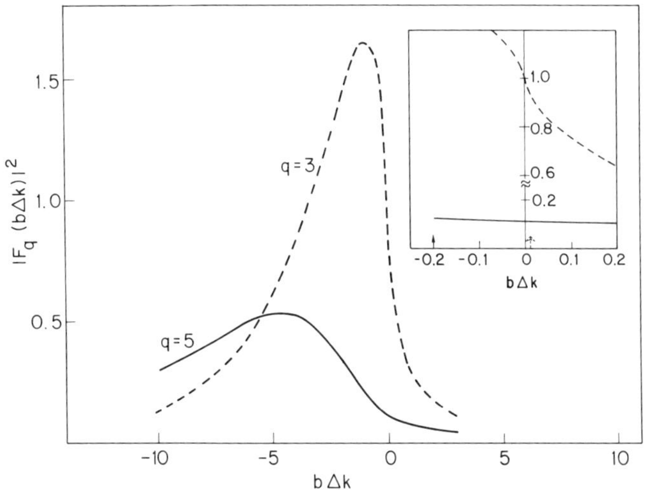

Fig. 27. Variation of the phase optimization integrals $|F_3|^2$ and $|F_5|^2$ as a function of $b\Delta k$ in a semi-infinite medium extending to one side of the beam waist. Solid arrow in the inset indicates the value of $b\Delta k_{15}$ that is appropriate for the experiments reported in Reintjes and She. Dotted arrow indicates the corresponding value for $b\Delta k_{13}$. (From Ref. 9.)

The expected ratio of the power in the two harmonics due to pump power, focused intensity, and the magnitude of the nonlinear susceptibilities is given by G_5/G_3. For the conditions that are again appropriate for the experiments of Ref. 9 ($P_1 = 12$ MW, $b = 0.5$ mm, $N = 6.34 \times 10^{17}$ cm^{-3}, $\chi_a^{(5)} = -3.39 \times 10^{-48}$ esu, $\chi_a^{(3)} = 3.86 \times 10^{-38}$ esu) this ratio has the value

$$G_5/G_3 = 83. \tag{83}$$

Combining Eqs. (83) and (82) we see that the power in the fifth harmonic is expected to exceed that in the third by the factor 10.4.

Measurements of the fifth- and third-harmonic power as a function of pump power at a helium pressure of 18 Torr are shown in Fig. 28. Also shown are measurements of the fifth harmonic at 25 Torr. For pump powers below about 12 MW, designated by the arrow, both the third- and fifth-harmonic signals follow their respective theoretical dependence on pump power. Their behavior in this region can be compared with the theory outlined earlier.

Extensive measurements of the type shown in Fig. 28 have indicated that the power at which the saturation of the fifth harmonic starts is independent of the helium pressure. We can therefore extrapolate the fifth-harmonic data at 18 Torr back to a pump power of 12 MW. At this power the ratio of the fifth-harmonic signal to the third-harmonic signal is 58. This value actually represents the ratio of the observed energies in the two harmonics. In order

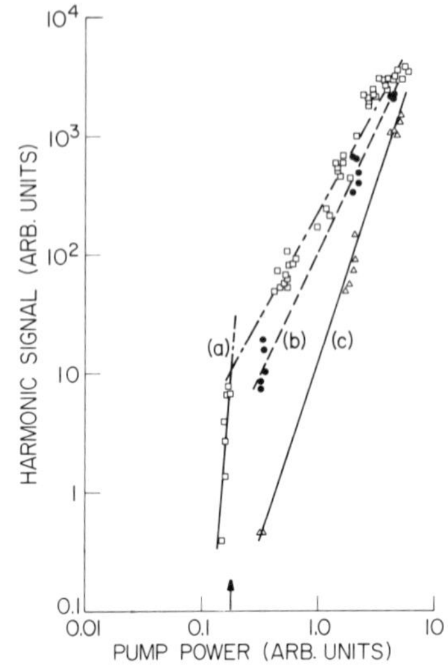

Fig. 28. Variation of the harmonic power with pump power. (a) Fifth harmonic, 25 Torr helium. (b) Fifth harmonic, 18 Torr helium. (c) Third harmonic, 18 Torr helium. (From Ref. 9.)

to compare it with the theory given earlier it must be corrected for the relative pulse duration of the two harmonics. The variation in spectral sensitivity of the detection system has already been included in the data in the figure. If we assume that the pump pulse had a Gaussian temporal and that the pump depletion is negligible, then the ratio of the duration of the fifth-harmonic pulse to that of the third-harmonic pulse is

$$T_5/T_3 = \sqrt{\tfrac{5}{3}}. \tag{84}$$

The measured signals must be corrected further for an interfering signal that was observed at the third-harmonic wavelength at low helium pressures. This signal was attributed to third-harmonic generation in the last absorption length of the CaF_2 lens. When these two corrections are made the experimental ratio of the power at the fifth harmonic to that at the third from generation in helium is 17, a value that is within a factor 2 of the one predicted by theory.

At pump powers above about 12 MW the fifth-harmonic signal saturates and no longer follows a fifth-power law dependence on the pump power. A proper description of this behavior requires a more complete theory which includes other nonlinear processes which affect the harmonic conversion process. A discussion of these effects will be given in Chapter 4. Here we can note that comparison of the saturation of the conversion to harmonics of different order can provide valuable information as to the mechanisms responsible for the saturation by indicating whether they involve effects that are common to both harmonics.

4.3.4b. Comparison of Fifth- and Seventh-Harmonic Conversion. The fifth- and seventh-harmonic signals obtained in helium at full laser power ($P_1 \times 330$ MW, $I_1 = 10^{15}$ W/cm^2) are shown in Fig. 29 as a function of gas pressure.[6] Both signals rise at low pressure as the square of the pressure and level off near 100 Torr. The strongest fifth-harmonic conversion was obtained with the 5-cm focal length focusing lens at a pressure of 200 Torr. A conversion efficiency of 10^{-5} was measured, corresponding to a peak intensity at 53.2 nm of about 3 kW. This conversion level is estimated to be

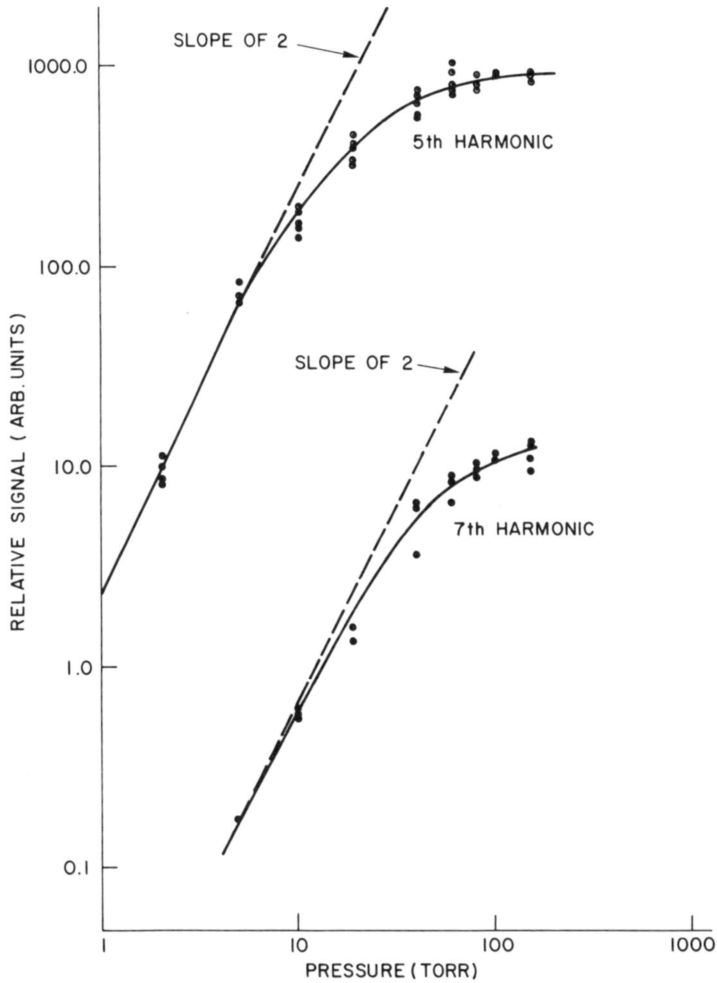

Fig. 29. Comparison of the fifth- and seventh-harmonic powers as a function of helium pressure. (From Ref. 6.)

accurate to within a factor of 3. The seventh-harmonic signals at the exit of the spectrometer were smaller than the fifth-harmonic signals by about a factor of 100. When these measurements were corrected for spectral variation in detector sensitivity and spectrometer efficiency, the maximum generated seventh-harmonic power is estimated to be about 10 times smaller than that of the fifth harmonic. Inasmuch as the seventh harmonic showed no signs of saturation (Fig. 29) it might be possible for the power in the seventh harmonic to exceed that in the fifth at higher levels of pump power. However, at the powers used in the experiments of Reintjes *et al.*[6] the fifth-harmonic power was always greater than the seventh-harmonic power.

The results of these comparisons are in general agreement with the considerations given in Section 3. For the pump and generated wavelengths involved the first resonance in helium occurs for the fifth-harmonic susceptibility, and indeed, conditions could be found for which the conversion to the fifth harmonic exceeded that to the third. No strong additional resonances are encountered in the seventh-harmonic susceptibility, and conversion to the seventh harmonic is less than conversion to the fifth. While it is possible that conversion to the fifth and seventh harmonics could become comparable at higher pump powers, it cannot be predicted from the present measurements whether additional saturation will become important at higher pump powers and prevent comparable conversion in the fifth and seventh harmonics.

4.3.5. *Higher-Order Processes With Harmonics of Nd:Glass Lasers*

Fifth-harmonic conversion of the second harmonic of a mode-locked Nd:glass laser has been reported by Mitev *et al.*[35] A similar interaction for the Nd:YAG laser was reported by Harris.[3] Mitev *et al.*[35] reported a conversion efficiency of 10^{-7} for measurements done in a phase-matched mixture of Na:Ar.

4.3.6 *Higher-Order Interactions with Rare-Gas Halide Lasers*

The rare-gas halide (RGH) lasers have been proposed as potential sources for pumping the higher-order interactions to provide a source of coherent XUV radiation.[11,13] As a class the RGH lasers are an efficient source of medium- to high-power radiation at several wavelengths in the near UV. When combined with radiation from dye lasers they can provide a source of narrow-band radiation that is tunable over virtually the entire range of wavelengths between 30 and 100 nm. An example of the tuning range is shown in Fig. 30 for various combinations of RGH lasers and dye lasers and for various nonlinear interactions involving four-, six-, and eight-wave mixing of the form $\omega_n = (n-1)\omega_{UV} \pm \omega_{dye}$, where n is the order of the interaction, ω_{UV} signifies a photon from a RGH laser, and ω_{dye} signifies a photon from a dye laser. In order to cover the entire range shown in Fig. 30

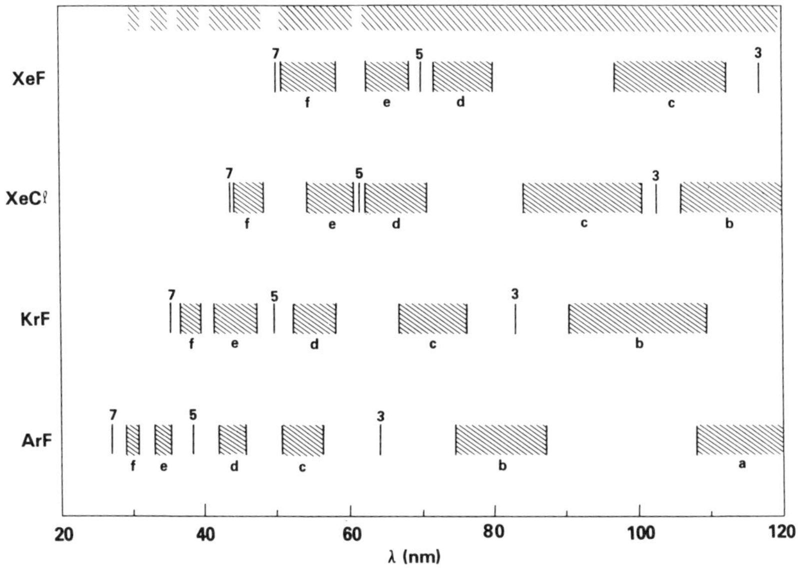

Fig. 30. Wavelength range covered by various nonlinear optical processes using radiation from RGH lasers and dye lasers extending from 300 to 900 nm. Each RGH laser is specified at left. Sharp lines indicate harmonic wavelengths with the order of the harmonic indicated. Shaded regions indicate spectral ranges covered by frequency mixing using the following processes: (a) $2\omega_{RGH} - \omega_{dye}$; (b) $2\omega_{RGH} + \omega_{dye}$; (c) $4\omega_{RGH} - \omega_{dye}$; (d) $4\omega_{RGH} - \omega_{dye}$; (e) $6\omega_{RGH} - \omega_{dye}$; (f) $6\omega_{RGH} + \omega_{dye}$. The composite tuning range is shown by the shaded region at the top. (From Ref. 13.)

dye lasers ranging from 330 to 900 nm were considered. The tuning range that can be obtained with a single dye varies from about 2.5 nm for the third-order processes to about 0.5 nm for the seventh-order interactions. The harmonics of the various RGH lasers are included as sharp lines, although some degree of tuning is possible with these processes as well.

All of these types of interactions that are included in Fig. 30 have been reported in the XUV with the exception of the eight-wave mixing interactions.[5–14] Shorter wavelengths than those shown here could be generated using interactions of still higher order. However, in view of the very low efficiency that has been reported for ninth-harmonic generation[15] (see Section 4.3.7) we have included interactions only up to order seven.

The third-order interactions contained in Fig. 30 were discussed in section 2-8.3. Fifth-harmonic conversion of XeCl laser radiation in Ar and Kr, producing radiation at 61.6 nm, has been reported by Reintjes *et al.*[14] The level diagrams for this process is shown in Fig. 31. The harmonic lies in the continuum for each of the gases. Although there are no published refraction data for these materials in this wavelength range the authors estimated that

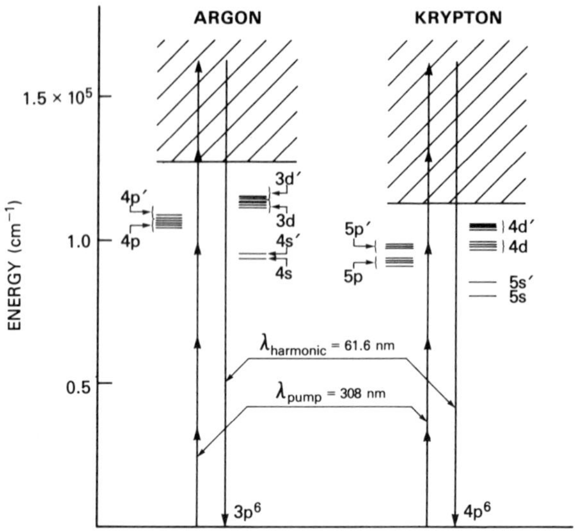

Fig. 31. Level diagram for fifth-harmonic conversion of XeCl laser radiation in Ar and Kr. (From Ref. 14.)

the dispersion for the fifth-harmonic process should be negative by comparison with the calculated behavior of the refractive index of neon in a similar region above its ionization threshold. There are no strong resonances in the susceptibility for either gas, although there may be some enhancement due to moderate three-photon resonances with the $4s'[\frac{1}{2}]^0$ level in Ar ($\Delta\tilde{\nu} = 2760\ \mathrm{cm}^{-1}$), or with the $4d[\frac{3}{2}]^0$ ($\Delta\tilde{\nu} = 285\ \mathrm{cm}^{-1}$), or the $4d[\frac{1}{2}]^0$ ($\Delta\tilde{\nu} = 316\ \mathrm{cm}^{-1}$) level in Kr.

Based on results obtained with higher-order harmonic conversion in nonresonant interactions with Nd:YAG lasers the authors estimated that a power level of 1 MW is needed for the third-order interactions, 10 MW for the fifth-order interactions, and 100 MW for the seventh-order interactions. In order to obtain the required power level for the fifth-harmonic process from an XeCl laser they used a Pockels cell shutter in the laser between an oscillator and an amplifier. This arrangement shortened the pulse duration from the laser from about 10 to about 0.25–0.5 nsec. The peak power was increased from 1 MW to about 20–40 MW and the peak focused intensity was of the order of 10^{14}–10^{15} W/cm^2.

Fifth-harmonic conversion was observed with the differential pumping arrangement shown in Fig. 17. Because the harmonic radiation lies in the absorption continuum of both gases harmonic signals were observed only when the position of the beam waist fell within a restricted range near the pinhole, as shown in Fig. 32. Also shown is the variation of the third harmonic with the position of the beam waist. Both harmonics drop off if the waist is

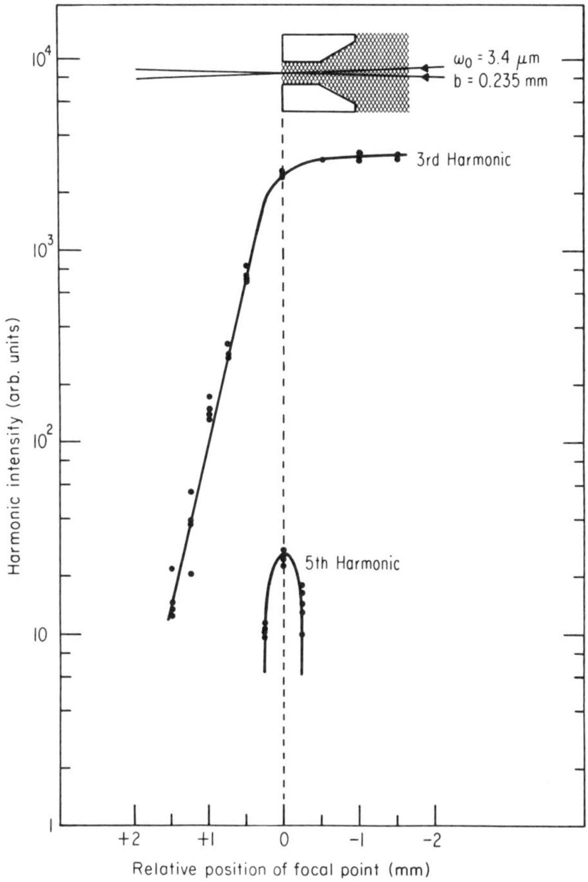

Fig. 32. Dependence of the third- and fifth-harmonic signals from XeCl laser radiation in Ar on the position of the beam waist relative to the pinhole used for differential pumping. Top: Pinhole geometry. Laser beam is incident from right. Shaded region indicates area occupied by the nonlinear medium. Bottom: Measured third- and fifth-harmonic signals. Positive distances correspond to beam waist positions inside the nonlinear medium.

positioned after the pinhole, providing a graphic demonstration of the drop in density of the nonlinear medium near the pinhole. As the beam waist moves into the nonlinear medium the third harmonic rises to a constant value, as would be expected for a negatively dispersive process in a transparent medium. The drop-off of the fifth harmonic as the beam waist moves into the nonlinear medium results from the absorption of the harmonic.

Measurements of the power dependence of the fifth harmonic and the variation with pressure made at the optimum focusing position are shown in Figs. 33a and 33b. The fifth harmonic shows good agreement with the fifth power law over most of the range of intensities that were used. Comparison

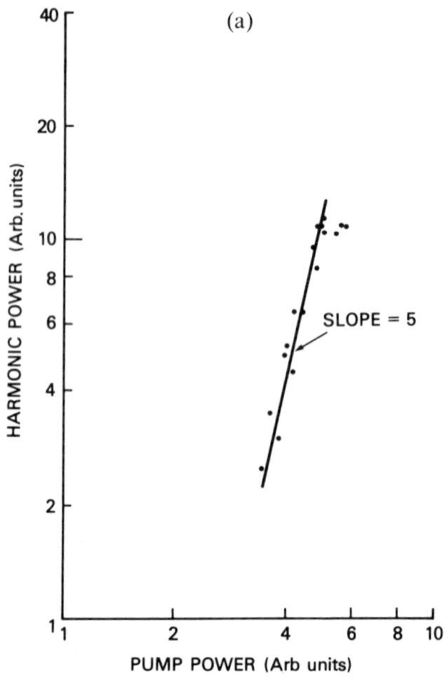

Fig. 33a. Measured dependence of the fifth-harmonic power on the pump power. The maximum pump power is of the order of 20–40 MW. The straight line is drawn through the data with a slope of 5. (From Ref. 14.)

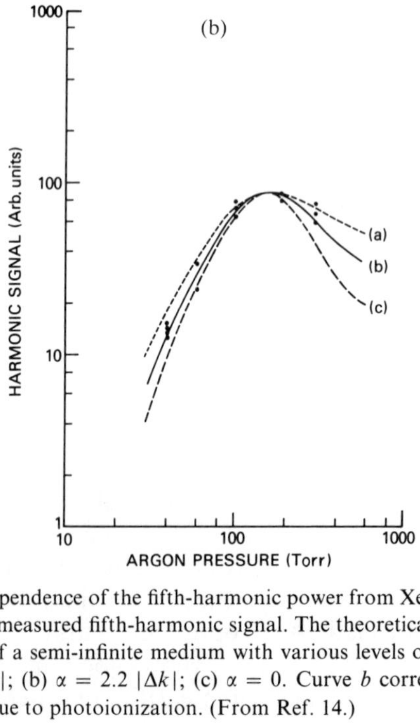

Fig. 33b. Pressure dependence of the fifth-harmonic power from XeCl laser radiation in Ar. The data points are the measured fifth-harmonic signal. The theoretical curves were calculated under the assumption of a semi-infinite medium with various levels of absorption at the harmonic: (a) $\alpha = 0.5\ |\Delta k|$; (b) $\alpha = 2.2\ |\Delta k|$; (c) $\alpha = 0$. Curve *b* corresponds to the expected amount of absorption due to photoionization. (From Ref. 14.)

TABLE 4
Plane-Wave Phase-Matching Ratios for Various Higher-Order Processes That Contribute to Ninth-Harmonic Generation of Nd:Glass Laser Radiation in Na:Ar Mixtures[a]

Interaction	Ratio $R = N_{Ar}/N_{Na}$
$\omega + \omega + \omega + \omega + \omega + \omega + \omega + \omega + \omega = 9\omega$	19.3
$7\omega + \omega + \omega = 9\omega$	5.2
$5\omega + \omega + \omega + \omega + \omega = 9\omega$	8.3
$3\omega + 3\omega + \omega + \omega + \omega = 9\omega$	2.4

[a] From Ref. 15.

of the pressure dependence of the fifth harmonic with theory indicates that the process is negatively dispersive and that the width of the phase optimization peak is larger than that of a transparent medium, but can be accounted for if the linear absorption of the harmonic is included in the analysis.

4.3.7. Ninth-Harmonic Conversion

Ninth-harmonic conversion of radiation from a mode-locked Nd:glass laser at 1.06 μm in mixtures of Na and Ar has been reported by Grozeva *et al.*[15] They have given calculated phase-matching ratios for the direct

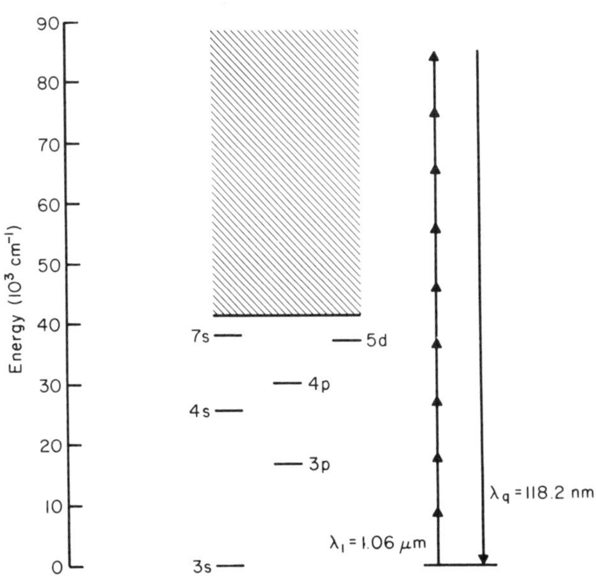

Fig. 34. Level diagram for the ninth-harmonic conversion of radiation from an Nd:glass laser in Na.

ninth-harmonic process and for three indirect processes for collimated beams. These are listed in Table 4. They reported observations of phase-matched conversion for the direct process and the first of the indirect processes. Level diagrams for these two interactions in Na and Ar are shown in Fig. 34. The experiments reported in Grozeva *et al.*[15] were done in focused beams for which the phase-matching ratio for the direct process 16.7 and that for the first indirect process was 4.73.

A conversion efficiency of about 10^{-16} was observed for the direct process at a Na pressure of 20 Torr, corresponding to a temperature of 550 K and a pump power of 250 MW. From their results Grozeva *et al.*[15] estimate that the effective ninth-order susceptibility, including the effects of the indirect processes listed in Table 4, is of the order of $\chi^{(9)}_{\mathrm{a,eff}} \approx 10^{-64}$ esu.

REFERENCES

1. S. A. Akhmanov, *Nonlinear Spectrosc., Corso Soc. Ital. Fis., 64th, Bologna* p. 239 (1977).
2. I. V. Tomov and M. C. Richardson, *IEEE J. Quantum Electron.* **QE-12**, 521 (1976).
3. S. E. Harris, *Phys. Rev. Lett.* **31**, 341 (1973).
4. D. I. Metchkov, V. M. Mitev, L. I. Pavlov, and K. V. Stamenov, *Opt. Commun.* **21**, 391 (1977).
5. J. Reintjes, R. C. Eckardt, C. Y. She, N. E. Karangelen, R. A. Andrews, and R. C. Elton, *Phys. Rev. Lett.* **37**, 1540 (1976).
6. J. Reintjes, C. Y. She, R. C. Eckardt, N. E. Karangelen, R. A. Andrews, and R. C. Elton, *Appl. Phys. Lett.* **30**, 480 (1977).
7. C. Y. She and J. Reintjes, *Appl. Phys. Lett.* **31**, 95 (1977).
8. J. Reintjes, C. Y. She, and R. C. Eckardt, *IEEE J. Quantum Electron.* **QE-14,** 581 (1978).
9. J. Reintjes and C. Y. She, *Opt. Commun.* **27**, 469 (1978).
10. J. Reintjes, *Philos. Trans. R. Soc. London, A* **298**, 273 (1980).
11. J. Reintjes, *Appl. Opt.* **19**, 3889 (1980).
12. C. Y. She and J. Reintjes, *in* "Adaptive Optics and Short Wavelength Sources" (S. F. Jacobs, M. Sargent, III, and M. O. Scully, eds.), p. 267. Addison-Wesley, Reading, Massachusetts, 1978.
13. J. Reintjes, *Proc. Soc. Photo-Opt. Instrum. Eng.* **279**, 2 (1981).
14. J. Reintjes, L. L. Tankersley, and R. Christensen, *Opt. Commun.* **39**, 334 (1981).
15. M. G. Grozeva, D. I. Metchkov, V. M. Mitev, L. I. Pavlov, and K. V. Stamenov, *Opt. Commun.* **23**, 77 (1977).
16. H. Puell, K. Spanner, W. Falkenstein, and W. Kaiser, *Phys. Rev. A* **14**, 2240 (1976).
17. J. Lukasik and J. Ducuing, *Phys. Rev. Lett.* **28**, 1155 (1972).
18. I. Chabay, G. Klauminzer, and B. S. Hudson, *Appl. Phys. Lett.* **28**, 27 (1976).
19. J. F. Ward and G. H. C. New, *Phys. Rev.* **185**, 57 (1969).
20. R. Fischer and L.-W. Wieczorek, *Ann. Phys. (Leipzig)* **35**, 389 (1978).
21. G. C. Bjorklund, *IEEE J. Quantum. Electron.* **QE-11**, 287 (1975).
22. V. G. Tunkin, T. Usmanov, and V. A. Slakirov, *Kvantovaya Elektron. (Moscow)* **2**, 117 (1972) [*Sov. J. Quantum Electron. (Engl. Transl.)* **2**, 487 (1973)].
23. C. C. Wang and E. L. Baardsen, *Appl. Phys. Lett.* **15**, 396 (1969).
24. E. Yablonovitch, C. Flytzanis, and N. Bloembergen, *Phys. Rev. Lett.* **29**, 865 (1972).

25. V. V. Rostovtseva, A. P. Sukhorukov, V. G. Tunkin, and S. M. Saltiel, *Opt. Commun.* **22**, 56 (1977).
26. S. A. Akhmanov, V. A. Martynov, S. M. Saltiel, and V. G. Tunkin, *Pis'ma Zh. Eksp. Teor. Fiz.* **22**, 143 (1975) [*JETP Lett.* (*Engl. Transl.*) **22**, 65 (1975)].
27. N. Bloembergen, "Nonlinear Optics." Benjamin, New York, 1965.
28. S. A. Akhmanov, A. N. Dubovik, S. M. Saltiel, N. B. Tomov, and V. G. Tunkin, *Pis'ma Zh. Eksp. Teor. Fiz.* **20**, 264 (1974) [*JETP Lett.* (*Engl. Transl.*) **20**, 117 (1974)].
29. V. L. Doitcheva, V. M. Mitev, L. I. Pavlov, and K. V. Stamenov, *Opt. Quantum Electron.* **10**, 131 (1978).
30. H. Eicher, *IEEE J. Quantum Electron.* **QE-11**, 121 (1975).
31. S. G. Dinev, O. R. Marazov, K. V. Stamenov, and I. V. Tomov, *Opt. Quantum Electron.* **12**, 183 (1980).
32. L. J. Zych and J. F. Young, *IEEE J. Quantum Electron.* **QE-14**, 147 (1978).
33. S. C. Wallace and G. Zdasiuk, *Appl. Phys. Lett.* **28**, 449 (1976).
34. R. T. Hodgson, P. P. Sorokin, and J. J. Wynne, *Phys. Rev. Lett.* **32**, 343 (1974).
35. V. M. Mitev, L. I. Pavlov, and K. V. Stamenov, *J. Phys. B* **11**, 819 (1978).

CHAPTER 4

LIMITATIONS ON CONVERSION EFFICIENCY DUE TO COMPETING EFFECTS

1. INTRODUCTION

In addition to producing nonlinear polarizations that radiate new optical fields, intense laser beams give rise to other nonlinear effects that can compete with and limit the efficiency of any of the frequency conversion processes that were discussed in the last two chapters. Some of the competing effects impose absolute limitations on the conversion efficiency, dependent only on the constants of the medium. Others merely place restrictions on the range of various parameters that can be used effectively. In some cases two or more effects, neither one of which is absolutely limiting by itself, can combine to form an absolute limit on the conversion efficiency.

The importance of these competing mechanisms in determining the ultimate efficiency of a given conversion process in a specific medium is determined by the magnitude of their effects relative to depletion of the pump wave. If the frequency conversion process is strong enough to deplete the pump wave before the conversion is limited by another mechanism, then the existence of the other mechanism is unimportant. If, on the other hand, the competing effects limit the conversion efficiency before pump depletion occurs, then they will play a major role in determining the maximum efficiency of the conversion process.

In some situations the conversion efficiency can be increased by the presence of a competing mechanism. For example, nonoptimum phase matching can be improved by a nonlinear index, or resonant enhancement of a susceptibility can be increased by a favorable Stark shift of an energy level. However, increased conversion can be expected to occur only for a limited range of operating parameters. For sufficiently high pump intensities all of the nonlinear competing process can be expected to limit the conversion efficiency.

Competing mechanisms can usually be neglected for second-harmonic generation or second-order frequency mixing in anisotropic solids. For this type of interaction conversion efficiencies of the order of 80–90%, with peak intensity conversion efficiency in excess of 99%, have been reported.[1–3] In gases and liquids, however, the nonlinear susceptibilities that determine the magnitude of the competing effects generally involve the same matrix elements that are contained in the susceptibilities that describe third- and higher-order frequency conversion. As a result, competing processes are usually more important in these media than they are in the anisotropic solids. Maximum measured conversion efficiencies for third- and higher-order precesses in gases and liquids have in general been much lower than for second-harmonic conversion in crystals, and significant pump depletion has been reported in only a few instances.[4,5] In most of the experiments that have been reported in gases and in liquids competing processes have limited the conversion efficiency to values ranging from 10^{-7} to 10^{-3}. These lower efficiencies were observed in many instances under conditions for which the theories of Chapters 2 and 3, which fail to account for the competing effects, predict effectively complete conversion of the pump wave energy.

Therefore any realistic theory that is used to describe or to predict the maximum realizable conversion efficiency for a given frequency conversion process in a gaseous or liquid medium must include the effects of any and all additional interactions that are present. The simple theories outlined in Chapters 2 and 3 will in general not be valid for quantitative predictions of efficiency except at extremely low pump intensities.

2. LIMITING MECHANISMS—GENERAL PROPERTIES

The role of competing mechanisms in limiting the efficiency of frequency conversion processes can be appreciated by considering the expression in Eq. (3-5) for the intensity generated in the qth harmonic:

$$I_q = \left[\frac{(2\pi)^{q+3}q^2}{n_q n_1^q \lambda_1^2 c^{q-1}}|\chi^{(q)}|^2\right]\left[\frac{L^2 \sin^2 \Delta kL/2}{(\Delta kL/2)^2}\right][I_1^q]. \tag{1}$$

This expression can be divided into three parts, as indicated in Eq. (1): one that depends on the nonlinear susceptibility, one that depends on the phase-

TABLE 1

Physical Processes That Limit Conversion Efficiency

Physical process	Effect on conversion efficiency
Linear absorption Nonlinear absorption Multiphon ionization Dielectric breakdown	Loss of energy at pump or generated frequency Saturation of the susceptibility Disturbance of phase-matching conditions Self-focusing, self-defocusing
Stark shifts	Saturation of the susceptibility Disturbance of phase-matching conditions Self-focusing, self-defocusing
Kerr effect	Disturbance of phase-matching conditions Self-focusing, self-defocusing

matching conditions, and one that depends on the pump intensity. A similar division can be made for the various expressions for conversion of focused beams. The competing processes can affect the conversion efficiency through any of these terms by causing: saturation of the susceptibility through changes in the population distribution or the position of the energy levels, changes of the coherence length through intensity-dependent phase changes, loss of optical energy through linear or nonlinear absorption, and changes in the beam intensity through self-focusing or defocusing. Except for linear absorption all of these mechanisms affect the power law dependence of the generated radiation on the pump radiation.

Some of the physical processes that are responsible for the effects that limit the conversion efficiency are listed in Table 1, along with the effects that they produce. All of these physical processes have been considered as sources of efficiency limitations in liquids and gases. Each of the physical processes in Table 1 can contribute to more than one of the effects on efficiency that were discussed earlier. For example, multiphoton absorption can cause energy loss at the fundamental or generated wavelengths, or breaking of phase matching or saturation of the susceptibility through changes in the population distribution. Conversely each of the effects on the conversion efficiency can be caused by more than one physical process. For example, breaking of phase matching can be caused by multiphoton absorption, multiphoton ionization, Stark shifts, or the Kerr effect.

This multivaluedness of the correspondence between the physical processes and the effects that they have on the conversion efficiency can make it difficult to determine exactly which of the many possible processes is responsible for observed limitations on efficiency. Ambiguity in interpretation of experimental data can result unless measurements are made of the dependence of the saturation of the conversion efficiency on such experimental parameters as gas pressure, detuning from resonance, strength of focusing, etc. Difficulties can also arise in theoretical predictions of the relative importance of the various competing processes unless all possible competing processes are considered and all the effects to which each gives rise are accounted for.

The relative importance of the various competing nonlinear effects depends on the particular combination of frequency conversion process, wavelength of pump and generated beams, and the nonlinear medium that is involved. Different effects can be important in limiting a given frequency conversion process in different media, or in a given medium under different operating conditions, such as pressure range or detuning from resonance. In some cases a single mechanism will be responsible for the efficiency limitation. More generally, however, two or more processes will be of equal importance.

The degree of resonant enhancement that is present in the nonlinear susceptibilities can determine to a certain extent which of the competing processes will be potentially most important. In nonresonant interactions breaking of phase matching through intensity-dependent phase changes can generally be expected to affect the conversion efficiency. In many instances this effect is, or can be made to be, the dominant limitation on conversion efficiency. If the nonlinear susceptibility is dominated by the detuning from a single resonance, even if the resonance condition is not very well satisfied, Stark shifts, which contribute to the nonlinear refractive index and cause saturation of the susceptibility, can also be important. At sufficiently high intensities, multiphoton ionization and dielectric breakdown can also affect the efficiency.

In interactions in which there is a strong resonant enhancement of the frequency-conversion susceptibility all of these effects will be enhanced. In addition, multiphoton absorption between discrete levels, leading to optical energy loss, saturation of the susceptibility, and additional contributions to the nonlinear index will be important.

In gases at low densities, typically of the order of 1–10 Torr, the primary effect of multiphoton absorption is in saturation of the susceptibility, while in gases at higher densities and in liquids the intensity-dependent phase changes caused by redistribution of the population can be more important. One exception to this last trend occurs for conversion in molecular liquids when resonant enhancement through vibrational energy levels is involved. Here the refractive index resulting from molecules in the excited state is almost the same as the index from molecules in the ground state and disturbance of the phase-matching conditions is relatively small.

For a given set of experimental conditions each nonlinear mechanism is characterized by a saturation intensity (or saturation power or energy density, depending on the exact situation) above which the conversion efficiency begins to be limited significantly. This saturation intensity generally varies inversely as some positive power of the susceptibility for the competing mechanism. Because the susceptibilities for the frequency conversion processes and the limiting processes tend to be related, large values of the susceptibility for the frequency conversion processes, such as would be obtained from resonant enhancement, for example, tend to be associated with low saturation intensities, while small values of the susceptibility tend to be associated with high saturation intensities.

Many types of competing mechanisms introduce a saturation term into the conversion efficiency of the form $1/[A + BI^{(q'-1)/2}]^2$, where q' is the order of the competing process, the constant A describes the behavior of the conversion in the absence of the competing process, and B is a constant determining the saturation characteristics. For this type of competing process the saturated efficiency for qth-order frequency conversion varies as the $(q - q')$ power of the pump intensity. A special case occurs when a qth-order frequency conversion process is limited by certain qth-order competing mechanisms, for example, third-harmonic generation limited by a third-order nonlinear index. In such a situation an absolute limit on conversion efficiency occurs, and the saturated efficiency is independent of the pump intensity. In situations in which two or more competing mechanisms are important, saturation terms of the type discussed above can occur in two or more of the factors in Eq. (1). An example of this is if the nonlinear susceptibility and the coherence length are both affected by the pump intensity. In some situations the effects of the two mechanisms can counteract each other, reducing the tendency for the conversion efficiency to saturate. For other conditions the effects of the two mechanisms can reinforce each other, increasing the saturation of the conversion efficiency. Absolute limitations on the efficiency of higher-order processes can also occur as the result of the effects of two or more lower-order processes.

The various mechanisms which affect the conversion efficiency depend differently on experimental parameters such as gas pressure, interaction length, peak pump intensity, pulse duration, etc. For example, saturation of the nonlinear susceptibility through nonlinear processes other than dielectric breakdown occurs on a per-atom or per-molecule basis and depends only on the peak pump intensity, or a combination of the peak intensity and peak energy density for transient effects. As a result, in the absence of depletion of the pump in the frequency conversion process, the pump intensity (or energy density where appropriate) at which the conversion saturates because of saturation of the susceptibility does not depend on gas pressure or interaction length. On the other hand, intensity-dependent phase changes do depend on the density of the medium and on the integral of the pump intensity through the nonlinear medium. Therefore the pump intensity at which the conversion efficiency saturates due to changes in the phase matching will depend on both the gas pressure and on the interaction length. These differences in dependence on experimental parameters can be used to determine which of several possible mechanisms are actually effective in limiting efficiency. A summary of these dependences on experimental parameters for various limiting mechanisms is given in Table 2.

TABLE 2

Experimental Parameters That Affect Limiting Mechanisms

Limiting mechanism	Experimental parameters	Scaling[a]
Energy loss		
Linear absorption	Density, interaction length	NL
Nonlinear absorption	Density, length, intensity	NLI^q
Saturation of the susceptibility	Peak intensity (except dielectric breakdown)	$(I/I_s)^q$
Third-order intensity-dependent phase changes	Density, intensity, interaction length	NIL (plane waves) NPb (focused beams)

[a] N—number density; L—length of medium; I—intensity; P—power.

The nonlinear polarizations that are responsible for the competing nonlinear processes can be transient when the time duration of the pump radiation is shorter than the response time of the material system, just as with the polarizations that give rise to the frequency conversion processes. Again, just as with the frequency conversion processes, the transient effects of the competing processes are most evident when they are resonantly enhanced. Generally speaking, the instantaneous competing processes tend to place limits on the peak pump intensity or pump power, while the transient competing processes also affect the total pump energy or energy density that can be used effectively. In addition, the transient response can change the relative importance of various competing mechanisms, especially when only one of the competing processes is transient.

3. PROPERTIES OF COMPETING NONLINEAR PROCESSES

Before examining the characteristics of the frequency conversion processes in the presence of the competing nonlinear effects let us discuss the nature of some of the physical processes that give rise to the competing mechanisms.

3.1. Multiphoton Absorption

There are many possible multiphoton absorption processes at both the pump and generated wavelengths that can occur, depending on the particular combination of pump and generated frequencies and nonlinear medium that is involved. Nonlinear absorption can proceed by absorption of an appropriate number of pump or generated photons, by simultaneous absorption of pump and generated photons, or, in media with suitable energy level structure, simultaneous absorption at one wavelength and emission at another (Raman processes). These are illustrated in Fig. 1.

Generally, multiphoton absorption that is induced by the pump intensity can be expected to be the most important since the pump waves are usually the strongest. However, in experiments in which there is significant conversion to the generated wave, or in which there are appropriate resonances in the nonlinear absorption susceptibility, nonlinear absorption induced by the generated intensity can also be important. Also, generally speaking, third-order (two-photon) absorption can be expected to be the most important unless a particular combination of pump and generated wavelengths and energy level positions provides preferential enhancement for higher-order absorption.

Multiphoton absorption in molecular liquids can involve transitions between either vibrational or electronic energy levels. Transitions between vibrational energy levels generally involve mid- to far-infrared radiation for sum- or difference-frequency generation, or near-infrared to visible radiation

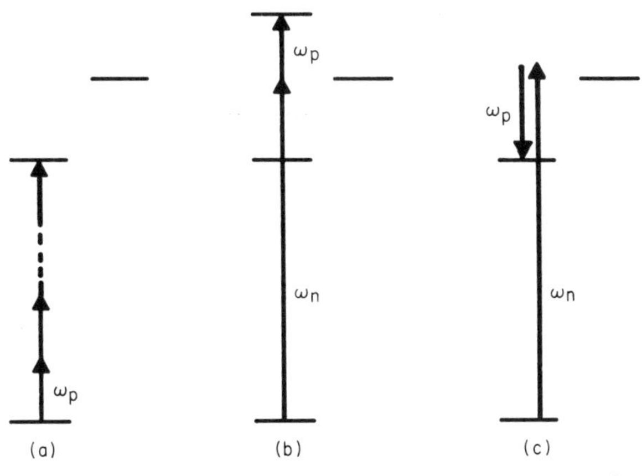

Fig. 1. Level diagrams showing various forms of multiphoton absorption. (a) n-photon absorption of pump radiation. (b) Two-photon absorption of one pump photon and one generated photon. (c) Raman processes involving absorption of one generated photon and emission of a pump photon.

for Raman-type interactions. They are generally important only when there is multiphoton enhancement of the nonlinear susceptibility for the frequency conversion process. Multiphoton transitions between electronic energy levels are most important for visible to ultraviolet radiation for which the transitions are usually of third or fifth order, depending on exactly where the lowest electronic energy level falls. As the wavelength of the radiation increases, the order of the nonlinear absorption process increases, and its strength generally decreases. Because the ultraviolet absorption region in liquids tends to be broadband, exact wavelength matching can be done over a wider range between narrow vibrational levels.

In gases multiphoton absorption can occur between discrete electronic or vibrational energy levels and also between the ground state and the ionization continuum. Transitions between discrete levels, as in the case of molecular liquids, will in general only be important when there is multiphoton enhancement of the frequency conversion susceptibility. Multiphoton transitions to the continuum will be considered in the next section.

Multiphoton absorption can give rise to steady-state or transient effects depending on the relation between the duration of the pump radiation and the relaxation times of the levels that are involved in the absorption process. Coherent propagation effects can also occur if the duration of the pump radiation is shorter than the dephasing time of the absorbing level.

Nonlinear absorption can affect the frequency conversion efficiency in four ways. It can result in optical energy losses at the pump or generated wavelengths, it can cause saturation of the nonlinear susceptibility, and it can cause intensity-dependent phase changes that result in breaking of phase matching or self-focusing or self-defocusing. Saturation of the susceptibility depends only on the peak pump intensity (or energy density) and so, for a given pump power, will be more important for beams that are tightly focused than for collimated beams.

3.2. Multiphoton Ionization

Multiphoton ionization is a special case of multiphoton absorption in which the final state lies in the ionization continuum. It is most commonly encountered in low-pressure gases where substantial free electron densities can be sustained without proceeding to avalanche breakdown. Multiphoton ionization can involve absorption of two or more photons from the pump or generated waves or simultaneous absorption of photons from both waves. Since the ionization threshold of most gases ranges from the visible to the vacuum ultraviolet, multiphoton ionization can be expected to be most important when the pump or generated radiation lies in the wavelength range from the near infrared to the ultraviolet, for which only a few photons are involved in the ionization process. Because the final state is in the continuum an exact combination of pump or generated wavelengths is not required, as it is for multiphoton absorption between discrete states. Rather it is only required that the combination of pump (or pump and generated) photon energies exceed the ionization threshold. The multiphoton ionization cross section can show interference maxima or minima near resonant intermediate states, and in some cases the effective order of the resonant interaction can be different from that of the nonresonant interaction.

Because it depends on the instantaneous intensity, the strength of the multiphoton ionization process depends on the statistical properties of the laser signal (see Section 2-6) and can be substantially greater for multimode laser radiation than for single-mode radiation. A similar effect can occur for the multiphoton absorption discussed in the last section.

Multiphoton ionization can be either transient or steady state, depending on the relative values of the pulse duration of the radiation and the response time of the nonlinear polarization of the medium. The response time of the polarization is determined by the inverse of the smallest detuning of the various intermediate virtual levels from the real intermediate levels, or, in the case of an exact multiphoton resonance, the inverse of the linewidth of the resonant level (see, e.g., Ref. 6). For interactions that are usually classed as nonresonant the response time ranges from values of the order of 10^{-13} sec for detunings from intermediate states of several hundred cm^{-1} to 10^{-15} sec or less for larger detunings. Such interactions will not be transient for all except the very shortest laser pulses that are currently available. For resonant interactions the response time can range from 10^{-11} sec for intermediate detunings of the order of several cm^{-1} to 10^{-9} sec or longer if the laser is tuned to an exact multiphoton resonance with a narrow level in a medium such as an atomic vapor.

Just as with multiphoton absorption, multiphoton ionization can affect the frequency conversion efficiency by saturation of the nonlinear susceptibility through depletion of the ground-state atoms or molecules, by causing a nonlinear refractive index leading to intensity-dependent phase mismatches, and by absorbing the radiation at the pump or generated wavelengths.

If transient effects due to the response time of the medium can be neglected, then the multiphoton ionization rate depends on an appropriate power of the pump intensity, or, when applicable, a product of powers of the pump and generated intensities. For a given pump power its effects will be greater for focussed beams than for collimated beams. Effects related to depletion of the ground state, for example, saturation of the susceptibility through ground-state depletion, will tend to be more important if all the photons that are involved in the ionization process come from the pump wave since the pump wave tends to be the most intense. However, effects arising from multiphoton ionization involving the generated radiation can be of significance if there is substantial conversion of the pump power or if the wavelength of the generated radiation lies considerably closer to the ionization threshold than does that of the pump radiation, allowing ionization of the medium to take place through a lower-order process.

Nonresonant multiphoton ionization by pump radiation with intensity I_1 is usually described by a rate equation of the form

$$\frac{dN_i(t)}{dt} = W^{(K_0)} N_0(t) I_1^{K_0}, \tag{2}$$

where $N_i(t)$ is the number density of ions, $N_0(t)$ is the number density of ground-state neutrals, $W^{(K_0)}$ is a rate constant that is usually termed the multiphoton ionization cross section, and K_0 is the smallest number of photons needed to reach the ionization continuum. The number densities for the ions and ground-state neutrals are subject to the condition

$$N_0(t) + N_i(t) = N_0(0).$$

Generally the ionization process described by Eq. (2) is the most important. However, higher-order processes can occur which involve the absorption of more than K_0 photons and excite the electron to higher states in the continum. These higher-order processes have been discussed by Agostini *et al.*[7] and by Kruit *et al.*[8] The formulas that are given in the succeeding sections will be based on the assumption that only the lowest order ionization process is significant.

Multiphoton ionization can result in attenuation of the pump radiation as described by the equation

$$dI/dz = -K_0 \hbar\omega W^{(K_0)} I^{K_0}. \tag{3}$$

If the generated frequency lies considerably closer to the continuum than the pump frequency, as it can for harmonic- or sum-frequency generation in rare gases or metal vapors using visible or ultraviolet radiation, ionization involving one pump and one generated photon can be significant. If the sum-frequency process is energetically allowed, then the ionization rate is

$$\frac{dN_i}{dt} = N_0 W^{(2)}(\omega, \omega_g, \omega_g + \omega) I_1 I_g, \tag{4}$$

where $W^{(2)}(\omega, \omega_g, \omega_g + \omega)$ is the sum-frequency ionization rate constant. The transmission of the medium at the generated frequency is then given by

$$\frac{dI_g}{dz} = -\hbar\omega_g N_0 W^{(2)} I_1 I_g. \tag{5}$$

Under some circumstances the induced absorption at the generated wavelength can be greater than that at the fundamental.

Depletion of the ground-state population is described by the equation

$$\frac{dN_0}{dt} = -\sum_{n,m} N_0 W^{(K_0-n+m)} I_1^{(K_0-n)} I_g^m, \tag{6}$$

where the sum is over all possible ionization processes involving combinations of $K_0 - n$ pump photons and m generated photons and it is assumed that the ground state is not repopulated during the laser pulse. If only the multiphoton ionization caused by the pump radiation is important, i.e., if $n = m = 0$, the ground-state population is given by

$$N_0(t) = N_0(0)\exp\left[-\int_{-\infty}^{t} W^{(K_0)} I^{K_0}(t')\,dt'\right]. \tag{7}$$

It is reduced to half of its initial value when

$$\int_{-\infty}^{t} W^{(K_0)} I^{K_0}(t')\,dt' = 0.69. \tag{8}$$

Contributions to the nonlinear refractive index come from both the change of the ground-state population and the plasma contribution of the electrons that are created in the ionization process. The contribution from the reduction of ground-state population is similar to that resulting from multiphoton absorption between discrete states and will be discussed in Section 4.3.4. The contribution from the free electrons is given by

$$\Delta n_e(\lambda) = -\frac{e^2\lambda^2}{2\pi mc^2} W^{(K_0)} \int N_0(t) I^{K_0}\,dt. \tag{9}$$

The resulting wave vector mismatch for qth-harmonic generation is

$$\Delta k_e = \frac{(q^2-1)}{q}\frac{e^2\lambda_1}{mc^2} W^{(K_0)} \int N_0 I^{K_0}\,dt. \tag{10}$$

Although the free electrons make a negative contribution to the refractive index, they make a positive contribution to the wave vector mismatch.

The multiphoton ionization rate constant $W^{(K)}$ is usually calculated using perturbation theory. Various approaches have been used to evaluate the sum over intermediate states, including phenomenological methods,[9–11] Green's functions,[12,13] and quantum defect theories.[14] Keldysh[15] has given a quasiclassical theory that reduces to multiphoton ionization for relatively large optical frequencies and to a tunneling theory at low optical frequencies. Multiphoton ionization in the presence of intermediate state resonances has been discussed by, for example, Mainfray and Manus,[6] Gontier and

Trahin,[16] and Lompré *et al.*[17] Extensive discussions of various aspects of multiphoton ionization have been given in review articles by Lambropoulos[18] and by Delone.[19]

Problems related to convergence of perturbation expansions can lead to difficulties in evaluating the multiphoton ionization cross section, especially near intermediate resonances. However, recent measurements of the resonant behavior of the multiphoton ionization cross section of Cs have shown excellent agreement with theory.[20]

Some measured and calculated values of the multiphoton ionization cross section at various wavelengths in selected gases are given in Table 3, along with the intensity required for depletion of the ground state to half of its initial value for a 1-nsec-duration pulse of constant intensity. Depending on the particular medium, the order of the ionization process and the degree of resonant enhancement ground-state depletion occurs in the range of intensities from 10^{10} to 10^{14} W/cm^2. These values will, of course, be higher if picosecond pulses are used. Products of the pump intensity, interaction length, and density for which a nonlinear phase change of π rad for third-harmonic generation results from the free electrons are also shown in Table 3, again assuming a pulse duration of 1 nsec. In some situations, especially when the initial ground-state density is high, substantial intensity-dependent phase changes can occur before there is significant depletion of the ground state.

Various aspects of the limitations on the efficiency of frequency conversion processes caused by multiphoton ionization have been considered by several authors. Georges *et al.*[21] have concluded, on the basis of calculations using theoretical values of the multiphoton ionization rate constant, that limitations on the efficiency of two-photon resonantly enhanced frequency conversion in metal vapors when the generated frequency lies in the ionization continuum are dominated by the depletion of ground state neutrals by multiphoton ionization. Miyazaki and Kashiwagi[22] have reported experiments of third-harmonic conversion in Na vapor, in which the generated wavelength was below the continuum, for which limitations were attributed to nonlinear index changes from the free electrons that were created in the multiphoton ionization process. The interplay between third-harmonic generation and coherent excitation of the medium when the laser is tuned to a three-photon resonance, and its effect on the multiphoton ionization process, has been discussed by Miller *et al.*,[23a] Payne *et al.*,[23b] and Jackson and Wynne.[23c]

A certain amount of caution should be exercised in arriving at conclusions regarding the importance of multiphotoionization in limiting conversion efficiency based on theoretical calculations of the amount of ionization. Since the multiphoton ionization rate depends critically on the magnitude of the rate constant, some ambiguity can result from the spread of several

TABLE 3

Multiphoton Ionization Parameters

Material	λ (nm)	Order (K_0)	W [(cm^2/W)^{K_0} sec^{-1}]	I_{dep} (W/cm^2)[a]	$N_0 I^{K_0} L$ (cm^{-2}[(W/cm^2)^{K_0}])[b]	Method	Reference
Li	347	2	2.26×10^{-13}	5.5×10^{10}	5.3×10^{38}	Theoretical	20a
	347	2	2.7×10^{-13}			Theoretical	12
	532	3	3.83×10^{-26}			Theoretical	12
Na	347	2	1.1×10^{-15}	7.9×10^{11}	1×10^{41}	Theoretical	20a
	347	2	1.99×10^{-15}			Theoretical	20b
	347	2	9.75×10^{-15}			Theoretical	12
	530	3	1.5×10^{-23}	3.5×10^{10}	5.3×10^{48}	Theoretical	12
	694.4	3	3×10^{-22}			Theoretical	12
	1059	5	2.3×10^{-47}			Theoretical	11
	1064.4	5	6.8×10^{-49}	2.5×10^{11}	5.7×10^{69}	Measured	22
Rb	514.5	2	8.3×10^{-13}			Theoretical	20b
	532	2	2.14×10^{-12}	1.8×10^{10}	3.7×10^{37}	Theoretical	12
	694.3	3	1.7×10^{-22}			Theoretical	12
	1059	4	1×10^{-38}	5.1×10^{11}	3.8×10^{63}	Theoretical	5
Cs	514.5	2	2.2×10^{-13}			Theoretical	20b
	528	2	4.7×10^{-13}	3.8×10^{10}	1.7×10^{38}	Experimental	20
	530	2	1.2×10^{-12}			Theoretical	12
	1056	4	5.9×10^{-34}	3.28×10^{10}	6.7×10^{58}	Experimental	20
He	694.3	14	5×10^{-179}	4.4×10^{12}	1.2×10^{204}	Theoretical	10
	1064.4	22	1.2×10^{-321}	8.78×10^{14}	3.3×10^{346}	Experimental	20c
Ne	694.3	13	1.4×10^{-158}	6.6×10^{12}	4.3×10^{183}	Theoretical	10
Ar	694.3	9	2.4×10^{-98}	6.7×10^{11}	2.5×10^{123}	Theoretical	10
Kr	694.3	8	3.5×10^{-85}	4.59×10^{11}	2.1×10^{111}	Theoretical	10
Xe	694.3	7	2.8×10^{-76}	1.13×10^{12}	2.1×10^{101}	Theoretical	10

[a] Intensity in a pulse of 1-nsec duration.

[b] Value required for π phase change assuming a constant intensity pulse of 1-nsec duration.

orders of magnitude in the values reported for the constants in various calculations and experiments. An example of such ambiguity can be found in the interpretation of the experiments reported by Puell *et al.*[5] There the authors concluded on the basis of constants given by Morton,[11] that almost complete ionization of the nonlinear medium occurred at their highest laser powers, while, for constants that they calculated using the theory of Keldysh,[15] a negligible amount of ionization occurred.

It can be anticipated that further calculations and experiments can contribute to a clearer understanding of the role of multiphoton ionization in limiting the efficiency of frequency conversion processes under various conditions.

3.3. Dielectric Breakdown

Dielectric breakdown can occur in both gases and liquids, but the breakdown intensities in liquids tend to be several orders of magnitude smaller than those in gases. The breakdown strength of a given medium is affected by a number of parameters, such as wavelength of the pump radiation, presence of resonances, linear and nonlinear absorption, size of focal volume for focused beams, and pulse duration for pulsed radiation. In gases it also depends on the pressure. At the subatmospheric pressures that are usually used for frequency conversion, the breakdown intensity is usually a decreasing function of gas pressure, and so breakdown becomes a more important effect as the pressure is increased. Dielectric breakdown will generally be more important for focused beams than for collimated beams, and will become increasingly more important as the beam is focused more tightly, although this trend can be violated if the focal volume becomes too small. Theories of dielectric breakdown have been discussed by several authors.[24–26] Some breakdown intensities for selected gases at specific wavelengths are listed in Table 4. Typical breakdown intensities are of the order of 10^8–10^{13} W/cm^2 depending on the wavelength, pulse duration, pressure, and gas.

The final state of the medium in the presence of dielectric breakdown is characterized by a very high electron density, the presence of multiply ionized species of the nonlinear medium, and almost complete destruction of the nonlinear medium. Gases and liquids do not suffer the permanent damage that solids do, but they can still be effectively destroyed during the pump pulse.

Dielectric breakdown can affect the efficiency of the frequency conversion process in all four ways that were discussed earlier for multiphoton absorption and ionization. In addition, if the electron density is high enough, then the medium can become opaque to the pump or generated radiation at the point in the pulse at which the breakdown occurs.

TABLE 4

Some Breakdown Intensities for Selected Gases

Gas	Pressure	Wavelength	Pulse duration	Intensity (W/cm^2)	Reference
He	2 atm	10.6 μm	200 nsec	2×10^9	24
	1000 Torr	694.3 nm	20 nsec	1.9×10^{11}	25
	1000 Torr	694.3 nm	50 psec	3×10^{13}	24
Ar	1 atm	10.6 μm	200 nsec	2×10^9	24
	1 atm	1.06 μm	40 psec	6×10^{12}	24
	1 atm	694.3 nm	18 psec	10^{14}	26
	1000 Torr	694.3 nm	50 psec	10^{13}	24
	1000 Torr	694.3 nm	20 nsec	6.6×10^{10}	25
	1000 Torr	1.06 μm	40 psec	4×10^{12}	24
	1000 Torr	1.06 μm	3 nsec	2×10^{11}	24
	1000 Torr	1.06 μm	50 nsec	8×10^{10}	24
N_2	1000 Torr	694.3 nm	20–100 psec	5×10^{13}	24
Air	1 atm	10.6 μm	200 nsec	4×10^9	25

3.4. Stark Shifts

Optical Stark shifts that are induced by either the pump or generated radiation affect the detuning denominators that appear in the susceptibilities. The lowest-order Stark shift of a level with an unshifted energy E_{i0} in the presence of radiation at frequency ω is

$$\Delta E_i = E_i - E_{i0} = \frac{1}{2\hbar} \sum \frac{|\mu_{ik}|^2 \omega_{ik}}{(\omega_{ik}^2 - \omega^2)} |A|^2, \tag{11}$$

where ω_{ik} is the frequency of transition between the unshifted levels i and k.

Optical Stark shifts can affect the frequency conversion process by saturating the nonlinear susceptibility and by changing the phase matching conditions. Generally speaking, Stark shifts are most important when near resonances in the susceptibility are involved. However, they can also be significant if the susceptibility is dominated by the detuning from a single resonance, even if the resonance condition is not well satisfied, or if very high intensities are used in nonresonant interactions.

In media with exact or nearly exact even photon resonances the Stark shift is closely connected with the redistribution of population between the resonant levels due to multiphoton absorption. In low-density media, in which the saturation of the susceptibility due to the resonance is usually more important than the intensity-dependent phase mismatches, the importance of the Stark shift relative to the population changes in saturating the susceptibility is determined by the ratio of appropriate matrix elements, and for steady-state interactions has a value determined only by the properties of the medium. In this regime stark shifts and population redistribution are usually of comparable importance. However, if the duration of the pump pulse is shorter than the relaxation time of the resonant level, then the effects of the Stark shift on the magnitude of the susceptibility can become more important than the redistribution of population. In higher-density media the intensity-dependent phase changes resulting from the population changes or Stark shifts can have a greater effect on the conversion efficiency than the saturation of the susceptibility.

The effects of Stark shifts on conversion efficiency have been discussed by several authors, and several experiments showing evidence of Stark shifts have been reported. Depending on the initial detuning from resonance, the Stark shift can either increase or decrease the nonlinear susceptibility. Increases in conversion due to Stark shifts have been discussed by Stappaerts[27] and by Anikin *et al.*[28] For sufficiently high pump intensities, however, conversion can be expected to be limited by the Stark shift of a near-resonant level. When saturation of the susceptibility due to both population changes and Stark shifts is present, the two effects can counteract each other to a certain extent. This effect has been discussed by Anikin *et al.*[28] and by Elgin and New.[29]

4. THE EFFECTS OF COMPETING PROCESSES ON CONVERSION EFFICIENCY

In this section we shall discuss the effects which the mechanisms that were just described have on the frequency conversion process. For purposes of illustration their effects on harmonic generation will be treated. In appropriate places qualitative comments will be made concerning their effects on other processes.

As was pointed out earlier a proper treatment of this subject requires consideration of all possible competing effects simultaneously. Such an analysis is very complex and generally requires numerical evaluation for individual systems. In this section we shall derive several expressions for the limiting conversion efficiency under various assumptions, each involving one, or at most a few, dominant competing processes. The reader should remember that the appropriate limiting form for a specific system depends on which mechanisms are dominant. We shall give rough estimates for the conditions under which the various types of competing processes are dominant in Sections 4.1.3 and 4.3.5. We shall also give references to reports in the literature that have analyzed more than one competing process for specific systems.

4.1. Limitations by Absorptive Energy Losses

4.1.1. *Linear Absorption*

Linear absorption in the generation of the qth harmonic can be accounted for by including appropriate absorption terms in Eq. (3-11) as follows:

$$\frac{1}{2ik_q}\nabla_\perp^2 A_q + \frac{\partial A_q}{\partial z} + \frac{\alpha_q}{2}A_q = i\kappa_q A_1^q e^{-i\Delta kz}, \tag{12}$$

$$\frac{1}{2ik_1}\nabla_\perp^2 A_1 + \frac{\partial A_1}{\partial z} + \frac{\alpha_1}{2}A_1 = i\frac{n_q}{n_1}\kappa_q A_q A_1^{q-1} e^{i\Delta kz}, \tag{13}$$

where

$$\kappa_q = q\pi^2\chi^{(q)}/2^{(q-3)}n_q\lambda_1 \tag{14}$$

and α_1 and α_q are the intensity absorption coefficients at the fundamental and harmonic wavelengths, respectively.

Analytic solutions for third-harmonic conversion of plane waves in the absence of pump depletion when only absorption at the harmonic is present have been given by Bey *et al.*[30] and have been discussed by Scheingraber *et al.*[31] Analytic solutions can also be obtained in the presence of pump depletion if equal absorption is present at both the fundamental and harmonic wavelengths. Solutions for second-harmonic generation under these conditions have been discussed by Akhmanov *et al.*[32]

4.1.1a. Absorption at the Harmonic. Linear absorption at only the harmonic wavelength can occur in gases when the generated wavelength lies in the ionization continuum or is tuned close to a dispersive level to satisfy phase-matching conditions. It can be encountered in liquids when the harmonic wavelength lies in the ultraviolet absorbing region. A similar case of absorption occurring only at a difference frequency can also be encountered if the generated radiation lies in an infrared absorbing region.

In the plane-wave approximation and in the absence of pump depletion through the harmonic process the harmonic intensity is given by

$$I_q = \frac{n_q}{n_1}K_q^2 I_{10}^q(1 + e^{-\alpha_q L} - 2e^{-\alpha_q L/2}\cos\Delta kL)/[(\alpha_q/2)^2 + \Delta k^2], \tag{15}$$

where

$$K_q^2 = \left(\frac{8\pi}{cn_1}\right)^{q-1}\kappa_q^2. \tag{16}$$

The spatial dependence of the qth harmonic is the same as that of the third harmonic described by Bey *et al.*[30]

At perfect phase matching (15) becomes

$$I_q = K_q^2 I_1^q (1 - e^{-\alpha_q L/2})^2/(\alpha_{q/2})^2. \tag{17}$$

For sufficiently large interaction lengths or absorption coefficients the harmonic intensity becomes independent of the interaction length:

$$\lim_{\alpha_q L \to \infty} I_q = 4K_q^2 I_1^q L_a^2, \tag{18}$$

where $L_a = \alpha^{-1}$ is the absorption length for the harmonic.

Linear absorption also reduces the oscillations that result from phase mismatches. In the limit of large distances or large absorption coefficients the oscillations disappear completely and the harmonic intensity again becomes independent of the interaction length, approaching the value

$$I_q = \frac{n_q K_q^2 I_1^q}{n_1 [(\alpha_q/2)^2 + (\Delta k)^2]}. \tag{19}$$

The dependence of the harmonic intensity on the phase mismatch in the presence of varying amounts of linear absorption has been studied by Scheingraber *et al.*[31] and is illustrated in Fig. 2. These calculations were done using parameters appropriate for third-harmonic conversion in Xe : Sr mixtures with the harmonic wavelength lying in the ionization continuum of Sr.

In a gaseous medium the absorption coefficient, the wavevector mismatch and the coupling parameter K_q are all proportional to the density and the limiting expressions for the harmonic intensity in (18) and (19) are independent of the gas density. In addition, if the harmonic wavelength is close enough to a dispersive level so that both the nonlinear susceptibility and the linear absorption coefficient are dominated by a term of the type $1/(\omega_{n_0} - \omega_q)$, then the limiting harmonic intensity will be independent of the detuning from this level at perfect phase matching. This independence of detuning from the near-resonance level will also be true for nonzero phase mismatches if the wave vector mismatch Δk is also dominated by the detuning from the same dispersive level that determines the harmonic absorption.

4.1.1b. Pump Depletion. When pump depletion must be considered, Eqs. (12) and (13) must be solved simultaneously. Solution for a general case usually requires numerical integration using specific parameters. When the absorption coefficients at the fundamental and harmonic wavelengths are equal, the qth harmonic field amplitude is given in the plane-wave

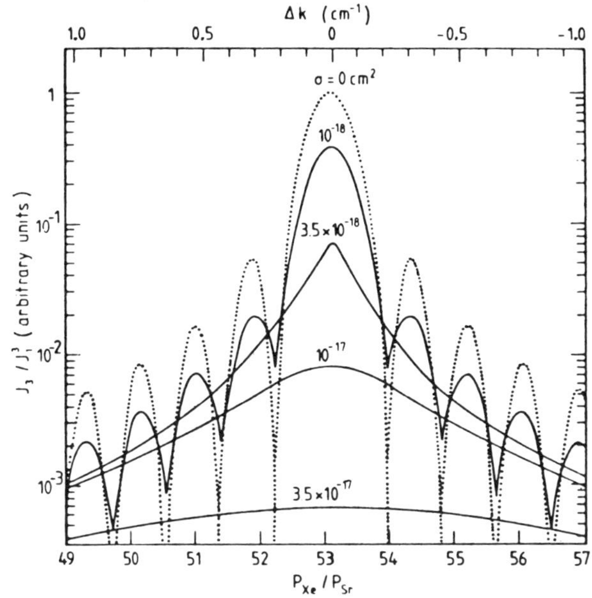

Fig. 2. Phase-matching curves for third-harmonic conversion in Sr:Xe mixtures showing effect of varying levels of absorption at the harmonic on the phase-matching interference structure for NL = 2.5×10^{18}-cm^{-2}. (From Ref. 31.)

approximation at perfect phase matching by the solution of the transcendental equation

$$\tan\theta \left\{ \frac{1}{(q-2)\cos^{q-3}\theta} + \frac{q-3}{(q-2)(q-4)\cos^{q-5}\theta} + \frac{(q-3)(q-5)}{(q-2)(q-4)(q-6)\cos^{q-7}\theta} + \cdots + \frac{(q-3)(q-5)\cdots(2)}{(q-2)(q-4)(q-6)\cdots(1)} \right\} = \kappa_q A_{10}^{q-1} z', \tag{20}$$

where

$$\frac{A_q}{A_{10}} = i\sin\theta\, e^{-\alpha z/2},$$

$$\frac{A_1}{A_{10}} = \cos\theta\, e^{-\alpha z/2},$$

$$z' = (1 - e^{-(q-1)\alpha z/2})/[(q-1)\alpha/2]$$

and κ_q is given by Eq. (14).

This equation is similar in form to (3-8) for the absorption-free case.

For third-harmonic generation, expressions for the fundamental and harmonic amplitudes can be obtained in closed form. The harmonic and fundamental intensities are given by

$$I_3 = \frac{(K_3^2/\alpha^2)(1 - e^{-\alpha z})^2 e^{-\alpha z} I_{10}^3}{1 + K_3^2(1 - e^{-\alpha z})^2 I_{10}^2/\alpha^2}, \tag{21a}$$

$$I_1 = \frac{I_{10} e^{-\alpha z}}{1 + K_3^2(1 - e^{-\alpha z})^2 I_{10}^2/\alpha^2}, \tag{21b}$$

where K_3 is given by Eq. (16) and (14).

The value of z at which optimum conversion occurs is shown in Fig. 3 in units of α^{-1} as a function of $(K_3 I_{10}/\alpha)^2$. Also shown are the conversion efficiency at the optimum value of z and the efficiency which would occur at the same value of z if the absorption were zero. For low values of $(K_3 I_{10}/\alpha)^2$ the optimum conversion occurs for $z \approx \alpha^{-1}$, while for larger values of $(K_3 I_{10}/\alpha)^2$ the optimum value of z lies between $(K_3 I_{10})^{-1}$ and α^{-1}.

The importance of linear absorption in harmonic generation is determined by the relation of the absorption length L_a to the effective interaction length L_{int}, which would dominate the frequency conversion process in the absence of absorption. The effective interaction length is the shortest of the length

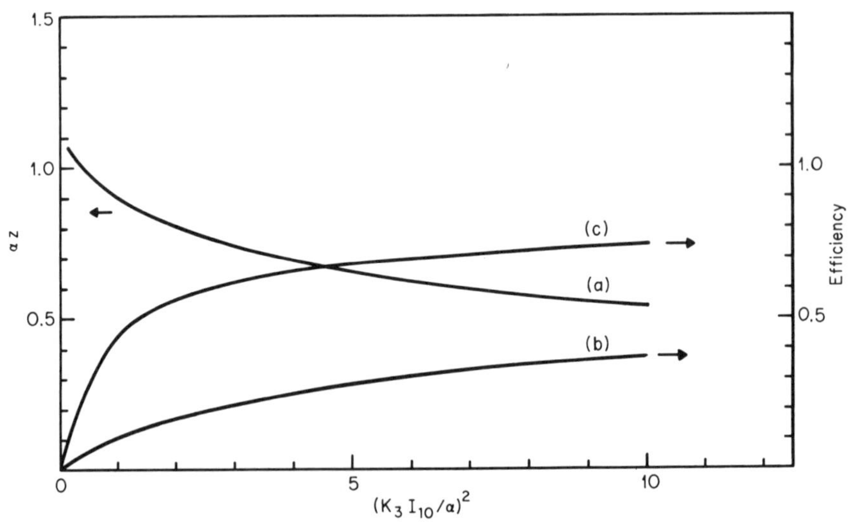

Fig. 3. (a) Value of z in units of α^{-1} at which the optimal third-harmonic conversion occurs as a function of the normalized incident pump intensity $(K_3 I_{10}/\alpha)^2$. (b) Efficiency at optimum value of z. (c) Corresponding value of efficiency at same value of z if the absorption were zero. The equations were solved including the effects of pump depletion and assuming that the pump and harmonic have the same absorption coefficient.

of the medium L_{med}, the coherence length L_{coh}, the pump depletion length L_{dep}, or the length determined by any of the nonlinear competing processes to be discussed in the next sections. If the absorption length is short compared to the interaction length, i.e., $L_a \ll L_{\text{int}}$, the harmonic intensity will be limited to a value given by (18) and will be substantially less than unity. If, on the other hand, $L_a \gg L_{\text{int}}$, then the linear absorption will not have a serious effect on the conversion efficiency.

Unlike the nonlinear competing effects that will be discussed in the following sections, linear absorption does not affect the power law dependence of the harmonic intensity on the pump intensity and as a result does not, by itself, impose an absolute limit on conversion efficiency. In some situations a certain degree of flexibility exists to allow reduction of the effects of linear absorption. For instance, increases in pump intensity can be used in place of increased density or interaction length to increase conversion efficiency. If sufficient pump intensity is available, and can be used, to reduce the pump depletion length below the absorption length, the effects of linear absorption can be made negligible. However, in many situations the required pump intensities are so high that one or more competing nonlinear effects become important and an absolute limit on the conversion efficiency is encountered.

4.1.2. *Nonlinear Absorption*

The effects of nonlinear absorption can be illustrated by considering the effects of two-photon absorption on third-harmonic generation. The plane-wave equations for third-harmonic generation at perfect phase matching in the presence of two-photon absorption are

$$\frac{dA_3}{dz} = i\kappa_3 A_1^3 - \beta_{13}|A_1|^2 A_3 - \beta_{33}|A_3|^2 A_3, \tag{22a}$$

$$\frac{dA_1}{dz} = i\kappa_3 A_3 A_1^{*2} - \beta_{11}|A_1|^2 A_1 - \beta_{31}|A_3|^2 A_1, \tag{22b}$$

where κ_3 is given by Eq. (14) and

$$\beta_{13} = \frac{18\pi^2}{n_3\lambda_1}\chi''(-\omega_3, \omega_1, -\omega_1, \omega_3),$$

$$\beta_{33} = \frac{9\pi^2}{n_3\lambda_1}\chi''(-\omega_3, \omega_3, -\omega_3, \omega_3),$$

$$\beta_{11} = \frac{3\pi^2}{n_1\lambda_1}\chi''(-\omega_1, \omega_1, -\omega_1, \omega_1),$$

$$\beta_{31} = \frac{6\pi^2}{n_1\lambda_1}\chi''(-\omega_1, \omega_3, -\omega_3, \omega_1).$$

The nonlinear absorption terms in the equation for the harmonic field are due to absorption induced by the fundamental intensity and the harmonic intensity, respectively. The first of these terms will usually be the more important unless there is significant conversion of the pump to the harmonic. The nonlinear absorption terms in the equation for the fundamental field likewise are due to absorption induced by the fundamental and harmonic intensities, respectively. The second of these terms can be positive or negative depending on whether the nonlinear absorption induced by the pump at the harmonic involves simultaneous absorption of the pump (sum-frequency absorption) or amplification of the pump (Raman processes).

The two-photon absorption processes considered here are assumed to arise from processes that do not involve resonant enhancement of the harmonic susceptibility. As a result we assume that the various absorption terms in Eqs. (22a) and (22b) are unrelated. Nonlinear absorption in resonantly enhanced conversion processes will be discussed later.

If we assume that there is little pump depletion in the harmonic process, then the last term in both equations and the first term in the second equation can be neglected. The field amplitudes are then given by

$$A_1 = \frac{A_{10}}{1 + \beta_{11}|A_{10}|^2 z}, \tag{23a}$$

$$A_3 = \frac{i\kappa_3 A_{10}}{\beta_{11}(\frac{1}{2} - \beta_{13}/\beta_{11})}\left[\frac{1}{(1 + \beta_{11}|A_{10}|^2 z)^{\beta_{13}/\beta_{11}}} - \frac{1}{(1 + \beta_{11}|A_{10}|^2 z)^{1/2}}\right], \tag{23b}$$

where A_{10} is the initial field amplitude of the pump wave. We can examine the behavior of this solution in several limits.

First let us assume that there is significant absorption only at the fundamental. Setting $\beta_{13} = 0$ we have

$$A_3 = i\frac{2\chi_3 A_{10}}{\beta_{11}}\left[1 - \frac{1}{(1 + \beta_{11}|A_{10}|^2 z)^{1/2}}\right]. \tag{24}$$

In the limit of large values of the product $\beta_{11}|A_{10}|^2 z$ the harmonic conversion efficiency approaches a constant value given by

$$\lim_{\beta_{11}|A_{10}|^2 z \to \infty} \eta_3 = I_3/I_1 = 4\kappa_3^2/\beta_{11}^2. \tag{25}$$

In some cases the nonlinear absorption can be expected to occur primarily at the harmonic wavelength. An example of such a situation is if multiphoton ionization involving one pump and one harmonic photon can occur but there no resonances that can contribute to multiphoton absorption at the fundamental. The harmonic field amplitude is then obtained by taking the limit of Eq. (23b) as $\beta_{11} \to 0$:

$$A_3 = i\frac{\kappa_3 A_{10}}{\beta_{13}}(1 - e^{-\beta_{13}|A_{10}|^2 z}). \tag{26}$$

In the limit of large nonlinear absorption the conversion efficiency again approaches a constant value

$$\lim_{\beta_{13}|A_{10}|^2 z \to \infty} \eta = I_3/I_1 = \kappa_3^2/\beta_{13}^2. \tag{27}$$

In each of these situations the limiting conversion efficiency is determined by the ratio of the nonlinear harmonic susceptibility to the appropriate nonlinear absorption coefficient but is independent of the interaction length and the initial pump intensity. This behavior can be contrasted to that experienced in linear absorption in which the limiting efficiency could be affected by choice of initial pump intensity.

In gases both the nonlinear susceptibility and the nonlinear absorption coefficient are proportional to the pressure. As a result the limiting conversion efficiencies in (25) or (27) are independent of the gas pressure. In some situations both coefficients are dominated by common terms such as detunings from resonances. Under such conditions the limiting efficiencies will also be independent of such detunings.

Attenuation through two-photon absorption will not impose an absolute limit on the conversion efficiency of the higher harmonics, but will weaken the power law dependence of the harmonic intensity on the pump intensity.

The assumption that was made at the beginning of this section that absorption of the pump or generated wave would occur before the pump wave is depleted in the harmonic conversion process can be expressed, in the case of pump absorption, as

$$L_{\mathrm{abs}} \ll L_{\mathrm{dep}} \tag{28}$$

or

$$\kappa_3/\beta_{11} \ll 1. \tag{29}$$

Comparison of Eqs. (25) and (29) shows that when this condition is satisfied, the conversion efficiency will be limited to a value considerably less than unity.

The second assumption involved the requirement that absorption of the pump occur before significant population changes could take place. This assumption places a restriction on the duration of a pulse of a given intensity. If we assume that the ground state is not repopulated during the time of the pump pulse by spontaneous radiation, by recombination, or by other processes, then the time required for the ground-state population to be reduced to half its initial value due to pump absorption is

$$t_{\text{depl}} = \frac{0.69cn\hbar\omega N_0}{8\pi\beta_{11}I_{10}^2} = \frac{1.4\hbar\omega L_{\text{abs}}N_0}{I_{10}}, \tag{30}$$

where $L_{\text{abs}} = (16\pi\beta_{11}I_{10}/cn)^{-1}$ is the initial two-photon absorption length. In order for the results of this section to be applicable the pulse duration must be shorter than this value:

$$t_{\text{pulse}} \ll t_{\text{depl}}. \tag{31}$$

This condition can also be expressed in terms of the required interaction length. In order for the results of this section to be applicable, the length of the nonlinear medium must reach an absorption length for the fundamental before significant ground-state depletion can occur. That is,

$$L_{\text{med}} > \frac{W_p t_{\text{depl}}}{1.4\hbar\omega N_0 t_p}, \tag{32}$$

where W_p is the initial energy density of the pump pulse (here assumed to be constant in time and of duration t_p) while at the same time

$$t_p < t_{\text{depl}}. \tag{33}$$

Condition (32) becomes easier to satisfy as the gas pressure increases and more molecules or atoms become available to absorb the pump radiation. It is also in general more easily met in liquids than in gases.

4.2. Limitation by Breaking of Phase Matching

The nonlinear refractive index can affect the frequency conversion process by introducing intensity-dependent phase shifts that alter phase mismatches and also by causing self-focusing or defocusing of the pump or generated beams. Because the frequency conversion process is so very sensitive to the phase-matching conditions, it can usually be expected that changes of phase matching due to the intensity-dependent phase shifts will be the more important effect. Most of the discussions in the literature have involved this aspect of the effects of the nonlinear index and is the one that we shall discuss here primarily.

Depending on the circumstances that are involved, the intensity-dependent phase shifts can either increase conversion efficiency by bringing the net phase mismatch closer to an optimum value or decrease conversion efficiency by moving the net phase mismatch away from the optimum value. Increases in the conversion efficiency can be expected only for a limited range of experimental parameters and ultimately the intensity-dependent phase shifts can be expected to limit the conversion efficiency.

The effects of intensity-dependent phase changes have been observed in a number of experiments involving both third- and fifth-order frequency conversion. The most important contributions to the intensity-dependent phase changes usually come from the third-order nonlinear refractive indices. The effects of intensity-dependent phase changes resulting from the Kerr effect and Stark shifts have been discussed theoretically and experimentally by several authors (Miles and Harris,[33] Puell *et al.*,[5] Zych and Young,[34] Reintjes *et al.*,[35] Stappaerts,[36] Puell and Vidal,[37] and Kildal and Brueck[38]). Puell *et al.*[5] have also discussed the effects of higher-order contributions to the nonlinear index. Intensity-dependent phase changes in resonant systems due to population changes will be treated in Sections 4.3.2 and 4.3.4.

The nonlinear index arises from terms in the nonlinear polarization that have the same frequency as the pump or generated waves. For third-harmonic generation there are in general four components of the nonlinear polarization that lead to an intensity-dependent refractive index. When the polarization responds fast enough to follow the pump pulse these components can be written as

$$P^{\mathrm{NL}}(\omega) = (3/4)\chi^{(3)}(-\omega, \omega, -\omega, \omega)A_1 A_1^* A_1 + (3/2)\chi^{(3)}(-\omega, 3\omega, -3\omega, \omega)A_3 A_3^* A_1, \tag{34a}$$

$$P^{\mathrm{NL}}(3\omega) = (3/2)\chi^{(3)}(-3\omega, \omega, -\omega, 3\omega,)A_1 A_1^* A_3 + (3/4)\chi^{(3)}(-3\omega, 3\omega, -3\omega, 3\omega)A_3 A_3^* A_3. \tag{34b}$$

The first terms in (34a) and (34b) describe a nonlinear polarization at the fundamental and harmonic frequencies, respectively, that is caused by the intensity of the fundamental wave. These terms generally make the dominant contribution to the nonlinear phase shifts. The second terms in these equations describe a nonlinear polarization at the fundamental and harmonic frequencies, respectively, that is caused by the intensity of the harmonic wave. In nonresonant interactions these terms will be significant only in situations in which there is substantial depletion of the pump wave in the harmonic conversion process.

The intensity-dependent phase shifts arise from the real parts of the susceptibilities in (34). In terms of the susceptibilities the nonlinear index changes and the contributions to the wave vector mismatch are

$$n(\omega) = n_0(\omega) + n_{21}(\omega)\langle A_1\rangle^2 + n_{23}(\omega)\langle A_3\rangle^2 = n_0(\omega) + \tfrac{1}{2}n_{21}(\omega)|A_1|^2 + \tfrac{1}{2}n_{23}(\omega)|A_3|^2, \tag{35}$$

where

$$n_{21}(\omega) = \frac{3\pi}{n_0(\omega)} \chi^{(3)'}(-\omega, \omega, -\omega, \omega),$$

$$n_{23}(\omega) = \frac{6\pi}{n_0(\omega)} \chi^{(3)'}(-\omega, 3\omega, -3\omega, \omega),$$

$$\begin{aligned} n(3\omega) &= n_0(3\omega) + n_{23}(3\omega)\langle A_3 \rangle^2 + n_{21}(3\omega)\langle A_1 \rangle^2 \\ &= n_0(3\omega) + \tfrac{1}{2}n_{23}(3\omega)|A_3|^2 + \tfrac{1}{2}n_{21}(3\omega)|A_1|^2, \end{aligned} \tag{36}$$

where

$$n_{23}(3\omega) = \frac{3\pi}{n_0(3\omega)} \chi^{(3)'}(-3\omega, 3\omega, -3\omega, 3\omega),$$

$$n_{21}(3\omega) = \frac{6\pi}{n_0(3\omega)} \chi^{(3)'}(-3\omega, \omega, -\omega, 3\omega),$$

and

$$\begin{aligned} \Delta k = \Delta k^{\mathrm{L}} + \Delta k^{\mathrm{NL}} &= \Delta k^{\mathrm{L}} + (k_{21}(3\omega) - 3k_{21}(\omega))|A_1|^2 \\ &+ (k_{23}(3\omega) - 3k_{23}(\omega))|A_3|^2, \end{aligned} \tag{37}$$

where

$$k_{21}(3\omega) = \frac{18\pi^2 \chi^{(3)'}(-3\omega, \omega, -\omega, 3\omega)}{\lambda_1 n_0(3\omega)},$$

$$k_{21}(\omega) = \frac{3\pi^2 \chi^{(3)'}(-\omega, \omega, -\omega, \omega, \omega)}{\lambda_1 n_0(\omega)},$$

$$k_{23}(3\omega) = \frac{9\pi^2 \chi^{(3)'}(-3\omega, 3\omega, -3\omega, 3\omega)}{\lambda_1 n_0(3\omega)},$$

$$k_{23}(\omega) = \frac{6\pi^2 \chi^{(3)'}(-\omega, 3\omega, -3\omega, \omega)}{\lambda_1 n_0(\omega)}.$$

All materials exhibit a nonlinear refractive index at sufficiently high intensities. The nonlinear index of most liquids at visible wavelengths is positive. In gases, and in molecular liquids at infrared wavelengths, the nonlinear index can be positive or negative, depending on the relation of the pump and generated frequencies to any resonances that may be present. In materials with discrete energy levels the third-order nonlinear indexes at the fundamental and harmonic wavelengths due to the pump intensity are given by

$$n_2(\omega, \omega) = n_{21}(\omega) = \frac{N\pi}{n_1 \hbar^3}\left\{\sum_{b\neq 0}\left[\frac{2|P_b^-|^2}{(\omega_{b0} - 2\omega)} + \frac{2|P_b^+|^2}{(\omega_{b0} + 2\omega)} + \frac{2|Q_b|^2}{\omega_{b0}}\right]\right.$$
$$\left. - 2Q_0 R_a + \frac{|P_0^+|^2}{\omega} - \frac{|P_0^-|^2}{\omega}\right\}, \tag{38a}$$

$$n_2(3\omega, \omega) = n_{21}(3\omega) = \frac{N\pi}{n_3 \hbar^3}\left\{\sum_{b\neq 0}\left[\frac{|T_b^-|^2}{\omega_{b0} - 2\omega} + \frac{T_b^- U_b^-}{\omega_{b0} - 4\omega} + \frac{|T_b^+|^2}{\omega_b + 2\omega}\right.\right.$$
$$\left. + \frac{T_b^+ U_b^+}{\omega_{b0} + 4\omega} + \frac{W_b Q_b}{\omega_{b0}}\right]$$
$$\left. - 2W_0 R_a - \frac{|T_0^-|^2}{2\omega} - \frac{T_0^- U_0^-}{4\omega} + \frac{|T_0^+|^2}{2\omega}\right\},$$
$$+ T_0^+ U_0^+/4\omega, \tag{38b}$$

where

$$P_b^\pm = \sum_a \frac{\mu_{0a}\mu_{ab}}{\omega_{a0} \pm \omega},$$
$$Q_b = \sum_a \mu_{0a}\mu_{ab}\left(\frac{1}{\omega_{a0} + \omega} + \frac{1}{\omega_{a0} - \omega}\right),$$
$$R_a = \sum_a |\mu_{0a}|^2 \left(\frac{1}{(\omega_{a0} + \omega)^2} + \frac{1}{(\omega_{a0} - \omega)^2}\right),$$
$$T_b^\pm = \sum_a \mu_{0a}\mu_{ab}\left(\frac{1}{\omega_{a0} \pm 3\omega} + \frac{1}{\omega_{a0} \mp \omega}\right),$$
$$U_b^\pm = \sum_a \mu_{0a}\mu_{ab}\left(\frac{1}{\omega_{a0} \pm 3\omega} + \frac{1}{\omega_{a0} \pm \omega}\right),$$
$$W_b = \sum_a \mu_{0a}\mu_{ab}\left(\frac{1}{\omega_{a0} + 3\omega} + \frac{1}{\omega_{a0} - 3\omega}\right).$$

Population changes of resonant even-photon levels and incoherent population changes of resonant single photon levels were neglected in deriving Eqs. (38a) and (38b) and must be added separately when they are important. Typical values at selected wavelengths of the coefficient of the nonlinear index n_2 for some liquids and values of n_2/N for some gases are given in Table 5.

A measure of the importance of the nonlinear index in limiting the frequency conversion process is given by the intensity-dependent phase shift

TABLE 5

Nonlinear Susceptibility Responsible for Nonlinear Refractive Index for Selected Materials

Material	Wavelength (nm)	Tensor component	Value (esu)	Reference
CS_2	1060	$\chi^{(3)}_{1111}(-\omega, \omega, -\omega, \omega)$	1.8×10^{-12}	38a, 38b
MBBA	694.3	$\chi^{(3)}_{1111}(-\omega, \omega, -\omega, \omega)$	1.96×10^{-9}	38c
Cs	1064	$\chi^{(3)}_{1111}(-\omega, \omega, -\omega, \omega)/N$	-2.74×10^{-30}	38d
Rb	1060	$\chi^{(3)}_{1111}(-\omega, \omega, -\omega, \omega)/N$	8×10^{-32}	5
	354.7	$\chi^{(3)}_{1111}(-\omega, \omega, -\omega, \omega)/N$	1.19×10^{-34}	5
	1060	$\chi^{(3)}_{1111}(-3\omega, \omega, -\omega, 3\omega)/N$	-1.66×10^{-34}	5
Mg	431	$\chi^{(3)}_{1111}(-\omega, \omega, -\omega, \omega)/N$	9.8×10^{-31} [a]	60
		$\chi^{(3)}_{1111}(-\omega, \omega, -\omega, \omega)/N$	2.56×10^{-34} [b]	60
		$\chi^{(3)}_{1111}(-3\omega, \omega, -\omega, 3\omega)/N$	9.1×10^{-34} [a]	60
		$\chi^{(3)}_{1111}(-3\omega, \omega, -\omega, 3\omega)/N$	6.2×10^{-36} [b]	60
Xe	354.7 nm	$[\chi^{(3)}_{1111}(-3\omega, \omega, -\omega, 3\omega) - \frac{1}{2}\chi^{(3)}_{1111}(-\omega, \omega, -\omega, \omega)]/N$	-8×10^{-35}	34

[a] Two-photon resonant contribution from the $3s^2$–$3s3d\,^1D$ transition with linewidth of 1 cm^{-1}.

[b] Nonresonant contribution.

that is accumulated in the interaction region. For the third-harmonic process this nonlinear phase shift is

$$\Delta\phi^{NL} = \frac{3\pi}{\lambda_1}\left\{(n_{21}(3\omega) - n_{21}(\omega)) \int |A_1(z)|^2\, dz \right.$$
$$\left. + (n_{23}(3\omega) - n_{23}(\omega)) \int |A_3(z)|^2\, dz \right\}. \tag{39}$$

As long as the nonlinear phase shift is small, the effects of the nonlinear index will be negligible. When the nonlinear phase shift becomes the order of π rad, the nonlinear indexes can affect the direction of power flow in a frequency conversion process that is phase matched at low intensities. Intensities at which the nonlinear phase shift is π rad for the various processes and materials listed in Table 5 are also shown.

Some of the properties of the nonlinear phase shifts that are important in determining their effect on a given frequency conversion process can be noted. First, the phase shift depends on the integral of the intensity through the interaction region. As a result its behavior is somewhat different from other nonlinear effects which depend only on the peak intensity. This difference can be especially important for focused beams (see Section 4.2.4). In gases the nonlinear phase shift depends on the density. It can therefore be distinguished from effects such as saturation of the susceptibility that do not depend directly on the density.

The equations for third-harmonic generation in the plane-wave approximation in the presence of nonlinear phase shifts are

$$\frac{dA_3}{dz} = i\kappa_3 A_1^3 e^{-i\Delta k^{\mathrm{L}} z} + ik_{21}(3\omega)|A_1|^2 A_3 + ik_{23}(3\omega)|A_3|^2 A_3, \quad (40a)$$

$$\frac{dA_1}{dz} = i\kappa_3 A_3 (A_1^*)^2 e^{i\Delta k^{\mathrm{L}} z} + ik_{21}(\omega)|A_1|^2 A_1 + ik_{23}(\omega)|A_3|^2 A_1, \quad (40b)$$

where κ_3 is given by Eq. (14) with $q = 3$.

Solution of these equations in the general case can require numerical computation. Here we shall discuss the properties of the solutions in specific circumstances. If the conversion to the harmonic is small, then the nonlinear indices of refraction due to the intensity at the harmonic can be neglected. The third-harmonic intensity is then

$$I_3 = \frac{n_3}{n_1} K_3^2 I_1^3 \frac{L^2 \sin^2[(\Delta k^{\mathrm{L}} + \Delta k_{21}|A_1|^2)L/2]}{[(\Delta k^{\mathrm{L}} + \Delta k_{21}|A_1|^2)L/2]^2}, \quad (41)$$

where K_3 is given by Eqs. (16) and (14) with $q = 3$, Δk^{L} is the phase mismatch at low intensities, and $\Delta k_{21} = k_{21}(3\omega) - 3k_{21}(\omega)$.

4.2.1. *Linear Phase Matching*

If the process is chosen to be phase matched at low intensities, i.e., $\Delta k^{\mathrm{L}} = 0$, the third-harmonic intensity is

$$I_3 = K_q^2 I_1^3 L^2 \frac{\sin^2[\Delta k_{21}|A_1|^2 L/2]}{[\Delta k_{21}|A_1|^2 L/2]^2}. \quad (42)$$

If the nonlinear indexes at the fundamental and harmonic wavelengths are equal, $\Delta k_{21} = 0$ and the process will remain phase matched at high intensities even though there may be a significant intensity-dependent phase shift at each wavelength. Under these conditions the nonlinear index will not affect the conversion process at low pump depletion although it may still be important at high levels of pump depletion.

In general, however, one can expect that the nonlinear indexes at the two wavelengths will be different and that the phase mismatch will be a function of pump intensity. As a function of distance the maximum conversion will occur when

$$L = L_{\mathrm{c}}^{\mathrm{NL}} = \frac{\pi}{\Delta k_{21}|A_1|^2}. \quad (43)$$

The quantity $L_{\mathrm{c}}^{\mathrm{NL}}$ can be considered as a nonlinear coherence length. When the optimum condition of (43) is satisfied, the harmonic intensity is

$$I_3 = \frac{4|\kappa_3|^2 I_1}{(\Delta k_{21})^2} = \frac{36\pi^4 |\chi_{\mathrm{h}}^{(3)}|^2 I_{10}}{n_1^2 \lambda_1^2 (\Delta k_{21})^2}. \quad (44)$$

Limitations on efficiency caused by nonlinear phase shifts in systems that are phase matched at low intensities have been discussed by Zych and Young.[34] At an optimum pump intensity determined by

$$I_1 = cn_1/8\Delta k_{21} L$$

the conversion efficiency $\eta = I_3/I_1$ is

$$\eta = \frac{1}{9n_1 n_3}\left|\frac{\chi_a^{(3)}(-3\omega, \omega, \omega, \omega)}{[\chi_a'(-3\omega, \omega, \omega, -3\omega)/n_3] - [\chi_a'(-\omega, \omega, \omega, -\omega)/2n_1]}\right|^2, \tag{45}$$

where the χ_a are the appropriate nonlinear susceptibilities per atom or molecule for the various nonlinear processes. The expression for the limiting efficiency in Eq. (45) has been given by Zych and Young in connection with third-harmonic conversion from 354.7 to 118.2 nm in Xe. As in the case of limitation by two-photon absorption, the limiting efficiency is independent of the pump intensity and interaction length and depends only on the constants of the medium. Furthermore, in gaseous media, it is independent of the density.

The importance of the nonlinear index in limiting the conversion efficiency in systems that are linearly phase matched is determined by the ratio of the nonlinear coherence length of Eq. (43) to the harmonic depletion lengths of Eqs. (3-10a)–(3-10d). For third-harmonic generation this ratio is given by

$$\frac{L_c^{NL}}{L_{depl}} = \frac{3\pi^3\chi_h^{(3)}}{n_3\lambda_1\Delta k_{21}}$$

$$= \frac{\pi}{6n_3}\frac{\chi_a(-3\omega, \omega, \omega, \omega)}{[\chi_a'(-3\omega, \omega, -\omega, 3\omega)/n_3] - [\chi_a'(-\omega, \omega, -\omega, \omega)/2n_1]}. \tag{46}$$

This ratio is independent of the pump power and depends only on the constants of the medium. The maximum conversion efficiency of Eq. (45) can be written in terms of this ratio as

$$\eta_{max} = \frac{4n_3}{n_1}\left(\frac{L_c^{NL}}{L_{depl}}\right)^2. \tag{47}$$

If the ratio of lengths in Eq. (46) is less than unity, then the intensity-dependent phase shifts associated with the nonlinear indices will limit the conversion efficiency to a value less than unity. If the ratio of lengths in Eq. (46) is greater than unity, then it is possible for significant conversion to the harmonic to take place. However, even under these conditions, nonlinear phase shifts can still prevent complete conversion to the harmonic if the nonlinear indices induced by the harmonic wave are different from those induced by the pump wave. In addition, conversion can be limited by other nonlinear mechanisms.

If the pump beam has a temporal or spatial profile, then additional effects can occur. For example, even though the maximum conversion efficiency of Eq. (45) is independent of the pump intensity, the optimum phase-matching condition does depend on intensity and cannot be maintained for all parts of the beam at all times. As a result the maximum energy conversion efficiency integrated over the beam in space and time will be less than that predicted by the constant intensity expression in (45). In addition, spatial and temporal modulation can develop if the maximum nonlinear phase shift is greater than π rad, leading to a loss of spatial and temporal coherence, increased beam divergence, and spectral broadening. These distortions are characteristic of the nonlinear index and are not introduced by saturation of the susceptibility or optical energy losses through linear or nonlinear absorption.

4.2.2. Linear Phase Mismatch

The limitations that are imposed on the conversion efficiency by the nonlinear phase shifts can be altered if the linear phase mismatch Δk^{L} is chosen to have a nonzero value such that optimum phase-matching conditions are met at a nonzero pump intensity. For the idealized assumptions of uniform pump intensity and negligible pump depletion of the preceding section, the optimum value of linear phase mismatch is given by

$$\Delta k^{\mathrm{L}} = -\Delta k_{21} |A_1|^2 \tag{48}$$

and the optimum conversion efficiency has the same form as the ideal formulas of Chapter 2:

$$\eta_{\mathrm{opt}} = \frac{n_3}{n_1} K_q^2 I_1^2 L^2. \tag{49}$$

Optimum conversion efficiency in this case is determined only by the harmonic conversion susceptibility and the length of the nonlinear medium.

The system just described can be optimized only for a single value of pump intensity. Any effect that changes the pump intensity, such as linear or nonlinear absorption or significant pump depletion, will prevent the optimum condition of (48) from being satisfied at all positions in the nonlinear medium. In addition, the idealized efficiency of (49) cannot be achieved at all parts of a pump beam that has a spatial or temporal profile.

As a function of pump intensity the harmonic intensity can grow faster than cubic for pump intensities below the optimum value $|A_1|^2_{\mathrm{opt}} = |\Delta k^{\mathrm{L}}/\Delta k_{21}|$, while it will vary more slowly than cubic for intensities above the optimum value. This type of behavior has been discussed by Marburger[39] in connection with second-harmonic generation in crystals.

When the pump beam has a temporal or spatial profile, the optimum phase-matching condition of (48) cannot be met for all parts of the beam at all times and an average linear phase mismatch must be used to maximize the total converted energy. If the maximum nonlinear phase shift that must be compensated at the peak of the fundamental pulse is too large, i.e., several π, the distortions that were discussed in the last section are almost sure to develop. Nevertheless use of a linear phase mismatch can result in improved total conversion efficiency. Calculations of such improvements have been made by Kildal and Brueck[38] for general two-photon resonant systems and by Puell *et al.*[40] for two-photon resonant conversion to the vacuum ultraviolet in Sr : Xe systems. These will be described in Section 4.3 in the discussion of two-photon resonant conversion.

If the linear phase mismatch is chosen to be non optimum, the optimum phase-matching ratios for component mixtures and the optimum focusing conditions for focused beams will be functions of pump intensity. This property is a unique feature of the nonlinear index and can be used to identify the presence of limitations due to nonlinear phase shifts.

4.2.3. *Pump Depletion*

We have seen in the previous section that compensation for nonlinear phase mismatches can lead to predictions of unlimited conversion efficiency for uniform beams of constant intensity provided that pump depletion is neglected. When significant pump depletion occurs, the nonlinear indices induced by the harmonic intensity must be accounted for. Apart from the unlikely event that the intensity-dependent phase mismatch induced by the harmonic is the same as that induced by the fundamental, the condition (48) will no longer be optimum at all distances into the nonlinear medium and full compensation cannot be made even for beams of uniform cross section. The effects of nonlinear phase mismatches in the presence of pump depletion were treated first by Armstrong *et al.*[41] and later by Puell and Vidal.[37] Puell and Vidal have calculated optimum conditions for conversion in terms of a nonlinear parameter β which, according to our definition of the nonlinear susceptibilities, is defined as

$$\beta = \frac{3}{4}\sqrt{\frac{n_1}{n_3}}\left\{\frac{(n_3/n_1)\chi^{(3)}(-\omega, \omega, -\omega, \omega) + (n_1/n_3)\chi^{(3)}(-3\omega, 3\omega, -3\omega, 3\omega) - 4\chi^{(3)}(-3\omega, \omega, -\omega, 3\omega)}{|\chi^{(3)}(-3\omega, \omega, \omega, \omega)|}\right\}. \tag{50}$$

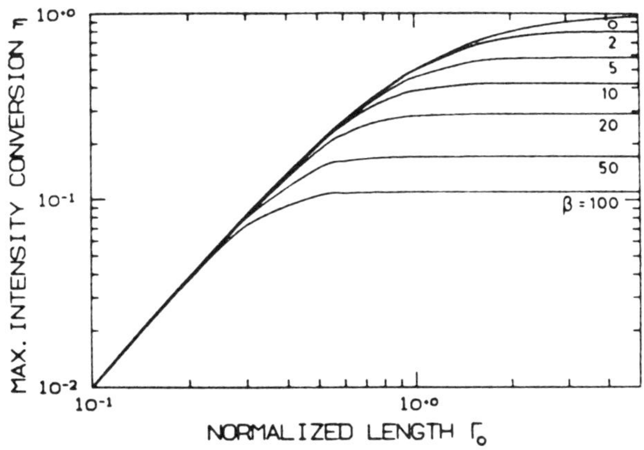

Fig. 4. Calculated third-harmonic conversion efficiency as a function of normalized propagation distance for different values of a nonlinear parameter β [see Eq. (50)] which describes the effects of the nonlinear refractive indices. (From Ref. 37, © 1978 IEEE.)

Their calculations for the maximum intensity conversion efficiency η as a function of the normalized interaction length Γ_0, which is defined in our notation as

$$\Gamma_0 = \frac{1}{2}\int_0^{\eta} \frac{d\eta}{\sqrt{\eta[(1-\eta)^3 - \eta(\alpha - \beta\eta)^2]}},$$

where α is a normalized phase mismatch given by

$$\alpha = \frac{1}{2}\left[\frac{\Delta k^{\mathrm{L}} L}{\Gamma_0} + \frac{[3\sqrt{n_3/n_1}\chi^{(3)}(-\omega, \omega, -\omega, \omega) - \sqrt{n_1/n_3}\chi^{(3)}(-3\omega, \omega, -\omega, 3\omega)]}{|\chi^{(3)}(-3\omega, \omega, \omega, \omega)|}\right],$$

are shown in Fig. 4 for various values of the parameter β. As originally pointed out by Armstrong *et al.*,[41] full conversion can occur only if $\beta = 0$. For nonzero values of β the maximum conversion efficiency is reduced.

4.2.4. *Focused Beams*

The effects of the nonlinear index on the conversion efficiency of focused beams are qualitatively similar to those on the conversion of plane waves, although evaluation of the detailed behavior is more difficult. A measure of the importance of the nonlinear index can be obtained by calculating the total on-axis phase shift caused by the nonlinear index. For a Gaussian beam in an infinite medium the on-axis phase shift is independent of the confocal parameter of the beam and depends only on the total power in the pump wave rather than on the focused intensity:

$$\Delta\phi^{\mathrm{NL}}(r = 0) = \Delta k_{21}|A_{10}|^2(b/2)\int_{-\infty}^{\infty}\frac{d\xi}{1+\xi^2}$$

$$= \frac{16\pi^2\Delta k_{21}}{cn_1\lambda_1}\mathscr{P}_1, \quad (51)$$

where $\mathscr{P}_1$ is the total pump power.

We recall from Chapter 2 that the third-harmonic power generated in a tightly focused beam is also dependent only on the total pump power. Thus the importance of the third-order nonlinear index in limiting third-harmonic conversion in focused beams cannot be changed by choosing a relatively weak focus to minimize the peak intensity in the focus or by choosing a tight focus to minimize the interaction length. However, its importance relative to other nonlinear effects that depend only on the intensity in the focus can be affected by choice of the confocal parameter.

Under some conditions the intensity-dependent phase changes can allow conversion in focused beams in a medium with positive dispersion by preventing complete cancellation of the radiation generated in the two halves of the focus. This effect was discussed by Anikin *et al.*[28] It can enlarge the selection of media that can be used for harmonic conversion.

4.2.5. *Limitations in Gases*

The nonlinear index has proved to be very important in limiting the conversion efficiency in many gases. Therefore let us discuss some of the behavior of the nonlinear index that is characteristic of gases with discrete energy levels. We consider the situation illustrated in Fig. 5. This would be

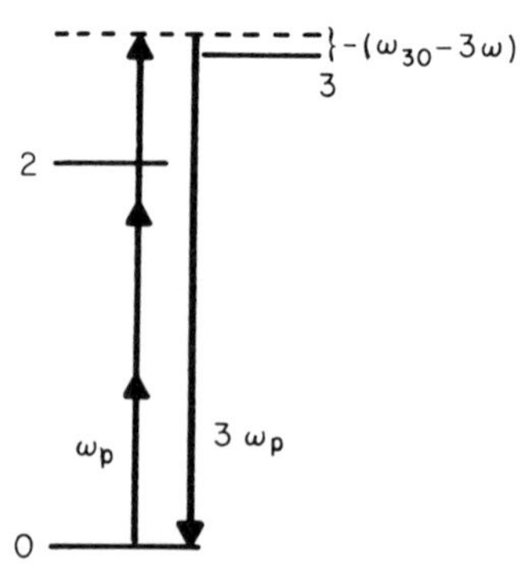

Fig. 5. Level diagram for third-harmonic generation pertinent to discussion of effects of nonlinear phase shift in gases.

typical of third-harmonic conversion in a negatively dispersive medium with no two-photon resonant enhancement, e.g., frequency tripling of radiation from an Nd laser in Rb vapor or tripling of near-UV radiation in rare gases. Two-photon resonant conversion will be discussed in the next section.

We have already pointed out that the nonlinear phase shift depends on the density of the gas [see Eq. (39)], making the effects that are due to the nonlinear indices differ from those that arise from other processes such as saturation of the susceptibility. We have also pointed out that the limiting efficiency of Eq. (45) is independent of the gas density. We can now consider effects that are characteristic of the detuning from the resonant levels. If the harmonic wavelength lies sufficiently close to the upper level in Fig. 5, then the nonlinear index at the harmonic wavelength will be larger than that at the fundamental and the intensity-dependent contribution to the wave vector mismatch will be dominated by the detuning from the upper level:

$$\Delta k^{(2)} = \Delta k_{23} - \Delta k_{21} \approx \frac{3\pi^2 N|\mu_{03}|^2}{n_3\lambda_1\hbar^3(\omega_{30} - 3\omega)^2} \times \sum_b |\mu_{3b}|^2\left(\frac{1}{\omega_{b0} - 2\omega} + \frac{1}{\omega_{b0} - 4\omega}\right). \tag{52}$$

If we assume that the third-harmonic susceptibility is also dominated by the detuning from the upper level, then the maximum conversion efficiency as determined by the intensity-dependent phase changes is

$$\eta = \frac{4}{n_1 n_3}(\omega_{30} - 3\omega)^2 \times \left|\frac{\sum_{a,b}\{\mu_{0a}\mu_{ab}\mu_{b3}/[(\omega_{b0} - 2\omega)(\omega_{a0} - \omega)]\}}{\mu_{30}\sum_b|\mu_{b3}|^2\{[1/(\omega_{b0} - 2\omega)] + [1/(\omega_{b0} - 4\omega)]\}}\right|^2. \tag{53}$$

This value increases with detuning from the upper level, but is independent of both the density and length of the medium.

4.2.6. *Higher-Order Harmonics*

The limitations on the efficiency of conversion to the higher-order harmonics by the third-order nonlinear index are not as severe as those on the third-harmonic efficiency. An analysis of plane-wave low-level qth-harmonic conversion similar to that which led to (41) gives for the qth-harmonic conversion efficiency

$$I_q = \frac{(2\pi)^{q+3}q^2|\chi^{(q)}|^2 I_1^q L^2}{n_q n_1^q \lambda_1^2 c^{q-1}} \frac{\sin^2[\Delta k^{\mathrm{L}} + \Delta k_{21}|A_1|^2 L/2]}{[\Delta k^{\mathrm{L}} + \Delta k_{21}|A_1|^2 L/2]^2}. \tag{54}$$

In the limit of high pump intensities the efficiency approaches the value

$$\lim_{I_1 \to \infty} \eta = \frac{I_q}{I_1} = \frac{(2\pi)^{q-3} I_1^{q-3}}{9 n_q n_1^{q-2} c^{q-3}} \times \left| \frac{\chi^{(q)}(-q\omega, \omega)}{[\chi^{(3)'}(-q\omega, \omega, -\omega, q\omega)/n_3] - [\chi^{(3)}(-\omega, \omega, -\omega, -\omega)/2n_1]} \right|^2 \tag{55}$$

The asymptotic efficiency is now not limited to a constant value but varies as the $q-3$ power of the pump intensity.

Similar comments apply to the conversion to higher-order harmonics in focused beams. Again the on-axis phase changes, similar to that in Eq. (51), give a measure of the importance of the effect. However, since the optimum conversion efficiency to the higher harmonics in focused beams increases as the beam is focused more tightly, while the limitation imposed by the third-order nonlinear index is independent of the confocal parameter, conversion to the higher-order harmonics can be favored relative to the limitations imposed by the third-order nonlinear index by tighter focusing.

4.3. Limitations in Resonantly Enhanced Frequency Conversion

It was pointed out in Chapter 2 that resonantly enhanced interactions have an important place in optical frequency conversion because they usually provide the largest nonlinear susceptibilities. They are also more strongly affected by the competing processes than are the nonresonant interactions. Different aspects of the saturation of resonantly enhanced processes have been discussed by several authors (Wang and Davis,[42] Ward and Smith,[43] Stappaerts,[27,36] Anikin *et al.*,[28,44] Harris,[45] Harris and Bloom,[46] Georges *et al.*,[47] Kildal and Brueck,[38] Scheingraber *et al.*,[31] Puell *et al.*,[40] Scheingraber and Vidal,[48] Elgin and New,[29] Elgin *et al.*,[49] New,[50] Grischowsky *et al.*,[51] Diels and Georges,[52] Diels,[53] and Poluektov and Nazarkin[54]), some of which will be treated in this section. Effects related to the saturation of the susceptibility are included under this heading because they are often, although not always, most prominent in resonant interactions. Saturation of resonant interactions usually involves more than one competing linear or nonlinear process, and so some of the effects that were discussed separately in the preceding sections will be discussed again in this section.

4.3.1. *Two-Photon Resonances*

Two-photon resonant interactions can be divided roughly into two classes, depending on whether significant population redistribution occurs. If no significant population changes are present, then saturation of the conversion can be caused primarily by the combined actions of two-photon absorption of the fundamental, Raman amplification of the fundamental with accompanying loss at the harmonic and intensity-dependent phase shifts associated with resonant or near-resonant Kerr effects. In some situations, e.g., transient interactions, Stark shifts can cause saturation of the susceptibility in the absence of significant population changes.

At exact two-photon resonance the resonant Kerr contributions to the intensity-dependent phase shifts are zero and the saturation of the conversion efficiency is dominated by the loss mechanisms of two-photon absorption and Raman scattering. At large detunings from the two-photon resonance the loss mechanisms become negligible and the limitations on conversion efficiency are dominated by the nonlinear phase shifts. At intermediate detunings, of the order of a few linewidths, both effects are important.

Stimulated Raman scattering that is induced by the harmonic and generates a Stokes wave at the difference frequency between the harmonic and the two-photon resonant level can also occur. At exact two-photon resonance it is identical to the Raman amplification of the pump, while at finite detunings it represents an additional loss mechanism for the harmonic. This last effect will not be treated in detail here, but the reader should keep in mind that this loss path can be significant if the stimulated Raman gain is high enough.

Under certain conditions the conversion in media at high density can be limited by nonlinear phase changes associated with population redistribution and Stark shifts, rather than absorptive energy losses and Raman scattering, even when the fractional population changes are small. The relative importance of these two types of effects will be discussed in Section 4.3.5.

Redistribution of the population can make added contributions to the nonlinear phase shifts and can cause saturation of the susceptibility as well as saturation of the effects discussed in the preceding paragraphs. In low-density gases frequency conversion is usually done using relatively high pump intensities. As a result, limitations on conversion efficiency due to saturation of the susceptibility generally outweigh the limitations caused by intensity-dependent phase changes or the multiphoton absorption processes. In these situations the effects of Stark shifts and power broadening are usually tightly coupled to those of population changes and all of these effects must be treated simultaneously. In some circumstances, for example, when short pulses are used, this coupling is relaxed and the Stark shift alone can affect the conversion efficiency of resonant processes.

In high-density gases and liquids, conversion can be done at lower intensity levels. For these conditions limitations on conversion efficiency due to multiphoton absorption and breaking of phase matching can occur without

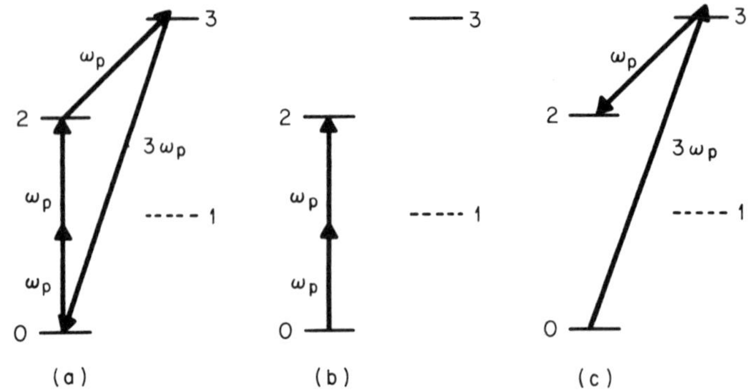

Fig. 6. (a) Level diagram for two-photon resonant third-harmonic generation. Competing processes of two-photon absorption (b) and Raman amplification of the pump (c) are also shown.

the significant population changes that are required for saturation of the nonlinear susceptibilities.

In this section we shall treat first the aspects of two-photon resonantly enhanced third-harmonic conversion that do not involve population changes or Stark shifts. These situations have been discussed by Stappaerts[36] and Kildal and Brueck.[38] We shall then discuss steady-state and transient aspects of two- and three-photon resonantly enhanced conversion when population changes, Stark shifts, and nonlinear phase shifts are involved.

A typical two-photon resonant level structure is shown in Fig. 6 along with the third-harmonic process and the competing processes of two-photon absorption and Raman amplification of the pump. In some systems level 1, one-photon resonant with the pump, is not present. Also, in some systems the detuning from the third level is so small that the system may be regarded as a three-level system to a high degree of accuracy.

The two-photon resonant contribution to the susceptibilities for third-harmonic generation, two-photon absorption at the pump, Raman scattering, and the intensity-dependent wave vectors are

$$\chi_{\text{h}}^{(3)} = \chi^{(3)}(-3\omega, \omega, \omega, \omega) = \frac{N}{\hbar^3(\omega_{20} - 2\omega - i/T_2)}\left(\sum_i \frac{\mu_{0i}\mu_{i2}}{\omega_{i0} - \omega}\right)$$

$$\times \left[\sum_j \mu_{2j}\mu_{j0}\left(\frac{1}{(\omega_{j0} - 3\omega)} + \frac{1}{\omega_{j0} + \omega}\right)\right], \tag{56}$$

$$\chi_{\text{abs}}^{(3)} = i \operatorname{Im} \chi^{(3)}(-\omega, \omega, -\omega, \omega) = i\,\frac{2N}{3\hbar^3}\left[\sum_i \frac{\mu_{0i}\mu_{i2}}{(\omega_{i0} - \omega)}\right]^2$$

$$\times \frac{1/T_2}{(\omega_{20} - 2\omega)^2 + (1/T_2)^2}, \tag{57}$$

$$\chi_{\mathrm{R}}^{(3)} = i\,\mathrm{Im}\,\chi^{(3)}(-3\omega, 3\omega, -\omega, \omega) = i\frac{N}{6\hbar^3}\left[\sum_i \mu_{0i}\mu_{i2} \right.$$
$$\left. \times\left(\frac{1}{\omega_{i0} - 3\omega} + \frac{1}{\omega_{i0} + \omega}\right)\right]^2 \frac{1/T_2}{(\omega_{20} - 2\omega)^2 + (1/T_2)^2}, \tag{58}$$

$$k_{21}(\omega) = \frac{2\pi^2 N}{n_1\lambda_1\hbar^3}\left[\frac{(\omega_{20} - 2\omega)}{(\omega_{20} - 2\omega)^2 + (1/T_2)^2}\right]\left[\sum_i \frac{\mu_{0i}\mu_{i2}}{(\omega_{i0} - \omega)}\right]^2, \tag{59a}$$

$$k_{21}(3\omega) = \frac{3\pi^2 N}{n_3\lambda_1\hbar^3}\left[\frac{\omega_{20} - 2\omega}{(\omega_{20} - 2\omega)^2 + (1/T_2)^2}\right]\left[\sum_i \mu_{0i}\mu_{i2} \right.$$
$$\left. \times\left(\frac{1}{\omega_{i0} - 3\omega} + \frac{1}{\omega_{i0} + \omega}\right)\right]^2, \tag{59b}$$

$$\chi_{\mathrm{Kerr}}(\omega) = \mathrm{Re}\,\chi^{(3)}(-\omega, \omega, -\omega, \omega) = \frac{2N}{3\hbar^3}\left[\sum_i \frac{\mu_{0i}\mu_{i2}}{(\omega_{i0} - \omega)}\right]^2$$
$$\times\left[\frac{\omega_{20} - 2\omega}{(\omega_{20} - 2\omega)^2 + (1/T_2)^2}\right], \tag{60a}$$

$$\chi_{\mathrm{Kerr}}^{(3)}(3\omega) = \mathrm{Re}\,\chi^{(3)}(-3\omega, \omega, -\omega, 3\omega) = \frac{N}{6\hbar^3}\left[\sum_i \mu_{0i}\mu_{i2} \right.$$
$$\left. \times\left(\frac{1}{\omega_{i0} - 3\omega} + \frac{1}{\omega_{i0} + \omega}\right)\right]^2\left[\frac{\omega_{20} - 2\omega}{(\omega_{20} - 2\omega)^2 + (1/T_2)^2}\right]. \tag{60b}$$

4.3.1a. Saturation in High-Density Media. If the nonresonant contributions to the susceptibilities and all effects due to population changes are neglected, then third-harmonic conversion is described by the equations

$$\frac{dA_3}{dz} = i\left(\frac{\kappa'}{\Delta - i\Gamma} A_1^3 e^{-i\Delta kz} + \frac{\kappa'}{\Delta - i\Gamma}\sqrt{R}A_3|A_1|^2\right), \tag{61a}$$

$$\frac{dA_1}{dz} = i\frac{n_3\kappa'}{3n_1}\left(\frac{2}{\Delta - i\Gamma} + \frac{1}{\Delta + i\Gamma}\right)A_3(A_1^*)^2 e^{i\Delta kz}$$
$$+ i\frac{2\kappa' n_3}{3n_1\sqrt{R}(\Delta - i\Gamma)}|A_1|^2 A_1 + i\frac{n_3\kappa'\sqrt{R}}{3n_1(\Delta + i\Gamma)}A_1|A_3|^2, \tag{61b}$$

where

$$\kappa' = \frac{3\pi\omega_1 N}{2n_3 c\hbar^3}\left[\sum_i \mu_{0i}\mu_{i2}\left(\frac{1}{\omega_{i0} - 3\omega} + \frac{1}{\omega_{i0} + \omega}\right)\right]\left[\sum_j \frac{\mu_{0j}\mu_{j2}}{\omega_{j0} - \omega}\right], \tag{62}$$

$$R = \left\{\sum_i \mu_{0i}\mu_{i2}\left(\frac{1}{\omega_{i0} - 3\omega} + \frac{1}{\omega_{i0} + \omega}\right)\Big/\left[\sum_j \frac{\mu_{0j}\mu_{j2}}{(\omega_{j0} - \omega)}\right]\right\}^2, \tag{63}$$

$$\Delta = \omega_{20} - 2\omega, \tag{64}$$

and $\Gamma = 1/T_2$ is the linewidth of the two-photon transition. In order for these equations to describe the harmonic conversion accurately the population changes must remain small. The conditions that must be satisfied to insure this will be discussed in Section 4.3.5.

Exact Resonance. At exact resonance the only competing resonant processes are two-photon absorption and Raman scattering. The equations (61) simplify to

$$\frac{d(A_3/A_{10})}{d\tilde{z}} = -(A_1/A_{10})^3 e^{-i\Delta kz} - \sqrt{R}(A_3/A_{10})|A_1/A_{10}|^2, \tag{65a}$$

$$\begin{aligned}\frac{d(A_1/A_{10})}{d\tilde{z}} &= -(1/3)(A_3/A_{10})(A_1^*/A_{10})^2 e^{i\Delta kz} \\ &\quad + (1/3)\sqrt{R}(A_1/A_{10})|A_3/A_{10}|^2 \\ &\quad - (2/3\sqrt{R})(A_1/A_{10})|A_1/A_{10}|^2,\end{aligned} \tag{65b}$$

where

$$\tilde{z} = \left(\frac{\kappa' A_{10}^2}{\Gamma}\right) z = \frac{8\pi}{cn}\frac{\kappa' I_{10}}{\Gamma} z \tag{66}$$

and A_{10} is the initial amplitude of the pump field.

The qualitative behavior of the solutions to these equations for phase-matched conditions is as follows. The harmonic intensity grows with the normalized distance $\tilde{z}$ until the condition

$$A_3(z)/A_{10} = -\,A_1(z)/\sqrt{R}A_{10} \tag{67}$$

is satisfied. When (67) is satisfied, the z derivatives of both the harmonic and fundamental field amplitudes are zero and both fields propagate without further change in amplitude regardless of the length of the nonlinear medium. The harmonic conversion process is then exactly balanced by the two-photon absorption and Raman processes. The difference in energy between that in the incident fundamental wave and that in the two optical waves when (67) is satisfied appears in the medium as excitation of the two-photon resonance. Note that when condition (67) is satisfied the two terms on the right-hand side of (65a) are of comparable magnitude, as are the three terms on the right-hand side of (65b). As a result, the effects of Raman amplification induced by the harmonic and the parametric back reaction cannot be neglected, even when the conversion efficiency is small.

In many situations the two-photon resonant harmonic process is conveniently characterized by the parameter R as pointed out by Stappaerts.[36] For large values of R the Raman process is stronger than the two-photon absorption and there is little depletion of the pump wave. In the limit of large R, $A_1(z)$ in Eq. (67) can be approximated by $A_1(z) \approx A_{10}$ and the maximum conversion efficiency approaches the value

$$\lim_{R\to\infty} \eta = 1/R, \tag{68}$$

or equivalently

$$\lim_{R\to\infty} \eta = \frac{1}{36}\frac{|\chi_{\mathrm{h}}|^2}{|\chi_{\mathrm{R}}|^2}. \tag{69}$$

For small values of R the two-photon absorption of the fundamental is stronger than the Raman process and the fundamental is substantially attenuated by the two-photon absorption process. Again there will be little harmonic generated and most of the energy ends up in excitation of the material system. For intermediate values of R, Eqs. (65a) and (65b) must be solved simultaneously.

Typical variation of the fundamental and harmonic intensities as a function of $\tilde{z}$ are shown in Fig. 7b for values of R greater than unity and in Fig. 7a for values of R less than unity. For large values of R the efficiency reaches its limiting value in normalized distances less than unity. For $R < 1$ the saturation of the conversion is more gradual and the saturation distance increases with decreasing R. For values of $R \gtrsim 0.01$ the limiting efficiency is effectively reached for $\tilde{z} \lesssim 30$.

The limiting conversion efficiency is shown as a function of R in Fig. 8. For large values of R the efficiency approaches the value $1/R$ as discussed earlier, whereas for small values of R the efficiency asymptotically approaches the value $\eta = 9R/4$, which can also be expressed as $\eta = |\chi_h|^2/|\chi_{\mathrm{abs}}|^2$. Substantial conversion is possible only for values of R near unity. A maximum limiting efficiency of $\eta \approx 0.31$ occurs for $R \approx 0.667$.

For detunings of several linewidths the absorptive terms become negligible and the dominant limitations arise from the intensity-dependent phase mismatches. When pump depletion is small the first term in Eq. (61b) can be neglected. Under the assumption of linear phase matching the solution of Eqs. (61a) and (61b) can be written as

$$\frac{I_3}{I_{10}} = \frac{64\pi^2}{n^2c^2}\frac{\kappa'^2L^2I_{10}^2}{\Delta^2}\frac{\sin^2\{(\kappa'/\Delta)|A_1|^2[(\sqrt{R}/2) - (1/\sqrt{R})]L\}}{\{(\kappa'/\Delta)|A_1|^2[(\sqrt{R}/2) - (1/\sqrt{R})]L\}^2}, \tag{70a}$$

$$= \frac{96\pi^2}{n^2c^2}k_{21}(\omega)k_{21}(3\omega)I_{10}^2L^2\frac{\sin^2\{[k_{21}(3\omega) - 3k_{21}(\omega)]|A_1|^2L/2\}}{\{[k_{21}(3\omega) - 3k_{21}(\omega)]|A_1|^2L/2\}^2}. \tag{70b}$$

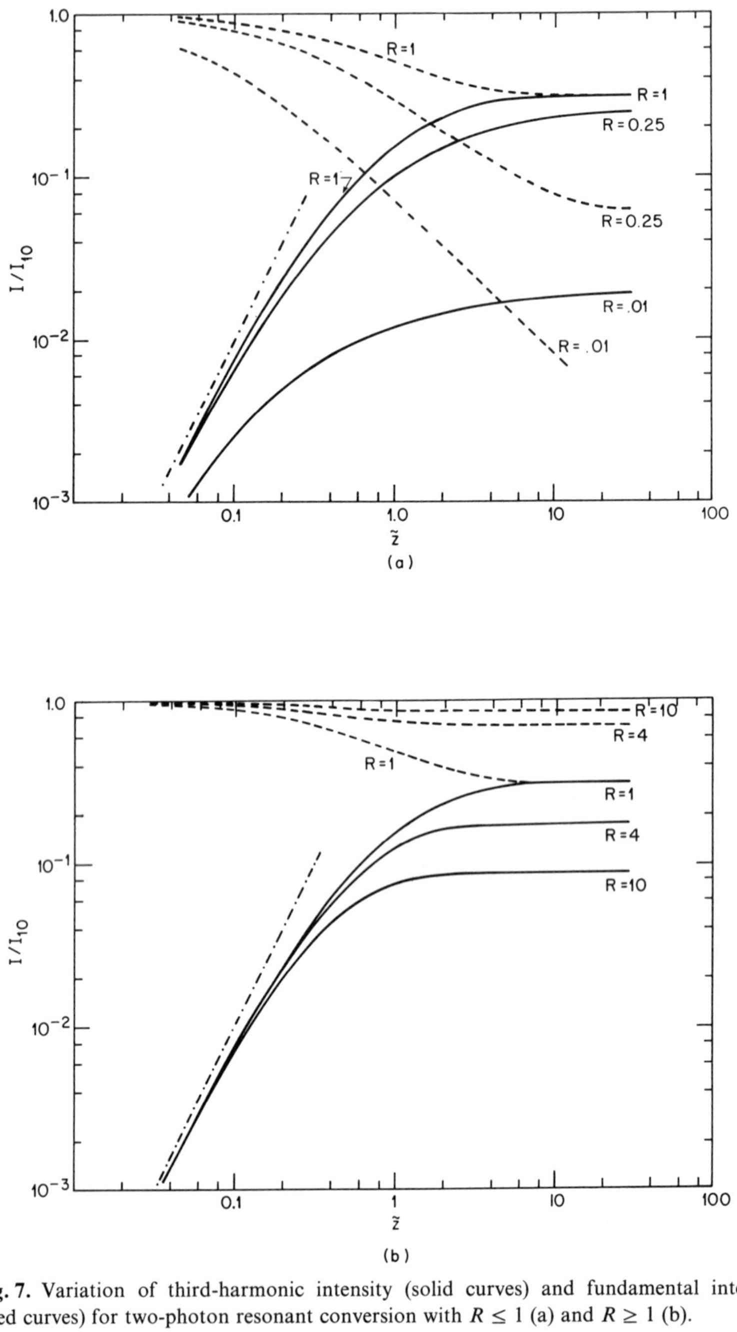

Fig. 7. Variation of third-harmonic intensity (solid curves) and fundamental intensity (dashed curves) for two-photon resonant conversion with $R \leq 1$ (a) and $R \geq 1$ (b).

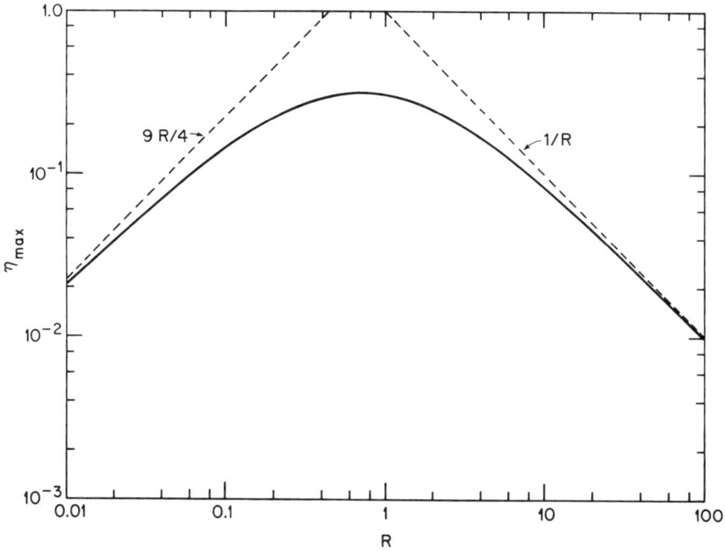

Fig. 8. Variation of maximum third-harmonic conversion efficiency with R.

The second form of this equation shows that conversion cannot be obtained for the near-two-photon resonant interaction without some degree of intensity-dependent wave vector mismatch at each of the frequencies involved. Limitations from the intensity-dependent refractive indices can be avoided only if the intensity-dependent phase shifts at the fundamental and harmonic frequencies cancel, i.e., if $k_{21}(3\omega) = 3k_{21}(\omega)$. Otherwise, as is usually the situation, the maximum harmonic intensity, obtained when

$$|A_1|^2[k_{21}(3\omega) - 3k_{21}(\omega)]L = \pi,$$

is

$$\eta = \frac{4R}{(R-2)^2}. \tag{71}$$

For large values of R the intensity-dependent phase changes at the harmonic dominates and the limiting efficiency is

$$\lim_{R\to\infty} \eta_{\max} = \frac{4}{R} = \frac{1}{9}\left|\frac{\chi_{\text{h}}^{(3)}}{\chi_{\text{Kerr}}^{(3)}(3\omega)}\right|^2.$$

For small values of R the intensity-dependent phase change at the fundamental dominates and the maximum conversion efficiency is

$$\lim_{R\to 0} \eta_{\max} = R = \frac{4}{9}\left|\frac{\chi_{\text{h}}}{\chi_{\text{Kerr}}(\omega)}\right|^2.$$

These expressions have the same functional dependence on R as the corresponding limiting efficiencies at exact resonance, but the proportionality factor is different. For each of these conditions the maximum conversion efficiency is determined by the magnitude of R and again substantial conversion to the harmonic is possible only for values of R near unity.

Stappaerts[36] has given an approximate analytic solution for the conversion efficiency as a function of detuning and two-photon transmission under the assumption that the pump depletion in the conversion process and the Raman amplification of the pump can be neglected. This solution predicts the peak conversion efficiency as obtained from solutions of Eqs. (65a) and (65b) and the required interaction length fairly well. It does not, however, give a good description of the behavior of the fundamental intensity for interactions with large values of R or for the behavior of either the fundamental or harmonic intensity at large distances.

Kildal and Brueck[38] have considered the effects of a linear phase mismatch for the special case of $R = 4$, for which the maximum conversion at exact resonance is $\eta = 0.17$. Their calculations show that at small detunings the use of an initial phase mismatch can increase the maximum conversion efficiency considerably. An example of this behavior is shown in Fig. 9.

They have also analyzed the effect of tight focusing on the two-photon resonant conversion process. Their results, shown in Fig. 10 for a focusing

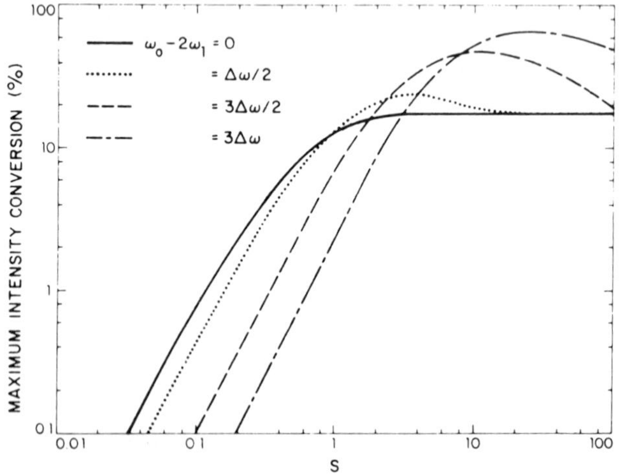

Fig. 9. Calculated third-harmonic conversion for various detunings from the two-photon resonance. The initial wavevector mismatch was optimized for each detuning. $s = \kappa' A_{10}^2 L/\Gamma$, $\Delta\omega = 2\Gamma$, $R = 4$. (From Ref. 38, © 1980 IEEE.)

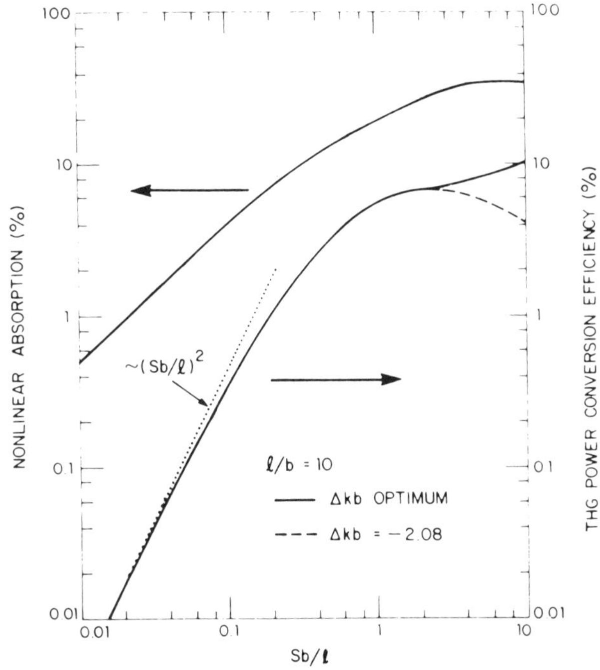

Fig. 10. Calculation of third-harmonic conversion for focused beam including effects of two-photon resonance. $s = \kappa' A_{10}^2 L/\Gamma$, $\Delta\omega = 2\Gamma$, $R = 4$. (From Ref. 38, © 1980 IEEE.)

parameter of $b/L = 0.1$, show that the maximum conversion efficiency that is attainable in a focused beam is somewhat less than that for a collimated beam because of the inability to maintain optimum phase-mismatch conditions in the presence of the distortions of the beam profile that develop as the pump beam is depleted.

4.3.1b. Population Changes and Stark Shifts. When the pump intensity is high enough, population redistribution and Stark shifting of the resonant levels can be significant. These processes can affect the conversion efficiency both by saturating the nonlinear susceptibility and changing the phase-matching conditions.

Saturation of the Susceptibility. We consider first the effects of population changes on conversion efficiency through saturation of the susceptibility. As discussed in Section 3.4 these effects are most likely to be encountered when conversion is done in low-density gases at high pump intensities.

The effects of population changes and Stark shifts have been considered experimentally and theoretically by several authors. Experimental observations of saturation of the susceptibility under two-photon resonant conditions have been reported by Wang and Davis[42] and Ward and Smith.[43] Elgin and New[29] and Anikin *et al.*[28] have discussed steady-state saturation of the susceptibility in the presence of both population changes and Stark shifts. Georges *et al.*[21,47] have treated transient aspects of the saturation of the susceptibility, including the effects of multiphoton ionization. Harris and Bloom[46] have considered the effects of transient limitations due to multiphoton absorption to two-photon resonant levels. Harris[45] has also discussed limitations due to n-photon absorption to dispersive levels. Stappaerts[27] has treated steady-state and transient limitations due to Stark shifts in multiphoton processes involving various multiphoton resonances. Limitations due to coherent population changes have been discussed by Grischowsky *et al.*[51] Coherent propagation effects have been discussed by Diels and Georges[52,53] and by Poluektov and Nazarkin.[54] Scheingraber and Vidal[48] and Puell *et al.*[40] have given numerical calculations specifically for conversion in Sr including effects of absorption of the harmonic wave, depletion of the fundamental wave in the conversion process, alteration of the phase-matching parameters associated with the population redistribution, and Stark shifts of the resonant levels in addition to the absorptive and Raman effects associated with the two-photon resonance that were discussed in the last section.

Several of these authors have pointed out the importance of including the Stark shift of the near-resonant level as well as its population in describing the saturation of the susceptibility. This is an example of a situation in which it is impossible to separate the effects of two different physical processes in limiting conversion efficiency.

Two-Photon Resonance. The equations describing third-harmonic generation in the plane-wave approximation in the presence of population changes associated with a two-photon resonance are[28]

$$\frac{\partial A_3}{\partial z} = i\,\frac{3(2\pi)^2}{n_3\lambda_1} A_1 \frac{N}{\hbar}\left[\sum \mu_{0i}\mu_{i2}\left(\frac{1}{(\omega_{i0} - 3\omega)} + \frac{1}{\omega_{i0} + \omega}\right)\right]\bar{\rho_{20}}, \tag{72a}$$

$$\frac{\partial \bar{\rho_{20}}}{\partial t} + i\left(\omega_{20} - 2\omega + \delta_{2,1}|A_1|^2 + \delta_{2,3}|A_3|^2 - \frac{i}{T_2}\right)\bar{\rho_{20}}$$
$$= \frac{i}{4\hbar^2}\left(\sum_i \frac{\mu_{0i}\mu_{i2}}{\omega_{i0} - \omega}\right)(A_1^2 + \sqrt{R}A_1^* A_3)(\rho_{00} - \rho_{22}), \tag{72b}$$

$$\frac{\partial}{\partial t}(\rho_{00} - \rho_{22}) + \frac{1}{T_1}(\rho_{00} - \rho_{22} - 1)$$
$$= \frac{i}{2\hbar^2}\left[\sum_j \frac{\mu_{0j}\mu_{j2}}{(\omega_{j0} - \omega)}\right]$$
$$\times [\rho_{20}^-(A_1^{*2} + \sqrt{R}A_1A_3^*) - \rho_{02}^+(A_1^2 + \sqrt{R}A_1^*A_3)], \quad (72c)$$

$$\rho_{02}^+ = (\rho_{20}^-)^*, \quad (72d)$$

$$\rho_{00} + \rho_{22} = 1, \quad (72e)$$

where R is given in Eq. (63) and the Stark shift parameters are given by

$$\delta_{2,n} = \frac{1}{2\hbar^2}\left[\sum_i \frac{|\mu_{2i}|^2\omega_{2i}}{\omega_{2i}^2 - (n\omega)^2} + \sum_i \frac{|\mu_{0i}|^2\omega_{i0}}{\omega_{i0}^2 - (n\omega)^2}\right]. \quad (73)$$

These equations include the effects of Stark shifts due to both the fundamental and harmonic fields, attenuation of the harmonic field due to Raman amplification of the pump, a nonlinear phase shift at the harmonic due to the fundamental intensity, saturation of the susceptibility due to two-photon transitions, and power broadening due to saturation of the two-photon transition. We have neglected intensity-dependent corrections to the lifetime T_1 and the dephasing time T_2. This approximation should be valid as long as the intensity does not become too large. The appropriate condition for the dephasing time is

$$|A_1|^2 \ll \frac{1}{2\hbar T_2}\left\{\sum_i \frac{|\mu_{2i}|^2}{[(\omega_{2i})^2 - \omega^2]T_{2i}} + \sum_i \frac{|\mu_{0i}|^2}{[(\omega_{i0})^2 - \omega^2]T_{0i}}\right\}, \quad (74)$$

where T_{mn} is the dephasing time for transitions between levels m and n. We have also neglected the linear phase mismatches and any nonlinear phase mismatches associated with the two-photon population changes. The effects of intensity-dependent phase shifts associated with the population redistribution will be discussed in Section 4.3.4.

Weak Pump Fields—Steady-State Solutions. Steady-state solutions of Eqs. (72) are appropriate for cw radiation or for pulsed radiation if the pulse duration is longer than T_1 or T_2. Analytic solutions can be obtained in the weak pump field limit for which the depletion of the pump field and the Stark shifts due to the harmonic field can be neglected. We will also neglect the attenuation of the harmonic field due to the Raman amplification of the pump on resonance and the nonlinear phase shift at the harmonic due to the fundamental intensity off resonance. In general the neglect of the linear and nonlinear phase mismatches will be valid as long as the nonlinear phase changes are small over the entire range of intensities that are used. Conditions under which this assumption is valid will be discussed in Section 3.3.5.

Under these assumptions the ρ_{20} component of the density matrix, the population changes, and the Stark shifts are determined only by the fundamental intensity. The steady-state solutions of Eqs. (72a)–(72e) are then

$$\rho_{20} = \frac{q_2 T_2[i + (\Delta_2' + \delta_2'|A_1|^2)]A_1^2}{4\hbar[1 + (\Delta_2' + \delta_2'|A_1|^2)^2 + (q_2^2 T_1 T_2 |A_1|^4/4\hbar^2)]}, \tag{75a}$$

$$\rho_{00} - \rho_{22} = \frac{1 + (\Delta_2' + \delta_2'|A_1|^2)^2}{1 + (\Delta_2' + \delta_2'|A_1|^2)^2 + q_2^2 T_1 T_2 |A_1|^4/4\hbar^2}, \tag{75b}$$

$$A_3 = i\frac{3\pi^2}{n_3\lambda_1} A_1^3 NL\left(\frac{q_2' q_2 T_2}{\hbar}\right) \times \frac{[i + (\Delta_2' + \delta_2'|A_1|^2)]}{[1 + (\Delta_2' + \delta_2'|A_1|^2)^2 + q_2^2 T_1 T_2 |A_1|^4/4\hbar^2]}, \tag{75c}$$

$$I_3 = \frac{576\pi^6}{n_3 n_1^3 \lambda_1^2 c^2} I_1^3 N^2 L^2 \left(\frac{q_2' q_2 T_2}{\hbar}\right)^2 \times \frac{[1 + (\Delta_2' + \delta_2'' I_1)^2]}{[1 + (\Delta_2' + \delta_2'' I_1)^2 + 16 q_2^2 T_1 T_2 I_1^2 \pi^2/\hbar^2 c^2 n_1^2]^2}, \tag{75d}$$

where

$$q_2 = \frac{1}{\hbar}\sum_i \frac{\mu_{2i}\mu_{i0}}{(\omega_{i0} - \omega)}, \tag{76a}$$

$$q_2' = \frac{1}{\hbar}\sum_j \mu_{0j}\mu_{j2}\left(\frac{1}{\omega_{j0} - 3\omega} + \frac{1}{\omega_{j0} + \omega}\right), \tag{76b}$$

$$\Delta_2 = \omega_{20} - 2\omega, \tag{77a}$$

$$\Delta_2' = \Delta_2 T_2, \tag{77b}$$

$$\delta_2' = \delta_{2,1} T_2, \tag{78a}$$

$$\delta_2'' = \frac{8\pi}{c n_1}\delta_2'. \tag{78b}$$

The conversion efficiency can be optimized either as a function of pump intensity for a constant detuning from the unshifted line or as a function of the initial detuning for a constant pump intensity. The behavior of the solutions for various initial conditions have been investigated by Anikin *et al.*[28] and by Elgin and New.[29]

The dependence of the conversion efficiency on the detuning from the unshifted line depends on the value of the pump intensity relative to the saturation intensity, defined as

$$I_s = \hbar c n_1 / 4\pi q_2 \sqrt{T_1 T_2}. \tag{79}$$

For pump intensities much less than the saturation intensity the maximum conversion efficiency occurs at the center of the Stark-shifted line:

$$\Delta_{\text{opt}} = -\delta_{2,1}|A_1|^2, \tag{80}$$

and has the value

$$\eta_{\max,1} = \frac{I_3}{I_1} = \frac{576\pi^6}{n_3 n_1^3 \lambda_1^2 c^2}\left(\frac{q_2' q_2 T_2}{\hbar}\right)^2 I_1^2 N^2 L^2 \tag{81a}$$

$$= \frac{36\pi^4 q_2'^2 N^2 L^2}{n_3 n_1 \lambda_1^2}\frac{T_2}{T_1} I_1^2/I_s^2. \tag{81b}$$

For pump intensities that are higher than the saturation intensity the optimum conversion efficiency occurs for two initial detunings that are symmetrically displaced from the center of the Stark-shifted line:

$$\Delta_{\text{opt}} = -\frac{8\pi}{c n_1}\delta_2 I_1 \pm \frac{1}{T_2}\sqrt{I_1^2/I_s^2 - 1}. \tag{82}$$

For the analysis given here, which neglected the contributions of the nonresonant terms in the susceptibility, the efficiency at either of the two optimum detunings of Eq. (82) is

$$\eta_{\max,2} = \frac{144\pi^6 N^2 L^2}{n_3 n_1 \lambda_1^2 c^2}\left(\frac{q_2' q_2 T_2}{\hbar}\right)^2 I_s^2 \tag{83a}$$

$$= 9\pi^4 N^2 L^2 q_2'^2 T_2 / n_3 n_1 \lambda_1^2 T_1. \tag{83b}$$

The remaining solution for the extrema as a function of frequency predicts a minimum at the center of the Stark-shifted line. When the nonresonant terms are included, asymmetry about the Stark-shifted line can occur. Such asymmetry is evident in the calculations given by Anikin *et al.*[28]

Typical examples of third-harmonic power as a function of detuning from resonance at a constant pump intensity as calculated by Elgin and New[29] are shown in Fig. 11. At low pump intensity the third-harmonic power shows typical resonant behavior with the peak centered near the unshifted two-photon frequency. At higher pump intensities the third harmonic shows the effects of Stark shifts and power broadening and eventually a dip at the

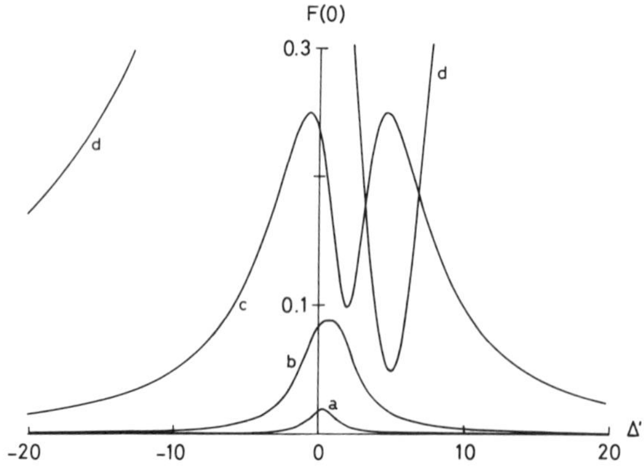

Fig. 11. Calculation of third-harmonic intensity for two-photon resonant conversion as a function of detuning from resonance. $\delta_2'' I_1 = 0.3$(a), $\sqrt{0.5}$(b), 2(c), and 5(d). (From Ref. 29.)

center of the Stark-shifted line appears. This dip at line center is a characteristic of the population changes and also appears when the limitations are due to intensity-dependent phase changes resulting from the population redistribution (see Section 4.3.4).

At a constant detuning from the unshifted resonant frequency the conversion efficiency has a maximum at a pump intensity given by

$$I_{\text{opt}} = \frac{I_s}{1 - (\delta_2'' I_s)^2} \left(\Delta_2' \delta_2'' I_s \pm \sqrt{1 + \Delta_2'^2 - (\delta_2'' I_s)^2}\right). \tag{84}$$

This maximum in intensity corresponds to the maximum in detuning at constant pump intensity as described by Eq. (82). At the optimum intensity of Eq. (84) the conversion efficiency has the value given in Eq. (83). When the initial detuning is such that the Stark shift brings the system into resonance, Eq. (84) predicts a single maximum near the positive root for I_{opt} when $(\delta_2'' I_s)^2 < 1$, two maxima when $1 + \Delta_2'^2 > (\delta_2'' I_s)^2 > 1$, and a single maximum near the center of the Stark-shifted line if $(\delta_2'' I_s)^2 > 1 + \Delta_2'^2$. Equation (84) predicts one maximum in efficiency when $(\delta_2'' I_s)^2 < 1$, but two maxima can occur if $1 + \Delta_2'^2 > (\delta_2'' I_s)^2 > 1$. In the range of pump intensities that satisfy the condition

$$I_1 \gg |\delta_2''^{-1}|, |\Delta_2'/\delta_2''|$$

but are still small enough so that intensity-dependent changes to T_1 and T_2 are not important the conversion efficiency approaches the value

$$\eta_\infty = \frac{36\pi^4 N^2 L^2 q_2'^2}{n_3 n_1 \lambda_1^2} \frac{T_2}{T_1} \frac{(I_{\text{St}}/I_s)^2}{[1 + (I_{\text{St}}/I_s)^2]^2}, \tag{85}$$

where $I_{St} = |\delta_2''|^{-1}$ is the intensity required for a Stark shift of T_2^{-1} sec^{-1}, as shown by Elgin and New. If the pump intensity becomes high enough for T_1 and T_2 to be affected, then the conversion efficiency can decrease with increasing pump intensity.

At either the optimum intensity of Eq. (83) or in the high-intensity limit, the conversion efficiency is independent of the pump intensity, just as it was when the conversion was limited by the third-order nonlinear index (see Section 4.2). Unlike under the limitations imposed by the nonlinear indices, however, the maximum efficiency allowed by the two-photon saturation and Stark shifts increases with the density of the medium and the interaction length.

The two-photon transition rate $W^{(2)}$ is defined by the equation

$$\frac{d}{dt}(\rho_{00} - \rho_{22}) + \frac{1}{T_1}(\rho_{00} - \rho_{22} - 1) = -W^{(2)}(\rho_{00} - \rho_{22}), \tag{86}$$

where

$$W^{(2)} = \frac{I_1^2}{T_1 I_s^2[1 + (\Delta_2' + I_1/I_{St})^2]}. \tag{87}$$

At the optimum intensity of Eq. (84), the steady-state value of $W^{(2)}$ is given by

$$W^{(2)}T_1 = 1, \tag{88}$$

as pointed out by Elgin and New. This is the optimum condition used by Harris and Bloom,[46] Harris,[45] and Stappaerts[36] in determining the limitations in resonant systems. Approximately one-quarter of the atoms or molecules are excited to the two-photon level under these conditions.

The ratio of the maximum conversion of (83) to the limiting efficiency at high intensities of (86) is determined by the relation between the population saturation and Stark shift parameters:

$$\frac{\eta_{\max, 2}}{\eta_\infty} = \frac{1}{4}\frac{[1 + (I_{St}/I_s)^2]^2}{[I_{St}/I_s]^2}. \tag{89}$$

For values of $(I_{St}/I_s)^2$ close to unity (Elgin and New[29] used a value of 2 for this parameter in their calculations shown in Fig. 11), saturation of the two-photon transition dominates the conversion process and prevents a significant increase in conversion from occurring as the system is brought into resonance by the Stark shift. For larger values of $(I_{St}/I_s)^2$ the limiting efficiency of (85) can be less than the maximum efficiency of (83) and increased conversion can result from a favorable Stark shift. Such behavior is evident in calculations given by Anikin *et al.*[28] for two-photon resonantly enhanced conversion in Na vapor. In either situation, however, the saturation of the two-photon resonance in the steady state will prevent the increases in efficiency of several orders of magnitude that otherwise would be expected when the Stark shift brings the system into resonance.

The Stark shift in turn plays an important role in preventing the complete saturation of the two-photon transition. This aspect of the interaction has been pointed out by both Elgin and New[29] and Georges *et al.*[47] Neglect of the Stark shift can result in a third harmonic intensity that ultimately decreases with increasing pump intensity [consider (75d) with the Stark-shift term set equal to zero]. This behavior is in contrast to the experimental observations of Wang and Davis.[42] By moving the system away from resonance at high intensities the Stark shift prevents complete saturation of the two-photon transition and actually increases conversion efficiency over what it would be otherwise.

Although the optimum conversion efficiency in the regime of strong saturation as given in Eq. (83) occurs at a different detuning for each value of pump intensity, the value of the optimum conversion efficiency is independent of the initial detuning and the pump intensity. The efficiency in the limit of high pump intensity is similarly independent of both the initial detuning and the pump intensity except that here the same efficiency is obtained for all values of initial detuning with no optimum being present. The maximum efficiency can therefore be obtained in a two-photon resonant interaction for any value of initial detuning, provided only that sufficient pump intensity be available to satisfy the conditions of Eq. (84) or (85). This property was pointed out by Harris and Bloom[46] in regard to frequency up-conversion by four-wave mixing. The chief value of the two-photon resonance then is in the increase in conversion efficiency that it provides at low pump intensities and in the fact that optimum conversion efficiency can be obtained at lower pump intensities for smaller detunings.

In many applications the total harmonic power is more important than the conversion efficiency in determining the optimum operation of the system. Under these conditions it is important to note that, although the conversion efficiency reaches a maximum value independent of the intensity and initial detuning of the pump, the harmonic power continues to grow in proportion to the pump power. Examples of the variation of the third-harmonic power with pump power for various initial detunings as calculated by Elgin and New[29] are shown in Fig. 12. The third-harmonic power goes from a cubic dependence on the pump power at low pump powers

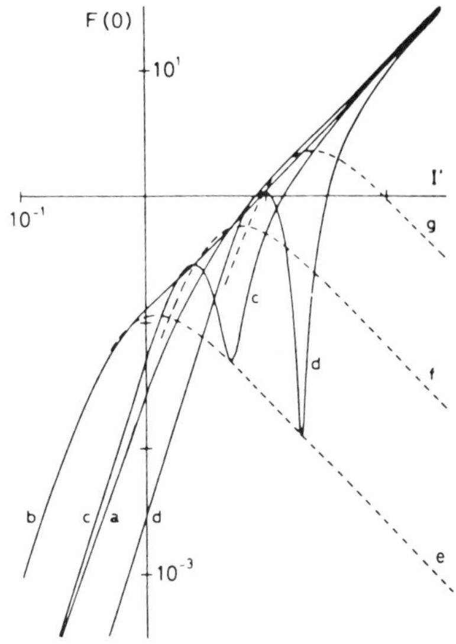

Fig. 12. Variation of third-harmonic intensity for two-photon resonant conversion as a function of pump intensity for various values of initial detuning. $I' = \delta_2'' I_1$; $\Delta' = 5$(a), 0(b), -5(c), -20(d). Without Stark shifts: $\Delta' = 0$(e), ± 5(f), ± 20(g). (From Ref. 29.)

to a linear dependence at high pump powers, consistent with the predictions of Eq. (86). In the transition range of intermediate intensities the dependence of the harmonic intensity on the pump intensity can go from its low-intensity behavior to its high-intensity behavior smoothly if the Stark shift moves the system farther from resonance, or it can show an oscillation if the Stark shift first moves the system closer to resonance. Ultimately, however, the harmonic intensity approaches the value $I_3 = \eta_\infty I_1$, with η_∞ given by Eq. (86), regardless of the sign or magnitude of the initial detuning. Whether higher harmonic intensity is generated for a given maximum pump intensity at the optimal detuning and pump intensity of Eqs. (82) and (84) or in the high-intensity limit depends on the value of the maximum available pump intensity and on the value of $(I_{St}/I_s)^2$. In general, however, it appears that the most harmonic power can be generated in the high-intensity limit, as is the case for the calculations shown in Fig. 12, if sufficient laser power is available.

One other consideration is that the limitation of the conversion efficiency in any real situation may be accompanied by other detrimental effects, such as pulse shape or beam distortion, or spectral broadening due to other competing effects that were not included in the present analysis. Under such conditions it may be advantageous to use a relatively large detuning, consistent with the constraints of the available laser power, to minimize the limitations and possible distortions of the beam due to the other processes.

4.3.2. *Three-Photon Resonances*

The equations describing third-harmonic conversion near a three-photon resonance, including population saturation and Stark shifts, are (see Anikin *et al.*[44])

$$\frac{\partial A_3}{\partial z} = i\frac{6(2\pi)^2}{n_3\lambda_1}\mu_{03}N\bar{\rho}_{30}e^{-i\Delta k_0 z}, \tag{90a}$$

$$\frac{\partial \bar{\rho}_{30}}{\partial t} + i(\Delta_3 + \delta_3|A_1|^2 - i/T_2)\bar{\rho}_{30} = i(\rho_{00} - \rho_{33})\left(\frac{q_3 A_1^3}{2^3\hbar} + \frac{\mu_{03}A_3}{2\hbar}e^{i\Delta k_0 z}\right), \tag{90b}$$

$$\frac{\partial}{\partial t}(\rho_{00} - \rho_{33}) + \frac{1}{T_1}(\rho_{00} - \rho_{33} - 1) = 2i\left\{\left[\left(\frac{q_3 A_1^{*3}}{2^3\hbar} + \frac{\mu_{30}}{2\hbar}A_3^* e^{-i\Delta k_0 z}\right)\bar{\rho}_{30}\right] - \left[\left(\frac{q_3 A_1^3}{2^3\hbar} + \frac{\mu_{30}}{2\hbar}A_3 e^{i\Delta k_0 z}\right)\overset{+}{\rho}_{03}\right]\right\}, \tag{90c}$$

where

$$q_3 = \frac{1}{\hbar^2}\sum_{i,j}\mu_{0i}\mu_{ij}\mu_{j3}/(\omega_{i0} - \omega)(\omega_{j0} - 2\omega),$$

$$\Delta_3 = \omega_{30} - 3\omega,$$

$$\delta_3 = \frac{1}{2\hbar^2}\left(\sum_i \frac{|\mu_{i3}|^2\omega_{3i}}{\omega_{3i}^2 - \omega^2} + \sum_j \frac{|\mu_{j0}|^2\omega_{j0}}{\omega_{j0}^2 - \omega^2}\right),$$

$$\overset{+}{\rho}_{03} = (\bar{\rho}_{30})^*.$$

The explicit phase mismatch Δk_0 includes the effect of the linear refractive index at the fundamental frequency, but not at the harmonic, in the resonant medium as well as any phase mismatch that may be due to a buffer gas. The linear refractive index at the harmonic frequency in the resonant medium is included in the equation for $\bar{\rho}_{30}$. In these equations we have assumed that only the 0–3 transition is nearly resonant, and therefore that only the Stark shift of the 0–3 transition need be considered. Stappaerts[27] has given formulas that include the Stark shifts of other resonant levels but neglect the population changes. Equations (90) are similar to Eqs. (72) for the two-photon resonance, except that they take into account the linear absorption at the harmonic by the 0–3 transition and they allow a phase mismatch due either to a linear mismatch or to the intensity-dependent refractive index at the harmonic due to depletion of the ground-state population or Stark shifts.

These terms in the nonlinear index will be larger than the corresponding terms for the fundamental frequency if the wave vector mismatch is determined by the dispersion in the refractive index at the harmonic frequency resulting from the three-photon resonance. We have also neglected the intensity dependence of the lifetimes T_1 and T_2 and the Stark-shift term due to the harmonic intensity. The first term in the first large parentheses on the right-hand side of Eq. (90c) involves three-photon absorption of the fundamental, while the second term involves single-photon absorption of the harmonic. Both terms also involve interference between the two fields through the product with ρ_{30}. The terms in the second large parentheses involve similar interactions.

In the steady-state these equations reduce to

$$\rho_{00} - \rho_{33} = \frac{1 + (\Delta_3' + \delta_3'|A_1|^2)^2}{\{1 + (\Delta_3' + \delta_3'|A_1|^2)^2 + 4T_1T_2[(q_3^2|A_1|^6/2^6\hbar^2) + (|\mu_{03}|^2/4\hbar^2)|A_3|^2 + (q_3\mu_{03}/2^4\hbar^2)(A_3A_1^{*}e^{i\Delta k_0 z} + A_3^{*}A_1^3 e^{-i\Delta k_0 z})]\}}, \tag{91a}$$

$$\frac{\partial A_3}{\partial z} + \frac{3(2\pi)^2 N|\mu_{03}|^2(\rho_{00} - \rho_{33})T_2}{n_3\lambda_1\hbar[1 + i(\Delta_3' + \delta_3'|A_1|^2)]}A_3 = -\frac{6(2\pi)^2 N\mu_{03}q_3A_1^3(\rho_{00} - \rho_{33})T_2 e^{-i\Delta k_0 z}}{2^3 n_3\lambda_1[1 + i(\Delta_3' + \delta_3'|A_1|^2)]}, \tag{91b}$$

where $\Delta_3' = \Delta_3 T_2$ and $\delta_3' = \delta_3 T_2$.

If saturation of the upper state population from linear absorption of the harmonic and interference between the harmonic and fundamental fields is important, then Eqs. (91a) and (91b) have to be solved numerically. If only saturation due to the three-photon absorption of the fundamental is important, however, then they can be solved analytically. Under this condition the harmonic intensity is given by

$$I_3 = \frac{n_3}{n_1}\left(\frac{8\pi}{cn_1}\right)^2 \frac{N^2 q_3^2 \sigma_{03}^2 I_1^3[1 + (\Delta_3' + \delta_3'' I_1)^2]}{16|\mu_{03}|^2\{[N\sigma_{03}]^2 + [\Delta k_0 + N\sigma_{03}(\Delta_3' + \delta_3'' I_1)]^2\}} \times \{1 + e^{-2N\sigma_{03}z} - 2e^{-N\sigma_{03}z}\cos[(\Delta k_0 + N\sigma_{03}(\Delta_3' + \delta_3'' I_1))z]\}, \tag{92}$$

where $\delta_3'' = \delta_3'(8\pi/cn_1)$ and

$$\sigma_{03} = \frac{(2\pi)^2|\mu_{03}|^2 T_2}{n_3\hbar\lambda_3[1 + (\Delta_3' + \delta_3'' I_1)^2 + I_1^3/I_{S,3}^3]} \tag{93}$$

is the saturated one-photon absorption cross section at the resonant level at the harmonic frequency and

$$I_{s,3} = \frac{cn_1}{8\pi} \sqrt[3]{16\hbar^2/q_3^2 T_1 T_2} \tag{94}$$

is the three-photon saturation intensity.

This solution has the same form as that given in Eq. (15) for third-harmonic conversion in the presence of linear absorption at the harmonic except that the absorption cross section now reflects the saturation of the 0–3 transition and the coherence length includes a nonlinear component that accounts for the effects of the nonlinear refractive index associated with the population saturation and Stark shift:

$$L_c = \pi[\Delta k_0 + N\sigma_{03}(\Delta_3' + \delta_3'' I_1)]^{-1}. \tag{95}$$

When the line-center absorption is large and the pump intensity is low enough so that the population changes and Stark shifts are small, the harmonic intensity at perfect phase matching has a minimum at line center and a maximum at a detuning that increases as the line-center absorption increases. This behavior is commonly encountered with three-photon resonant interactions. When the pump intensity is strong enough for significant population changes to occur, somewhat different behavior can be encountered.

The effect of the population changes on the three-photon resonant conversion process depends on whether the interaction length is determined by the length of the sample, the linear or nonlinear coherence length, or the saturated absorption length at the third harmonic. If the sample length is shorter than the linear and nonlinear coherence lengths and the saturated absorption length, i.e., if $L \ll N\sigma_{03}, L_c$, the interaction is effectively linearly phase matched and the harmonic intensity is given by

$$I_3 = \frac{576\pi^6 N^2 L^2 I_1^3}{n_3 n_1^3 \lambda_1^2 c^2} \left(\frac{\mu_{03} q_3 T_2}{\hbar}\right)^2 \frac{[1 + (\Delta_3' + \delta_3'' I_1)^2]}{[1 + (\Delta_3' + \delta_3'' I_1)^2 + (I_1/I_{s,3})^3]^2}, \tag{96a}$$

$$= \frac{4n_3 \pi^2 q_3^2 I_1^3}{n_1^3 c^2 |\mu_{03}|^2} (N\sigma_{03} L)^2 \{1 + (\Delta_3' + \delta_3'' I_1)^2\}. \tag{96b}$$

This expression has the same form as that for the harmonic intensity in the presence of a saturated two-photon resonance [see Eq. (75d)], except that the saturation term now depends on I^3. Note that in order for the approximation that the excitation of the three-photon resonant level by single-photon absorption of the harmonic radiation be small compared to the excitation resulting from three-photon absorption of the fundamental radiation the quantity $(N\sigma_{03} L)^2[1 + (\Delta_3' + \delta_3'' I_1)^2]$ must be small compared to unity.

Within this approximation the conditions for maximum third-harmonic intensity near the three-photon resonance are similar to those for the maximum efficiency near the two-photon resonance. If the pump intensity is less than the saturation intensity, then the maximum third-harmonic intensity is generated at the center of the Stark-shifted line:

$$\Delta'_{3,\,\mathrm{opt},\,1} = -\,\delta''_3 I_1, \tag{97}$$

and the maximum intensity is

$$I_{3,\,\max,\,1} = \frac{576\pi^6 N^2 L^2}{n_3 n_1^3 \lambda_1^2 c^2} \left(\frac{\mu_{03} q_3 T_2}{\hbar}\right)^2 \frac{I_1^3}{[1 + (I_1/I_{\mathrm{S},\,3})^3]^2}. \tag{98}$$

When the pump intensity exceeds the three-photon saturation intensity, the maximum harmonic intensity for a given pump intensity occurs at a detuning given by

$$\Delta'_{3,\,\mathrm{opt},\,2} = -\,\delta''_3 I_1 \pm \sqrt{(I_1/I_{\mathrm{s},\,3})^3 - 1} \tag{99}$$

and has the value

$$I_{3,\,\max,\,2} = \frac{144\pi^6 N^2 L^2}{n_3 n_1^3 \lambda_1^2 c^2} \left(\frac{\mu_{03} q_3 T_2}{\hbar}\right)^2 I_{\mathrm{s},\,3}^3 \tag{100a}$$

$$= \frac{9c\pi^3 N^2 L^2 \mu_{03}^2 T_2}{2n_3 \lambda_1^2 T_1}. \tag{100b}$$

For a given initial detuning from the unshifted resonance frequency the optimum pump intensity is determined by the solution of Eq. (99) for the intensity. At the optimum detuning and intensity the condition

$$W^{(3)} T_1 = 1 \tag{101a}$$

is satisfied. Here $W^{(3)}$, defined by

$$W^{(3)} = \frac{(q_3)^2 |A_1|^6 T_2}{16\hbar[1 + (\Delta'_3 + \delta'_3 |A_1|^2)^2]}, \tag{101b}$$

is the three-photon transition rate for the 0–3 transition. This condition, and corresponding ones for higher-order processes were used by Harris[45] and by Stappaerts[36] for estimating the limitations in qth-order frequency conversion processes involving multiphoton absorption to dispersive levels.

Just as for the maximum efficiency in the two-photon resonant situation the maximum harmonic intensity for the three-photon resonant interaction occurs at a different detuning for each value of pump intensity. However, the value of the maximum harmonic intensity is independent of both the optimum initial detuning and the pump intensity at which it occurs.

In the limit of high pump intensities the harmonic intensity becomes a decreasing function of pump intensity:

$$\lim_{I_1 \to \infty} I_3 = \frac{18\pi^3 N^2 L^2 c |\mu_{03}|^2 T_2}{n_3 \lambda_1^2 T_1} \frac{I_s^3}{I_1 I_{St}}. \tag{102}$$

Unlike the case for the two-photon resonance, the Stark shift of the three-photon resonant level does not prevent saturation of the population in the upper level from dominating the conversion efficiency. It does, however, weaken the saturation effects which would otherwise occur.

Higher conversion efficiency can be obtained if conditions are adjusted so that the effective interaction length is determined by an appropriate absorption or coherence length rather than an arbitrary length of the nonlinear medium. When this is the case, the nature of the solution will depend on the magnitude of Δk_0, on the detuning from resonance, and on the magnitude of the Stark shift relative to the initial detuning. In general the applicability of Eq. (92) in this regime can be limited because excitation of the three-photon resonance by single-photon absorption of the harmonic field can become comparable to that resulting from three-photon absorption of the fundamental field. However, as we shall see, the limiting efficiency will be independent of the degree of saturation of the resonance.

In interactions in which Δk_0 has a significant value (for example, in interactions involving phase matching by a buffer gas), and if conversion is done in one coherence length and absorption at the harmonic frequency is negligible over this length, the harmonic intensity as given by Eq. (92) is

$$I_3 = \frac{16 n_3 q_3^2 I_1^3 N^2 \sigma_{03}^2 L_c^2 [1 + (\Delta_3' + \delta_3'' I_1)^2]}{n_1^3 c^2 |\mu_{03}|^2}. \tag{103}$$

In order for excitation of the resonance by single-photon absorption of the harmonic be negligible we must again have the condition

$$N^2 \sigma_{03}^2 L_c^2 [1 + (\Delta_3' + \delta_3'' I_1)^2] \ll 1.$$

In a general case the intensity given in this approximation is reduced from that in the phase-matched interaction by the factor $(L_c^2/\pi)^2$ [see Eq. 96b)].

We can recover the functional form discussed qualitatively in Section 4.2.5 for three-photon resonant media by assuming, in accordance with the approximations of Section 4.2.1, that Δk_0 is adjusted to compensate for the linear refractive index at the harmonic frequency in the resonant medium, that conversion is done in one nonlinear coherence length, that absorption over this length can be neglected, that the Stark shift is small compared to the initial detuning, and that saturation of the resonance is negligible. In this situation the dominant resonant term in the wave-vector mismatch due to the quadratic Kerr effect [see Eq. (38b)] is the same as that given by the Stark shift. The appropriate coherence length is

$$L_c = \frac{\pi \Delta_3'^2}{N\sigma_{03}'(0)\delta_3'' I_1},$$

where $\sigma_{03}'(0)$ is the unsaturated absorption coefficient at the harmonic frequency at line center. The resulting harmonic intensity as obtained from Eq. (103) is

$$I_3 = \frac{16 n_3 \pi^2 q_3^2 \Delta_3'^2 I_1}{n_1^3 c^2 |\mu_{03}|^2 \delta_3''^2}. \tag{104a}$$

Here we see the same behavior as was described in Section 4.2.5, namely that the conversion efficiency reaches a limiting value that is independent of the initial detuning and the density of the medium and increases as the square of the initial detuning from resonance.

In order for Eq. (104a) to be valid it is necessary to satisfy simultaneously the conditions that the absorption of the harmonic be small over the effective coherence length and that the Stark shift be small compared to the initial detuning. A sufficient condition that allows these requirements to be satisfied and still permits the interaction length to be adjusted to a coherence length is

$$1 \ll \delta_3'' I_1 \ll \Delta_3'. \tag{104b}$$

If Δk_0 is small, as, for example, if conversion is done in a single-component medium with no single-photon resonances at the fundamental frequency, or if absorption over the coherence length is large, conditions can readily be encountered that violate the assumptions leading to Eq. (92). In such a situation Eqs. (91a) and (91b) must be solved numerically to determine the detailed behavior of the harmonic intensity. We can, however, determine the limiting behavior of the harmonic intensity from the structure of Eq. (91b).

We consider first the situation in which the detuning from the Stark shifted line is zero. In this situation we can expect that absorption of the harmonic will be important in limiting the harmonic conversion. A qualitative description of the behavior of the harmonic intensity is as follows. The harmonic field will grow 180° out of phase with the fundamental field, until the condition

$$A_3(z) = -\frac{q_3 A_1^3(z)}{4\mu_{03}} \tag{105}$$

is satisfied. At this point the excitation of the resonance by three-photon absorption of the fundamental radiation is exactly balanced by an opposite excitation through single-photon absorption at the harmonic. As a result no further excitation of the resonance takes place and the fields will propagate without further change. This type of behavior has been discussed by Jackson and Wynne[23c] in connection with three-photon resonantly enhanced multiphoton ionization. It is also exactly the same kind of behavior that was encountered in the two-photon resonant interaction in Section 4.3.1a.

When depletion of the pump wave in the harmonic conversion process can be neglected we can replace $A_1(z)$ in Eq. (105) by A_{10}, and the limiting harmonic intensity is given by

$$I_3 = \frac{4n_3\pi^2 q_3^2 I_{10}^3}{n_1^3 c^2 |\mu_{03}|^2}. \tag{106}$$

Comparison of this result with that of Eq. (96b) shows that the harmonic intensity generated under these conditions can be higher because of the restriction on the value of $(N\sigma_{03}L)^2[1 + (\Delta_3' + \delta_3'' I_1)^2]$ in Eq. (96b).

When the pump intensity is small compared to the three-photon saturation intensity, the approximations leading to Eq. (92) are valid and analysis shows that Eq. (106) will be satisfied for distances that are large compared with the linear absorption length at the harmonic. When saturation of the resonance is important, a comparison of Eq. (105) with Eq. (91a) shows that the terms involving A_3 in the last set of brackets in the denominator of the expression for $\rho_{00} - \rho_{33}$ can no longer be ignored. As a result we cannot simply replace the unsaturated absorption length with its saturated value as the condition for satisfying Eq. (105). The limiting intensity in Eq. (106) will, however, continue to be valid regardless of the degree of saturation because any decrease in nonlinear susceptibility due to saturation of the resonance is offset by an increase in the effective interaction length.

When the system is detuned by several linewidths from the center of the Stark-shifted line the primary limitation on conversion will arise from intensity-dependent phase shifts at the harmonic frequency associated with the three-photon resonance. In the absence of saturation the limiting behavior can be obtained from Eq. (92) with σ_{03} replaced by its unsaturated value. If conversion is done in one coherence length the limiting harmonic intensity is given by

$$I_3 = \frac{16n_3\pi^2 q_3^2 I_1^3}{n_1^3 c^2 |\mu_{03}|^2}. \tag{107}$$

When saturation of the resonance is important we must again consider the effects of the harmonic field on the excitation of the resonant medium. Equations (91a) and (91b) should then be solved together. We can, however, obtain an analytic solution when pump depletion can be neglected and Δk_0 is negligible. Under these conditions the solution of Eq. (91b) is

$$A_3 = -\frac{q_3 A_1^3}{4\mu_{03}}\left\{1 - \exp\left[-i\frac{N\sigma_{03}^{(0)}(\rho_{00} - \rho_{33})z}{\Delta_3' + \delta_3'|A_1|^2}\right]\right\},$$

where $\sigma_{03}^{(0)}$ is the unsaturated, line-center, single-photon absorption coefficient at the harmonic frequency. Substitution of this result in Eq. (91a) yields for the population difference

$$\rho_{00} - \rho_{33} = \frac{1 + (\Delta_3' + \delta_3'' I_1)^2}{1 + (\Delta_3' + \delta_3'' I_1)^2 + I_1^3/I_s^3}.$$

The population difference can be seen to be independent of the harmonic intensity. In this situation the excitation of the resonance through single-photon absorption of the harmonic radiation exactly cancels the excitation due to interference of the harmonic and fundamental fields, leaving only the excitation through multiphoton absorption of the fundamental radiation.

Optimum conversion under these conditions can be obtained by adjusting the system so the medium is one nonlinear coherence length long, satisfying the condition

$$L = \frac{\pi(\Delta_3' + \delta_3'' I_1^2)}{N\sigma_{03}^{(0)}(\rho_{00} - \rho_{33})}.$$

When this condition is met the maximum harmonic intensity is again given by Eq. (107).

The limiting intensity that is obtained in a single coherence length off-resonance is then four times the value that can be obtained on resonance. When conversion is done off-resonance the maximum harmonic intensity is obtained at only one value of interaction length that depends on the medium density, the pump intensity and the degree of saturation. In contrast the maximum harmonic intensity on line center can be obtained for all distances greater than some minimum value. In either situation the optimum harmonic intensity is generated at only one value of the pump intensity for a given experimental arrangement because of either the need to maintain resonance with the Stark-shifted line or to keep the interaction length equal to a single coherence length. As a result optimum conversion will not in general be possible at all times for pulsed excitation or at all radial positions for a pump beam with a transverse profile with a varying intensity.

The limiting harmonic intensity for either the on-resonant or off-resonant interactions is independent of the density of the nonlinear medium, the initial detuning (for the off-resonant interactions) and the interaction length, regardless of the degree of saturation of the resonance. This behavior reflects the fact that the effective interaction length that is required to reach the limiting harmonic intensity depends on the density, the initial detuning, and the degree of saturation and automatically adjusts to compensate for changes in any of these parameters. It can also be noted that the limiting harmonic intensity in either Eq. (106) or Eq. (107) increases as the third power of the pump intensity even in the presence of saturation of the resonance. This again is a reflection of the fact that the effective interaction length increases as the resonance saturates, offsetting the decrease in the nonlinear susceptibility.

It is of interest to consider the limiting behavior of the harmonic intensity for off-resonant conversion since it is very different in the quadratic Kerr regime [Eq. (104a)] and in the regime in which full saturation of the resonance is taken into account [Eq. (107)]. In either case the limiting intensity is reached when the density and length of the nonlinear medium is adjusted to provide a single coherence length. Which of the two limiting regimes is applicable depends on the magnitude of the initial detuning from the unshifted line and on the size of the Stark shift relative to the initial detuning. When $\Delta_3' \gg 1$, condition (104b) can be satisfied and the maximum harmonic intensity will saturate as the pump intensity is increased, taking on a linear dependence on the pump intensity. In this regime the pump intensity will be small enough that the Stark shift is small compared to the initial detuning. As the intensity is increased further the Stark shift can eventually become larger than the initial detuning and the limiting behavior of Eq. (107) will be observed, with the harmonic intensity again increasing as the third power of the pump intensity. In this regime the increasing Stark shift has saturated the resonant term in the third-order Kerr effect that is responsible for the limitations at lower intensities. This type of behavior has been reported by Puell *et al.*[5] in third-harmonic conversion of Nd : glass laser radiation in Rb vapor, although the saturation of the nonlinear index was attributed in their work to another mechanism. It can also be noted that if $\Delta_3' \not\gg 1$ it is possible that the conditions required for limitation by the quadratic Kerr effect cannot be satisfied. In such a situation the saturation of the harmonic intensity as a function of pump intensity may not be observed.

In a real system the behavior described here must be modified to take into account the presence of other levels or nonresonant terms in both the harmonic conversion process and the quadratic Kerr effect. For example, in the limit of very large intensities the Stark shift will eventually move the system sufficiently far from resonance that the resonant approximation is no longer valid. In this limit the conversion efficiency can be expected to be limited by the nonresonant Kerr effect, although at a higher value than that given by Eq. (104a), according to the discussion of Section 4.2, and the harmonic process will be described by a nonresonant susceptibility. A sufficiently large Stark shift can also bring other levels into resonance, introducing new resonant terms.

4.3.3. *Transient Effects*

Incoherent transient effects, arising from time-dependent incoherent population changes in the resonant level, can occur when the condition $T_2 < t_p < T_1$, where t_p is the duration of the pump pulse and T_2 and T_1 are the dephasing and population decay times of the resonant level, is satisfied. This situation can arise, for example, in gaseous media when collision broadening is dominant and the pulse duration is shorter than the population lifetime of the resonant level. If the pulse duration is shorter than T_2, then the coherent response of the medium must be taken into account.

4.3.3a. INCOHERENT TRANSIENT EFFECTS. In the incoherent transient regime the time-dependent equations for the population of the resonant level [Eq. (72c) for two-photon resonances or Eq. (90c) for three-photon resonances] must be solved. In the extreme incoherent transient regime $T_2 \ll t \ll T_1$ and for a pulse of constant intensity that is turned on at $t = 0$ the population differences at time t for conditions under which the population of the resonant level is determined only by the pump intensity are

$$\rho_{00}(t) - \rho_{22}(t) = \exp\left(-\frac{\int_{-\infty}^{t} I_1^2(t')\,dt'}{I_{s,2}^2 T_1[1 + (\Delta_2' + \delta_2'' I_1(t))^2]}\right) \tag{108a}$$

for the two-photon resonance and

$$\rho_{00}(t) - \rho_{33}(t) = \exp\left(-\frac{\int_{-\infty}^{t} I_1^3(t')\,dt'}{I_{s,3}^3 T_1[1 + (\Delta_3' + \delta_3'' I_1(t))^2]}\right) \tag{108b}$$

for the three-photon resonance. Here $I_{s,2}$ and $I_{s,3}$ are the two- and three-photon saturation intensities given by Eqs. (79) and (94), respectively, and Δ_i' and δ_i'' are the normalized detunings and Stark-shift parameters as given in Eqs. (77b), (78b), (91c), (91d), and (93a).

For two-photon resonant interaction when pump depletion and linear and nonlinear phase mismatches can be neglected the harmonic intensity is given by

$$I_3 = \frac{576\pi^6 N^2 L^2}{n_3 n_1^3 \lambda_1^2 c^2}\left(\frac{q_2' q_2 T_2}{\hbar}\right)^2 \frac{I_1^3 \exp\left(-\dfrac{2\int_{-\infty}^{t} I_1^2(t')\,dt'}{I_{s,2}^2 T_1[1 + (\Delta_2' + \delta_2'' I_1(t))^2]}\right)}{[1 + (\Delta_2' + \delta_2'' I_1(t))^2]}. \tag{109}$$

When the condition

$$\int_{-\infty}^{t} I_1^2(t')\,dt' < I_{s,2}^2 T_1/2 \tag{110}$$

is satisfied the optimum intensity conversion efficiency occurs at the center of the Stark-shifted line. When condition (110) is not satisfied the optimum detuning is given by

$$\Delta'_{2,\,\mathrm{opt}}(t) = -\delta''_2 I_1(t) \pm \left(\frac{2\int_{-\infty}^{t} I_1^2(t')\,dt'}{I_{\mathrm{s},2}^2 T_1} - 1\right)^{1/2}. \tag{111}$$

The intensity conversion efficiency at the optimum detuning is

$$\eta_{\mathrm{opt}}(t) = \frac{576\pi^6 N^2 L^2}{n_3 n_1^3 \lambda_1^2 c^2}\left(\frac{q'_2 q_2 T_2}{\hbar}\right)^2 \frac{I_1^2(t) I_{\mathrm{s},2}^2 T_1 e^{-1}}{2\int_{-\infty}^{t} I_1^2(t')\,dt'} \tag{112a}$$

$$= \frac{18\pi^4 N^2 L^2 q_2'^2 T_2 e^{-1} I_1^2(t)}{n_3 n_1 \lambda_1^2 \int_{-\infty}^{t} I_1^2(t')\,dt'}. \tag{112b}$$

The time dependence of the optimum detuning implies that for a pump pulse with a fixed frequency the optimum efficiency cannot be maintained once saturation of the resonance becomes important.

In the limit of large intensities the conversion efficiency is given by

$$\eta_{\infty}(t) = \frac{576\pi^6 N^2 L^2}{n_3 n_1^3 \lambda_1^2 c^2}\left(\frac{q'_2 q_2 T_2}{\hbar}\right)^2 I_{\mathrm{St}}^2 \exp\left(-\frac{2I_{\mathrm{St}}^2 \int_{-\infty}^{t} I_1^2(t')\,dt'}{I_{\mathrm{s},2}^2 I_1^2(t) T_1}\right). \tag{113}$$

The ratio of the intensity conversion efficiency at the optimum detuning of Eq. (111), or equivalently at an optimum intensity for a given detuning, to that in the limit of large pump intensities is

$$\frac{\eta_{\mathrm{opt}}(t)}{\eta_{\infty}(t)} = \frac{I_1^2 I_{\mathrm{s},2}^2 T_1 \exp\left(-1 + \dfrac{2I_{\mathrm{St}}^2 \int_{-\infty}^{t} I_1^2(t')\,dt'}{I_1^2(t) I_{\mathrm{s},2}^2 T_1}\right)}{2I_{\mathrm{St}}^2 \int_{-\infty}^{t} I_1^2(t')\,dt'}. \tag{114}$$

We recall that this ratio was typically of the order of unity in the steady state [see Eq. (89)]. In the transient regime, however, it can become large when $T_1 \gg t_p$. In such a situation a favorable Stark shift can be much more effective in enhancing the intensity conversion efficiency in a transient interaction than it is in the steady state. A calculation of four-photon resonantly enhanced fifth-harmonic conversion in Li vapor showing the effects of increased efficiency at the Stark-shifted frequency in the absence of population saturation has been given by Stappaerts.[27]

Similar results can be obtained for three-photon resonant conversion. If we again assume that the interaction is effectively phase matched over the range of parameters considered, that pump depletion and absorption of the harmonic can be neglected, and that saturation of the resonant level proceeds only through three-photon absorption of the pump radiation the harmonic intensity in the extreme transient limit is given by

$$I_3 = \frac{576\pi^6 N^2 L^2}{n_3 n_1^3 \lambda_1^2 c^2} \left(\frac{\mu_{03} q_3 T_3}{\hbar}\right)^2 \frac{I_1^3 \exp\left(-\dfrac{2\int_{-\infty}^{t} I_1^3(t')\,dt'}{I_{s,3}^3 T_1[1 + (\Delta_3' + \delta_3'' I_1(t))^2]}\right)}{[1 + (\Delta_3' + \delta_3'' I_1^3(t))^2]}. \tag{115}$$

Again the maximum conversion efficiency occurs at the center of the Stark-shifted line when

$$\int_{-\infty}^{t} I_1^3(t')\,dt' < I_{s,3}^3 T_1/2, \tag{116}$$

whereas for larger times the optimum detuning is given by

$$\Delta_{3,\text{opt}}'(t) = -\delta_3'' I_1(t) \pm \left(\frac{2\int_{-\infty}^{t} I_1^3(t')\,dt'}{I_{s,3}^3 T_1} - 1\right)^{1/2}. \tag{117}$$

At the optimum detuning of Eq. (117) the harmonic intensity is given by

$$I_{3,\text{opt}} = \frac{576\pi^6 N^2 L^2}{n_3 n_1^3 \lambda_1^2 c^2} \left(\frac{\mu_{03} q_3 T_2}{\hbar}\right)^2 \frac{I_1^3(t) I_{s,3}^3 T_1 e^{-1}}{2\int_{-\infty}^{t} I_1^3(t')\,dt'} \tag{118a}$$

$$= \frac{9\pi^3 c N^2 L^2 |\mu_{03}|^2 e^{-1} I_1^3(t) T_2}{n_3 \lambda_1^2 \int_{-\infty}^{t} I_1^3(t')\,dt'}. \tag{118b}$$

In the limit of large intensities at a given initial detuning the harmonic intensity is

$$I_{3,\infty} = \frac{576\pi^6 N^2 L^2}{n_3 n_1^3 \lambda_1^2 c^2} \left(\frac{\mu_{03} q_3 T_2}{\hbar}\right)^2 I_{St}^2 I_1(t) \exp\left(-\frac{2 I_{St}^2 \int_{-\infty}^{t} I_1^3(t')\,dt'}{I_1^2(t) I_{s,3}^3 T_1}\right). \tag{119}$$

Just as in the steady state the harmonic intensity in the incoherent transient regime goes to zero in the limit of large I_1.

In many situations involving pulsed radiation the total generated energy and the energy conversion efficiency are more important than the instantaneous harmonic intensity. The generated energy can be obtained by integrating the intensity in Eq. (109) or (115) over the pump pulse duration. An analytic solution can be obtained for an idealized pump pulse of intensity I_{10} that is turned on at $t = 0$. In this situation the generated energy for either the two- or three-photon resonant conversion can be written in the form

$$\mathscr{W}_3(t) = \mathscr{W}_{3,\max}(1 - e^{-2W^{(q)}t}), \tag{120}$$

where the maximum energy $\mathscr{W}_{3,\max}$ is given by

$$\mathscr{W}_{3,\max} = \frac{18\pi^4 N^2 L^2 q_2'^2 I_{10} T_2}{n_3 n_1 \lambda_1^2} \tag{121a}$$

for the two-photon resonant interaction and

$$\mathscr{W}_{3,\max} = \frac{9\pi^3 c N^2 L^2 |\mu_{03}|^2 T_2}{n_3 \lambda_1^2} \tag{121b}$$

for the three-photon resonant process. In these expressions $W^{(q)}$ is the appropriate multiphoton transition probability as given in Eq. (87) or (101b). For each process the generated energy increases monotonically with time, approaching the value $\mathscr{W}_{3,\max}$ as $t \to \infty$. There is no optimum pulse duration for the generation of harmonic energy.

As a function of detuning from the resonance for a pulse of given intensity and duration the harmonic energy is always maximum at the center of the Stark-shifted line for both types of resonance and for all pump intensities regardless of the degree of saturation term of the resonance. This behavior can be contrasted to that of the intensity in the steady-state regime, for which an optimum detuning from the Stark-shifted line exists when saturation is important.

In the limit of large t_p the energy generated in a two-photon resonant process takes on a linear dependence on the pump intensity but becomes independent of the pulse duration and the initial detuning. The energy in a three-photon resonant interaction approaches a constant value independent of the pump intensity, the pulse duration, and the initial detuning in the limit of large t_p.

The independence of the third-harmonic energy on the initial detuning was pointed out by Harris,[45] Harris and Bloom,[46] and Stappaerts.[27] Harris[45] and Stappaerts[27] have also shown a similar independence on detuning for higher-order frequency conversion processes involving limitations by multiphoton absorption.

The energy conversion efficiency for the two types of processes can be expressed as

$$\eta_{\mathscr{W}}(t) = \eta_{\mathscr{W},\max}\left(\frac{1 - e^{-2W^{(q)}t}}{2W^{(q)}t}\right), \tag{122}$$

where

$$\eta_{\mathscr{W},\max} = \frac{576\pi^6 N^2 L^2}{n_3 n_1^3 \lambda_1^2 c^2}\left(\frac{q_2' q_2 T_2}{\hbar}\right)^2 \frac{I_{10}^2}{[1 + (\Delta_2' + \delta_2'' I_1)^2]} \tag{123a}$$

for the two-photon process and

$$\eta_{\mathscr{W},\max} = \frac{576\pi^6 N^2 L^2}{n_3 n_1^3 \lambda_1^2 c^2}\left(\frac{\mu_{03} q_3 T_2}{\hbar}\right)^2 \frac{I_{10}^2}{[1 + (\Delta_3' + \delta_3'' I_1)^2]} \tag{123b}$$

for the three-photon process.

The behavior of both the harmonic energy and the energy conversion efficiency for the idealized square pulse are shown in Fig. 13. The condition $W^{(q)}t = 1$ has been used by some authors (Stappaerts,[27] Harris,[45] and Harris and Bloom[46]) as an estimate of the optimum condition for transient conversion in analogy with the optimum condition for the steady-state intensity conversion [see Eq. (88)]. The point corresponding to the condition is indicated in Fig. 13. At this point the harmonic energy has reached about 86% of its final value but the energy conversion efficiency is only 43% of its maximum value.

For the idealized square pulse treated here the energy conversion efficiency is maximum at the beginning of the pulse and decreases as t increases. For a more realistic pulse shape we might expect that the maximum conversion efficiency would occur somewhat later in the pulse. However, the maximum energy should still be generated at the end of the pulse.

4.3.3b. Coherent Transient Effects. Coherent transient effects involve a coherent transfer of energy from the radiation fields to the material system and back again. Analysis of these effects requires solution of the time-dependent equations for both the diagonal and off-diagonal elements of

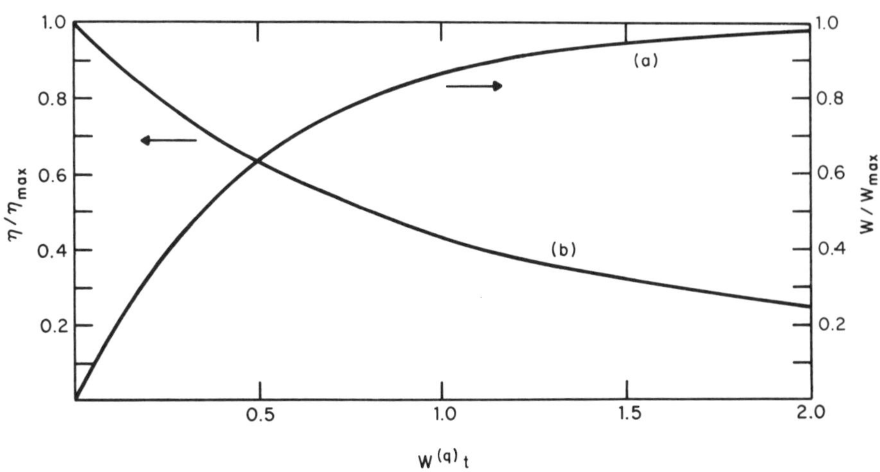

Fig. 13. Dependence of the normalized third-harmonic energy (a) and energy conversion efficiency (b) on time for two- or three-photon resonant conversion in the incoherent transient regime. $W^{(q)}$ is the appropriate two- or three-photon absorption probability, and the pump pulse was assumed to be constant in time starting at $t = 0$.

the density matrix [e.g., Eqs. (72b) and (72c)]. Generally these analyses are given for the extreme transient regime in which $t_p \ll T_1$, T_2 and the effects of the relaxation times can be neglected. The coherent effects usually involve some degree of resonant enhancement of the nonlinear susceptibility. If there is no incoherent absorption, e.g. due to finite relaxation times or to continuum states, the interactions can be analyzed within the framework of the vector model as discussed in Section 2-6.2.2.

Near-Resonant Interactions. Four-wave sum-frequency mixing of the form $\omega_4 = 2\omega_1 + \omega_2$ with a near-two-photon resonance has been discussed by Grischowsky *et al.*[51] within the approximations of the adiabatic following model (see Section 2-6.2.2). Anikin *et al.*[28] have given a similar discussion for third-harmonic generation. The nonlinear polarization for third-harmonic generation under conditions of two-photon adiabatic following is

$$P^{\mathrm{NL}}(3\omega) = \frac{1}{4}\frac{Nq_2 q_2' A_1^3}{[(\Delta_2 + \delta_2|A_1|^2)^2 + q_2^2|A_1|^4/4\hbar^2]^{1/2}} \tag{124}$$

and the third-harmonic intensity for phase-matched plane waves is

$$I_3 = \frac{576\pi^6 N^2 L^2}{n_1^3 n_3 \lambda_1^2 c^2}\left(\frac{q_2 q_2' T_2^2}{\hbar^2}\right)\frac{I_1^3}{[(\Delta_2' + \delta_2'' I_1)^2 + (I_1^2/I_s^2)(T_2/T_1)]}, \tag{125}$$

where q_2, q_2', Δ_2, Δ_2', δ_2, δ_2'', and I_S are given in Eqs. (76a), (76b), (77a), (77b), (73), (78a), (78b), and (79). Although the details of the saturation behavior in this regime are different from those in the steady state, the harmonic intensity again becomes proportional to the pump intensity at high pump intensities:

$$\lim_{I_1\to\infty} I_3 = \frac{36\pi^4 q_2'^2 N^2 L^2 I_1}{n_1 n_3 \lambda_1^2}\frac{I_{\mathrm{St}}^2/[I_s^2 T_1/T_2]}{[1 + I_{\mathrm{St}}^2/[I_s^2 T_1/T_2]]} \tag{126}$$

and the efficiency approaches a value that is independent of the pump intensity just as it does in the steady state [see Eq. (86)].

Transient Coherent Effects at Exact Resonance. Transient coherent effect under conditions of exact resonance have been discussed by New[50] and by Georges *et al.*[21,47] At exact resonance, neglecting Stark shifts, pump depletion, and relaxation effects and assuming perfect phase matching, Eqs. (72) become

$$\frac{dA_3}{dz} = i\frac{3(2\pi)^2}{n_3\lambda_1} Nq_2' A_1 \rho_{\bar{2}0}, \tag{127a}$$

$$\frac{\partial \rho_{\bar{2}0}}{\partial t} = i\frac{q_2}{4\hbar} A_1^2(\rho_{00} - \rho_{22}), \tag{127b}$$

$$\frac{\partial}{\partial t}(\rho_{00} - \rho_{22}) = -\frac{q_2}{\hbar}\,\mathrm{Im}(\rho_{\bar{2}0} A^{*2}), \tag{127c}$$

$$\rho_{00} + \rho_{22} = 1, \tag{127d}$$

where q_2 and q'_2 are given in Eqs. (76a) and (76b), respectively. New has treated this interaction within the framework of the vector model as discussed in Section 2-6.2.2. The solution for the nonlinear polarization at the third harmonic, which in this situation is given by $P_{\mathrm{NL}}^{3\omega} = q'_2 A_1 \rho_{\bar{2}0}$, is

$$P_{\mathrm{NL}}^{3\omega}(t) = iq'_2 \frac{A_1^2(t)}{A_1^*(t)} \sin\theta(t), \tag{128}$$

where $\theta(t)$ is the two-photon area of the pulse defined as

$$\theta(t) = \frac{q_2}{2\hbar}\int_{-\infty}^{t} |A_1(t')|^2\,dt'. \tag{129}$$

The corresponding time-dependent third-harmonic intensity for phase-matched plane waves is given by

$$I_3(t) = \frac{36\pi^4 N_0^2 L^2 q_2'^2}{n_1 n_3 \lambda_1^2} I_1(t)\sin^2\theta(t). \tag{130}$$

Optimum intensity conversion efficiency occurs when the two-photon area of the pulse is $\theta = (2m+1)\pi/2$. Because of the integrating nature of the nonlinear polarization the optimum efficiency can occur at a point where there is relatively little intensity if the pulse reaches the optimum area near its end. Optimum conversion efficiency will occur at the center of symmetric pulses of total area $(2m+1)\pi$. The optimum efficiency as determined from Eq. (130) with $\theta = (2m+1)\pi/2$ is again independent of the pump intensity.

Because of the integrating nature of the nonlinear polarization, much of the pump pulse is wasted as far as the frequency conversion process is concerned for the interactions discussed in the previous paragraph. New has pointed out that the sum-frequency process $\omega_4 = 2\omega_1 + \omega_2$ can be used to obtain more efficient conversion of the pulse at ω_2 if it is delayed in time to allow the coherent population to build up during the pulse at ω_1. This effect was discussed in Section 2-6.2.1, and an example of the improved conversion of the pulse at ω_2 is shown in Fig. 2-26. Nonsynchronous excitation has also been discussed in connection with the measurement of coherent relaxation times by Matsuoka *et al.*[55] and by Royt and Lee.[56]

Georges *et al.*[21,47] have treated transient third-harmonic generation when the generated radiation lies in the photoionization continuum. This situation is typical of many experiments involving the generation of ultraviolet and vacuum ultraviolet radiation in metal vapors. Under these conditions the equation for ρ_{22} carries an extra term describing the photoionization process.

In addition, Eq. (127d) is no longer satisfied. As a result the vector model cannot be used and the equations must be integrated numerically.

Georges *et al.*[47] have given numerical results for the specific case of third-harmonic conversion of ruby laser radiation in Cs vapor. In addition to the effects discussed before, they included the nonlinear phase shifts associated with the population changes along with the saturation of the susceptibility. Their results for the total third-harmonic energy as a function of the incident pump intensity are shown in Fig. 14.

Consideration of the nonlinear phase changes associated with the decrease in the ground-state population led to an increase in the calculated third-harmonic energy because the large wave vector mismatch that was present in the unperturbed system was reduced. Georges *et al.*[47] have pointed out that when photoionization is important the process never reaches steady state for pulse durations shorter than the recombination time. As a result

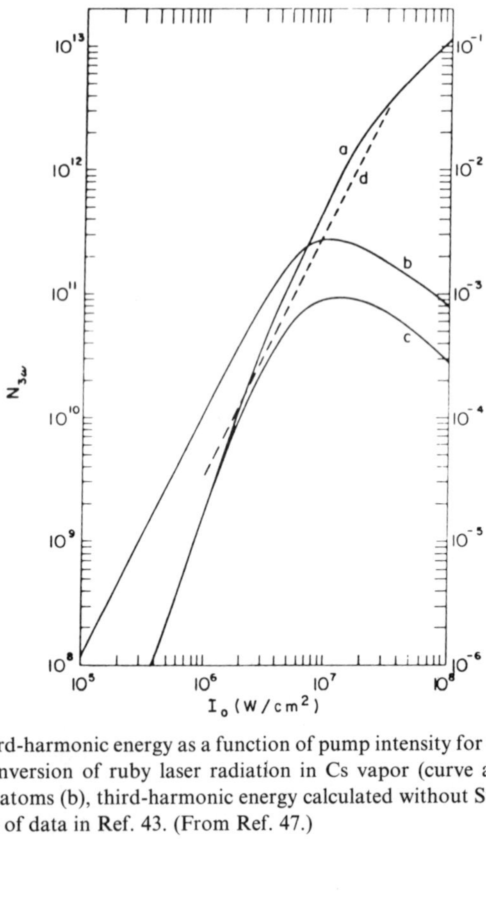

Fig. 14. (a) Third-harmonic energy as a function of pump intensity for two-photon resonant third-harmonic conversion of ruby laser radiation in Cs vapor (curve a). Other curves give fraction of excited atoms (b), third-harmonic energy calculated without Stark shift of resonant level (c), and slope of data in Ref. 43. (From Ref. 47.)

they have concluded that the third-harmonic conversion of ruby laser radiation in Cs vapor may well be limited by the almost complete depletion of the ground-state population by multiphoton ionization and suggest that this process may be important in other systems involving two-photon resonant conversion in metal vapors.

Coherent Propagation Effects. When the pulse duration is shorter than the dephasing time of a two-photon resonant transition, coherent propagation effects can occur. These effects usually involve pulse reshaping due to the interchange of energy between the material system and the optical fields as the pulse propagates. Their effects on third- and fifth-harmonic conversion have been analyzed by Diels and Georges.[52,53] Poluektov and Nazarkin[54] have treated third-harmonic conversion with coherent propagation.

When coherent propagation effects are present the time-dependent partial derivative must be retained in the equation for the optical fields in addition to the time derivatives in the equations for the elements of the density matrix. The equations describing third-harmonic generation now become

$$\frac{\partial A_3}{\partial z} + \frac{n_3}{c}\frac{\partial A_3}{\partial t} = i\frac{3(2\pi)^2}{n_3 \lambda_1} N q_2' A_1 \bar{\rho}_{20}, \tag{131a}$$

$$\frac{\partial A_1}{\partial z} + \frac{n_1}{c}\frac{\partial A_1}{\partial t} = i\frac{(2\pi)^2}{n_3 \lambda_1} N[q_2' \overset{+}{\rho}_{02} A_3 + 2q_2 \bar{\rho}_{20} A_1^*], \tag{131b}$$

$$\frac{\partial \bar{\rho}_{20}}{\partial t} + i(\Delta_2 + \delta_2 |A_1|^2 - i/T_2)\bar{\rho}_{20} = \frac{i}{4\hbar}[q_2 A_1^2 + q_2' A_1^* A_3], \tag{131c}$$

$$\frac{\partial}{\partial t}(\rho_{00} - \rho_{22}) + \frac{1}{T_1}(\rho_{00} - \rho_{22} - 1) = -\mathrm{Im}\left(\frac{q_2}{\hbar}\bar{\rho}_{20} A_1^{*2} + \frac{q_2'}{\hbar}\overset{+}{\rho}_{02} A_1 A_3^*\right), \tag{131d}$$

$$\overset{+}{\rho}_{02} = (\bar{\rho}_{20})^*. \tag{131e}$$

These equations include the effects of two-photon absorption and Raman amplification of the pump as well as nonlinear phase changes at the pump and harmonic.

Using equations equivalent to (131a)–(131e), and also including linear absorption at both frequencies, Diels and Georges[52,53] considered conversion of pulses with various two-photon coherent areas [see Eq. (129)]. Their analysis was directed expressly toward conversion in two-photon resonant metal vapors, for which in general $q_2' > q_2$ and for which the absorption arose from photoionization from the ground and excited states. Their calculations showed that improved conversion can be obtained through the use of a pump pulse that has a 90° phase shift halfway through. Such a pulse compensates for the change in sign of the nonlinear polarization as population is transferred from the ground state to the two-photon resonant level. An example of the calculated third-harmonic power as a function of propagation distance is shown in Fig. 15. Pulse reshaping at both the fundamental and harmonic frequencies due to the coherent interaction as well as phase modulation are evident in their calculations.

As a result of their numerical analysis Diels[53] arrived at the empirical result that the optimum third-harmonic energy conversion efficiency is given by

$$\eta_{\text{opt}} = \left[\sum \mu_{0i}\mu_{i2}\left(\frac{1}{\omega_{i0} - 3\omega} + \frac{1}{\omega_{i0} + \omega}\right)\right] \Bigg/ \sum \frac{\mu_{0i}\mu_{i2}}{\omega_{i0} - \omega} = R^{1/2} \quad (132)$$

for materials in which $R < 1$ (compare with the results of Section 4.31a in which the limiting intensity efficiency for steady-state interactions was shown to be approximately equal to R when $R \ll 1$). They predict on the basis of numerical analysis that a third-harmonic energy conversion efficiency of 6.5% for a process that is two-photon resonant with the 2s–4s transition in Li, producing radiation at 190 nm, should be possible. Still larger conversion efficiencies might be obtained in media with a larger value of R.

Pulse reshaping is also evident in the calculations of Poluektov and Nazarkin.[54] These authors analyzed the numerical solutions of the coherent propagation equations over a wide range of material parameters and concluded that conversion efficiencies above 20% should be possible for pump pulses with a two-photon area greater than 4π.

4.3.4. *Effects of Wave Vector Mismatch in Resonant Interactions*

The effects that lead to saturation of the nonlinear susceptibility can also cause intensity-dependent phase changes that affect the conversion efficiency. As was pointed out earlier the effects related to the saturation of the susceptibility are more likely to be dominant in interactions involving relatively low pressure gases and high intensities for which the transition probability to the resonant level is relatively high. The effects related to changes of the coherence length, on the other hand, are more likely to be encountered in interactions in relatively high pressure gases or liquids at lower intensities for which the transition probabilities to the resonant levels remain relatively small.

Expressions for the nonlinear refractive indices at the fundamental and harmonic wavelengths in a medium with discrete energy levels were given in Eqs. (38a) and (38b). The contributions of the incoherent population changes were not included in these expressions and must be added separately.

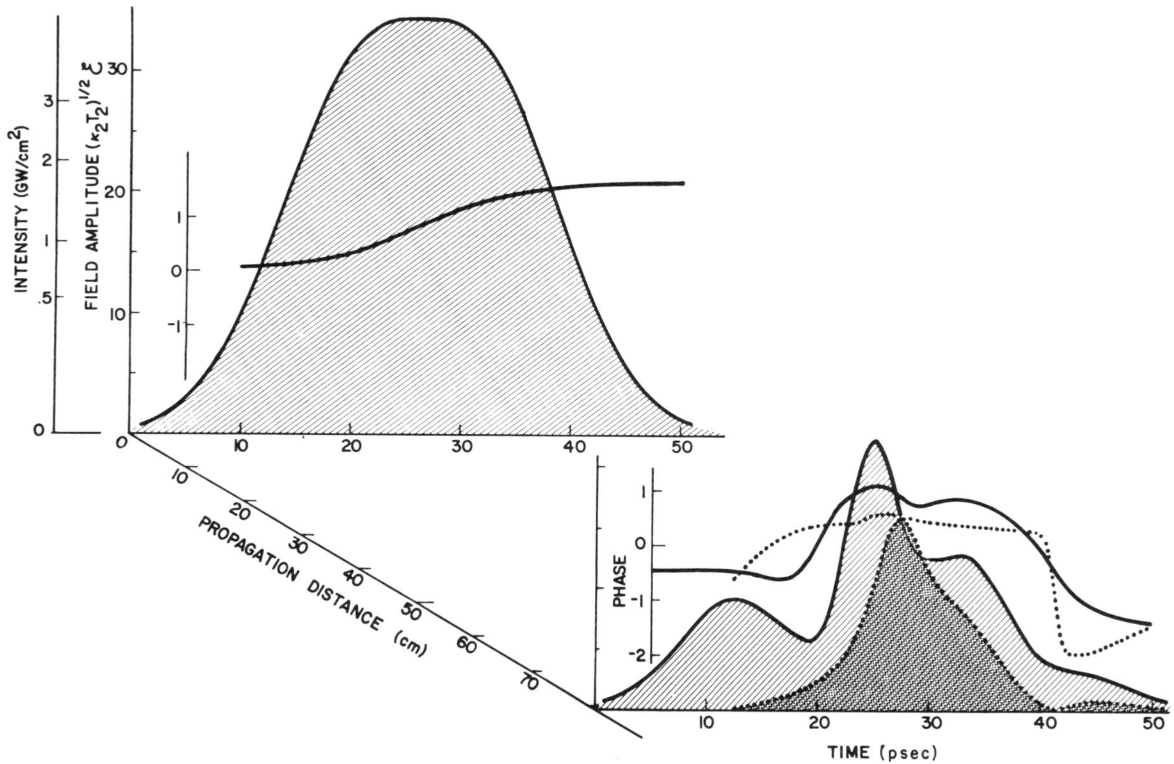

Fig. 15. Coherent propagation effects in third-harmonic generation. Initial pulse has a two-photon area of 4.89 and a 90° phase shift. ——, fundamental; · · ·, third harmonic. (From Ref. 52.)

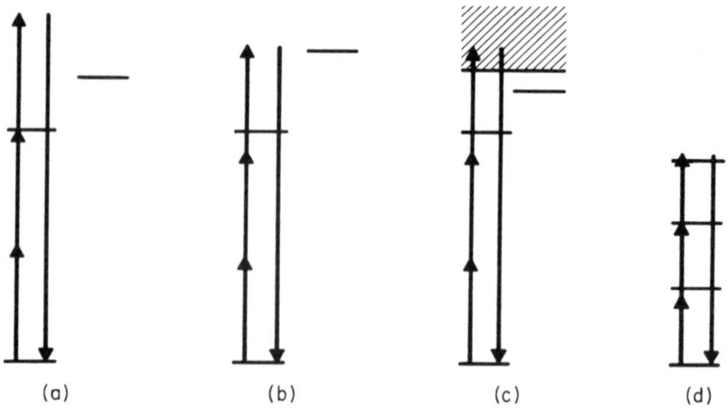

Fig. 16. Level diagrams for various types of interactions that lead to saturation of the susceptibility. (a) Two-photon resonant third-harmonic generation in a negatively dispersive medium. Resonance is with electronic levels. Saturation occurs because of two-photon absorption. (b) Three-photon resonant third-harmonic conversion. Saturation occurs because of multiphoton absorption to the same level that determines the linear phase mismatch. (c) Saturation due to multiphoton ionization. (d) Two-photon resonant conversion and saturation in a molecular vibrational system.

An example of the effect of the nonlinear refractive index resulting from incoherent population changes on the coherence length was given in Section 4.3.2 for three-photon resonant interactions. In general the magnitude of the effect of the population changes on the wave vector mismatch depends on the particular material and the wavelengths that are involved. As a result each particular system must be evaluated individually. In this section we shall give some general trends for particular types of interactions involving population redistribution. Since only general properties are considered Stark shifts will not be included in the discussion.

Four types of interaction that can lead to saturation of the susceptibility are illustrated in Fig. 16. In Fig. 16a third-harmonic generation occurs in a negatively dispersive medium and is enhanced by a two-photon resonance with an electronic level. Saturation of the susceptibility is assumed to occur by two-photon absorption to the resonant level. The situation on Fig. 16b is similar except that the two-photon resonance is not necessary and the saturation of the susceptibility occurs through multiphoton absorption to the same level that determines the linear coherence length. In Fig. 16c the limitation on the susceptibility is assumed to arise from multiphoton ionization. The situation in Fig. 16d involves levels that are almost equally spaced such as might be encountered in two-photon resonantly enhanced third-harmonic conversion of infrared radiation using the vibrational levels of molecules.

The general trends of the intensity-dependent phase changes associated with the population redistribution in these situations are as follows. In all four situations the ground-state population is reduced when saturation of the susceptibility occurs. This will reduce the magnitude of the quantity $[n(\omega) - 1]$ at both the fundamental and third harmonic. Because the intial dispersion was chosen to be negative, this leads to a positive contribution to the wave vector mismatch.

Population in the final state also makes a contribution to the wave vector mismatch. In the two-photon resonant process of Fig. 16a the contribution to the refractive index at the third-harmonic frequency from the two-photon population is likely to be negligible compared to the near-resonant contribution of the gound state population, at least at the optimum values of the populations as determined in Section 4.3.1b. In this case the entire intensity-dependent refractive index at the harmonic due to the population changes arises from the depletion of the ground state. The population in the two-photon level can make a substantial contribution to the refractive index at the fundamental, however, because of the near-one-photon resonance with the 2–3 transition. For the energy level structure of Fig. 16a this contribution is negative, leading to another positive contribution to the wave vector mismatch which reinforces the effect of the gound-state depletion. If sufficient population in the two-photon level is produced, then the sign of the dispersion can be reversed.

The change in the difference in the refractive indices at the harmonic and fundamental due to the population redistribution can be expressed as

$$\delta(\Delta n) = N[(n_2(3\omega) - n_2(\omega)) - (n_0(3\omega) - n_0(\omega)]\rho_{22} = N\{\Delta n\}\rho_{22}, \quad (133)$$

where $n_i(\omega_j)$ is the contribution of the population in level i to the refractive index at ω_j. If we assume that the dominant contributions to the refractive indexes at both frequencies comes from terms that are resonant with level 3 and that the population distribution is given by the optimum values as determined by Eqs. (75b) and (82). This change can be expressed as

$$\{\Delta n\} = -\frac{\pi N}{4\hbar(\omega_{30} - 3\omega)}(|\mu_{30}|^2 + |\mu_{23}|^2). \quad (134)$$

The corresponding change in the wave vector mismatch is

$$\begin{aligned} \Delta k_{\text{two-photon}} &= \Delta k_{\text{final}} - \Delta k_{\text{initial}} \\ &= \frac{3\pi^2 N}{\lambda_1 \hbar(3\omega - \omega_{30})}(|\mu_{30}|^2 + |\mu_{23}|^2). \end{aligned} \quad (135)$$

The net change in the dispersion for this process is illustrated schematically in Fig. 17a.

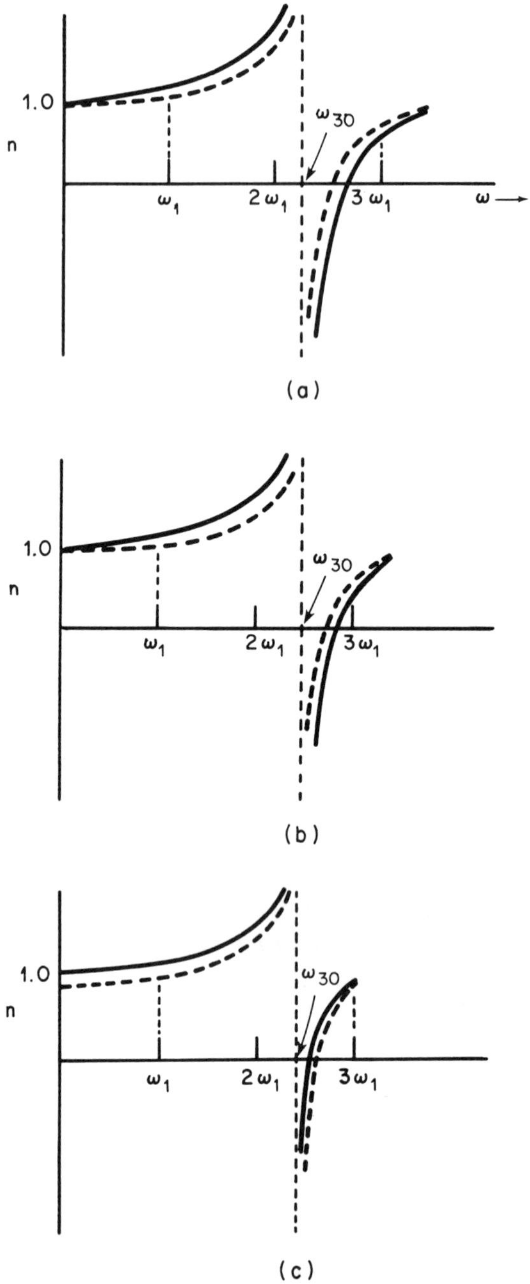

Fig. 17. (a) Change in dispersion due to two-photon population changes for system shown in Fig. 16a. (b) Change in dispersion due to population changes caused by multiphoton absorption to level that determines the linear dispersion. (c) Change in dispersion due to free electrons resulting from multiphoton ionization.

For the process shown in Fig. 16b the population in the upper level will contribute primarily to the refractive index at the harmonic wavelength, leading to a further reduction in the magnitude of the quantity $n_3 - 1$ and an additional positive contribution to the wave vector mismatch. If there is no two-photon resonant level that develops significant population, then the dominant change in the refractive index at the fundamental will come from the reduction in the ground-state population. The contribution to the wave vector mismatch from the change in population is

$$\Delta k_{\text{three-photon}} = \frac{|2\pi^2|\mu_{30}|^2\omega_{30}N}{\lambda_1\hbar}\left(\frac{1}{\omega_{30}^2 - (3\omega)^2} - \frac{1}{\omega_{30}^2 - \omega^2}\right) \times (\rho_{00} - \rho_{33} - 1). \quad (136)$$

If the wave vector mismatch is determined entirely by the detuning of the harmonic from the upper level, then this expression becomes at the condition of optimum population change [see Eqs. (82) and (75b)]

$$\Delta k_{\text{three-photon}} = -\frac{6\pi^2|\mu_{30}|^2 N}{\lambda_1\hbar(\omega_{30} - 3\omega)}. \quad (137)$$

These changes are illustrated in Fig. 17b.

If there is a two-photon resonant level that develops significant population in addition to the population in the three-photon dispersive level, then the refractive index at the fundamental will receive a near-resonant contribution, as was discussed in connection with the process in Fig. 16a. In such a situation the net contribution to the wave vector mismatch of the population redistribution will depend on the relative population in the 2 and 3 levels and on the relative oscillator strengths of the 3–0 and 3–2 transitions.

In the process shown in Fig. 16c the refractive indices at the fundamental and harmonic wavelengths receive contributions from the free electrons that are created by the multiphoton ionization process in addition to the changes caused by the redistribution of population among the discrete levels. An expression for the contribution of the free electrons to the refractive indices was given in Eq. (9). The contribution of the free electrons to the refractive index at both wavelengths is negative, of the same sense as the contribution of the reduction of the ground-state population at the fundamental wavelength but of opposite sense to that at the harmonic wavelength. The contributions of the ions to the refractive indexes is usually negligible in comparison to that of the electrons. The changes in dispersion of the refractive index for this process are illustrated in Fig. 17c. It was assumed for purposes of illustration that the negative contribution of the free electrons to the refractive index at the harmonic outweighed the positive contribution caused by the reduction in ground-state population. Actual changes in the refractive index at the harmonic can be either positive or negative depending on the details of the situation. Qualitatively similar changes is disposition can be expected from dielectric breakdown.

The situation shown in Fig. 16d is slightly different from the others in that the nearly resonant vibrational levels are more nearly equally spaced then are the electronic levels. The change in the wave vector mismatch because of the transfer of population from the ground state to the two-photon resonant level at the optimum fractional population transfer of $\frac{1}{4}$ is given by

$$\Delta k = \frac{12\pi^2 N}{\lambda_1}\left[\frac{|\mu_{21}|^2\omega_{21}}{2\hbar(\omega_{21}^2 - (3\omega)^2)} - \frac{|\mu_{10}|^2\omega_{10}}{\hbar(\omega_{10}^2 - (3\omega)^2)} - \frac{|\mu_{21}|^2\omega_{21}}{2\hbar(\omega_{21}^2 - \omega^2)} + \frac{|\mu_{10}|^2\omega_{10}}{\hbar(\omega_{10}^2 - \omega^2)}\right]. \tag{138}$$

Using the relation between the transition moments of the first and second vibrational levels, $|\mu_{21}|^2 = 2|\mu_{10}|^2$, and the fact that $\omega_{21} \approx \omega_{10}$, it can be seen that the change in wave vector mismatch due to population changes among the vibrational levels is considerably smaller than those resulting from population changes among electronic levels. Since the resonant Kerr contribution to the nonlinear index is zero at exact resonance, the effects of the intensity-dependent phase changes for conversion that is enhanced by vibrational resonances can be less important relative to the two-photon absorption and Raman processes than they are in media which have electronic resonances. This characteristic was pointed out by Stappaerts[36] and appears to be substantiated by the observations of Kildal and Brueck[38] on the third-harmonic conversion of CO_2 laser radiation in liquid CO.

The effects which these intensity-dependent phase changes have on the efficiency of frequency conversion depends on the conditions under which the conversion is done. If the conversion process is phase matched at low intensities, then the nonlinear phase changes associated with population redistribution will lead to limitations of the type discussed in Section 4.2. If the process is appropriately phase mismatched at low intensities, then the nonlinear index can improve the phase matching, thus increasing the conversion efficiency. If conversion is done in one coherence length, as, for example, in a collimated beam at an optimum interaction length or in a focused beam at an optimum phase mismatch, then the coherence length can be increased by the nonlinear phase changes associated with the population redistribution. This in turn leads to an increase in the conversion efficiency over that which would be obtained if only saturation of the susceptibility occurred. This type of behavior was evident in the analysis of three-photon resonantly enhanced conversion in Section 4.3.2. It must be remembered, however, that in some situations, these increases may only partially offset the limiting effects of saturation of the susceptibility that also accompanies the population changes.

The simultaneous effects of saturation of the susceptibility and changes in the coherence length due to population redistribution have been considered by a few authors. We have already pointed out the effects that were calculated by Georges *et al.*[47] for third-harmonic conversion of ruby laser radiation in Cs vapor (see Section 4.3.3b). Puell *et al.*[40] have analyzed two-photon resonant third-harmonic conversion including the effects of two-photon absorption of the fundamental, linear absorption at both wavelengths due to photoionization, repumping of the harmonic into the fundamental through Raman scattering, and intensity-dependent phase shifts due to the two-photon population changes. Their analysis in Puell *et al.*[40] neglected the Stark shifts of the resonant level, implying that these results are appropriate in general for pump intensities that are lower than the saturation intensity of the two-photon transition for which the fractional population changes are small. Stark shifts were included in the analysis given in Scheingraber and Vidal.[48]

They have shown that significant nonlinear phase changes can develop for population changes that are too small to cause saturation of the susceptibility and that these effects are more likely to be encountered in high-density media than in low-density ones. Their analysis shows that when the two-photon absorption is important, the harmonic intensity develops a minimum at the center of the Stark-shifted line just as it does when the conversion is limited by two-photon saturation of the susceptibility (see Section 4.3.1b). Their calculations also indicate that for a given pump intensity there is an optimum linear phase mismatch, and therefore an optimum product for the density and path length, for which conversion can be optimized at high intensities. An example of their calculated third-harmonic intensity is shown as a function of the pump frequency for various pump intensities in Fig. 18. These calculations were done for two-photon resonant conversion in Sr vapor, producing radiation at ~ 192 nm. The column density was optimized for each value of pump intensity and frequency. Strong resonant behavior is evident at the low pump intensities, but at high pump intensities the harmonic intensity is relatively independent of the detuning from the two-photon resonance. Puell *et al.*[40] have also examined the improvements to be gained from use of a linear phase mismatch. An example of their calculations, along with some experimental results, are shown in Fig. 19. Here the degree of linear phase mismatch was controlled by the ratio of Sr to Xe. Although each curve shows saturation at high intensities, the curve with the greater linear phase mismatch ultimately produces the higher harmonic intensity.

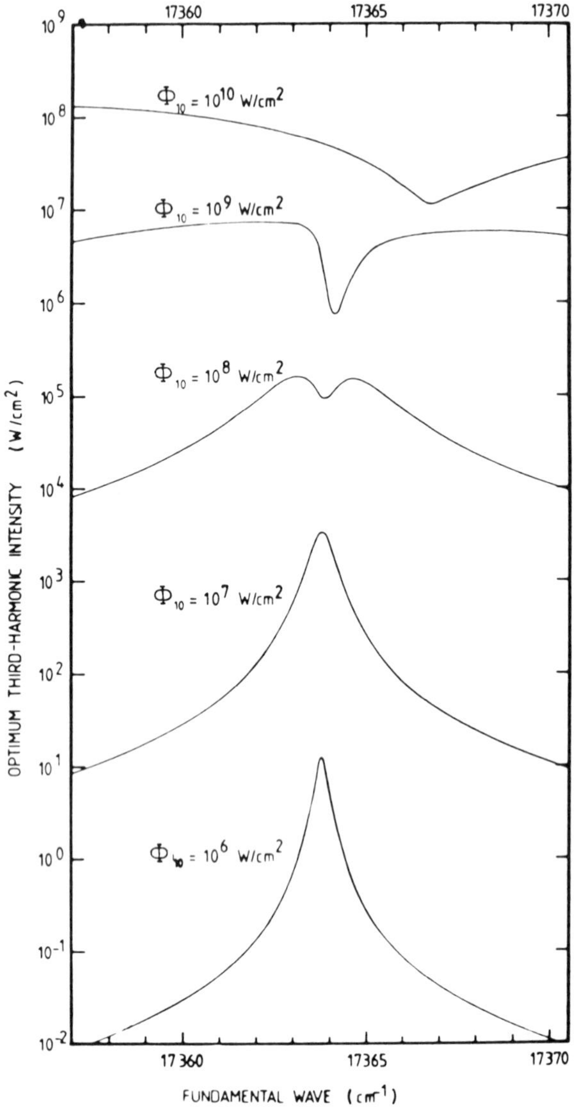

Fig. 18. Calculated optimum third-harmonic intensity for two-photon resonant conversion in Sr vapor. Calculations took into account phase mismatch due to two-photon population and Stark shift of resonant level. Column density was adjusted to provide an optimum linear phase mismatch for each incident pump intensity. (From Ref. 48.)

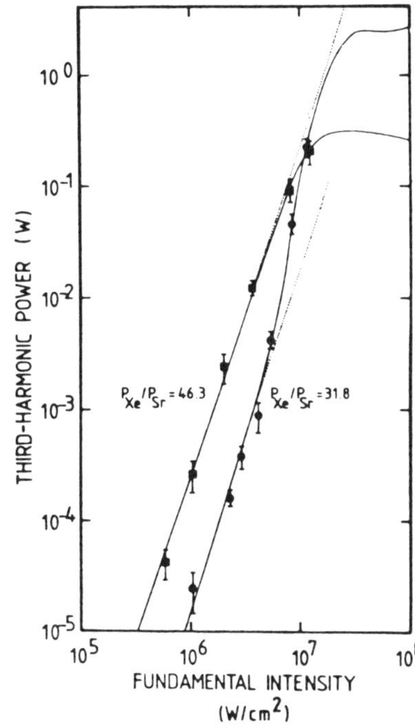

Fig. 19. Third-harmonic conversion in Sr:Xe as a function of incident pump intensity for Xe pressures above (square points) and below (dots) the linear phase-matching value. The effects of saturation of the two-photon resonance are included in the solid curve but are neglected in the dotted curve. (From Ref. 40.)

4.3.5. *Relative Importance of Wave Vector Mismatches, Absorption, and Saturation of the Susceptibility in Resonant Interactions*

The discussions in the preceding sections have treated limitations on third-harmonic generation in resonant systems due to various competing processes such as multiphoton absorption, Raman amplification of the pump, breaking of phase matching due to near-resonant Kerr effects or multiphoton population changes, and saturation of the susceptibility. In general, all of these processes can be expected to contribute to the limitations on conversion efficiency in a given situation. However, under certain conditions that depend on parameters such as the density of the nonlinear medium, the interaction length, the pump intensity, the pulse duration, and the detuning from exact resonance, it is possible for the limitations on conversion efficiency to arise primarily from one type of competing process. The results of the preceding sections for specific competing processes are summarized in Table 6. In this section we shall examine the conditions under which the various competing processes can be expected to dominate the limitations on conversion efficiency, quantifying some of the qualitative statements that were made earlier.

TABLE 6 *Summary of Maximum Conversion*

Limiting process	Efficiency	Intensity/Energy
Nonresonant interactions		
Linear absorption (*q*th-harmonic generation)	$\dfrac{4K_{a,q}^2 I_1^{(q-1)}}{\alpha_{a,q}^2}$	
Two-photon absorption (third-harmonic generation)	$\dfrac{4\kappa_3^2}{\beta_{11}^2}$ $\dfrac{\kappa_3^2}{\beta_{13}^2}$	
Third-order nonlinear refractive index (*q*th-harmonic generation)	$\dfrac{(2\pi)^{q-3} q^2 I_1^{q-3}}{81 n_q n_1^{q-2} c^{q-3}} \times \left\| \dfrac{\chi_a^{(q)}(-q\omega, \omega, \omega, \ldots, \omega)}{\dfrac{\chi_a'(-q\omega, \omega, -\omega, q\omega)}{n_q} - \dfrac{\chi_a'(-\omega, \omega, -\omega, \omega)}{2n_1}} \right\|^2$	
Resonant interactions		
Two-photon absorption, Raman amplification	$1/R$ $9R/4$	
Two-photon resonant Kerr effect	$\dfrac{R}{(R/2 - 1)^2}$	
Population redistribution, Stark shifts	$\dfrac{9\pi^4 N^2 L^2 q_2'^2 T_2}{n_1 n_3 \lambda_1^2 T_1}$	

Conditions	Comments
$\alpha_{a,q} \gg \Delta k, K_q^{-1}(L_{dep}^{-1})$	Absorption at harmonic; no pump depletion or pump absorption
$\beta_{11}^2 \lvert A_{10}\rvert^2 \gg 1/L, 1/L_{dep}$ $I_{10} t_{pulse} \ll 1.4\hbar\omega N_0 L_{abs}$	Two-photon absorption at fundamental; phase-matched plane waves
$\beta_{13}^2 \lvert A_{10}\rvert^2 \gg 1/L, 1/L_{dep}$	Two-photon absorption at harmonic induced by intensity at fundamental
$\Delta k_{21} L = \pi$ $L_c \ll L_{dep}$	Plane waves, linear phase matching, no pump depletion
$R \gg 1;\ \tilde{z} \gtrsim 3/\sqrt{R}$ $I_1 \ll I_s, R\pi\hbar c/6\{\Delta n\}T_1$ $\Delta_2 = 0$	Plane waves, linear phase matching throughout interaction, negligible population changes
$R \ll 1;\ \tilde{z} \gtrsim 3/\sqrt{R}$ $I_1 \ll I_s, 2\pi\hbar c/3\{\Delta n\}T_1$ $\Delta_2 = 0$	Plane waves, linear phase matching throughout interaction, negligible population changes
$\Delta k_{21} L = \pi$ $R \ll 1$ or $R \gg 1;\ \Delta_2 T_2 \gg 1;\ I_1 \ll I_s$ $I_1 < \dfrac{n_1 \hbar c \Delta_2 T_2 (R-2)}{4\{\Delta n\} n_3 T_1}$	Plane waves, linear phase matching, negligible population changes
$\Delta_2 = \left[-\delta_2' I \pm \sqrt{(I_1^2/I_{s,2}^2 - 1)}\right] T_2^{-1}$ $I_1 > I_{s,2}$	Two-photon resonance, plane waves, no pump depletion, linear phase matching, maximum efficiency at optimal detuning

(continued)

TABLE 6 *(continued)*

Limiting process	Efficiency	Intensity/Energy
	$\frac{36\pi^4 N^2 L^2 q_2'^2 T_2}{n_1 n_3 \lambda_1^2 T_1} \frac{(I_{\mathrm{St}}/I_{\mathrm{S},2})^2}{[1 + (I_{\mathrm{St}}/I_{\mathrm{S},2})^2]^2}$	
		$\frac{9\pi^3 c N^2 L^2 \lvert\mu_{03}\rvert^2 T_2}{2n_3 \lambda_1^2 T_1}$
		$\frac{16 n_3 \pi^2 q_3^2 I_1 \Delta_3'^2}{n_1^3 c^2 \lvert\mu_{03}\rvert^2 \delta_3''^2}$
		$\frac{16\pi^2 n_3 q_3^2 I_1^3}{n_1^3 c^2 \lvert\mu_{03}\rvert^2}$
		$\frac{4\pi^2 n_3 q_3^2 I_1^3}{n_1^3 c^2 \lvert\mu_{03}\rvert^2}$
Transient effects Population changes, Stark shifts	$\frac{576\pi^6 N^2 L^2}{n_3 n_1^3 \lambda_1^2 c^2} \left(\frac{q_2' q_2 T_2}{\hbar}\right)^2 I_1^2 \times \exp\left(-2 \frac{\int_{-\infty}^{t} I_1^2(t')\,dt'}{I_{\mathrm{s},2}^2 T_1}\right)$	
	$\frac{18\pi^4 N^2 L^2 q_2'^2 T_2}{n_3 n_1 \lambda_1^2} \frac{e^{-1} I_1^2}{\int_{-\infty}^{t} I_1^2(t')\,dt'}$	
		$\frac{576\pi^6 N^2 L^2}{n_3 n_1^3 \lambda_1^2 c^2} \left(\frac{\mu_{03} q_3 T_2}{\hbar}\right)^2 I_1^3 \exp\left(-\frac{2\int_{-\infty}^{t} I_1^3(t'}{I_{\mathrm{s},3}^3 T_1}\right.$

Conditions	Comments
$I_1 \gg \lvert(\delta_2'')^{-1}\rvert, \lvert\Delta_2'/\delta_2''\rvert$ $I_1 > I_{s,2}$	Two-photon resonance, plane waves, no pump depletion, linear phase matching, high-intensity limit
$\Delta_3 = [-\delta_3' I_1 \pm \sqrt{(I_1^3/I_s^2 - 1)}]T_2^{-1}$ $L \ll (N\sigma_{03})^{-1}, L_c$	Three-photon resonance, plane waves, linear phase matching maximum intensity at optimum detuning
$1 \ll \lvert\delta_3'' I_1\rvert \ll \lvert\Delta_3'\rvert$	Three-photon resonance, plane waves, no pump depletion, near-resonant interaction, conversion done in one nonlinear coherence length, Stark shift small compared to initial detuning
$\lvert\delta_3'' I_1\rvert \gg \lvert\Delta_3'\rvert$	Three-photon resonance, plane waves, maximum intensity for conversion in one nonlinear coherence length, no pump depletion, near-resonant interaction, Stark shift large compared to initial detuning
$N\sigma_{03}L \gg 1$ $\delta_3'' I_1 = -\Delta_3'$	Three-photon resonance, plane waves, no pump depletion, interaction resonant with Stark-shifted line, maximum harmonic intensity
$T_2 < t < T_1$; $\Delta_2' = -\delta_2'' I_1; \int_{-\infty}^{t} I_1^2(t')\,dt' < I_{s,2}^2 T_1/2$	Two-photon resonance, phase-matched plane waves, incoherent transient regime, short-time behavior, intensity conversion efficiency
$T_2 < t < T_1$ $\Delta_2' = -\delta_2'' I_1 \pm \left(\frac{2\int_{-\infty}^{t} I_1^2(t')\,dt'}{I_{s,2}^2 T_1} - 1\right)^{1/2}$ $\int_{-\infty}^{t} I_1^2(t')\,dt' > I_{s,2}^2 T_1/2$	Two-photon resonant interaction, phase-matched plane waves, incoherent transient regime, maximum intensity conversion efficiency for long times
$T_2 < t < T_1; \Delta_3' = -\delta_3'' I_1$ $\int_{-\infty}^{t} I_1^3(t')\,dt' < I_{s,3}^3 T_1/2$	Three-photon resonance, linear phase matching, plane waves, incoherent transient regime, short time behavior, maximum harmonic intensity

(*continued*)

TABLE 6 *(continued)*

Limiting process	Efficiency	Intensity/Energy
		$\frac{9\pi^3 cN^2L^2\lvert\mu_{03}\rvert^2 e^{-1}}{n_3\lambda_1^2}\,\frac{I_1^3T_2}{\int_{-\infty}^{t} I_1^3(t')\,dt'}$
		$\frac{18\pi^4N^2L^2q_2'^2T_2I_{10}}{n_3n_1\lambda_1^2}$
		$\frac{9\pi^3cN^2L^2\lvert\mu_{03}\rvert^2T_2}{n_3\lambda_1^2}$
	$\frac{576\pi^6N^2L^2}{n_3n_1^3\lambda_1^2c^2}\left(\frac{q_2'q_2T_2}{\hbar}\right)^2 I_1^2$	
	$\frac{576\pi^6N^2L^2}{n_3n_1\lambda_1^2c^2}\left(\frac{\mu_{03}q_3T_2}{h}\right)^2 I_1^2$	
Coherent transients	$\frac{36\pi^4q_2'^2N^2L^2}{n_1n_3\lambda_1^2}\,\frac{I_{St}^2/(I_{S,2}^2T_1/T_2)}{[1+I_{St}^2/(I_{S,2}^2T_1/T_2)]}$	
	$\frac{36\pi^4q_2'^2N^2L^2}{n_1n_3\lambda_1^2}$	
	$R^{1/2}$	

[a] Key: $K_{a,q} = \left(\frac{8\pi}{cn_1}\right)^{(q-1)/2} q\pi^2 x_a^{(q)}/2^{q-3}n_q\lambda_1$

$\kappa_3 = 3\pi^2 N x_a^{(3)}/n_3\lambda_1$

$$R = \left(\frac{\sum_i \mu_{0i}\mu_{i2}\left(\frac{1}{\omega_{i0}-3\omega}+\frac{1}{\omega_{i0}+\omega}\right)}{\sum_j \frac{\mu_{0j}\mu_{j2}}{(\omega_{j0}-\omega)}}\right)^2$$

$$\tilde{z} = \left(\frac{8\pi}{cn_1}\right)\frac{3\pi\omega_1 N}{2n_3 c\hbar} q_2^2 R^{1/2} T_2 I_{10} z$$

$$q_2 = \frac{1}{\hbar}\sum_i \frac{\mu_{2i}\mu_{i0}}{(\omega_{i0}-\omega)}$$

$$q_2' = \frac{1}{\hbar}\sum_i \mu_{0i}\mu_{i2}\left(\frac{1}{\omega_{i0}-3\omega}+\frac{1}{\omega_{i0}+\omega}\right)$$

$$I_{s,2} = \hbar c n_1/4\pi q_2\sqrt{T_1T_2}$$

$$I_{St} = \lvert\delta_2''^{-1}\rvert$$

$$\delta_j'' = \frac{8\pi}{cn}\left\{\frac{1}{2\hbar^2}\left[\sum_i \frac{\lvert\mu_{ji}\rvert^2\omega_{ji}}{\omega_{ji}^2-\omega^2}+\sum_i\frac{\lvert\mu_{0i}\rvert^2\omega_{i0}}{\omega_{i0}^2-\omega^2}\right]\right\}T_2$$

$$I_{s,3} = \frac{cn_1}{8\pi}\sqrt[3]{16\hbar^2/q_3^2T_1T_2}$$

$$q_3 = \frac{1}{\hbar^2}\sum_{i,j}\mu_{0i}\mu_{ij}\mu_{j3}/(\omega_{i0}-\omega)(\omega_{j0}-2\omega)$$

Conditions	Comments
$_2 < t < T_1; \Delta_3' = -\delta_3'' I_1 \pm \left(\frac{2\int_{-\infty}^{t} I_1^3(t')\,dt'}{I_{s,3}^3 T_1} - 1\right)^{1/2}$ $\int_{-\infty}^{t} I_1^3(t')\,dt' > I_{s,3}^3 T_1/2$	Three-photon resonance, linear phase matching, plane waves, incoherent transient regime, maximum harmonic intensity for long times
$T_2 < t < T_1; I_1(t) = I_0\Theta(t)$ $I_1^2 t/I_S^2 T_1 \gg 1 + (\Delta_2' + \delta_2'' I_1)^2$	Two-photon resonant interaction, incoherent transient regime, maximum harmonic energy in large time limit, pump pulse of constant intensity
$T_2 \ll t \ll T_1; I_1(t) = I_0\Theta(t);$ $I_1^3 t/I_{S,3}^3 T_1 > 1 + (\Delta_3' + \delta_3'' I_1)^2$	Three-photon resonant interaction, incoherent transient regime, maximum harmonic energy in large time limit, pump pulse of constant intensity
$T_2 < t < T_1; I_1(t) = I_0\Theta(t);$ $\Delta_2' = -\delta_2'' I_0; \frac{I_1^2 t}{I_{s,2}^2 T_1} \ll 1$	Two-photon resonant interaction, maximum energy conversion efficiency at short times, pump pulse of constant intensity
$T_2 < t < T_1, I_1(t) = I_0\Theta(t),$ $\Delta_3' = -\delta_3'' I_0; \frac{I_1^3 t}{I_{s,3}^3 T_1} \ll 1$	Three-photon resonant interaction, maximum energy conversion efficiency at short times pump pulse of constant intensity
$t \ll T_2, T_1$ Adiabatic following conditions	Two-photon resonant interaction, phase-matched plane waves, two-photon adiabatic following approximation
$t \ll T_2, T_1$ $\frac{q_2}{2\hbar}\int_{-\infty}^{t} \lvert A_1(t')\rvert^2\,dt' = \pi/2$ $\Delta_2 = 0$	Two-photon resonant interaction, phase-matched plane waves, Stark shifts neglected
$R < 1$	Two-photon coherent propagation, empirical result

$\Delta_2 = \omega_{20} - 2\omega, \Delta_2' = \Delta_2 T_2, \Delta_3 = \omega_{30} - 3\omega; \Delta_3' = \Delta_3 T_2$

$$_{03} = \frac{3(2\pi)^2 \lvert \mu_{03}\rvert^2 T_2}{n_3 \hbar \lambda_1 [1 + (\Delta_3' + \delta_3'' I_1)^2 + I_1^3/I_{s,3}^3]^2}$$

$$(t) = \begin{cases} 1, & t > 0 \\ 0, & t < 0 \end{cases}$$

$$_{21} = k_{21}(3\omega) - 3k_{21}(\omega)$$
$$= \frac{18\pi^2}{\lambda_1}\left[\frac{\chi(-q\omega, \omega, -\omega, q\omega)}{n_3} - \frac{\chi(-\omega, \omega, -\omega, \omega)}{2n_1}\right]$$

As was pointed out earlier, conversion in relatively high density media can be done at relatively low pump intensities. The fractional population changes can then be kept small and saturation of the susceptibility can be avoided. The conversion efficiency in two-photon resonant media under these conditions will be limited either by the intensity dependent phase changes associated with the population changes that do occur, as discussed in Section 4.3.4, or by the combined effects of pump absorption Raman amplification, and the nonlinear index changes due to the near-resonant Kerr effect, as discussed in Section 4.3.1a. The relative importance of these mechanisms in limiting the efficiency depends on the distance required for significant nonlinear phase changes due to the population redistribution to occur compared to the distance required for limitation of the efficiency due to the multiphoton absorption and Raman processes (see Fig. 7).

We consider first the conditions that must be met so that the absorption and Raman processes dominate the limitations on conversion efficiency. These effects were derived in Section 4.3.1a under the assumptions that the population changes and the nonlinear phase changes were small and that the system was phase matched. The condition that the fractional population changes be small can be expressed as

$$\rho_{22} \ll 1. \tag{139}$$

This is equivalent to the condition that the pump intensity be less than the saturation intensity [see Eq. (75b)], which can be expressed as

$$I_{10} \ll \hbar c n_1 / 4\pi q_2 \sqrt{T_1 T_2} \tag{140}$$

in the notation of Section 4.3.1b or, in the notation of Section 4.3.1a, as

$$I_{10} \ll \left(\frac{3\sqrt{R}\,\hbar \omega c n_1^2}{32\pi \kappa_a' T_1 T_2 n_3} \right)^{1/2}, \tag{141}$$

where $\kappa_a' = \kappa'/N$ and κ' is given in Eq. (62).

The nonlinear phase shift associated with the population changes can be expressed as

$$\Delta\phi^{NL} = (6\pi/\lambda_1)\{\Delta n\} N_0 \int \rho_{22}(z)\, dz, \tag{142}$$

where $\{\Delta n\}$ is given in Eq. (133). The integral in Eq. (142) must be taken over the distance required for maximum conversion subject to the limitations of the competing effects. Inspection of the curves in Fig. 7 shows that when conversion is limited by two-photon absorption and Raman amplification, the harmonic intensity reaches 90% of its final value for normalized interactions lengths $\tilde{z}$ [see Eq. (66)] greater than $3/\sqrt{R}$ for $R \gtrsim 0.01$. The nonlinear phase shift can then be expressed as

$$\Delta\phi^{\mathrm{NL}} = \frac{6\pi}{\lambda_1}\{\Delta n\}N_0 \int_0^{3/\sqrt{R}} \rho_{22}(\tilde{z}) \frac{cn\Gamma}{8\pi\kappa_a' I_{10} N}\, d\tilde{z}. \tag{143}$$

Further simplification requires knowledge of $\rho_{22}(\tilde{z})$ throughout the nonlinear medium. Approximate analytic expressions can be derived in the limit of small and large R.

In media with small R the two-photon absorption is stronger than the Raman amplification of the pump and the two-photon population is determined almost entirely by the pump intensity. Using the condition that the fractional population changes are small, along with Eqs. (75b) and (72e), we can write

$$\rho_{22}(\tilde{z}) = I_1^2(\tilde{z})/2I_s^2. \tag{144}$$

In media with small R the variation of the pump intensity with $\tilde{z}$ can be approximated by

$$I_1(\tilde{z}) = \frac{I_{10}}{1 + 4\tilde{z}/3\sqrt{R}}. \tag{145}$$

This expression is not perfectly accurate because it neglects both the Raman amplification of the pump and the back reaction of the harmonic field on the pump intensity. However, it is a fairly good approximation for the normalized distances required for the harmonic field to reach about 90% of its final value.

When Eqs. (145) and (144) are substituted into Eq. (143), it is evident that the upper limit may be replaced with $\tilde{z} = \infty$ when R is small. The nonlinear phase shift is then given by

$$\Delta\phi_{\mathrm{NL}} = 3\{\Delta n\}I_{10}T_1/2\hbar c. \tag{146}$$

In order that the nonlinear phase shifts associated with the population redistribution be small, the condition $\Delta\phi_{\mathrm{NL}} \ll \pi$ must be satisfied, leading to the following condition on the incident intensity:

$$I_{10} \ll \frac{2\pi\hbar c}{3\{\Delta n\}T_1}. \tag{147}$$

At the same time the incident intensity must also satisfy the condition

$$I_{10} \geq \frac{3cn_1}{N\kappa_a' T_2 L 8\pi\sqrt{R}} \tag{148}$$

to ensure that the interaction length L is long enough to give maximum conversion.

When R is large the Raman amplification of the pump beam is stronger than the two-photon absorption and the pump beam is not attenuated significantly. It can be shown that under these conditions the steady-state value of $\rho_{22}(z)$ is given by

$$\rho_{22}(z) = \frac{q_2^2 T_1 T_2}{8\hbar^2}[|A_1|^4 + R|A_1|^2|A_3|^2 + \sqrt{R}(A_1^3 A_3^* + A_3 A_1^*)] \quad (149)$$

when $\rho_{22} \ll 1$. When the harmonic intensity is large enough to satisfy the limiting condition (67), $\rho_{22}(z) = 0$. The nonlinear phase shift therefore accumulates only over the distance required to reach the limiting efficiency. We can obtain an approximation to the value of $\rho_{22}(z)$ by setting the terms in the brackets in Eq. (149) equal to $|A_{10}|^4$ and integrating Eq. (143) between the limits $\tilde{z} = 0$ and $\tilde{z} = 3/\sqrt{R}$. With these approximations the nonlinear phase shift is given by

$$\Delta\phi_{\mathrm{NL}} = 6\{\Delta n\}\frac{I_{10}T_1}{\hbar c R}. \quad (150)$$

In order for the nonlinear phase shifts to be small when R is large, the incident intensity must satisfy the condition

$$I_{10} \ll \pi R\hbar c/6\{\Delta n\}T_1 \quad (151)$$

while at the same time satisfying the relation (148) to ensure sufficient interaction length to attain full conversion.

The various conditions that must be satisfied so that the conversion is limited by absorption and Raman scattering are summarized in Table 7. Basically the incident pump intensity must be kept small enough to satisfy the more restrictive of conditions (147) or (151) and (140).

Neither the condition on the fractional population change (140) nor that on the nonlinear phase shift (147) or (151) depends on the interaction length or on the density of the medium. This is to be expected for the condition on the fractional population change but is somewhat surprising for the condition on the nonlinear phast shift, which is usually expected to build up over the interaction length and therefore should be proportional to the product $N_0 L$. In the present situation this behavior is a result of the fact that the distance over which the nonlinear phase shift builds up scales as $1/N_0$. As a result the condition for having a small nonlinear phase shift due to population changes can be satisfied by choosing I_{10} to be sufficiently small. The product of the density and the interaction length can then be chosen to satisfy condition (148).

When condition (147) or (151) is not satisfied, the nonlinear phase shift associated with the two-photon population changes can be significant. If

$N_0\{\Delta n\}$ is sufficiently large, then this can happen under conditions of small population changes, i.e., $\rho_{22} \ll 1$. In this situation the saturation of the susceptibility will be negligible and the main effect of the population redistribution will be in the nonlinear phase changes. If $\Delta\phi_{\text{NL}} > \pi$ before $\tilde{z}$ reaches the value $3/\sqrt{R}$ required for maximum conversion based on the limitations imposed by the effects of two-photon absorption and Raman amplification, then the nonlinear phase changes due to the two-photon population changes will dominate the limitations on conversion efficiency.

In this situation the integral in Eq. (143) is taken over the entire interaction length L and the pump intensity is assumed to be constant. In order for the nonlinear phase shift to be greater than π the incident intensity must satisfy the condition

$$I_{10}^2 \geq \frac{\lambda_1 I_s^2}{3\{\Delta n\} N_0 L}. \tag{152}$$

At the same time the incident intensity must satisfy the condition

$$I_{10}^2 \ll I_S^2 \tag{153}$$

to ensure that the fractional population changes are small and the condition

$$I_{10} \ll \frac{3cn_1}{8\pi\sqrt{R}\,\kappa_a' T_2 N_0 L} \tag{154}$$

to ensure that the two-photon absorption and Raman amplification processes are not important. The condition on the magnitude of the nonlinear phase shift [inequality (152)] now depends on the product $N_0 L$. This will be true as long as $z \ll 3/\sqrt{R}$ and the nonlinear phase shift builds up over the entire interaction length. The conditions that must be met so that the nonlinear phase shifts associated with the two-photon population changes dominate the limitations on conversion efficiency are also summarized in Table 7.

Whether the efficiency is limited by the nonlinear phase shifts or by the two-photon absorption and Raman processes when the fractional population changes are small is determined by the ratio

$$D = \left(\frac{3cn_1}{8\pi\sqrt{R}\,\kappa_a' T_2}\right)^2 \frac{3\{\Delta n\}}{\lambda_1 I_S^2 N_0 L}, \tag{155}$$

which depends only on the constants of the medium and the product of the density and interaction length. If this ratio is greater than unity, then conditions (152) and (154) can be satisfied simultaneously and the experimental conditions can be arranged so that the nonlinear phase shifts are the dominant limiting mechanisms, provided I_{10} is subject to conditions (152)–(154).

TABLE 7

Conditions Required for the Various Effects Related to a Two-Photon Resonance to Dominate the Limitations on Conversion Efficiency

Dominant limiting process	Condition	Effect
Two-photon absorption, Raman amplification	$I_{10} \ll I_S$	Small population changes
	$I_{10} \ll 2\pi\hbar c/3\{\Delta n\}T_1,\ R \ll 1$ $I_{10} \ll \pi R\hbar c/6\{\Delta n\}T_1,\ R \gg 1$	Small nonlinear phase shift from population change
	$I_{10} \geq \dfrac{3cn_1}{8\pi N\kappa'_{a,3}T_2L\sqrt{R}}$	Maximum conversion limited by two-photon absorption or Raman amplification
Phase shifts from near-resonant Kerr effect	$\Delta T_2 \gg 1$	Detuning large enough for Kerr effect to dominate absorptive processes
	$I_1 \ll I_S$	Small population changes
	$\|D'\| < 1$	Phase changes from Kerr effect larger than those from population changes
	$N_0L = \dfrac{4n_3\lambda_1\Delta T_1T_2I_S^2}{3n_1\hbar cI_1(R-2)}$	Interaction length sufficient to obtain optimal conversion
Phase shifts associated with population changes	$I_{10} \ll I_S$	Small population change
	$I_{10}^2 > \lambda_1 I_S^2/6\{\Delta n\}N_0L$	A Significant index change from population change
	$I_{10} \ll \dfrac{3cn_1}{8\pi\kappa'_{a,3}T_2L\sqrt{R}}$	B Small effect due to two-photon attenuation and Raman amplification
	$\|D\| > 1$	Requirement so that A and B can be satisfied simultaneously on resonance

Dominant limiting process	Condition	Effect
	$\lvert D' \rvert > 1$	Requirement so that phase shift from population change can dominate Kerr effect phase shift for near-resonant interactions
Saturation of susceptibility from two-photon absorption and Stark shifts	$I_{10} \gg I_S$	Large population changes
	$D'' \ll 1$	Saturation of susceptibility occurs before nonlinear phase shifts become important

$$I_S = \hbar c n_1 / 4\pi q_2 \sqrt{T_1 T_2}$$

$$D = \left(\frac{3cn_1}{8\pi\sqrt{R}\,\kappa_a T_2} \right)^2 \frac{3\{\Delta n\}}{\lambda_1 I_S^2} \frac{1}{N_0 L}$$

$$D' = 4\{\Delta n\} n_3 T_1 I_1 / n\hbar c \Delta T_2 (R - 2)$$

$$D'' = 6\pi\{\Delta n\} N_0 L / \lambda_1$$

$$\kappa'_{a,3} = \frac{3\pi\omega_1}{2n_3 c\hbar} R^{1/2} q_2^2$$

$$R = \left[\sum \mu_{0i}\mu_{i2} \left(\frac{1}{\omega_{i0} - 3\omega} + \frac{1}{\omega_{i0} + \omega} \right) \Big/ \sum \frac{\mu_{0i}\mu_{i2}}{\omega_{i0} - \omega} \right]^2$$

$$q_2 = \frac{1}{\hbar} \sum \mu_{0i}\mu_{i2} \left(\frac{1}{\omega_{i0} - \omega} \right)$$

$$\Delta = \omega_{20} - 2\omega$$

$$\{\Delta n\} = [(n_2(3\omega) - n_2(\omega)) - (n_0(3\omega) - n_0(\omega))]$$

When $D < 1$, conditions (152) and (154) cannot be met simultaneously and significant nonlinear phase changes will not be encountered before the two-photon absorption and Raman effects become important. The condition

$$D > 1 \tag{156}$$

becomes harder to satisfy as the product $N_0 L$ increases. Thus for a given medium the two-photon absorption and Raman processes can be made to dominate the limitations on conversion efficiency by choosing I_{10} small enough to satisfy inequalities (140) and (147) or (151), provided the product $N_0 L$ can be made large enough to satisfy the condition

$$D < 1. \tag{157}$$

When the detuning from the two-photon resonance is larger than a few linewidths, the effects of the nonlinear phase changes associated with the near-resonant Kerr effect are larger than those resulting from two-photon absorption and Raman amplification of the pump. The intensity-dependent phase changes caused by the population redistribution must then compete with those caused by the near-resonant Kerr effect. If we assume that the pump intensity is less than the saturation intensity and that the detuning is larger than the Stark shift and the linewidth, then we can express the nonlinear phase changes due to the population redistribution as

$$\Delta\phi_{\text{pop}}^{\text{NL}} = \frac{3\pi}{\lambda_1}\{\Delta n\}\frac{NLI_1^2}{I_s^2\Delta^2 T_2^2}. \tag{158}$$

The nonlinear phase change due to the Kerr effect is given by

$$\Delta\phi_{\text{Kerr}}^{\text{NL}} = \frac{3\pi}{4\lambda_1}\frac{n_1\hbar c}{n_3\Delta T_1 T_2}\frac{N_0 L I_1}{I_s^2}(R - 2) \tag{159}$$

for $R \gg 1$ or $R \ll 1$.

Comparison of these two expressions shows that the phase changes due to the population redistribution are more important than those due to the near-resonant Kerr effect when

$$|D'| = \left|\frac{4\{\Delta n\}n_3 T_1 I_1}{n_1\hbar c\Delta T_2(R - 2)}\right| > 1. \tag{160}$$

Condition (160) is independent of the density and length of the nonlinear medium. It becomes easier to satisfy at larger values of pump intensity and smaller values of detuning.

When condition (160) is not satisfied the phase changes due to the Kerr effect will outweigh those due to the population redistribution. Just as for the on-resonant interaction, limitations due to the Kerr effect can be made to dominate if the pump intensity is chosen to be sufficiently small, this time satisfying the condition

$$I_1 < \frac{n_1 \hbar c \Delta T_2 (R - 2)}{4\{\Delta n\} n_3 T_1}. \tag{161}$$

The product of the density and interaction length can then be chosen to satisfy the condition

$$N_0 L = \frac{4 n_3 \lambda_1 \Delta T_1 T_2 I_s^2}{3 n_1 \hbar c I_1 (R - 2)} \tag{162}$$

in order to ensure that the interaction length is sufficient to obtain optimum conversion.

When condition (140) is not satisfied saturation of the susceptibility must be accounted for. As discussed in Section 4.3.1b the Stark shift of any near-resonant levels must usually be included in the analysis at the same time. Along with saturation of the susceptibility which governs the frequency conversion process, changes in the linear susceptibilities that govern the phase-matching conditions will occur. An example of the interplay between these two effects was given in Section 4.3.2.

Just as with limitations due to the nonlinear phase shifts associated with the population redistribution, limitations due to saturation of the susceptibility can dominate if $\tilde{z} < 3/\sqrt{R}$. Under this condition the relative importance of the nonlinear phase changes and the saturation of the susceptibility is determined by the quantity

$$D'' = 6\pi\{\Delta n\} N_0 L/\lambda_1. \tag{163}$$

If $D'' \gg 1$, then the nonlinear phase shift associated with the population redistribution will become significant before the susceptibility saturates. If $D'' \ll 1$, then the saturation of the susceptibility can dominate the limitations on efficiency, as discussed in Section 4.3.1b, before the nonlinear phase changes become significant. As pointed out earlier, saturation of the susceptibility is more likely to dominate in low-density media, while the nonlinear phase shifts will be more important in high-density media. These conditions are also summarized in Table 7.

5. EXPERIMENTAL RESULTS

Experimentally the presence of competing nonlinear processes is usually indicated by a deviation of the intensity in the generated beam from its ideal power law dependence on the pump intensity. Other experimental indications include deviations from the theoretical phase-matching conditions or from the theoretical pressure or interaction-length scalings. Limitations due to linear absorption are indicated by a reduction in the phase-matching peak, a loss of contrast in phase-matching oscillations, or a failure to maintain the theoretical scaling with interaction length or medium density. Conversion efficiency that is significantly less than the theoretical estimates can also be an indication of limitations due to competing processes, but this is not always a reliable guide because of the difficulty in being sure that all the assumptions of the theory are met in the experiment, e.g., the phase-matching conditions, accurate measure of beam profiles, absolute intensities, and the magnitude of the nonlinear susceptibilities.

Observations of limitations due to competing mechanisms have been made under a wide range of experimental conditions (resonant, nonresonant, phase matched, non-phase matched, collimated beams, and focused beams) and in a wide variety of materials (rare gases, metal vapors, molecular vapors, and molecular liquids). The limitations generally become evident at peak pump intensities ranging from 10^6–10^9 W/cm^2 [40,42,43,57,58] for resonant conversion processes to intensities of the order of 10^{10}–10^{12} W/cm^2 [5,34,35] for nonresonant processes.

The observed limitations in nonresonant conversion in various systems have been attributed by various authors to intensity-dependent phase changes associated with the Kerr effect and to saturation of the susceptibility due to breakdown and Stark shifts. Limitations in resonant conversion in various systems have been ascribed to saturation of the susceptibility by population changes, Stark shifts, power broadening, multiphoton ionization, and dielectric breakdown, to competition between the harmonic conversion and two-photon absorption and Raman scattering, and to intensity-dependent phase changes associated with near-resonant Kerr effects and population redistribution.

In some cases the proper limiting processes can be identified beforehand by a theoretical analysis. The accuracy of such an approach depends on just how many of the possible limiting mechanisms are included in the analysis and on the accuracy of the numerical values that are used for the various susceptibilities.

Experimental studies, on the other hand, can provide valuable insight into the limiting mechanisms that are important both in specific circumstances and in general classes of interactions. In particular, studies of the saturation of the conversion efficiency as a function of such experimental parameters as pump power, intensity, and pulse duration, detuning from resonance, the confocal parameter of focused beams, the interaction length, medium density, and phase-matching conditions can serve to distinguish among effects that depend only on pump intensity, those that depend on a combination of pump intensity and medium density, and those that depend on the integral of the pump intensity over the interaction length. Measurements of the interference peaks associated with phase matching can give information on the importance of linear absorption.

Currently the role of some of the competing processes has been unambiguously identified in several specific experiments. It can be anticipated that in the future continued theoretical and experimental study will yield a clear picture of the circumstances under which the various competing effects will be important in limiting conversion efficiency. We shall now review some specific experimental results.

5.1. Limitations in Nonresonant Harmonic Generation

Several authors have reported experiments in which limitations were observed in nonresonant harmonic conversion in metal vapors and rare gases. Zych and Young[34] have studied third-harmonic conversion from 354.7 to 118.2 nm in Xe and phase-matched mixtures of Xe and Ar. They observed that the conversion efficiency for tightly focused beams in pure Xe reached a constant value of about 2×10^{-4}, becoming independent of the pump power for pump intensities above 5×10^{12} W/cm^2, as shown in Fig. 20. At the same time they observed that the Xe pressure required for phase optimization in the tight focus decreased with increasing pump power, indicating the presence of intensity-dependent phase changes. Their analysis showed that the largest contribution to the nonlinear index in this system comes from the mixed-frequency Kerr susceptibility $\chi^{(3)}(-3\omega, \omega, -\omega, 3\omega)$, as can be expected in an interaction which involves only a near-three-photon resonance as discussed in Section 4.2.

Their calculations indicated that the maximum conversion efficiency for this process when limited by the nonlinear phase changes should be of the order of 9×10^{-3}. A comparison of the measured conversion efficiency with the calculated value indicated that yet another limiting mechanism was present. By using mixtures of Xe and Ar they could change the peak intensity at which the nonlinear index dominated the limitation of the conversion. In this manner efficiencies up to 2×10^{-3} were obtained. In all cases the efficiency was limited at a peak intensity of 5×10^{12} W/cm^2. These observations confirmed the presence of limitations due to the nonlinear index and also indicated that the second mechanism is one that depends only on the peak pump intensity. Dielectric breakdown was suggested for the second mechanism but its identity was not conclusively established.

Reintjes *et al.*[35] have observed similar limitations in experiments on nonresonant fifth-harmonic conversion of 266.1-nm radiation in helium. A maximum conversion efficiency of 10^{-5} was measured in these experiments at a pump power of 100 MW and a peak focused intensity of 10^{15} W/cm^2

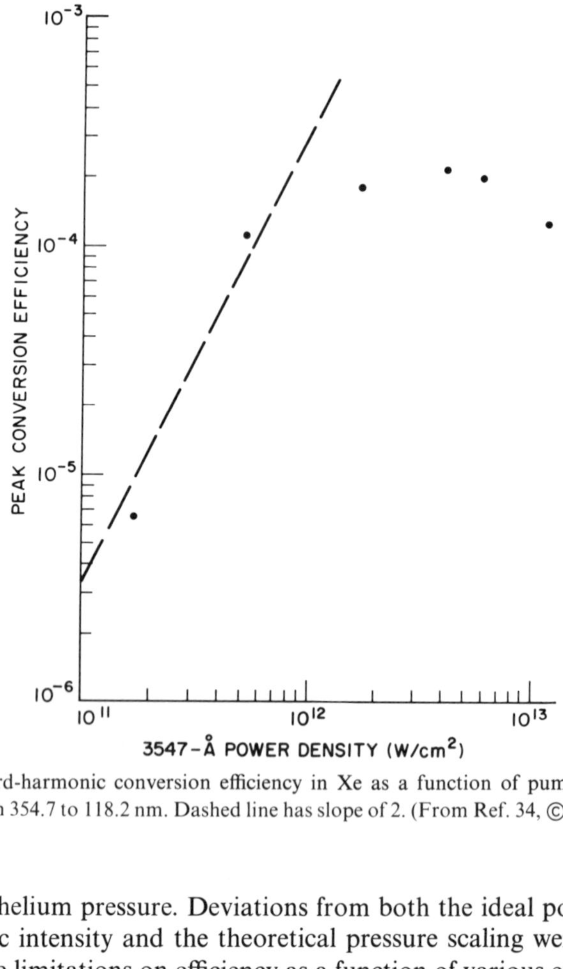

Fig. 20. Third-harmonic conversion efficiency in Xe as a function of pump intensity for conversion from 354.7 to 118.2 nm. Dashed line has slope of 2. (From Ref. 34, © 1978 IEEE.)

at 100 Torr helium pressure. Deviations from both the ideal power law for the harmonic intensity and the theoretical pressure scaling were observed. Studies of the limitations on efficiency as a function of various experimental parameters indicated that both intensity-dependent phase changes and saturation of the susceptibility were important.

The variation of the fifth-harmonic power as a function of the pump power for various helium pressures is shown in Fig. 21. The presence of competing processes is indicated by the change in slope of the curves at a pump power near 12 MW. The fact that the pump power at which the change of slope occurs is independent of pressure indicates that the competing mechanism that is responsible for this behavior causes saturation of the susceptibility. For pump powers above 12 MW the variation of the harmonic power with pump power changes from a cubic dependence at 7.5 Torr to an approximately 1.5-power dependence at 120 Torr. This behavior indicates the

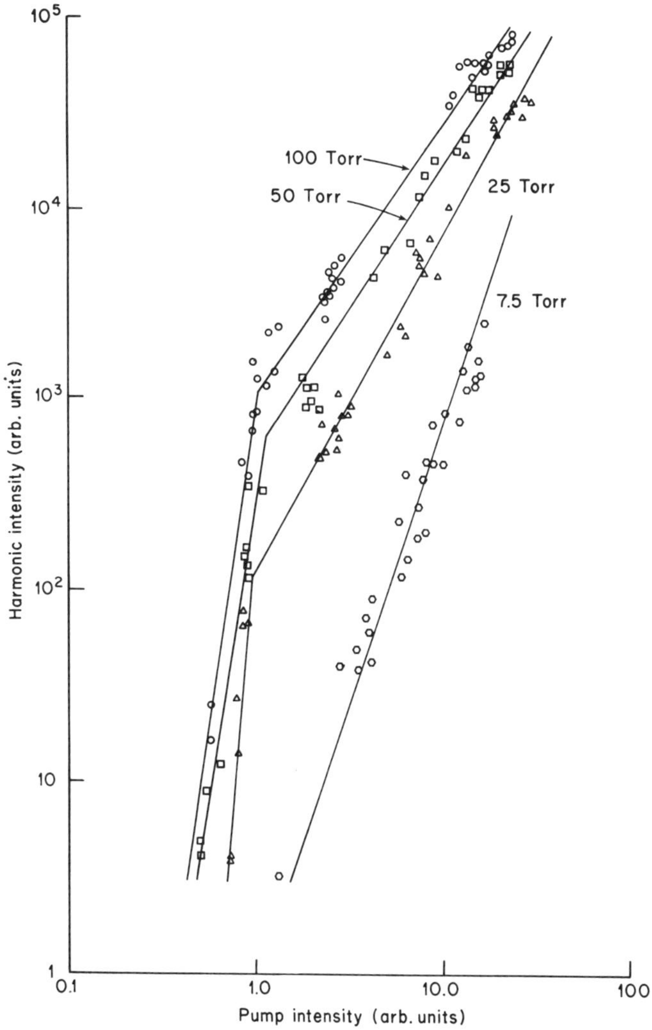

Fig. 21. Fifth-harmonic intensity as a function of pump intensity for various helium pressures for conversion from 266.1 to 53.2 nm.

presence of a second competing mechanism, one whose effects do depend on gas pressure.

Similar behavior is evident in the variation of the harmonic power with gas pressure, an example of which is shown in Fig. 22. Here the vertical scale for the measured values was determined by comparing the harmonic power generated at the highest pump power and gas pressure with that

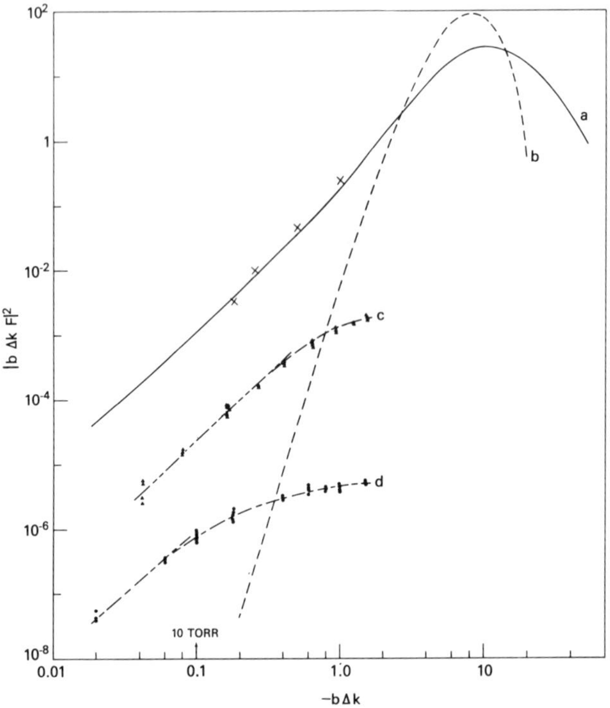

Fig. 22. Fifth-harmonic power as a function of helium pressure (proportional to $b\Delta k$) for various pump powers: 12 MW (crosses), 75 MW (triangles, curve c), and 300 MW (solid circles, curve d). The curves were calculated for an infinite (solid curve) and a semi-infinite medium (dashed curve). Data was adjusted to match observed conversion efficiency at the maximum value of curve d. Experimental points are plotted with the fifth-power dependence divided out. (From Ref. 57.)

predicted from the theory of Chapter 3 for a semi-infinite medium using measured beam parameters and the calculated fifth-order susceptibility. The expected fifth-power dependence on the pump power has been divided out of the experimental data so that data taken at different pump powers can be compared with the same theoretical curve. The two theoretical curves were calculated for infinite (curve a) and semi-infinite (curve b) media using the theory of Chapter 3 with no adjustable parameters.

At the lowest pump power the agreement between theory and experiment for the semi-infinite medium is quite good. At the higher pump powers two kinds of deviation from the theory are observed. At low gas pressures the square law scaling with pressure is followed but the measured harmonic power falls increasingly farther from the theoretical curve as the pump power is increased, indicating a deviation form the fifth-power law. In this region the conversion is being affected by a saturation of the susceptibility. At higher gas pressures the data start deviating from the theoretical pressure scaling at a pressure that decreases as the pump power is increased. This behavior indicates the presence of an additional effect that depends on the pressure as well as the pump power. Further measurements of this type using different focal length lenses showed that the pressure-dependent effect depends on the product of the total pump power and the gas pressure but not independently on the peak pump intensity or the confocal parameter. This behavior is characteristic of limitations due to intensity-dependent phase changes. Perturbation calculations show that the largest contribution to the nonlinear index comes from the mixed Kerr susceptibility $\chi^{(3)}(-5\omega, \omega, -\omega, 5\omega)$, causing a nonlinear index at the harmonic due to the intensity at the fundamental. This is the same kind of term that was responsible for the saturation of the third-harmonic conversion in Xe : Ar mixtures.

Further insight into the limiting mechanisms was obtained by comparing the fifth- and third-harmonic conversion.[58] An example of this comparison was given in Fig. 3-28, where it was pointed out that saturation was not observed in the third harmonic. This result shows that phase shifts at the fundamental are not important, confirming the conclusion about the dominance of the mixed frequency Kerr susceptibility. In addition, it shows that the saturation of the susceptibility is not due to depletion of the ground-state population, ruling out effects such as multiphoton absorption, multiphoton ionization, or dielectric breakdown. The result strongly indicates that the limiting mechanism involves a resonant enhancement at the fifth harmonic.

An example of an effect which could give rise to the observed behavior is a Stark shift of the 3P level in helium due to the intensity of the fundamental wave. A Stark shift of this level would introduce a saturation of the susceptibility of the form $\chi^{(5)} \propto 1/(\omega_{3p,0} - 5\omega + \delta I)$ and a wave vector mismatch of the form $\Delta k = \Delta k^0 + \Delta k_{21} I_1$. Calculated dependence of the fifth-harmonic power on the gas pressure for different pump powers using saturation and phase mismatch terms of these types shows qualitative agreement with the experimental results (see Fig. 23).[57] Confirmation of this interpretation, however, requires observation of substantial level shifts during the pump pulse.

Efficiency limitations have also been reported in non-resonant third-harmonic conversion in metal vapors. Puell *et al.*[5] observed such limitations in the efficiency of conversion of Nd:YAG laser radiation in Rb:Xe mixtures using both 300-psec- and 7-psec-duration pulses. For the longer pulses the conversion was observed to saturate when the pump intensity was above about 10^{10} W/cm^2 as shown in Fig. 24. From the numbers given in Puell *et al.*[5] the maximum intensity conversion efficiency is estimated to be

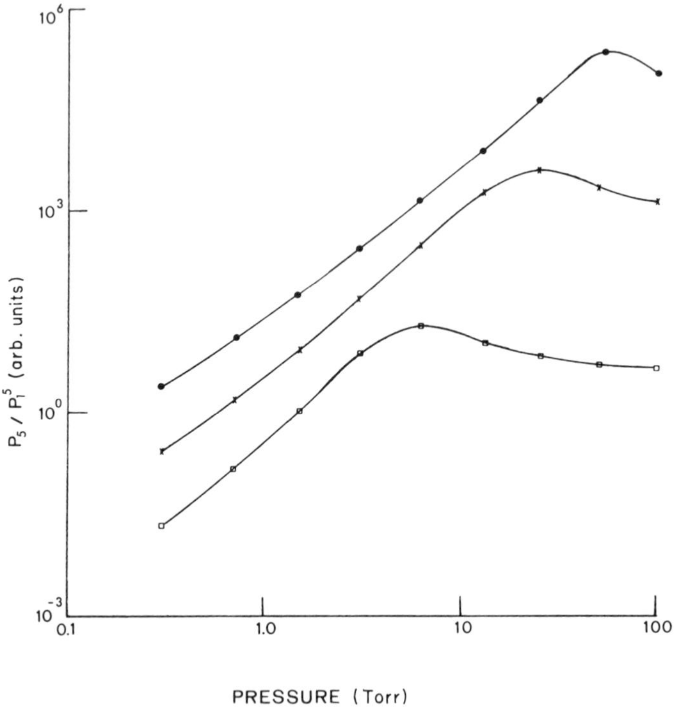

Fig. 23. Calculated pressure dependence of the fifth-harmonic power for various pump powers including the effects of a nonlinear refractive index ($n_2 = -4 \times 10^{-34}$ esu) and saturation of the fifth-order susceptibility according to the form $\chi^{(5)}/(1 + I_1/I_S)$. •, 23 MW; ×, 75 MW; □, 300 MW.

about 0.014. Measurements of the saturation behavior at different gas pressures showed that the maximum efficiency was independent of the medium density, suggesting that the limitations were due to a nonlinear index. This interpretation was shown to be consistent with calculations of the third-order Kerr effect.

Measurements with the 7-psec pulses showed saturation of the efficiency for pump intensities above 5×10^{10} W/cm^2 as shown in Fig. 25. Again the limitations were attributed to the intensity-dependent phase changes associated with the Kerr effect. For the shorter pulses the efficiency was observed to start increasing again for pump intensities above about 10^{11} W/cm^2. This behavior was attributed to saturation of the third-order Kerr effect due to a fifth-order term. A maximum intensity conversion efficiency of 0.1 and a maximum energy efficiency of 0.028 was observed for the 7-psec pulses at a pump intensity of 2×10^{11} W/cm^2.

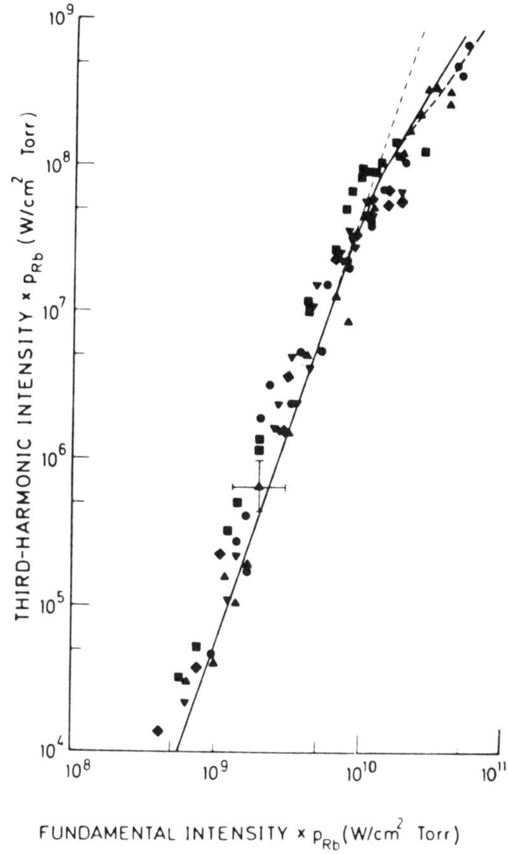

Fig. 24. Third-harmonic intensity as a function of pump intensity for conversion of 1.06-μm radiation in Rb: Xe mixtures. The pulse duration was 300 psec. ●, 1.0:355 Torr; ▲, 0.57:200 Torr; ■, 0.36:121 Torr; ▼, 0.29:100 Torr; ◆, 0.25:88 Torr. (From Ref. 5.)

Experiments reported on the same system by Bloom *et al.*[4] produced somewhat higher conversion efficiencies (10%) with lower intensities, higher pressures, and a geometry that more nearly approximated a collimated beam. A comparison of these two sets of experiments appears to indicate that benefits may result from using as low a pump intensity as possible and from using collimated beams rather than beams that are even weakly focused.

Miyazaki and Kashiwagi[22] have reported saturation of third-harmonic conversion of Nd:YAG laser radiation in Na vapor. In this interaction the third harmonic lies below the ionization threshold but the sodium can be

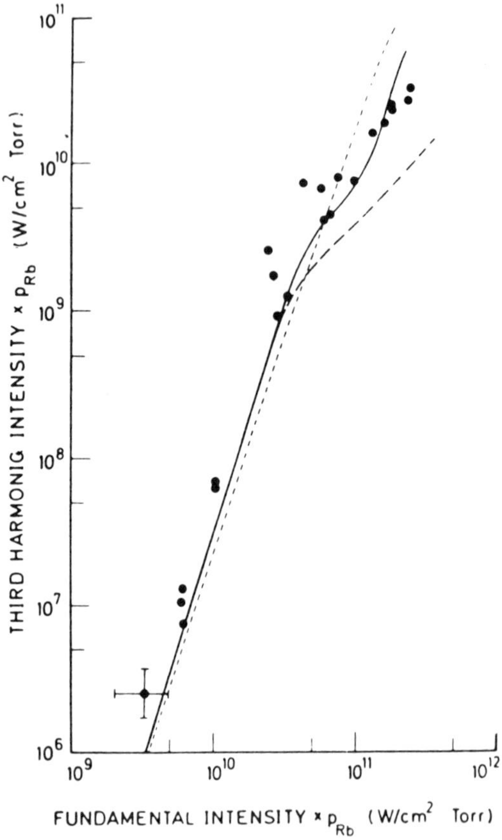

Fig. 25. Third-harmonic intensity as a function of pump intensity for conversion of 1.06-μm radiation in a 1:372 Rb:Xe mixture. Pulse duration was 7 psec. The dashed curve accounts for saturation due to third-order effects, whereas the solid curve includes fifth-order effects. (From Ref. 5.)

ionized by a five-photon process. Miyazaki and Kashiwagi observed that the third-harmonic intensity increased as the cube of the fundamental intensity up to pump intensities of the order of 5×10^{11} W/cm^2. Above this intensity the third-harmonic intensity increased more slowly than linearly with the pump intensity (Fig. 26). The authors interpreted this effect as being due to a nonlinear phase shift resulting from the free electrons that are created in the five-photon ionization process. Using their own measured value of the five-photon ionization cross section they obtained good agreement between the form of the experimental dependence of the harmonic intensity on the fundamental intensity and the calculated dependence including the effects of the free electrons on the phase mismatch, as shown in Fig. 26.

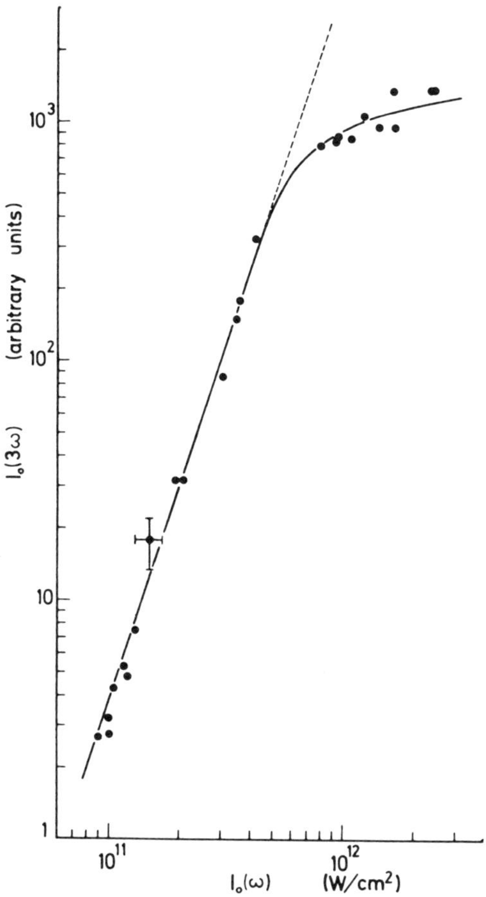

Fig. 26. Third-harmonic intensity as a function of pump intensity for conversion of Nd : YAG laser radiation in Na. Curve is calculated taking into account breaking of phase matching due to the free electrons created by multiphoton ionization. (From Ref. 24.)

5.2. Limitations in Two-Photon Resonant Interactions

Limitations on two-photon resonant third-harmonic conversion have been observed by various authors in experiments involving both low- and high-density gases and liquids. Depending on the circumstances the observed limitations have been attributed to saturation of the susceptibility combined with Stark shifts (see Section 4.3.1b), refractive index changes associated with the two-photon population (see Section 4.3.4), and two-photon absorption and Raman amplification of the pump (see Section 4.3.1a).

Saturation of two-photon resonant third-harmonic conversion has been observed in Na vapor by Ward and Smith[43] and in Cs vapor by Wang and Davis.[42] Ward and Smith observed that the pump power at which saturation occurred decreased as the resonance was approached, but, as has been discussed earlier, this behavior is characteristic of all the limiting mechanisms. Evidence of power broadening of the resonant level was also reported. Ward and Smith observed a dip in the harmonic power at the center of the resonance, indicating the presence of multiphoton population changes. Stark shifts were also evident in their measurements. Analysis by Elgin and New[29] has indicated that these results can be explained in terms of two-photon population saturation and Stark shifts. Georges *et al.*[21,47] have given a similar but not identical analysis including transient effects and multiphoton ionization as well as nonlinear phase changes due to ground state depletion (see Fig. 14) and have obtained reasonable numerical agreement with the experimental results.

Limitations on two-photon resonant third-harmonic conversion of dye laser radiation in Mg vapor were reported by Wallace and Zdasiuk[59] and in Mg:Kr mixtures by Junginger *et al.*[60] The presence of competing processes was identified in each case by a deviation of the dependence of the harmonic power from the ideal cubic-law variation with pump power at pump intensities above about 10^7 W/cm^2. The results of Junginger *et al.*[60] were accounted for by a theory that took into account transient population changes that caused transient intensity-dependent phase changes of the type discussed in Section 3.4. A maximum power conversion efficiency of 1.2×10^{-4} was observed in these experiments, while Wallace and Zdasiuk[59] reported a conversion efficiency of 10^{-3}. The difference was attributed[60] to the possibility of increased conversion that can occur at small detunings from exact resonance under conditions of two-photon saturation of the resonant level (see Section 4.3.1b).

Saturation of third-harmonic conversion of CO_2 laser radiation in liquid CO–O_2–SF_6 mixtures has been studied by Kildal and Brueck.[38,61] In this interaction radiation from the R(6) line of the CO_2 laser is nearly two-photon resonant with the vibrational mode of the CO molecule [$\Delta_2' = (\omega_{20} - 2\omega)T_2 = 0.0667$]. In their early experiments, which were done in a tight focus geometry ($5 \lesssim L/b \lesssim 30$) with pulse durations of 120 nsec, Kildal and Brueck observed that conversion was limited by dielectric breakdown. Conversion efficiency was observed to increase as the confocal parameter increased and the peak pump intensity decreased. A maximum conversion efficiency of 0.02 at an incident pump power of 600 kW, and a peak focused intensity of about 5×10^8 W/cm^2, was reported.

In subsequent experiments[61] done with shorter pulses (8 nsec) and a tight-focused geometry ($b = 1$ cm, $L = 10$ cm) the conversion efficiency was

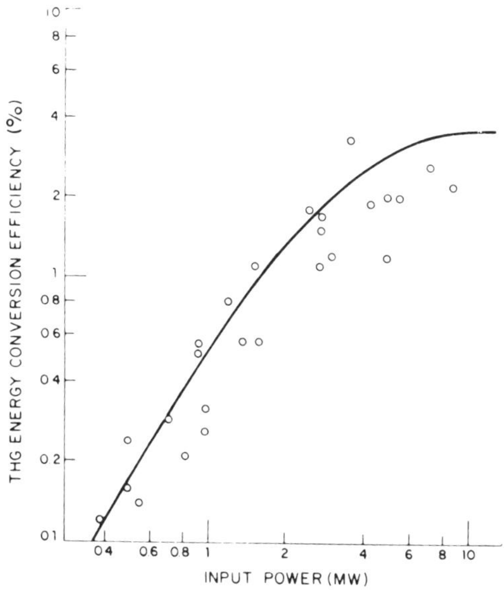

Fig. 27. Dependence of third harmonic of CO_2 laser radiation on pump power. Conversion was done in liquid $CO–O_2–SF_6$ mixture. Calculation includes effects of two-photon absorption and Raman amplification of pump beam. (From Ref. 38, © 1980 IEEE.)

observed to follow its theoretical variation as the square of the pump power for pump powers up to 4 MW, above which the efficiency became independent of pump power. Kildal and Brueck[38] obtained good agreement between the observed behavior and the predictions of the theory outlined in Section 4.3.1a which accounted for two-photon attenuation and Raman amplification of the pump and intensity-dependent phase shifts due to the near-resonant Kerr effect (see Fig. 27). A maximum conversion efficiency of 0.018 in a single-pass geometry and 0.04 in a double-pass geometry[61] (see Section 2-8.4a) was reported at a maximum incident pump power of 10 MW and a peak pump intensity of about 3×10^{10} W/cm^2. Somewhat higher conversion efficiencies were predicted for collimated beams.

5.3. Limitations From Linear Absorption of the Harmonic

Limitations on third-harmonic efficiency due to linear absorption at the harmonic have been investigated by Scheingraber *et al.*[31] for two-photon resonant conversion Sr:Xe mixtures. In this system the absorption arose from the photoionization continuum of Sr. As was pointed out in Section 4.1.1 linear absorption at the harmonic does not change the third power law scaling of the harmonic intensity on the pump intensity. Apart from a

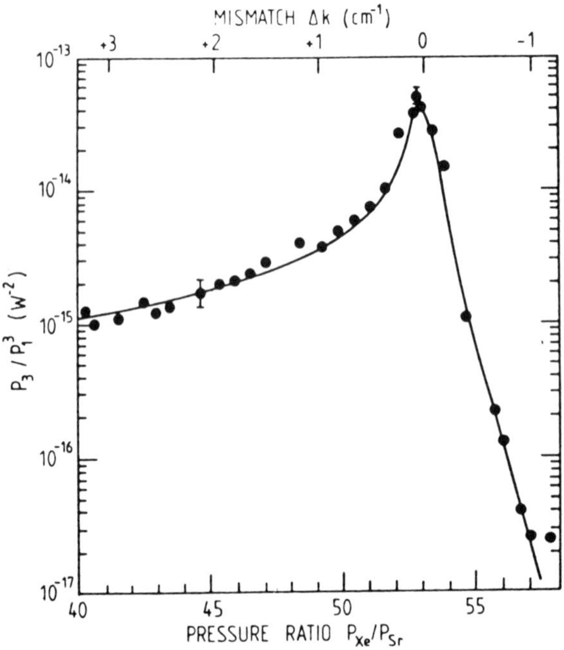

Fig. 28. Third-harmonic conversion efficiency as a function of wave vector mismatch for conversion in a Sr:Xe mixture. Absorption at the harmonic was present because of photoionization in the Sr. The curve was calculated with the theory of Ref. 31. $\sigma = 3.5 \times 10^{-18}\ \mathrm{cm}^2$; $\alpha = 0.6\ \mathrm{cm}^{-1}$. (From Ref. 31.)

reduction of the total harmonic intensity over that which would be obtained in the absence of the absorption, its primary effect is in the area of phase matching. An example of the reduced contrast of the phase-matching structure due to absorption was shown in Fig. 2. Absorption of the harmonic also limits the maximum increase in conversion efficiency that can be obtained by optimizing the phase mismatch. An example of the phase-matching peak in an Sr:Xe mixture is shown in Fig. 28. While phase matching for the conditions under which these data were taken still gives a maximum in conversion efficiency, the contrast with non-phase-matched efficiencies is considerably smaller than if the absorption were not present (see Fig. 2-1).

REFERENCES

1. J. Reintjes and R. C. Eckardt, *Appl. Phys. Lett.* **30**, 91 (1977).
2. Yu. A. Matveets, D. N. Nikogosyan, V. Kabelka, and A. Piskarskas, *Kvantovaya Elektron.* (*Moscow*) **5**, 664 (1978) [*Sov. J. Quantum Electron.* (*Engl. Transl.*) **8**, 386 (1978)].
3. W. Seka, S. D. Jacobs, J. E. Rizzo, R. Boni, and R. S. Craxton, *Opt. Commun.* **34**, 469 (1980).

4. D. M. Bloom, G. W. Bekkers, J. F. Young, and S. E. Harris, *Appl. Phys. Lett.* **26**, 12 (1975).
5. H. Puell, K. Spanner, W. Falkenstein, W. Kaiser, and C. R. Vidal, *Phys. Rev. A* **14**, 2240 (1976).
6. G. Mainfray and C. Manus, *Appl. Opt.* **19**, 3934 (1980).
7. P. Agostini, F. Fabre, G. Mainfray, G. Petite, and N. K. Rahman, *Phys. Rev. Lett.* **42**, 1127 (1979).
8. P. Kruit, J. Kinman, and M. J. van der Wiel, *J. Phys. B* **14**, 597 (1981).
9. A. Gold and H. B. Bebb, *Phys. Rev. Lett.* **14**, 60 (1965).
10. H. B. Bebb and A. Gold, *Phys. Rev.* **143**, 1 (1966).
11. V. M. Morton, *Proc. Phys. Soc., London* **92**, 301 (1967).
12. E. J. McGuire, *Phys. Rev. A* **23**, 186 (1981).
13. J. Morellec, D. Normand, and G. Petite, *Phys. Rev. A* **14**, 300 (1976).
14. A. Rachman, G. Laplanche, and M. Jaouen, *Phys. Lett. A* **68**, 433 (1978).
15. L. V. Keldysh, *Zh. Eksp. Teor. Fiz.* **47**, 1945 (1964) [*Sov. Phys.—JETP* (*Engl. Transl.*) **20**, 1307 (1965)].
16. T. Gontier and M. Trahin, *Phys. Rev. A* **19**, 264 (1979).
17. L. A. Lompré, G. Mainfray, and C. Manus, *J. Phys. B* **13**, 85 (1980).
18. P. Lambropoulos, *Adv. At. Mol. Phys.* **12**, 87 (1976).
19. N. B. Delone, *Usp. Fiz. Nauk* **115**, 361 (1975) [*Sov. Phys.—Usp.* (*Engl. Transl.*) **18**, 169 (1975)].
20. D. Normand and J. Morellac, *J. Phys. B* **13**, 1551 (1980); J. Morellac, D. Normand, G. Mainfray, and C. Manus, *Phys. Rev. Lett.* **14**, 1394 (1980).
20a. J. Mizuno, *J. Phys. B* **6**, 314 (1973).
20b. P. Lambropoulous and M. R. Teague, *J. Phys. B* **9**, 587 (1976).
20c. L. A. Lompré, G. Mainfray, C. Manus, S. Repoux, and J. Thebault, *Phys. Rev. Lett.* **36**, 949 (1976).
21. A. T. Georges, P. Lambropoulos, and J. H. Marburger, *Phys. Rev. A* **15**, 300 (1977).
22. K. Miyazaki and H. Kashiwagi, *Phys. Rev. A* **18**, 635 (1978).
23a. J. C. Miller, R. N. Compton, M. G. Payne, and W. R. Garrett, *Phys. Rev. Lett.* **45**, 114 (1980).
23b. M. G. Payne, W. R. Garrett, and H. C. Baker, *Chem. Phys. Lett.* **75**, 468 (1980); M. G. Payne and W. R. Garrett, *Phys. Rev. A* **26**, 356 (1982).
23c. D. J. Jackson and J. J. Wynne, *Phys. Rev. Lett.* **49**, 543 (1982); erratum, *Phys. Rev. Lett.* **49**, 1057 (1982).
24. C. Grey Morgan, *Rep. Prog. Phys.* **38**, 625 (1975).
25. D. C. Smith and R. G. Meyerland, Jr., *in* "Principles of Laser Plasmas" (G. Bekefi, ed.), p. 457. Wiley, New York, 1977.
26. R. J. Dewhurst, *J. Phys. D* **10**, 283 (1977).
27. E. A. Stappaerts, *Phys. Rev. A* **11**, 1664 (1975).
28. V. I. Anikin, V. D. Gora, K. N. Drabovich, and A. N. Dubovik, *Kvantovaya Elektron.* (*Moscow*) **3**, 330 (1976) [*Sov. J. Quantum Electron.* (*Engl. Transl.*) **6**, 174 (1976)].
29. J. N. Elgin and G. H. C. New, *Opt. Commun.* **16**, 242 (1976).
30. P. P. Bey, J. F. Giuliani, and H. Rabin, *Phys. Rev. Lett.* **19**, 819 (1967).
31. H. Scheingraber, H. Puell, and C. R. Vidal, *Phys. Rev. A* **18**, 2585 (1978).
32. S. A. Akhmanov, A. I. Kovrygin, and A. P. Sukhorukov, *in* "Quantum Electronics" (H. Rabin and C. L. Tang, eds.), Vol. 1, Part B, p. 516. Academic Press, New York, 1975.
33. R. B. Miles and S. E. Harris, *IEEE J. Quantum Electron.* **QE-9**, 470 (1973).
34. L. J. Zych and J. F. Young, *IEEE J. Quantum Electron.* **QE-14**, 147 (1978).
35. J. Reintjes, C. Y. She, and R. C. Eckardt, *IEEE J. Quantum Electron.* **QE-14**, 581 (1978).

36. E. A. Stappaerts, *IEEE J. Quantum Electron.* **QE-15**, 110 (1979).
37. H. P. Puell and C. R. Vidal, *IEEE J. Quantum Electron.* **QE-14**, 364 (1978).
38. H. Kildal and S. R. J. Brueck, *IEEE J. Quantum Electron.* **QE-16**, 566 (1980).
38a. E. R. Bliss, D. R. Speck, and W. W. Simmons, *Appl. Phys. Lett.* **25**, 728 (1974).
38b. M. J. Moran, C. Y. She, and R. L. Carman, *IEEE J. Quantum Electron.* **QE-11**, 259 (1975).
38c. G. K. L. Wong and Y. R. Shen, *Phys. Rev. A* **10**, 1277 (1974).
38d. R. H. Lehmberg, J. Reintjes, and R. C. Eckardt, *Phys. Rev. A* **13**, 1095 (1976).
39. J. H. Marburger. Unpublished results.
40. H. Puell, H. Scheingraber, and C. R. Vidal, *Phys. Rev. A* **22**, 1165 (1980).
41. J. A. Armstrong, N. Bloembergen, J. Ducuing, and P. Pershan, *Phys. Rev.* **127**, 1918 (1962).
42. C. C. Wang and L. I. Davis, Jr., *Phys. Rev. Lett.* **35**, 650 (1975).
43. J. F. Ward and A. V. Smith, *Phys. Rev. Lett.* **35**, 653 (1975).
44. V. I. Anikin, S. A. Akhmanov, K. N. Drabovich, and A. N. Dubovik, *Kvantovaya Elektron.* (*Moscow*) **3**, 2014 (1976) [*Sov. J. Quantum Electron.* (*Engl. Transl.*) **6**, 1096 (1976)].
45. S. E. Harris, *Phys. Rev. Lett.*
46. S. E. Harris and D. M. Bloom, *Appl. Phys. Lett.* **24**, 229 (1974).
47. A. T. Georges, P. Lambropoulos, and J. H. Marburger, *Opt. Commun.* **18**, 509 (1976).
48. H. Scheingraber and C. R. Vidal, *Opt. Commun.* **38**, 75 (1981).
49. J. N. Elgin, G. H. C. New, and K. E. Orkney, *Opt. Commun.* **18**, 250 (1976).
50. G. H. C. New, *Opt. Commun.* **19**, 177 (1976).
51. D. Grischowsky, M. M. T. Loy, and P. F. Liao, *Phys. Rev. A* **12**, 2514 (1975).
52. J.-C. Diels and A. T. Georges, *Phys. Rev. A* **19**, 1589 (1979).
53. J.-C. Diels, *Proc. Vavilov Conf. Coherence Nonlinear Opt., 6th, Novosibirsk, SSSR*, p. 270 (1979).
54. I. A. Poluektov and A. V. Nazarkin, *Kvantovaya Elektron.* (*Moscow*) **6**, 2525 (1979) [*Sov. J. Quantum Electron.* (*Engl. Transl.*) **9**, 1495 (1979)].
55. M. Matsuoka, H. Nakatsuka, and J. Okada, *Phys. Rev. A* **12**, 1062 (1975).
56. T. R. Royt and C. H. Lee, *Appl. Phys. Lett.* **30**, 332 (1977).
57. C. Y. She and J. Reintjes, *in* "Picosecond Phenomena" (C. V. Shank, E. P. Ippen, and S. L. Shapiro, eds.), p. 266. Springer-Verlag, Berlin and New York, 1978.
58. J. Reintjes and C. Y. She, *Opt. Commun.* **27**, 469 (1978).
59. S. C. Wallace and G. Zdasiuk, *Appl. Phys. Lett.* **28**, 449 (1976).
60. H. Junginger, H. B. Puell, H. Scheingraber, and C. R. Vidal, *IEEE J. Quantum Electron.* **QE-16**, 1132 (1980).
61. H. Kildal and S. R. J. Brueck, *IEEE J. Quantum Electron.* **QE-16**, 566 (1980).

CHAPTER 5

SELF-ACTION EFFECTS AND DEGENERATE FOUR-WAVE MIXING

1. INTRODUCTION

In the previous chapters we were primarily concerned with nonlinear polarizations that were created at frequencies other than those in the incident laser beams. Nonlinear polarizations at the incident frequencies were considered only in the context of their effect on the frequency conversion processes. In this chapter we shall consider in more detail the nonlinear polarizations that are at the same frequencies as those that are present in the incident beams. The effects of these polarizations can be divided into two broad classes: self-action effects, in which the nonlinear polarization created by an incident beam affects the propagation or other properties of the same beam, and a process usually termed degenerate four-wave mixing, in which the nonlinear polarization created by an incident beam or beams affect another beam that is at the same frequency but is otherwise distinguishable, e.g., by its direction of propagation or polarization.

2. NONLINEAR POLARIZATIONS

The most common interactions involving the nonlinear polarization at the incident frequencies arise from the third-order terms. For steady-state interactions in isotropic media the third-order polarization amplitude can be expressed as[1]

$$P^{\mathrm{NL}}(\omega) = \tfrac{1}{4}[C(A \cdot A^*)A + \tfrac{1}{2}D(A \cdot A)A^*], \tag{1}$$

where $A = \sum A_i$ is the vector sum of the incident field amplitudes and we have allowed for the incident fields to be polarized in different directions. The coefficients C and D are related to the tensor components of the third-order susceptibility: $C = 3(\chi^{(3)}_{1122} + \chi^{(3)}_{1212})$, $D = 6\chi^{(3)}_{1221}$. If there is only one incident wave, then Eq. (1) yields for the polarization amplitude resulting from linearly polarized light

$$P^{\mathrm{NL}}(\omega) = \tfrac{1}{4}(C + \tfrac{1}{2}D)|A_1|^2 A_1 = \tfrac{3}{4}\chi^{(3)}_{1111}(-\omega, \omega, -\omega, \omega)|A_1|^2 A_1. \quad (2)$$

The most commonly encountered effects arising from these nonlinear polarizations are multiphoton absorption, self-focusing, self-defocusing, self-phase modulation, and degenerate four-wave mixing. The first of these effects comes from the imaginary part of the two-photon resonant susceptibility in (1) and involves energy transfer to the medium. It will not be treated in detail in this chapter. The remaining effects, which will be covered here, can arise from either the real or imaginary parts of the susceptibility.

In discussions of the self-action effects it is common to express the real part of the susceptibility in (2) in terms of a nonlinear dielectric constant, ϵ_2 or a nonlinear refractive index n_2. If the dielectric constant is defined as

$$\begin{aligned} \epsilon \equiv \epsilon^{\mathrm{L}} + \delta\epsilon &= \epsilon^{\mathrm{L}} + \epsilon_2 \langle A_1 \rangle^2 \\ &= \epsilon^{\mathrm{L}} + \tfrac{1}{2}\epsilon_2 |A_1|^2, \end{aligned} \quad (3)$$

then the coefficient ϵ_2 is given in terms of the real part of the third-order susceptibility as

$$\epsilon_2 = 6\pi\chi^{(3)\prime}(-\omega, \omega, -\omega, \omega). \quad (4)$$

The corresponding expressions for the nonlinear refractive index are, for small nonlinear contributions to the refractive index,

$$n \equiv \sqrt{\epsilon} = n^{\mathrm{L}} + \delta n = n^{\mathrm{L}} + n_2 \langle A_1 \rangle^2 = n^{\mathrm{L}} + \tfrac{1}{2} n_2 |A_1|^2,$$

$$n_2 = \frac{3\pi\chi(-\omega, \omega, -\omega, \omega)}{n^{\mathrm{L}}}. \quad (5)$$

These quantities are sometimes defined in the literature as coefficients of the intensity instead of the square of the field strength:

$$\epsilon = \epsilon^{\mathrm{L}} + \epsilon_2(I) I, \quad (6a)$$

$$n = n^{\mathrm{L}} + n_2(I) I. \quad (6b)$$

The intensity coefficients are related to the field coefficients by the following formulas:

$$\epsilon_2(I) = \frac{c(\epsilon^{\mathrm{L}})^{1/2}}{4\pi}\, \epsilon_2(A), \quad (7a)$$

$$n_2(I) = \frac{cn^{\mathrm{L}}}{4\pi}\, n_2(A). \quad (7b)$$

3. MATERIALS

It was pointed out in Section 4-4.2 that all materials exhibit a nonlinear index at sufficiently high optical intensities. The origin of the nonlinear index depends on the type of medium involved and on the characteristics of the optical radiation. In liquids with anisotropic molecules, orientation of the molecules in response to the incident fields makes a major contribution to the nonlinear index. In describing the nonlinear index due to molecular motion in response to pulsed fields it is often necessary to take into account the finite response time of the molecular motion.

The change in refractive index δn in such materials can usually be described by a diffusion equation of the form

$$\tau \frac{d\delta n(t)}{dt} + \delta n(t) = \tfrac{1}{2} n_2 |A_1(t)|^2. \tag{8}$$

Here τ is the response time of the molecules to the incident radiation and n_2 is the steady-state coefficient of (5). Typical response times for liquids are of the order of several picoseconds to hundreds of picoseconds.

In addition to orientational motion, librations have been proposed to contribute to the nonlinear index in certain circumstances. The appropriate equation for this situation is

$$\tau_2^2 \frac{d^2\delta n}{dt^2} + \tau_1 \frac{d\delta n}{dt} + \delta n = \tfrac{1}{2} n_2 |A_1|^2. \tag{9}$$

Here τ_2 is determined by the librational frequency and τ_1 is the damping time.

The orientational effect always makes a positive contribution to the nonlinear refractive index.

Other types of molecular or atomic motion can result in nonlinear refractive indices. Electrostriction, whereby atoms or molecules are drawn to regions of high intensity, generally results in a positive contribution to the nonlinear refractive index. The time constant for this effect involves the time it takes for a molecule to move a distance comparable to the beam size and is typically of the order of nanoseconds or longer.

Optical absorption can result in local heating, leading to a decrease in density in regions where the beam is most intense, resulting in a negative contribution to the refractive index. This effect gives rise to a form of self-defocusing known as thermal blooming. The time constant associated with it is of the order of microseconds to milliseconds. It is thus not generally important for Q-switched or mode-locked laser pulses, which typically have durations less than 20 nsec, but it can be significant for pulse durations of the order of microseconds or longer. It is especially important in the propagation of high-power infrared pulses in the atmosphere.

A nonlinear refractive index can also arise from the electronic response of atoms and molecules. An expression for the nonlinear index in media with isolated electronic energy levels was given in Eq. (4-38a). The response time of the electronic nonlinear index is usually instantaneous on the time scale of available laser pulses. However, a transient response can be significant if exact, or nearly exact, one- or two-photon resonances are involved. The electronic contributions are present in all materials and can make the dominant contribution to the nonlinear index in media that do not contain anisotropic molecules, such as atomic gases and liquids, liquids with symmetric molecules, and many solids.

The nonresonant contributions of these terms are usually positive and have been shown to make a significant contribution to the positive n_2 in materials such as laser glass[2] and liquids with symmetric molecules such as CCl_4.[3a] Resonant contributions, which can be encountered in, for example, gases, can be positive or negative, depending on the relation of the incident frequency or its multiples to the appropriate energy levels.

Many of the effects that cause a nonlinear index of refraction, for example, molecular orientation or molecular clustering, also cause optically induced birefringence, resulting in an optical Kerr effect. The effective Kerr constant depends on the mechanism responsible for the nonlinear index. For example, for molecular orientation the induced birefringence is related to the nonlinear susceptibility by

$$\delta n_{\parallel} - \delta n_{\perp} = \frac{9\pi\chi^{(3)}_{1111}}{4n^{\mathrm{L}}} |A_1|^2, \tag{10}$$

while for nonresonant electronic distortions of spherically symmetric molecules the relation is

$$\delta n_{\parallel} - \delta n_{\perp} = \frac{\pi\chi^{(3)}_{1111}}{n^{\mathrm{L}}} |A_1|^2. \tag{11}$$

TABLE 1

Nonlinear Refractive Indexes and Relaxation Times of Some Liquids

Liquid	n_2 (esu)	τ	Reference
CS_2	1.1×10^{-11}	2.3 psec	3b
Toluene		7 psec	
Nitrobenzene		20 psec	
MBBA	$\dfrac{2 \times 10^{-9}}{T(°\mathrm{K}) - 314.7}$	$\dfrac{9.1 \times 10^{-11}\, e^{2800/T(°\mathrm{K})}\ \mathrm{sec}^{a}}{T(°\mathrm{K}) - 314.7}$	3c

[a] $T > 314.7$°K.

Here the designations $\parallel$ and $\perp$ refer to directions parallel and perpendicular to the direction of the polarization of the incident light.

The optically induced birefringence can be used as a diagnostic for identifying various properties of the nonlinear index. It can also give rise to rotation of the sense of polarization of elliptically polarized light [1–3a] and forms the basis for certain forms of nonlinear spectroscopy.

A list of values of n_2 along with appropriate relaxation times is given for various materials in Table 1.

4. SELF-ACTION EFFECTS

Self-action effects—self-focusing, self-defocusing, and self-phase modulation—arise from a combination of a nonlinear index and intensity variations in the incident radiation which result in intensity-dependent contributions to the phase of the incident beam that vary in space or time. Self-focusing and self-defocusing are associated with spatial intensity variations of the incident beam. Since the intensity of most laser beams varies across the diameter due to mode structure, diffraction, or other causes, the spatial self-action effects will generally be encountered in all materials at high enough intensities. Self-phase modulation is associated with temporal intensity variations. It gives rise to frequency broadening, the extent of which is determined by a combination of the effective pulse duration, the magnitude of n_2, and the propagation length.

Because spatial phase variations are converted to intensity variations through linear diffraction and temporal phase variations are changed to temporal intensity variations through dispersion, it is usually impossible to separate the effects of self-focusing or defocusing from those of self-phase modulation. In some instances, however, especially if the nonlinear phase changes are kept small, it is possible for either the spatial or temporal self-action effects to dominate the interaction. In this chapter we shall treat the spatial and temporal effects separately, but the reader should keep in mind that in most cases they will occur together.

The self-action effects have been studied for many years and have been the subject of numerous review articles and book chapters.[4–7] Therefore only the basic principles of these interactions will be described here. For further details the reader is referred to the appropriate references at the end of the chapter.

4.1. Self-Focusing

Self-focusing occurs as the result of a combination of a positive nonlinear index and a spatial variation of laser intensity in which the beam is more intense in the center than at the edges, as shown in Fig. 1. In the presence of

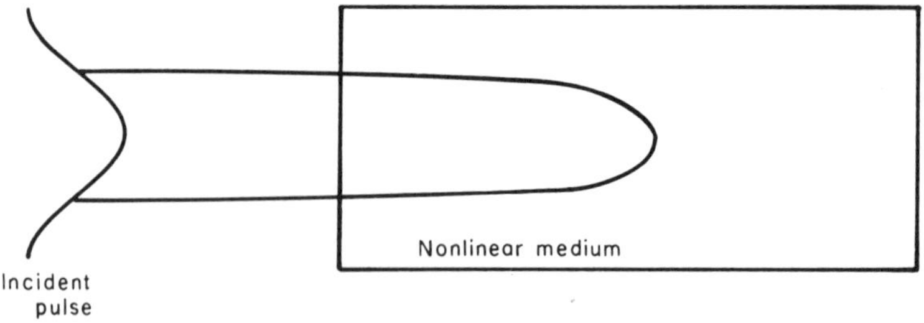

Fig. 1. Schematic illustration of catastrophic focusing associated with self-focusing.

the optical field the increase in refractive index follows the intensity profile of the incident beam, larger at the center and smaller at the edges. This profile corresponds to a positive lens and the beam focuses. The focusing of the beam increases the power of the effective nonlinear lens, causing the beam to focus yet further. As a result the beam focuses catastrophically, rather than gently, as is usual in focusing with a linear lens. This behavior is illustrated by the contour lines in Fig. 1.

Self-focusing has been observed in many kinds of materials and with many types of laser sources. It is commonly observed in Kerr active liquids, that is liquids with a large optical Kerr effect, with both mode-locked and *Q*-switched laser pulses. Self-focusing of mode-locked pulses in non-Kerr liquids, such as liquid argon,[8] has also been reported. Self-focusing of mode-locked laser pulses in moecular gases has been observed.[9] Self-focusing of pulsed dye laser radiation in gases due to resonant interactions has been reported.[10] Finally, self-focusing of cw dye laser radiation in Na vapor has been observed.[11] Self-focusing of pulsed laser radiation also occurs in many transparent solids at high intensities and often results in damage of these materials.

The qualitative behavior of a self-focused beam depends on a number of factors, such as the total power in the beam relative to a critical power for the onset of self-focusing (see Section 4.1.1), whether the self-focal point lies inside the nonlinear medium, as shown in Fig. 1, or outside the medium, and on the presence of any spatial structure on the beam profile (see Section 4.1.5).

Whole beam self-focusing, in which the entire incident beam self-focuses as a unit, is usually observed only under special conditions. For beams whose total power is only slightly greater than the critical power, whole beam self-focusing can be observed when the focal point lies within the nonlinear medium. For beams whose power greatly exceeds the critical power, whole beam self-focusing can usually be observed only if the beam profile is extremely smooth and if the total nonlinear phase shift is kept small, of the order of a few π. For these conditions the catastrophic self-focal point usually lies outside the nonlinear medium and will never be reached. Instead, the medium acts as an intensity-dependent positive lens. Distortion of the beam can result if the whole beam self-focusing is strong enough to introduce non-spherical wavefronts. If the incident radiation is in the form of a pulse, the focal length of the nonlinear lens will vary in time.[12]

When the total beam power greatly exceeds the critical power, it is more common for the beam to break up into many self-focal points each of which carries only one or a few critical powers, an amount that is far less than that contained in the entire beam. This beam breakup is usually termed small-scale self-focusing. The diameter of the individual focal points is of the order of a few micrometers to a few tens of micrometers for a wide range of materials and input powers.[13] Similar behavior is observed in the small-scale self-focusing of mode-locked and Q-switched laser pulses with regard to beam breakup and focal spot size.

4.1.1. Self-Focusing Equations

Steady-state self-focusing is described by the equation

$$\nabla_\perp^2 A + 2ik\frac{\partial A}{\partial z} = -\frac{n^{\mathrm{L}}\omega^2}{c^2} n_2 |A|^2 A. \tag{12}$$

This equation has been analyzed by many authors. Numerical solutions of Eq. (12) for a beam with an initially Gaussian profile were given first by Kelley.[14] Dawes and Marburger[15] have analyzed the behavior of a beam in a nonlinear medium using various beam shapes and have included the effects of linear and nonlinear absorption. Similar analyses have been given by Goldberg *et al.*[16] and Wagner *et al.*[17]

The solution of Dawes and Marburger shows that the behavior of an initially Gaussian beam with the profile

$$A(r) = A_0 e^{-r^2/2a^2} \tag{13}$$

depends on the total power in the beam relative to two critical powers, $\mathscr{P}_1$ and $\mathscr{P}_2$, which are determined by the nonlinear index of the medium and the vacuum wavelength of the radiation, λ_0, and using our definition of n_2 are given by

$$\mathscr{P}_2 = (1.22\lambda_0)^2 c/128 n_2, \tag{14a}$$

$$\mathscr{P}_1 = 0.273\mathscr{P}_2. \tag{14b}$$

For incident powers that satisfy

$$\mathscr{P} > \mathscr{P}_2 \tag{15}$$

the beam propagates to a catastrophic self-focus at a distance z_f given by

$$z_f = 0.369ka^2/(\sqrt{\mathscr{P}/\mathscr{P}_2} - 0.858). \quad (16)$$

At the self-focal point the calculated on-axis intensity tends to infinity and very few numerical calculations have followed the beam through the catastrophic self-focal point.

The detailed behavior of a beam undergoing self-focusing depends on the exact beam shape. Dawes and Marburger have shown that for non-Gaussian beams the form of (16) for z_f is valid in the asymptotic limit $\mathscr{P} \gg \mathscr{P}_2$ but the constants depend on the beam profile. Calculations by Dawes and Marburger[15] and by other authors[16,17] predict the occurrence of multiple focal points and the development of ring structure when $\mathscr{P} > 2\mathscr{P}_2$.

For incident powers satisfying the condition

$$\mathscr{P}_1 < \mathscr{P} < \mathscr{P}_2$$

the on-axis intensity first increases due to weak self-focusing and then decreases as the self-focusing is overcome by diffraction. For incident powers satisfying the condition

$$\mathscr{P} < \mathscr{P}_1$$

no self-focusing is predicted.

4.1.2. *Limiting Spot Sizes in Self-Focusing*

It should be noted that the requirement for self-focusing is on the total incident beam power and not on the incident intensity. As a result, once the critical power is exceeded and the beam starts to focus, the steady-state theory outlined above predicts that the beam diameter will become arbitrarily small, and the intensity arbitrarily large, unless limited by another process. Several processes have been considered for limitation of self-focusing—saturation of the nonlinear index,[11,15,18] stimulated Raman scattering,[7,19,22] dielectric breakdown,[20] and transient response of the nonlinear index.[21,22]

Saturation of the nonlinear index has been observed to limit the self-focusing process in a cw resonant interaction in Na vapor.[11] In this experiment the light was confined to a narrow channel after the self-focusing point. The tendency of the light to diffract was offset by the self-focusing process, while further self-focusing was prevented by the saturation of the nonlinear index.

However, in Kerr active liquids the maximum index change at the minimum diameters that are observed is only of the order of 10^{-4}, too small for saturation of the nonlinear index to limit the self-focusing process in these materials. Saturation of the index nonlinearity would require index changes of the order of 0.1–0.5 and minimum spot diameters that are over an order of magnitude smaller than those that are actually observed. Other mechanisms are thus expected to play a role in determining the minimum size of the self-focal spots in these materials. For example, in many of the liquids the maximum optical fields are comparable to the breakdown fields, and dielectric breakdown has been suggested as a mechanism to explain the smallest diameters of the self-focal spots in many liquids.[20] Dielectric breakdown can be expected to occur in most materials at sufficiently high intensities, and it can be expected to affect the self-focusing process if some other effect does not limit the process at lower intensities.

A transient response of the nonlinear index is important in describing the self-focusing of mode-locked laser pulses in many media, and the effect of a transient response has been analyzed by several authors[21,22] (see Section 4.1.4). For a discussion of various limiting processes the reader is directed to Refs. 5–7, 19–22, and the references contained therein.

4.1.3. *Moving Self-Foci*

If the incident radiation is pulsed with a duration long compared to the relaxation time of the nonlinear index, then the steady-state theory outlined in the previous section is applicable, except that the position of the self-focal point in (9) varies with time. This situation describes, for example, self-focusing of Q-switched laser pulses with durations typically of the order of 10–20 nsec in most Kerr active liquids or the self-focusing of Q-switched or mode-locked laser pulses in media with fast responding electronic nonlinearities.

The existence of the moving self-focal points was proposed by Dyshko *et al.*[23] and Lugovoi *et al.*[24] and was later explored in detail by Loy and Shen[25] and Korobkin *et al.*[26]

The self-focusing of an incident pulse gives rise to two focal points whose trajectories are determined by the time-dependent form of Eq. (16), and are illustrated in Fig. 2. The first focal point comes from radiation on the leading edge of the pulse and occurs at a distance indicated by the point D in the figure.

From this position two focal points emerge, one moving toward the end of the cell and caused by the radiation at earlier times on the leading edge of the pulse and the other moving backward toward the front of the cell and caused by radiation occurring later on the rising edge. The backward moving focal point reaches a minimum distance from the entrance of the cell given by

$$z_{\min} = \frac{0.369ka^2}{\sqrt{\mathscr{P}_0/\mathscr{P}_2} - 0.858}, \tag{17}$$

where $\mathscr{P}_0$ is the maximum power in the pulse, and then turns around and moves toward the end of the cell during the second half of the pulse.

The moving self-focal points have been used to explain the apparently continuous streaks of light that are observed from the side of a cell during the self-focusing process. Such streaks were originally interpreted in terms of a trapped filament. The trapping picture is accurate for the cw radiation used in the self-focusing experiments in Na vapor for which the filament diameter was stabilized by saturation of the nonlinear index. However, as pointed out earlier, such an effect is not likely to be significant in typical Kerr active liquids. The appearance of constant diameter streaks are now believed to arise from scattering from a succession of self-focal points that follow the trajectories illustrated in Fig. 2.

The duration of the radiation in a single self-focused spot can be considerably shorter than that of the incident pulse. For example, in many Kerr active liquids pulse durations ranging from several picoseconds to hundreds of picoseconds are observed in the self-focused regions generated by incident pulses of the order of 10 nsec. The dynamics of the radiation in the self-

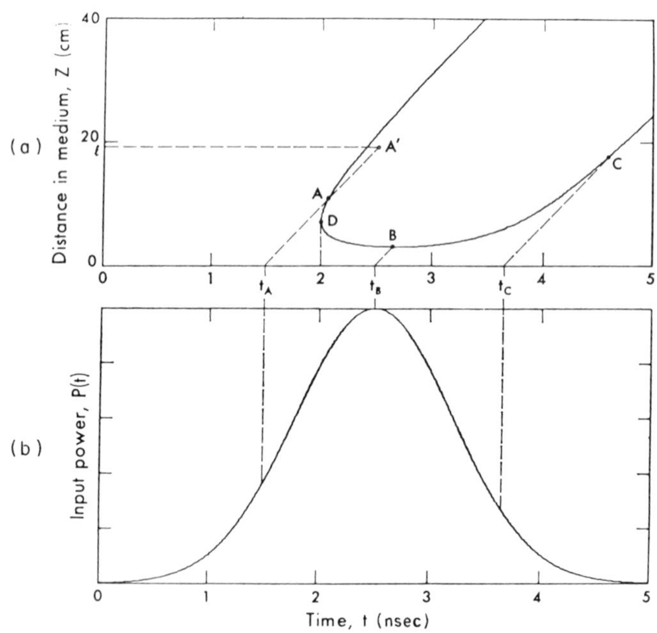

Fig. 2. (a) Trajectory of moving focal point in self-focusing. (b) Incident intensity as a function of time. (From Ref. 21, © 1973 IEEE.)

focused region will be governed by a transient theory if the minimum duration of the focused radiation becomes comparable to or shorter than the relaxation time of the nonlinear index.

4.1.4. *Transient Effects*

If the duration of the laser pulse is comparable to or shorter than the response time of the nonlinear index, then the nonlinear polarization cannot respond instantaneously on the time scale of the radiation. This situation can occur, for example, when a mode-locked laser pulse self-focuses in a medium which exhibits an orientational Kerr effect. It can also be appropriate for describing the details of the behavior of the self-focused region of longer pulses as pointed out in the previous section.

Under these conditions Eq. (12) describing self-focusing must be replaced by one in which the nonlinear polarization is time dependent. The nonlinear polarization in turn is determined by an appropriate differential equation. For example, for self-focusing in a medium in which the nonlinear index is described by a simple relaxation equation, the time-dependent self-focusing equations are

$$\nabla_{\perp}^2 A + 2ik\frac{\partial A}{\partial z} + \frac{2i\omega}{c}\frac{\partial A}{\partial t} = -\frac{2n_0\omega^2}{c^2}\,\delta n(t)A, \tag{18a}$$

$$\tau\frac{\partial \delta n}{\partial t} + \delta n = \tfrac{1}{2}n_2|A|^2. \tag{18b}$$

The transient response of the nonlinear index can have a qualitative effect on the self-focusing process. This effect can be explained physically in terms of a "horn" model, which has been investigated theoretically and experimentally by many authors.[21,22,27] The horn model is illustrated in Fig. 3. As the beam propagates through the cell the front of the pulse does not focus because the nonlinear index cannot respond fast enough. The leading edge does set up, however, a nonlinear lens that causes later sections of the pulse to focus. Eventually the diameter of the beam takes on the appearance of a horn, with the self-focused region confined to the rear of the pulse.

As the pulse propagates through the nonlinear medium the front of the pulse diffracts. Eventually the nonlinear lens set up by the front of the pulse becomes too weak to focus the rear of the pulse and the self-focused region disappears. This behavior is consistent with the observations of Reintjes *et al.*,[28] who observed that self-focused regions from mode-locked pulses in CS_2 existed at distances from 1 up to about 10 cm from the entrance of a cell but disappeared at longer distances.

Since the diffraction of the self-focused region at the rear of the pulse is controlled by the larger diameter at the leading edge, it can propagate for a

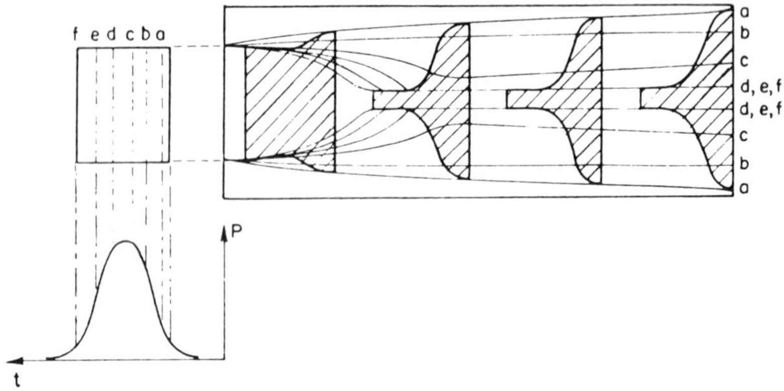

Fig. 3. Development of pulse according to horn model of transient self-focusing. (Reprinted with permission from Y. R. Shen, *Prog. Quantum Electron.* **4**, copyright 1977, Pergamon Press, Ltd.)

distance that is many times its own diffraction length without changing size significantly. This behavior has given rise to the name dynamic self-trapping for this process. The calculations of Shimizu[22] have shown that the transient response, coupled with diffraction of the leading edge of the pulse, can result in a self-focused region near the tail of the pulse with a diameter that undergoes stable oscillations about an equilibrium value. The minimum diameter of the self-focused region was very sensitive, however, to the total input power and for many conditions was predicted to be smaller than the values that are observed. Shimizu has thus postulated the presence of additional mechanisms to account for the limiting spot size in transient focusing.

4.1.5. Beam Breakup—Spatial-Frequency Analysis

The theories given in the preceding sections described the formation of a single self-focal spot from an incident beam. As pointed out earlier, however, when the power in the beam greatly exceeds the critical power, it is more likely that many self-focal spots will be formed. The breakup of the beam into these multiple focal spots is related to the growth of components within the beam with nonzero transverse spatial frequencies. The growth of spatial modulation on a beam was originally described by Bespalov and Talanov.[29] It was later studied theoretically by Suydam[30] and Glass[31] and experimentally by Campillo *et al.*[32a,32b] and by Bliss *et al.*[33]

To derive a growth rate for spatial modulation on a beam we express the complex amplitude of the incident wave as the product of a real amplitude and a phase factor:

$$A(r, z) = A_0(r, z)e^{i\phi}. \tag{19}$$

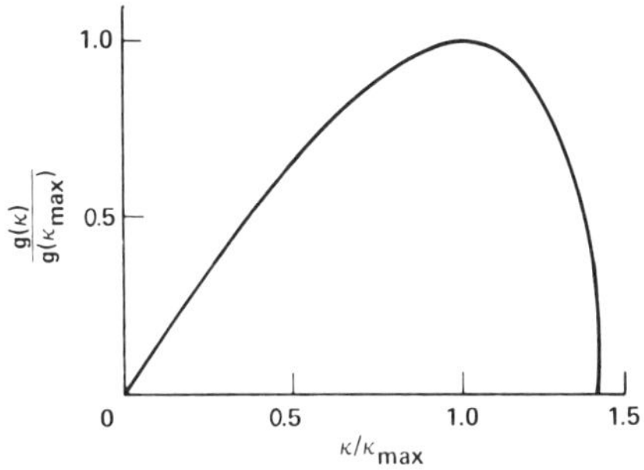

Fig. 4. Exponential growth rate of modulation in beam as a function of transverse spatial frequency due to nonlinear refractive index. (From Ref. 31.)

We assume that the amplitude and phase carry a weak modulating term with a distribution of transverse spatial frequencies k_x and k_y:

$$A_0(r, z) = A_0 + \psi_k(z)e^{i(k_x x + k_y y)}, \tag{20a}$$

$$\phi = \phi_0 + \tilde{\psi}(z)e^{i(k_x x + k_y y)}. \tag{20b}$$

Using Eq. (7), and linearizing in the weak functions ψ_k and $\tilde{\psi}$, we find for the modulation components the solutions

$$\psi_{K_\perp}(z) = \psi_{K_\perp}(0)e^{gz}, \tag{21a}$$

$$\tilde{\psi}_{K_\perp}(0) = \tilde{\psi}_{K_\perp}(0)e^{gz}, \tag{21b}$$

where the exponential growth rate is given by

$$g = (K_\perp/2k)(2n_0\omega^2 n_2 A_0^2/c^2 - K_\perp^2)^{1/2}, \tag{22}$$

and $K_\perp = (k_x^2 + k_y^2)^{1/2}$. The exponential growth rate is shown in Fig. 4 as a function of spatial frequency $K_\perp$. A maximum growth rate of

$$g_{\max} = \frac{\omega^2 n_2 A_0^2}{2kc^2} \tag{23}$$

occurs for an optimal spatial frequency

$$K_{\text{opt}} = (n_0 n_2 \omega^2 A_0^2/c^2)^{1/2}. \tag{24}$$

Slower growth rates are experienced by spatial frequencies larger and smaller than K_{opt}, while no growth is experienced for spatial frequencies above

$$K_{\max} = (2n_0\omega^2 n_2 A_0^2/c^2)^{1/2} = \sqrt{2}K_{\text{opt}}. \tag{25}$$

The exponential growth rate has been measured as a function of spatial frequency by Bliss *et al.*[33] They used wedged glass plates to introduce modulation with varying spatial frequency onto a plane wave. Their results are shown

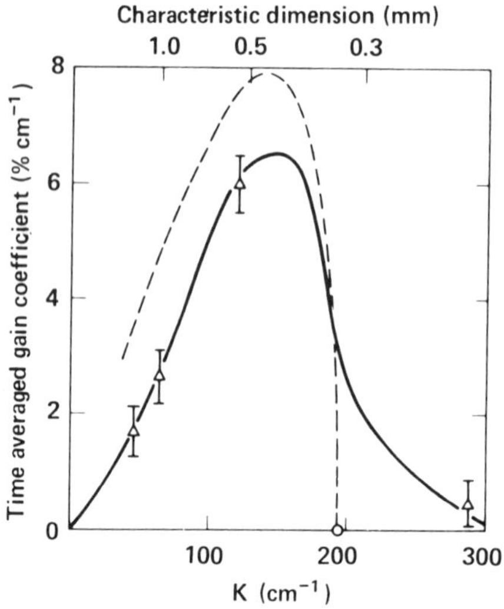

Fig. 5. Experimental demonstration of exponential growth of periodic modulation on laser beam. (From Ref. 33.)

in Fig. 5. The dotted curve is calculated from the linearized theory outlined earlier, while the solid curve is calculated from a more complete analysis that accounted for the phase and amplitude of the initial modulation. The more complete analysis does not predict a sharp cutoff for spatial frequencies above K_{max} and is in good agreement with the observations.

Campillo *et al.*[32a,32b] have investigated the relation between the growth of spatial frequencies and the breakup of a beam due to self-focusing. They have shown that because of the exponential growth of the transverse spatial frequencies, a beam has a tendency to break up into a periodic array of cells of dimension $a \sim K_{opt}^{-1}$ each of which eventually forms a separate self-focal point. Although the initial growth rate of a component with a given value of K is independent of the values of k_x and k_y, they have shown that the most favorable distribution for beam breakup occurs when $k_x = k_y$, yielding an array of symmetric cells. The power contained in each cell is of the order of a few critical powers, with the exact value depending on the profile of the main beam. The breakup of a beam which initially had a ring pattern due to Fresnel diffraction into a circular array of focal points is shown in Fig. 6.

Spatial modulation is usually present on all laser beams to some degree due to such sources as propagation through inhomogeneous media, from dust or imperfections on optical surfaces or from diffraction from the edges of apertures. If the incident beam carries modulation at a well-defined spatial frequency, then the beam breakup can be very regular, as shown by the pictures in Fig. 6. If, however, the strong modulating components are at fre-

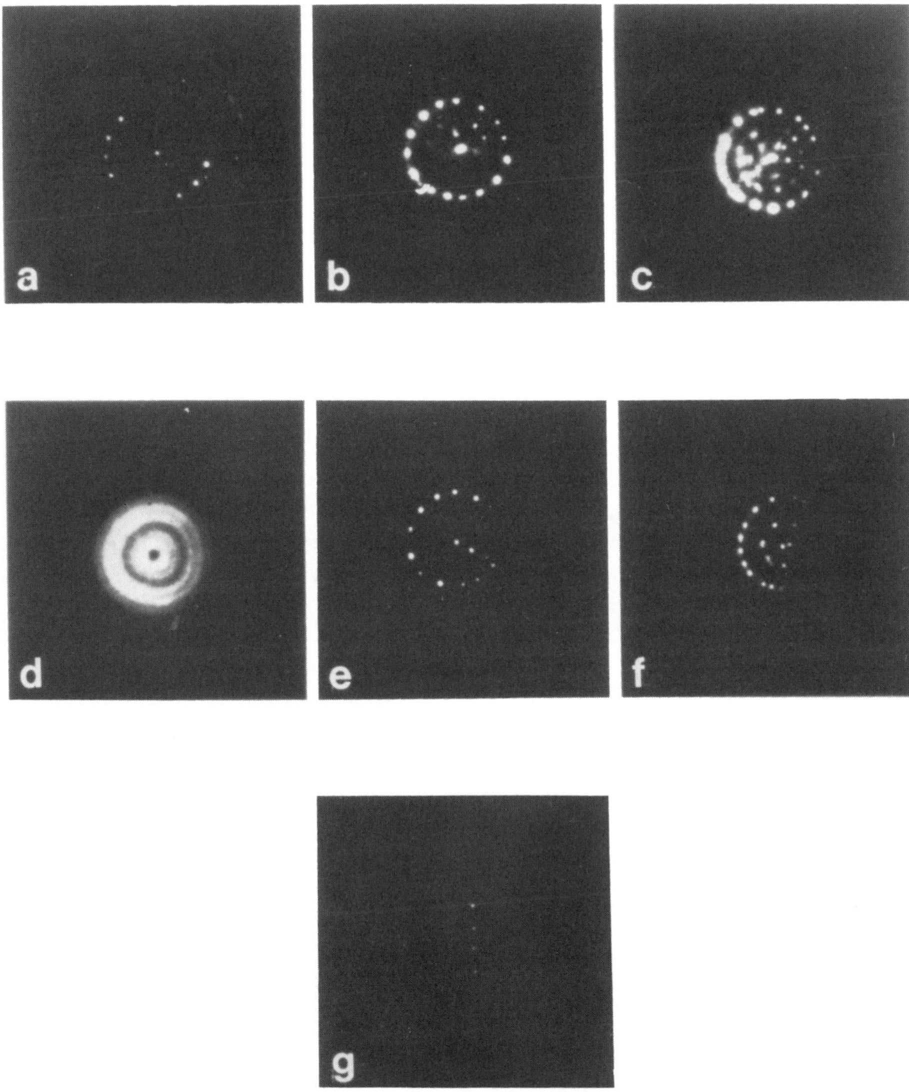

Fig. 6. Illustration of beam breakup due to periodic modulation. (From Ref. 32a.)

quencies other than K_{opt}, then the focal spot pattern can be unrelated to the optimum spatial frequency. If the initial modulation is random, then the resulting focal spot pattern will also be random. This last behavior is commonly observed in the self-focusing of laser beams that greatly exceed the critical power. The exponential growth of components with spatial frequencies in the range $K_\perp \lesssim K_{\perp,\max}$ explains why it is so difficult to avoid beam breakup in self-focusing media once the critical power is exceeded.

4.2. Self-Defocusing

Self-defocusing occurs as a result of the combination of a negative nonlinear index and a beam with an intensity profile that is larger in the center than at the edges. The refractive index at the center of the beam is now less than at the edges and the nonlinear lens that is formed in the medium is negative, causing the beam to defocus. When the degree of defocusing is small, the beam shape changes little as it defocuses. If the degree of defocusing is large, however, then the beam flattens, reducing the tendency for the beam to undergo further defocusing. Thus the self-defocusing process has a form of self-limitation rather than the catastrophic behavior that was present in self-focusing.

Self-defocusing from electronic nonlinearities is usually observed only under special circumstances. The expression for the nonlinear index from isolated resonances [Eq. (4-38a)] shows that n_2 is negative only for detunings that are just below single-photon resonances or just above two-photon resonances. Self-defocusing in atomic vapors under each of these conditions has been reported.[34,35] Thermal self-defocusing or thermal blooming can be observed in media that are slightly absorbing. Thermal blooming in liquids[36] and in the atmosphere[37,38] has been reported.

4.2.1. *Self-Defocusing Equations*

Self-defocusing is described by the same equations that describe self-focusing, except for the change in sign of n_2. A discussion of the effective focal lengths of the nonlinear lens in absorbing media that exhibit thermal blooming has been given by Akhmanov *et al.*[5]

4.2.2. *Spatial Frequencies*

An analysis of the behavior of spatial frequencies modulating an incident beam, similar to that given in Section 4.1, leads to a prediction of oscillation of the amplitudes of the spatial frequency components, rather than attenuation as might be expected. The amplitude of the spatial frequency components obtained from solution of Eq. (7) using (13) and a negative n_2 is

$$\psi_{K_\perp}(z) = \psi_K(0)e^{i2\pi z/L_{osc}}, \tag{26}$$

where the spatial oscillation distance is

$$L_{osc} = \frac{4\pi k}{K_\perp(2n_0 n_2 \omega^2 A_0^2/c^2 + K_\perp^2)^{1/2}}. \tag{27}$$

Self-defocusing will tend to wash out local intensity structure as the spatial frequencies that form the structure become dephased with distance. However, the fact that the amplitudes of the spatial frequency components are not attenuated must be kept in mind if it is attempted to compensate for the effects of self-focusing and beam breakup with a self-defocusing medium.

4.2.3. Experimental Demonstrations of Electronic Self-Defocusing

4.2.3a Single-Photon Resonances. An example of self-defocusing due to a coherent interaction near a single-photon resonance has been given by Grischowsky and Armstrong.[34] Self-defocusing of dye laser radiation in Rb vapor was observed when the dye laser was tuned to the long-wave side of the $^2P_{1/2}$ resonance line as shown in the level diagram in Fig. 7. For these experiments a magnetic field was used to split the resonance line so that the incident radiation was resonant with only one upper level, allowing accurate use of a two-level model for the system.

The experiments were done with pulse durations of 10 nsec, which was significantly shorter than the dephasing time of the upper level. The response of the atom was thus described by the adiabatic following model (see Section 2-6.2.2). Under these conditions the nonlinear index is given by

$$n_2 = -2\pi N|\mu_{12}|^4/n_0[\hbar(\omega_{i0} - \omega)]^3. \tag{28}$$

An example of the measured self-defocused beam profiles, along with the calculated ones, for two different circular polarizations is shown in Fig. 8.

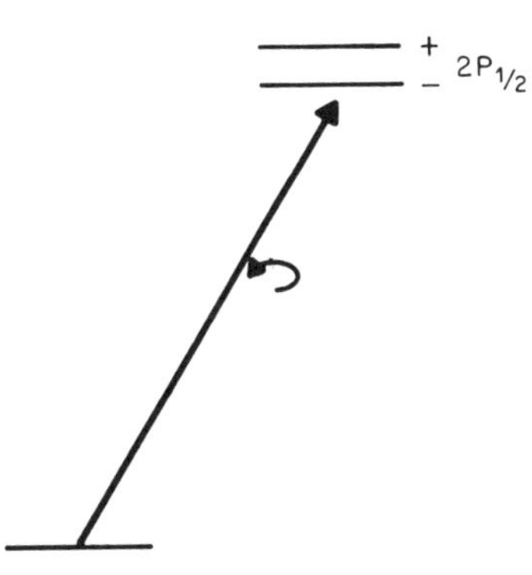

Fig. 7. Level diagram for self-defocusing of dye laser radiation in Rb.

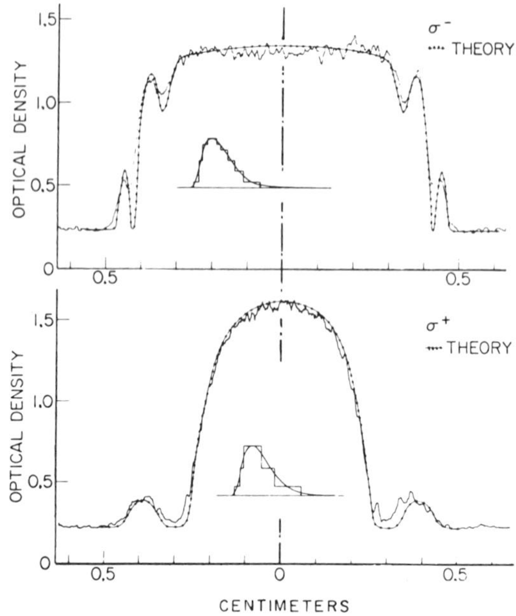

Fig. 8. Self-defocused beam profiles in Rb. Curve was calculated from theory using adiabatic following model. (From Ref. 34.)

4.2.3b. TWO-PHOTON RESONANCE. Self-defocusing from a two-photon resonance has been reported by Lehmberg *et al.*[35a,35b] They used 30-psec pulses from a mode-locked Nd:YAG laser at a wavelength of 1.064 μm and observed self-defocusing in Cs vapor. The level diagram is shown in Fig. 9. The two-photon energy of the laser lies 250 cm^{-1} above the 7s level, giving a two-photon enhancement and a negative contribution to the nonlinear index. An additional negative contribution to the nonlinear index comes from the single-photon resonance with the 6p levels. This detuning is greater than the two-photon detuning, about 2500 cm^{-1}, and the single-photon enhancement of the nonlinear index is not as great as the two-photon enhancement. However, for circularly polarized light the two-photon enhancement between the s levels is not observed. An example of the self-defocused profiles that were observed at various laser powers for a Cs density of 0.32×10^{17} cm^{-3} is shown in Fig. 10. From defocusing measurements made at an input intensity of ~ 6 GW/cm^2 for which beam distortion is negligible, a value of $n_2/N = -2.84 \pm 0.2 \times 10^{-30}$ esu was obtained. This value is within a factor 2 of the value predicted from a first principle calculation of the nonlinear susceptibility. The values of n_2 that were observed for linearly and circularly polarized light are shown as a function of Cs density in Fig. 11.

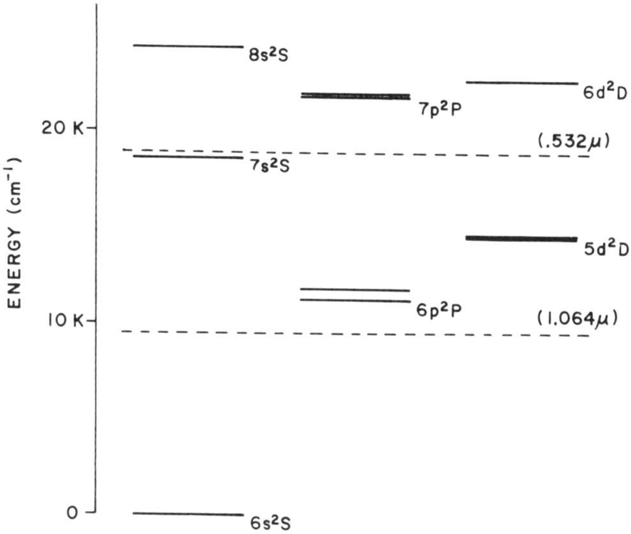

Fig. 9. Level diagram for self-defocusing of Nd:YAG laser radiation in Cs vapor. (From Ref. 35b.)

The ratio of the values of $(n_2)_{\text{linear}}/(n_2)_{\text{circular}}$ that was obtained from these measurements is in excellent agreement with the relative value of 5 that is predicted from theory.

4.3. Self-Phase Modulation

Self-phase modulation results from a combination of a nonlinear index and a temporal variation of the laser intensity. While self-phase modulation can occur for a laser pulse of any duration, its primary manifestation is in a broadening of the laser spectrum that is usually observed with laser pulses with durations typically of the order of 100 psec or less. Such pulses are produced, for example, with mode-locked lasers or in the self-focused components of longer-duration Q-switched lasers.

A qualitative picture of self-phase modulation is illustrated in Fig. 12. In the presence of a nonlinear index the phase ϕ of an intense laser pulse with a field, which is defined by

$$\mathscr{E}(r, z, t) = \tfrac{1}{2}[A(r, z, t)e^{-i(\omega t - kz + \phi)} + \text{c.c.}], \tag{29}$$

becomes time dependent. If the nonlinear index can follow the temporal variations of the laser intensity, then the phase takes on the same temporal variation as the laser pulse. This situation is illustrated in Fig. 12b for a laser pulse with a bell-shaped intensity profile in time.

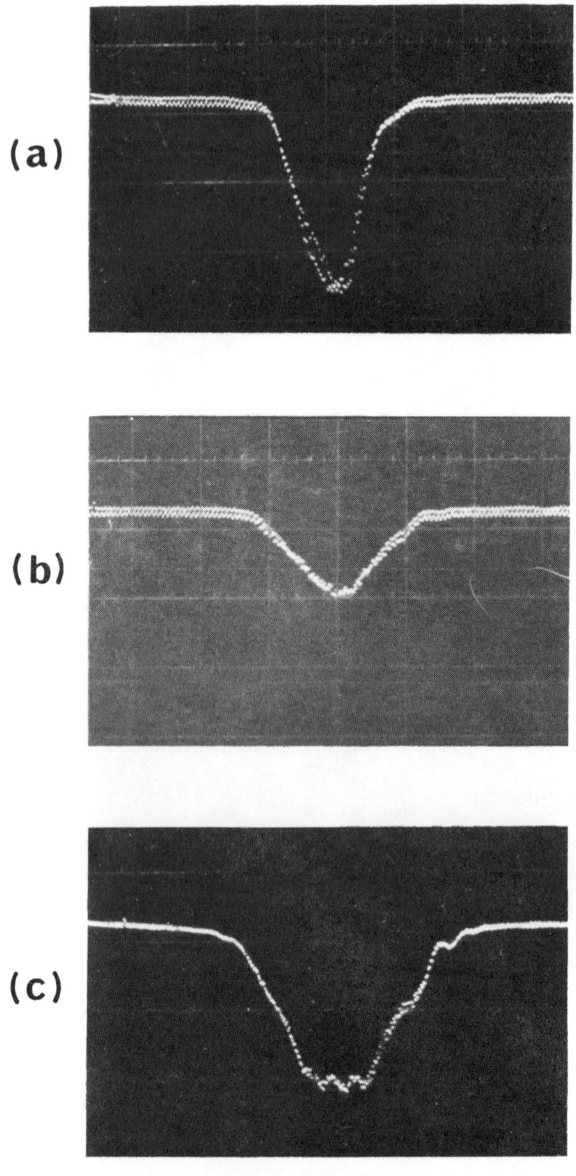

Fig. 10. Profiles of self-defocused beam for different incident laser powers. (a) 0.27×10^{10} W/cm^2; (b) 0.64×10^{10} W/cm^2; (c) 1.9×10^{10} W/cm^2. (From Ref. 35b.)

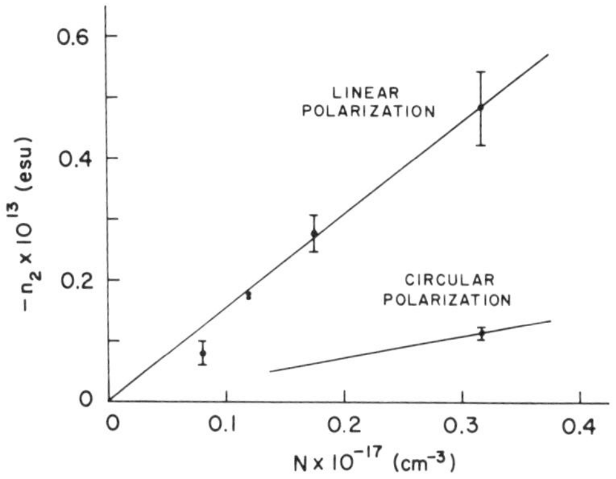

Fig. 11. Measured values of n_2 in Cs at 1.06 μm as a function of density for linear and circular polarization. (From Ref. 35b.)

4.3.1. Frequency Broadening

The instantaneous frequency of the pulse relative to the incident frequency ω_0 is defined as the time derivative of the phase ϕ,

$$\omega(t) = d\phi/dt, \tag{30}$$

and is as shown in Fig. 12c. It is a double-valued function of time with its greatest excursions from the unshifted frequency occurring at the times of greatest slope of the laser pulse.

The frequency spectrum as measured, for example, in a spectrometer shows interference between components of the pulse that have the same frequency but originate at different times in the pulse. Calculated spectra for pulses with different degrees of phase modulation are shown in Fig. 13. The spacing of the interference maxima is determined by the laser pulse duration and can be used as a measure of the effective duration of the self-focused part of the pulse. The number of interference maxima is determined by the maximum nonlinear phase shift. For the conditions shown in Fig. 12 with a positive n_2 the unshifted part of the spectrum is determined by the second half of the pulse and the downshifted part of the spectrum is controlled by the leading half of the pulse.

The symmetry of the spectrum is affected both by the symmetry of the laser pulse and by the response characteristics of the nonlinear index. If the laser pulse is asymmetric, then the spectrum will be asymmetric in extent, in spacing of the interference maxima, and in energy content. However, the number of interference maxima in the upshifted and downshifted parts of the spectrum will be the same. An example of an asymmetric spectrum caused by laser pulse asymmetry with an instantaneous nonlinear index is shown in

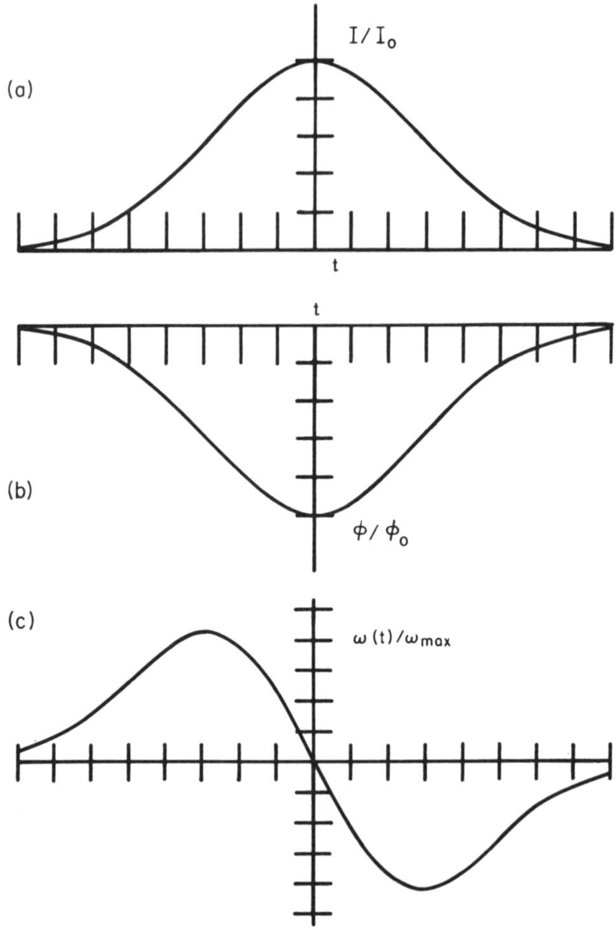

Fig. 12. Schematic illustration of frequency broadening due to self-phase modulation. (a) Laser intensity as a function of time. (b) Instantaneous phase. (c) Instantaneous frequency. Illustrations in (b) and (c) are for a medium with a positive n_2.

Fig. 14. The spectrum extends farther to the side controlled by the steeper part of the pulse, but the energy content of this side is smaller.

If the nonlinear index cannot respond instantaneously on the time scale of the laser pulse, then the nonlinear phase shift lags the laser pulse in the plane wave approximation as shown in Fig. 15. The spectrum is now asymmetric, with the upshifted part of the spectrum containing less energy, extending to a lesser degree, and containing fewer interference peaks than the downshifted part of the spectrum. The degree of symmetry of the broadened spectrum

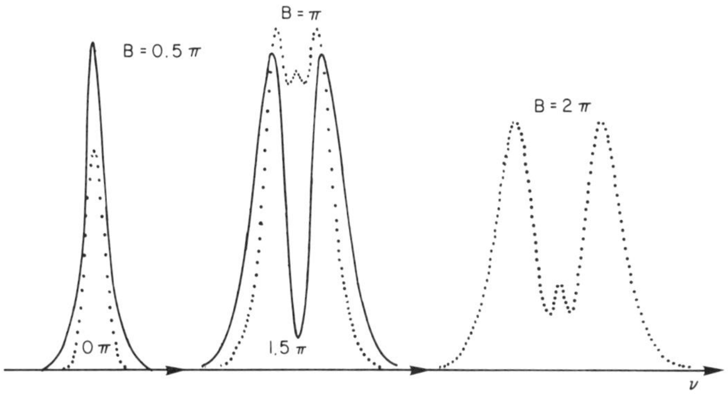

Fig. 13. Calculated spectra for different amounts of peak self-phase modulation. (From Ref. 55.)

together with its extent and number and spacing of its interference maxima can provide valuable information about the details of the self-phase modulation interaction.

The above considerations apply for a medium with a positive n_2. In media with a negative n_2 the sense of the nonlinear phase shift in Fig. 12 is reversed. The form of the spectrum for a medium with an instantaneous response remains the same except that the upshifted part of the spectrum is determined by the leading edge of the pulse and the down shifted part of the spectrum is determined by the trailing edge of the pulse. In media with relaxation effects the spectrum will be predominantly upshifted.

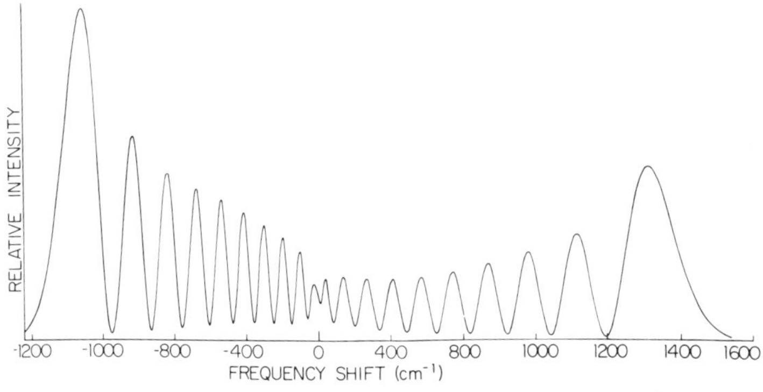

Fig. 14. Asymmetric self-phase-modulated spectra resulting from asymmetric pulses.

$$E = E_0 e^{-t/T_1}, \qquad t < 0,$$
$$= E_0 e^{-t/T_2}, \qquad t > 0,$$

$$T_2 = 0.7T_1.$$

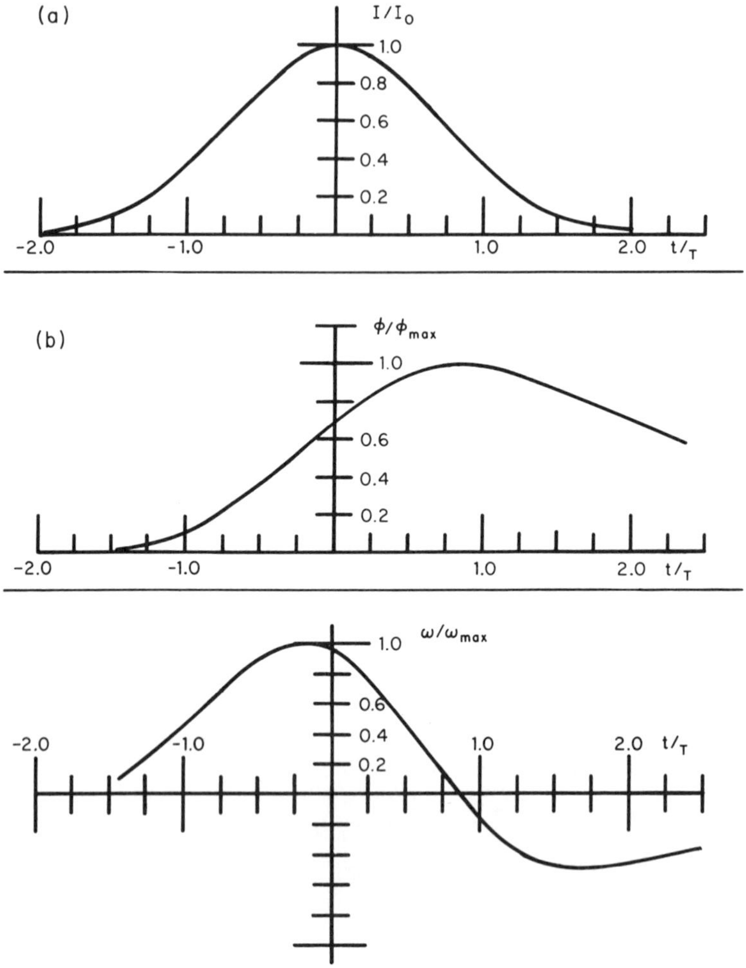

Fig. 15. Schematic development of self-phase-modulated spectra in a medium with a finite response time. (a) Laser intensity as a function of time. (b) Instantaneous phase. (c) Instantaneous frequency. Laser intensity had the form $I = I_0 \exp[-(t/T)^2]$ and the relaxation time was $\tau = 3T$.

Experimental observation of the characteristic interference oscillations in the spectrum depends on the severity of the spatial effects of self-focusing or defocusing. In order for a clean spectrum to be observed it is important that the beam intensity vary smoothly across its diameter. Clean spectra are usually observed only for pulses that have undergone only small amounts of self-focusing or defocusing or that have self-focused to isolated focal spots.

Self-Phase Modulation Equations. In the plane-wave approximation, neglecting the spatial effects of self-focusing, self-phase modulation is described by

$$\frac{\partial}{\partial z} A(z, t) + \frac{n_0}{c} \frac{\partial}{\partial t} A(z, t) = i \frac{2\pi\omega}{n_0 c} P^{\mathrm{NL}}, \tag{31}$$

where the nonlinear polarization P^{NL} is given by an appropriate equation as discussed in Section 3. The simplest case is for an instantaneous response for which the solution of (31) is

$$A(z, t) = A_0(t') e^{-i\phi(z', t')}, \tag{32}$$

where z' and t' are coordinates in a reference frame moving with the pulse:

$$z' = z, \tag{33}$$

$$t' = t - z/c, \tag{34}$$

and the nonlinear phase is given by

$$\phi(L, t') = -\frac{\pi n_2}{\lambda} \int_0^L |A_0(t')|^2 \, dz'. \tag{35}$$

If relaxation of the nonlinear index is important, then the solution of (32) is still given by (33), but the nonlinear phase shift is now given by

$$\phi(L, t') = -\frac{\pi n_2}{\lambda} \int_0^L \int_{-\infty}^{t'} |A_0(t'')|^2 e^{-(t' - t'')/\tau} \, dt'' \, dz. \tag{36}$$

The spectrum of the phase-modulated pulse is given by the Fourier transform of the field amplitude:

$$I(\omega) = \frac{c}{8\pi} \left| \int e^{i\omega t} A_0(t) e^{-i(\omega_0 t + \phi(t))} \, dt \right|^2. \tag{37}$$

For an incident beam with a Gaussian envelope

$$A(t) = A_0 e^{-2t^2/t_p^2}, \tag{38}$$

where t_p is the full width of the intensity envelope at the $1/e$ point, the maximum frequency shift for an instantaneous response is

$$|\omega_{\max}| = 1.7\pi n_2 L A_0^2/\lambda t_p. \tag{39}$$

The number of interference peaks is $1/2\pi$ times the maximum nonlinear phase shift and is given by

$$N = \phi_{\max}/2\pi = \tfrac{1}{2} n_2 A_0^2 L/\lambda = 0.093 t_p |\omega_{\max}|. \tag{40}$$

Spectral broadening of intense laser pulses has been observed in many materials, including molecular liquids with strong orientational Kerr effects;[21,39,40] liquids with symmetric molecules, and therefore no orientational Kerr effect, such as CCl_4;[28] liquids such as H_2O that have a small n_2;[41–51] atomic liquids such as liquid Ar;[52] and solids.[53] In materials with a large n_2 the spectral broadening has been attributed to self-phase modulation,[39] and in many cases a clean set of interference maxima has been observed in the spectrum to confirm this interpretation.[39,40] In media with small values of n_2 it has been suggested that four-wave mixing[43,45,46–48] (see Sections 4.3.2 and 2-2.2) is responsible in whole or in part for the observed spectral extent.

Spectral broadening can be observed in media with positive or negative n_2 but has been studied most extensively in media with a positive n_2. In such media it is possible to observe self-phase modulation without catastrophic self-focusing or beam breakup under properly chosen experimental conditions. Generally this requires that the beam diameter be relatively large and smooth. In addition, the total nonlinear phase shift must be kept small, of the order of a few π. Under these conditions the plane-wave theory outlined earlier describes the spectral broadening fairly well.

Spectral broadening in media with positive n_2 is more typically observed accompanying the catastrophic beam breakup associated with small-scale self-focusing. Such spectral broadening in Kerr active liquids has been observed with both Q-switched and mode-locked laser pulses.[39,40]

Spectra from Q-switched laser pulses typically extend to frequencies of the order of several hundred cm^{-1} from the initial laser frequency. They are usually asymmetrically broadened, extending predominantly to the low-frequency side, as shown in Fig. 16. The spacing of the interference maxima is typical of pulses of duration of the order of several picoseconds, thereby providing an estimate of the duration of the radiation in the self-focused region of the incident pulse. These observations have been reasonably well explained within the framework of the moving focal point theory by taking into account the spatial and temporal variation of the refractive index within the moving focal point as well as the velocity of the self-focal point.[21]

Spectral broadening with mode-locked laser pulses is also observed in Kerr active liquids and typically extends from several hundred to thousands of cm^{-1} from the initial laser frequency, as shown in Fig. 17. Spectra observed with mode-locked pulses are typically more symmetric than those observed with Q-switched laser pulses. The symmetry of these spectra has been discussed by many authors. Polloni *et al.*[40] have suggested the existence of a fast molecular response associated with librations to provide the up-shifted spectrum, but no evidence for a fast-responding component has been observed in these materials. Bloembergen and Yablonovitch[20] have

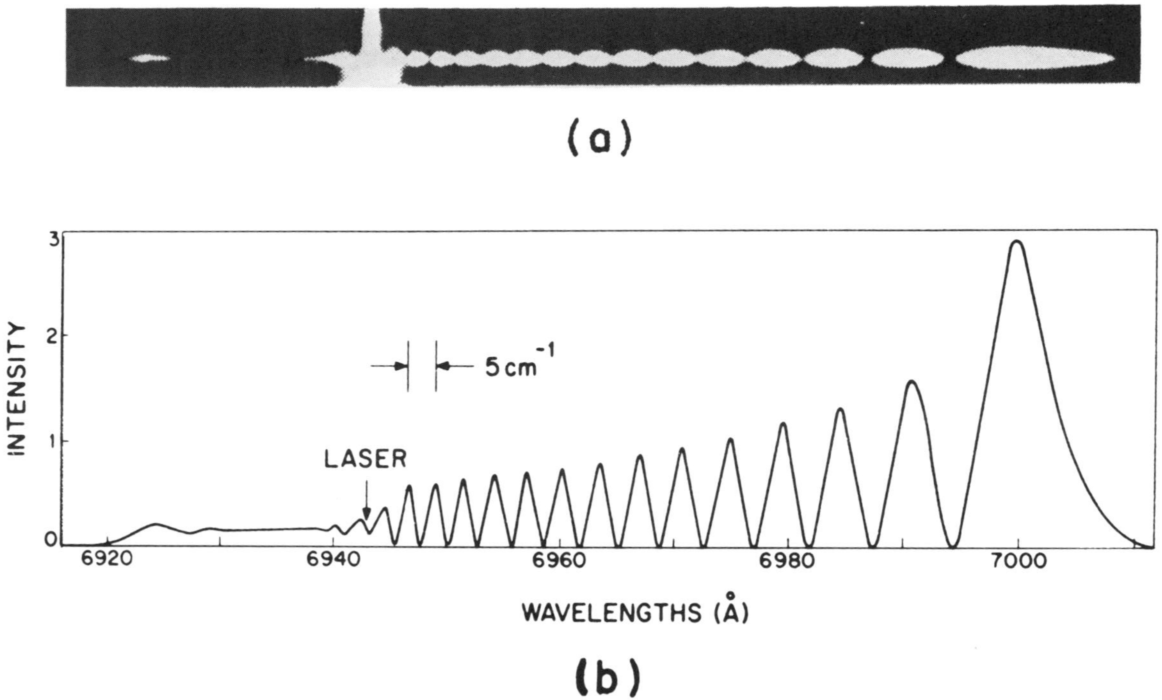

Fig. 16. Typical self-phase-modulated spectra from Q-switched laser pulses. (From Ref. 39.)

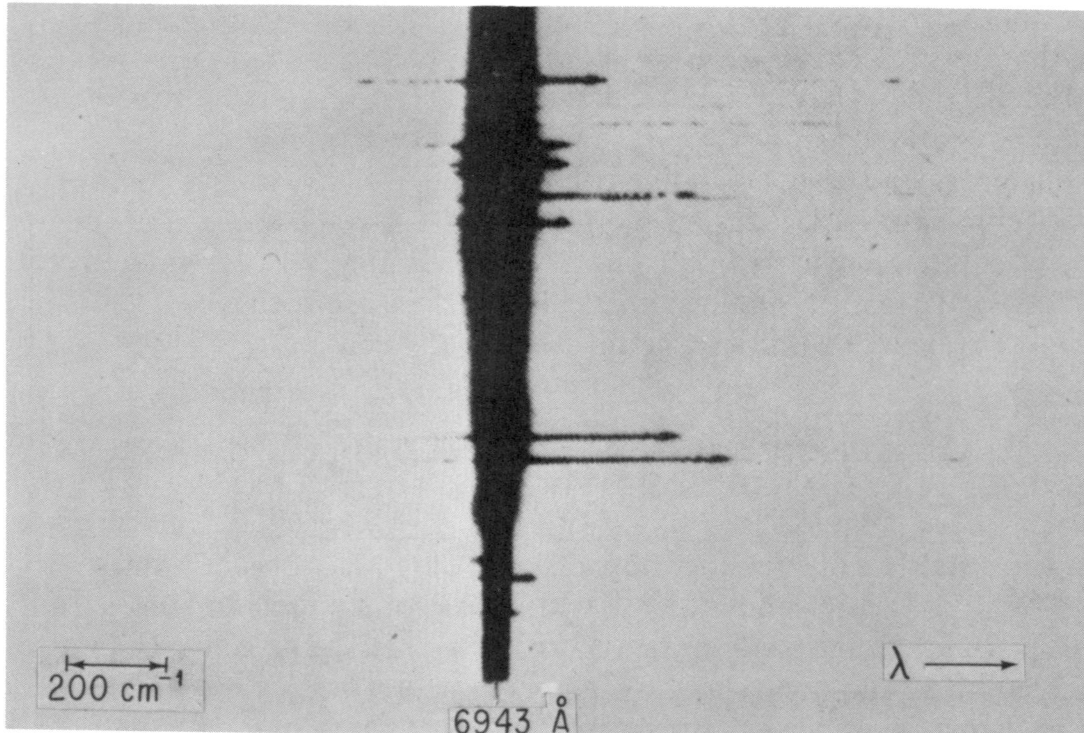

Fig. 17. Typical spectra from self-phase-modulated mode-locked laser pulses. (From Ref. 40.)

suggested that the upshifted components result from the rapid changes in refractive index that accompany the plasma formation at the end of the pulse that is implicated in limiting the size of the beam in the self-focused region. Numerical calculations done by Shimizu using the horn model of transient self-focusing and the retarded nonlinear index given by Eq. (8) predict spectra with a degree of symmetry that is comparable to that observed experimentally.[22] Similar results were obtained by Loy and Shen[21] who showed that the upshifted components can arise from phase variations associated with the rapid change in beam diameter near the horn of the transiently self-focused mode-locked pulse.

4.3.2. *Superbroadened Spectra and Picosecond Continua*

Considerably more extensive spectral broadening is observed with mode-locked laser pulses in certain materials with a low value of n_2, including H_2O, D_2O, CCl_4, isopropyl alcohol, liquid and solid argon, glasses, and other solids. Typically this superbroadening (SB) extends over thousands to tens of thousands of cm^{-1} and can range from the infrared to the ultraviolet, depending on the initial laser wavelength. The superbroadened spectra have been generated in the various liquids with picosecond pulses from mode-locked Nd:YAG and Nd:glass lasers and their second harmonics,[41,43–48] mode-locked ruby lasers[42] and mode-locked dye lasers.[54]

Among the liquids the broadest spectra are usually observed in H_2O and D_2O.[41–43] When very intense pulses are used the intensity of the spectrum from D_2O has been observed to be up to 250 times as intense as the spectrum from H_2O,[43] although the intensities of the spectra from the two liquids are more nearly comparable when the laser pulses are weaker.[48]

The superbroadened spectra from H_2O and D_2O are typically smooth and featureless when pulse durations of 10–14 psec and relatively high intensities are used,[43] although structure in the continuum has been reported for lower intensities[48] and longer pulses.[44] Generally the on-axis spectrum extends well beyond the Raman–Stokes and anti-Stokes frequencies and the upshifted spectrum usually extends beyond the second harmonic of the incident laser as well. Enhanced emission near the Raman–Stokes frequencies and decreased emission near the Raman anti-Stokes frequencies has been reported. Off-axis rings near the Raman frequencies arising from four-wave mixing have also been reported.[42,44]

For a particular system the extent of the spectrum and its symmetry about the initial laser wavelength are influenced by absorption in the medium. For example, if a mode-locked laser pulse from an Nd:YAG laser at 1.06 μm is used, the anti-Stokes spectrum from D_2O has been observed to extend throughout the near infrared and visible and into the ultraviolet, from about 800 to 380 nm.[43] The Stokes broadening, on the other hand, is not observed

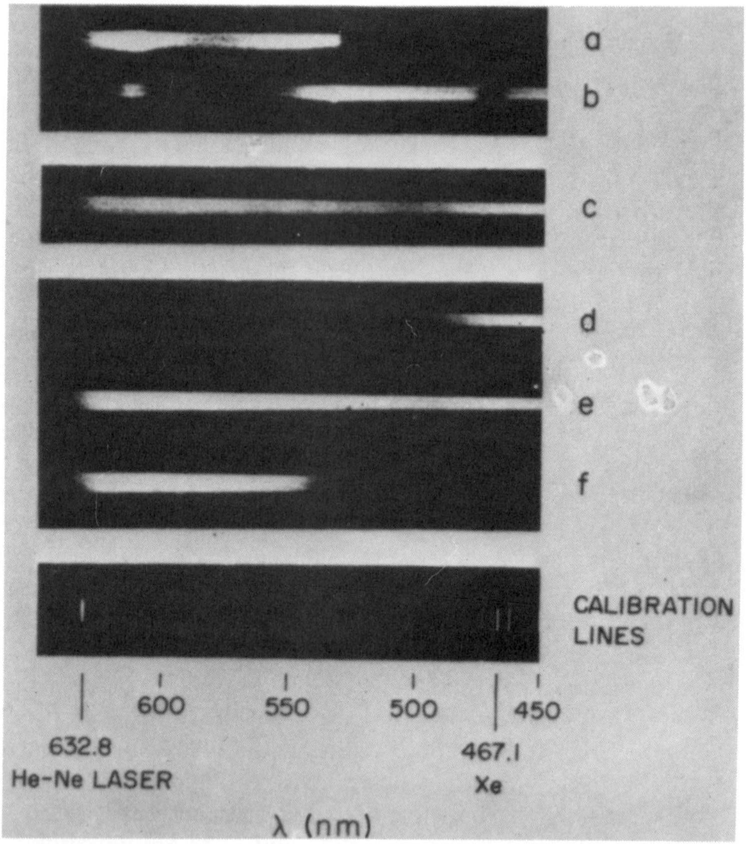

Fig. 18. Superbroadened spectra in D_2O using 0.53-μm light (a) and (b) and 1.06-μm light (c)–(f) and various optical filters: (a) OG 550; (b) BG-12; (c) KG 3; (d) KG 3 + BG 12; (e) KG 3; (f) KG 3 + BG 12; (f) calibration lines. (a)–(c): 50-μm slits, (d)–(f): 100-μm slits. (From Ref. 43.)

to be as extensive, presumably due to IR absorption. If the second harmonic of the Nd:YAG laser at 532 nm is used, then the spectrum shows more or less symmetric broadening about the initial laser wavelength. The visible portion of typical spectra from D_2O obtained with both 1.06-μm and 532-nm exciting light are shown in Fig. 18. The 1.06-μm pulses had a peak power of the order of 10^9 W and were focused into a 25-cm-long cell of D_2O with a 25-cm focal length lens. The power in the 532-nm pulse was about 7% of the power in the 1.06-μm pulse.

Measurements of the temporal properties of the broadened radiation[43,49] indicate that its duration is slightly shorter than that of the initial laser pulse, providing a picosecond or subpicosecond continuum for time-resolved spectroscopic measurements.[49–51,54] Conversion efficiencies of up to 10% in the entire continuum have been reported.

Although the properties of the superbroadened spectra have been studied quite extensively, the details of the formation mechanism are not yet fully resolved. The superbroadened spectra are typically observed in materials that have a small nonlinear index. This has led many authors[43,45,46–48] to suggest that these spectra are the result of a four-wave mixing process of the form $2\omega_1 = \omega_2 + \omega_3$ where ω_1 is the initial laser frequency and ω_2 and ω_3 are generated in the nonlinear interaction. The appropriate phase-matching condition $2\vec{k_1} = \vec{k_2} + \vec{k_3}$ determines which pair of frequencies is generated. Off-axis or intensity-dependent phase matching can allow effective phase matching over a wide range of generated frequencies. This model can explain the presence of the superbroadened spectra in materials with a small nonlinear index and also the absence of the interference maxima that are characteristic of the self-phase-modulated spectra. The four-wave mixing model does imply, however, that the spectral intensity should drop off for frequencies higher than the second harmonic of the initial laser frequency, a behavior in contradiction with the experimental observations.

Smith *et al.*[44] have reported observations of self-focusing and dielectric breakdown accompanying the superbroadening of 30-psec-long pulses in CS_2 and have suggested that self-phase modulation may provide seed radiation, allowing the parametric interactions to extend the broadened radiation to wavelengths shorter than the laser second harmonic. Breakdown was not observed in measurements done with shorter-duration pulses,[45] however, raising the possibility that the breakdown occurs near the end of the longer pulses, perhaps after the broadening process is initiated.

4.3.3. *Nonlinear Phase Compensation*

Lehmberg *et al.*[55] have investigated the use of a medium with a negative nonlinear index to compensate for the nonlinear phase changes that are introduced to a beam by a medium with a positive nonlinear index. Such a technique can prove useful in high-power solid-state lasers operating at short pulse durations (e.g., less than 100 psec) for which the self-phase modulation and self-focusing arising in the laser amplifiers can limit the power that can be obtained from the laser.

An example of nonlinear phase compensation using pulses from a mode-locked Nd:YAG laser is shown in Fig. 19a,b. The positive nonlinear phase changes were caused partly by the YAG amplifiers and partly by cells of CS_2 that were placed in the beam after the last amplifier as shown in the schematic of the experiment in Fig. 19a. Phase compensation was done in

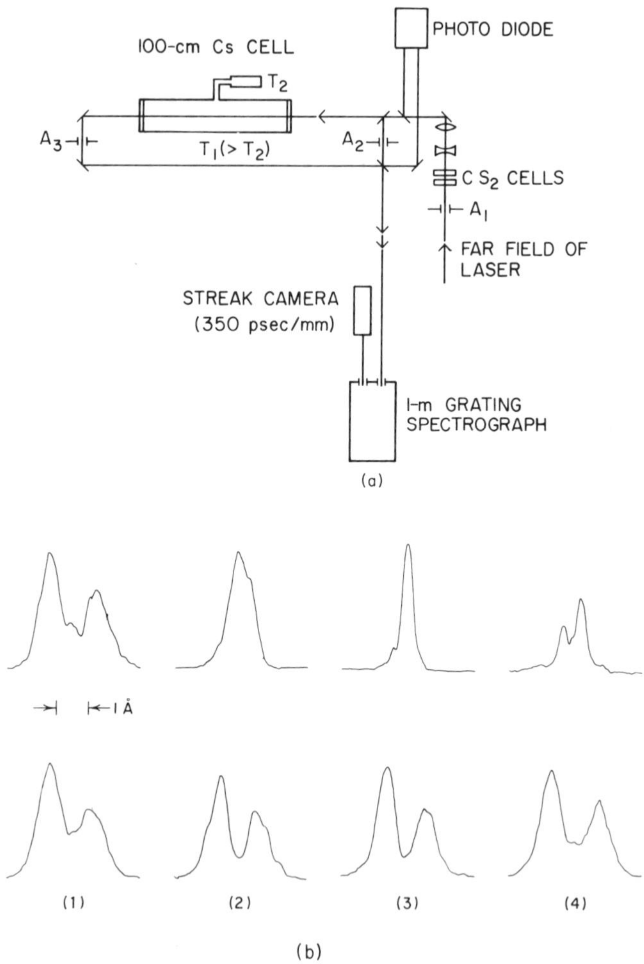

Fig. 19. (a) Schematic illustration of apparatus used to demonstrate nonlinear phase compensation in Cs vapor. (b) Densitometer traces of incident spectra (lower) and transmitted spectra (top) for cesium densities of (a) $N = 0$, (b) $N = 4.6 \times 10^{16}\ \text{cm}^{-3}$, (c) $7.6 \times 10^{16}\ \text{cm}^{-3}$, and (d) $1.13 \times 10^{17}\ \text{cm}^{-3}$. (From Ref. 55.)

a 1-m-long cell of Cs vapor, and the presence of the nonlinear phase changes was detected by observing the spectral broadening with a spectrometer. Compensation of the nonlinear phase changes is evident in the sequence of pictures in Fig. 19b. At low Cs densities very little compensation is present and the spectrum is relatively broad. As the Cs density increases, the spectrum narrows. It reaches a width that is almost the value determined by the inverse of the pulse duration, corresponding to full compensation, at a Cs density of

7.6×10^{16} cm^{-3}. At higher Cs densities, the spectrum broadens again due to overcompensation. Compensation of nonlinear phase changes up to 3π were observed in this manner.

In order for the compensation to be accurate the effects of the nonlinear index, both temporal and spatial, must remain in the phase and not be transferred to the amplitude by dispersion or by diffraction. This technique works best if the nonlinear phase change is corrected before it can build to a large value. It is therefore ideally suited for use in laser oscillators in which the nonlinear phase change can be compensated on each cavity transit.

5. DEGENERATE FOUR-WAVE MIXING

Degenerate four-wave mixing (DFWM) is a term used to describe the nonlinear interaction of waves that are at a common optical frequency but are otherwise distinguishable, for example by their direction of propagation or polarization. DFWM can take place in any medium that exhibits a third-order nonlinear susceptibility and can be caused by, for example, a nonlinear refractive index, two-photon absorption, or saturation of linear absorption. DFWM has to date been reported for a variety of conditions involving transparent and absorbing media, liquids, gases, and solids, resonant and nonresonant interactions, and pulsed and cw radiation. DFWM gives rise to several different effects, such as amplification of a forward directed probe wave, amplified reflection, phase conjugation, spontaneous oscillation, holographic image reconstruction, and other forms of optical signal processing. A related effect that has been termed nondegenerate four-wave mixing involves the nonlinear coupling of two waves at a common frequency to two other waves that are oppositely offset in frequency from the first pair. DFWM is also related to a more general process of dynamic holography which can involve waves at more than one frequency.

These various effects have been considered theoretically by many authors. Gerritsen[56] discussed DFWM in an aborbing medium for use in real time holography in 1967. Following a proposal by Yariv[57] for phase conjugation using three-wave mixing in anisotropic crystals, Hellwarth[58] proposed use of DFWM with counterpropagating pump waves in a transparent medium for generation of phase conjugate waves for correction of phase aberrations. Yariv[59] and Yariv and Pepper[60] have analyzed several aspects of DFWM in transparent media, including backward wave generation, amplified reflection, forward wave amplification, oscillation, and phase conjugation. Yariv[61] and Pepper *et al.*[62] have explored the used of DFWM for real time holography and other optical processing techniques, such as spatial convolution and optical correlations.

The analysis of DFWM has been extended to include arbitrary polarization combinations (Steel *et al.*,[63] Zel'dovich and Shkunov,[64] Martin *et al.*,[65] and Bol'shov *et al.*[66]), the effects of nonlinear phase shifts (Marburger and Lam[67]), pump depletion (Marburger and Lam,[68] Hsu,[69] and Lazaruk and Rubanov[70]), saturable absorbers (Abrams and Lind,[71] Apanasevich *et al.*,[72] Watkins *et al.*,[73] and Brown[74]), Doppler broadening and atomic motion (Wandzura,[75] Nilsen and Yariv,[76] and Humphrey *et al.*[77]), thermal effects (Key *et al.*,[78] Eichler *et al.*,[79] Heilweil *et al.*,[80] Andrews and Hochstrasser,[81] and Tocho *et al.*[82,83]), frequency offsets and spectral bandpass (Pepper and Abrams,[84] Fu and Sargent,[85] Wandzura,[75] Nilsen and Yariv,[76] and Pilipetsky and Shkunov[86]), two-photon resonant media (Fu and Sargent[87]), transient responses (Rigrod *et al.*[88]), and DFWM in plasmas (Steel and Lam[89]). Phase conjugation in optical waveguides has been discussed by Yariv *et al.*,[90] Jensen and Hellwarth,[91] AuYeung *et al.*,[92] and Hellwarth.[93] Phase conjugation of multiline laser signals has been discussed by Steel and Lam[94] and Depatie and Haueisen.[95] The properties of optical resonators with phase-conjugating mirrors have been investigated by Bel'dyugin *et al.*,[96] Bel'dyugin and Zemskov,[97] AuYeung *et al.*,[98] Belanger *et al.*,[99] Lam and Brown,[100] Reznikov and Khizhnyak,[101] Feinberg and Hellwarth,[102] Hardy,[103] and Lind *et al.*[104] The use of phase conjugate mirrors in interferometers has been discussed by Hopf.[105] The reversal of temporal phase structure has been discussed by Marburger,[106] and compensation of group velocity dispersion has been treated by Yariv *et al.*[107] The use of DFWM for producing two-photon coherent states of the radiation field has been discussed by Yuen and Shapiro.[108]

The early experiments in this area were described in terms of dynamic holography, that is, scattering from a grating formed in the nonlinear medium by the interference of two intense laser beams. Carman *et al.*[109] reported observations of a process they termed light-by-light scattering resulting from such an interaction in nitrobenzene illuminated with two beams from a ruby laser. Gerritsen[56] later demonstrated reconstruction of real time holographic images formed by ruby laser radiation in cryptocyanine. Stepanov *et al.*[110] used a ruby laser for real time holographic reconstruction of images and also for correction of phase aberrations on the laser beam using DFWM with counterpropagating beams in cryptocyanine. Subsequently, many aspects of the DFWM process using counterpropagating beams, including backward and forward wave amplification and amplified reflection,[59,63,73,77,81–83,89,92,95,111–129] oscillation,[59,102,111,112,122] reconstruction of phase-distorted images,[65,130,131] and correction of phase distortions on laser beams.[59,82,91,94,115–117,120,132–135] The use of phase conjugate mirrors in laser resonators,[98,102,104] and high-resolution holographic imaging[65,136,137] have also been demonstrated experimentally in various media.

Reviews on nonlinear optical phase conjugation have been written by Yariv[59a] and by Pepper.[138] Yariv[59b] has also given a chronological bibliography of many of the earlier papers on DFM, dynamic holography, and wavefront reconstruction. Hellwarth[93] has given a review paper on DFWM in optical waveguides. Reviews have also been given of DFWM in resonant media (Lind *et al.*[139]), wavefront reconstruction and signal processing (White and Yariv[140]), and of applications of wavefront reconstruction (Giuliano[141]).

5.1. DFWM Geometries

DFWM can be accomplished using various geometries. Two of these are illustrated in Fig. 20 along with appropriate wave vector diagrams. The arrangement shown in Fig. 20a involves two beams propagating at a small forward angle to each other. This is the geometry used by Carman *et al.*[109] and later by Heer and Griffen.[142] In this arrangement three of the four waves involved in the four-wave mixing interaction are supplied by the two incident beams and a fourth wave is generated at an angle as shown in the figure. This geometry is inherently phase mismatched, with a wave vector defect $\Delta k = \theta^2 k$, as indicated in the figure. This process can therefore be expected to be weak unless $\theta < (\lambda/2L)^{1/2}$. Higher-order diffraction beams can also be observed for sufficiently intense pump beams.

An alternative geometry that has been used more extensively involves counterpropagating pump waves, as shown in Fig. 20b. In this interaction E_1 and E_2 are strong pump waves at frequency ω incident on a nonlinear medium from opposite directions. A probe wave E_p at frequency ω is incident

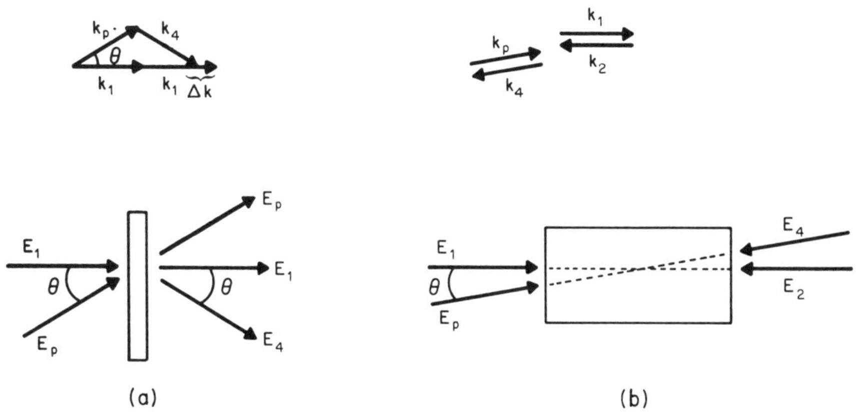

Fig. 20. Schematic illustration of arrangements used to observe phase conjugation by degenerate four-wave mixing. (a) Forward wave interaction. (b) Backward interaction.

at an angle θ to the pump waves. The nonlinear interaction couples these three waves to a fourth wave E_4, also at frequency ω, which propagates in a direction opposite to that of E_p. This geometry has an advantage over that shown in Fig. 20a in that the phase matching requirements are automatically satisfied regardless of the angle θ, as will be discussed later.

The angle between the probe wave and the pump waves is more or less arbitrary for many situations, although under some conditions the efficiency of the DFWM process arising from certain interactions depends on the magnitude of θ, as will be discussed later. Common experimental arrangements have the probe wave propagating collinearly with, at a small angle to, or at 90° to the pump waves. Propagation collinearly with the pump waves or at small angles to them provides the longest interaction paths and hence the largest nonlinear interactions. When the pump and probe waves are exactly collinear they must be orthogonally polarized in order for them to be distinguishable. For other geometries the pump and probe waves may be linearly polarized either parallel or perpendicular to one another. The effects of various polarization combinations will be discussed in Section 5.2.1. For the remainder of this chapter we shall consider in detail only the DFWM interaction involving counterpropagating pump waves.

5.2. DFWM Theory

The basic effects arising from DFWM with counter-propagating pump waves can be illustrated by considering the geometry shown in Fig. 20b. We assume for the forward and backward pump waves fields of the form

$$\mathscr{E}_1(z', t) = \tfrac{1}{2}(E_1(z')e^{-i\omega t} + \text{c.c.}) = \tfrac{1}{2}(A_1 e^{-i(\omega t - k_1 z')} + \text{c.c.}), \tag{41a}$$

$$\begin{aligned}\mathscr{E}_2(z', t) &= \tfrac{1}{2}(E_2(z')e^{-i\omega t} + \text{c.c.}) = \tfrac{1}{2}(A_2 e^{-i(\omega t - k_2 z')} + \text{c.c.}) \\ &= \tfrac{1}{2}(A_2 e^{-i(\omega t + k_1 z')} + \text{c.c.}).\end{aligned} \tag{41b}$$

The fields of the probe and backward generated wave are

$$\mathscr{E}_p(r, z, t) = \tfrac{1}{2}(E_p(r, z)e^{-i\omega t} + \text{c.c.}) = \tfrac{1}{2}(A_p(r, z)e^{-i(\omega t - k_p z)} + \text{c.c.}), \tag{41c}$$

$$\begin{aligned}\mathscr{E}_4(r, z, t) &= \tfrac{1}{2}(E_4(r, z)e^{-i\omega t} + \text{c.c.}) = \tfrac{1}{2}(A_4(r, z)e^{-i(\omega t - k_4 z)} + \text{c.c.}) \\ &= \tfrac{1}{2}(A_4(r, z)e^{-i(\omega t + k_p z)} + \text{c.c.}).\end{aligned} \tag{41d}$$

Here the direction z' is used for the direction of propagation of the pump waves, z for the direction of propagation of the probe and generated waves, and r for the transverse coordinate of the probe and generated waves. The pump waves are assumed to be plane waves.

Components of the nonlinear polarization at frequency ω arising from the interactions of these four waves can be obtained explicitly from Eq. (1). For example, for geometries in which the pump and probe waves are not collinear the amplitude of the polarization component proportional to e^{-ikz}, which describes the propagation of the backward wave E_4 can be written in a general form as

$$\begin{aligned}
P_4(\omega) = {} & \tfrac{1}{4}\{C'(A_1 \cdot A_p^*)A_2 + C''(A_2 \cdot A_p^*)A_1 + D(A_1 \cdot A_2)A_p^*\} \\
& + \tfrac{1}{4}\{C(|A_1|^2 + |A_2|^2 + |A_3|^2 + |A_4|^2)A_4 \\
& + C''(A_4 \cdot A_1^*)A_1 + C'(A_4 \cdot A_2^*)A_2 + C(A_4 \cdot A_3^*)A_3 \\
& + D(A_1 \cdot A_4)A_1^* + D(A_2 \cdot A_4)A_2^* + D(A_3 \cdot A_4)A_3^* \\
& + \frac{D}{2}|A_4|^2 A_4\}.
\end{aligned} \tag{42}$$

The terms in the first pair of curly brackets describe the generation of the backward wave by DFWM. The remaining terms describe the effects of nonlinear phase shifts and have usually been neglected in elementary descriptions of DFWM. Similar terms can be written down for the polarization components proportional to e^{ikz}, $e^{ikz'}$, and $e^{-ikz'}$, which describe the propagation of the probe wave and the two pump waves.

In addition to the components given before, the nonlinear polarization also contains components of the form

$$\begin{aligned}
P = {} & \tfrac{1}{4}\{C'''(A_1 \cdot A_2^*)A_4 + (D/2)(A_1 \cdot A_1)A_p^*\}e^{i2kz'-kz} \\
& + \tfrac{1}{4}\{C'''(A_2 \cdot A_1^*)A_p + (D/2)(A_2 \cdot A_2)A_4^*\}e^{-i2kz'+kz}.
\end{aligned} \tag{43}$$

These terms can make phase-matched contributions to the probe and backward waves when collinear geometries are used with the probe and pump waves orthogonally polarized. For noncollinear interactions the terms in Eq. (43) describe the generation of waves in directions other than those of the probe or backward waves. For this situation these terms produce a non-phase-matched component of the nonlinear polarization for the backward wave and can usually be neglected. For collinear interactions, on the other hand, they must in general be retained.

Finally the polarization contains terms that are proportional to $e^{\pm i3kz}$, $e^{\pm i3kz'}$, $e^{\pm i(2kz+kz')}$, and $e^{\pm i(2kz'+kz)}$. For many interactions these terms are also not phase matched and can be neglected. However, in certain circumstances, such as in interactions involving saturation of a resonant transition these terms can take a nonzero contribution to the nonlinear polarization.

Each of the terms describing the generation of the backward wave in Eq. (42) can be given a physical interpretation. The first two terms in the first pair of square brackets represent interference of the probe wave with the forward and backward pump waves, respectively. This interference gives

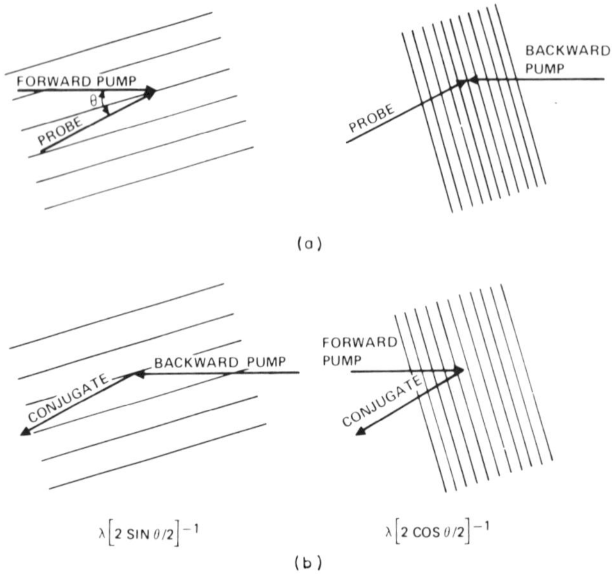

Fig. 21. Schematic illustrations of gratings formed by interference of incident probe wave with forward (left) and backward (right) pump waves: (a) formation, (b) readout. (From Ref. 139.)

rise to a spatially periodic modulation of the polarization of the medium that amounts to a volume diffraction grating or hologram, as illustrated in Fig. 21. It can be either a phase or amplitude grating, or both, depending on the nature of the nonlinear interaction. The spacing of the planes of the gratings formed by interference with each of the pump beams is different, as shown in the figure. The grating formed by the interference of the probe wave with the forward pump wave has a spacing given by

$$\Lambda_f = \lambda/(2 \sin \theta/2). \tag{44a}$$

This spacing is relatively wide and decreases as the angle θ increases. The spacing of the grating formed by the interference of the probe wave with the backward pump wave is given by

$$\Lambda_b = \lambda/(2 \cos \theta/2). \tag{44b}$$

This spacing is relatively narrow and increases as the angle θ increases. For 90° geometries the spacings of the two gratings are equal. The generation of the backward wave by these terms results from the scattering from each volume grating of the pump wave that is not involved in the formation of that grating. The condition for Bragg scattering of a pump wave from the appropriate grating is identical to the phase-matching condition $\Delta k = 0$.

In deriving the forms of both Eqs. (42) and (43) we have allowed the terms that involve the formation of spatial gratings through the interference of different combinations of waves to have different coefficients. The various constants $C^{(n)}$ involve, in addition to the nonlinear susceptibility, effects related to smearing of the grating due, for example, to atomic motion (see Section 5.2.7) or thermal diffusion (see Section 5.2.6b). These effects are sensitive to the spacing of the gratings and so, in general, affect the forward and backward grating differently.

The third term in the first pair of curly brackets in Eq. (42) describes a standing-wave modulation of the polarization that oscillates at a frequency 2ω. This standing-wave modulation scatters A_p and A_4^* into each other in an interaction that is automatically phase matched because of the cancellation of the pump wave k-vectors in the interference of E_1 and E_2.

The total backward wave is given by the coherent sum of the components generated by each of the terms in the first pair of curly brackets in Eq. (42). The relative importance of the terms depends on the nature of the nonlinear interaction, the choice of polarization for the pump and probe waves, and the relation between the polarization of the waves and the plane containing the k-vectors of the pump and probe waves. Some specific examples will be given here.

In transparent, nonresonant media, for which the nonlinear interaction occurs through the nonresonant Kerr effect, all three terms can in general contribute. In this situation all the constants $C^{(n)}$ are equal. The relative importance of the three terms in the first curly brackets in Eq. (42) is then determined by the properties of the medium, which determine the relative values of C and D, and the polarization of the various waves.

In absorbing media, for which the nonlinear interaction occurs as a result of saturable absorption or thermal gradients, on the other hand, the first two terms in the first set of curly brackets in Eq. (42) make the dominant contribution. The relative importance of these terms is determined by the direction of polarization of the various waves and any effects due to grating washout that may be present.

Finally, in two-photon resonant media the third term in the first pair of curly brackets in Eq. (42) makes the dominant contribution.

Even though a particular term has a nonzero coefficient according to the discussion of the preceding paragraphs, it can still be suppressed by an appropriate choice of pump- and probe-wave polarizations. These restrictions are easily determined by inspection of the terms in the first pair of curly brackets in Eq. (42). For example, if the pump waves are orthogonally polarized, then the last term cannot contribute. Furthermore, if the probe wave is polarized parallel to one of the pump waves. then only one of the first two terms will be nonzero. A second example is one in which the pump waves are polarized parallel to each other but perpendicular to the probe wave. In this situation the first two terms are zero and only the third term can make a contribution. On the other hand, if all three waves are polarized parallel to one another, then there are no restrictions imposed by the polarizations. However, there may still be effects related to the polarization that depend on the DFWM geometry.

In determining a component of nonlinear polarization to use in the equations for the backward wave it is necessary to take into account the projection of the total nonlinear polarization onto a direction perpendicular to the direction of propagation of the backward wave. This component of the nonlinear polarization can depend not only on the directions of polarization of the pump and probe waves but also on how the probe wave is rotated with respect to the pump waves. To illustrate these dependences we consider first a situation in which the probe and pump waves are polarized parallel to one another and the probe wave is incident collinear with the forward pump wave. If the propagation direction of the probe wave is now rotated in the plane perpendicular to the one containing the polarization vectors, then, if effects of grating washout can be neglected, the nonlinear polarization that generates the backward wave will be independent of the angle θ. If, on the other hand, grating washout is important, the nonlinear polarization will depend on θ. If the propagation direction of the probe wave is rotated in the plane containing the polarization vectors the nonlinear polarization will depend on θ, varying from a value given by

$$P = \tfrac{1}{4}(C' + C'' + D)A_1 A_2 A_p^*$$

at $\theta = 0°$ to a value of

$$P = \tfrac{1}{4}DA_1 A_2 A_p^*$$

at $\theta = 90°$.

A similar variation can exist if the pump waves are orthogonally polarized. To illustrate this we assume that the probe wave is polarized parallel to one of the pump waves, taken to be E_1. If the probe-wave propagation direction is rotated in the plane containing E_1 or E_2, then the component of nonlinear polarization that generates the backward wave will vary from

$$P = (C'/4)A_1 A_2 A_p^*$$

at $\theta = 0°$ to

$$P = 0$$

at $\theta = 90°$.

5.2.1. Polarization of the Backward Wave

The direction of polarization of the backward wave can be determined for specific polarization combinations by inspection of the terms in the first pair of curly brackets in Eq. (42). The last term in the first pair of curly brackets generates a component of the backward wave that has the same polarization as the probe wave. The polarization of the component of the backward wave that is generated by the first two terms, however, depends on the directions of polarization of the pump and probe waves. The overall polarization of the backward wave, then, depends on the relative polarization of the probe and pump waves, the experimental geometry, and the nature of the nonlinear interaction. For certain specific choices of polarizations, or in media in which one type of term in the first pair of curly brackets in Eq. (42) makes a negligible contribution, the probe wave polarization can be reproduced in the backward wave. In general, however, the backward wave polarization will not be a faithful replica of the probe-wave polarization.

In one of the more common experimental geometries the pump waves are linearly polarized parallel to each other. If the probe wave is linearly polarized in a direction either parallel or perpendicular to that of the pump radiation, then the backward wave will be linearly polarized parallel to the probe wave. If the probe wave is linearly polarized at an arbitrary angle to the pump waves or is elliptically or randomly polarized, however, then the polarization of the backward wave will not in general be the same as that of the probe wave. One exception to this last condition is for nonlinear media that have susceptibilities satisfying the condition $C' = C'' = 3(\chi_{1111} - \chi_{1221}) = 0$. In this case only the last term in the first pair of curly brackets in Eq. (42) contributes and the backward wave polarization is identical to that of the probe wave regardless of its orientation.

In general, the direction of polarization of the backward wave will be independent of the angle θ, although the amplitude of the effective component of the nonlinear polarization may depend on θ for certain geometries. One exception to this behavior occurs when the probe wave is polarized at a random angle to the pump waves and the probe-wave k-vector is rotated in the plane containing the pump-wave polarization vectors. In this situation both the direction and magnitude of the nonlinear polarization are a function of the angle θ.

If the pump waves are linearly polarized perpendicular to each other, then the last term in the first pair of curly brackets in Eq. (42) does not contribute. If the first two terms in the first pair of curly brackets make equal contributions to the nonlinear polarization, then the x and y components of the probe-wave polarization are interchanged. In particular, for any linearly polarized probe wave the backward wave will have an orthogonal polarization. In media for which the first two terms in the first pair of curly brackets in Eq. (42) do not make an equal contribution, a probe wave polarized at an arbitrary angle to the pump waves will not result in an orthogonally polarized backward wave. However, even with unequal contributions from the first two terms, a probe wave polarized parallel to one of the pump waves will produce a backward wave polarized along the other pump wave. Some of these polarization properties have been demonstrated by Steel *et al.*[63]

It is evident from this discussion that if linearly polarized pump waves are used, then the polarization of an arbitrarily polarized probe wave can be reproduced in the backward wave only under special conditions. For some applications it is desirable to be able to reproduce an arbitrary probe-wave polarization under more general conditions. The reproduction of arbitrary probe-wave polarization for a general interaction has been discussed by Zel'dovich and Shkunov,[64] Martin *et al.*,[65] and Bol'shov *et al.*[66] One method of accomplishing this task is to use pump waves with opposite circular polarization, as proposed by Zel'dovich and Shkunov.[64] When this combination is used the first two terms in the first pair of curly brackets in Eq. (42) combine to give an approximate replica of the probe-wave polarization for small values of θ. This arrangement can be used to reproduce an arbitrary probe-wave polarization using nonlinear interactions in which the first two terms in the first pair of curly brackets in Eq. (42) contribute equally. Martin *et al.*[65] have reported an experimental demonstration of this technique. Phase conjugation using elliptically polarized pump waves has been discussed by Bol'shov *et al.*[66]

The various states of polarization of the backward wave for specific combinations of polarization of the pump and probe waves are summarized in Table 2. The appropriate geometry is indicated at the top of the table. The probe wave is assumed to be incident with its k-vector lying in the y'–z' plane of the pump waves. Expressions for the amplitude of the appropriate component of the nonlinear polarization for some of the combinations are also given.

5.2.2. *Backward Wave Generation*

In illustrating the effects associated with backward wave generation in DFWM it is simplest to use the geometry shown in Fig. 20b for which the angle θ is nonzero. We shall assume that the medium is transparent and that all the waves are linearly polarized in the same direction. We shall also assume that the plane containing the propagation vectors of the probe and forward pump waves is perpendicular to the plane containing the polarization vectors. This system has been analyzed by a number of authors. Here we follow the treatment of Yariv,[59a] allowing for the difference in designation of the probe and backward waves. If the nonlinear phase shifts are neglected, then the field amplitudes are described by the equations

$$\frac{dA_4}{dz} = -i\frac{6\pi^2}{n\lambda}\chi A_1 A_2 A_p^*, \tag{45a}$$

$$\frac{dA_p}{dz'} = -i\frac{6\pi^2}{n\lambda}\chi A_1 A_2 A_4^*, \tag{45b}$$

$$\frac{dA_1}{dz'} = i\frac{6\pi^2}{n\lambda}\chi A_4 A_p A_2^*, \tag{45c}$$

$$\frac{dA_2}{dz'} = -i\frac{6\pi^2}{n\lambda}\chi A_4 A_p A_1^*, \tag{45d}$$

where

$$3\chi = 3\chi_{1111}(-\omega, \omega, \omega, -\omega) = C + \tfrac{1}{2}D$$

is the same susceptibility that is responsible for the nonlinear refractive index.

The phase-matching requirement $\Delta k = k_4 - (k_1 - k_2 - k_p) = 0$ which was so important in other nonlinear interactions is automatically satisfied for the backward wave in the arrangement of Fig. 20b regardless of the angle θ because the two pump waves are oppositely directed ($k_1 = -k_2$) and all waves are at the same frequency ($k_4 = -k_p$). Exact phase matching for a backward wave at a slightly offset frequency $\omega_4 = \omega_1 + \delta$ can be obtained by using a probe beam at an oppositely offset frequency $\omega_p = \omega - \delta$ and arranging the pump beams to propagate at a small angle. The relation between the frequency offset and the angle between the pump beams has been analyzed by Pilipetzky and Shkunov[86] for various situations.

If we neglect pump depletion and assume that the pump amplitudes are uniform over the interaction path of the probe beam, then the solution of Eqs. (45a) and (45b) is

$$A_p(z) = A_p(0)\frac{\cos[\kappa|A_1A_2|(z-L)]}{\cos[\kappa|A_1A_2|L]} + iA_4^*(L)\frac{\sin[\kappa|A_1A_2|z]}{\sin[\kappa|A_1A_2|L]}, \tag{46a}$$

$$A_4(z) = A_4(L)\frac{\cos[\kappa|A_1A_2|z]}{\cos[\kappa|A_1A_2|L]} - iA_p^*(0)\frac{\sin[\kappa|A_1A_2|(z-L)]}{\cos[\kappa|A_1A_2|L]}, \tag{46b}$$

where

$$\kappa = 6\pi^2\chi/\lambda n \tag{47}$$

and L is the length of interaction between the pump and probe beams along z.

Many of the effects of DFWM that were mentioned in the previous section can be derived as special cases of these solutions. The characteristics of the

TABLE 2

Polarization Combinations and θ Dependence of the Nonlinear Polarization in DFWM[a]

Geometry

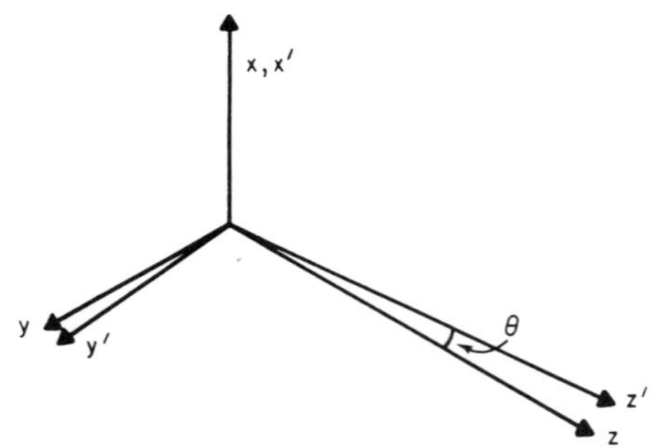

Direction of E-field polarization				θ dependence of nonlinear polarization	
				Direction of polarization	Amplitude
E_1	E_2	E_p	E_4		
$\hat{x}'$	$\hat{x}'$	$\hat{x}$	$\hat{x}$	None	$\frac{C' + C'' + D}{4} A_1 A_2 A_p^*$
$\hat{x}'$	$\hat{x}'$	$\hat{y}$	$\hat{y}$	None	$\frac{D}{4} A_1 A_2 A_p^*$
$\hat{y}'$	$\hat{y}'$	$\hat{x}$	$\hat{x}$	None	$\frac{D}{4} A_1 A_2 A_p^*$

$\hat{y}'$	$\hat{y}'$	$\hat{y}$	$\hat{y}$	None	$\frac{(C' + C'')\cos^2\theta + D}{4} A_1 A_2 A_p^*$
$\hat{x}'$	$\hat{y}'$	$a\hat{x} + b\hat{y}$ $(a^2 + b^2) = 1$	$\hat{x}\left(\frac{a(C' + C'' + D)}{\sqrt{a^2(C' + C'' + D)^2 + b^2D^2}}\right) + \hat{y}\left(\frac{bD}{\sqrt{a^2(C' + C'' + D)^2 + b^2D^2}}\right)$	None	Independent of θ
$\hat{y}'$	$\hat{y}'$	$a\hat{x} + b\hat{y}$	$\hat{x}\left(\frac{aD}{\sqrt{a^2D^2 + b^2[C' + C'')\cos^2\theta + D]^2}}\right) + \hat{y}\left(\frac{b[(C' + C'')\cos^2\theta + D]}{\sqrt{a^2D^2 + b^2[(C' + C'')\cos^2\theta + D]^2}}\right)$	Yes	Depends on θ
$\hat{x}'$	$\hat{y}'$	$a\hat{x} + \mathrm{b}\hat{y}$	$\frac{bC''\hat{x} + aC'\hat{y}}{\sqrt{b^2C''^2 + a^2C'^2}}$	Yes $(C' \neq C'')$ None $(C' = C'')$	$\frac{1}{4}\sqrt{a^2C'^2 + b^2C''^2}\cos\theta\, A_1 A_2 A_p^*$
$\frac{\hat{x}' + i\hat{y}'}{\sqrt{2}}$	$\frac{\hat{x}' - i\hat{y}'}{\sqrt{2}}$	$a\hat{x} + b\hat{y}$	$\hat{x}\left(\frac{a}{\sqrt{a^2 + b^2\cos^2\theta}}\right) + \hat{y}\left(\frac{b\cos^2\theta}{\sqrt{a^2 + b^2\cos^2\theta}}\right)^b$	Yes	Depends on θ

[a] For transparent nonresonant media, $C' = C'' = 3(\chi_{1122} + \chi_{1212})$, $D = 6\chi_{1221}$.
[b] For $C' = C''$.

solutions that are important for the discussions that follow are the following: (a) A backward wave A_4 can be generated even in the absence of an input signal for this wave, i.e., if $A_4(L) = 0$. (b) When there is no input for A_4 the backward wave amplitude is the complex conjugate of the probe beam amplitude. (c) When the condition $\kappa|A_1A_2|L = \pi/2$ is satisfied the system can oscillate, producing radiation in waves A_p and A_4 even with no input in either wave.

5.2.2a. DFWM AMPLIFICATION. A DFWM system as illustrated in Fig. 20b can act as either a backward or forward wave amplifier if there is input at only one of the waves A_p or A_4. For example, if $A_4(L) = 0$, then the probe and backward wave amplitudes are given by

$$A_p(z) = A_p(0)\frac{\cos[\kappa|A_1A_2|(z - L)]}{\cos[\kappa|A_1A_2|L]}, \tag{48a}$$

$$A_4(z) = -iA_p^*(0)\frac{\sin[\kappa|A_1A_2|(z - L)]}{\cos[\kappa|A_1A_2|L]}. \tag{48b}$$

Under these conditions the probe beam at the exit of the cell is amplified with the amplitude gain factor

$$G_A = A_p(L)/A_p(0) = \sec(\kappa|A_1A_2|L), \tag{49}$$

while the backward wave is generated according to (41) and has the value

$$A_4(0) = +iA_p^*(0)\tan(\kappa|A_1A_2|L) \tag{50}$$

at the entrance to the nonlinear medium. Both waves gain energy at the expense of energy in the pump waves.

5.2.2b. CONJUGATE REFLECTION. Most of the attention in studies of DFWM with counterpropagating pump waves has been paid to the backward wave. Its amplitude is proportional to the complex conjugate of the amplitude of the incident probe beam, as shown in Eq. (50). The implications of this property will be discussed in a later section. Under the conditions appropriate to the solutions (48a) and (48b) the DFWM system acts as a conjugate mirror. The reflectivity of the mirror, defined as $R = I_4(0)/I(0)$, is

$$R = \tan^2(\kappa|A_1A_2|L). \tag{51}$$

The reflectivity is not restricted to values less than or equal to unity but can exceed unity substantially if the interaction is strong enough. For sufficiently strong interactions the power in the backward wave will be limited to a value comparable to that in one of the pump waves and the assumption of negligible pump depletion that led to Eq. (48) will no longer be valid.

For the transparent media considered in this section under the assumptions that pump depletion and the nonlinear phase shifts induced by the pump waves can be neglected, the reflectivity and gain coefficients depend only on the product $|A_1A_2|^2$. For other systems and other conditions the dependence of these quantities on the pump amplitudes can be more complex.

5.2.3. *DFWM Oscillation*

The result of Eq. (50) shows that when the condition

$$\kappa|A_1A_2|L = \pi/2 \tag{52}$$

is satisfied, the reflected wave amplitude goes to infinity, implying oscillation, or the presence of a finite amplitude at A_4 even when the input amplitude at A_p is zero. Similarly the amplitude $A_p(L)$ diverges for the same condition [see Eq. (48a)]. An example of the calculated field distribution inside the nonlinear medium at the oscillation threshold is illustrated in Fig. 22.

The minimum oscillation condition of (52) is usually met first for waves that are collinear with the pump waves for which the interaction path is longest. An alternative geometry, shown in Fig. 23, uses a mirror placed at an angle to the pump waves. The extra mirror defines an axis for the oscillation and reduces the threshold condition to[59a]

$$\kappa|A_1A_2|L = \tan^{-1}(1/|r|), \tag{53}$$

where $|r|^2$ is the mirror reflectivity.

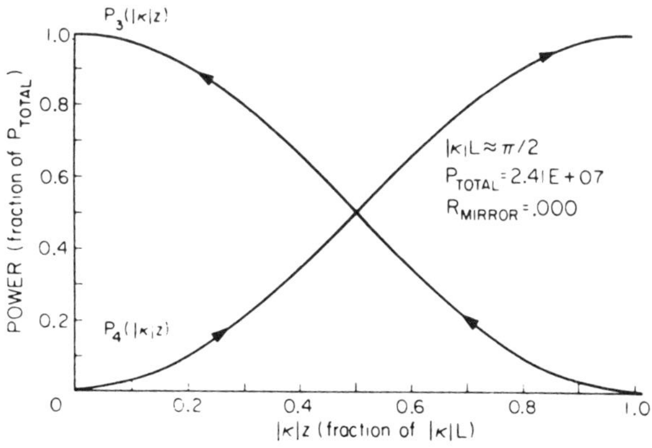

Fig. 22. Calculated field distribution inside nonlinear medium at DFWM oscillation threshold. (From Ref. 59a, © 1978 IEEE.)

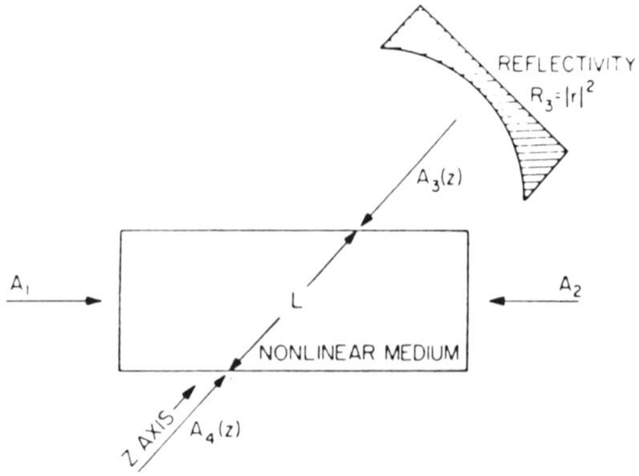

Fig. 23. An alternative geometry for DFWM oscillation using an external mirror to define an axis of oscillation and to reduce the threshold. (From Ref. 59a, © 1978 IEEE.)

5.2.4. *Nonlinear Phase Shifts*

Marburger and Lam[67] have extended the analysis of DFWM in transparent media to include the effects of the nonlinear phase shifts introduced by the pump beams. These effects arise from the terms in the nonlinear polarization in Eq. (42) that were left out in deriving Eqs. (45). The equations for the field amplitudes when these terms are included are

$$\frac{dA_1}{dz'} = i\frac{3\pi\omega}{2nc}\chi A_1(|A_1|^2 + 2|A_2|^2), \tag{54a}$$

$$\frac{dA_2}{dz'} = -i\frac{3\pi\omega}{2nc}\chi A_2(|A_2|^2 + 2|A_1|^2), \tag{54b}$$

$$\frac{\omega}{\omega_3}\frac{dA_\mathrm{p}}{dz} = i\frac{3\pi\omega}{nc}\chi\{(|A_1|^2 + |A_2|^2)A_\mathrm{p} + |A_1A_2|A_4^* e^{i(\phi_1+\phi_2-\Delta kz)}\}, \tag{54c}$$

$$\frac{\omega}{\omega_4}\frac{dA_4}{dz} = -i\frac{3\pi\omega}{nc}\chi\{(|A_1|^2 + |A_2|^2)A_4 + |A_1A_2|A_\mathrm{p}^* e^{i(\phi_1+\phi_2-\Delta kz)}\}, \tag{54d}$$

where χ is the appropriate susceptibility for the experimental geometry.

Here we have assumed for the incident fields the forms

$$\mathscr{E}_1 = \tfrac{1}{2}\{A_1 e^{-i(\omega t - k_1 z)} + \text{c.c.}\} = \tfrac{1}{2}\{|A_1|e^{i\phi_1}e^{-i(\omega t - k_1 z)} + \text{c.c.}\}, \tag{55a}$$

$$\mathscr{E}_2 = \tfrac{1}{2}\{A_2 e^{-i(\omega t + k_1 z)} + \text{c.c.}\} = \tfrac{1}{2}\{|A_2|e^{i\phi_2}e^{-i(\omega t + k_1 z)} + \text{c.c.}\}, \tag{55b}$$

$$\mathscr{E}_{\mathrm{p}} = \tfrac{1}{2}\{A_3(z)e^{-i(\omega_3 t - k_3 z)} + \text{c.c.}\}, \tag{55c}$$

$$\mathscr{E}_4 = \tfrac{1}{2}\{A_4(z)e^{-i(\omega t + k_4 z)} + \text{c.c.}\}, \tag{55d}$$

where

$$\omega_3 = \omega - \delta, \tag{56a}$$

$$\omega_4 = \omega + \delta, \tag{56b}$$

and have allowed for a frequency offset between the pump and probe waves. The phase mismatch Δk is given by

$$\Delta k = -k_4 + k_3 = -2\delta/c. \tag{57}$$

In deriving these equations it has also been assumed that only the nonlinear phase shift induced by the pump beams is important and that pump depletion can be neglected.

Equations (54) can be solved for the DFWM reflectivity, yielding the result

$$R = \frac{\tan^2 \kappa' L}{1 + [(\Delta k - \Delta k_m)^2 \sec^2 \kappa' L]/4\alpha^2}, \tag{58}$$

where

$$\alpha^2 = (\pi\omega/nc)^2(3\chi)^2|A_1 A_2|^2, \tag{59a}$$

$$4\kappa'^2 = 4\alpha^2 + (\Delta k - \Delta k_m)^2, \tag{59b}$$

$$\Delta k_m/k = 3\pi\chi(|A_1|^2 - |A_2|^2)/2n^2. \tag{59c}$$

These solutions are identical to the ones given by Marburger and Lam[67] when the difference in definition of the field amplitudes is accounted for.

When $(\kappa' L)^2 \ll 1$ the reflectivity is small and reduces to the same value given earlier in Eq. (51). For this condition the DFWM reflectivity is not affected by the nonlinear phase shifts. For larger reflectivities the nonlinear phase shifts cause the DFWM reflectivity to be reduced from the value of Eq. (51) by a factor that depends on the forward–backward pump wave asymmetry. Within the approximations that led to Eq. (58) the effects of a forward–backward pump wave asymmetry can be compensated for by introducing a finite detuning such that the phase mismatch satisfies the condition $\Delta k = \Delta k_m$.

The results of Marburger and Lam[67] are important near the oscillation condition of (52). They have shown that in the presence of nonlinear phase shifts oscillation in a system with no auxiliary mirrors is possible at the pump frequency only for equal pump intensities. For unequal pump intensities oscillation can only take place at a shifted frequency whose value is determined by the forward–backward pump asymmetry.

For collinear geometries Marburger and Lam[67] have also shown that the polarization components resulting from the nonlinear phase shift terms in (43) and from the terms in (44) generate a component of the backward wave that is not conjugate to the input probe wave even when the amplitudes of the forward and backward waves are equal. The amplitude equations for the probe and backward waves for this situation, assuming input waves of the form given in Eq. (55) and equal forward and backward pump wave amplitudes are

$$\frac{dA_{\mathrm{p}}}{dz} = i\frac{\pi\omega}{2nc} D|A_1|^2\left[\frac{2C}{D} A_{\mathrm{p}} + A_4^* + \tfrac{1}{2}\left(\frac{2C}{D} A_4 + A_{\mathrm{p}}^*\right)e^{i2\phi_1}\right], \quad (60a)$$

$$\frac{dA_4}{dz} = -i\frac{\pi\omega}{2nc} D|A_1|^2\left[\frac{2C}{D} A_4 + A_{\mathrm{p}}^* + \tfrac{1}{2}\left(\frac{2C}{D} A_{\mathrm{p}} + A_4^*\right)e^{-i2\phi_1}\right]. \quad (60b)$$

Here we have assumed that the probe and pump beams are orthogonally polarized.

The solution to these equations derived by Marburger and Lam[67] shows that the ratio of the power in the conjugate component of the backward wave to the power in the nonconjugate component varies from $(D/C)^2 = [2\chi_{1122}/(\chi_{1111} - \chi_{1122})]^2$ for weak interactions, i.e., when

$$[12(C/D + 1)\pi\omega D A_1^2 L/nc \ll 1],$$

and therefore low reflectivities, to a value of unity when the interaction approaches the oscillation condition. For finite detunings of the probe- and pump-wave frequencies the conjugate component of the backward wave is produced at a different frequency than the nonconjugate component.

5.2.5. *Pump Depletion*

The effects of pump depletion on the backward wave reflectivity have been considered by Marburger and Lam[68] and Hsu.[69] Hsu treated DFWM in a transparent medium using the full set of equations in (45) and assumed equal pump-wave amplitudes. Marburger and Lam included the effects of nonlinear phase shifts through the terms in Eq. (42) and allowed for unequal pump intensities. Hsu's results predict that full conversion of the power in one of the pump waves to the backward wave is possible for any value of input probe intensity. They also predict that oscillation as the pump intensity approaches the condition in (52) is suppressed if the input probe intensity has a nonzero value. Marburger and Lam obtained solutions for a collinear geometry in which the pump waves were polarized parallel to each other, as were the probe and backward generated waves, but the probe wave was polarized perpendicular to the pump wave (the configuration on line 2 in Table 2). Their solutions show that, in general, the intensity in the generated wave is less than or equal to the incident intensity in the backward pump wave. Full conversion of the backward pump wave to the backward generated wave was predicted to occur only if the incident forward and backward pump intensities were equal and if the incident probe wave amplitude was small compared to a quantity which, in our notation, is $16I_1(0)/(2C + D)$. Both Hsu and Marburger and Lam predict oscillating behavior for an incident probe wave with a finite amplitude, with the power in the probe and backward waves being returned to the pump waves at higher pump powers than that required for maximum backward wave intensity.

5.2.6. *DFWM in Absorbing Media*

DFWM can also take place in absorbing media. In many cases of interest the DFWM interaction arises from the absorption processes itself, either as a saturation of linear absorption or from thermal index changes. In other situations the absorption process may not be directly involved in the DFWM interaction, but it can still affect the results through attenuation of the various waves. The equations for DFWM in these media must be modified to take into account the absorption of the waves. In saturable absorbers the DFWM gain and the losses due to absorption are connected, and so some of the behavior present in transparent media will undergo qualitative changes as well as quantitative ones.

DFWM in absorbing media has been discussed by several authors. For example, DFWM in saturable absorbers has been analyzed by Abrams and Lind,[71] Apanasevich *et al.*,[72] Dunning and Steel,[143] and by Brown.[74,144] A discussion of various effects in these media has been given by Lind *et al.*[139] Bragg scattering from thermal gratings has been discussed by Key *et al.*[78] and by Eichler *et al.*[79] Thermal effects in the context of DFWM and phase conjugation have been discussed by Heilweil *et al.*,[80] Andrews and Hochstrasser,[81] Tocho *et al.*, [82,83] Martin and Hellwarth,[145] and Caro and Gower.[126–128] Caro and Gower[128] have given a general treatment covering some aspects of both thermal media and saturable absorbers. Here we shall treat the two types of interaction separately.

5.2.6a. SATURABLE ABSORBERS. DFWM in saturable absorbers involves a single-photon resonance and can be expected in general to be effective at relatively low pump powers. Abrams and Lind considered a system in which the nonlinear interaction results only from saturation of the single-photon absorption, while Apanasevich *et al.*[72] also included contributions from the third-order susceptibility that do not involve saturation of the absorption. Both of these treatments neglected attenuation of the pump radiation either from absorption or from conversion in the nonlinear interaction. These latter two effects were discussed in the treatment by Brown.

In media with saturable absorption the spatial part of the total nonlinear polarization in the steady state is given by

$$\mathscr{P} = \frac{1}{2}\left[i\frac{\alpha_0 ncE}{2\pi\omega(1 + i\Delta')(1 + |E|^2/A_s^2)} + \chi^{(3)}E^2E^* + \text{c.c.}\right]. \tag{61}$$

Here $\alpha_0 = (2\pi\omega/cn)|\mu_{10}|^2 T_2 N/\hbar$ is the unsaturated field attenuation coefficient at line center for a transition between levels 0 and 1,

$$A_s = [|\mu_{10}|^2 T_1 T_2/\hbar^2(1 + \Delta'^2)]^{-1/2} \tag{62}$$

is the frequency-dependent saturation field amplitude, and $\Delta' = (\omega_{10} - \omega)T_2$ is the normalized detuning from resonance. The field components are given by

$$E = \Sigma E_i = \Sigma A_i e^{ik_i z}.$$

The nonlinear interaction arising from saturation of the absorption at the resonant frequency is given by the first term in Eq. (61). The second term contains the effects of all the remaining interactions that result in a nonlinear polarization at frequency ω. These include, for example, the nonlinear susceptibility arising from other levels that may be present in the system. All of the latter interactions involve terms that are proportional to the ground-state population, and some contain the population in level 1 as well as resonant terms in the detuning Δ'. Thus even though these terms do not directly involve the saturation of the absorption at the resonant frequency they must be suitably modified from the form of the nonresonant susceptibility to take into account the redistribution of population between levels 0 and 1 when the absorption is saturated.

If we assume that the probe and backward wave amplitudes are everywhere small compared to the sum of the pump wave amplitudes, that all waves are polarized in the same direction, and consider only the nonlinear effects due to saturation of the absorption, then we can expand the denominator of the first term in Eq. (62), giving for the spatial part of the total polarization

$$\mathscr{P} = i\frac{\lambda\alpha_0 \Sigma E_i}{8\pi^2(1 + i\Delta')D'} \times \left(1 - \frac{[(E_3 + E_4)(E_1^* + E_2^*) + (E_3^* + E_4^*)(E_1 + E_2)]/A_s^2}{D'}\right) + \text{c.c.}, \tag{63}$$

where

$$D' = 1 + |A_1 e^{ik_1 z'} + A_2 e^{-ik_1 z'}|^2/A_s^2$$
$$= 1 + |A_1/A_s|^2 + |A_2/A_s|^2 + 2|A_1 A_2/A_s^2|\cos(2kz' + \phi_1 - \phi_2),$$

and ϕ_1 and ϕ_2 are the phases of A_1 and A_2, respectively.

Because of the coupling of the forward and backward waves by the oscillating term in the denominator in Eq. (63) we must obtain the nonlinear polarizations that drive each of the waves by taking the appropriate Fourier component of the total polarization. We will illustrate the types of terms that are obtained in this manner for the polarization that drives the forward pump wave and the backward generated wave.

The nonlinear polarization amplitude with the same spatial Fourier component as E_1 is given by

$$P_1 = \frac{i\lambda\alpha_0}{4\pi^2(1 + i\Delta')}\left(\frac{1}{2\pi}\int\frac{A_1(z')\,dz'}{D'(z')} + \frac{1}{2\pi}\int\frac{A_2(z')e^{-i2kz'}\,dz'}{D'(z')}\right). \tag{64}$$

In the slowly varying envelope approximation the integrals can be evaluated over a distance of several wavelengths and the field amplitudes can be treated as constants. The first integral always makes a nonzero contribution to the polarization for E_1. The second term makes a nonzero contribution because of the coupling of the forward and backward waves through the interference term in the denominator. In interactions that do not involve saturation of a resonant transition or in ones in which the pump waves are orthogonally polarized the second term would not make a phase matched contribution to the polarization for E_1.

The nonlinear polarization amplitude with the same Fourier component as E_4 has four types of terms:

$$\begin{aligned} P_4 = {} & \frac{i\lambda\alpha_0}{8\pi^3(1 + i\Delta')}\left(\int\frac{A_4\,dz}{D'(z')}\right.\\ & \left. - \int\frac{A_4(|A_1|^2 + |A_2|^2)/A_s^2 + 2A_1A_2A_3^*/A_s^2}{(D'(z'))^2}\,dz\right)\\ & - \frac{i\lambda\alpha_0}{8\pi^3(1 + i\Delta')}\left(\int\frac{(A_3^*A_1^2 + A_4A_1A_2^*)/A_s^2 e^{i2kz'}}{(D'(z'))^2}\,dz\right.\\ & \left. + \int\frac{(A_3^*A_2^2 + A_4A_1^*A_2)/A_s^2 e^{-i2kz'}}{(D'(z'))^2}\,dz\right)\\ & + \frac{i\lambda\alpha_0}{8\pi^3(1 + i\Delta')}\left(\int\frac{A_2e^{ik(z-z')}}{D'(z')}\,dz\right.\\ & \left. - \int\frac{(A_4^*A_2^2 + A_3A_1^*A_2)/A_s^2 e^{i2k(z-z')}}{(D'(z'))^2}\,dz\right)\\ & + \frac{i\lambda\alpha_0}{8\pi^3(1 + i\Delta')}\left(\int\frac{A_1e^{ik(z'+z)}}{D'(z')}\,dz + \int\frac{A_3e^{i2kz}}{D'(z')}\,dz\right.\\ & - \int\frac{(A_4^*A_1^2 + A_3A_1A_2^*)/A_s^2 e^{i2k(z'+z)}}{(D'(z'))^2}\,dz\\ & \left. - \int\frac{(2A_4^*A_1A_2 + A_3(|A_1|^2 + |A_2|^2))/A_s^2 e^{i2kz}}{(D'(z'))^2}\,dz\right). \end{aligned} \tag{65}$$

The terms in the first parentheses make the usual phase-matched contribution to the polarization for the backward generated wave. The terms in the second parentheses make a nonzero contribution to the polarization for the backward wave in saturable resonant media because of the interference term in the denominator. The terms in the first two sets of parentheses will be used to derive equations for the field amplitudes in the saturable absorber. The remaining terms have a different period for the exponential factor and the interference term in the denominator. In general they do not make a phase-matched contribution to the polarization for the backward wave. They can, however, make a nonzero contribution if the angle between the probe wave and the forward pump wave is less than a value given by

$$\theta_0 = (\lambda/L_c)^{1/2},$$

where L_c is the characteristic distance over which the field amplitudes vary. In this situation all of the terms in the second and third sets of parentheses in Eq. (65) will make a nonzero contribution to the backward wave polarization. Note that even in the absence of the interference term in the denominator the terms in the second parentheses contribute when $\theta < \theta_0$. These contributions can be especially important for collinear interactions. When $\theta \gg \theta_0$ the contributions of the terms in the second and third sets of parentheses will oscillate with distance and average to zero over the entire interaction length.

If we assume that $\theta > \theta_0$ and yet small enough so that the difference between z and z' can be neglected in the first and second sets of terms in Eq. (65), then we can derive the following set of equations for the amplitudes of the four waves:

$$\frac{dA_1}{dz} = -\bar{\alpha}_1 A_1, \tag{66a}$$

$$\frac{dA_2}{dz} = \bar{\alpha}_2 A_2, \tag{66b}$$

$$\frac{dA_3}{dz} = -\bar{\alpha}_3 A_3 + \bar{\kappa}' A_4^*, \tag{66c}$$

$$\frac{dA_4}{dz} = \bar{\alpha}_3 A_4 - \bar{\kappa}' A_3^*, \tag{66d}$$

$$\bar{\alpha}_1 = \alpha_0 \left[\frac{1}{[D(z)]^{1/2}} - \frac{1}{2A_1^2(z)/A_s^2}\left(1 - \frac{1 + (A_1^2(z) + A_2^2(z))/A_s^2}{[D(z)]^{1/2}}\right)\right], \tag{67a}$$

$$\bar{\alpha}_2 = \alpha_0 \left[\frac{1}{[D(z)]^{1/2}} - \frac{1}{2A_2^2(z)/A_5^2}\left(1 - \frac{1 + [A_1^2(z) + A_2^2(z)]/A_s^2}{[D(z)]^{1/2}}\right)\right], \tag{67b}$$

$$\bar{\alpha}_3 = \alpha_0\left(\frac{1 + [A_1^2(z) + A_2^2(z)]/A_s^2}{[D(z)]^{3/2}}\right) \tag{67c}$$

$$= (\bar{\alpha}_3(z))_R + i(\bar{\alpha}_3(z))_I,$$

$$\bar{\kappa}' = \alpha_0\left(\frac{2A_1(z)A_2(z)/A_s^2}{[D(z)]^{3/2}}\right), \tag{67d}$$

$$D(z) = 1 + 2(A_1^2(z) + A_2^2(z))/A_s^2 + (A_1^2(z) - A_2^2(z))^2/A_s^4. \tag{67e}$$

In obtaining these equations we have assumed that the system is at exact resonance and that all the fields are real.

In the treatment of Abrams and Lind[71] the attenuation of the pump waves due to absorption or to depletion in the nonlinear interaction was neglected. Under these conditions only Eqs. (66c) and (66d) need to be considered and the pump waves in Eqs. (67c) and (67d) can be treated as constants.

Abrams and Lind have used these approximations along with the assumption that the incident pump wave amplitudes were equal to obtain the following expression for the conjugate reflectivity:

$$R = \frac{|\bar{\kappa}' \sin wL|^2}{|w \cos wL + (\bar{\alpha}_3)_R \sin wL|^2}, \tag{68a}$$

where

$$w = (|\bar{\kappa}'|^2 - (\bar{\alpha}_3)_R^2)^{1/2}. \tag{68b}$$

These expressions were generalized by Dunning and Steel[143] to account for unequal pump wave amplitudes.

Evaluation of these solutions by Abrams and Lind[71] for various conditions has shown that the reflectivity increases with increasing pump intensity when $I_{1,2} < I_S$, but falls off for pump intensities greater than the single-photon saturation intensity because of the decreasing nonlinear susceptibility. They also predict that the maximum reflectivity that can be obtained for nonzero detunings is greater than the maximum reflectivity at exact resonance.

The solutions given before neglected attenuation of the pump intensity due either to absorption or to depletion in the four-wave interaction. Qualitatively we can expect that the neglect of pump absorption should become increasingly less valid as the absorption increases unless the pump intensity greatly exceeds the saturation intensity. Experimental observation of deviation from the theoretical forms given before has been reported by Watkins *et al.*[73] for $\alpha_0 L = 3$ and a pump intensity that was greater than 10 times the saturation intensity.

Solutions for the reflectivity under conditions when the pump waves are attenuated due to absorption or nonlinear depletion have been discussed by

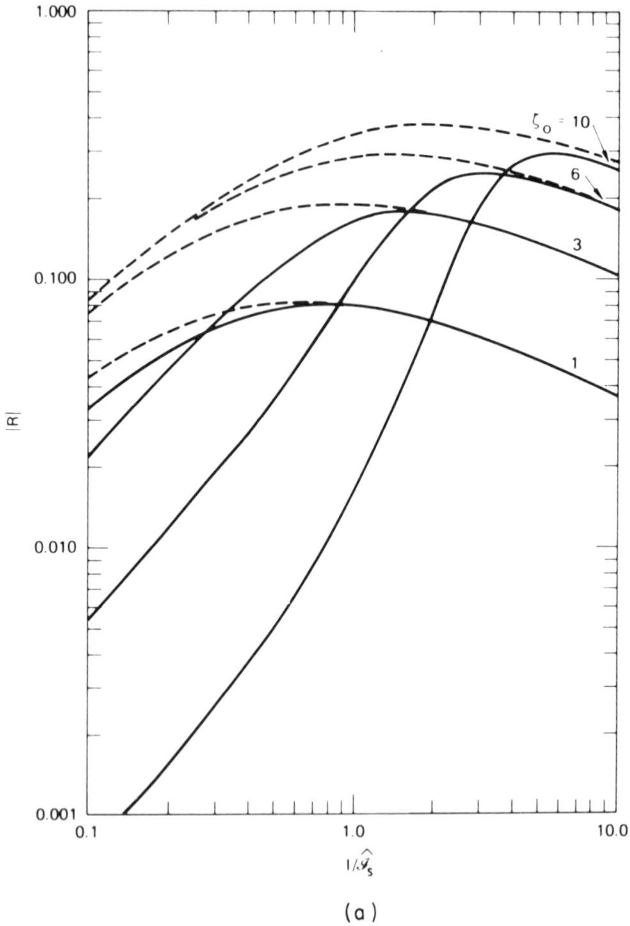

Fig. 24a. DFWM field amplitude reflectivity in a saturable absorber as a function of incident pump intensity. The curves are parametrized by the small-signal intensity absorption coefficient-length product and the effects of pump absorption are included. The dashed curves are calculated reflectivities for corresponding small-signal absorption without pump attenuation. All waves polarized parallel, $\Delta' = 0$. (From Ref. 74.)

Brown.[74,144] When only pump absorption is important the equations for the pump waves decouple from those for the probe and backward wave as in Eq. (62), but now all four equations must be solved simultaneously.

Numerical solutions given by Brown[74,144] and shown in Fig. 24a show that agreement with the constant pump solutions is reasonably good either for $\alpha_0 L < 1$ or for $I_1 \gg I_s$, but that deviation from the former solutions becomes increasingly important for $\alpha_0 L > 1$ and $I_1 \lesssim I_s$. For small values

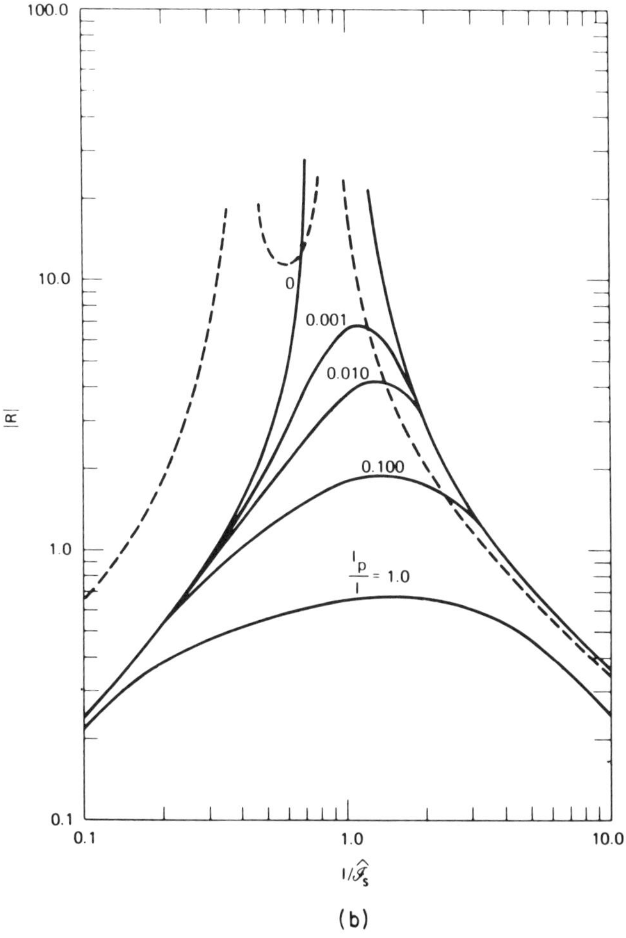

Fig. 24b. Calculated field amplitude reflectivity for a saturable absorber as a function of the normalized pump intensity for various values of the ratio of the incident probe intensity to the incident pump intensity. The dashed curves are for the constant pump approximation. The normalized detuning is $\Delta' = 15$ and the intensity attenuation coefficient-length product $2\alpha L = 2$. Polarizations: $E_1 \parallel E_4$, $E_2 \parallel E_3$, $E_1 \perp E_2$. (From Ref. 74.)

of $\alpha_0 L$ the optimum pump intensity is approximately equal to the saturation intensity, but increases as $\alpha_0 L$ increases. The value of the maximum reflectivity is generally less than unity on resonance and approaches unity only in the limit of large $\alpha_0 L$. In addition, no oscillation is possible on resonance for saturable absorbers. Larger maximum reflectivities can be obtained off resonance. Under these conditions the optimum pump intensity is determined by the frequency-dependent saturation intensity, which increases with detuning from resonance. In addition, oscillation is possible off resonance because of the increasing importance of the dispersive part of the nonlinear susceptibility.

When both the pump absorption and nonlinear depletion must be taken into account the full nonlinear polarization of Eq. (61) must be used and the equations for the pump waves do not decouple from those for the probe and backward waves. Under these circumstances Brown has given solutions that predict that approximately one-third of the energy in one of the incident pump waves can be transferred to the backward wave as shown in Fig. 24b. In addition, when a finite incident probe wave amplitude is accounted for, the oscillation that is predicted for off-resonant interactions is suppressed.

5.2.6b. THERMAL EFFECTS. Phase gratings can also be created in absorbing media due to thermal index changes. These arise as a result of the interference between the probe and pump waves through the first two terms in the first pair of curly brackets in Eq. (42). They can be encountered, for example, in absorbing liquids or in solutions of absorbing materials such as dyes in relatively transparent liquids. In the latter situation the energy is usually absorbed by the solute, but the thermal pattern is developed in the solvent after the energy is thermalized and transferred to the solvent.

Caro and Gower[128] have given the following expression for the DFWM reflectivity in an absorbing medium allowing for attenuation of the pump waves due to absorption but not due to depletion in the four-wave mixing interaction, and with the assumption that the absorption of one pump wave is not affected by the presence of the other:

$$R = \left| \frac{2Q \sin(HL/2)}{H \cos(HL/2) + \alpha \sin(HL/2)} \right|^2. \tag{69}$$

Here Q is related to the nonlinear susceptibility and the product of the pump wave amplitudes and for thermal media is given by

$$|Q| = \frac{dn}{dT} \phi \left(\frac{\omega}{\rho c C_\mathrm{p}} \right) t_\mathrm{p} \alpha I_1 e^{-\alpha L'}, \tag{70}$$

where dn/dT is the coefficient of variation of the refractive index with temperature, I_1 is the incident pump intensity, ϕ is the fraction of absorbed radiation that is turned into heat, ρ is the density, C_p is the specific heat at constant pressure, L' is the interaction distance measured along the direction of the probe wave, t_p is the duration of the laser pulse, α is the field absorption coefficient, and $H = (4|Q|^2 - \alpha^2)^{1/2}$.

Thermal diffusion tends to wash out the thermal gratings by equilibrating the temperature in the high and low intensity regions of the interference pattern. This washout is similar to that due to atomic motion in gaseous media, which will be discussed in the next section. The effects of thermal diffusion give the gratings a characteristic lifetime that is related to the time it takes for heat to diffuse a distance corresponding to the grating spacing. Eichler *et al.*[79] have given the following expression for the lifetime of thermal gratings due to thermal diffusion:

$$\tau = \left(\frac{\rho C}{\lambda_w}\right)\left(\frac{\Lambda}{2\pi}\right)^2. \tag{71}$$

Here λ_w is the thermal diffusivity, C is the specific heat, ρ is the density, and Λ is the grating spacing as given in Eq. (45). For a given material the time constant is a function of the angle between the pump and probe waves. For interactions with a small angle between the pump and probe waves the lifetime of the grating formed by the interference of the probe wave with the forward pump wave is greater than that of the grating involving the backward pump wave. For typical materials and visible wavelengths the lifetime of the forward grating can vary in magnitude from values of the order of tens of nanoseconds for angles θ near 90° to values of the order of hundreds of microseconds for near-collinear interactions.

The time constants associated with thermal gratings can be used to distinguish interactions involving thermal effects from those that involve saturation of the absorption. In addition, the relatively long time constants associated with the thermal gratings make the thermal interactions more effective than faster responding mechanisms in phase conjugating relatively long pulses. This can be especially true of the forward grating in near-collinear geometries.

Eichler *et al.*[79] have shown that in the limit of low reflectivities the reflectivity from a thermal grating in the presence of washout effects is given by

$$R = I_0^2 T(1 - T)^2\left[\frac{dn}{d\theta}\left(\frac{2\pi}{\lambda}\right)\left(\frac{\tau}{\rho C_p}\right)\right]$$
$$\times\ [1 + (2\tau/t_p)e^{-t_p/\tau} - (\tau/2t_p)e^{-t_p/\tau} - 3\tau/t_p], \tag{72}$$

where I_0 is the incident pump intensity, T is the transmission of the medium, $dn/d\theta$ is the temperature of the refractive index, τ is the decay time of the grating as given by Eq. (71), ρ is the density of the medium, C_p is the specific heat, and t_p is the duration of the incident pulse.

Because the reflectivity from thermal gratings is not intimately connected with saturation of the nonlinear susceptibility, it can be higher in some situations than that obtained from saturable absorbers. It can be especially important for relatively long pulses that allow the thermal gradients to be established, and for high incident pump intensities for which the reflectivity from saturable absorbers tends to saturate. Thermal gratings have proved to give higher reflectivity in the ultraviolet than either gratings due to saturable absorption or the Kerr effect.

5.2.7. *Effects of Atomic Motion*

The effects of atomic motion on the DFWM reflectivity in single-photon resonant interactions have been discussed by Wandzura[75] and by Nilsen and Yariv.[76] Both groups have treated Doppler broadening, while Wandzura has also discussed homogeneously broadened systems. In both types of system atomic motion tends to reduce the DFWM reflectivity and introduce a dependence of the reflectivity on the angle θ between the pump and probe waves. These effects have been attributed to a washing out of the dynamic grating by the atomic motion. Atomic motion can also affect the spectral bandpass of the DFWM interaction, as will be discussed in the next section.

Calculations given by Wandzura for homogeneously broadened atomic media shown in Fig. 25a indicate that the reflectivity decreases when the lifetime T_1 becomes comparable to the time required for an atom traveling at thermal speeds to move a distance comparable to the spacing of the dynamic gratings. For collinear interactions only the grating formed by the probe and the backward pump wave is affected. The reflectivity approaches one-half of its zero-lifetime value as T_1 approaches infinity and the backward grating is completely washed out. For nonzero values of θ both gratings are

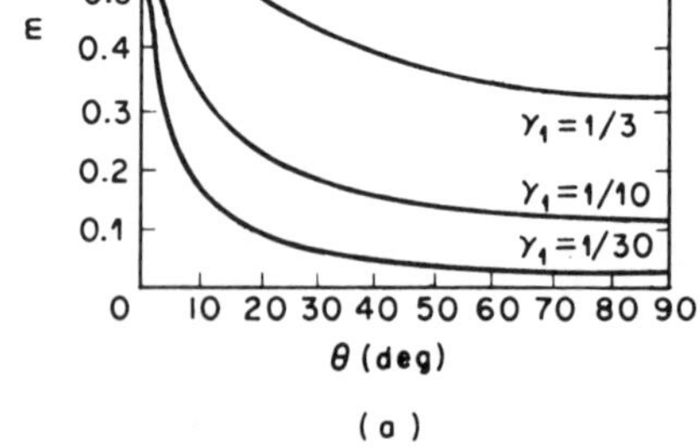

Fig. 25a. Normalized amplitude of the generated field in a homogeneously broadened medium as a function of the angle between the probe wave and the forward pump wave. The curves are parametrized by the ratio of the time required for an atom to move approximately one wavelength at the average thermal velocity to the atomic lifetime. (From Ref. 75.)

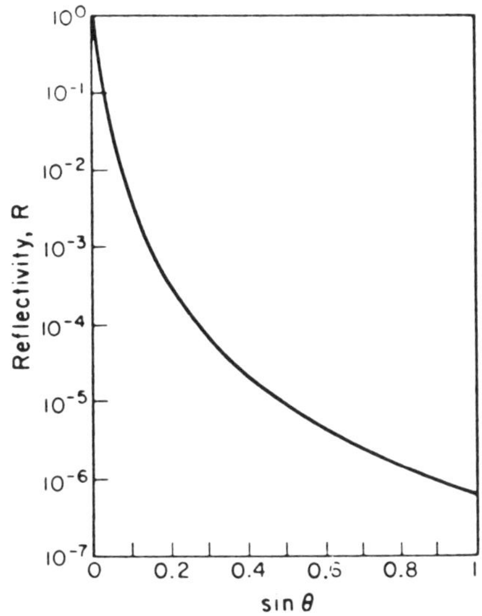

Fig. 25b. Dependence of the DFWM reflectivity for a Doppler-broadened medium as a function of the angle between the pump and probe beams. The detuning for the pump and probe frequencies and the pump frequency from the atomic resonance frequency were both zero. (From Ref. 76.)

affected. The reflectivity decreases with increasing angle when T_1 is large enough for grating washout effects to be important.

In Doppler-broadened media the reflectivity is also reduced by atomic motion and is also a decreasing function of the angle θ. Both Wandzura[75] and Nilsen and Yariv[76] have shown that, for large values of θ, the reflectivity falls off as $|u_N \sin\theta|^{-4}$, where θ is the angle between k_1 and k_{probe} as defined in Fig. 20a and $u_N = v_{th}\omega_0 T_2/c$ is a normalized thermal velocity. Nilsen and Yariv's[76] results, shown in Fig. 25b, indicate that for small angles the angular width of the DFWM reflectivity is inversely proportional to u_N. As a result the field of view of the DFWM interaction in Doppler-broadened media will be limited. The angular width increases as the pump frequency is detuned from the center of the single-photon resonance, but this increase is accompanied by a decrease in the maximum reflectivity.[76] Humphrey *et al.*[77] have discussed DFWM in Doppler-broadened media in which optical pumping effects are important. They have shown that under these conditions the reflectivity falls off with angle as $(\sin\theta)^{-2}$.

5.2.8. *Spectral Bandpass*

When the frequency of the probe beam is detuned from that of the pump waves ($\omega_p = \omega_1 - \delta$), the nonlinear interaction couples the probe wave to a fourth wave that is detuned in an opposite sense ($\omega_4 = \omega_1 + \delta$). If the pump waves remain exactly counterpropagating, then a phase mismatch given by

$$\Delta k = k(\omega_1 - \delta) - k(\omega_1 + \delta)$$

will be introduced, which for small detunings is

$$\Delta k = -2n\delta/c.$$

The presence of the phase mismatch results in a narrow-bandpass spectral filter. In transparent media for geometries for which the reflectivity has no dependence on the angle between the pump and probe beams, the DFWM process can provide a narrow-band filter with a wide field of view. The reflectivity and bandpass of such a filter have been analyzed by Pepper and Abrams.[84] They show that in the presence of a phase mismatch the reflectivity is given by

$$R = \left|\frac{A_4(0)}{A_{\text{probe}}(0)}\right|^2 = \frac{|\kappa A_1 A_2 L|^2 \tan^2 \beta L}{|\kappa A_1 A_2 L|^2 + (\Delta k L/2)^2 \sec^2 \beta L},$$

where $\beta = [|\kappa A_1 A_2|^2 + (\Delta k/2)^2]^{1/2}$ and κ is given by (47). The reflectivity is shown in Fig. 26 as a function of the normalized detuning $\psi = \Delta k L/2\pi = nL\delta/\lambda\omega$ for various values of the nonlinear gain $\kappa A_1 A_2 L$ with the peak reflectivity normalized to unity for each value of $\kappa A_1 A_2 L$. The spectral

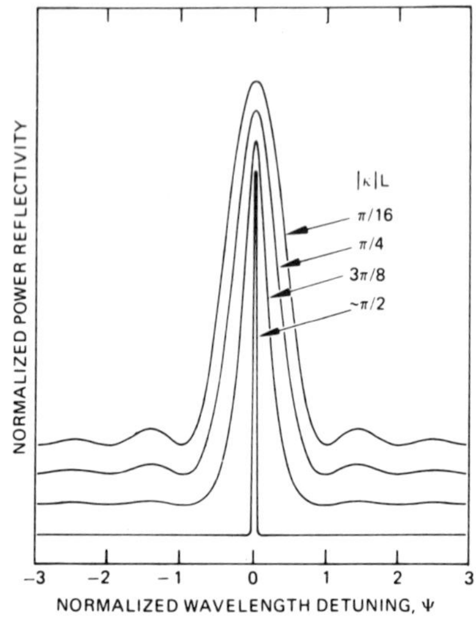

Fig. 26. Dependence of the power reflectivity for DFWM as a function of normalized detuning $\psi = nL\delta/\lambda\omega$ for various values of the nonlinear gain $\kappa A_1 A_2 L$. The peak value of the reflectivity for each curve is normalized to unity. (From Ref. 84.)

bandpass becomes narrower as the nonlinear gain increases, approaching the width of the pump waves as $\kappa A_1 A_2 L$ approaches $\pi/2$.

In single-photon resonant media frequency dependent reflectivity can occur either in degenerate four-wave mixing as the frequency of all the waves is tuned about the resonant frequency or in nearly degenerate four-wave mixing as the frequency of the probe wave is tuned relative to that of the pump waves. The spectral properties of DFWM in single photon resonant media in the absence of atomic motion have been discussed by Fu and Sargent.[85] Spectral properties in single-photon resonant media with atomic motion have been treated by Wandzura[75] and by Nilsen and Yariv.[76] Fu and Sargent have shown that the spectral width of the DFWM reflectivity as a function of detuning of the probe wave from the pump waves decreases as the ratio T_1/T_2 increases.

Nilsen and Yariv have given curves for the reflectivity as a function of probe-wave detuning for various pump-wave detunings from resonance. The reflectivity peak becomes lower and broader as the pump waves are detuned from resonance, as shown in Fig. 27. Both Wandzura and Nilsen and Yariv have examined the resonant peak that occurs in exactly degenerate four-wave mixing at the resonant frequency. Both groups have shown that the width of the peak is Doppler-free for collinear interactions, being determined entirely by the homogeneous width T_1^{-1}. Wandzura has also shown that the spectral width of the resonant peak increases as the angle θ increases.

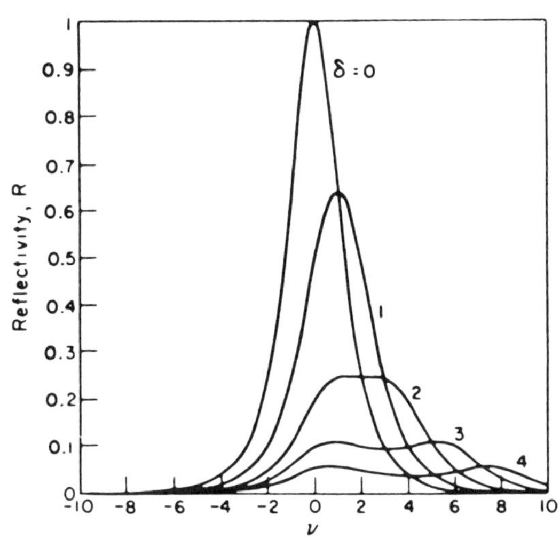

Fig. 27. Reflectivity in a Doppler-broadened medium as a function of the detuning of the probe frequency from the pump-wave frequency for various values of the detuning of the pump-wave frequency from the atomic resonance. (From Ref. 76.)

In the treatment of Nilsen and Yariv,[76] third-order perturbation theory was used, implying $I_{1,2} < I_s$, and the reflectivity was always maximum at the center of the resonant line. Analysis by Agrawal *et al.*[146] has shown that at higher pump intensities the resonant peak exhibits saturation splitting when $|I_1 - I_2| > I_s$ with a local minimum developing at line center. Similar results have also been calculated by Nilsen and Yariv.[147]

5.2.9. *Linewidth Restrictions*

The analysis of DFWM in the preceding sections assumed that all waves were monochromatic. In practice we must be concerned with the interactions of waves that carry varying amounts of spectral bandwidth. In a general interaction we can expect such bandwidth to affect both the DFWM reflectivity and the fidelity of the conjugation. The effects of the bandwidth of the various waves, in general, depend on the geometry of the interaction, the nature of the source of the bandwidth of the radiation, and the response time of the nonlinearity giving rise to the phase conjugation.

When the effects of bandwidth are considered we must include the time dependence of the field envelopes. In the plane wave approximation for an interaction in a medium with an instantaneous response the backward generated wave amplitude is proportional to the integral over the interaction length of the product of appropriately retarded probe and pump wave amplitudes:

$$A_4(0, t') = -\frac{2\pi^2}{\lambda}(C + D/2)\int_0^L A_1\left(0, t' - \frac{z' + z}{c}\right)A_2\left(0, t' + \frac{z' - z}{c}\right) \times A_3^*\left(0, t' - \frac{2z}{c}\right)dz,$$

where t' is the time coordinate for a reference frame that moves with the generated wave.

Generally speaking the results for the DFWM reflectivity that were presented in the preceding sections will be valid if the contributions to the backward wave at time t' that are radiated from various positions along the interaction path arrive in phase at the front of the cell. The minimum conditions for this requirement to be met depend on the interaction geometry. We consider first the geometry shown in Fig. 20b with a small angle between the probe wave and the forward pump wave. In this situation $z' \approx z$ and the difference of the phases of the waves A_1 and A_3 at a given position t' in the envelope remains constant over the interaction path, although the phase difference can be different at different times in the envelope. In addition, the radiation in A_4 at a given time t' arises from radiation in the envelope A_2 at approximately the same time t', regardless of the position in the medium from which it is radiated.

The most general case to consider is one in which the probe wave and the pump waves are not mutually coherent. Such a situation arises, for example, if the probe and pump waves are derived from different lasers, or if the probe wave is derived from the pump wave, but is delayed relative to the pump wave by more than their common coherence lengths. If A_1 and A_3 carry phase modulation that satisfies the conditions of stationarity, the difference in the phases of the two waves will depend only on the time difference t' between the parts of the envelopes of A_1 and A_3 over which interference can take place. A sufficient condition for the results of the preceding section to be valid is for the waves to be self-coherent over twice the interaction length:

$$\Delta\omega_{1,\mathrm{p}} < 4\pi c/L.$$

At the same time the backward pump wave propagates almost coherently with the backward generated wave. A sufficient condition for the bandwidth of the backward pump wave, in order that the results of the previous sections for the DFWM reflectivity remain valid under the conditions being considered here, is

$$\Delta\omega_2 < 4\pi c/L\theta^2.$$

The bandwidth of the backward wave is allowed to increase as the angle θ becomes smaller, and, for exactly collinear interactions, there is no restriction on the backward pump wave bandwidth. Any variations in phase that are present in the backward pump wave will, however, be transferred to the conjugate wave. If the phase variations are uniform across the backward pump beam, they will not, however, degrade the spatial fidelity of the conjugation.

The relaxed requirements on the bandwidth of the backward wave are, in many cases, academic, since the backward pump wave is usually derived from the same source as the forward pump wave and therefore will satisfy the more restrictive requirements of that wave. They do, however, allow the transfer of temporal information from the backward pump wave to the backward generated wave without loss of efficiency in the DFWM process.

If the probe wave is coherent with the forward pump wave, as, for example, if it is derived from the forward pump wave and delayed by a distance less than the coherence length, then the phase difference of the probe and forward pump waves will be approximately constant regardless of their bandwidth or the parts of the envelopes that interfere in the formation of the backward wave. In the small angle approximation the requirement on bandwidth can be expressed as

$$\Delta\omega_{1,\mathrm{p}} < 4\pi c/L\theta^2.$$

Now for small angles the allowed bandwidth for the forward waves is much larger and there is no restriction for exactly collinear interactions.

A second common geometry is for the probe and pump waves to propagate at right angles. In this situation the requirement for the results of the previous sections to be valid is again for the components of the backward conjugate wave to be in phase at the entrance to the nonlinear medium. Now the probe wave and the forward pump wave will not in general stay in phase unless the waves are self-coherent for twice the interaction length. In addition, this condition now applies also to the backward pump wave. The interaction path is usually the cross section of the pump waves and is usually much smaller than for small angle interactions. As a result, the bandwidths allowed for right angle interactions are usually larger than those allowed for small angle interactions when the pump and probe waves are not mutually coherent.

When the nonlinear medium has a finite response time the interference gratings will integrate over a certain amount of the probe and pump wave envelopes. This situation can arise, for example, in interactions involving resonant enhancement (saturable absorbers or saturable amplifiers) or thermal interactions. In order for phase variations in the various waves not to affect the contrast of the gratings it becomes necessary for the waves to remain in phase for the lifetime of the grating. This is especially important for the forward grating in gaseous saturable absorbers or thermal media, which can have a relatively long lifetime.

Various aspects of the coherence requirements of the pump and probe beams, including effects related to the response time of the nonlinear medium, have been discussed by Grossman and Shemwell.[148]

5.3. Experimental Observations of DFWM

5.3.1. Experimental Geometries

Several pump beam geometries have been used in DFWM experiments. One of the simplest involves retroreflecting the forward pump beam after it has passed through the nonlinear medium, as shown in Fig. 28a. This geometry is well suited for experiments that involve transparent nonlinear media and that do not involve pump depletion. A variation on this geometry, which has been used by several authors, uses the standing waves that are present in linear laser cavities to provide the oppositely directed pump waves, as shown in Fig. 28b. In this situation the nonlinear medium can either be the laser medium itself or another medium introduced into the cavity. A third geometry that has been used involves splitting the pump beam before the nonlinear medium and bringing the backward pump wave around the nonlinear medium, as shown in Fig. 28c. In this geometry each pump beam makes only

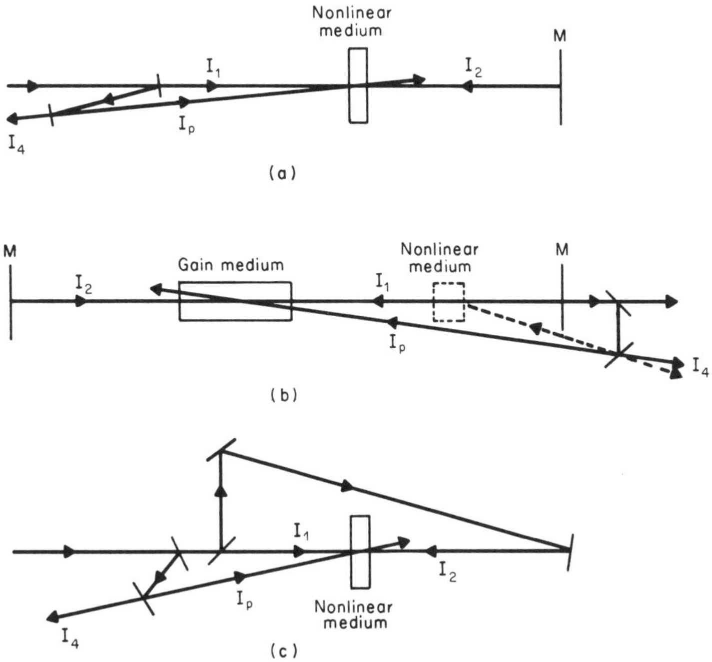

Fig. 28. Typical experimental geometries used for observations of DFWM. (a) Retroreflection of pump wave after nonlinear medium. This method is of most use in transparent media. (b) Pump waves are obtained from counterpropagating waves in a linear laser cavity. The nonlinear medium can either be a separate element in the cavity or the laser gain medium. (c) The pump waves are obtained by splitting the forward pump wave before the nonlinear medium. This method is useful for absorbing media. In (a) and (c) the probe wave is obtained by splitting off a portion of the forward pump wave before the nonlinear medium. In (b) the probe wave is obtained from the output of the laser.

one pass through the nonlinear medium. Although this method reduces the intensity of each pump beam by a factor 2 over that which would be obtained with the geometry of Fig. 28a, it can provide better performance in experiments which involve absorbing media or pump depletion. In each of these geometries the probe beam is usually split off from one of the pump beams, as indicated in Fig. 28a, and then is brought into the nonlinear medium at the desired angle.

5.3.2. Backward Wave Generation and Amplified Reflection

Experimental measurements of backward wave generation and amplified reflection have been reported by many authors. Bloom *et al.*[111] have reported backward wave generation using pulsed dye laser radiation tuned near the D lines of atomic Na. They used an arrangement in which the backward pump wave was obtained by retroreflecting the forward pump wave after the nonlinear medium. The probe beam, obtained by splitting off part of the forward pump wave, propagated at an angle of 0.5° to the direction of the forward pump wave. The nonlinearity of this system was attributed by the authors to a nonlinear index resulting from an optical Stark shift of the resonance levels. As a function of detuning from the resonance lines the DFWM reflectivity peaked at a finite detuning above or below each D line but was minimum at the center of each line. The authors measured reflectivities up to the order of 100 (10^4 %), as shown in their results reproduced in Fig. 29, for a detuning of 1.25 cm^{-1} on the low-frequency side of the D_1 line. Saturation of the reflectivity was observed at pump intensities above about 30 kW/cm^2 and was attributed to the combined effects of pump depletion and self-focusing or defocusing, depending on which side of the nearest line the dye laser was tuned.

Grischowsky *et al.*[113] reported reflectivity measurements using dye laser radiation in atomic Rb vapor. Saturation of the reflectivity was again observed at higher pump intensities, as shown in their results in Fig. 30. The authors accounted for their observations reasonably well by the predictions of the adiabatic following model (see Section 2-6.2.2).

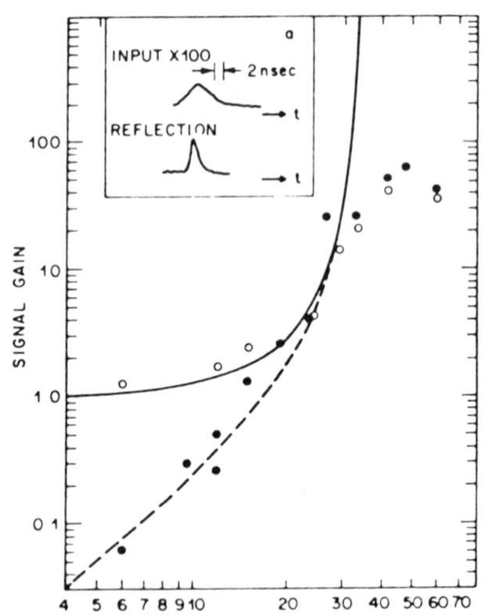

Fig. 29. Measurements of DFWM in Na vapor as a function of the pump-wave intensity. The open circles are the measured gain of the forward probe wave; the solid circles are the measured reflectivity of the backward wave. Solid lines are calculated from theory. Inset shows incident probe pulse (top) and backward wave pulse (bottom). (From Ref. 111.)

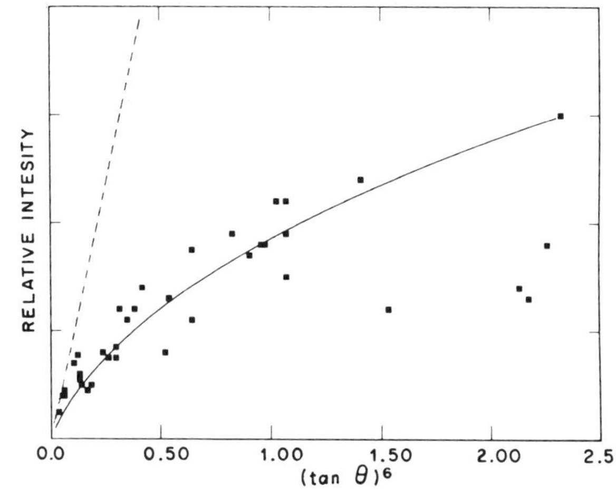

Fig. 30. Measurements of DFWM reflectivity in the adiabatic following regime in Rb. Solid line is calculated from the adiabatic following approximation; dashed line is calculated from the low-intensity theory. The horizontal axis is a measure of the pump-wave intensity. (From Ref. 113.)

Pepper *et al.*[112] have reported backward wave generation in CS_2 using light from a ruby laser. In these experiments the probe beam traveled exactly parallel to the pump beams, as shown schematically in Fig. 31, but was separated from the pump beams by its orthogonal polarization. Reflectivities of 2–3 were observed for pump intensities of the order of 2×10^7 W/cm^2 and an interaction length of 40 cm.

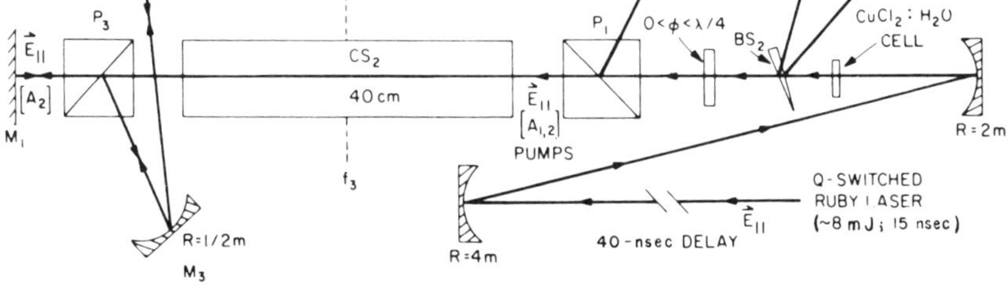

Fig. 31. Experimental apparatus used to measure DFWM reflectivity in CS_2 with ruby laser radiation. (From Ref. 112.)

Moses and Wu[115] have observed backward wave generation in the saturable absorber BDN using a Q-switched Na:YAG laser. Reflectivities as high as 6 were reported. Saturation of the reflectivity was observed at constant pump intensity as the incident probe beam intensity increased. DFWM reflectivity at 1.06 μm has also been observed in Si with reflectivities of over 1.8.[119]

Lind *et al.*[116] have reported backward wave generation of CO_2 laser radiation using a saturable absorption interaction in SF_6. Reflectivities of 0.005 were observed using a pump intensity of the order of 50 kW/cm^2, an interaction length of 2 cm, an SF_6 pressure of about 5 Torr, and a geometry in which the backward pump wave was formed by retroreflecting the forward pump wave after it passed through the nonlinear medium. Higher reflectivity, up to 0.07, was obtained by splitting the pump wave before the cell into two separate beams, each of which entered the cell from opposite sides. Reflectivity was observed to rise at SF_6 pressures below 5 Torr but to fall off for higher pressures. The falloff was attributed to increasing pump absorption at the higher pressures. Fisher and Feldman[117] later observed DFWM reflections with a CO_2 laser using the nonlinearity arising from the saturated gain of the inverted CO_2 within the cavity. DFWM reflectivity of less than 0.01 was observed, but the net reflectivity was raised to about 2.5 by multipass amplification in the inverted CO_2. DFWM of CO_2 laser radiation in Ge with reflectivities of 2% has been reported by Bergmann *et al.*[133] and with reflectivities up to 8 (800%) by Watkins *et al.*[122]

Depatie and Haueisen[95] have demonstrated DFWM of radiation from a DF laser in Ge. A reflectivity of 1.4×10^{-3} was reported at a pump intensity of 12 MW/cm^2 for single-line radiation, and a reflectivity 3×10^{-4} was observed for multiline radiation with a total pump intensity of 15 MW/cm^2.

DFWM in the ultraviolet has been observed with radiation from the fourth harmonic of a Nd:YAG laser (Feldman *et al.*[123]) and radiation from a KrF laser (Bigio *et al.*,[125] Caro and Gower[126–128]). In their studies of DFWM at 266 nm, Feldman *et al.* used dilute solutions of either Kerr active media (CS_2, DMSO, and β-carotene) or saturable absorbers (Fluorescein, Rhodamine 6G, Kodak A9740 dye, and 2-naphthol 6-sulfonate) in various ultraviolet-transmitting solvents (cyclohexane, ethyl alcohol, 1,2-dichloroethane, N-butanol, or water). The pump waves carried an energy of about 400 μJ in pulses about 15 psec long. The CS_2 had to be diluted to reduce the absorption and DFWM reflectivities up to 10^{-3} were observed with a 60% by volume solution of CS_2 in hexane. Similar reflectivities were measured for CS_2 in other solvents. Weaker DFWM reflectivity was observed for the other Kerr materials and the saturable absorbers.

Caro and Gower[126–128] have studied DFWM of radiation from a KrF laser at 249 nm in a number of solutions containing organic dyes or CS_2.

Reflectivities up to 3 (300%) were observed in a solution of Rhodamine 6G in ethanol. Similar reflectivities were observed with a pump intensity of ~ 12 MW/cm^2 in solutions of CS_2 in hexane appropriately diluted to give reasonable transmission at 248 nm.[128] In these experiments the high reflectivities were attributed to a thermal interaction and comparison with the theory outlined in Section 5.2.6b was good.

Tocho *et al.*[82,83] have reported DFWM reflectivity measurements in the saturable dye DODCI in several solvents using light from a frequency-doubled Nd:YAG laser at 532 nm. Reflectivity was measured for both picosecond- and microsecond-duration pulses. Reflectivities up to 0.5 (50%) were reported for the picosecond pulses, while reflectivities up to 4 (400%) were observed with the longer pulses.[83] The observed integration of the reflectivity signal during the microsecond pulses along with the relatively high reflectivity indicated that these observations were due to thermal gratings.

DFWM reflectivity in liquid crystals has been reported by Fekete *et al.*[124] and by Garibyan *et al.*[135] Energy reflectivities of the order of 0.65 were reported by Fekete *et al.*[124] for measurements in MBBA for pump pulses with energies of the order of 3 mJ and durations of 20 nsec. Garibyan *et al.*[135] observed phase conjugate reflectivities of the order of 0.01 with pump powers of the order of 10.

3. *DFWM Oscillation*

DFWM oscillation was first reported by Bloom *et al.*[111] using a pulsed dye laser and atomic Na. An auxiliary mirror was used in these experiments to reduce the oscillation threshold in accordance with Eq. (53). Pepper *et al.*[112] have reported oscillation using the apparatus illustrated in Fig. 32. Here all waves propagate collinearly and the generated wave was separated from the pump waves by its orthogonal polarization. In the geometry of Fig. 32 the generated wave is reflected by mirror M1 and the oscillation condition of (53) is appropriate. An oscillation threshold of 8.8 MW/cm^2 was reported. DFWM oscillation has also been reported in Watkins *et al.*[122]

5.4. Phase Conjugation

One of the most interesting aspect of the DFWM interaction is the conjugate relation between the probe beam and the backward beam amplitudes. This conjugate relation has been emphasized by a number of authors. Hellwarth[58] and Yariv and Pepper[60] have shown that within the approximations leading to (50) the conjugate relation holds for a probe beam amplitude that is an arbitrary function of the transverse coordinate r, i.e.,

$$A_4(r, 0) = +i \tan(KL)\, A_p^*(r, 0).$$

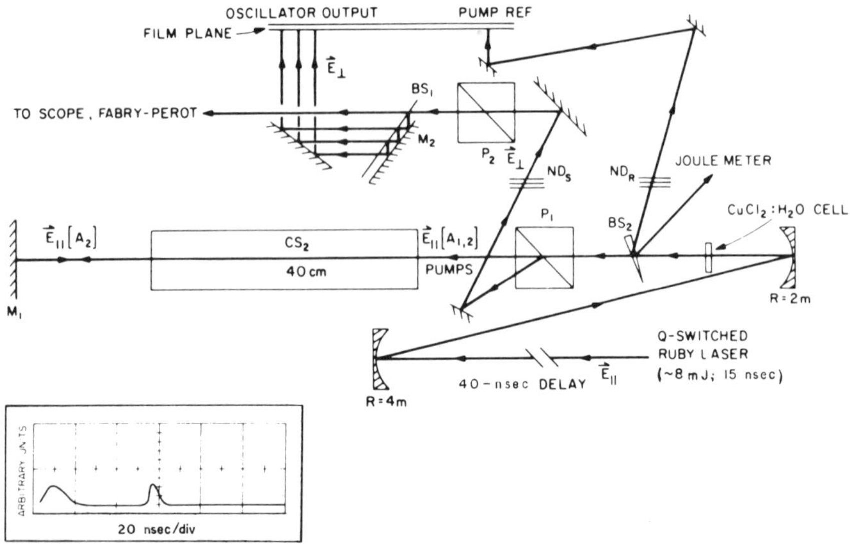

Fig. 32. Schematic diagram of apparatus used to observe DFWM oscillation in CS_2 with ruby laser radiation. Mirror M1 not only reflects the pump wave but also forms a one-sided cavity for the oscillation signal. (From Ref. 59a, © 1978 IEEE.)

Thus the transverse spatial variation of the backward wave amplitude is proportional to the complex conjugate of the probe-wave amplitude for input waves of arbitrary complexity. The generation of a conjugate wave has a number of practical applications in the areas of image reconstruction, the correction of optical distortions, and in various forms of optical signal processing.

Conjugate waves can be generated by other nonlinear interactions in addition to DFWM with counterpropagating waves. Numerous authors have reported phase conjugation using various forms of stimulated scattering, including stimulated Brillouin,[149–155] Raman,[156,157] and Rayleigh[158] scattering. Phase conjugation using a three-wave mixing interaction in a nonlinear crystal has been reported by Avizonis *et al.*[159] Heer and Griffen[142] have generated a forward conjugate wave in thin nonlinear optical media using the DFWM geometry shown in Fig. 20a. Compared to these other techniques phase conjugation by backward DFWM can generate a true conjugate wave (at least to the extent that the effects discussed in Section 4.2.4 can be neglected) as opposed to the approximate conjugate that is usually generated in stimulated scattering.[160] The backward DFWM interaction is also automatically phase matched, as opposed to the restrictive phase matching requirements that are usually encountered in three-wave interactions in anisotropic crystals[159] or in the forward DFWM interaction of Heer and Griffen.[142] This automatic phase matching regardless of the angle between the probe and pump beams greatly increases the spatial frequencies, and therefore the spatial complexity and field of view, of the wave that can be conjugated, unless there are restrictions on angular acceptance imposed by atomic motion (see Section 5.2.7) or choice of the polarization of the incident waves (see Section 5.2.1). In addition, for appropriate conditions the conjugate wave is generated at exactly the same frequency as the probe wave, a property that can be important if distortions in a medium with sharp spectral structure near the probe-wave frequency are being corrected. Finally, the backward DFWM interaction allows conjugation of each spectral component of a multiline probe beam, an important consideration for the conjugation of signals from lasers which operate on more than one spectral line.

5.4.1. Correction of Phase Distortions

One of the more important applications of a phase conjugate wave is in the correction of phase distortions on an optical beam. Such distortions can occur, for example, as a result of propagation through distorting optical components, nonideal media, nonlinear media, or multimode fibers. The use of phase conjugation to correct phase distortions introduced by propagation through arbitrary linear media has been discussed by a number of authors.[57–60,149,153,160] Yariv *et al.*[90] and Hellwarth[93] have considered the use of phase conjugation to correct distortions caused by mode disperison in optical waveguides. Pepper and Yariv[161] and Bol'shov *et al.*[66,162] have discussed the correction of phase distortions caused by a nonlinear index, such as might occur in a medium that exhibits thermal blooming. Pepper and Yariv have shown that the nonlinear phase distortions can be corrected in principle when the reflectivity of the phase conjugate mirror is unity. The efficiency of various types of correction systems has been discussed by Lukin.[163] In the following discussions of phase distortion correction we follow the treatment of Yariv.

The correction of phase distortions by phase conjugation can be seen by considering a wave with amplitude $\psi_R(r, z)$ that propagates through a distorting medium from left to right, as illustrated in Fig. 33. The medium is assumed to have a dielectric constant that can be a function of position. The equation describing the propagation of the incident wave is

$$\nabla_\perp^2 \psi_R + [\epsilon(r, z)\omega^2/c^2 - k^2]\psi_R + 2ik\frac{\partial \psi_R}{\partial z} = 0. \tag{73}$$

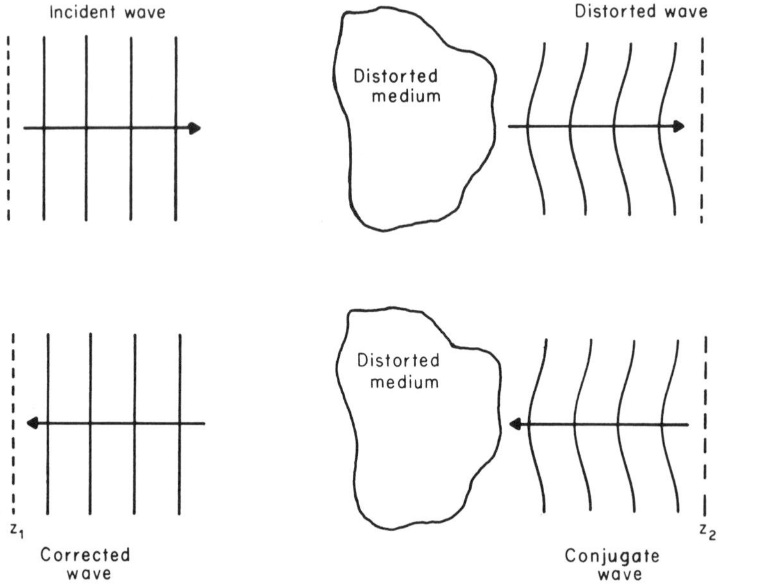

Fig. 33. Schematic diagram illustrating the correction of phase distortions with a conjugate reflection. A plane wave, propagating from left to right, is incident on a distorting medium. After passing through the distorting medium it is reflected from a conjugate mirror, and the sense of the phase distortions are reversed relative to the direction of propagation (bottom). After propagating through the same medium from right to left the wave fronts of the beam are again plane.

We next consider a wave ψ_{L}, propagating from right to left through the same medium, which is described by the equation

$$\nabla_{\perp}^2 \psi_{\mathrm{L}} + [\epsilon(r, z)\omega^2/c^2 - k^2]\psi_{\mathrm{L}} - 2ik \frac{\partial \psi_{\mathrm{L}}}{\partial z} = 0. \tag{74}$$

If $\epsilon(r, z)$ is real, then Eq. (74) has the form of the complex conjugate of Eq. (73). Thus if the wave traveling from left to right is generated with an amplitude that is the conjugate of the amplitude of the wave traveling from right to left at a plane $z = z_2$ to the right of the distorting medium as illustrated in Fig. 33, i.e., $\psi_{\mathrm{L}}(r, z) = \psi_{\mathrm{R}}^*(r, z)$, the transverse variation of the backward wave in an arbitrary plane z will be the complex conjugate of the spatial variation of the incident wave in the same plane, i.e., $\psi_{\mathrm{L}}(r, z) = \psi_{\mathrm{R}}^*(r, z)$. In particular, if the incident beam has a plane wave front or a simple spherical wave front at a plane z_1 to the left of the distorting medium, then the conjugate wave will also have an undistorted plane or spherical wave front at position z_1, regardless of the distortions imposed by the medium.

This correction of distortions is illustrated for a simple system in Fig. 34, where a plane parallel plate is introduced into part of the path of the incident wave, causing a phase delay for part of the wave front as shown. Upon reflection from the phase conjugate mirror the sense of the phase disturbance is reversed and it now amounts to a phase advance. After a second traversal of the distorting medium the distorted part of the beam has its phase delayed again and the exit wave front is plane. Also shown is the corresponding situation for reflection from a normal mirror. In this situation the sense of the phase disturbance is not changed by the reflection and the net phase distortion is doubled after the second traversal of the distorting medium.

Conceptually the phase conjugate mirror can be thought of as having a contour that matches the phase front of the probe wave. The appropriate shape of the mirror for the situation in Fig. 34 is illustrated in Fig. 35. Perhaps a more familiar example of a conjugate mirror is that of a spherical mirror placed in a beam in such a way that the curvature of the mirror matches the curvature of the beam. In this situation the incident beam is focused back to the point from which it has diverged. For more complicated phase contours the curvature of the mirror must be adjusted so that it matches the curvature of the beam everywhere.

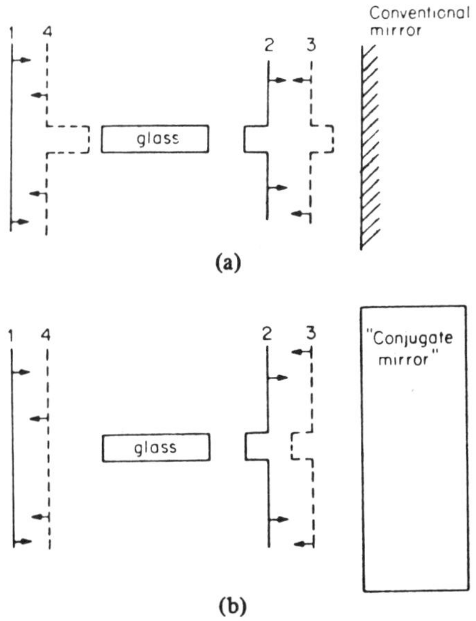

Fig. 34. Use of simple system to illustrate the correction of phase distortions with a conjugate reflector. Top: phase distortion at center of beam is doubled after reflection from a conventional mirror. Bottom: phase distortion of beam is removed after reflection from a conjugate mirror. (From Ref. 59a, © 1978 IEEE.)

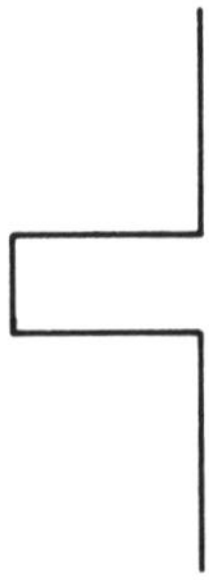

Fig. 35. Shape of conjugate mirror for system in Fig. 34.

It was pointed out in Section 5.2 that the DFWM phase conjugation process can be thought of as arising from a dynamic grating formed by the interference of the probe and pump waves. When the probe wave carries phase distortions, the contours of the grating are distorted to match. This situation is illustrated in Fig. 36 for the distortion of Fig. 34. It can be seen that the countour of the grating that reflects the second pump wave has exactly the form of the conceptual conjugate mirror of Fig. 35.

If the dielectric constant of the distorting medium is complex, then the identification of the conjugate of Eq. (73) with the equation that describes the propagation of a backward wave will not be exact. This means, for example, that the correction of distortions in media that show either loss or gain, such as laser amplifiers, will not in general be exact, even in principle. Saturation of the gain in amplifiers can also degrade the correction fidelity as

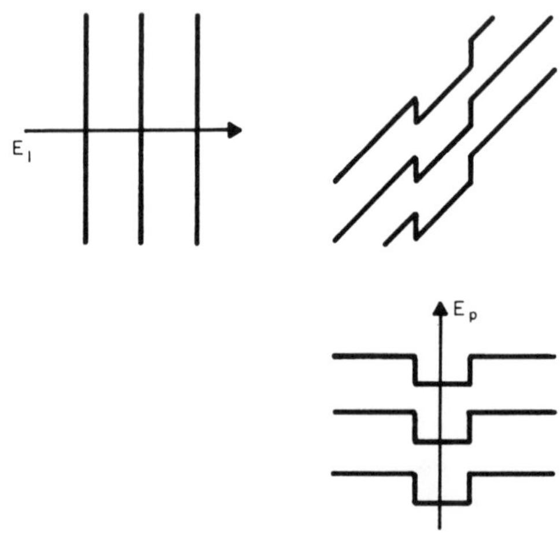

Fig. 36. Formation of contours in the dynamic grating to provide conjugate reflection for the system in Fig. 34.

discussed by O'Meara.[164] Under some circumstances, however, wavefront correction in distorting media with gain or loss can be expected. Intuitively one can expect, for example, that correction of distortions from media with linear gain or loss that is distributed uniformly across the beam profile should be possible with only a uniform change in the amplitude of the backward wave. Kochemasov and Nikolaev[165] have given an analysis that shows that wavefront correction should be possible in laser amplifiers that act as thin lenses, but that correction may not be possible if the amplifier acts as a thick lens. Experimentally, wavefront correction has been demonstrated with ruby (Nosach *et al.*[150]), Nd:glass (Basov *et al.*[166]), Nd:YAG (Zubarev *et al.*[154]), and KrF (Gower[167]) amplifiers.

5.4.2. Experimental Demonstrations—Spatial Phase Corrections

Correction of spatial phase distortions has been investigated experimentally by several authors. Experimental demonstration of phase correction is usually accomplished by introducing a phase-aberrating plate (for example, a wavy or etched piece of glass) into an otherwise plane wave in front of a phase-conjugating mirror. Such a system is shown schematically in Fig. 37. Phase correction is demonstrated by photographing the reflected beam after it has made two passes through the distorting plate. Several criteria have been used to establish the presence of phase conjugate reflections, including the resolution of an image viewed through the distorting plate or the size of the far-field spot after propagation through the distorting medium. Alternatively the insensitivity of the direction of the reflected beam to adjustment of turning mirrors in the probe beam or the reflection of the backward beam through a pinhole through which the probe beam is focused, as shown in Fig. 37c, has been used as evidence of conjugate reflection. The last configuration has the added advantage of suppressing reflections that are not conjugates of the incident probe beam, since these will not return through the pinhole, but it can also distort the relative amounts of power that are carried in the conjugate and nonconjugate components of the backward beam.

Bloom and Bjorklund[130] have demonstrated image reconstruction using DFWM by photographing the image of a resolution chart using the apparatus shown in Fig. 38. For these experiments pump powers of the order of 1.6×10^7 W/cm^2 were used and reflected powers into the image wave were of the order of 2×10^{-3} of the power in the probe wave. The nonlinear medium was CS_2. Typical results are shown in Fig. 39. The left-hand pictures were taken without the phase-aberrating plate, while the pictures at the right were taken with the aberrating plate in place. The picture at the top right was taken with the phase-conjugating system perfectly aligned and shows a corrected resolution of about 30 lp/mm. The picture at the lower

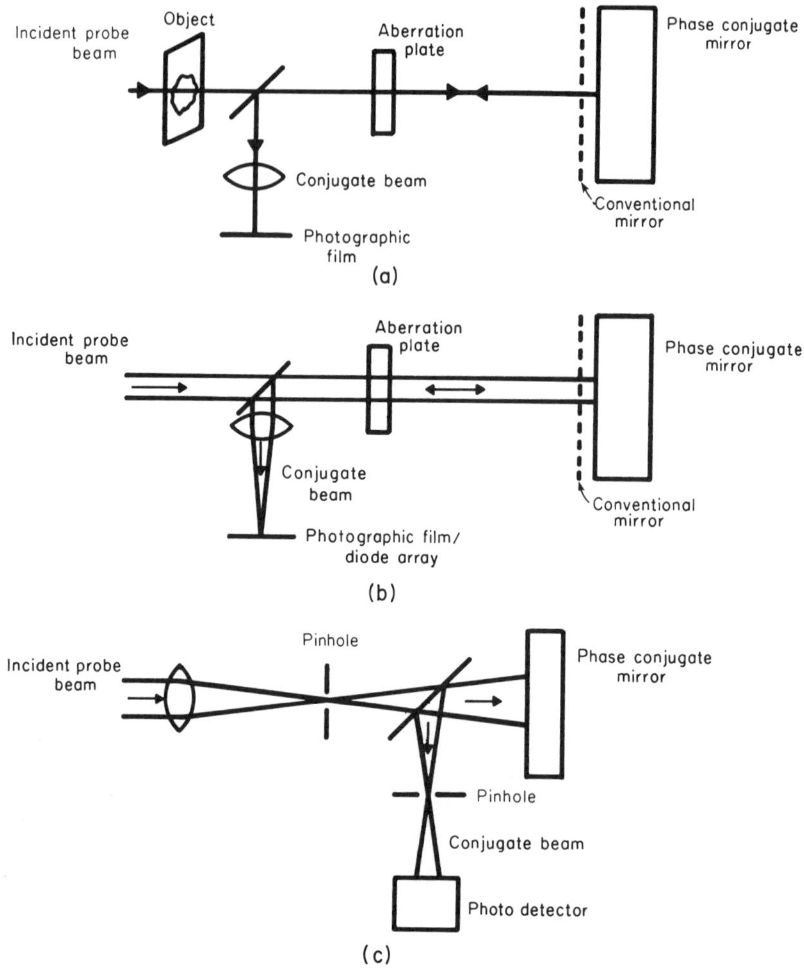

Fig. 37. Schematic diagram of methods typically used to demonstrate correction of phase distortions by conjugate reflection. (a) Photographing an image through a distorting plate. (b) Photographing the far-field distribution of the reflected beam. (c) Focusing of the conjugate reflection through a pinhole by focusing the incident probe beam through a similar pinhole. The removal of phase distortions in (a) and (b) is usually demonstrated by comparing the pattern observed with an aberrating plate in place using the conjugate mirror with that obtained by substituting a plane mirror in front of the conjugate mirror.

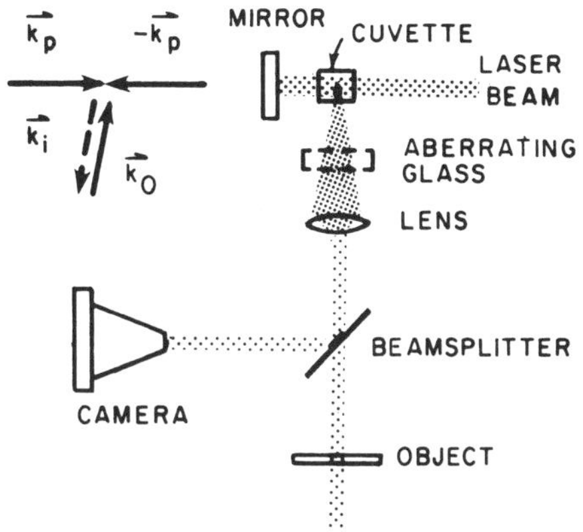

Fig. 38. Schematic diagram of apparatus used by Bloom and Bjorklund to observe phase corrections in image of resolution chart. (From Ref. 130.)

right shows the degradation of the image due to the phase aberrations when the phase conjugation system is misaligned by 0.25 mrad. Bloom *et al.*[111] later reported amplified DFWM images obtained using a pulsed dye laser at powers up to 40 kW/cm^2 tuned to the D lines in atomic Na vapor. Liao *et al.*[131] have obtained reconstructed images using a cw dye laser in atomic Na vapor. For these experiments a pump of 15 mW was used and conversion efficiency of the pump to the object wave was of the order of 2×10^{-3}.

Phase conjugation can also be used to correct distortions on a laser beam profile, restoring the beam to a diffraction-limited divergence. The use of DFWM for this application was first demonstrated by Stepanov *et al.*[110] They used DFWM in cryptocyanine to remove phase distortions from a ruby laser beam that propagated at a small angle to the pump beams. Similar results were later reported by Jensen and Hellwarth,[132] who used CS_2 for the nonlinear medium. For these experiments the probe beam carrying the phase distortions propagated at a small angle to the pump beams and the far-field profile of both the probe and conjugated beams were photographed with and without an aberrating plate. The authors reported that phase corrections for the ruby laser–CS_2 system could be obtained for phase conjugate reflectivities up to a value of about 0.5.

The correction of phase distortions on laser beams has since been reported by many authors using many different laser sources and nonlinear media. The correction of phase distortions in an Nd:YAG resonator using saturation of the gain of the YAG rod to provide the nonlinear interaction was demonstrated by Tomita.[134] Lind *et al.*[116] have shown correction of aberrations on a CO_2 laser beam using a single-photon resonant interaction

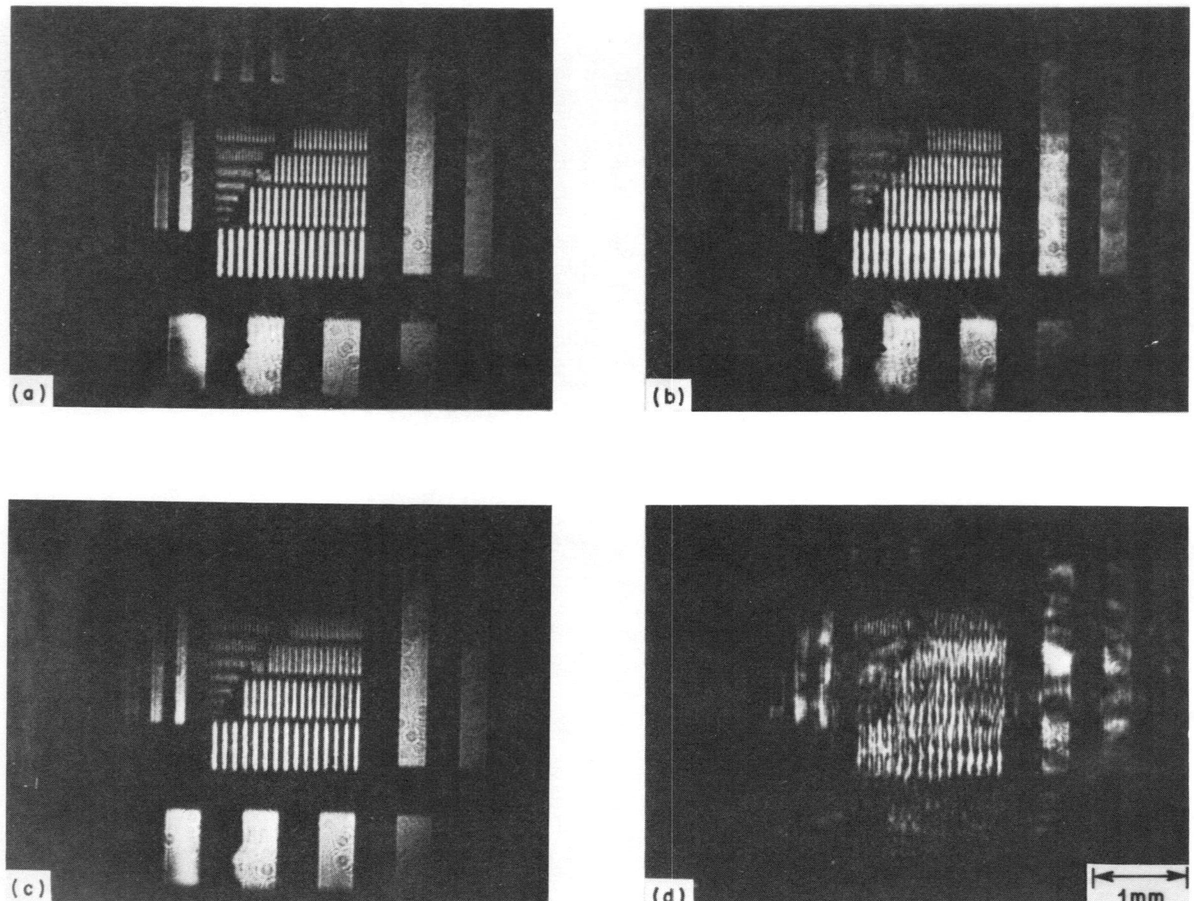

Fig. 39. Images of resolution chart obtained with apparatus in Fig. 37: (a) no aberrating glass; (b) aberrating glass in place; (c) no aberrating glass, rear mirror misaligned; (d) aberrating glass, rear mirror misaligned. (From Ref. 130.)

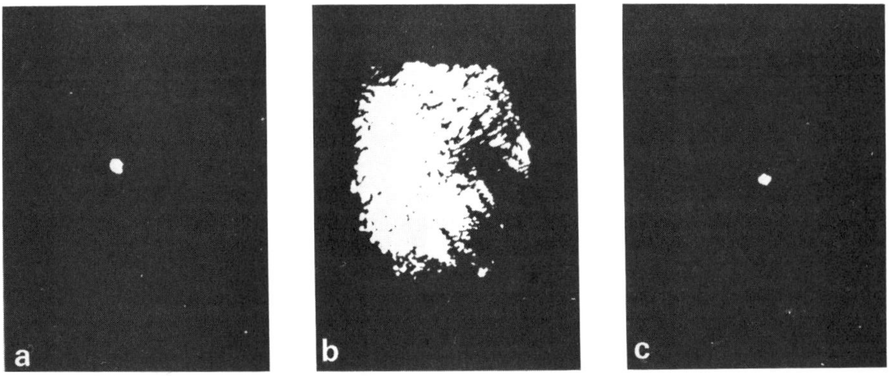

Fig. 40. Correction of phase distortions on a laser beam. (a) Far field pattern of unaberrated probe beam after double pass with normal mirror (laser input signal). (b) Far field pattern with aberrator and normal mirror (aberrated input signal). (c) Far field pattern with aberrator and conjugating mirror (corrected signal). The radiation was from a ruby laser and the nonlinear material was either Schott glass or cryptocyanine. (From Ref. 144.)

in SF_6. The correction of phase aberrations on a CO_2 laser beam using intracavity DFWM in Ge has also been reported.[122] Steel and Lam[94] reported correction in SF_6 of beam distortions of a CO_2 laser operating on two rotational lines. Multiline phase correction of a DF laser using Ge for the nonlinear medium was reported by Depatie and Haueisen.[95] Jensen and Hellwarth[91] have demonstrated the correction of phase distortions resulting from propagation in a waveguide. The correction of phase distortions for various laser systems have also been reported in Refs. 82, 115–117, and 120.

An example of the correction of phase distortions on a laser beam is shown in Fig. 40. For these experiments the probe beam entered the nonlinear medium after passing through an aberrating plate, and the far field pattern of the return beam was photographed after a second pass through the aberrating plate in a manner similar to that shown in Fig. 37b. These experiments were done using a ruby laser and either Schottky glass or cryptocyanine for the nonlinear medium. The far field distribution shown in Fig. 40a was obtained without the aberrating plate and using a reference mirror in place of the conjugating medium. The pattern in Fig. 40b was obtained with the aberrating plate in place and with the reference mirror, whereas the pattern in Fig. 40c was obtained with both the aberrating plate and the conjugating mirror. The far field pattern was reduced with the conjugating mirror from nearly 11 times diffraction limited to a distribution close to the unaberrated pattern.

5.4.3. Holographic Image Reconstruction

DFWM can be used for real time holographic image reconstruction. An analogy between the DFWM and the usual holographic process can be drawn by considering the dynamic grating discussed in Section 5.2. In this picture the probe beam and one of the pump beams interfere to form a dynamic hologram in the nonlinear medium. The second pump beam serves as a readout beam. A real time holographic image will be formed in the backward wave of any structure that is present in the incident probe beam. The image will be formed at a distance from the conjugate mirror equal to the distance of an object placed in the probe beam, as indicated in Fig. 41. This process is a special case of a more general technique for real time holography using four-wave mixing in which the readout beam may be at a different frequency than the beams that form the hologram. In the general case the image will be formed if the angle of the readout beam is adjusted to Bragg scatter off of the dynamic grating formed by the other two beams. Magnification or demagnification of the image will occur depending on the relative wavelengths of the pump and readout beams. Note that in this technique no lens is used in contrast to the arrangement shown in Fig. 38.

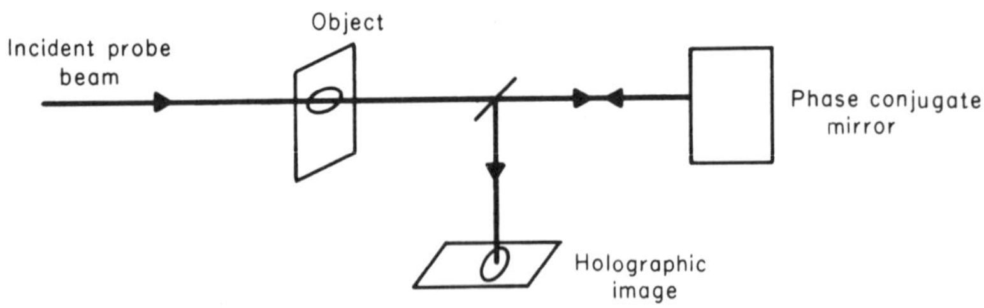

Fig. 41. Schematic illustration of formation of holographic image using DFWM.

Many of the early experiments in DFWM were directed toward observation of holographic imaging.[56,110] Yariv[59,61] has given theoretical analyses of the real time holographic properties of DFWM. More recently Levenson,[136] and Levenson *et al.*[137] have demonstrated high-resolution imaging with this technique. Using the apparatus shown schematically in Fig. 42 Levenson obtained images with a resolution of 500 lp/mm with an illumination wavelength of $\lambda = 514.5$ nm from an argon laser, corresponding to an effective aperature of 0.48. Somewhat higher resolution (800 lp/mm) was obtained by Levenson *et al.*[137] with a similar apparatus using $LiNbO_3$ as the phase conjugating material. An example of these images is shown in Fig. 43.

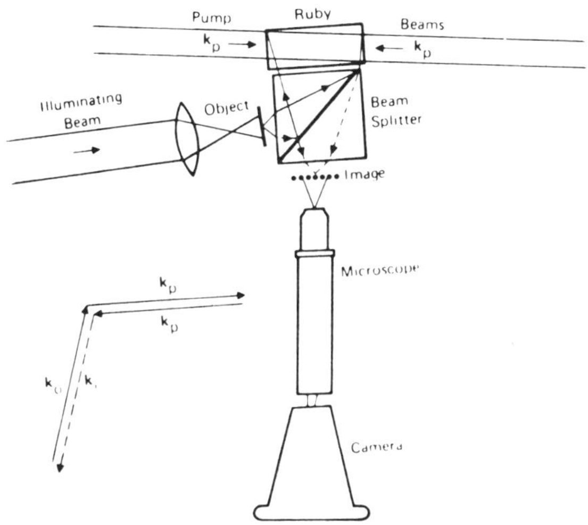

Fig. 42. Schematic diagram of apparatus used to obtain holographically reconstructed images. (From Ref. 136.)

5.4.4. Resonators with Phase Conjugate Mirrors

The use of phase conjugate mirrors (PCM) on laser resonators has been discussed by Bel'dyugin *et al.*,[96] Bel'dyugin and Zemskov,[97] AuYeung *et al.*,[98] Belanger *et al.*,[99] Lam and Brown,[100] Reznikov and Khizhnyak,[101] Feinberg and Hellwarth,[102] Hardy,[103] and Lind *et al.*[104] An example of a cavity of this type, using one normal mirror and one PCM is shown schematically in Fig. 44. The attractions of this arrangement are that the cavity can be autoaligning and the PCM can correct for distortions caused by the gain medium or by optics in the cavity. The mode structure of resonators with a single PCM have been investigated by Bel'dyugin *et al.*, Bel'dyugin and Zemskov, AuYeung *et al.*, Belanger *et al.*, Lam and Brown, Reznikov and Khizhnyak, and Lind *et al.* When the PCM is operated at exact degeneracy the oscillating frequency is determined only by the frequency of the pump waves and not by the length of the cavity. This lack of longitudinal mode structure at exact degeneracy occurs because the phase shift accumulated during a one-way trip through the cavity is subtracted on the way back as a result of the phase reversal at the PCM. The round trip phase shift is therefore independent of both the frequency of the radiation and the length of the cavity. The lack of longitudinal modes allows the cavity to be operated at a single narrow frequency without the complications of mode hopping. If the frequency of the pump waves is tuned, then the frequency of the PCM resonator will also tune continuously without mode structure.

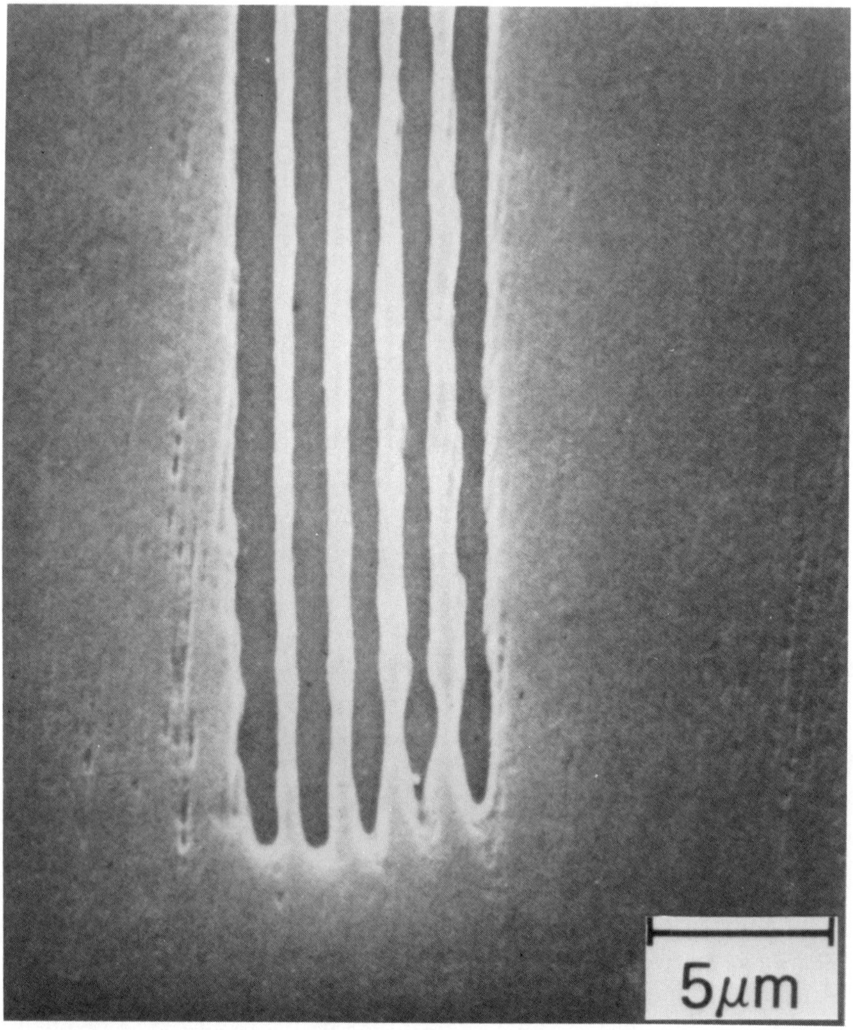

Fig. 43. Holographically reconstructed image of lines from a test chart using an apparatus similar to the one in Fig. 41. The lines were 0.75 μm wide, spaced by 0.5 μm, corresponding to a spatial frequency of 800 line pairs per millimeter. The illumination wavelength was 413 nm and the nonlinear material in the conjugator was $LiNbO_3$. (From Ref. 137.)

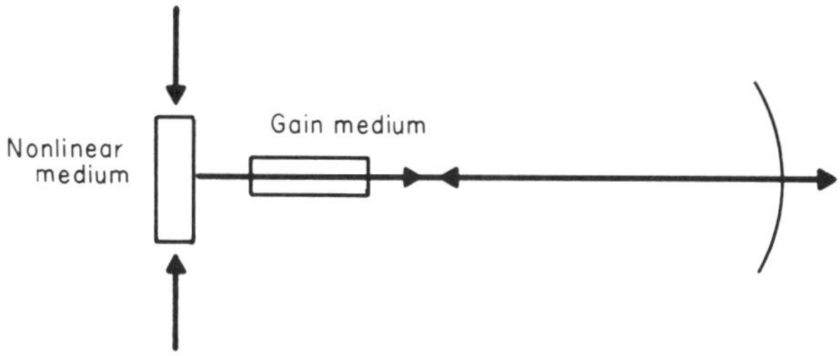

Fig. 44. Schematic diagram of a laser cavity with one-phase conjugate mirror.

Under conditions that are usually encountered the narrow bandwidth of the PCM allows only degenerate operation. If the bandwidth is broadened, however, e.g., by mistuning from an atomic resonance or by saturation of the DFWM reflectivity, nondegenerate operation of the PCM can be obtained. Under these conditions the cavity will contain longitudinal mode structure because of the coupling of radiation at $\omega = \omega_{\text{pump}} \pm \delta$, which occurs in nondegenerate frequency mixing. The mode structure is centered on the pump frequency and each mode consists of radiation at the pair of frequencies $\omega = \omega_{\text{p}} \pm \pi c/2L$. The spacing of the frequencies of the components of the modes on either side of the pump frequency is one-half of the spacing of the modes in a cavity with normal reflectors.

The transverse mode structure of a cavity with a PCM has also been investigated. It has been shown that any intensity distribution with a curvature that matches the curvature of the normal mirror is a mode of a cavity with one PCM and an infinite transverse dimension. In addition, all the transverse modes are degenerate in frequency. When the cavity has a finite transverse extent, either because of a finite mirror diameter or an intracavity aperature, the lowest-order mode becomes a well-defined Gaussian with its diameter determined by the cavity geometry.

The effect of phase aberrations on the quality of the output beam has been analyzed by Hardy.[103] His calculations show that a PCM of a given diameter can prevent degradation of the far-field pattern of the resonator output provided that the phase aberration is not too great. Some loss of correctability is predicted for large phase aberrations and is attributed to diffraction of the aberrated beam around the PCM. Even with this loss of correctability, however, substantial improvements in the far-field intensity were predicted for most conditions over that which could be achieved with normal mirrors.

Resonators using PCMs have been investigated experimentally by Feinberg and Hellwarth[102] and by Lind *et al.*[104]. Feinberg and Hellwarth used an arrangement similar to the one shown in Fig. 23 in which the gain for the cavity was provided by the PCM. They showed that aberrations due to intracavity optics could be corrected on the output beam. Lind *et al.*[104] used a separate gain medium in a cavity similar to the one shown in Fig. 44. They investigated the longitudinal mode structure under degenerate and nearly degenerate conditions and found results that were in excellent agreement with theory. They also demonstrated the correction of aberrations on the output beam.

5.4.5. *Quality of Spatial Phase Correction*

An important question in the subject of phase conjugation concerns the degree to which the backward wave is a true conjugate of the incident probe wave. A related question concerns the quality of correction of phase aberrations that can be achieved or the resolution of the holographic images that can be produced. Within the simple approximations that have been made in most of this chapter, i.e., plane pump waves, no pump depletion, and no nonlinear phase shifts, there is no provision for a component of the backward wave to be generated that is not conjugate to the incident probe wave. Under these conditions the quality of the phase correction or the resolution of the holographic images should be determined entirely by the effective aperture of the optical system.

This limitation can take any of several forms. For example, an aperture placed between the object and the holographic image plane or between the phase aberrator and the nonlinear medium can limit the resolution of the system by preventing some of the information-bearing or aberrated rays from reaching the DFWM medium. The system aperture can also be limited by the diameter of the DFWM interaction region. Levenson[136] and Hellwarth[58] have discussed limitations due to effective apertures in the optical system. The effects of a finite diameter of the DFWM interaction region have been discussed in Ref. 144. The limitation on the resolution of the corrected images reported by Bloom *et al.*[111] was attributed to the finite diameter of the nonlinear interaction region. Similar limitations are apparent in the calculations of phase correction in a phase conjugate mirror resonator by Hardy.[103] These effects can be ameliorated by a certain extent by focusing or imaging the aberrated beam into the conjugate mirror, as discussed in Ref. 144.

Another type of limitation can exist for double-pass phase corrections of the type illustrated in Fig. 32 if the conjugate wave does not retrace the path of the probe wave exactly or if the aberrating medium changes during the time between the forward pass of the probe wave and the return pass of the conjugate wave. The effect of misalignment of the conjugate wave was evident in the incompletely corrected images of Bloom and Bjorklund.[130] The effects of a changing medium can limit the quality of correction in aberrating media such as the atmosphere.

The quality of phase conjugation can also be affected by all of the processes that were left out of the analysis of this chapter. Nonlinear phase shifts can produce a nonconjugate component in the backward wave in collinear interactions, as discussed in Section 5.2.4. Nonlinear phase shifts can also lead to self-focusing or self-defocusing or ellipse rotation. Self-focusing or self-defocusing, along with pump depletion, was suggested as a cause of the saturation of the DFWM reflectivity observed by Bloom *et al.*,[111] but the effect on the phase conjugation was not examined. A weak nonconjugate component was observed in the backward wave by Pepper *et al.*[112] and was attributed to ellipse rotation in the nonlinear medium. Residual distortions of the order of $\lambda/2$ in the corrected beam have been reported by Hopf *et al.*[168] in experiments in DFWM in Si at 1.06 μm. These distortions were attributed to self-focusing. Degradation of the phase correction of a distorted laser beam has been observed[144] at sufficiently high intensities in saturable absorbers and has been attributed to self-focusing or thermal blooming.

Distortion of the conjugate wave can also arise from distortion or structure of the pump waves. Variation of the pump-wave amplitude across the diameter can create structure in the backward wave that was not present in the incident probe wave, for example. This effect was observed in the imaging experiments of Bloom and Bjorklund.[130] Pump-wave structure can also be used to impart desirable spatial properties or information to the conjugate wave. These effects have been discussed in Ref. 144. DFWM with randomly structured pump waves has been discussed by Blaschuk *et al.*[169] Yariv and Koch[170] have described a technique for transferring information to the backward generated wave from a structured pump wave.

Saturation of the DFWM interaction, either as a result of pump depletion or saturation of the nonlinear susceptibility, can also affect the quality of the phase conjugation. The saturation of the reflectivity observed in Bloom *et al.*[111] has already been mentioned. Grischowsky *et al.*[113] has also observed saturation of the DFWM reflectivity with increasing pump intensity and attributed this effect to saturation of the nonlinear susceptibility. Tomita[133] observed saturation of the DFWM reflectivity with increasing probe-wave intensity. The effect of the saturation of the DFWM reflectivity on the phase conjugation was not examined in any of these experiments. The effects of pump depletion on the DFWM reflectivity have been investigated theoretically by Marburger and Lam,[68] Hsu,[69] and Lazaruk and Rubanov,[70] but again, the effects on phase conjugation were not treated.

5.4.6. *Temporal Phase Correction—Chirp Reversal*

In addition to spatial phase variations, a laser beam can carry temporal phase disturbances. Such disturbances can arise, for example, from propagation of a pulse through a medium with a nonlinear refractive index and can cause spectral broadening, as was discussed in Section 4.3. Marburger[106] has shown theoretically that such a frequency sweep can be effectively reversed using a DFWM interaction in an optically thin medium. For thicker media the reflected-wave amplitude is proportional to the integral of the probe beam amplitude and the reversal of the frequency sweep will not be accurate.

REFERENCES

1. R. W. Hellwarth, *Prog. Quantum Electron.* **5**, 1 (1979).
2. A. Owyoung, *IEEE J. Quantum Electron.* **QE-9**, 1064 (1973).
3a. R. W. Hellwarth, A. Owyoung, and N. George, *Phys. Rev. A* **4**, 2342 (1971).
3b. M. J. Moran, C. Y. She, and R. L. Carman, *IEEE J. Quantum Electron.* **QE-11**, 259 (1975).
3c. G. K. L. Wong and Y. R. Shen, *Phys. Rev. A* **10**, 1277 (1974).
4. S. A. Akhmanov, A. P. Sukhorukov, and R. V. Khokhlov, *Usp. Fiz. Nauk* **93**, 19 (1967) [*Sov. Phys.—Usp.* (*Engl. Transl.*) **10**, 109 (1968)].
5. S. A. Akhmanov, R. V. Khokhlov, and A. P. Sukhorukov, *in* "Laser Handbook" (F. T. Arecchi and E. O. Schulz-duBois, eds.), Vol. 2, Chap. E3, p. 1151. North-Holland Publ., Amsterdam, 1972.
6. Y. R. Shen, *Prog. Quantum Electron.* **4**, 1 (1977).
7. J. H. Marburger, *Prog. Quantum Electron.* **4**, 35 (1977).
8. J. P. McTague, C. H. Lin, T. K. Gustafson, and R. Y. Chiao, *Phys. Lett. A* **32A**, 82 (1969).
9. M. E. Mack, R. L. Carman, J. Reintjes, and N. Bloembergen, *Appl. Phys. Lett.* **16**, 209 (1970).
10. D. Grischowsky, *Phys. Rev. Lett.* **24**, 866 (1970).
11. J. E. Bjorkholm and A. Ashkin, *Phys. Rev. Lett.* **32**, 129 (1974).
12. F. S. Felber, *Appl. Phys. Lett.* **36**, 723 (1980); P. D. McWane, *Nature* (*London*) **211**, 1081 (1966); A. E. Kaplan, *Izv. Vyssh. Uchebn. Zaved., Radiofiz.* **12**, 869 (1969) [*Radiophys. Quantum Electron.* (*Engl. Transl.*) **12**, 692 (1969)].
13. R. G. Brewer and C. H. Lee, *Phys. Rev. Lett.* **21**, 267 (1968).
14. P. L. Kelley, *Phys. Rev. Lett.* **15**, 1005 (1965).
15. E. L. Dawes and J. H. Marburger, *Phys. Rev.* **179**, 862 (1969).
16. V. N. Goldberg, V. I. Talanov, and R. E. Erm, *Izv. Vyssh. Uchebn. Zaved. Radiofiz.* **10**, 674 (1967) [*Radiophys. Quantum Electron.* (*Engl. Transl.*) **10**, 368 (1967)].
17. W. G. Wagner, H. A. Haus, and J. H. Marburger, *Phys. Rev.* **175**, 256 (1968).
18. J. H. Marburger and E. L. Dawes, *Phys. Rev. Lett.* **21**, 556 (1968).
19. O. Rahn and M. Maier, *Phys. Rev. Lett.* **29**, 558 (1972).
20. E. Yablonovitch and N. Bloembergen, *Phys. Rev. Lett.* **29**, 907 (1972); N. Bloembergen, *Opt. Commun.* **8**, 288 (1973).
21. M. M. T. Loy and Y. R. Shen, *IEEE J. Quantum Electron.* **QE-9**, 409 (1973).
22. F. Shimizu, *IBM J. Res. Dev.* **17**, 286 (1973).
23. A. L. Dyshko, V. N. Lugovoi, and A. M. Prokhorov, *Zh. Eksp. Teor. Fiz., Pis'ma Red.* **6**, 655 (1967) [*JETP Lett.* (*Engl. Transl.*) **6**, 146 (1967)].
24. V. N. Lugovoi and A. M. Prokhorov, *Zh. Eksp. Teor. Fiz., Pis'ma Red.* **7**, 153 (1968) [*JETP Lett.* (*Engl. Transl.*) **7**, 117 (1968)].
25. M. M. T. Loy and Y. R. Shen, *Phys. Rev. Lett.* **22**, 994 (1969); M. M. T. Loy and Y. R. Shen, *Phys. Rev. Lett.* **25**, 1333 (1970).

26. V. V. Korobkin, A. M. Prokhorov, R. V. Serov, and M. Y. Schelev, *Pis'ma Zh. Eksp. Teor. Fiz.* **9**, 108 (1969) [*JETP Lett.* (*Engl. Transl.*) **9**, 61 (1969)].
27. S. A. Akhmanov, A. P. Sukhorukov, and R. V. Khokhlov, *Zh. Eksp. Teor. Fiz.* **51**, 296 (1966) [*Sov. Phys.—JETP* (*Engl. Transl.*) **24**, 198 (1967)].
28. J. Reintjes, R. L. Carman, and F. Shimizu, *Phys. Rev. A* **8**, 1486 (1973).
29. V. I. Bespalov and V. I. Talanov, *Zh. Eksp. Teor. Fiz., Pis'ma Red.* **3**, 471 (1966) [*JETP Lett.* (*Engl. Transl.*) **3**, 307 (1966)].
30. B. R. Suydam, *in* "Laser Induced Damage in Optical Materials" (A. J. Glass and A. Guenther, eds.), NBS Spec. Publ., No. 387, p. 42. U.S. Gov. Print. Off., Washington, D.C., 1973; B. R. Suydam, *IEEE J. Quantum Electron.* **QE-10**, 837 (1974).
31. A. J. Glass, *in* "Laser Induced Damage in Optical Materials" (A. J. Glass and A. Guenther, eds.), NBS Spec. Publ., No. 387, p. 36. U.S. Gov. Print. Off., Washington, D.C., 1973.
32. A. J. Campillo, S. L. Shapiro, and B. R. Suydam, *Appl. Phys. Lett.* **23**, 628 (1973).
32b. A. J. Campillo, S. L. Shapiro, and B. R. Suydam, *Appl. Phys. Lett.* **24**, 178 (1974).
33. E. S. Bliss, D. R. Speck, J. F. Holtzrichter, J. H. Erkkila, and A. J. Glass, *Appl. Phys. Lett.* **25**, 448 (1974).
34. D. Grischowsky and J. A. Armstrong, *Phys. Rev. A* **6**, 1566 (1972).
35a. R. H. Lehmberg, J. Reintjes, and R. C. Eckardt, *Appl. Phys. Lett.* **25**, 374 (1974).
35b. R. H. Lehmberg, J. Reintjes, and R. C. Eckardt, *Phys. Rev. A* **13**, 1095 (1976).
36. R. J. Hull, P. L. Kelley, and R. L. Carman, *Appl. Phys. Lett.* **17**, 539 (1970).
37. J. L. Walsh and P. B. Ulrich, *in* "Laser Beam Propagation in the Atmopshere" (J. W. Strohbehn, ed.), p. 223. Springer-Verlag, Berlin and New York, 1978.
38. D. C. Smith, *Proc. IEEE* **65**, 1679 (1976).
39. T. K. Gustafson, J.-P. Taran, H. A. Haus, J. R. Lifsitz, and P. L. Kelley, *Phys. Rev.* **177**, 306 (1969).
40. R. Polloni, C. A. Sacchi, and O. Svelto, *Phys. Rev. Lett.* **23**, 690 (1969).
41. N. N. Il'ichev, V. V. Korobkin, V. A. Korshunov, A. A. Malyukin, T. K. Okroashvili, and P. P. Pushinin, *Pis'ma Zh. Eksp. Teor. Fiz.* **15**, 191 (1972) [*JETP Lett.* (*Engl. Transl.*) **15**, 133 (1972)].
42. W. Werncke, A. Lau, M. Pfeiffer, K. Lenz, H.-J. Weigmann, and C. D. Thuy, *Opt. Commun.* **4**, 413 (1972).
43. D. Sharma, R. W. Yip, D. F. Williams, and S. E. Sugamori, *Chem. Phys. Lett.* **41**, 460 (1976).
44. W. L. Smith, P. Liu, and N. Bloembergen, *Phys. Rev. A* **15**, 2396 (1977).
45. D. K. Sharma and R. W. Yip, *Opt. Commun.* **30**, 113 (1979).
46. A. Penzkofer, A. Laubereau, and W. Kaiser, *Phys. Rev. Lett.* **31**, 863 (1973).
47. A. Penzkofer, *Opt. Commun.* **11**, 265 (1974).
48. A. Penzkofer, A. Seilmeir, and W. Kaiser, *Opt. Commun.* **14**, 363 (1975).
49. G. E. Busch, R. P. Jones, and P. M. Rentzepis, *Chem. Phys. Lett.* **18**, 178 (1973).
50. C. A. G. O. Varma and P. M. Rentzepis, *Chem. Phys. Lett.* **19**, 162 (1973).
51. A. Penzkofer, W. Falkenstein, and W. Kaiser, *Appl. Phys. Lett.* **28**, 319 (1976).
52. R. R. Alfano and S. L. Shapiro, *Phys. Rev. Lett.* **24**, 1217 (1970).
53. R. R. Alfano and S. L. Shapiro, *Phys. Rev. Lett.* **24**, 592 (1970).
54. E. P. Ippen, C. V. Shank, J. M. Wiesenfeld, and A. Migus, *Philos. Trans. R. Soc. London, Ser. A* **298**, 225 (1980).
55. R. H. Lehmberg, J. Reintjes, and R. C. Eckardt, *Appl. Phys. Lett.* **30**, 487 (1977).
56. H. J. Gerritsen, *Appl. Phys. Lett.* **10**, 239 (1967).
57. A. Yariv, *J. Opt. Soc. Am.* **66**, 301 (1976).
58. R. W. Hellwarth, *J. Opt. Soc. Am.* **67**, 1 (1977).
59a. A. Yariv, *IEEE J. Quantum Electron.* **QE-14**, 650 (1978).

59b. A. Yariv, *IEEE J. Quantum Electron.* **QE-15**, 256 (1979).
60. A. Yariv and D. M. Pepper, *Opt. Lett.* **1**, 16 (1977).
61. A. Yariv, *Opt. Commun.* **25**, 23 (1978).
62. D. M. Pepper, J. AuYeung, D. Fekete, and A. Yariv, *Opt. Lett.* **3**, 7 (1978).
63. D. G. Steel, R. C. Lind, J. F. Lam, and C. R. Giuliano, *Appl. Phys. Lett.* **35**, 376 (1979).
64. B. Ya. Zel'dovich and V. V. Shkunov, *Sov. J. Quantum Electron.* **9**, 379 (1979).
65. G. Martin, L. K. Lam, and R. W. Hellwarth, *Opt. Lett.* **5**, 185 (1980).
66. L. A. Bol'shov, F. V. Bunkin, and D. V. Vlasov, *Sov. J. Quantum Electron.* **10**, 1197 (1980).
67. J. H. Marburger and J. F. Lam, *Appl. Phys. Lett.* **35**, 249 (1979).
68. J. H. Marburger and J. F. Lam, *Appl. Phys. Lett.* **34**, 389 (1979).
69. H. Hsu, *Appl. Phys. Lett.* **34**, 855 (1979).
70. A. M. Lazaruk and A. S. Rubanov, *Sov. J. Quantum Electron.* **10**, 1148 (1980).
71. R. L. Abrams and R. C. Lind, *Opt. Lett.* **2**, 94 (1978); erratum, *Opt. Commun.* **3**, 205 (1978).
72. P. A. Apanasevich, A. A. Afanas'ev, and S. P. Zhvaryi, *Sov. J. Quantum Electron.* **10**, 906 (1980).
73. D. E. Watkins, J. F. Figueira, and S. J. Thomas, *Opt. Lett.* **5**, 169 (1980).
74. W. P. Brown, *J. Opt. Soc. Am*, **73**, 629 (1983).
75. S. M. Wandzura, *Opt. Lett.* **4**, 208 (1979).
76. J. Nilsen and A. Yariv, *J. Opt. Soc. Am.* **71**, 180 (1981).
77. L. M. Humphrey, J. P. Gordon, and P. F. Liao, *Opt. Lett.* **5**, 56 (1980).
78. P. Y. Key, R. G. Harrison, V. I. Little, and J. Katzenstein, *IEEE J. Quantum Electron.* **QE-6**, 641 (1970).
79. H. Eichler, G. Enterlein, F. Glozbach, J. Munschau, and H. Stahl, *Appl. Opt.* **11**, 372 (1972).
80. F. J. Heilweil, R. M. Hochstrasser, and H. Souma, *Opt. Commun.* **35**, 227 (1980).
81. J. R. Andrews and R. M. Hochstrasser, *Chem. Phys. Lett.* **76**, 213 (1980).
82. J. O. Tocho, W. Sibbett, and D. J. Bradley, *in* "Picosecond Phenomena II" (R. M. Hochstrasser, W. Kaiser, and C. V. Shank, eds.), p. 372. Springer-Verlag, Berlin and New York, 1980.
83. J. O. Tocho, W. Sibbett, and D. J. Bradley, *Opt. Commun.* **37**, 67 (1981).
84. D. M. Pepper and R. L. Abrams, *Opt. Lett.* **3**, 212 (1978).
85. T. Fu and M. Sargent, III, *Opt. Lett.* **4**, 366 (1979).
86. N. F. Pilipetsky and V. V. Shkunov, *Opt. Commun.* **37**, 217 (1981).
87. T. Fu and M. Sargent, III, *Opt. Lett.* **5**, 433 (1980).
88. W. W. Rigrod, R. A. Fisher, and B. J. Feldman, *Opt. Lett.* **5**, 105 (1980).
89. D. G. Steel and J. F. Lam, *Opt. Lett.* **4**, 363 (1979).
90. A. Yariv, J. AuYeung, D. Fekete, and D. M. Pepper, *Appl. Phys. Lett.* **32**, 635 (1978).
91. S. M. Jensen and R. W. Hellwarth, *Appl. Phys. Lett.* **32**, 166 (1978).
92. J. AuYeung, D. Fekete, D. M. Pepper, A. Yariv, and R. K. Jain, *Opt. Lett.* **4**, 42 (1979).
93. R. W. Hellwarth, *IEEE J. Quantum Electron.* **QE-15**, 101 (1979).
94. D. G. Steel and J. F. Lam, *Opt. Lett.* **5**, 297 (1980).
95. D. Depatie and D. Haueisen, *Opt. Lett.* **5**, 252 (1980).
96. I. M. Bel'dyugin, M. G. Galushkin, and E. M. Zemskov, *Kvantovaya Elektron.* (*Moscow*) **6**, 38 (1979) [*Sov. J. Quantum Electron.* (*Engl. Transl.*) **9**, 20 (1979)].
97. I. M. Bel'dyugin and E. M. Zemskov, *Kvantovaya Elektron.* (*Moscow*) **6**, 2036 (1979) [*Sov. J. Quantum Electron.* (*Engl. Transl.*) **9**, 1198 (1979)].
98. J. AuYeung, D. Fekete, D. M. Pepper, and A. Yariv, *IEEE J. Quantum Electron.* **QE-15**, 1180 (1979).
99. P. A. Belanger, A. Hardy, and A. E. Siegman, *Appl. Opt.* **19**, 602 (1980).

100. J. F. Lam and W. P. Brown, *Opt. Lett.* **5**, 61 (1980).
101. M. G. Reznikov and A. I. Khizhnyak, *Kvantovaya Elektron.* (*Moscow*) **7**, 1105 (1980) [*Sov. J. Quantum Electron.* **10**, 633 (1980)].
102. J. Feinberg and R. W. Hellwarth, *Opt. Lett.* **5**, 519 (1980).
103. A. Hardy, *IEEE J. Quantum Electron.* **QE-17**, 158 (1981).
104. R. C. Lind, D. G. Steel, and J. F. Lam, *CLEO '81, Washington, D.C.* Pap. FP4 (1981); R. C. Lind and D. G. Steel, *Opt. Lett.* **6**, 554 (1981).
105. F. A. Hopf, *J. Opt. Soc. Am.* **70**, 1320 (1980).
106. J. H. Marburger, *Appl. Phys. Lett.* **32**, 372 (1978).
107. A. Yariv, D. Fekete, and D. M. Pepper, *Opt. Lett.* **4**, 52 (1979).
108. H. P. Yuen and J. H. Shapiro, *Opt. Lett.* **4**, 334 (1979).
109. R. L. Carman, R. Y. Chiao, and P. L. Kelley, *Phys. Rev. Lett.* **17**, 1281 (1966).
110. B. I. Stepanov, E. V. Ivakin, and A. S. Rubanov, *Dokl. Akad. Nauk SSSR* **196**, 567 (1971) [*Sov. Phys.—Dokl.* (*Engl. Transl.*) **16**, 46 (1971)].
111. D. M. Bloom, P. F. Liao and N. P. Economou, *Opt. Lett.* **2**, 58 (1978).
112. D. M. Pepper, D. Fekete, and A. Yariv, *Appl. Phys. Lett.* **33**, 41 (1978).
113. D. Grischowsky, N. S. Shiren, and R. J. Bennett, *Appl. Phys. Lett.* **33**, 805 (1978).
114. P. F. Liao and D. M. Bloom, *Opt. Lett.* **3**, 4 (1978).
115. E. I. Moses and F. Y. Wu, *Opt. Lett.* **5**, 64 (1980).
116. R. C. Lind, D. G. Steel, M. B. Klein, R. L. Abrams, C. R. Giuliano, and R. K. Jain, *Appl. Phys. Lett.* **34**, 457 (1979).
117. R. A. Fisher and B. J. Feldman, *Opt. Lett.* **4**, 140 (1979).
118. R. K. Jain and M. B. Klein, *Appl. Phys. Lett.* **35**, 454 (1979).
119. R. K. Jain, M. B. Klein, and R. C. Lind, *Opt. Lett.* **4**, 328 (1979).
120. H. Mandelberg, *Opt. Lett.* **5**, 258 (1980).
121. N. Tan-no, T. Hoshimiya, and H. Inaba, *IEEE J. Quantum Electron.* **QE-16**, 147 (1980).
122. D. E. Watkins, C. R. Phipps, Jr., and S. J. Thomas, *Opt. Lett.* **6**, 76 (1981).
123. B. J. Feldman, R. A. Fisher, and S. L. Shapiro, *Opt. Lett.* **6**, 84 (1981).
124. D. Fekete, J. AuYeung, and A. Yariv, *Opt. Lett.* **5**, 51 (1980).
125. I. J. Bigio, B. J. Feldman, R. A. Fisher, and M. Slatkine, *CLEO '81, Washington, D.C.* postdeadline Pap. THS3 (1981).
126. R. G. Caro and M. C. Gower, *Opt. Lett.* **6**, 557 (1981).
127. R. G. Caro and M. C. Gower, *Appl. Phys. Lett.* **39**, 855 (1981).
128. R. G. Caro and M. C. Gower, *IEEE J. Quant. Electron.* **QE-18**, 1375 (1982).
129. D. S. Hamilton, D. Heiman, J. Feinberg, and R. W. Hellwarth, *Opt. Lett.* **4**, 124 (1979).
130. D. M. Bloom and G. C. Bjorklund, *Appl. Phys. Lett.* **31**, 592 (1977).
131. P. F. Liao, D. M. Bloom, and N. P. Economou, *Appl. Phys. Lett.* **32**, 813 (1978).
132. S. M. Jensen and R. W. Hellwarth, *Appl. Phys. Lett.* **32**, 166 (1978).
133. E. E. Bergmann, I. J. Bigio, B. J. Feldman, and R. A. Fisher, *Opt. Lett.* **3**, 82 (1978).
134. A. Tomita, *Appl. Phys. Lett.* **34**, 463 (1979).
135. O. V. Garibyan, I. N. Kompanets, A. V. Perfyonov, N. F. Pilipetsky, V. V. Shkunov, A. N. Sudarkin, A. V. Sukhov, N. V. Tabiryan, A. A. Vasiliev, and B. T. Zel'dovich, *Opt. Commun.* **38**, 67 (1981).
136. M. D. Levenson, *Opt. Lett.* **5**, 182 (1980).
137. M. D. Levenson, K. M. Johnson, V. C. Hanchett, and K. Chiang, *J. Opt. Soc. Amer.* **71**, 737 (1981).
138. D. M. Pepper, *Opt. Eng.* **21**, 156 (1982).
139. R. C. Lind, D. G. Steel, and G. J. Dunning, *Opt. Eng.* **21**, 190 (1982).
140. J. O. White and A. Yariv, *Opt. Eng.* **21**, 224 (1982).
141. C. R. Giuliano, *Phys. Today* Apr., p. 27 (1981).

142. C. V. Heer and N. C. Griffen, *Opt. Lett.* **4**, 239 (1979).
143. G. J. Dunning and D. G. Steel, *IEEE J. Quantum Electron.* **QE-18**, 3 (1982).
144. "Correction of Phase Distortions by Nonlinear Optical Techniques," Tech. Rep., Hughes Research Laboratories, Hughes Aircraft Co., Malibu California, 1981.
145. G. Martin and R. W. Hellwarth, *Appl. Phys. Lett.* **34**, 371 (1979).
146. G. P. Agrawal, A. Van Lerberghe, P. Aubourg, and J. L. Boulnois, *Opt. Lett.* **7**, 540 (1982).
147. J. Nilsen and A. Yariv, *IEEE J. Quantum Electron.* **QE-18**, 1947 (1982).
148. W. M. Grossman and D. M. Shemwell, *J. Appl. Phys.* **51**, 914 (1980).
149. B. Y. Zel'dovich, V. I. Popovichev, V. V. Ragul'skii, and F. S. Faizullov, *Pis'ma Zh. Eksp. Teor. Fiz.* **15**, 160 (1972) [*JETP Lett.* (*Engl. Transl.*) **15**, 109 (1972)].
150. O. Y. Nosach, V. I. Popovichev, V. V. Ragul'skii, and F. S. Faizullov, *Pis'ma Zh. Eksp. Teor. Fiz.* **16**, 617 (1972) [*JETP Lett.* (*Engl. Transl.*) **16**, 435 (1972)].
151. G. G. Kochemasov and V. D. Nikolaev, *Kvantovaya Elektron.* (*Moscow*) **4**, 115 (1977) [*Sov. J. Quantum Electron.* (*Engl. Transl.*) **7**, 60 (1977)].
152. B. Ya. Zel'dovich and V. V. Shkunov, *Kvantovaya Electron.* (*Moscow*) **5**, 36 (1978) [*Sov. J. Quantum Electron.* (*Engl. Transl.*) **8**, 15 (1978)].
153. V. Wang and C. R. Giuliano, *Opt. Lett.* **2**, 4 (1978).
154. I. G. Zubarev, A. B. Mironov, and S. I. Mikhailov, *Kvantovaya Elektron.* (*Moscow*) **7**, 2035 (1980) [*Sov. J. Quantum Electron.* (*Engl. Transl.*) **10**, 1179 (1980)].
155. R. H. Lehmberg, *Phys. Rev. A* **22**, 2156 (1980).
156. B. Ya. Zeldovich and V. V. Shkunov, *Kvantovaya Elektron.* (*Moscow*) **4**, 1090 (1977) [*Sov. J. Quantum Electron.* (*Engl. Transl.*) **7**, 610 (1977)].
157. B. Ya. Zel'dovich, N. A. Mel'nikov, N. F. Pilipetzky, and V. V. Ragul'skii, *Pis'ma Zh. Eksp. Teor. Fiz.* **25**, 41 (1977) [*JETP Lett.* (*Engl. Transl.*) **25**, 36 (1977)].
158. A. D. Kudriavtseva, A. I. Sokolovskaia, J. Gazengel, N. Phu Xuan, and G. Rivoire, *Opt. Commun.* **26**, 446 (1978).
159. P. V. Avizonis, F. A. Hopf, W. D. Bomberger, S. F. Jacobs, A. Tomita, and K. H. Womack, *Appl. Phys. Lett.* **31**, 435 (1977).
160. A. Yariv, *Opt. Commun.* **21**, 49 (1977).
161. D. M. Pepper and A. Yariv, *Opt. Lett.* **5**, 59 (1980).
162. L. A. Bol'shov, D. V. Vlasov, M. A. Dykhne, V. V. Korobkin, K. S. Saidov, and A. N. Starostin, *Pis'ma Zh. Eksp. Teor. Fiz.* **31**, 311 (1980) [*JETP Lett.* (*Engl. Transl.*) **31**, 286 (1980)].
163. V. P. Lukin, *Opt. Lett.* **4**, 15 (1979).
164. T. R. O'Meara, *Opt. Eng.* **21**, 243 (1982).
165. G. G. Kochemasov and V. D. Nikolaev, *Kvantovaya Elektron.* (*Moscow*) **6**, 864 (1979) [*Sov. J. Quantum Electron.* (*Engl. Transl.*) **9**, 514 (1979)].
166. N. G. Basov, V. F. Efimkov, I. G. Zubarev, A. V. Kotov, A. B. Mironov, S. I. Mikhaolov, and M. G. Smirnov, *Kvantovaya Elektron.* (*Moscow*) **6**, 765 (1979) [*Sov. J. Quantum Electron.* (*Engl. Transl.*) **9**, 455 (1979)].
167. M. C. Gower, *Opt. Lett.* **7**, 423 (1982).
168. F. A. Hopf, A. Tomita, and T. Liepmann, *Opt. Commun.* **37**, 72 (1981).
169. V. N. Blaschuk, B. Ya. Zel'dovich, and V. V. Shkunov, *Kvantovaya Elektron.* (*Moscow*) **7**, 2559 (1980) [*Sov. J. Quantum Electron.* (*Engl. Transl.*) **10**, 1494 (1980)].
170. A. Yariv and T. L. Koch, *Opt. Lett.* **7**, 113 (1982).

CHAPTER 6

APPLICATIONS

1. INTRODUCTION

In the previous chapters the nonlinear interactions were discussed from the viewpoint of nonlinear generation of radiation fields, either at the same frequencies as those in the incident waves or at different frequencies. Nonlinear interactions can also be used in various types of applications in which the primary emphasis is not on the generation of radiation fields. Generally speaking these applications fall into two broad categories—materials studies, in which the nonlinear interactions are used as a diagnostic tool to obtain information about the properties of the nonlinear medium, and various device applications in which a nonlinear interaction is used to control, manipulate, or otherwise affect the transfer of information.

Some applications of each type have already been mentioned in earlier chapters. For example, spectroscopic studies of autoionizing levels of metal vapors and of even-parity electronic states of molecules using four-wave sum-frequency mixing was discussed in Chapter 2. The use of degenerate four-wave mixing to correct phase distortions was treated in Chapter 5. In this chapter other applications will be touched on.

In selecting a set of applications for discussion a compromise must be made between a complete exposition of all possible uses of nonlinear optics and the need to keep the discussion to a finite length. In addition, the selection of particular applications reflects of necessity the interests and background of the author, and so may not overlap with similar restricted lists drawn up by individual readers. Thus the applications discussed here are meant to be examples of the possible uses of nonlinear optics rather than an exhaustive list. Indeed, it can be expected that new and different applications will constantly become available in the future.

2. MATERIALS STUDIES

Nonlinear optical materials studies include various forms of nonlinear spectroscopy and measurement of kinetics of excited states, including lifetimes, decay times of coherence, excited-state absorption, or the identification of decay routes. In addition, the nonlinear effects themselves are the most widely used interaction for measuring nonlinear susceptibilities, especially in media for which techniques for first principle calculation of the susceptibilities have not yet been established.

2.1. Nonlinear Spectroscopy

Nonlinear spectroscopy makes use of the multiphoton resonances that occur in the nonlinear susceptibility in a manner analogous to the use of single-photon resonances in linear spectroscopy. The desired nonlinear spectroscopic information is provided by the variation of the end quantity of the nonlinear interaction (the power generated in a four-wave mixing experiment, the excited-state population in a multiphoton absorption measurement, the current produced in a multiphoton ionization experiment, the polarization change in a measurement involving an optical Kerr effect, or the induced gain or loss in a measurement involving saturation spectroscopy) as the harmonics or sum- or difference-frequency combinations of the incident waves are tuned through an appropriate resonance. Nonlinear spectroscopy can thus be differentiated from applications of frequency mixing to linear spectroscopy in which a radiation field is created through a nonlinear interaction in one medium and is subsequently used as an illumination source for another material.

Nonlinear spectroscopic studies can provide information on energy levels or states in the medium that are not usually accessible in linear spectroscopy because of parity restrictions or because they lie in an absorbing region of the spectrum. Various forms of nonlinear spectroscopy can also provide increases in signal levels, signal-to-noise ratios, or spectral resolution over linear spectroscopic techniques under certain conditions.

All of the nonlinear interactions that were discussed in previous chapters can be used for nonlinear spectroscopy and many have been applied in this manner. Some of the processes that have been used for nonlinear spectroscopy are indicated in Table 1, along with the quantities that are varied and measured and the type of information that can be obtained. Each of the nonlinear interactions has its own characteristics and its own set of conditions for optimum application.

There have been numerous discussions of nonlinear spectroscopic techniques in various publications. A review of the nonlinear spectroscopic applications utilizing Raman resonances can be found in Tolles *et al.*[37] These and other nonlinear spectroscopic techniques are discussed in Akhmanov and Koroteev.[38] In the following sections some general aspects of some of these techniques will be treated.

2.1.1. *Spectroscopy Using Four-Wave Mixing*

Various forms of four-wave mixing have been used for nonlinear spectroscopy. One of the most widely used of these methods is coherent anti-Stokes–Raman scattering (CARS), which uses an interaction of the form $\omega_{AS} = 2\omega_1 - \omega_2$. Two-photon Raman resonances of the form $\omega_R = \omega_1 - \omega_2$ are typically studied with this technique.

CARS can be used for obtaining the Raman spectrum of any material that is used in spontaneous Raman scattering. CARS spectroscopy has been done in liquids, solids, and gases and has been used for study of various types of Raman resonances, including vibrational and rotational modes of molecules in gases and liquids and polariton modes and lattice vibrations in solids. CARS has been used for measurement of such quantities as the position of energy levels, the concentration of components in mixtures, density and temperature distributions in flames and other media, lifetimes and coherent decay times of excited levels, and the magnitudes of nonlinear susceptibilities. Some of the reported applications of CARS are summarized in Table 1.

2.1.2. *CARS Phase Matching*

Phase matching for CARS in liquids is most commonly achieved using noncollinear pump waves, as shown by the vector diagram in Fig. 1a. In this geometry the anti-Stokes–Raman wave ω_{AS} is generated in a direction different from the pump waves, allowing spatial separation of the generated wave and aiding in detection. The angles between the pump waves and between wave ω_1 and the anti-Stokes wave are determined by the refractive indexes of the material and by the wavelengths of the waves that are used according to the formulas of Eq. (2-49). For typical organic liquids, phase-matching angles are of the order of 30–60 mrad for visible radiation and Raman shifts of the order of 2000–3000 cm^{-1}. A collinear geometry as shown in Fig. 1b can be used in situations for which the dispersion is small, e.g., for CARS in gases, especially for rotational Raman shifts, or in liquids when the interaction length is less than the coherence length, which is typically of the order of 100 μm. Some other phase-matching geometries that have been used for CARS studies are indicated in Figs. 1c[39] and 1d.[40]

TABLE 1

Applications of Nonlinear Optics to Spectroscopy

Nonlinear interaction	Quantity varied	Quantity measured	Information obtained	References
Multiphoton absorption				
Atom + $2\omega_{laser} \rightarrow$ atom*	ω_{laser}	Fluorescence from excited level	Energy levels of states with same parity as ground state	1a–1e
			Sub-Doppler resolutions and linewidths	
Multiphoton ionization				
Atom + $2\omega_{laser} + \mathscr{E}^a \rightarrow$ atom$^+$	ω_{laser}	Ionization current	Rydberg energy levels	2–4
Atom (molecule) + $(\omega_1 + \omega_2)$ $\rightarrow$ atom$^+$ (molecule$^+$)	ω_2	Ionization current	Even-parity autoionizing levels	5, 6
Four-wave mixing				
Sum-frequency mixing				
$\omega_4 = 2\omega_1 + \omega_2$	ω_2	Optical power at ω_4	Energy structure of levels near ω_4, e.g., autoionizing levels, matrix elements, line shapes	7
Third-harmonic generation				
$\omega_3 = 3\omega_1$	ω_1	Optical power at ω_3	Energy levels near $2\omega_1$ with same parity as ground state	8, 9
Difference-frequency mixing				
$\omega_4 = 2\omega_1 - \omega_2$ (CARS)	ω_2	Optical power at ω_4	Raman energy levels	
			Solids	
			Polaritons	10
			Lattice vibrations	11
			Gases	12

			Measurement of nonlinear susceptibilities	
			Solids	13
			Liquids	14–16
			Concentrations of liquids in solutions	12, 17, 18
			Concentrations of gases in mixtures	19
			Temperature measurements in gases	20, 21
			Measurements in flames, combustion diagnostics	19–22
			Interference of nonlinear susceptibilities	23–25
			Time-resolved measurements	
			Lifetimes, dephasing times	26, 27
			CARS background suppression	28
Four-frequency CARS				
$\omega_4 = \omega_1 + \omega_2 - \omega_3$	$\omega_1 - \omega_3$ $\omega_2 - \omega_3$	Optical power at ω_4	Same as CARS, CARS background suppression	
Raman-induced Kerr effect				
$\omega_{2,y} = \omega_1 - \omega_1 + \omega_{2,x}$	ω_2	Polarization changes at ω_2	Same as CARS, CARS background suppression	29–31
Higher-order Raman processes				
$\omega_4 = 3\omega_1 - 2\omega_2$	ω_2	Optical power at ω_4	Same as CARS	32
Coherent Stokes–Raman spectroscopy				
$\omega_4 = 2\omega_1 - \omega_2\ (\omega_1 < \omega_2)$	ω_2	Optical power at ω_4	Same as CARS	33
Coherent Stokes scattering				
$\omega_S = \omega_1(t) - \omega_1(t) + \omega_S(t) + \omega_1(t + \Delta t) - \omega_1(t + \Delta t)$	Δt	Optical power at $\omega_S(t + \Delta t)$	Lifetimes, dephasing times, resonant contributions to $\chi^{(3)}$	27, 34
Raman gain spectroscopy				
$\omega_S = \omega_L - \omega_L + \omega_S$	ω_S	Gain or loss at ω_S	Raman energy levels	35
Saturation spectroscopy		Induced gain or loss	High-resolution spectra	36

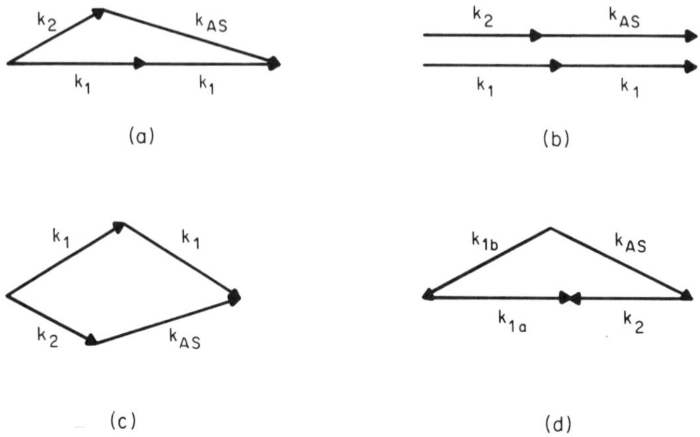

Fig. 1. Phase-matching arrangements for CARS. (a) Noncollinear, normally used in liquids. (b) Collinear, can be used in low-dispersion samples such as gases or very thin liquid samples. (c) BOXCARS. (d) ARCS.

2.1.3. *CARS Intensities*

The anti-Stokes intensity in a plane-wave phase-matched interaction without pump depletion is given by (see Section 2-2.2)

$$I_{\mathrm{AS}} = (KNL)^2 I_1^2 I_2, \tag{1}$$

where

$$K^2 = (24\pi^3/\lambda_{\mathrm{AS}} c)^2 |\chi(-\omega_{\mathrm{AS}}, \omega_1, \omega_1, -\omega_2)|^2 / n_{\mathrm{AS}} n_1^2 n_2,$$

N is the number density of the Raman active medium, L is the interaction length, n_i is the refractive index for the ith wave, and χ is the appropriate third-order susceptibility per molecule. If all the waves have the same cross-sectional area A, the total anti-Stokes power is given by

$$\mathscr{P}_{\mathrm{AS}} = (KNL)^2 I_1^2 \mathscr{P}_2 = (KNL/A)^2 \mathscr{P}_1^2 \mathscr{P}_2. \tag{2}$$

The power generated in one coherence length is

$$\mathscr{P}_{\mathrm{AS}} = (2KNL_{\mathrm{coh}}/\pi A)^2 \mathscr{P}_1^2 \mathscr{P}_2.$$

Corresponding formulas for focused beams have been discussed by Shaub *et al.*[41]

CARS offers advantages over spontaneous Stokes–Raman scattering for spectroscopic studies of certain materials. Under certain conditions the signal levels and signal-to-noise ratios in a CARS interaction can exceed those in spontaneous Raman scattering. In addition, the CARS interaction produces radiation in a region of the spectrum that is different from that of any spontaneous fluorescence which can occur in some organic samples and obscure weak Raman modes.

A comparison of the signal levels that can be obtained in CARS and spontaneous Raman scattering can be made by considering the spontaneous Stokes power that is generated in a 90° scattering geometry:

$$\mathscr{P}_{\text{Stokes}} = NL\Omega(d\sigma/d\Omega)\mathscr{P}_{\text{laser}}, \tag{3}$$

where Ω is the solid angle collected by the detector and $d\sigma/d\Omega$ is the differential Raman cross section per molecule, related to the CARS susceptibility at exact resonance by

$$|3\chi_{\text{AS}}|^2 = (2c^4\ d\sigma/d\Omega/\hbar\omega_{\text{S}}^4\Gamma)^2, \tag{4}$$

where Γ is the Raman linewidth.

Higher power can be generated at the anti-Stokes wavelength by a CARS process than at the Stokes wavelength in spontaneous scattering when

$$\left(\frac{2\lambda s^4}{hc\lambda_{\text{AS}}\Gamma}\right)^2 \frac{NL\ d\sigma/d\Omega\ I_1^2}{\Omega n_{\text{AS}} n_1^2 n_2} > 1, \tag{5}$$

assuming that the power in the pump wave at ω_2 is equal to the power used for the laser in spontaneous scattering.

This condition is typically met for interactions involving samples of relatively high density and long interaction length and is most commonly satisfied when the material under study is a pure material or a major constituent of a mixture. The properties of CARS spectroscopy in mixtures will be discussed in more detail later. The nonlinear dependence of the CARS intensity on the pump intensity gives added control over the ratio of the CARS power to the spontaneous Raman power. A discussion of the relative properties of spontaneous Raman scattering and CARS can be found in Tolles and Turner.[42]

2.1.4. *Interference with Nonresonant Background*

In addition to the resonant Raman contribution the wave at $\omega_{\text{AS}} = 2\omega_1 - \omega_2$ receives contributions from interactions involving nonresonant levels, such as electronic states or nearby Raman levels. These nonresonant contributions create a background signal against which the resonant CARS signal must be measured. The background signal can obscure weak Raman modes or can place a lower limit on the density of a solute that can be measured by CARS in a mixture. In addition, the interference between the fields created in the resonant and nonresonant interactions can result in a distortion of the Raman line shape.

In general the nonresonant susceptibility will have a complex value. In addition, the resonant CARS susceptibility will also have both real and imaginary parts for finite detunings from the center of the Raman line. The total susceptibility per molecule for the general case can be expressed as

$$\chi_{\mathrm{T,m}} = \frac{\chi'_{\mathrm{B}}[1 + (\delta/\Gamma)^2] + (\delta/\Gamma)\chi''_{\mathrm{R}}(0) + i\{\chi''_{\mathrm{B}}[1 + (\delta/\Gamma)^2] + \chi''_{\mathrm{R}}(0)\}}{1 + (\delta/\Gamma)^2}, \quad (6)$$

where χ'_{B} and χ''_{B} are the real and imaginary parts of the background susceptibility, $\chi''_{\mathrm{R}}(0)$ is the value of the CARS susceptibility at the center of the Raman line, and $\delta/\Gamma = [\omega_{\mathrm{R}} - (\omega_1 - \omega_2)]/\Gamma$ is the normalized detuning from the Raman frequency.

Interference between the resonant CARS signal and the nonresonant electronic background signal has been discussed by Wynne,[10] Regnier and Taran,[19] Levenson *et al.*,[11] and Levenson and Bloembergen.[13,14] For this situation the background susceptibility is real and the third-order susceptibility near a single Raman resonance has the form[14]

$$\chi_{\mathrm{T,m}}(-\omega_{\mathrm{AS}}, \omega_1, \omega_1, -\omega_2) = \chi_{\mathrm{B}} + \frac{\omega_{\mathrm{R}}|\alpha_{\mathrm{R}}|^2/6\hbar}{\omega_{\mathrm{R}}^2 - (\omega_1 - \omega_2)^2 - 2i(\omega_1 - \omega_2)\Gamma}$$

$$\approx \chi_{\mathrm{B}} + \frac{|\alpha_{\mathrm{R}}|^2/12\hbar}{\omega_{\mathrm{R}} - (\omega_1 - \omega_2) - i\Gamma}, \quad (7)$$

where $|\alpha_{\mathrm{R}}|^2$ is related to the spontaneous Raman cross section $d\sigma/d\Omega$ by

$$|\alpha_{\mathrm{R}}|^2 = \frac{8c^4}{\omega_{\mathrm{S}}^4}\, d\sigma/d\Omega \quad (8)$$

and χ_{B} is the background susceptibility.

The anti-Stokes intensity is proportional to $|\chi^{(3)}_{\mathrm{T,m}}|^2$, which is given by

$$|\chi^{(3)}_{\mathrm{T,m}}|^2 = |\chi_{\mathrm{B}}|^2\left\{\frac{(\omega_{\mathrm{R}} - (\omega_1 - \omega_2) + |\alpha_{\mathrm{R}}|^2/12\hbar\chi_{\mathrm{B}})^2 + \Gamma^2}{[\omega_{\mathrm{R}} - (\omega_1 - \omega_2)]^2 + \Gamma^2}\right\}. \quad (9)$$

For Raman resonances that are strong enough to satisfy the condition

$$\frac{|\alpha_{\mathrm{R}}|^2}{12\hbar\chi_{\mathrm{B}}} \gg 4\Gamma^2 \quad (10)$$

a resonant peak in the CARS signal occurs at the position of the Raman frequency and an interference minimum appears on the high-frequency side. A typical CARS spectrum for benzene near the 992-cm^{-1} Raman mode is shown in Fig. 2a.

When interference occurs between Raman resonances and electronics states involving two-photon absorption or between nearby Raman resonances

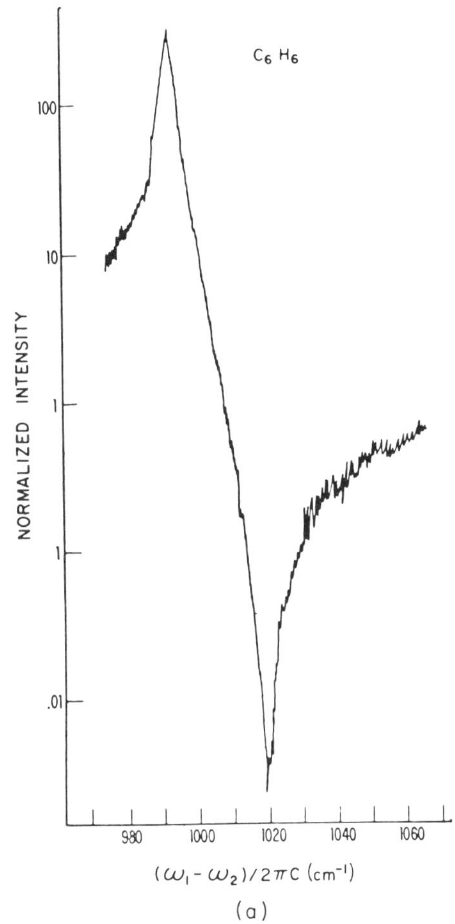

Fig. 2a. Line shapes in CARS. Interference with electronic susceptibility for 992-cm^{-1} mode of benzene. [From Ref. 14.]

in different media, additional terms must be included in Eqs. (7)–(9) to account for the imaginary part of χ_B. Interference between nearby Raman modes of different materials using two pump waves has been discussed by Begley *et al.*[17] and Levenson and Bloembergen.[14] Similar effects using three pump waves in the interaction $\omega_4 = \omega_1 + \omega_2 - \omega_3$ have been described by Lynch *et al.*[23] and Lotem *et al.*[25] Depending on the relation of the pump frequencies to the Raman frequencies, the CARS spectrum can have a bell shape with a single peak or a dispersionlike shape. The interference minimum can occur on either side of the resonant peak, depending on the sign of χ_{NR}. An example of the various line shapes that can occur

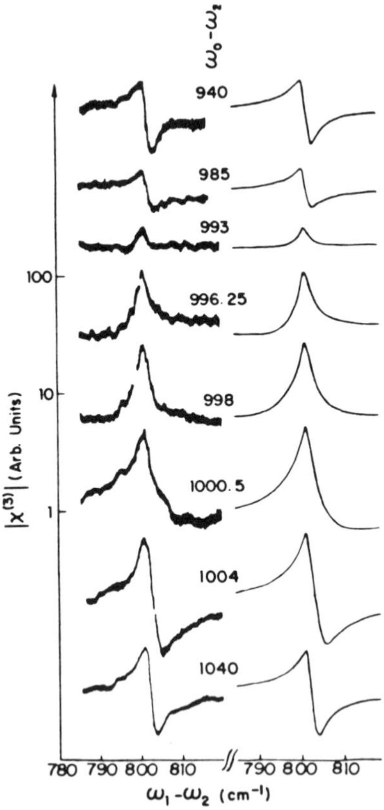

(b)

Fig. 2b. Line shapes in CARS. Line shapes in four-frequency CARS. Interference is between Raman modes of cyclohexane and benzene in a mixture of materials. [From Ref. 25.]

is shown in Fig. 2b. Interferences between Raman resonances and the background susceptibility due to two-photon absorption have been discussed by Levenson and Bloembergen.[13,14] Lotem and Lynch[24] have reported observations of destructive interference in the imaginary part of the susceptibility due to interference between Raman and two-photon absorption resonances.

2.1.5. *Measurement of Nonlinear Susceptibilities*

The interference of the resonant and nonresonant signals provides a method of measuring the nonresonant third-order susceptibility that does not rely on accurate measurements of interaction parameters or the power levels of the pump or scattered radiation. Errors in the determination of these quantities are often the largest source of inaccuracies in absolute measurements of the nonlinear susceptibilities.

The interference technique can be used to determine the magnitude of the nonresonant susceptibility $\chi_{NR}(-\omega_4, \omega_1, \omega_1, -\omega_2)$ relative to the peak Raman cross section, a quantity that is usually known accurately from linear-scattering measurements. Measurements can also be made of the two-photon absorption cross section or nonlinear refractive index relative to the Raman cross section in the same material. In addition, the nonresonant susceptibility of one material can be measured relative to the Raman cross section in another material if mixed samples are used.

The interference technique is based on measurements of the frequency separation of the maximum and minimum in the CARS spectrum. For a single medium involving interference between a Raman mode and a nonresonant electronic background this frequency difference, as obtained from Eq. (9), is given by

$$\omega_{max} - \omega_{min} = -\{[(\alpha_R)^2/12\hbar\chi_{NR}]^2 + 4\Gamma^2\}^{1/2}. \tag{11}$$

If the Raman resonance is strong enough to satisfy condition (10), then the frequency separation of the maximum and minimum is given approximately by

$$\omega_{max} - \omega_{min} \approx -\frac{|\alpha_R|^2}{12\hbar\chi_{NR}}. \tag{12}$$

For weak Raman modes both the position of the resonant maximum and the frequency separation of the maximum and minimum shift.

Levenson and Bloembergen have used the interference method to measure the nonresonant susceptibility of many liquids and solids.[11,13,43] Lotem and Lynch[24] have used the technique to measure the two-photon absorption cross section of several materials. Wiener-Avenear *et al.*[15] have used the interference between resonant and nonresonant contributions in liquid solutions of varying concentration to determine the nonresonant third-order susceptibility without the need for exact line shape fitting. Measurement of nonlinear refractive indexes by this technique has been discussed by Levenson *et al.*[11]

2.1.6. *Limitations of Density Measurements in Mixtures*

The nonresonant four-wave mixing signal from electronic states or nearby Raman modes of a solvent can place a lower limit on the concentration of a solute that can be detected using CARS. These effects have been discussed by Begley *et al.*,[17] Levenson and Bloembergen,[14] Regnier and

Taran,[19] Chabay *et al.*,[32] and Carreira *et al.*[18] The anti-Stokes intensity from a solution is determined by the magnitude of the total nonlinear susceptibility, which is given by

$$|\chi_T|^2 = |N_1\chi_1 + N_2\chi_2|^2, \tag{13}$$

where N_i and χ_i are the number density and susceptibility per molecule of the ith component. In general the susceptibilities of both the solvent and solute will be complex and the total susceptibility will have the form

$$|\chi_T|^2 = (N_1\chi_1' + N_2\chi_2')^2 + (N_1\chi_1'' + N_2\chi_2'')^2. \tag{14}$$

If the susceptibility of one component is real while that of the other is imaginary, then the susceptibilities of the components will add in quadrature and the anti-Stokes intensity will depend on the square of the density of each component. If, however, one component carries both real and imaginary parts, then the anti-Stokes intensity can have a region in which it varies linearly with the density of the solute.

The susceptibility of the solvent will be dominantly real if the frequency difference between the pump waves is detuned from the nearest Raman resonance of the solvent by more than a few linewidths and if there are no electronic resonances in the solvent susceptibility. For this situation the total nonlinear susceptibility in the vicinity of a Raman resonance of the solute can be written as

$$|\chi_T|^2 = N_1^2|\chi_1|^2\left(\left|1 + \frac{N_2}{N_1}\frac{\chi_2''/\chi_1}{\Delta\omega/\Gamma + i}\right|^2\right), \tag{15}$$

where χ_2'' is the susceptibility of the solute at the center of the Raman line, $\Delta\omega = \omega_R - (\omega_1 - \omega_2)$ is the detuning, and Γ is the Raman linewidth. As the density of the solvent is decreased, the resonant contribution of even a strong Raman mode can become comparable to or less than that of a weaker background susceptibility in the more concentrated solvent.

At solute concentrations great enough to satisfy the condition

$$N_2 \gg \frac{N_1\chi_1}{\chi_2''} \tag{16}$$

the maximum anti-Stokes intensity occurs for approximately zero detuning ($\Delta\omega = 0$) and varies as the square of the solute density. For lower concentrations, such that (16) is not satisfied, the maximum anti-Stokes intensity occurs at a detuning that is concentration dependent and is given by

$$\Delta\omega = \Gamma\left(-\frac{b}{2} + \frac{1}{2}\sqrt{b^2 + 4}\right), \tag{17}$$

where $b = N_2\chi_2''/N_1\chi_1$.

In the limit of small concentrations the total nonlinear susceptibility at the optimum detuning is given approximately by

$$|\chi_T|^2 = N_1^2 |\chi_1|^2 + N_1 N_2 \chi_1 \chi_2''. \tag{18}$$

In this limit the maximum resonant contribution to $|\chi_T|^2$ varies linearly with solute density, although the resonant contribution at zero detuning continues to vary as N_2^2. If the solvent susceptibility has an imaginary part due to Raman or electronic resonances, then the anti-Stokes intensity at the center of the Raman line will also have a component that varies linearly with solute density at low concentrations.

The minimum concentration that can be detected is determined by the ratio χ_1/χ_2'' and is obviously smaller for solvents with smaller nonresonant susceptibilities. Minimum detectable concentrations are also dependent to a certain extent on specific experimental arrangements. Chabay *et al.*[32] have reported measured concentrations of diphenyloctatetraene in benzene of 10^{-3} *M* using a carefully controlled experimental arrangement. They estimated that concentrations as low as 10^{-5} *M* could be detected with their apparatus.

The range of detectable concentrations can be extended by using electronic resonances to enhance the Raman susceptibility of the solute. Such a technique was used by Carreira *et al.*[18] to measure concentrations of β-carotene in paradioxane down to values as low as 5×10^{-5} *M*.

2.1.7. *Suppression of the CARS Background*

Several alternative interactions that can reduce the background signal from nonresonant levels have been proposed and studied. Some of these are listed in Table 1. The Raman-induced Kerr effect (RIKE)[30–32] makes use of the change in polarization of a light wave at ω_2 that can be caused by the birefringence induced by a wave at another frequency ω_1 in Kerr active media. The magnitude of the change in polarization is resonantly enhanced whenever the frequency difference $\omega_1 - \omega_2$ approaches a Raman frequency. When the pump wave ω_1 is circularly polarized, optically induced birefringence occurs only from the Raman resonances, reducing the effects of the nonresonant background. This process does not need to be phase matched and so is easier to use than CARS. It can be affected, however, by any birefringence that is present in the sample. The effects of such background birefringence can be reduced by using circular polarization for both waves at ω_1 and ω_2.

Various four-wave mixing interactions of the form $\omega_4 = \omega_1 + \omega_2 - \omega_3$ have also been used for background suppression. One of these techniques makes use of the destructive interference that occurs near the Raman resonance for one pair of frequencies to reduce the background for the

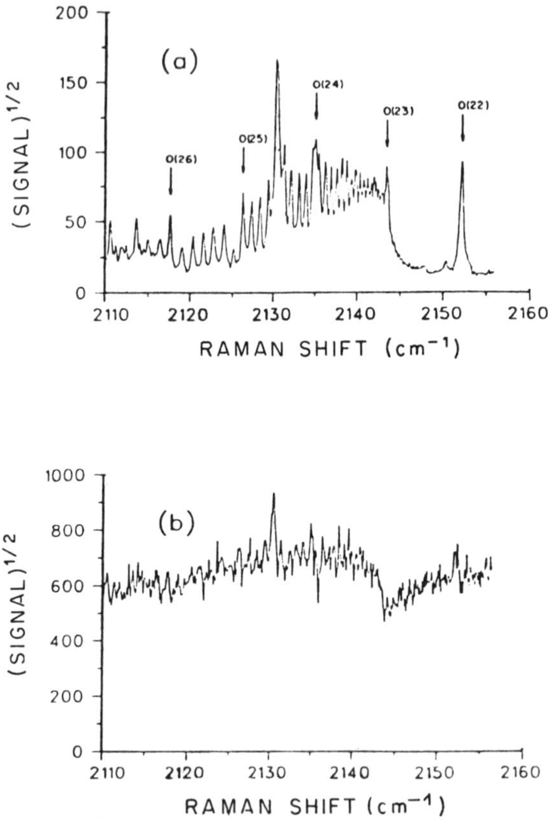

Fig. 3. CARS spectrum of CO in a methane–air flame. (a) Background suppression by selection of polarization angles of pump beams. (b) Conventional CARS spectrum. (From Ref. 22.)

Raman spectrum that is obtained by varying the third frequency.[32] Another technique involves a choice of polarizations of the three pump waves and of the detected wave at ω_4 to suppress the nonresonant background.[22,31,32,44] An example of background suppression in the CARS spectrum of CO in a methane–air flame using this technique is shown in Fig. 3.[22] Raman modes of desired symmetry can be rejected or selected by varying the orientation of a polarizer placed in the generated beam. Background suppression can also be obtained by using circularly polarized pump waves.[32] Finally, a properly sequenced set of pulses with duration shorter than the Raman dephasing time can allow background suppression by making use of the longer dephasing time of the resonant interaction.[29]

2.1.8. *Lifetime Measurements*

CARS has also been used to measure the dephasing times of Raman transitions.[26,27,28] For liquids and solids these times are usually of the order of subpicoseconds to tens of picoseconds. The information obtained in this manner previously could be obtained only by inference from spontaneous-scattering linewidths.

3. INFRARED UP-CONVERSION

Frequency mixing can be used to transfer radiation from one region of the spectrum to another. In the discussions of previous chapters this property was used to provide laser-like radiation in regions of the spectrum that are otherwise inaccessible to coherent radiation. Frequency shifting of radiation from one region of the spectrum to another can also be done to make the radiation more easily detectable. This application is encountered, for example, in the conversion of infrared (IR) radiation to the visible, where the detectors can be faster and more sensitive.

Infrared up-conversion in gases and liquids can be accomplished using any of several four-wave mixing interactions. The process $\omega_{vis} = 2\omega_1 + \omega_{IR}$, where ω_1 is a strong pump wave, usually in the visible or near IR, and ω_{IR} is the infrared wave to be converted, has been used for IR signal detection[45] and IR image up-conversion.[46] The process $\omega_{vis} = \omega_1 + \omega_2 - \omega_{IR}$, where ω_1 is a strong visible wave, ω_2 an infrared pump wave, and ω_{IR} the image-bearing IR wave, has been used for IR image conversion to the visible using dynamic holography.[47] The interaction $\omega_{vis} = \omega_1 - \omega_2 + \omega_{IR}$, where ω_1 is a visible wave, ω_2 an IR pump wave, and ω_{IR} the IR wave to be converted to the visible, has been used for IR absorption spectroscopy.[48,49]

Generally, considerations of efficiency of the IR up-conversion process are concerned with the efficiency of conversion of the IR wave and not primarily with efficient use of the pump radiation. Indeed, in many applications of the type considered here the primary emphasis is on obtaining any information that may be carried by the IR wave, so a faithful reproduction in the visible of the IR radiation with respect to temporal, spatial, or spectral variation is important. If the conversion process so efficiently uses the pump radiation that pump depletion occurs, distortion of the information on the converted signal may result. In many cases, then, we may assume that the pump wave is strong compared to the IR wave and remains undepleted during the up-conversion process. Overall efficiency of a particular system can be important, however, when size or power restrictions are present.

Infrared up-conversion through the process $\omega_{vis} = 2\omega_1 + \omega_{IR}$ in a nonresonant system is described in the plane-wave approximation at perfect phase matching by the equations

$$\frac{dA_{vis}}{dz} = i\frac{3\pi\omega_{vis}}{2n_{vis}c}\chi(-\omega_{vis}, \omega_1, \omega_1, \omega_{IR})A_1^2 A_{IR}, \tag{19a}$$

$$\frac{dA_{IR}}{dz} = i\frac{3\pi\omega_{IR}}{2n_{IR}c}\chi(-\omega_{IR}, -\omega_1, -\omega_1, \omega_{vis})A_{vis}(A_1^*)^2, \tag{19b}$$

$$\frac{dA_1}{dz} = i\frac{3\pi\omega_1}{n_1 c}\chi(-\omega_1, \omega_{vis}, -\omega_{IR}, -\omega_1)A_{vis}A_{IR}^* A_1^*. \tag{19c}$$

If we neglect depletion of the pump wave ω_1, then the intensities of the visible and IR waves are given by

$$I_{vis} = I_{IR}(0)\frac{\omega_{vis}}{\omega_{IR}}\sin^2(K\sqrt{\omega_{vis}\omega_{IR}/n_{IR}n_{vis}}|A_1|^2 L), \tag{20a}$$

$$I_{IR} = I_{IR}(0)\cos^2(K\sqrt{\omega_{vis}\omega_{IR}/n_{IR}n_{vis}}|A_1|^2 L), \tag{20b}$$

where $K = 3\pi\chi(-\omega_{vis}, \omega_1, \omega_1, \omega_{IR})/2$, $I_{IR}(0)$ is the incident IR intensity, and L is the interaction length. In this approximation the generated signal is proportional to the incident IR intensity regardless of the degree of depletion of the IR wave. This is a consequence of the assumption that the pump wave is undepleted and will not necessarily be true if significant pump depletion occurs.

Total conversion of the IR wave occurs when

$$K(\omega_{vis}\omega_{IR}/n_{vis}n_{IR})^{1/2}|A_1|^2 L = \pi/2. \tag{21}$$

For larger values of the quantity on the left-hand side of Eq. (21) the IR conversion drops as the reverse parametric process $\omega_{IR} = -2\omega_1 + \omega_{vis}$ converts the visible radiation back into the infrared. The generated wave at ω_{vis} can carry more power than that contained in the incident IR wave with the excess power supplied by the pump wave. Thus the IR up-converter can also act as a power amplifier with a maximum power amplification given by ω_{vis}/ω_{IR}.

For weak interactions the visible intensity takes the form

$$I_{vis} = \left(\frac{16\pi^2}{\lambda_{vis}}\right)^2 \frac{K^2 L^2 I_1^2 I_{IR}(0)}{n_1^2 n_{vis} n_{IR}}, \tag{22}$$

in agreement with the formulas given in Chapter 2 for sum-frequency mixing. In systems that are not perfectly phase matched the intensity that is generated in one coherence length is

$$I_{vis} = \left(\frac{32\pi^2}{\lambda_{vis}\Delta k}\right)^2 \frac{K^2 I_1^2 I_{IR}(0)}{n_1^2 n_{vis} n_{IR}} = \left(\frac{32\pi}{\lambda_{vis}}\right)^2 \frac{K^2 L_c^2 I_1^2 I_{IR}(0)}{n_1^2 n_{vis} n_{IR}}. \tag{23}$$

All of the considerations that were discussed in Chapter 4 concerning effects that limit the conversion efficiency of frequency-mixing processes apply to IR up-conversion and can prevent the system from reaching the condition for complete conversion. One of the more common interactions in which the competing processes are important involves a two-photon resonance with the pump wave ω_1. In this situation we must take into account the effects of changes in the ground-state population, loss of pump power due to two-photon absorption, and loss of visible power due to Raman scattering from the visible to the infrared. As discussed in Chapter 4 the relative importance of these effects is determined by the degree of change in the level populations that is induced in the medium. In relatively low density media, such as low-pressure gases, efficient conversion requires relatively high pump intensities and population changes can be significant. This situation has been treated by Harris and Bloom[50] and by Stappaerts.[51] In relatively high density media, such as high-pressure gases or liquids, lower pump intensity is required and the effects of optical absorption and Raman scattering can cause the dominant limitation. This situation has been discussed by Kildal and Brueck.[52]

In low-density media we may consider that the absorptive losses at the pump, IR, and visible frequencies are negligible and further that the process can be phase matched over the range of experimental parameters that is commonly encountered. Two-photon resonant conversion is then described by the same type of analysis as given in Chapter 4 for third-harmonic generation. If depletion of the IR wave is neglected, then the visible field amplitude is described at perfect phase matching by the equations

$$\frac{dA_{\text{vis}}}{dz} = i\frac{(2\pi)^2 N}{n_{\text{vis}}\lambda_{\text{vis}}}\left[\sum \frac{\mu_{0i}\mu_{i2}}{\hbar}\left(\frac{1}{\omega_{i0} - \omega_{\text{vis}}} + \frac{1}{\omega_{i0} + \omega_2}\right)\right]\rho_{20}^{(-2\omega_1)}A_{\text{IR}}, \tag{24a}$$

$$\frac{\partial}{\partial t}\rho_{20}^{(-2\omega_1)} + i(\omega_{20} - 2\omega_1 + \delta|A_1|^2 - i/T_2)\rho_{20}$$

$$= i(\rho_{00} - \rho_{22}) \times \frac{1}{4\hbar^2}\sum \frac{\mu_{2i}\mu_{i0}}{\omega_{i0} - \omega} A_1^2, \tag{24b}$$

$$\frac{\partial}{\partial t}(\rho_{00} - \rho_{22}) + \frac{1}{T_1}(\rho_{00} - \rho_{22} - 1)$$

$$= \frac{i}{2\hbar^2}\left(\sum_i \frac{\mu_{0i}\mu_{i2}}{(\omega_{i0} - \omega_1)}\right)(\rho_{20}^{(-2\omega_1)}A_1^{*2} - \rho_{02}^{(+2\omega_1)}A_1^2), \tag{24c}$$

where

$$\delta - \frac{1}{2\hbar}\left(\sum_j \frac{|\mu_{2j}|^2\omega_{2j}}{\omega_{j2}^2 - \omega_1^2} + \sum \frac{|\mu_{0j}|^2\omega_{j0}}{\omega_{j0}^2 - \omega_1^2}\right). \tag{25}$$

In the steady state ρ_{20} is given by Eq. (4-75a) and the steady-state visible field amplitude is given by

$$A_3 = i\frac{\pi^2 Nq'qT_2[i + (\Delta + \delta|A_1|^2)T_2]}{n_{\text{vis}}\lambda_{\text{vis}}\hbar[1 + (\Delta + \delta|A_1|^2)^2T_2^2 + q^2T_1T_2|A_1|^4/4\hbar^2]}, \tag{26}$$

where

$$\Delta = \omega_{20} - 2\omega_1, \tag{27}$$

$$q = \frac{1}{\hbar}\sum\frac{\mu_{2i}\mu_{i0}}{\omega_{i0} - \omega_1}, \tag{28}$$

$$q' = \frac{1}{\hbar}\sum\mu_{0i}\mu_{i2}\left(\frac{1}{\omega_{i0} - \omega_{\text{vis}}} + \frac{1}{\omega_{i0} + \omega_{\text{IR}}}\right). \tag{29}$$

The conditions for optimum conversion are the same as those discussed in Section 4-4.3 for third-harmonic generation. If we assume that the up-conversion remains phase matched at its low-intensity conditions as the two-photon transition saturates, then the optimum conversion occurs at a detuning given by

$$\Delta = -\delta I_1 \pm (1/T_2)(I_1^2/I_s^2 - 1)^{1/2} \tag{30}$$

for pump intensities above the two-photon saturation intensity given by

$$I_{\text{sat}} = \hbar c n_1/4\pi q(T_1T_2)^{1/2}. \tag{31}$$

The optimal conversion efficiency is the same as that for third-harmonic generation:

$$\eta_{\text{opt}} = \frac{\pi^4 q'^2 L^2 N^2}{n_1 n_{\text{vis}}\lambda_{\text{vis}}^2}\frac{T_2}{T_1}. \tag{32}$$

Two-photon resonant up-conversion of CO_2 laser radiation at various wavelengths near 9.6 and 10.6 μm to the wavelength range near 332 nm has been reported by Bloom *et al.*[45] The pump radiation was obtained from a parametric oscillator tuned to the 3s–3d transition in Na at 685.6 nm. At a pump power of 3 kW (focused to a peak intensity of 10 MW/cm^2) power conversion efficiencies ranged from 7.9% for the 10.6-μm line to 1620% for the 9.26-μm line.

3.1. Infrared Image Up-Conversion

The fact that the phase-matching condition for four-wave mixing in gases does not depend on the angle between the waves makes the process ideal for converting images from the IR to the visible where they can be more easily detected or photographed. Such an application has been demonstrated

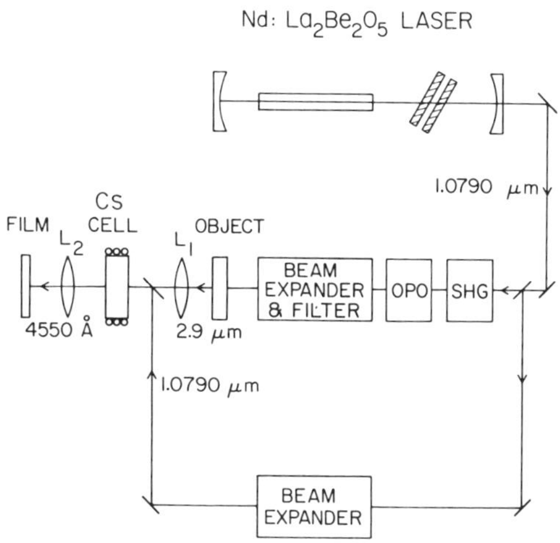

Fig. 4. Schematic diagram of apparatus used for IR image up-conversion. (From Ref. 46.)

by Stappaerts *et al.*[46] using the process $\omega_{vis} = 2\omega_1 + \omega_{IR}$. They converted image-bearing radiation from a parametric oscillator at 2.9 μm to the visible at 0.455 μm with the apparatus shown schematically in Fig. 4. For the pump wave ω_1 they used radiation from an Nd:BEL (lanthanum beryllate) laser at 1.079 μm for which there is a two-photon resonance with the $6s^2\,{}^1S_0$–$7s^2\,{}^1S_0$ transition in Cs. Visible images were reported that had approximately 1000 resolvable spots. A photon conversion efficiency from the IR to the visible of 20% (130% power conversion efficiency) was reported with a pump intensity of 8×10^4 W/cm^2 and a Cs density of about 10^{17} cm^{-3}.

3.2. Infrared Image Up-Conversion with Dynamic Holography

Infrared image up-conversion using the interaction $\omega_{vis} = \omega_1 - \omega_2 + \omega_3$ has been reported by Martin and Hellwarth[47] using the arrangement shown in Fig. 5. Here ω_2 and ω_3 are infrared beams of the same frequency, with the image carried by the beam at ω_2, and ω_1 is a visible pump beam. The image is converted to the visible by the interaction shown in the figure. The authors have described the image conversion mechanism as resulting from holographic imaging similar to the image reconstruction that can be done with DFWM. In this situation a diffraction grating bearing the image information is created by the interference of beams 2 and 3. The image information is read out by beam 1, which is Bragg scattered by the grating to form beam 4, and the image is produced at the position of the real holographic image.

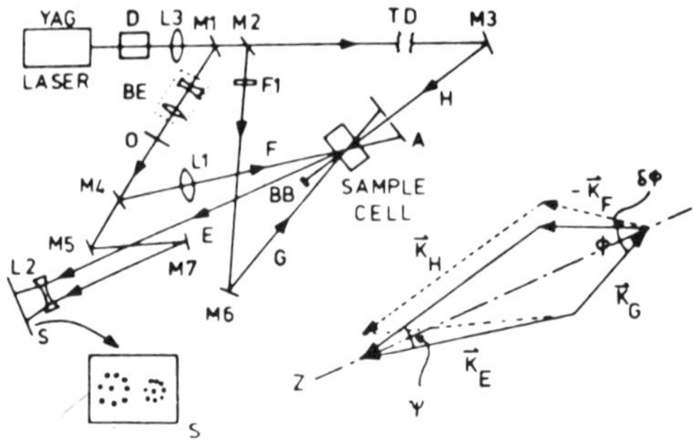

Fig. 5. Schematic diagram of apparatus used for observation of up-converted phase conjugate images. (From Ref. 47.)

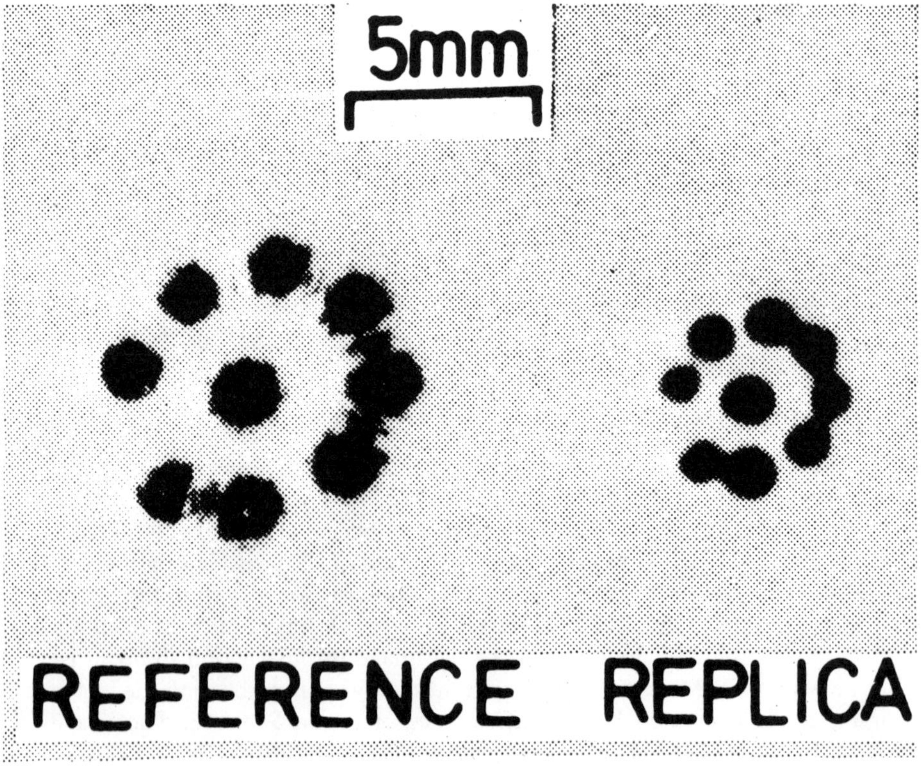

Fig. 6. Up-converted phase conjugate images obtained with apparatus in Fig. 5. Incident pattern is on the left; upconverted pattern is on the right. (From Ref. 47.)

Martin and Hellwarth have reported obtaining images with about 10^3 resolution elements from several organic liquids in this manner. For these materials the interaction involved the formation of thermal gratings by absorption of the IR waves. An example of such an image is shown in Fig. 6. For this picture, radiation from an Nd:YAG laser at 1.06 μm was used for beams 2 and 3 and the second harmonic of the YAG laser at 0.532 μm was used for beam 1. A maximum conversion efficiency of 5×10^{-2} was reported using an IR absorbing solution of dichloroethane and Kodak 14015 *Q*-switching dye.

3.3. Time-Resolved Infrared Spectroscopy

Infrared up-conversion can also be used to convert IR spectral information to the visible where it can be detected with greater sensitivity. Such conversion can make it possible to record time-resolved IR spectra. A technique for doing this has been reported by Bethune *et al.*[48,49] They have recorded IR absorption spectra over the range 4000–2600 cm^{-1} using the apparatus shown schematically in Fig. 7. The IR radiation was converted in an alkali

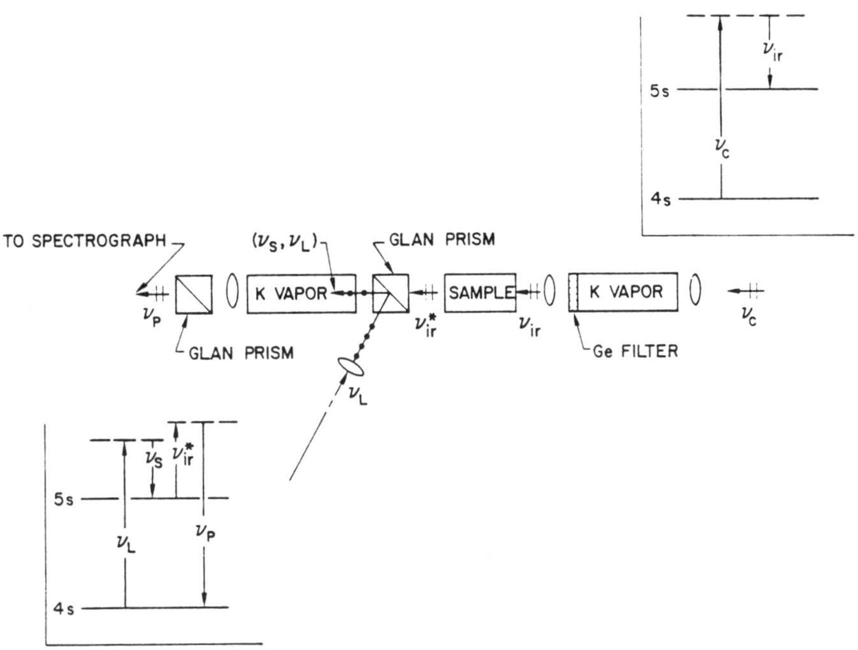

Fig. 7. Schematic diagram of apparatus used for time-resolved IR spectroscopy. (From Ref. 48.)

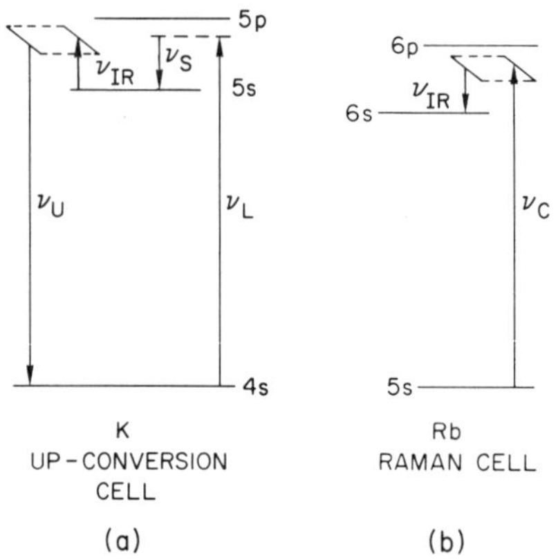

Fig. 8. (a) Level diagram for IR up-conversion used for time-resolved spectroscopy. (b) Level diagram of process used to create the IR continuum. (From Ref. 49. Copyright 1979 by International Business Machines Corporation; reprinted with permission.)

metal vapor to the visible through the interaction $\omega_{vis} = \omega_1 - \omega_s + \omega_{IR}$, where ω_{IR} is a broad-band IR illumination source used to obtain the spectral information, ω_1 is a narrow-band visible wave, and ω_s is a narrow-band Raman–Stokes beam that is generated from ω_1 by stimulated Raman scattering in the alkali metal vapor, as shown in Fig. 8a. Because of the use of the Raman–Stokes frequency for ω_s, the four-wave mixing process is automatically two-photon resonantly enhanced.

The IR continuum was generated from a visible continuum by stimulated electronic Raman scattering in a suitable metal vapor (Fig. 8b). The visible continuum in turn is generated by amplifying superfluorescent radiation from a dye laser pumped by a suitable pulsed laser.

The authors have observed that the conversion from the IR to the visible is effectively uniform over a relatively wide spectral range, indicating that the frequency dependence of the phase-matching conditions is not significant in this interaction. The up-converted signal thus provides a faithful reproduction of the spectral intensity. By choosing various combinations of laser dyes and alkali metal vapors the region of IR coverage can be adjusted. Approximately 800-cm^{-1} spectral width in the IR is covered with one combination, limited by the fluorescence of the laser dye. Visible spectra are recorded on photographic film and time resolution is provided by the shorter of the IR pulses or the visible dye laser pulse.

The authors have reported absorption spectra from several materials, e.g., CO_2 vapor, H_2O vapor, and various materials with C–H absorption bands. Spectra showing time-dependent cooling, deduced from the spectral width and distribution of the rotational absorption spectrum, in CH_3NC following a controlled explosion have also been reported (Fig. 9).

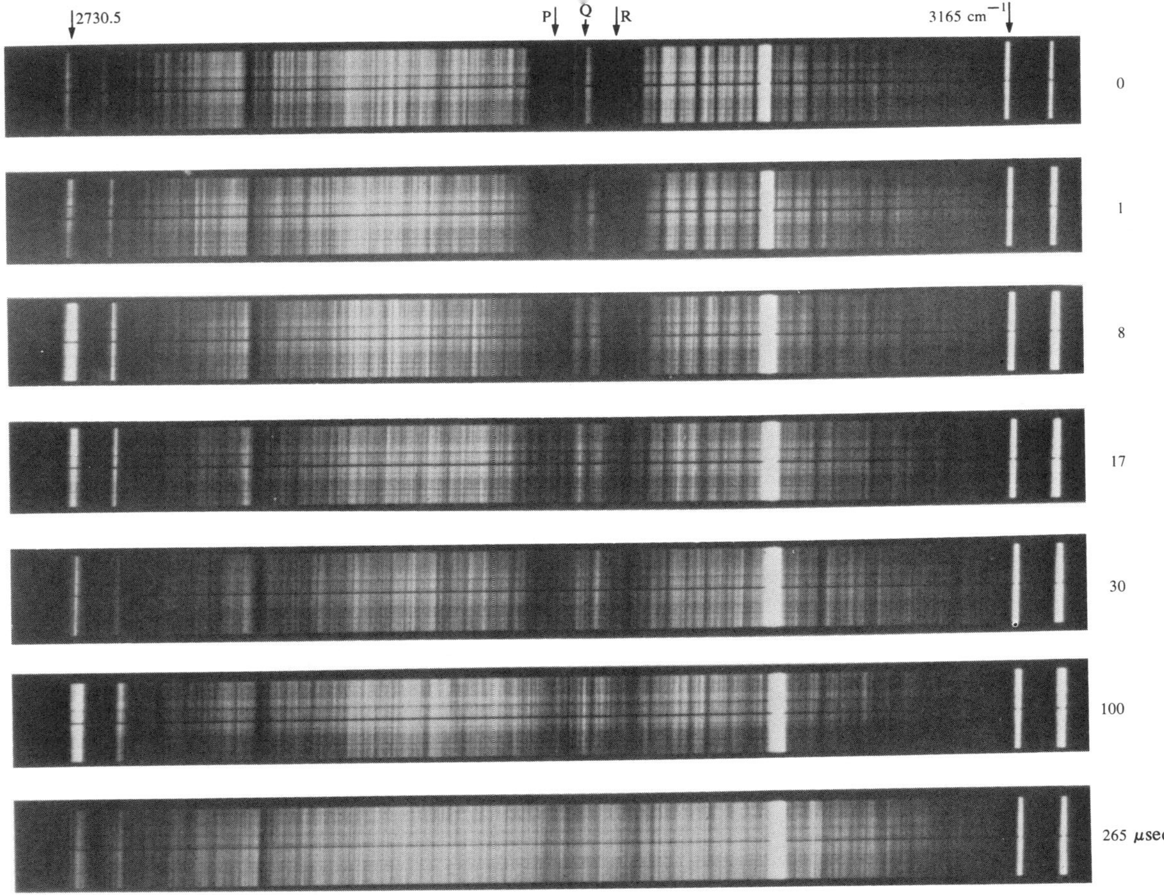

Fig. 9. Photograph of up-converted infrared spectrum of CH_3NC as it thermally explodes at various times following the application of a CO_2 laser pulse. (From Ref. 49. Copyright 1979 by International Business Machines Corporation; reprinted with permission.)

4. OPTICAL BISTABILITY AND RELATED EFFECTS

Several optical devices have been shown to exhibit optical bistability and related effects such as differential gain, optical switching, clipping, limiting, and regenerative pulsing. The first class of device that was shown to exhibit these effects was a nonlinear Fabry–Perot (FP) resonator, that is, a FP resonator that is filled or partially filled with a nonlinear medium, as shown in Fig. 10a. Although nonlinear FP resonators using saturable absorbers were proposed first, and have been studied extensively theoretically, most of the experimental work has been done on reactive FPs, that is, FP resonators with a medium whose nonlinear effects arise from a nonlinear refractive index.

The properties of absorptive FP resonators using saturable absorbers were discussed by Seidel,[53] Szoke *et al.*,[54] and McCall.[55] A steady-state analysis of optical bistability due to cooperative effects in a two-level saturable absorber has been developed by Bonifacio *et al.*[56,57] using a mean field theory. Gronchi and Lugiato[58] have considered the effects of inhomogeneous broadening and refractive index changes in saturable absorbers. Willis[59] has developed a quantum mechanical master equation for the two-level saturable absorber in the steady state, and Willis and Day[60] have extended this theory to include the effects of refractive index changes in the two-level system.

Nonlinear FP resonators using a reactive medium have been discussed in the steady state by Felber and Marburger.[61,62] A theoretical description that includes both absorptive and reactive effects has been given approximately by Gibbs *et al.*[63] and in more detail by Agrawal and Carmichael.[64]

Other aspects of optical bistability have been investigated by many authors. The transient response of absorptive optical bistable devices has been discussed by Bonifacio and Meystre,[65] and the transient response of reactive devices has been treated by Bischofberger and Shen,[66–68] Grischowsky,[69] Hopf *et al.*,[70,71] and Goldstone and Garmire.[72] Critical slowing down near the transition point has been treated by Bonifacio and

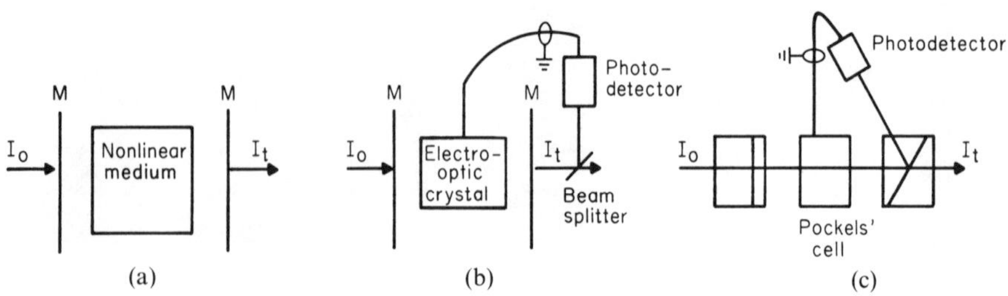

Fig. 10. Schematic diagrams of nonlinear bistable devices. (a) All-optical device. (b) Hybrid electro-optical device. (c) Mirrorless bistable device.

Meystre[73] and by Goldstone and Garmire.[72] The spectrum of the radiation that is transmitted by an absorptive bistable device has been discussed by Bonifacio and Lugiato[74] and Lugiato.[75] Regenerative pulsing has been discussed by McCall,[76] Bonifacio *et al.*,[77] Okada and Takizawa,[78] Lugiato,[79] and Benza *et al.*[80] Space and time effects have been treated by Abraham *et al.*;[81] effects due to fluctuations of a two-level saturable absorber have been discussed by Schenzle and Brand,[82] and effects related to noise on the laser source have been treated by Chrostowski and Zardecki[83] and by Ritze and Bandilla.[84] Optical bistability in laser resonators has been described by Levin and Tang[85] and by Haag *et al.*[86]

Experimental demonstrations of several steady-state optical effects using both reactive and absorptive FPs have been given by Szoke *et al.*[54] and Gibbs *et al.*[63] Transient effects have been demonstrated by Bischofberger and Shen[66–68] and by Grischowsky.[69]

A second class of FP device has also been shown to exhibit the property of optical bistability and many of the other effects mentioned. This type of device, shown in Fig. 10b, consists of an electrooptical (EO) crystal in an FP resonator. The nonlinear behavior needed for performing the various optical functions mentioned is obtained by controlling the properties of the EO crystal with an electrical signal that is derived from the transmitted optical intensity. This form of hybrid optical-electrical FP resonator has been discussed by Smith *et al.*[87,88]

It was later realized that a resonant cavity was not necessary for optical bistability. A mirrorless bistable device based on a Pockels cell between crossed polarizers is shown in Fig. 10c. This configuration was first proposed by Garmire *et al.*[89] and later discussed by Feldman.[90,91] The transient response of this type of device has been explored by Garmire *et al.*[92] and by Goldstone *et al.*[93] Various other configurations that exhibit optical bistability have also been proposed.[94,95]

The hybrid optical-electrical devices of Figs. 10b and 10c generally require lower optical intensities for operation than does the all-optical device of Fig. 10a. In addition, the mirrorless device of Fig. 10c can operate with broad-band incoherent light. The hybrids have been constructed in miniaturized form compatible with fiber optical and integrated optical formats. The properties of such integrated optical configurations have been studied by Smith *et al.*[96,97] and by Garmire *et al.*[98]

Optical bistability due to other nonlinear processes is also possible. For example, bistable effects in degenerate four-wave mixing have been discussed by Winful and Marburger[99] and by Agrawal and Flytzanis.[100]

Optical bistability has been the subject of a special issue of the *IEEE Journal of Quantum Electronics*,[101] and a more extensive list of references can be found there.

4.1. Reactive FP Resonators

Reactive FP resonators use a medium with a nonlinear refractive index. The nonlinear resonator behavior results from shifts in the transmission resonances caused by intensity-dependent phase changes in the nonlinear medium. Marburger and Felber[62] have shown that when the intensity-dependent contribution to the refractive index is small compared to the linear index, the transmission of a plane parallel FP, defined as the ratio of the transmitted intensity to the incident intensity I_t/I_0, is given by

$$T = [1 + F \sin^2(\delta + \alpha I_t)]^{-1}. \tag{33}$$

Here $F = 4R/(1 - R)^2 = 4\mathscr{F}^2/\pi^2$, where $\mathscr{F}$ is the finesse of the FP cavity, R is the mirror reflectivity, $\delta = 2\pi n_0 D/\lambda$ is the single-pass phase shift in the cavity at low intensities, and D is the spacing of the FP mirrors. The initial detuning of the FP from the nearest transmission resonance is given by $\phi = \delta - M\pi$, where M is the largest integer required to make ϕ fall between $-\pi/2$ and $+\pi/2$. The effects of the nonlinear index, defined as $n = n_0 + n_2|A|^2$, are contained in the parameter $\alpha = 6\pi D n_2(1 + R)/n_0(1 - R)$.

If we consider only the effects due to the nearest transmission resonance, regions of initial detuning for optical limiting, stable gain, and optical bistability can be determined from Eq. (33), as discussed by Gibbs *et al.*[102] and by Agrawal and Carmichael.[64] For initial detunings that satisfy

$$0 > \phi > -\sqrt{3}\pi/2\mathscr{F}, \tag{34}$$

stable gain occurs. For detunings in this range the maximum gain is given by

$$G_{\max} = (1 - 4\delta^2\mathscr{F}^2/3\pi^2)^{-1}. \tag{35}$$

For initial detunings in the range

$$\phi_1 < \phi < -\sqrt{3}\pi/\mathscr{F} \tag{36}$$

optical bistability occurs, while for initial detunings that satisfy

$$\phi_2 > \phi \geq 0 \tag{37}$$

optical limiting occurs. Optical limiting will also occur in the high-intensity regimes of either the stable gain or bistable operation.

Condition (34) is equivalent to condition (7.22) of Ref. 64 when differences in notation are accounted for. The maximum detunings ϕ_1 and ϕ_2 in Eqs. (36) and (37) are determined by the finesse of the FP resonator.

A graphical construction for the solution of Eq. (33). similar to the one described by Felber and Marburger,[61] is shown in Fig. 11. Here the transmission resonances corresponding to the right-hand side of Eq. (33) are shown as a function of $(\alpha I_t + \delta)$ for a finesse of 5. Also shown are straight

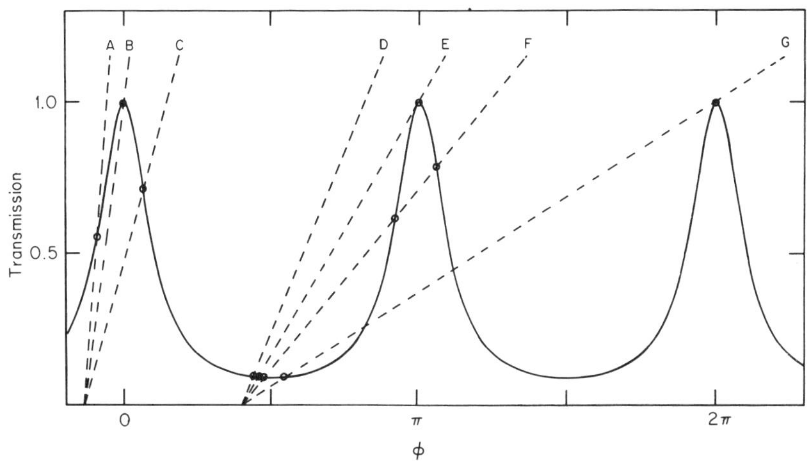

Fig. 11. Graphical construction of transmission of reactive nonlinear Fabry–Perot cavity. The cavity resonances are indicated by the solid line. The dashed lines have slope $(\alpha I_0)^{-1}$ and are drawn for different detunings, corresponding to different regimes of operation: *A–C*, stable gain; *D–F*, bistable; *G*, multistable.

lines with slope $(\alpha I_0)^{-1}$, corresponding to different values of incident intensity. The intersections of the two types of curves show possible allowed values of the transmission.

Sets of curves are shown for two different detunings, displayed for clarity on two separate transmission resonances. Lines *A–C* are for a relatively small initial detuning ($\delta = -0.15\pi$) satisfying condition (34). Each of the curves has only one intersection with the FP transmission curve and therefore only one value of transmission. Line *A* corresponds to a region of stable gain. Line *B*, at the peak of the transmission curve, marks the division between gain and optical limiting, which occurs for line *C*. Higher intensities would show regions of multistability which are shown in more detail in the remaining lines, *D–G*.

Lines *D–G* correspond to a larger initial detuning ($\delta = -0.4\pi$) for which bistable and multistable regions occur. Line *D* has one intersection, corresponding to a low value of transmission. Line *E* has two intersections and marks the onset of the bistable region. Line *F*, at higher incident intensity, has three intersections, two of which correspond to stable transmission values, one high and the other low, and the third to an unstable solution of Eq. (33). Higher incident intensities, e.g., line *G*, result in regions of multistability, as the nonlinear phase shift causes the FP to tune through multiple transmission resonances.

From these curves a graph of transmitted intensity versus incident intensity can be constructed for a given value of initial detuning. Samples of such curves are shown in Fig. 12a–c, corresponding to the three regions defined by conditions (34), (36), and (37), respectively. Stable gain is illustrated in Fig. 12a for the detuning of lines *A–C* in Fig. 11. As the incident intensity is increased from a low value, the nonlinear phase shift resulting from the increasing intracavity intensity causes the FP resonator to move closer to resonance. As a result the transmitted intensity increases faster than the incident intensity, resulting in the gain shown in the figure. The sharpness of the transition between the low- and high-transmission regions is determined by the maximum gain, which is related to the initial detuning relative to the width of the transmission resonance, as given by Eq. (35). As the incident intensity increases above the region in which gain is observed, the nonlinear phase shift causes the FP to move away from resonance, resulting in optical limiting.

A graph of transmitted versus incident intensity in the bistable region is shown in Fig. 12b for the detuning of lines *D–G* in Fig. 11. As the incident

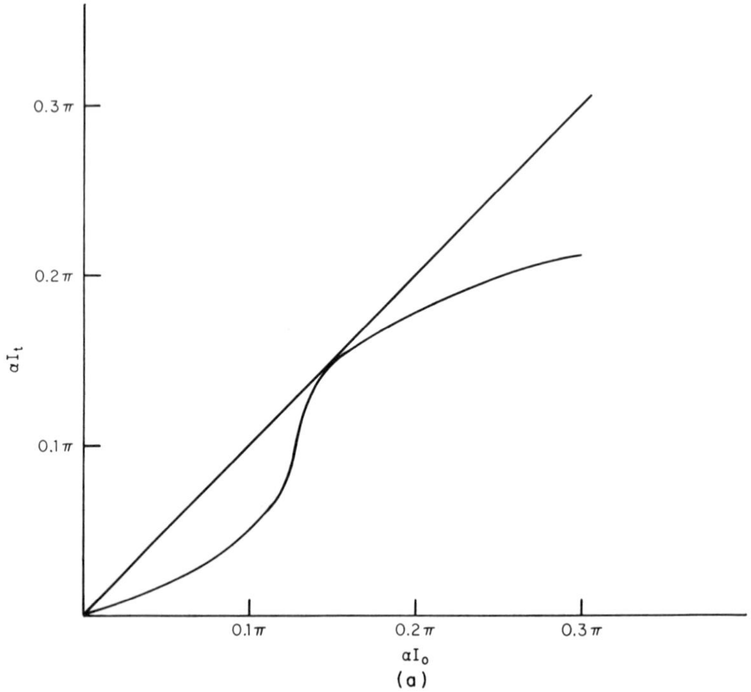

Fig. 12. (a) Transmitted intensity as a function of incident intensity for stable gain region *A–C* of Fig. 11. (b) transmitted intensity in bistable region *D–F*. (c) Optical limiting, which occurs for positive detunings from an FP resonance.

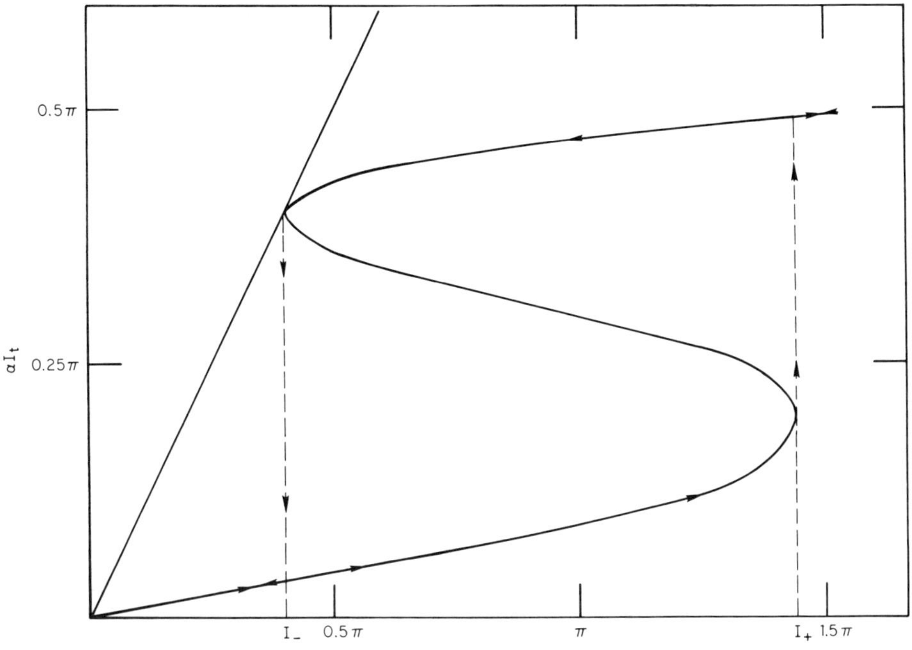

0.5π
0.25π
αI_t
I_-
0.5π
π
I_+
1.5π
αI_o

(b)

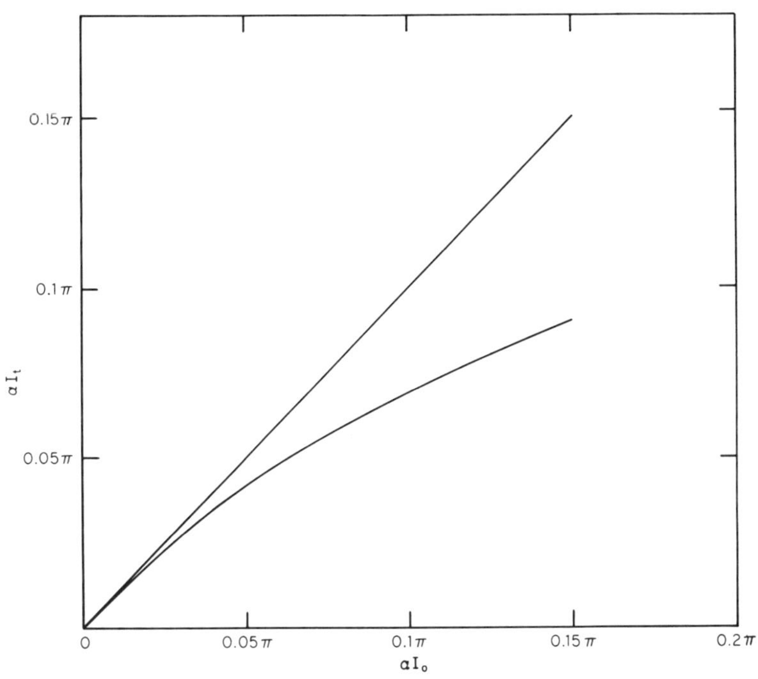

0.15π
0.1π
0.05π
αI_t
0
0.05π
0.1π
0.15π
0.2π
αI_o

(c)

intensity is increased from a low value the transmitted intensity remains low until a switching intensity I_+ is reached. When the incident intensity exceeds I_+ the transmission switches from a low value to a high value, as indicated by the dotted lines in Fig. 12b. Again at higher incident intensities the nonlinear phase shift results in optical limiting.

As the incident intensity is decreased from a region of high transmission the FP moves closer to resonance and the transmitted intensity remains high until the incident intensity reaches a maintaining value I_- for which the transmission is unity. When the incident intensity is decreased below I_- the transmission switches to a low value. The maintaining intensity is less than the switching intensity I_+ and the graph of transmitted versus incident intensity exhibits hysteresis for detunings satisfying (36). The dependence of I_+ and I_- on initial detuning has been studied by Felber and Marburger. The nonlinear FP can be used in the bistable mode for any of the functions that bistable electrical circuits are used for, such as memory applications.

The stable gain that is encountered for detunings that satisfy (34) is not the same as that encountered in laser amplifiers with population inversions. Rather it is similar to the gain in active electronic circuits. A technique for observing such gain is illustrated in Fig. 13. Here a signal beam is used to modulate the refractive index of the nonlinear medium. The modulation appears on the more intense transmitted main beam, amplified by the gain of the device.

As the transition between stable gain and bistable operation is approached the slope of the transmission curve becomes very steep. In this region the device can be used for optical switching by biasing it just below its transition and using a signal beam to switch the transmission from its low to its high

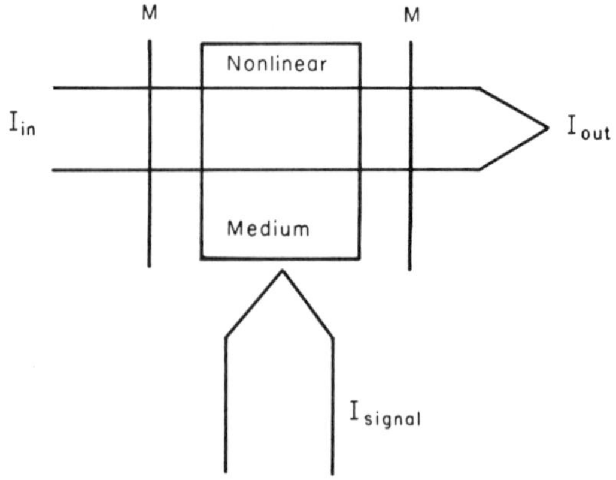

Fig. 13. Schematic diagram of arrangement in which the nonlinear FP is used as an optical switch.

value. The amount of optical power needed for switching depends on the steepness of the transition region and decreases as the transition region becomes steeper. Generally minimum switching requirements are controlled by fluctuations in the incident laser intensity, which limit how close to the transition region the device can be biased.

If the initial detuning is chosen to be zero or positive, then effects of limiting and clipping can be observed without differential gain or bistability. For this situation the transmission of the FP cavity decreases monotonically as the incident intensity increases, as shown in Fig. 12c.

4.2. Absorptive Nonlinear FP Resonators

In FP resonators that use saturable absorbers the nonlinear behavior results from intensity-dependent changes in the amount of feedback in the resonator as the intracavity fields bleach the absorbing medium. No shift of the resonant frequencies is involved.

To describe the operation of such a device qualitatively, we consider an FP cavity tuned to resonance. At low intensity most of the radiation is reflected because the absorption of the medium within the cavity prevents resonant transmission from occurring. As the incident intensity is increased the intracavity absorption is bleached, allowing the intracavity intensity to build up and cause further bleaching of the nonlinear medium. This regenerative operation eventually results in a transition from a condition of low transmission to one of high transmission. Depending on the relation of the initial absorption to the empty cavity finesse the absorptive FP can exhibit gain, bistability, optical switching, or regenerative pulsing.

In the region above the transition from low to high transmission the transmitted intensity is proportional to the incident intensity, and as a result the absorptive FP does not exhibit limiting or clipping. It also does not exhibit multistability.

In the bistable mode hysteresis occurs in the graph of transmitted intensity versus incident intensity because as the incident intensity decreases in a region of high transmission the intracavity intensity maintains the saturated condition of the nonlinear absorber until the incident intensity drops below the value at which the transmission switches from low to high as the intensity is increased.

The transmission of the absorptive FP, as shown by several authors, (see, for example, Agrawal and Carmichael[64]), is given by

$$T = I_t/I_0 = [1 + 2C/(1 + I_t')]^{-2}, \tag{38}$$

where $C = \alpha LR/2(1 - R)$, $I_t' = 2I_t/I_s(1 - R)$, L is the length of the FP cavity, R is the mirror reflectivity, I_s is the saturation intensity, and α is the

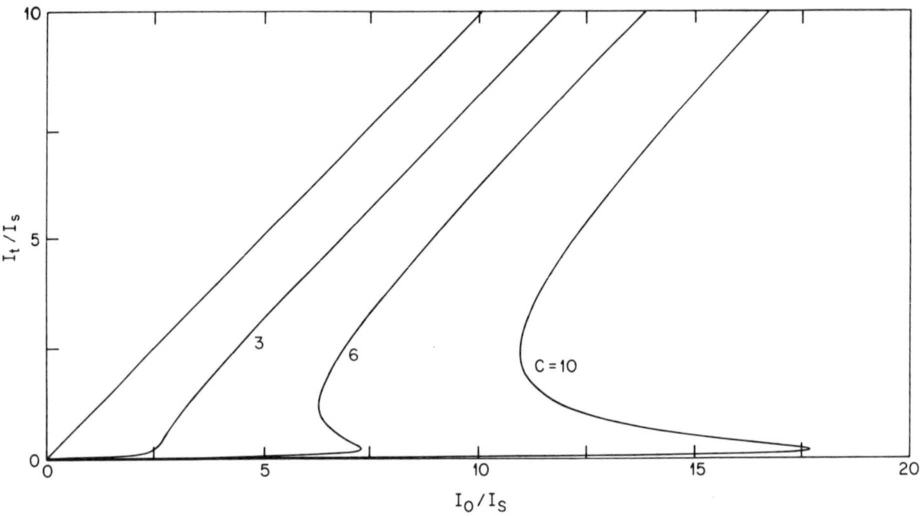

Fig. 14. Transmission of absorptive FP cavity for various values of the parameter C.

linear absorption coefficient. For initial absorption coefficients and mirror reflectivities such that $C > 4$ bistability occurs, while for absorptions such that $C < 4$ stable gain is obtained. Curves for the transmitted intensity as a function of incident intensity similar to the ones obtained for the reactive FP can be constructed. An example of such curves for an absorptive nonlinear FP with a finess of 9.2 is shown in Fig. 14 for various values of C. For a given initial absorption the steepness of the transition and the existence of stable gain or bistability can be controlled by the mirror reflectivity.

4.3. Hybrid Electrical-Optical Bistable Devices

The hybrid device of Fig. 10b is similar to the reactive nonlinear FP in that the nonlinear cavity behavior results from an intensity-dependent shift of the cavity transmission resonances. In this case the nonlinear phase shifts occur in the EO crystal and are controlled by the electrical signal derived from the transmitted intensity. The effective size of the nonlinearity can be controlled by the size of the signal used to drive the crystal. This device will exhibit the same effects as those encountered in the reactive FP, namely, bistability, switching limiting, clipping, differential gain, and regenerative pulsing.

The hybrid mirrorless device of Fig. 10c uses the electrical signal produced by the transmitted beam to change the polarization of the beam that strikes the second polarizer. Again the size of the nonlinearity can be controlled by the magnitude of the electrical signal and this device also exhibits the same effects as the reactive FP.

4.4. Experimental Results

Several modes of operation of a nonlinear FP were demonstrated by Gibbs *et al.*[63] They used an FP with a mirror reflectivity of 0.9 and a finesse of 30. Na vapor was used for the nonlinear medium and operating wavelengths were chosen in the vicinity of the D lines. Depending on the exact relation of the frequency of the incident radiation to the D doublet the dominant response of the Na vapor can be reactive, absorptive, or a mixture of both. The authors observed bistable operation, differential gain, and optical clipping. Stable gains of the order of 2 were observed in the amplification mode. For most of their operation they attributed their observations to the reactive part of the Na nonlinearity. At three specific frequencies, however, the reactive nonlinearity was small and effects due to saturation of the absorption could be observed.

A transient response of a nonlinear FP can occur as a result of either a finite buildup time for the fields in the cavity or a finite response time of the nonlinear medium. Transient responses can limit switching speed and can cause ringing and overshoot as the FP switches from one transmission value to another.

The transient characteristics of reactive FPs were investigated by Bischofberger and Shen.[66–68] They used cavities filled with either a liquid crystal, for which the response time of the nonlinear medium was the dominant source of the transient response, or with CS_2 or nitrobenzene, for which cavity buildup effects were important. They observed transient responses in the form of distortion of the transmitted pulse for operation in the stable gain, bistable, and limiting regions. Their results indicate that even when the response of the nonlinear medium is effectively instantaneous the minimum switching time will be limited by the finite buildup time in the cavity. An example of their results is shown in Fig. 15 for three modes of operation.

Grischowsky[69] has investigated transient aspects of a reactive FP in which the nonlinear index resulted from adiabatic following in a nearly resonant alkali metal vapor. In addition to limitations in switching speed imposed by cavity buildup times, he observed effects on performance which he attributed to four-wave mixing.

The properties of hybrid FP resonators have been investigated by Smith *et al.*[96,97] Optical limiting, bistability, and stable gain all have been reported. Optical switching with only microwatts of optical power at the detector

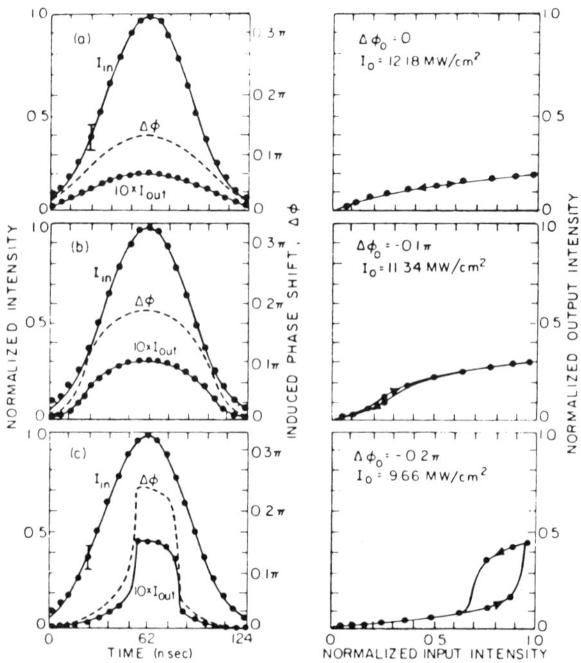

Fig. 15. Experimental observations of transmission of nonlinear FP cavity. (a) Optical limiting. (b) Stable gain. (c) Bistable. The cavity buildup time was 0.55 nsec, longer than the response time of the nonlinear medium. Time-dependent behavior is shown at left. (From Ref. 67.)

was observed. Overshoot and critical slowing down due to the transient response in mirrorless bistable devices has been reported by Garmire *et al.*[92] and by Goldstone *et al.*[93]

REFERENCES

1a. L. S. Vasilenko, V. P. Chebotaev, and A. V. Shisaev, *Pis'ma Zh. Eksp. Teor. Fiz.* **12**, 161 (1970) [*Sov. Phys.—JETP Lett.* (*Engl. Transl.*) **12**, 1131 (1970)].

1b. B. Cagnac, G. Grynberg, and F. Biraben, *J. Phys.* (*Orsay, Fr.*) **34**, 845 (1973).

1c. D. Pritchard, J. Apt, and T. W. Ducas, *Phys. Rev. Lett.* **32**, 641 (1974).

1d. F. Biraben, B. Cagnac, and G. Grynberg, *Phys. Rev. Lett.* **32**, 643 (1974).

1e. M. D. Levenson and N. Bloembergen, *Phys. Rev. Lett.* **32**, 645 (1974).

2. J. A. Armstrong, P. Esherick, and J. J. Wynne, *J. Opt. Soc. Am.* **69**, 211 (1979).
3. P. Esherick, *Phys. Rev. A* **15**, 1920 (1977).
4. K. C. Harvey and B. P. Stoicheff, *Phys. Rev. Lett.* **38**, 537 (1977).
5. J. J. Wynne and J. P. Hermann, *Opt. Lett.* **4**, 106 (1979).
6. J. Sugar, T. B. Lucatorto, T. J. McIlrath, and A. W. Weiss, *Opt. Lett.* **4**, 109 (1979).

7. J. A. Armstrong and J. J. Wynne, *Phys. Rev. Lett.* **33**, 1183 (1974).
8. S. C. Wallace and K. K. Innes, *J. Chem. Phys.* **72**, 4805 (1980).
9. C. Tai, F. W. Talby, and G. L. Giles, *Phys. Rev. A* **20**, 233 (1979).
10. J. J. Wynne, *Phys. Rev. Lett.* **29**, 650 (1972).
11. M. D. Levenson, C. Flytzanis, and N. Bloembergen, *Phys. Rev. B* **6**, 3962 (1972).
12. B. Attal, O. O. Schneff, and J.-P. Taran, *Opt. Commun.* **24**, 77 (1978).
13. M. D. Levenson and N. Bloembergen, *Phys. Rev. B* **10**, 4447 (1974).
14. M. D. Levenson and N. Bloembergen, *J. Chem. Phys.* **60**, 1323 (1974).
15. E. Wiener-Avnear, S. Chandra, and A. Compaan, *Appl. Phys. Lett.* **32**, 286 (1978).
16. S. Saikan and G. Morowsky, *Opt. Commun.* **26**, 466 (1978).
17. R. F. Begley, A. B. Harvey, and R. L. Byer, *Appl. Phys. Lett.* **25**, 387 (1974).
18. L. A. Carreira, L. P. Goss, and T. B. Malloy, Jr., *J. Chem. Phys.* **69**, 855 (1978).
19. P. R. Regnier, F. Moya, and J.-P. Taran, *AIAA J.* **12**, 826 (1974); P. R. Regnier and J.-P. Taran, *Appl. Phys. Lett.* **23**, 240 (1973).
20. F. Moya, S. A. J. Druet, and J. P. E. Taran, *Opt. Commun.* **13**, 169 (1975).
21. R. J. Hall and A. C. Eckbreth, *Opt. Eng.* **20**, 494 (1981).
22. L. A. Rahn, L. J. Zych, and P. L. Mattern, *Opt. Commun.* **30**, 249 (1979).
23. R. T. Lynch, Jr., S. D. Kramer, H. Lotem, and N. Bloembergen, *Opt. Commun.* **16**, 372 (1976).
24. H. Lotem and R. T. Lynch, Jr., *Phys. Rev. Lett.* **37**, 334 (1976).
25. H. Lotem, R. T. Lynch, Jr., and N. Bloembergen, *Phys. Rev. A* **14**, 1748 (1976).
26. R. R. Alfano and S. L. Shapiro, *Phys. Rev. Lett.* **26**, 1247 (1971).
27. A. Lauberau and W. Kaiser, *Rev. Mod. Phys.* **50**, 607 (1978).
28. F. M. Kamga and M. G. Sceats, *Opt. Lett.* **5**, 126 (1980).
29. D. Heiman, R. W. Hellwarth, M. D. Levenson, and G. Martin, *Phys. Rev. Lett.* **36**, 189 (1976).
30. J. J. Song, G. L. Eesly, and M. D. Levenson, *Appl. Phys. Lett.* **29**, 567 (1976).
31. M. D. Levenson, J. J. Song, R. W. Hellwarth, and D. Heiman, *Opt. Commun.* **18**, 133 (1976).
32. I. Chabay, G. Klauminzer, and B. S. Hudson, *Appl. Phys. Lett.* **28**, 27 (1976).
33. R. T. Lynch, Jr., H. Lotem, and N. Bloembergen, *J. Chem. Phys.* **66**, 4250 (1977).
34. W. Zinth, A. Laubereau, and W. Kaiser, *Opt. Commun.* **26**, 457 (1978).
35. A. Owyoung, *IEEE J. Quantum Electron.* **QE-14**, 192 (1978).
36. V. S. Letokhov, *in* "High Resolution Spectroscopy" (K. Shimoda, ed.), p. 95. Springer-Verlag, Berlin and New York, 1976; C. E. Wagstaff and M. H. Dunn, *Opt. Commun.* **35**, 353 (1980); B. Couillaud, L. A. Bloomfield, J. E. Lawler, A. Siegel, and T. W. Hansch, *Opt. Commun.* **35**, 359 (1980).
37. W. M. Tolles, J. W. Nibler, J. R. McDonald, and A. B. Harvey, *Appl. Spectrosc.* **31**, 253 (1977).
38. S. A. Akhmanov and N. I. Koroteev, *Usp. Fiz. Nauk* **123**, 405 (1977) [*Sov. Phys.—Usp.* (*Engl. Transl.*) **20**, 899 (1977)].
39. A. C. Eckbreth, *Appl. Phys. Lett.* **32**, 421 (1978).
40. G. Laufer and R. B. Miles, *Opt. Commun.* **28**, 250 (1979).
41. W. M. Shaub, A. B. Harvey, and G. C. Bjorklund, *J. Chem. Phys.* **67**, 2547 (1977).
42. W. M. Tolles and R. D. Turner, *Appl. Spectrosc.* **31**, 96 (1977).
43. M. D. Levenson, *IEEE J. Quantum Electron.* **QE-10**, 110 (1974).
44. J. L. Oudar, R. W. Smith, and Y. R. Shen, *Appl. Phys. Lett.* **34**, 758 (1979).
45. D. M. Bloom, J. T. Yardley, J. F. Young, and S. E. Harris, *Appl. Phys. Lett.* **24**, 427 (1974).
46. E. A. Stappaerts, S. E. Harris, and J. F. Young, *Appl. Phys. Lett.* **29**, 669 (1976).
47. G. Martin and R. W. Hellwarth, *Appl. Phys. Lett.* **34**, 371 (1979).

48. D. S. Bethune, J. R. Lankard, and P. P. Sorokin, *Opt. Lett.* **4**, 103 (1979).
49. D. S. Bethune, J. R. Lankard, M. M. T. Loy, and P. P. Sorokin, *IBM J. Res. Dev.* **23**, 556 (1979).
50. S. E. Harris and D. M. Bloom, *Appl. Phys. Lett.* **24**, 229 (1974).
51. E. A. Stappaerts, *IEEE J. Quantum Electron.* **QE-15**, 110 (1979).
52. H. Kildal and S. R. J. Brueck, *IEEE J. Quantum Electron.* **QE-16**, 566 (1980).
53. H. Seidel, U.S. Patent No. 3610731 (1971).
54. A. Szoke, V. Daneu, J. Goldhar, and N. A. Kurnit, *Appl. Phys. Lett.* **15**, 376 (1969).
55. S. L. McCall, *Phys. Rev. A* **9**, 1515 (1974).
56. R. Bonifacio and L. A. Lugiato, *Opt. Commun.* **19**, 172 (1976).
57. R. Bonifacio, L. A. Lugiato, J. D. Farina, and L. M. Narducci, *IEEE J. Quantum Electron.* **QE-17**, 357 (1981).
58. M. Gronchi and L. A. Lugiato, *Opt. Lett.* **5**, 108 (1980).
59. C. R. Willis, *Opt. Commun.* **23**, 151 (1977); *Opt. Commun.* **26**, 62 (1978).
60. C. R. Willis and J. Day, *Opt. Commun.* **28**, 137 (1979).
61. F. S. Felber and J. H. Marburger, *Appl. Phys. Lett.* **28**, 731 (1976).
62. J. H. Marburger and F. S. Felber, *Phys. Rev. A* **17**, 335 (1978).
63. H. M. Gibbs, S. L. McCall, and T. N. C. Venkatesan, *Phys. Rev. Lett.* **36**, 1135 (1976).
64. G. F. Agrawal and H. J. Carmichael, *Phys. Rev. A* **19**, 2074 (1979).
65. R. Bonifacio and P. Meystre, *Opt. Commun.* **27**, 147 (1978).
66. T. Bischofberger and Y. R. Shen, *Appl. Phys. Lett.* **32**, 156 (1978).
67. T. Bischofberger and Y. R. Shen, *Opt. Lett.* **4**, 40 (1979).
68. T. Bischofberger and Y. R. Shen, *Phys. Rev. A* **19**, 1169 (1979).
69. D. Grischowsky, *J. Opt. Soc. Am.* **68**, 641 (1978).
70. F. A. Hopf, P. Meystre, P. D. Drummond, and D. F. Walls, *Opt. Commun.* **31**, 245 (1979).
71. F. A. Hopf and P. Meystre, *IEEE J. Quantum Electron.* **QE-17**, 366 (1981).
72. J. A. Goldstone and E. M. Garmire, *IEEE J. Quantum Electron.* **QE-17**, 366 (1981).
73. R. Bonifacio and P. Meystre, *Opt. Commun.* **29**, 131 (1979).
74. R. Bonifacio and L. A. Lugiato, *Phys. Rev. Lett.* **40**, 1023 (1978).
75. L. A. Lugiato, *Nuovo Cimento Soc. Ital. Fiz. B* **50**, 89 (1979).
76. S. L. McCall, *Appl. Phys. Lett.* **32**, 284 (1978).
77. R. Bonifacio, M. Gronchi, and L. A. Lugiato, *Opt. Commun.* **30**, 129 (1979).
78. M. Okada and K. Takizawa, *IEEE J. Quantum Electron.* **QE-16**, 770 (1980).
79. L. A. Lugiato, *Opt. Commun.* **33**, 108 (1980).
80. V. Benza, L. A. Lugiato, and P. Meystre, *Opt. Commun.* **33**, 113 (1980).
81. E. Abraham, R. K. Bullough, and S. S. Hassan, *Opt. Commun.* **29**, 109 (1979).
82. A. Schenzle and H. Brand, *Opt. Commun.* **27**, 485 (1978).
83. J. Chrostowski and A. Zardecki, *Opt. Commun.* **29**, 230 (1979).
84. H.-H. Ritze and A. Bandilla, *Opt. Commun.* **29**, 126 (1979).
85. K. H. Levin and C. L. Tang, *Appl. Phys. Lett.* **34**, 376 (1979).
86. G. Haag, M. Munz, and G. Morowsky, *IEEE J. Quantum Electron.* **QE-17**, 349 (1981).
87. P. W. Smith and E. H. Turner, *Appl. Phys. Lett.* **30**, 280 (1977).
88. P. W. Smith, E. H. Turner, and B. B. Mumford, *Opt. Lett.* **2**, 55 (1978).
89. E. Garmire, J. H. Marburger, and S. D. Allen, *Appl. Phys. Lett.* **32**, 320 (1978).
90. A. Feldman, *Appl. Phys. Lett.* **33**, 243 (1979).
91. A. Feldman, *Opt. Lett.* **4**, 115 (1979).
92. E. Garmire, J. H. Marburger, S. D. Allen, and H. G. Winful, *Appl. Phys. Rev.* **34**, 374 (1979).
93. J. A. Goldstone, P.-T. Ho, and E. Garmire, *Appl. Phys. Lett.* **37**, 126 (1980).

94. U. K. Sengupta, V. H. Gerlach, and S. A. Collins, *Opt. Lett.* **3**, 199 (1978).
95. A. Schnapper, M. PaPuchon, and C. Puech, *Opt. Commun.* **29**, 364 (1979).
96. P. W. Smith, I. P. Kaminov, P. J. Maloney, and L. N. Stulz, *Appl. Phys. Lett.* **33**, 24 (1978).
97. P. W. Smith, I. P. Kaminov, P. J. Maloney, and L. N. Stulz, *Appl. Phys. Lett.* **34**, 62 (1979).
98. E. Garmire, S. D. Allen, and J. H. Marburger, *Opt. Lett.* **3**, 69 (1978).
99. H. G. Winful and J. H. Marburger, *Appl. Phys. Lett.* **36**, 613 (1980).
100. G. P. Agrawal and C. Flytzanis, *IEEE J. Quantum Electron.* **QE-17**, 374 (1981).
101. *IEEE J. Quantum Electron.* **QE-17** (1981).
102. H. M. Gibbs, S. L. McCall, and T. N. C. Venkatesan, *in* "Coherence in Spectroscopy and Modern Physics" (F. T. Arecchi, R. Bonifacio, and M. O. Scully, eds.), p. 111. Plenum, New York, 1978.

INDEX

D

E

F

G

H

S

Social Media, Criminal Law and Legality

Utilising Lon Fuller's conception of legality, this book argues that current legal provisions often used to control online abuse aided by social media do not conform to the basic principles of legality in the criminal law, in turn, threatening freedom of expression. How we regulate inappropriate behaviour online, often referred to as online abuse, particularly online abuse aided by social media, is a contemporary concern for governments across the globe. Tragedies, such as the death of a celebrity following a campaign of online abuse, often hit the headlines, followed by the same echo: there should be a law against this. Yet, in England and Wales, numerous laws exist to control, prosecute, and convict individuals who use the likes of social media to harass, intimidate, and abuse others online. So why is the law failing to keep pace with modern technology? This monograph critically examines this fundamental question, from the perspective of legality. Applying criminal law to three growing areas of concern, it covers: (1) racist speech, (2) cyberharassment/cyberstalking, and (3) the sending of abusive messages online. It then turns to examine the latest attempts by UK officials to tackle these issues through the implementation of the Online Safety Act 2023 and France's, Germany's, and India's attempts to regulate social media. The book will be of interest to researchers in the field of criminal law and cyber law, as well as online abuse, harassment, and discrimination.

Laura Higson-Bliss is a Lecturer in Law at Keele University, United Kingdom, joining Keele School of Law in 2021. Her interests, broadly conceived, are in the interaction between social media and the criminal law, legality, freedom of expression and law and new technologies. Laura is an author of several blogs and a regular contributor to wider discussions on social media regulation on the radio and television, both nationally and internationally.

Routledge Research in the Law of Emerging Technologies

Regulating Artificial Intelligence
Binary Ethics and the Law
Dominika Ewa Harasimiuk and Tomasz Braun

Cryptocurrencies and Regulatory Challenge
Allan C. Hutchinson

Regulating Artificial Intelligence in Industry
Edited by Damian M. Bielicki

The Law of Global Digitality
Edited by Matthias C. Kettemann, Alexander Peukert and Indra Spiecker gen. Döhmann

Internet of Things and the Law
Legal Strategies for Consumer-Centric Smart Technologies
Guido Noto La Diega

Regulating the Metaverse
A Critical Assessment
Ignas Kalpokas and Julija Kalpokienė

The Regulation of Digital Technologies in the EU
Act-ification, GDPR Mimesis and EU Law Brutality at Play
Vagelis Papakonstantinou and Paul de Hert

Privacy, Data Protection and Data-driven Technologies
Edited by Martin Ebers and Karin Sein

Social Media, Criminal Law and Legality
Laura Higson-Bliss

For more information about this series, please visit: https://www.routledge.com/Routledge-Research-in-the-Law-of-Emerging-Technologies/book-series/LAWTECHNOLOGY

Social Media, Criminal Law and Legality

Laura Higson-Bliss

Routledge
Taylor & Francis Group
LONDON AND NEW YORK

First published 2025
by Routledge
4 Park Square, Milton Park, Abingdon, Oxon OX14 4RN

and by Routledge
605 Third Avenue, New York, NY 10158

Routledge is an imprint of the Taylor & Francis Group, an informa business

British Library Cataloguing-in-Publication Data
A catalogue record for this book is available from the British Library

ISBN: 978-1-032-30155-6 (hbk)
ISBN: 978-1-032-30156-3 (pbk)
ISBN: 978-1-003-30368-8 (ebk)

DOI: 10.4324/9781003303688

Typeset in Times New Roman
by Deanta Global Publishing Services, Chennai, India

This one's for you Dad. We miss you.

Also, to Walt, who has heard me read this monograph out loud dozens of times during the proofreading process. You are loved beyond words.

Contents

Figures

Acknowledgements

This monograph has been a long time coming and would never have been made possible without the support of so many people. I would firstly like to thank Routledge for their support throughout the writing process and for being understanding when asking for extensions following the many occasions the Online Safety Act 2023 was delayed during its passing through Parliament!

This monograph is loosely based on my PhD thesis, submitted to Edge Hill University in 2019. I would like to thank my examiners, Professor Richard Parrish, and Dr Holly Powell-Jones for their constructive and supportive feedback during the viva process.

I am also grateful to my colleagues at Keele School of Law for their support and advice, with particular thanks to Dr Emma Allen, Dr Stella Coyle, Dr Bleddyn Davies, Dr Elizabeth Faulkner, Eve Lakin, Lisa Mountford, Professor Yossi Nehushtan, Dr Abigail Pearson, Dr Sotirios Santatzoglou, and Professor Anthony Wrigley. A special thank you to my mentors, Dr Kathryn Simpson and Professor Alannah Tomkins for all their support whilst I have been at Keele University. I am also grateful for the support I have received from the Foundation Year Team, alongside all of our admin staff at Keele, and the support of so many colleagues during my teaching fellowship at the University of Birmingham.

This monograph would not have been possible without the continued support of my PhD supervisor Adam Pendlebury – he really has gone above and beyond as a PhD supervisor. I am also incredibly lucky that I had Dr Grace Robinson walking the long journey of completing a PhD by my side. Having you made those days that felt never-ending so worthwhile. You truly are the most wonderful person. It takes a lot to get inside the trenches of grief when you are at your happiest moment. But you did that for me, and that I will always be thankful for.

I have also been heavily supported by my friends. Kelly Wilson, Janette McMillan, Steph Farish, and Charna Paton, you really have been my lifelines since school. We may all have grown up and have families of our own, but I know you will always be there for me. Michelle Cliffe, Liz Attwood, and Nic Chatterley – I owe all three of you so much. The next Saturday coffee after the gym is on me.

To James Bliss, my not-so-little brother. You have grown into one of the most wonderful human and your strength of character makes me so proud every single day. We lost our Dad in 2013, and things were difficult for us both.

But none of this would have been possible without my husband, Jonathan Higson-Bliss. You are my rock, and I am forever grateful to have you next to me. I am excited for our next big adventure as parents to Walt, who is already loved so much, following the loss of our twins in December 2023 – we carry them both with us in our hearts.

Finally, to my Dad – everything I do is for you. I love you. I miss you. Always.

Cases

Legislation

Part I

Setting the Scene

This section provides the contextualisation of the key arguments made throughout this monograph. It is broken down into three chapters. The first chapter provides an introductory overview of the monograph, highlighting the dominance of social media today. Chapter 2 discusses at length the right to freedom of expression, an important democratic right, alongside that of the right to privacy. The chapter highlights why freedom of expression is important in the digital world and why we need to be careful in using the coercive nature of the criminal law to limit our online expressions. Chapter 3 provides the theoretical positioning which will be used to critique the criminal law in later chapters, legality. The key arguments put forward in these chapters, the dominance of social media, the importance of freedom of expression in a digital world and the significance of ensuring the principle of legality is maintained in the criminal law, provides the grounding for future arguments made in this monograph.

DOI: 10.4324/9781003303688-1

1 Introduction

1.1 Introduction

The ‘harms’ associated with the online world have become the moral panic of the digital era. It is very rare for a week to go by without a story emerging in the mainstream media discussing a negative event that has taken place online, both inside and outside of the United Kingdom. From cyberbullying, to harassment, to suicides which can be linked back to a person’s interactions with content on the internet, the online world is often portrayed as the ‘wild west’ which needs to be tamed.[1] But the idea that the internet is beyond regulation is untrue. The law can, and does, apply to conduct carried out online. The issue remains, however, in how far the law should intervene. This becomes an even more prominent question when it comes to communications sent *via* the use of social media platforms.

Today anyone in the world can become a publisher. Indeed, any social media user who actively posts online, is a publisher, just without the legal training. The ease of access to the online world means that communications sent online are rising year-on-year, resulting in more content being published than ever before. The key difference today is that when a comment is made online it can be recorded in a form that can be searched and examined long after the statement was made, even if the message only took seconds to write.[2] It can also reach millions of people from across the globe in a matter of seconds. Nothing has caused as much of a headache for governments globally, then attempting to make the online world a safer environment.

At the point of writing, the regulation of the online world, in particular social media platforms, is a top priority for governments across the globe, with the United Kingdom leading the way in attempting to become ‘one of the safest places to go online’ through the creation and implementation of the Online Safety Act 2023

1 David Omand, ‘The dark net: Policing the internets underworld’ (2016) Winter 2015/16 World Policy Journal <https://worldpolicy.org/2015/12/09/the-dark-net-policing-the-internets-underworld/> accessed 13 January 2020.

2 Jacob Rowbottom, ‘To Rant, Vent and Converse: Protecting Low Level Digital Speech’ (2012) 71(2) C.L.J. 355, 367.

DOI: 10.4324/9781003303688-2

(OSA).[3] At the same time, pressure is being felt across police forces as social media related complaints are on the rise.[4] Not only are the police being asked to patrol the physical streets, there is now a growing expectation that the virtual streets also need to be monitored. As eloquently put by Rowbottom:

> [T]wenty years ago where a person made a racist remark in a social setting, the chances of the police ever hearing about it were small. Now a recipient can direct the police to the statement made online and allow them to witness it first hand.[5]

We are now witnessing a growing trend in the application of the criminal law framework to conduct carried out on the likes of Facebook, X (formally Twitter),[6] Instagram, TikTok, Snapchat, and YouTube, to name but a few. This is especially true for England and Wales, which will be the primary jurisdiction explored in this monograph.

Yet the application of the criminal law to conduct carried out on social media platforms does not necessarily adhere to the basic principle of legality in the criminal law—an important underlying principle which should underpin any legal provision, as set out in Chapter 3. This will be the thesis of this monograph—the criminal law framework, as it currently stands, when applied to certain conduct carried out on social media does not uphold the very basics of legality. In turn, the right to freedom of expression, as protected under both national and international law, as discussed at length in Chapter 2 is under threat. The discussion of legality and freedom of expression form Part One of this monograph.

To illustrate this thesis, Part Two focusses on three key 'harms' associated with the online world and social media:

- Racist comments made online, focussing primarily on s.18 and 19 of the Public Order Act 1986 (POA), alongside discussing s.66 of the Sentencing Act 2020 (SA) and s.31 and 32 of the Crime and Disorder Act 1988 (CDA) in Chapter 4;
- Cyberharassment and cyberstalking with reference to the Protection from Harassment Act 1997 (PfHA) in Chapter 5; and
- More broadly what can be defined as communication offences, as prohibited under both the Malicious Communications Act 1988 (MCA) and s.127 of the Communications Act 2003 (CA) in Chapter 6. In addition, following

3 Technology Secretary Michelle Donelan. See, Department for Science, Innovation and Technology, 'UK children and adults to be safer online as world-leading bill becomes law' (*Gov.uk,* 26 October 2023) <https://www.gov.uk/government/news/uk-children-and-adults-to-be-safer-online-as-world-leading-bill-becomes-law> accessed 1 November 2023.

4 The BBC, 'Teenager's life "ruined" by Live.me and Twitter "trolls"' *The BBC* (London, 24 October 2017) <http://www.bbc.co.uk/news/uk-england-41693437> accessed 30 January 2018.

5 Rowbottom (n 2).

6 Throughout this monograph, where reference is made to Twitter, this is because the content of the discussion relates to something that happen prior to the Twitter rebrand. Otherwise, X will be used.

the passing of the OSA in October 2023, some of the 'new' communication offences will be discussed in Chapter 7.

Each of these chapters will set out the law, focussing on the *actus reus* (the criminal conduct), the *mens rea* (the mental element of the offence) and where appropriate defences, before turning to examine their application in a social media context against the principle of legality and the right to freedom of expression.

Whereas Part Two focuses on the application of the criminal law to certain types of conduct carried out on social media, Part Three turns to examine the regulation of social media companies themselves. Chapter 8 starts by outlining some of the key elements of the OSA against the principle of legality, before turning to examine in Chapter 9, how France, Germany, and India have attempted to regulate social media platforms. This allows for comparisons to be made between different jurisdictions. The final chapter will draw the arguments made in this monograph to a close. Section 1.4 below, provides an overview of each chapter.

However, before setting out each chapter, it is important to contextualise the arguments made in this monograph. To do this, the following discussion will provide a brief overview of the rise in social media platforms. First, what is meant by the term social media will be discussed, before turning to provide a brief overview of the creation and dominance of the World Wide Web (WWW), commonly referred to as the internet. Next, the rise in social media will be outlined. Finally, a discussion will take place exploring the 'good' and the 'bad' when it comes to social media.

1.2 The Development of Social Media

It is hard to imagine a world today without social media. In fact, we are now living in a time where some individuals do not know a world without the internet, let alone social media. These individuals are commonly referred to as digital natives.[7] But like many of the abstract concepts that will be explored in this monograph, such as online abuse, harassment, or offensive commentary, there is no one true definition of social media. The term social media can cover a range of online interactions, for instance, blogging, video sharing, and virtual world reality games can all be considered a form of social media.[8] Many of us, when considering what social media is, will think of Facebook, owned by Meta. Less will consider WhatsApp (another platform owned by Meta) as a traditional form of social media, but it is, for the purposes of regulation, considered a social media platform.

Despite there being no true definition of social media, it has come to be accepted that social media is 'the ability to share, to co-operate, with one another, and to take collective actions, all outside the framework of traditional institutions and

7 Marc Prensky, 'Digital Natives, Digital Immigrants' 2001 9(5) Horiz 1.
8 Laura Scaife, *Handbook of Social Media and the Law* (Routledge 2015) 8–9.

organisations'.[9] As outlined by the Law Commission during their 2018 scoping report into the criminal law and online abuse, there are several fundamental characteristics associated with social media,

> including users being invited to provide data to generate a 'profile', and to share information such as status updates, videos, photographs, and music, so as to facilitate interaction with other users.[10]

Indeed, the OSA does not define the term social media, nor mentions the term social media. This is despite discussions inside and outside of Parliament, at its time of enactment, primarily focussing on regulating social media platforms, in a bid to stop them from 'marking their own homework'.[11] Instead, such platforms are termed user-to-user services, defined as,

> an internet service by means of which content that is generated directly on the service by a user of the service, or uploaded to or shared on the service by a user of the service, may be encountered by another user, or other users, of the service.[12]

For the purposes of this monograph, the term social media is used to refer to platforms which allow users to create their own profiles (either in their own name or under a pseudonym), interact with others (either as a lurker or someone who actively engages with the platform), generate or view content and are created and run by private organisations for the purposes of generating a profit.

1.2.1 The World Wide Web

Surprisingly, the internet, which can be defined as 'a telecommunications system for computer networks',[13] was created in response to the first man-made object entering outer space in 1957. The announcement by the Soviet Union that they had achieved what was considered the impossible in launching an object into space came as a shock to those living in Western society at the time. As a result, the then President of the United States Dwight Eisenhower, in a bid not to be taken by surprise at the emergence of new technology in the future, created the

9 Clay Shirky, *Here Comes Everyone* (Penguin 2008) 20. For a discussion on different definitions of social media see, Christian Fuchs, *Social Media a Critical Introduction* (Sage Publications 2014) 35–37.

10 Law Commission, *Abusive and Offensive Online Communications: A Scoping Report* (Law Com No 381, 2018) [2.32].

11 Lorna Woods, William Perrin and Maeve Walsh, 'It's (nearly) here: a short guide to the Online Safety Act' (*Carnegie UK*, 19 September 2023) <https://carnegieuktrust.org.uk/blog-posts/its-nearly-here-a-short-guide-to-the-online-safety-act/> accessed 27 September 2023.

12 Online Safety Act s.3(1).

13 Andrew Murray, *Information Technology Law: The Law and Society* (5th edn, Oxford University Press 2023) 21.

Advanced Research Projects Agency.[14] It was this research body that created the world's first successful computer network, known as ARPANET (the Advanced Research Projects Agency Network) in the late 1960s.[15] Following the success of ARPANET, further networks were manufactured including the Transmission Control Protocol, which allowed networks to be created between local computers.[16] In 1988 computers based in Canada and France were linked to networks created in the United States, paving the way for the creation of the WWW.[17]

The WWW consists of millions of interlinked web pages which use the internet to connect. The foundations for the WWW were created in 1989 by Sir Tim Berners-Lee.[18] Berners-Lee envisioned a database where users would be able to obtain any information they desired with minimum fuss. In 1990 the first website went 'live'.[19] By 1993, the WWW or Web 1.0 as it is commonly referred to, was made available to the wider public. The purpose of Web 1.0 was to allow users to actively search between static websites.[20] Web 1.0 was, to put it simply, an information database, where users were considered consumers.[21]

By the turn of the next millennium, the advancement of changing technology had altered Web 1.0 completely. Internet users went from consumers to content creators.[22] By 2005 it had been accepted that a new version of the WWW had been created, Web 2.0:

> The term Web 2.0 […] is said to describe the period in which websites became more interactive, collaborative, and social. This is typically contrasted with more passive website interactions, where users simply viewed or downloaded content from websites.[23]

Web 2.0 unlike its predecessor, was much more interactive, dynamic, and user-driven.[24] The expansion of the WWW changed the job market, altered how individuals communicated, and has become integral to how society functions. At the time of writing, there is an ongoing debate in the wider literature that we are

14 Ibid., 22.
15 Ibid., 32.
16 Jeffrey C. Jackson, *Web Technologies: A Computer Science Perspective* (Pearson Education 2007) 2.
17 Ibid., 3.
18 Kevin M. Rogers, *The Internet and the Law* (Macmillan International Higher Education 2011) 3.
19 Madhumita Murgia, 'The world's first website went online 25 years ago today' *The Telegraph* (London, 21 December 2015) <https://www.telegraph.co.uk/technology/internet/12061803/The-worlds-first-website-went-online-25-years-ago-today.html> accessed 2 November 2018.
20 Rogers (n 18) 213.
21 Matthew Allen, 'What was Web 2.0? Versions as the Dominant Mode of Internet History' (2012) 15(2) New Media Soc. 260, 263.
22 Ibid.,
23 Law Com No 381 (n 10) [2.29].
24 Rogers (n 18) 213.

perhaps moving away from Web 2.0 towards Web 3.0,[25] driven by the creation of the Metaverse.[26]

Since its creation, the WWW and the internet have grown in terms of websites and online users.[27] In the United Kingdom in 2018, 90% of adults were regular internet users, an increase of 1% on the previous year. By 2020, this figure had risen to 92%.[28] Of those aged between 16 and 34, 99% were regular users of the internet.[29] Similarly, internet usage has increased across the world. In 2017 it was reported that 3.58 billion people worldwide were users of the internet, an increase of 190 million on the previous year.[30] By July 2023, this figure reached 5.19 billion, of which 4.88 billion were also social media users.[31]

The WWW has therefore gone from static websites, whereby users were able to use the internet as an information database, to interactive websites, which allow for, though not limited to, instant communication, political debate, and the selling of goods online.[32] Today, the internet can be considered 'messy, dynamic, and constantly changing'[33] whilst facilitating the exchange of information on a scale unprecedented in human history.[34] Consequently, the purpose of the internet has expanded, as illustrated in the work of Bernal. Bernal suggests that the purpose of the internet can be split into six categories: the use of the internet as an information database, the use of the internet as a business platform, the use of the internet as a political platform, the internet as a public space, the internet as a communications

25 'Web 3.0 entails an integrated Web experience where the machine will be able to understand and catalogue data in a manner similar to humans'. See, Riaan Rudman and Rikus Bruwer, 'Defining Web 3.0: opportunities and challenges' (2016) 34(1) Electron. Libr. 132.

26 The Metaverse is considered 'the idea of integrating real and digital worlds ever more seamlessly, using technologies such as virtual reality (VR) and augmented reality (AR)'. See, Marcus Carter and Ben Egliston, 'Facebook relaunches itself as "Meta" in a clear bid to dominate the metaverse' (*The International Forum for Responsible Media Blog*, 20 October 2021) <https://inforrm.org/2021/10/30/facebook-relaunches-itself-as-meta-in-a-clear-bid-to-dominate-the-metaverse-marcus-carter-and-ben-egliston/> accessed 27 September 2023.

27 Websites have since expanded from 130 in 1993 to 17,087,182 by the end of 2000. See, Internet Live Stats, 'Total number of websites' (*Internet Live Stats*, 2018) <http://www.internetlivestats.com/total-number-of-websites/> accessed 2 November 2018.

28 Office of National Statistics, 'Internet users, UK: 2020' (*Office of National Statistics*, 6 April 2021) <https://www.ons.gov.uk/businessindustryandtrade/itandinternetindustry/bulletins/internetusers/2020> accessed 27 September 2023. At the time of writing, there are no updated national statistics on internet access in the United Kingdom.

29 Ibid.

30 Statista, 'Number of internet users worldwide from 2005 to 2017 (in millions)' (*Statista*, 2018) <https://www.statista.com/statistics/273018/number-of-internet-users-worldwide/> accessed 2 November 2018.

31 Statista, 'Number of internet and social media users worldwide as of July 2023 (in billions)' (*Statista*, 22 September 2023) <https://www.statista.com/statistics/617136/digital-population-worldwide/> accessed 27 September 2023.

32 Note, this is not a definitive list.

33 Paul Bernal, *The Internet, Warts and All: Free Speech, Privacy and Truth* (Cambridge University Press 2018) 2.

34 Law Com No 381 (n 10) [1.41].

device, and the use of the internet in being integral to society.[35] It is the use of the internet as a communications device, to which this chapter will now turn to examine.

1.2.2 The Early Days of Online Communications and Social Media

The expansion of changing technologies has not only changed how we obtain information, adapted political campaigns, or indeed altered how we shop, but it has also transformed how members of a society interact with each other:

> Technological advancement in the 19th century sparked a revolution in the speed of communication. The invention of the telegraph was the first form of electrical telecommunication that had this effect. Subsequent innovations such as the telephone, radio, television, the internet, and most recently, the emergence of social media, have radically transformed the way we communicate.[36]

In 1971 the first email between two computers was transmitted.[37] Since then, technological advances have continued to revolutionise how we communicate. We have gone from being able to send written or verbal communications to one another, to being able to share pictures, videos, memes, emojis, newspaper articles etc., instantly, at the click of a button, with social media playing a pivotal role in the revolution of online communications.

The first form of social media as society knows it today emerged in 1997, 'SixDegrees.com', in which users of the site could connect with friends and family.[38] Following the initial success of SixDegrees, which boosted around 3.5 million users, other websites started to emerge based on a similar format, such as 'Friendster', 'LinkedIn', and 'MySpace', which all launched in 2002.[39] By 2015 it was estimated that there were over 200 different social media platforms across the globe.[40] However, the biggest turning point for the revolution of social media was the creation of 'TheFacebook' in 2004, which was rebranded as Facebook in August 2005. And despite it being considered 'old', Facebook still, to this day, is the biggest social media platform worldwide.[41]

35 Bernal (n 33) 5–19.
36 Law Com No 381 (n 10) [2.29].
37 Centre for Computing History, '1971: First Network Email Sent by Ray Tomlinson' (*Centre for Computing History*, 2016) <http://www.computinghistory.org.uk/det/6116/First-e-mail-sent-by-Ray-Tomlinson/> accessed 27 November 2018.
38 Scaife (n 8) 4.
39 Broadband Search, 'History of social media (It's younger than you think)' (*Broadband Search*, 10 May 2023) <https://www.broadbandsearch.net/blog/complete-history-social-media> accessed 27 September 2023.
40 Scaife (n 8) 4.
41 Josh Constine, 'Facebook now has 2 billion monthly users … and responsibility' (*Tech Crunch*, 2017) <https://techcrunch.com/2017/06/27/facebook-2-billion-users/> accessed 29 November 2018.

1.2.3 Social Media as We Know It Today and Its Global Reach

On 4 February 2004 Mark Zuckerberg, a college student at Harvard University, alongside several other students, launched Facebook. The initial purpose of Facebook was to create a communications mechanism for students at Harvard University, with the site later being made available to all higher education students in the Boston area.[42] In 2006, Facebook was released to the world, with anyone who claimed[43] to be over 13 years of age being able to create a Facebook page.

Facebook, 'operates as a social networking site based on interconnection[s] with other users to generate content'.[44] Here users must register with the site and build personal profiles of themselves, which can include, likes and dislikes, a picture of themselves, their date of birth, and other unique information such as where they were educated. In fact, Facebook is built on the premise of getting its users to reveal private information about themselves.

By 2016 Facebook had expanded to become the sixth-most valuable public company in the world.[45] In June 2017 Mark Zuckerberg announced, *via* Facebook, that Facebook had 2 billion users, averaging around 580,000 new users each day.[46] In the United Kingdom alone, 48.5 million citizens are active Facebook users.[47] Meta, who owns Facebook, have also extended the number of other social media platforms in its portfolio including, WhatsApp (with a usership worldwide of 2 billion in 2023);[48] Instagram (with a usership worldwide of 1.21 billion in 2021);[49] and most recently, Threads (which surpassed 100 million users in its first 5 days of launch)[50] as a direct response to the decline of X, which until recently was one of its biggest competitors.[51] Other platforms have emerged since the launch of Facebook, but none have surpassed its reach, as illustrated in Figure 1.1.

42 Sarah Phillips, 'A brief history of Facebook' *The Guardian* (London, 25 July 2007) <https://www.theguardian.com/technology/2007/jul/25/media.newmedia> accessed 26 November 2018.

43 Very little is done to verify the age of users on Facebook. However, under the Online Safety Act 2023 platforms will have to do more to ensure those who should not be on the platform, cannot access it through age-verification and age-estimation. See Chapter 8.

44 Scaife (n 8) 9.

45 Ursula Smartt, *Media & Entertainment Law* (Taylor & Francis 2017) 79.

46 James Titcomb, 'Facebook now has 2 billion users, Mark Zuckerberg announces' *The Telegraph* (London, 27 June 2017) <https://www.telegraph.co.uk/technology/2017/06/27/facebook-now-has-2-billion-users-mark-zuckerberg-announces/> accessed 29 November 2018.

47 Statista, 'Distribution of Facebook users in the United Kingdom as of April 2023, by age group' (*Statista*, 22 May 2023) <https://www.statista.com/statistics/1030055/facebook-users-united-kingdom/> accessed 27 September 2023.

48 Statista, 'WhatsApp – Statistics & Facts' (*Statista*, 30 August 2023) <https://www.statista.com/topics/2018/whatsapp/#topicOverview> accessed 27 September 2023.

49 Statista, 'Number of Instagram users worldwide from 2020 to 2025 (in billions)' (*Statista*, 29 August 2023) <https://www.statista.com/statistics/183585/instagram-number-of-global-users/> accessed 27 September 2023.

50 The BBC, 'Threads app signs up 100m users in less than a week' *The BBC* (London, 10 July 2023) <https://www.bbc.co.uk/news/technology-66153244> accessed 27 September 2023.

51 Josh Taylor and Josh Nicolas, 'Twitter traffic sinks in wake of changes and launch of rival platform Threads' *The Guardian* (London, 10 July 2023) <https://www.theguardian.com/technology/2023

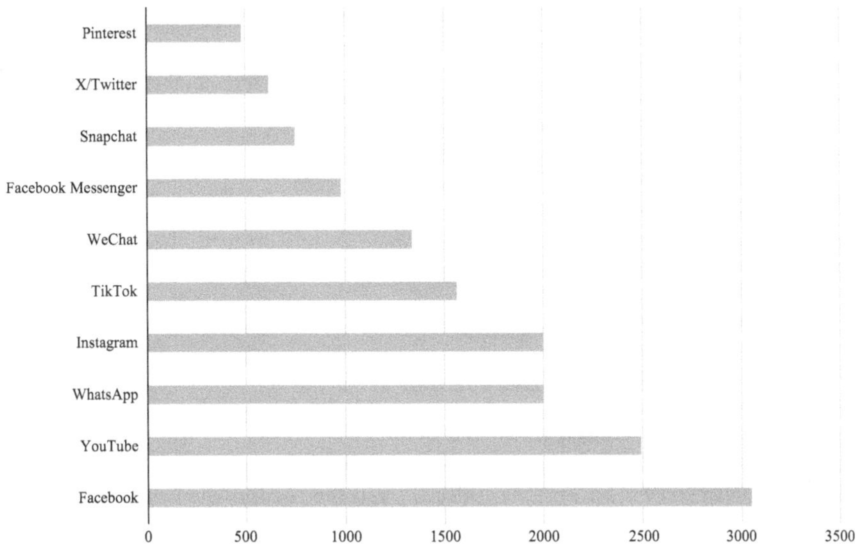

Figure 1.1 Most popular social networks worldwide as of January 2024, ranked by number of monthly active users (in millions).

Note: Statista, 'Most popular social networks worldwide as of January 2024, ranked by number of monthly active users (in millions)' (*Statista*, 31 January 2024) <https://www.statista.com/statistics/272014/global-social-networks-ranked-by-number-of-users/> accessed 1 March 2024.

To put the dominance of social media into context, in 2023:[52]

- An estimated 4.9 billion people globally used at least one form of social media;
- The average user had between six and seven social media platforms, many of which were easily accessed *via* a mobile phone at the click of a button;
- The average person spent around 145 minutes a day on social media;
- 84% of those aged 18 to 29 and 81% of people between 30 to 49 actively used at least one social media site; and
- 77% of businesses worldwide used at least one form of social media.

Whilst in the United Kingdom, in January 2023:[53]

/jul/10/twitter-traffic-sinks-in-wake-of-changes-and-launch-of-rival-platform-threads-elon-musk> accessed 23 September 2023.

52 Belle Wong, 'Top social media statistics and trends of 2023' (*Forbes,* 18 May 2023) <https://www.forbes.com/advisor/business/social-media-statistics/> accessed 27 September 2023.

53 Simon Kemp, 'Digital 2023: Global overview report' (*Datareportal*, 26 January 2023) <https://datareportal.com/reports/digital-2023-global-overview-report> accessed 23 September 2023.

- 57.10 million individuals were social media users (84.4% of the total population);
- The average UK citizen spent 116 minutes each day on social media;
- The average user in the United Kingdom had six social media platforms; and
- 51.6% of UK social media users were female.

Social media therefore dominates much of everyday life, with its usage increasing dramatically during the global COVID-19 pandemic.[54] In turn, the way in which humans interact with each other, has also changed, both for the better and for the worse.

1.3 Social Media: The Good and the Bad

There is no doubt that social media has changed many aspects of everyday life—both positively and negatively. However, it is the negative side of social media that is often raised by government officials and the media when discussing such forms of communication. It is this continued rhetoric around the 'darker' side of the online world which has traditionally led to calls for stronger regulation, to control and punish those who use such platforms to abuse others. Though as noted by Rowbottom:

> [t]he legal responses can, however, seem heavy-handed for what might have been a statement made with little thought while the speaker was sat at a desk at home. Words typed in seconds followed by hitting the enter key can lead to a criminal record or costly civil litigation.[55]

Recently, as discussed in Chapter 8, there has been a change in direction, with emphasis being placed on regulating social media platforms.

There is no doubt that social media can be used to abuse others, with regular research being published highlighting the 'darker' side of social media. From research exposing the use of such platforms to target those in the public domain, including athletes,[56] politicians,[57] and celebrities,[58] to research indicating a rise in

54 Stacey Jo Dixson, 'Social media use during COVID-19 worldwide – Statistics & Facts' (*Statista*, 31 August 2023) <https://www.statista.com/topics/7863/social-media-use-during-coronavirus-covid-19-worldwide/#topicOverview> accessed 23 September 2023.

55 Rowbottom (n 2) 356.

56 Professional Footballers' Association, 'AI research study on online abuse' (*Professional Footballers' Association*, 2021) <https://www.thepfa.com/news/2020/10/21/pfa-charity-ai-research-study> accessed 21 September 2023.

57 Laura Bliss, 'Abuse of women MPs is not just a scandal – it's a threat to democracy' *The Conversation* (London, 17 July 2017) <https://theconversation.com/abuse-of-women-mps-is-not-just-a-scandal-its-a-threat-todemocracy-80781> accessed 9 December 2018.

58 Christopher Hand, 'Why is celebrity abuse on Twitter so bad? It might be a problem with our empathy' *The Conversation* (London, 4 March 2021) <https://theconversation.com/why-is-celebrity-abuse-on-twitter-so-bad-it-might-be-a-problem-with-our-empathy-154970> accessed 23 September 2023.

the use of such platforms to target children.[59] Indeed, anyone can become subjected to horrific forms of abuse online. And although this monograph does not justify the abuse that some individuals experience online, it does dispute the use of the criminal law in this area, particularly criminal law provisions which are incompatible with the principle of legality when applied in a social media context.

Further, as will be noted throughout Chapters 4, 5, and 6, the statistics provided to justify why we need stronger criminal legal provisions to tackle the 'harms' associated with the online world, need to be considered with caution. Many studies use vague definitions of the 'harm'. For instance, as discussed in Chapter 5, cyberharassment and cyberstalking are often used interchangeably and even when applied correctly, studies define such behaviours differently. Additionally, research conducted by The Alan Turing Institute, as noted in Chapter 6, estimates that only 0.001% to 1% of content online can be considered so highly abusive, that further action should be taken.[60] Indeed, research by OFCOM (the UK communications regulator) with adults on 'harmful' content they have seen online, found those who took part in the study, had witnessed in the four weeks prior:

- 'Alternative viewpoints' (14%);
- 'Offensive/upsetting language' (10%); and
- 'Offensive/upsetting videos/images' (7%).

Those who took part in the study considered these behaviours to be 'harmful'.[61] Yet, such content, as discussed in Chapter 2, falls within the right to freedom of expression.

It is studies like these that are often used to justify the paternalistic approach undertaken by governments to regulate individuals' conduct online, where it has come to be accepted that self-regulation is no longer working (if it ever did) and therefore the law must intervene. As highlighted by Gillespie, it is not a defence to say, 'it's only Twitter'.[62] But there is a 'brighter' side to social media, and it's this 'brighter' side that is often missed in conversations. For instance, as noted in Chapter 2, social media can be used as a modern form of protest to challenge state coercion or abuse; it allows individuals to keep updated with loved ones;

59 Nick Stripe, 'Online bullying in England and Wales: year ending March 2020' (*Office of National Statistics*, 16 November 2020) <https://www.ons.gov.uk/peoplepopulationandcommunity/crimeandjustice/bulletins/onlinebullyinginenglandandwales/yearendingmarch2020> accessed 23 September 2023.

60 Bertie Vidgen, Helen Margetts, and Alex Harris, 'How much online abuse is there? A systematic review of evidence for the UK. Policy Briefing – Full Report' (*The Alan Turing Institute*, 2020) 4 <https://www.turing.ac.uk/sites/default/files/2019-11/online_abuse_prevalence_full_24.11.2019_-_formatted_0.pdf> accessed 3 May 2023.

61 OFCOM, 'Ofcom pilot online harms survey 2020/21' (*OFCOM*, 2021) 18 <https://www.ofcom.org.uk/__data/assets/pdf_file/0014/220622/online-harms-survey-waves-1-4-2021.pdf> accessed 23 September 2023. The sample range in total was 4,242 internet users aged 13–84 in the United Kingdom.

62 Alisdair A. Gillespie, 'Twitter, Jokes and the Law' (2012) 76(5) J. crim. law 364 (note).

it can bring people together; and for some, it's the only place they can truly be themselves,[63] but to name just a few examples. In fact, further research by OFCOM around young people's perceptions of the online world and social media, directly opposes much of the current rhetoric that the online world is 'bad' for children:

> Specifically, in relation to social media, a majority of children aged 8 to 17 using these apps or sites were able to identify the benefits of these apps or sites, *all or most of the time.* About seven in ten said they felt safe using these types of apps (73%). There was a positive year-on-year shift in the proportion of children aged 8–17 claiming that these platforms made them feel happy (67%, up from 59% in 2021), while 66% thought that it helped them stay close to their friends (up from 61% in 2021).[64]

And although those who took part in the study acknowledged that there were 'downsides' to being online, overall the good outweighed the bad.[65]

It is vital therefore that we strike a balance when it comes to regulating individuals' conduct online, in which the criminal law should only intervene in the most extreme cases and only when such legal provisions are compatible with the principle of legality. This is because 'for democracy to flourish, we need informed citizens, accountability, transparency, [and] inclusive public debate'.[66] It is only through having difficult conversations, that change can be enacted. We need to be careful to ensure we are not shutting down the voices of the minority to suit the needs of the majority. To put it bluntly 'there [will] always [be] stupid people and now they are all in one place'.[67] Arguably, many of the 'negative' aspects of social media are inherent within society, and social media is just a magnification of conversations that happen every day in the physical world, which can now be documented and searched for, long after the exchange took place. We therefore need to ensure that the law is only applied in extreme circumstances and does not become the go-to option for every little dispute that takes place online. And where such provisions are applied, we need to ensure legality is maintained—something we are currently failing to adhere to.

63 Daria Kuss, 'How social media affects children at different ages – and how to protect them' (*LSE Blog*, 4 May 2022) <https://blogs.lse.ac.uk/parenting4digitalfuture/2022/05/04/social-media-at-different-ages/#:~:text=Engaging%20with%20social%20media%20can,health%20and%20wellbeing%20as%20well> accessed 23 September 2023.

64 OFCOM, 'Children and parents: Media use and attitudes' (*OFCOM,* 29 March 2023) 34 <https://www.ofcom.org.uk/__data/assets/pdf_file/0027/255852/childrens-media-use-and-attitudes-report-2023.pdf/> accessed 23 September 2023.

65 Ibid., 35.

66 Select Committee on Democracy and Digital Technologies, *Digital Technology and the Resurrection of Trust* (HL 2019-21, 77) [10].

67 City University, 'City law school webinar: The law of Facebook' (*YouTube*, 21 April 2021) <https://www.youtube.com/watch?v=zjaOFPmm8wo> accessed 18 October 2023.

1.4 Chapter Overview

The proceeding chapters are split into three parts. Part One sets the contextual background for later discussions in Part Two of this monograph and consists of this chapter and two other chapters: Chapter 2 – Freedom of Expression: The Online Context and Chapter 3 – Regulating Social Media: Theoretical Positioning.

Chapter 2 provides a discussion on the right to freedom of expression and the right to privacy in the online world. First, Article 19 of the Universal Declaration of Human Rights is outlined, before turning to examine aspects of the European Convention on Human Rights which by way of the Human Rights Act 1998 has been incorporated into the legal framework of the United Kingdom. Of focus are Articles 10 (the right to freedom of expression), 8 (the right to privacy), and 17 (the prohibition of abuse rights). Next, why freedom of expression is essential in a digital world is outlined, focussing on the 'good' that can come from social media usage, which is often neglected in wider discussions; before turning to examine the often-misunderstood concept of privacy online, and how, despite what some may think, privacy does exist even in the virtual world. Chapter 2 concludes by reenforcing the importance of protecting freedom of expression online and how ultimately the use of the law to govern inappropriate communications on social media sites, will not change social norms and instead, such use merely threatens the right to freedom of expression.

The overall premise of Chapter 3 is to provide the framework which will be used to critically examine legal provisions and their use to curtail social media abuse in Part Two of this monograph—the principle of legality. To do this, the first half of Chapter 3 explores four theoretical concepts of online regulation: (1) cyber-libertarianism; (2) cyber-paternalism; (3) network communitarianism; and (4) the governance model. It is argued that regardless of the theoretical positioning you take to the regulation of social media platforms, legality needs to be maintained. The second half of Chapter 3 introduces the concept of legality outlining with reference to legal provisions and case law examples, legality in international criminal law, followed by a further in-depth discussion on legality from a European Human Rights perspective. The final section of this chapter explores Lon Fuller's[68] eight principles of legality, with reference to other scholars in the field, such as Joseph Raz,[69] Cian Murphy,[70] and Tom Bingham.[71] It is Fuller's conception of legality which is used to critique the law in the proceeding chapters.

Part Two provides the substantive discussions for this monograph and its split into four chapters; each exploring a different area of the criminal law and its application to a social media context: Chapter 4 – Online Racism and the Criminal Law;

68 Lon L. Fuller, *The Morality of Law* (Yale University Press 1964).

69 Joseph Raz, 'The Rule of Law and its Virtue' (1977) 93 L.Q.R. 195.

70 Cian C. Murphy, 'The Principle of Legality in Criminal Law under the European Convention on Human Rights' (2010) 2 Eur. Hum. Rights Law Rev. 192.

71 Tom Bingham, *The Rule of Law* (Penguin 2011). See also, Lord Bingham, 'The Rule of Law' (2007) 66(1) Camb. Law. J. 67.

Chapter 5 – Harassment and Stalking Online; Chapter 6 – Abusive Communications Online: The Development of Communication Law; and Chapter 7 – The Future of Communication Law.

Chapter 4 starts by outlining the rise in online hate crimes, before turning to examine the use of such sites to send racial abuse—the hate characteristic which will be the focus of this chapter. Reference will be made to the growing concern that social media sites are increasingly being used to spread racial hatred. Following on from this, Chapter 4 examines how the criminal justice system is attempting to tackle the rise in racial abuse on social media sites by examining three legal provisions which prohibit such behaviour: s.66 of the SA, s.31 and 32 of the CDA and s.18 and 19 of the POA. Using a narrative review and case law examples, each provision is taken in turn, looking specifically at the language used within these provisions and their application to a social media context. It highlights, with reference to the Crown Prosecution Service guidelines on hate crimes, the failure of the law to keep pace with digital technology. The final section of Chapter 4 contextualises and demonstrates how the application of these provisions, to a social media context, breaches the fundamental principles of legality.

Cyberharassment and cyberstalking are the focus of Chapter 5. This chapter provides a narrative review, with reference to case law examples and statistics, on the application of the PfHA to cyberharassment and/or cyberstalking. Taking a similar structure to Chapter 4, it starts by contextualising the growth in the use of social media sites to harass and stalk individuals online, before turning to examine the PfHA. Taking harassment and stalking in turn, the *actus reus* and *mens rea* of each offence is critically examined, looking at the language used within the Act, before turning to look at the PfHA and the principle of legality. Chapter 5 concludes by demonstrating how the Act, when applied in a social media context, does not uphold the principle of legality and in turn, freedom of speech is under threat.

Whereas Chapters 4 and 5 focus on substantive criminal offences, Chapter 6 turns to examine communication offences in the criminal law framework of England and Wales. Until very recently, two communication offences existed: the MCA and s.127(1) and (2) of the CA. Chapter 6 starts by briefly outlining the historical development of these two Acts, before discussing the rise in online abuse—a term that is often thrown around and used to justify the coercive nature of the criminal law but lacks any definitive definition. Next, like that of the previous chapters, the MCA and s.127(1) and (2) of the CA are outlined, examining the offences which were originally prohibited under these provisions. Of focus will be the vague terminology used in both provisions—highlighting significant issues with legality, and showing why these provisions were, and still are, considered the most controversial approach to regulating speech. Therefore allowing, in Chapter 7, direct comparisons between the 'new' communication offences and the 'old' offences.

Chapter 7 explores the future of communication law, following the Law Commission in 2021 making several radical recommendations to revamp this area of the criminal law. Many of these recommendations were included in the OSA, others were not. The passing of the OSA has made some small amendments to both the MCA and s.127(2) of the CA. Of note and as discussed in Chapter 7 are offences related to the sending of harmful, false, and threatening communications. Each is taken in turn, discussing the *actus reus* and *mens rea* of each offence, before turning to examine the right to freedom of expression and the principle of legality in the criminal law. This chapter will conclude by demonstrating how these 'new' offences have not addressed previous issues with the application of communication law to conduct carried out on social media platforms.

Whereas Part Two of this monograph focussed primarily on three significant 'harms' associated with the online world from a criminal law perspective, Part Three turns to examine the regulation of social media platforms themselves as opposed to individuals. Part Three is made up of two substantive chapters, followed by the conclusive chapter: Chapter 8 – Regulating Online Behaviour in the United Kingdom: Beyond the Criminal Law; and Chapter 9 – International Perspectives on Social Media and the Law.

Chapter 8 explores and critically examines the United Kingdom's OSA. Under the Act, platforms which fall within its scope, including most social media companies, are legally obliged to do more to tackle the 'harms' associated with the digital world. Including, though not limited to, doing more to reduce the behaviours discussed in Chapters 4, 5, 6, and 7 on their platforms. Chapter 8 starts by outlining the background to the OSA, before turning to critically discuss the Act from the perspective of legality. The rationale for including a separate chapter looking at the regulatory context, beyond just the criminal law, is to situate the discussions and arguments made in previous chapters, within the wider political, ethical, and legal context of online regulation.

The final substantive chapter, Chapter 9, provides a comparative chapter exploring how other countries have attempted to regulate online (mis)behaviours including, France, Germany, and India. Each State is examined in turn, with a narrative review of the key provisions used by each jurisdiction to regulate both individual behaviours online, and social media companies themselves. Through a narrative review of these legal provisions, issues with legality and freedom of expression are exposed. In fact, in France, legislation enacted to tackle online hate has previously been struck down by the French Constitutional Court for being incompatible with the right to freedom of expression. The rationale for focussing on these States, as opposed to others, is due to the similarities in approaches, but the outcomes have varied. The purpose of Chapter 9 is to demonstrate how other countries have attempted to tackle online abuse to aid further discussions, particularly in the conclusive chapter, Chapter 10, on how we can better regulate online behaviours, whilst upholding the principle of legality and protecting freedom of expression.

Reference list

Allen M, 'What was Web 2.0? Versions as the dominant mode of internet history' (2012) 15(2) New Media Soc 260

Bernal P, *The Internet, Warts and All: Free Speech, Privacy and Truth* (Cambridge University Press 2018)

Bingham T, *The Rule of Law* (Penguin 2011)
Bliss L, 'Abuse of women MPs is not just a scandal – it's a threat to democracy' *The Conversation* (London, 17 July 2017) https://theconversation.com/abuse-of-women-mps-is-not-just-a-scandal-its-a-threat-todemocracy-80781
Broadband Search, 'History of Social Media (It's Younger Than You Think)' (*Broadband Search*, 10 May 2023) https://www.broadbandsearch.net/blog/complete-history-social-media
Carter M & Egliston B, 'Facebook Relaunches Itself as "Meta" in a Clear Bid to Dominate the Metaverse' (*The International Forum for Responsible Media Blog*, 20 October 2021) https://inforrm.org/2021/10/30/facebook-relaunches-itself-as-meta-in-a-clear-bid-to-dominate-the-metaverse-marcus-carter-and-ben-egliston/
Centre for Computing History, '1971: First Network Email sent by Ray Tomlinson' (*Centre for Computing History*, 2016) http://www.computinghistory.org.uk/det/6116/First-e-mail-sent-by-Ray-Tomlinson/
City University, 'City Law School Webinar: The Law of Facebook' (*YouTube*, 21 April 2021) https://www.youtube.com/watch?v=zjaOFPmm8wo
Constine J, 'Facebook Now Has 2 Billion Monthly Users … and Responsibility' (*Tech Crunch*, 2017) https://techcrunch.com/2017/06/27/facebook-2-billion-users/
Department for Science, Innovation and Technology, 'UK Children and Adults to Be Safer Online as World-Leading Bill Becomes Law' (*Gov.uk,* 26 October 2023) https://www.gov.uk/government/news/uk-children-and-adults-to-be-safer-online-as-world-leading-bill-becomes-law
Dixson S J, 'Social Media Use During COVID-19 Worldwide - Statistics & Facts' (*Statista*, 31 August 2023) https://www.statista.com/topics/7863/social-media-use-during-coronavirus-covid-19-worldwide/#topicOverview
Fuchs C, *Social Media a Critical Introduction* (Sage Publications 2014)
Fuller L L, *The Morality of Law* (Yale University Press 1964)
Gillespie A A, 'Twitter, Jokes and the Law' (2012) 76(5) J Crim Law 364 (note)
Hand C, 'Why is celebrity abuse on Twitter so bad? It might be a problem with our empathy' *The Conversation* (London, 4 March 2021) https://theconversation.com/why-is-celebrity-abuse-on-twitter-so-bad-it-might-be-a-problem-with-our-empathy-154970
HM Government, 'Internet Safety Strategy – Green Paper' (*Gov.uk*, October 2018) https://assets.publishing.service.gov.uk/government/uploads/system/uploads/attachment_data/file/650949/Internet_Safety_Strategy_green_paper.pdf
Internet Live Stats, 'Total Number of Websites' (*Internet Live Stats*, 2018) http://www.internetlivestats.com/total-number-of-websites/
Jackson J C, *Web Technologies: A Computer Science Perspective* (Pearson Education 2007)
Kemp S, 'Digital 2023: Global Overview Report' (*Datareportal*, 26 January 2023) https://datareportal.com/reports/digital-2023-global-overview-report
Kuss D, 'How Social Media Affects Children at Different Ages – and How to Protect Them' (*LSE Blog*, 4 May 2022) https://blogs.lse.ac.uk/parenting4digitalfuture/2022/05/04/social-media-at-different-ages/#:~:text=Engaging%20with%20social%20media%20can,health%20and%20wellbeing%20as%20well
Law Commission, *Abusive and Offensive Online Communications: A Scoping Report* (Law Com No 381, 2018)
Lord Bingham, 'The Rule of Law' (2007) 66(1) Cam Law J 67
Murgia M, 'The world's first website went online 25 years ago today' *The Telegraph* (London, 21 December 2015) https://www.telegraph.co.uk/technology/internet/12061803/The-worlds-first-website-went-online-25-years-ago-today.html
Murphy C C, 'The Principle of Legality in Criminal Law Under the European Convention on Human Rights' (2010) 2 Eur Hum Rights Law Rev 192
Murray A, *Information Technology Law: The Law and Society* (5th edn, Oxford University Press 2023)

OFCOM, 'Children and Parents: Media Use and Attitudes' (*OFCOM,* 29 March 2023) https://www.ofcom.org.uk/__data/assets/pdf_file/0027/255852/childrens-media-use-and-attitudes-report-2023.pdf/

OFCOM,'Ofcom Pilot Online Harms Survey 2020/21' (*OFCOM*, 2021) https://www.ofcom.org.uk/__data/assets/pdf_file/0014/220622/online-harms-survey-waves-1-4-2021.pdf

Office of National Statistics, 'Internet Users, UK: 2020' (*Office of National Statistics*, 6 April 2021) https://www.ons.gov.uk/businessindustryandtrade/itandinternetindustry/bulletins/internetusers/2020

Omand D, 'The Dark Net: Policing the Internets Underworld' (2016) Winter 2015/16 World Policy Journal https://worldpolicy.org/2015/12/09/the-dark-net-policing-the-internets-underworld/

Phillips S, 'A brief history of Facebook' *The Guardian* (London, 25 July 2007) https://www.theguardian.com/technology/2007/jul/25/media.newmedia

Prensky M, 'Digital Natives, Digital Immigrants' 2001 9(5) *Horiz* 1

Professional Footballers' Association, 'AI Research Study on Online Abuse' (*Professional Footballers' Association*, 2021) https://www.thepfa.com/news/2020/10/21/pfa-charity-ai-research-study

Raz J, 'The Rule of Law and its Virtue' (1977) 93 LQR 195

Rogers K M, *The Internet and the Law* (Macmillan International Higher Education 2011)

Rowbottom J, 'To Rant, Vent and Converse: Protecting Low Level Digital Speech' (2012) 71(2) CLJ 355

Rudman R & Bruwer R, 'Defining Web 3.0: Opportunities and Challenges' (2016) 34(1) Electron Libr 132

Scaife L, *Handbook of Social Media and the Law* (Routledge 2015)

Select Committee on Democracy and Digital Technologies, *Digital Technology and the Resurrection of Trust* (HL 2019-21, 77)

Shirky C, *Here Comes Everyone* (Penguin 2008)

Smartt U, *Media & Entertainment Law* (Taylor & Francis 2017)

Statista, 'Most Popular Social Networks Worldwide as of January 2024, Ranked by Number of Monthly Active Users (in Millions)' (*Statista*, 31 January 2024) https://www.statista.com/statistics/272014/global-social-networks-ranked-by-number-of-users/

Statista, 'Number of Internet and Social Media Users Worldwide as of July 2023 (in Billions)' (*Statista*, 22 September 2023) https://www.statista.com/statistics/617136/digital-population-worldwide/

Statista, 'WhatsApp - Statistics & Facts' (*Statista*, 30 August 2023) https://www.statista.com/topics/2018/whatsapp/#topicOverview

Statista, 'Number of Instagram Users Worldwide from 2020 to 2025 (in Billions)' (*Statista*, 29 August 2023) https://www.statista.com/statistics/183585/instagram-number-of-global-users/

Statista, 'Distribution of Facebook users in the United Kingdom as of April 2023, by Age Group' (*Statista*, 22 May 2023) https://www.statista.com/statistics/1030055/facebook-users-united-kingdom/

Statista, 'Number of Internet Users Worldwide from 2005 to 2017 (in Millions)' (*Statista*, 2018) https://www.statista.com/statistics/273018/number-of-internet-users-worldwide/

Stripe N, 'Online bullying in England and Wales: Year Ending March 2020' (*Office of National Statistics*, 16 November 2020) https://www.ons.gov.uk/peoplepopulationandcommunity/crimeandjustice/bulletins/onlinebullyinginenglandandwales/yearendingmarch2020

Taylor J & Nicolas J, 'Twitter traffic sinks in wake of changes and launch of rival platform Threads' *The Guardian* (London, 10 July 2023) https://www.theguardian.com/technology/2023/jul/10/twitter-traffic-sinks-in-wake-of-changes-and-launch-of-rival-platform-threads-elon-musk

The BBC, 'Threads app signs up 100m users in less than a week' *The BBC* (London, 10 July 2023) https://www.bbc.co.uk/news/technology-66153244

The BBC, 'Teenager's life "ruined" by Live.me and Twitter "trolls"' *The BBC* (London, 24 October 2017) http://www.bbc.co.uk/news/uk-england-41693437

Titcomb J, 'Facebook now has 2 billion users, Mark Zuckerberg announces' *The Telegraph* (London, 27 June 2017) https://www.telegraph.co.uk/technology/2017/06/27/facebook-now-has-2-billion-users-mark-zuckerberg-announces/

Vidgen B, Margetts H & Harris A, 'How Much Online Abuse Is There? A Systematic Review of Evidence for the UK. Policy Briefing – Full Report' (*The Alan Turing Institute*, 2020) https://www.turing.ac.uk/sites/default/files/2019-11/online_abuse_prevalence_full_24.11.2019_-_formatted_0.pdf

Wong B, 'Top Social Media Statistics and Trends of 2023' (*Forbes,* 18 May 2023) https://www.forbes.com/advisor/business/social-media-statistics/

Woods L, Perrin W & Walsh M, 'It's (Nearly) Here: A Short Guide to the Online Safety Act' (*Carnegie UK*, 19 September 2023) https://carnegieuktrust.org.uk/blog-posts/its-nearly-here-a-short-guide-to-the-online-safety-act/

2 Freedom of Expression
The Online Context

2.1 Introduction

As potently put by Newman:

> Words can be hurtful. They can antagonize [*sic*], they can wound. But words also inform and educate. They can dispel prejudice and reveal unanticipated affinities. We need strong laws to protect us from violence and discrimination, but laws that protect us from hurtful words tend to silence us as well as our antagonists, because they close down rather than open conversations and because they teach good citizens to call the police rather than do the hard work of combating hateful ideas.[1]

Indeed, you cannot discuss the limits of online communications without first discussing the right to freedom of expression. It is a right that intertwines with several fundamental human rights, including, thought, association, and assembly, with these rights often being described as 'closely related civil and political rights'.[2] And '[a]lthough the right to freedom of expression is not in every case the ace of trumps, it is a powerful card'.[3] In fact, for Bernal, freedom of expression is prized above all other freedoms.[4] It is both an individual right, in that citizens have the freedom to speak their own mind and a collective right, in that we have the right to exchange ideas and listen to others' viewpoints. It can be traced back to ancient Greece and Socrates.[5] Since, elements of freedom of expression have been found in various legal instruments in one way or another, including the British Bill of Rights written in 1689 which guaranteed freedom of expression within Parliament.

1 Stephen L. Newman, 'Finding the Harm in Hate Speech: An Argument against Censorship' (2017) 50(3) CJPS 679, 693.

2 Kevin Boyle, 'Thought, Expression, Association, and Assembly' in Daniel Moeckli, Sangeeta Shah, and Sandesh Sivakumaran (eds), *International Human Rights Law* (Oxford University Press 2010) 257.

3 *Douglas v Hello! Ltd* [2001] QB 967 per Brooke L.J. at 982.

4 Paul Bernal, *The Internet, Warts and All: Free Speech, Privacy and Truth* (Cambridge University Press 2018) 102.

5 For an in-depth discussion on the history of freedom of expression, see, Jacob Mchangama, *Free Speech: A Global History from Socrates to Social Media* (Basic Books 2022).

DOI: 10.4324/9781003303688-3

The right to speak freely, without interference, plays a pivotal role in any democratic society.[6] It allows us to challenge state authority, aids in the discovery of truth and can be considered a necessary element of self-fulfilment. There are some states, in which the right to freedom of expression does not exist, but these are states in which many of us would not want to live. But the right to freedom of expression is not absolute, nor has it ever been, with human rights provisions allowing governments to restrict the right to freedom of expression when certain criterions are met, discussed further in Section 2.3. Even as early as 1857 there are examples of individuals having their speech restricted in the United Kingdom, such as the prosecution of a person for writing offensive words relating to Christianity on a gate.[7] Indeed, up until 2008, blasphemy was a criminal offence in England and Wales. In fact, it was not until 2021 that blasphemy was decriminalised in Scotland.

Despite the weight given to freedom of expression, there are clearly limits. The issue lies in what constitutes an exception and it is evidently difficult to find the line between protected and non-protected speech. Nothing has demonstrated this more than the increased usage of social media sites, where internet users have gone from consumers to content creators.[8] Today we live in a world where we share our most personal and intimate information online—but at what point does our behaviour on social media sites go beyond what is legally acceptable, and therefore within the boundaries of freedom of expression, to wholly unacceptable and therefore warrants state interference? Indeed, '[t]he internet is the conduit for communications in the digital age, making it the heart of any system of free expression'.[9] Today, anyone with an internet connection can 'become a town crier with a voice that resonates farther than it could from any soapbox'.[10]

For the UK House of Lords Select Committee on Freedom of Expression, abusing another online was considered to fall outside the scope of freedom of expression.[11] Yet for the UK's Divisional Court 'free speech encompasses the right to offend, and indeed to abuse another'.[12] At the same time, it has come to be accepted that although the online world is inherently public, privacy does exist online—an expansive right that encompasses more than just a private life away from the public domain. It includes the right to personal autonomy.[13] A right that is infringed when a person is subjected to online abuse. As potently put by Bernal '[t]he problem with the internet is that the boundaries between what is public and what is private have

6 *Handyside v The United Kingdom* (1976)1 EHRR 737.

7 John Stuart Mill, *On Liberty* (first published 1859, Enhanced Media 2016) 25.

8 Newman (n 1) 680.

9 Emily B. Laidlaw, *Regulating Speech in Cyberspace: Gatekeepers, Human Rights and Corporate Responsibility* (Cambridge University Press 2015) 280.

10 *Reno v ACLU*, 521 U.S. 884, 896-97 (1997).

11 Communications and Digital Committee, *Free for All? Freedom of Expression in the Digital Age* (2021–22 HL 54) 3.

12 *Scottow v CPS* [2020] EWHC 3421 (Admin) per Justice Warby [43].

13 *Pfeifer v Austria App* no 125561/03 [2007] ECTHR 935 [33].

been more than just blurred; they have been all but obliterated'.[14] A balance must therefore be struck, even in the digital world, between freedom of expression and the right to privacy.

The following chapter will outline how freedom of expression is currently protected through international treaties, focussing on Article 19 of the Universal Declaration of Human Rights (UDHR) and regional human rights instruments, with emphasis on Article 10 of the European Convention on Human Rights (ECHR). The right to privacy as protected under Article 8 of the ECHR will also be explored, alongside Article 17. By way of the Human Rights Act 1998 (HRA), the ECHR has since been incorporated into the legal system of the United Kingdom. The final parts of this chapter will explore the balance between freedom of expression and the right to privacy in a digital age—in which it will be argued throughout this monograph that current approaches to online regulation are a serious threat to freedom of expression. Indeed, if we truly want to eradicate online abuse, societal norms need to be altered first, before legal norms will take effect.

2.2 Article 19 and the Universal Declaration of Human Rights

In 1948 the UDHR was adopted by the United Nations General Assembly to create a unilateral understanding of all aspects of human rights. It was drafted as a response to the travesties of the Second World War and set out for the first time on an international scale, fundamental rights to be universally protected, with 48 states signing the declaration. Though not legally binding on states, it has been incorporated into several further legal provisions and is widely considered as paving the way for other human rights treaties, including, the International Covenant on Civil and Political Rights; the International Convention on the Elimination of all forms of Racial Discrimination; the Convention on the Rights of the Child'; and the Convention on the Rights of Persons with Disabilities—each of which protects freedom of expression.

Under Article 19(1) and (2) of the UDHR:

> Everyone shall have the right to hold opinions without interference. Everyone shall have the right to freedom of expression; this right shall include freedom to seek, receive, and impart information and ideas of all kinds, regardless of frontiers, either orally, in writing, or in print, in the form of art, or through any other media of his choice.

Article 19 of the UDHR is embodied further in Article 19 of the ICCPR—in which the United Kingdom is party to and ratified in 1976—with the United Nations clarifying when this right maybe restricted: '(a) For respect of the rights or reputations of others; (b) For the protection of national security or of public order (ordre public), or for public health or morals'.[15] For the United Nations General

14 Bernal (n 4) 18.
15 International Covenant on Civil and Political Rights Article 19(3).

Assembly, such weight is given to the right to speak freely as it contributes 'to the full development of a person', it 'is essential for any society' and can be considered the 'foundation stone for every free and democratic society'.[16] Indeed, freedom of expression is now considered to be a norm of customary international law and is further instilled in Article 10 of the ECHR—in which its creation was heavily influenced by the UDHR.[17]

2.3 The European Convention on Human Rights

The ECHR came into force on 21 January 1959 and was signed by all 47 Member Sates of the European Council, including the United Kingdom. The ECHR itself consists of several Articles protecting what is considered basic human rights, including the prohibition of torture, the right to life, freedom of expression and the right to privacy. As noted above, by way of the HRA, the ECHR has been incorporated into UK law, taking legal effect in October 2010.

Under the HRA, all public bodies in the United Kingdom are under a lawful obligation to protect human rights, with section 6(3) of the Act defining public bodies as: '(a) a court or tribunal, and (b) any person certain of whose functions are functions of a public nature'. Here, public bodies within the criminal justice system, including the courts, the police, and the Crown Prosecution Service, are under a legal duty to consider human rights when coming to a decision. For example, if a matter were to go before a court of law concerning if a particular message sent online was contrary to law, the court must also consider the right to freedom of expression, alongside other rights contained in the ECHR, such as that of privacy. Further, the courts of the United Kingdom must consider the jurisprudence of the European Court of Human Rights (ECtHR) when coming to a judgment as set out in s.2 of the HRA. However, it does not necessarily mean the court will follow the jurisprudence of Strasbourg, as exceptions have been created through the Common Law.[18]

All rights contained in the ECHR can be split into three types: absolute, qualified, and limited. Absolute rights are considered those basic human rights which the state cannot infringe under any circumstances. So, for instance, the prohibition of torture is an absolute right, which cannot be breached even in times of national emergency or during times of war.[19] Whereas Article 5, the right to liberty, is a limited right whereby it can be restricted under the exceptions contained within

16 Human Rights Committee, 'United Nations International Covenant on Civil and Political Rights: General comment No. 34 – Article 19: Freedoms of opinion and expression' (*United Nations*, 2011) [2] <https://www2.ohchr.org/english/bodies/hrc/docs/gc34.pdf> accessed 8 July 2022.

17 Laidlaw (n 9) 87.

18 The English courts may decline to follow the jurisprudence of the European Court of Human Rights if it is clear that the European Court did not sufficiently appreciate the particular nature of the English Legal System (as per *R v Horncastle* [2009] UKSC 14) or if they are bound by the doctrine of precedent (*Leeds v Price* [2005] W.L.R. 1825).

19 The European Convention on Human Rights Article 15(2).

the Article itself. For instance, an individual can be deprived of their liberty when detained following a court conviction.[20]

The right to freedom of expression and the right to privacy are qualified rights:

> A public authority can sometimes interfere with your rights if it's in the interest of the wider community or to protect other people's rights. These rights are qualified. Qualified rights may need to be balanced against other people's rights or the rights of the wider community to achieve a fair outcome.[21]

All qualified rights can be restricted when three criterions are met: the restriction is prescribed by law (the legality element), the restriction fulfils a legitimate aim (contained in the second paragraph of the Article), and the restriction can be considered as necessary in a democratic society (in which the least intrusive method must be taken).

2.3.1 Prescribed by Law

At its very basic, the concept prescribed by law means that there must be a legal rule in place prohibiting the conduct in question and includes both Common Law provisions and Acts of Parliament. However, it is not simply a case of pointing to a legal provision and stating, 'there's the law'. For a provision to be considered as prescribed by law, it needs to encompass key elements of the principle of legality, a fundamental principle of due process which is discussed at length in the following chapter. The ECtHR has identified four key questions in determining if a legal provision can be said to be prescribed by law:[22]

1. Does the domestic legal system sanction the infraction?
2. Is the relevant legal provision accessible to the citizen?
3. Is the legal provision sufficiently precise to enable the citizen reasonably to foresee the consequences which a given action may entail?
4. Does the law provide adequate safeguards against arbitrary interference with the respective substantive rights?

2.3.1.1 Does the Domestic Legal System Sanction the Infraction?

The ECtHR has given a wide definition to sanction and includes Acts of Parliament, the Common Law, international legal obligations, and secondary sources, such as

20 Ibid., Article 5(1)a.

21 Citizens Advice, 'When can a public authority interfere with your human rights?' (*Citizens Advice*, 2018) <https://www.citizensadvice.org.uk/law-and-courts/civil-rights/human-rights/when-can-a-public-authority-interfere-with-your-human-rights/> accessed 5 September 2018.

22 *Huvig v France*, [1990] ECHR, application no. 11105/84, judgment of 24 April 1990; *Kruslin v France*, [1990] ECHR, application no. 11801/85, judgment of 24 April 1990.

royal decrees. In essence, there must first be a legal provision in place as affirmed in *Norman v The United Kingdom*.[23]

2.3.1.2 *Is the Relevant Legal Provision Accessible to the Citizen?*

The issue of accessibility was discussed at length by the ECtHR in *Sunday Times v The United Kingdom*.[24] *The Sunday Times* had run several newspaper articles detailing an ongoing legal dispute between users of the pharmaceutical drug Thalidomide, and the manufacturer of the drug Distillers. Thalidomide had been given to pregnant women in the late 1950s and early 1960s to help with morning sickness. However, it was later discovered that the drug caused significant deformities in the foetus.

The newspaper articles included arguments that Distillers should not mount a legal defence to the allegations against them. In addition, the articles also included suggestions on some of the evidence that may be presented before the Court. Consequently, the Attorney General was granted an injunction by the domestic courts of the United Kingdom for fear that the reports may affect the outcome of the Distillers trial, under the Common Law offence of Contempt of Court. The Sunday Times challenged the decision before the ECtHR on the grounds that the injunction breached their right to freedom of expression. For the ECtHR, freedom of expression could be limited to maintain the authority of the judiciary, however the law limiting Article 10 needs to be accessible: 'the law must be adequately accessible: the citizen must be able to have an indication that is adequate in the circumstances of the legal rules applicable to a given case'.[25] In essence, citizens must be able to understand the legal provision—a concept that underpins legality as discussed further in Chapter 3. An element the ECtHR concluded was missing in the Common Law provision of Contempt of Court. As a result, the ECtHR concluded the legal provisions governing Contempt of Court at the time were not sufficiently clear, breaching the applicant's right to freedom of expression.[26]

2.3.1.3 *Is the Legal Provision Sufficiently Precise to Enable the Citizen to reasonably foresee the Consequences which a Given Action may Entail?*

For a legal provision to be considered compatible with the ECHR, and a principle which also underpins our understanding of legality, as discussed further in the following chapter, it must be 'action-guiding':

> [A] norm cannot be regarded as a 'law' unless it is formulated with sufficient precision to enable the citizen to regulate his conduct: he must be able—if

23 *Norman v The United Kingdom* (Application no. 41397/17), 6 July 2021.

24 *Sunday Times v The United Kingdom* (1979) 2 EHRR 245.

25 Ibid., [49].

26 The decision of the European Court of Human Rights paved the way for the creation of the Contempt of Court Act 1981.

> need be with appropriate advice—to foresee, to a degree that is reasonable in the circumstances, the consequences which a given action may entail.[27]

Though it is accepted by the ECtHR, and indeed, in many legal philosopher's conceptions of legality, that complete precision would be unattainable and therefore some vagueness is inherent in the law, it cannot be so inherently vague that it would be unforeseeable that a person's actions would invoke the law, or it imposes significant discretion by the state on how the legal provision should be applied.[28] A concept upheld by the ECtHR in *Cantoni v France.*[29]

2.3.1.4 Does the Law Provide Adequate Safeguards against arbitrary Interference with the Respective Substantive Rights?

The foundations of the 'prescribed by law' element in the ECHR is to protect individuals from arbitrary interference of their rights by a state. The law should therefore not confer such discretion on members of the executive or, for instance, actors within the criminal justice system, without providing adequate safeguards from abuse:

> [I]t would be contrary to the rule of law for the legal discretion granted to the executive to be expressed in terms of an unfettered power. Consequently, the law must indicate the scope of any such discretion conferred on the competent authorities and the manner of its exercise with sufficient clarity, having regard to the legitimate aim of the measure in question, to give the individual adequate protection against arbitrary interference.[30]

In essence, safeguards need to be in place to stop arbitrary interference by the state with respective substantive rights, not only in theory but also in reality.[31] Indeed, in the joined cases of *Huvig v France*[32] and *Kruslin v France*,[33] the ECtHR found that despite seventeen safeguards cited by the French government as protecting citizens from the abuse of phone tapping by public authorities, these safeguards were not adequate and therefore the actions of the state were incompatible with the ECHR.

2.3.2 Legitimate Aims

If the legal provision under scrutiny can be considered as prescribed by law, next the court must consider if the restriction placed on the substantive right, for instance,

27 *Sunday Times* (n 24) [49].
28 See for example, *Malone v the United Kingdom*, [1984] ECHR, application no. 8691/79, judgment of 2 August 1984.
29 *Cantoni v France*, no. 7862/91, 12 April 1995.
30 *Malone* (n 28) [68].
31 *Gillan v The United Kingdom* (2010) 50 EHRR 45.
32 *Huvig* (n 22).
33 *Kruslin* (n 22).

a legal provision restricting the right to freedom of expression, fulfils a legitimate aim. Under each qualified right within the ECHR a list of aims whereby a right can be restricted is laid out. The legitimate aims differ depending on the right under question. For example, the right to freedom of expression can be restricted when pursuing one of the following aims[34] (alongside fulfilling the criterions set out in Section 2.3.1 above and 2.3.3 below):

1. National security;
2. Territorial integrity or public safety;
3. The prevention of disorder or crime;
4. The protection of health or morals;
5. The protection of the reputation or rights of others;
6. Preventing the disclosure of information received in confidence; or
7. For maintaining the authority and impartiality of the judiciary.

Failure in a restriction to fulfil a legitimate aim will render the restriction unlawful.[35] As discussed in Section 2.3.6 similar aims can be found in Article 8: the right to privacy.

However, it is not simply a case of stating, for example, that by law we will prohibit anti-vaxers from suggesting that vaccinations are being used to track citizens, under the presumption that this would protect the health of those living in a society. Or indeed, we may not mention another person online without permission to protect the rights and freedoms of others (a legitimate aim under Article 8(2)). Instead, the legal provision restricting the substantive right not only needs to fulfil a legitimate aim, but it also needs to be necessary in a democratic society.

2.3.3 Necessary in a Democratic Society

For a restriction to be considered as necessary in a democratic society, it is for the courts to establish:

> whether: (i) the legislative objective is sufficiently important to justify limiting a fundamental right; (ii) the measures designed to meet the legislative objective are rationally connected to it; and (iii) the means used to impair the right or freedom are no more than is necessary to accomplish the objective.[36]

In essence, the state must take the least intrusive method possible to restrict a substantive Article right to render it necessary in a democratic society and in turn, to be

34 The European Convention on Human Rights Article 10(2).

35 *Handyside* (n 6) [49].

36 *de Freitas v Permanent Secretary of Ministry of Agriculture, Fisheries, Lands and Housing* [1999] 1 A.C. 69 (PC) 80.

compatible with the ECHR, as reinforced by the ECtHR in *Dudgeon v The United Kingdom.*[37]

Dudgeon concerned the lawfulness of a legal provision contained in the law of Northern Ireland which criminalised homosexuality. For the applicant who sought leave before the ECtHR, the law breached Article 8(1) of the Convention: the right to privacy. For the ECtHR, the law in question (section 11 of the Criminal Law Amendment Act 1885) was sufficiently clear and pursued one of the legitimate aims contained under Article 8(2) (protection of morals). However, issues arose in relation to the restriction being necessary in a democratic society:

> It cannot be maintained in these circumstances that there is a 'pressing social need' to make such acts criminal offences, there being no sufficient justification provided by the risk of harm to vulnerable sections of society requiring protection or by the effects on the public. On the issue of proportionality, the Court considers that such justifications as there are for retaining the law in force unamended are outweighed by the detrimental effects which the very existence of the legislative provisions in question can have on the life of a person of homosexual orientation like the applicant. Although members of the public who regard homosexuality as immoral may be shocked, offended or disturbed by the commission by others of private homosexual acts, this cannot on its own warrant the application of penal sanctions when it is consenting adults alone who are involved.[38]

Essentially, it was found that criminalising homosexual acts between two consenting adults was not considered necessary in a democratic society therefore breaching Article 8 of the ECHR.[39] As eloquently put in *Olsson v Sweden,* any 'inference [must] correspond to a pressing social need and, in particular, that it is proportionate to the legitimate aim pursued'.[40]

2.3.4 Summary

So long as the three elements discussed above, prescribed by law, fulfilment of a legitimate aim and necessity are met, a state can lawfully infringe on a qualified right protected under the ECHR. It is also important to note that upon making a judgment as to whether a state has impinged upon a substantive right under the ECHR, the ECtHR must have regard to the margin of appreciation whereby the ECtHR often asserts that states are, in principle, in a better position than the courts

37 *Dudgeon v The United Kingdom* (1981) 4 EHRR 149.

38 Ibid., [60].

39 Later parts of this chapter will examine in detail the right to privacy as guaranteed under Article 8 of the European Convention of Human Rights.

40 *Olsson v Sweden* (A/250) (1994) 17 EHRR 134 [67].

to assess the needs of its citizens.[41] Consequently, a state is provided with a degree of discretion on how it fulfils its obligations under the ECHR—including how it protects and balances, the right to freedom of expression and the right to privacy, though the ECHR makes it clear, under Article 17, that you cannot use one right, such as the right to freedom of expression, to disregard another's right, such as the right to privacy.

2.3.5 Article 10

The right to freedom of expression goes beyond just a right to speak freely without state interference. It can include any form of communication, including, though not limited to, art, radio, books, and dance. As outlined at the start of this chapter it is often considered the touchstone of all rights and plays a fundamental role in any democratic society:

> Not only is freedom of expression inseparable from freedom of thought, association, and assembly, it is essential for the enjoyment of all rights, including economic, social, and cultural rights. It is a vital freedom for development, the functioning of a democracy, and modern economies.[42]

The right to freedom of expression is protected under Article 10(1) of the ECHR:

> Everyone has the right to freedom of expression. This right shall include freedom to hold opinions and to receive and impart information and ideas without interference by public authorities and regardless of frontiers.

In essence, every citizen has the right to express an opinion or thought under Article 10 and this does not stop at the borders of a state (the regardless of frontiers element of the right). In fact, in today's digital world, the protection of frontiers would include the prohibition of blocking access to the internet.[43] But as previously acknowledged, the right to freedom of expression can be limited under certain circumstances, though needs to be carefully monitored 'because if wrongly or excessively invoked, they can "chill" the freedom or even eliminate it'.[44] The question remains, at what point does a person's communication or conduct go beyond the right to freedom of expression to warrant state interference?

41 The doctrine of the margin of appreciation has been developed by the European Court of Human Rights as a way to ensure that the Court recognises that the Convention is an international instrument covering 47 countries who may differ in significant ways with regard to their cultural and societal norms and that, therefore, it would not be appropriate for the Court too readily rule against a signatory state for a breach of a Convention right.

42 Boyle (n 2) 266.

43 *Akdeniz v Turkey* (dec.) (no. 20877/10, 11 March 2014).

44 Boyle (n 2).

The jurisprudence of both the ECtHR and the domestic courts of the United Kingdom supports the idea that different types of speech incur different types of protection:

> There are undoubtedly different types of speech, just as there are different types of private information, some of which are more deserving of protection in a democratic society than others. Top of the list is political speech. The free exchange of information and ideas on matters relevant to the organisation of the economic, social and political life of the country is crucial to any democracy. Without this, it can scarcely be called a democracy at all.[45]

Whereas political speech is given the highest form of protection, speech which can be labelled as mere gossip is not necessarily shielded from the law, even in today's digital world.[46] Yet, despite the notion that mere gossip may not be protected under the law, the ECtHR has upheld that the right to freedom of expression includes the right to be offensive in *Handyside v The United Kingdom.*[47]

In the early 1970s, Richard Handyside was prosecuted and convicted under the Obscene Publications Act 1959 and 1964, for the distribution of 'The Little Red School Book'. The book, which was written by two Danish school teachers in 1969, contained several pages on sex, drugs, and alcohol—with the United Kingdom courts upholding that the distribution of the book was contrary to law, resulting in Handyside lodging an application before the ECtHR, claiming his right to freedom of expression had been breached.

Though Handyside's appeal was unsuccessful, with the ECtHR concluding that the restriction did not breach Article 10, the judges did support the concept that freedom of expression encompasses the right to be offensive:

> Freedom of expression constitutes one of the essential foundations of such a society, one of the basic conditions for its progress and for the development of every man. Subject to Article 10(2), it is applicable not only to 'information' or 'ideas' that are favourably received or regarded as inoffensive or as a matter of indifference, but also to those that offend, shock or disturb the state or any sector of the population.[48]

A concept that is further supported by the United Nations who maintain that there needs to be a 'genuine and serious incitement to extremism, as opposed to ideas that simply offend, shock or disturb others'[49] to restrict freedom of expression.

But despite the notion that we do have the right to be offensive under the ECHR, there are limits for the ECtHR, including, speech which can be labelled as holocaust

45 *Campbell v MGN Ltd* [2004] UKHL 22, [2004] 2 A.C. 457 per Lady Baroness-Hale [148].
46 Jacob Rowbottom, 'To Rant, Vent and Converse: Protecting Low Level Digital Speech' (2012) 71(2) CLR 355, 357.
47 *Handyside* (n 6).
48 Ibid., [49].
49 Laura Scaife, *Handbook of Social Media and the Law* (Routledge 2015) 42.

denial,[50] hate speech,[51] and defamatory statements,[52] to name but a few restrictions. In fact, the ECtHR has even upheld that the publication of photographs of Princess Caroline of Monaco was not protected under Article 10 and was an infringement upon her right to privacy, as protected under Article 8.[53]

2.3.6 Article 8

Under Article 8 '[e]veryone has the right to respect for his private and family life, his home and his correspondence'. As discussed earlier, like freedom of expression, the right to privacy is a qualified right and can be restricted when prescribed by law, where it is necessary in a democratic society, and the restriction fulfils one or more of the following legitimate aims:[54]

1. National security;
2. Public safety or the economic well-being of the country;
3. The prevention of disorder or crime;
4. The protection of health or morals; or
5. The protection of the rights and freedoms of others.

But the right to privacy encompasses more than just a right to a private life beyond the public domain, it is one of the most expansive rights under the ECHR and has been interpreted to include, for instance, the protection of reproductive rights,[55] the rights of one's image,[56] police surveillance,[57] and gender identity.[58]

In fact, privacy:

> extends to aspects relating to personal identity, such as a person's name or picture, and furthermore includes a person's physical and psychological integrity; the guarantee afforded by Article 8 of the Convention [ECHR] is primarily intended to ensure the development, without outside interference, of the personality of each individual in his relations with other human beings. There is therefore a zone of interaction with others, even in a public context, which may fall within the scope of 'private life'.[59]

Privacy therefore includes the right for a person not to have their physical or psychological integrity infringed. With Brownsword arguing that, 'privacy is the most

50 *Lehideux and Isorni v France* ECHR 23 September 1998.
51 *Glimmerven en Hagenbeek v Netherlands* [1979] ECTHR 8.
52 *Lingens v Austria* (1986) 8 EHRR 407.
53 *Von Hannover v Germany* (2005) 40 EHRR 1.
54 The European Convention on Human Rights Article 8(2).
55 *A, B and C v Ireland* (2010) Application no. 25579/05.
56 *López Ribalda and others v Spain* (2019) Applications nos.1874/13 and 8567/13.
57 *Shimovolos v Russia* (2011) Application no. 30194/09.
58 *Hämäläinen v Finland* (2014) Application no. 37359/09.
59 *Pfeifer* (n 13).

protean of concepts, with [...] both spatial and informational articulations'.[60] The concept of privacy therefore underpins free speech,[61] complete free speech would limit the speech of the minority,[62] whilst also having significant psychological effects on those subjected to it. Indeed, a right to privacy, even on the internet, does exist—as the Samaritans discovered following the creation, and ultimate backlash, in its 'Samaritans Rader App'.

In 2014 the Samaritans, a UK-based mental health charity, released an App aimed at reducing suicide. The App, once downloaded onto a person's smart device, would link to the users Twitter profile to scan tweets in the user's homepage. Using a list of predetermined keywords, the App would then alert the user if someone who they 'followed' on Twitter indicated possible suicidal tendencies, even if the other person did not have access to the App. Following its launch, the Samaritans came under heavy criticism for breaching Twitter user's privacy. The original stance taken by the Samaritans was to justify the App's usage by arguing that comments made on Twitter were in the public domain, and as a result, there was no expectation to privacy. Two months after the App was launched, the Samaritans suspended its usage.[63]

A balance must therefore be struck between freedom of expression and privacy. In fact:

> any restriction of the right to freedom of expression must be subjected to very close scrutiny. But so too must any restriction of the right to respect for private life. Neither Article 8 nor Article 10 has any pre-eminence over the other in the conduct of this exercise. As Resolution 1165 of the Parliamentary Assembly of the Council of Europe (1998), para 11, pointed out, they are neither absolute nor in any hierarchical order, since they are of equal value in a democratic society.[64]

A concept reinforced in Article 17 of the ECHR.

2.3.7 Article 17

Article 17(1) states:

> Nothing in this Convention may be interpreted as implying for any State, group or person any right to engage in any activity or perform any act aimed at the destruction of any of the rights and freedoms set forth herein or at their limitation to a greater extent than is provided for in the Convention.

60 Roger Brownsword, *Law, Technology and Society: Re-imaging the Regulatory Environment* (Routledge 2019) 303 as inspired by Graeme Laurie, *Genetic Privacy* (Cambridge University Press 2002).

61 Bernal (n 4) 145.

62 Ibid., 106.

63 Samaritans, 'Samaritans rader' (*The Samaritans*, 10 March 2015) <https://www.samaritans.org/how-we-can-help-you/supporting-someone-online/samaritans-radar> accessed 26 October 2018.

64 *Campbell* (n 45) per Lord Hope of Craighead [115].

Put simply, a person cannot rely on the protection of a given Article within the ECHR to the detriment of another right. So, for example, a person cannot spread racial hatred as part of their right to freedom of expression if it is considered to significantly affect other rights contained in the ECHR as illustrated in *Glimmerven en Hagenbeek v Netherlands.*[65]

In *Glimmerven en Hagenbeek* the ECtHR was asked to consider if the Netherlands had breached the applicant's right to freedom of expression by prohibiting them from distributing leaflets. Glimmerven and Hagenbeek, which had previously been declared as a prohibited organisation under the Civil Code of the Netherlands, had distributed leaflets advocating the removal of 'all Surinamers, Turks and other so-called guest workers from the Netherlands'. Before the ECtHR the Dutch Government accepted that they had infringed the applicant's right to free speech, however the Government argued that Article 17(1) prohibited Glimmerven and Hagenbeek from exploiting Article 10 to spread racial hatred:

> The Netherlands' authorities in allowing the applicants to proclaim freely and without penalty their ideas would certainly encourage the discrimination prohibited by [these] provisions of the Convention [… such activities being] contrary to the text and spirit of the Convention.[66]

The arguments put forward by the Dutch government's legal team were accepted by the ECtHR, where the court held that the applicant's right to freedom of expression was lawfully infringed by the state.

2.3.8 Summary

It is evident that both Articles 10 and 8 can, at times, come into conflict with one another. Though in theory, neither provision is more important than the other. However, for Foster:

> [i]nstead of assuming free speech to be good in itself, and only capable of restriction in the most exceptional cases, we are in danger of reversing that assumption and challenging the speaker to prove the overriding good that speech will prevail.[67]

This is even more apparent when it comes to the regulation of the digital world, in particular, the regulation of speech on social media platforms. In recent years, we have seen a drive towards regulating not only social media platforms themselves, through Acts such as the United Kingdom's Online Safety Act 2023 (OSA), but also a continued emphasis being placed on the criminal law to tackle abusive communications on social media sites. Indeed, as will be reflected throughout

65 *Glimmerven en Hagenbeek* (n 51).

66 Ibid., [196].

67 Steve Foster, 'Do We Want Free Speech or Not? Modern Challenges to Free Speech' (2019) 24(2) Cov. L. J 70, 81.

discussions in this monograph, we are at huge risk of restricting freedom of expression and in turn, overcriminalising speech, even though such use of the law can be considered a fundamental breach of the principle of legality. For instance, as will be discussed in Chapter 6, we have seen an alarming increase in the use of s.127(1) of the Communications Act 2003 and the Malicious Communications Act 1988 in criminalising communications, despite the vague terminology used in both provisions. In fact, when it comes to discussions on freedom of expression and social media, society seems to heavily focus on the 'bad' rather than the 'good'.

2.4 Freedom of Expression v the Right to Privacy: The Online Context

The balancing of Articles 8 and 10 is not unique to the digital age. Much of the jurisprudence relating to these two rights stems from traditional forms of media such as the press.[68] In *Campbell v MGN Limited*, the House of Lords had to directly address the issue of freedom of expression versus the right to privacy:

> The present case concerns one aspect of invasion of privacy: wrongful disclosure of private information. The case involves the familiar competition between freedom of expression and respect for an individual's privacy. Both are vitally important rights. Neither has precedence over the other. The importance of freedom of expression has been stressed often and eloquently, the importance of privacy less so.[69]

The continued expansion of the internet and the sheer dominance of social media today, have created further complexities when it comes to balancing freedom of expression and privacy, as the lines between what is considered public and private have been heavily distorted.

As internet usage has grown, how we communicate our thoughts, anger, frustrations, or feelings has evolved, with social media playing a pivotal role. Whereas once-upon-a-time, we would have vented our frustrations to our loved ones, or even our pets, today we can now vent these frustrations to the world at large, in a matter of seconds—with detrimental costs. For example, in 2015 an individual tweeted the following: 'Going to Africa. Hope I don't get AIDS [*sic*]. Just kidding. I'm white!' to her 170 followers just as she got on a plane from JFK Airport to Cape Town. By the time she arrived at her destination, eleven hours later, her name was trending on Twitter, she was making headline news across the world and in the process of her flight, she was dismissed from her employment and was met at Cape Town airport by armed police.[70] Yet, these are extreme cases and it is the extreme

68 Lorna Woods, 'Social Media: It is not just about Article 10' in David Mangan and Lorna E. Gillies (eds), *The Legal Challenges of Social Media* (Edward Elgar Publishing 2017) 109. For Bernal, the use of traditional media law is not necessarily suitable for social media. See, Bernal (n 4) 25.

69 *Campbell* (n 45) per Lady Baroness-Hale [148] and per Lord Nicholls of Birkenhead [12].

70 Jon Ronson, 'How one stupid tweet blew up Justine Sacco's life' *The New York Times Magazine* (New York, 12 February 2015) <https://www.nytimes.com/2015/02/15/magazine/how-one-stupid-tweet-ruined-justine-saccos-life.html> accessed 5 February 2019.

cases that tend to focus the public's attention, resulting in the cry that being able to say such things online 'should be criminal'. There will of course always be cases which should invoke the criminal law, so long as those legal provisions criminalising such conduct comply with the principle of legality, as discussed in the following chapter. But if we focus primarily on the extreme cases, the cases which catch the attention of the media, and use these as the rationale for increasing the criminal law in prohibiting certain types of speech, we run the risk of 'borderline' speech also being caught up in such an approach. We also run the risk of Government's implementing legislation which can be abused for their own political agenda, such as the OSA, which puts pressure on social media companies to remove content that illustrates those crossing the channel in a 'positive' light.[71] In fact, despite, the 'negative' side of social media, which is often the focus of such discussions, freedom of expression on social media platforms plays a significant role within society, and an important one at that.

2.4.1 Freedom of Expression Online

The ability to express oneself freely without fear of being censored is the bedrock of any democratic society. 'If competing views are publicly debated and exposed to public scrutiny the good will, over time, drive out the bad'[72] with social media playing an important role in amplifying the 'wrongs' within a given society. There is no doubt that social media has changed every aspect of everyday life, but this is not necessarily always a bad 'thing'.

Politicians can now directly communicate with their constituents on mass about important matters. Likewise, constituents can contact their local MP with a 'click-of-a-button'. The police can use social media to help locate a person who has been reported missing. Citizens can also use social media to draw attention to campaigns, call for change in society or even hold politicians to account. There are ample examples of how social media has been pivotable in drawing people's attention to social injustice, two of which, #MeToo and Black Lives Matter are explored below. Indeed, for Laidlaw '[w]here the technology of the internet is particularly facilitative is a platform for mobile licence social and political change'.[73]

2.4.1.1 #MeToo

#MeToo is a political movement that started in 2006 on the now debunked social media platform MySpace but gained significant attention in 2017, following the exposure of sexual-abuse allegations against a former Hollywood movie director,

71 See, OFCOM, 'Protecting people from illegal harms online. Volume 2: The causes and impacts of online harms' (*OFCOM*, 9 November 2023) 168 – 178 <https://www.ofcom.org.uk/__data/assets/pdf_file/0019/271243/volume-2-illegal-harms-consultation-1.pdf> accessed 4 January 2024.

72 *Campbell* (n 45) per Lord Bingham [148].

73 Laidlaw (n 9) 25.

Harvey Weinstein. The hashtag, which trended on several occasions in 2017, is used to expose the continued sexual harassment and sexual assault experienced by individuals, in particular women, within society: 'If all the women who have been sexually harassed or assaulted wrote "Me too" as a status, we might give people a sense of the magnitude of the problem'.[74] Here, using the hashtag, individuals spoke publicly on platforms such as Twitter to tell their stories of sexual harassment and assault, allowing those who had been subjected to this abuse, having their voices heard—and in many cases, having their voices heard for the first time.

But the #MeToo movement did not just stop at social media sites, it did create significant changes within society. Widespread coverage of the movement, particularly in Hollywood, led to high-profile dismissals and adaptions of the hashtag emerging to call out sexual harassment in schools, sports, and politics. Through social media, voices could be heard on mass putting pressure on organisations, governments, and the criminal justice system to do more to tackle sexual harassment. Indeed, we have seen the use of social media sites such as Facebook and X used to highlight the everyday sexism that exists within society with #everydaysexism;[75] the use of social media to organise protests to challenge the 'sexism, racism and xenophobia of the Trump Administration';[76] and as a space for women to challenge stereotypical conceptions of rape.[77] Social media has been pivotal in drawing attention to how women are treated within society—it has been a force for good.

Though of course, it must not be dismissed that social media has also been used to target women, as discussed in Section 2.4.2.2 below. However, if we always focus on the 'bad' we risk losing the 'good'—such as the use of social media platforms to evoke wider societal change.

2.4.1.2 Black Lives Matter

Black Lives Matter (BLM) is a movement advocating for non-violent civil disobedience in protest against incidents of police brutality and all racially motivated violence against African-American people. It was founded in 2013 but gained significant attention in 2020, following the death of George Floyd on 25 May 2020

74 Nadja Sayei, 'Alyssa Milano on the #MeToo movement: "We're not going to stand for it any more"' *The Guardian* (London, 1 December 2017) <https://www.theguardian.com/culture/2017/dec/01/alyssa-milano-mee-too-sexual-harassment-abuse> accessed 6 July 2022.

75 Laura Bates, 'The everyday sexism project' (*Everydaysexism*, 2019) <https://everydaysexism.com/> accessed 6 February 2019.

76 Kaitlynn Mendes, Jessica Ringrose and Jessalynn Keller, *Digital Feminist Activism: Girls and Women Fight Back Against Rape Culture* (Oxford University Press 2019) 4.

77 Jessamy Gleeson, '"(Not) working 9–5": The consequences of contemporary Australian-based online feminist campaigns as digital labour' (2016) 16(1) MIA 77 <http://journals.sagepub.com/doi/pdf/10.1177/1329878X16664999> accessed 13 January 2020.

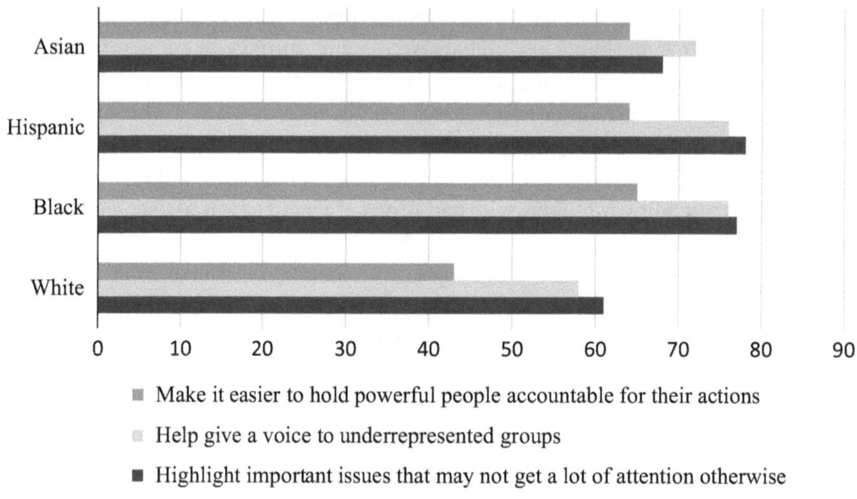

Figure 2.1 Percentages of social media users by race who say the following statements describe social media very or somewhat well.

Note: Brooke Auxier, 'Social media continue to be important political outlets for Black Americans' (*Pew Research Centre*, 11 December 2020) <https://www.pewresearch.org/fact-tank/2020/12/11/social-media-continue-to-be-important-political-outlets-for-black-americans/> accessed 18 July 2022.

at the hands of a Minneapolis police officer.[78] His death made headline news, with video footage of the incident being actively shared across social media sites.[79]

Though the tragic events of that day were also shared through traditional forms of media, such as the press and radio, social media, like that of the #MeToo movement discussed above, played an important role in bringing the events of George Floyd's death to the attention of the public worldwide. In fact, in the following days after the death of George Floyd, nearly 8.8 million tweets contained the hashtag: #BlackLivesMatter.[80] Indeed, social media has played an important role in giving a voice to marginalised groups, with research undertaken by the Pew Research Centre, as shown further in Figure 2.1, finding that for those Black Americans who took part in the study, social media was an 'important political outlet':

> Black social media users have been particularly likely to say that these sites are personally important to them for getting involved with issues they care

78 Kim Parker, Juliana Menasce and Monica Anderson, 'Amid protests, majorities across racial and ethnic groups express support for the Black Lives Matter Movement' (*Pew Research Centre*, 12 June 2020) <https://www.pewresearch.org/social-trends/2020/06/12/amid-protests-majorities-across-racial-and-ethnic-groups-express-support-for-the-black-lives-matter-movement/> accessed 27 July 2022.

79 Brooke Auxier, 'Social media continue to be important political outlets for Black Americans' (*Pew Research Centre*, 11 December 2020) <https://www.pewresearch.org/fact-tank/2020/12/11/social-media-continue-to-be-important-political-outlets-for-black-americans/> accessed 18 July 2022.

80 Ibid.,

about or finding like-minded people. They are also likely to express positive views about the impact of these platforms for holding powerful people accountable for their actions and giving a voice to underrepresented groups.[81]

2.4.1.3 Summary

Freedom of expression on social media sites must be protected from arbitrary government interference. Its use allows users to call out social injustice and express a need for change within society and this should not be forgotten—as noted throughout the discussion above if we always focus on the 'bad' we risk eroding the 'good':

> First, if any opinion is compelled to silence, that opinion may, for aught we can certainly know, be true. To deny this is to assume our own infallibility. Secondly, though the silenced opinion be an error, it may, and very commonly does, contain a portion of truth [...] it is only by collusion of adverse opinions, that the remainder of the truth has any chance of being supplied.[82]

But it is also important to remember that complete free speech online does not exist because complete free speech would, in turn, reduce the speech of the minority and directly affect another's right to privacy.

2.4.2 Privacy Online

There is a 'darker' side to social media, where voices of the minority can be shut down and behaviours such as racism are rife. As evidenced in Article 17 of the ECHR, freedom of expression cannot be used to justify such behaviour(s). In essence, we cannot, and should not, use our own right to freedom of expression to limit another's right to free speech or indeed, to infringe upon another's right to privacy. Social media now forms part of everyday life which cannot be separated from that of the real world.[83] Or as potently put by Bernal, '[t]he internet is now integral to the way society operates'.[84] And even in the digital world, privacy does exist online.

As outlined in Section 2.3.6 privacy is more than just a right to have your private information kept out of the public domain. It encompasses a range of expansive rights, including the right to bodily autonomy, with the ECtHR finding that a person's body is an intimate aspect of a person's private life[85] and includes protection of a person's mental health.[86] When it comes to online abuse, it is clear, that such

81 Ibid.,
82 Mill (n 7) 40.
83 Barry Wellman and Caroline Haythornthwaite (eds), *The Internet in Everyday Life* (John Wiley & Sons 2008) 25.
84 Bernal (n 4) 19.
85 *Y.F. v Turkey* (24209/94, 22 October 2003).
86 *Bensaid v the United Kingdom* (no. 44599/98, 6 May 2001).

behaviours can have a significant effect on a person's mental wellbeing and in turn, breaches that person's right to privacy:

> We have seen that specific harms resulting from being the recipient of abusive and offensive communication[s] online can include: (1) psychological effects and emotional harms; (2) physiological harms; including suicide and self-harm; (3) exclusion from public online space and corresponding feelings of isolation; (4) economic harms; and (5) wider societal harms.[87]

Indeed, like that of freedom of expression, there are ample examples of how a person's right to privacy is infringed in the online world, two examples, cyberbullying and violence against women and girls, are explored in Sections 2.4.2.1 and 2.4.2.2 below.

2.4.2.1 Cyberbullying

Not only has the internet created new and unique ways to taunt another, but behaviours which were once confined to the physical world have now emerged online, with cyberbullying being a prime example.[88] Cyberbullying takes the form of traditional bullying behaviour but is conducted online and unlike bullying, which is commonly associated with the playground, it can occur around the clock.

Each year Ditch the Label, with the help of educational institutions, conducts a survey exposing the extent of bullying in the United Kingdom. In recent years they have included the concept of cyberbullying. In 2016 they found that 65% of participants had experienced some form of cyberbullying, an increase of 3% on the previous year.[89] By 2018, this figure had risen to 66%.[90] However, it is important to note that in the 2020 bullying study by Ditch the Label, this figure had dropped to 27%.[91] Whilst on the other hand, research by OFCOM into the online habits of users in the United Kingdom found that '[a]mong children aged 8–15 who have experienced bullying, more than eight in ten (83%) had experienced it through a

87 Law Commission, *Abusive and Offensive Online Communications: A Scoping Report* (Law Com No 381, 2018) [3.30].

88 Ibid., [1.33].

89 Ditch the Label, 'The annual bullying survey 2016' (*Ditch the Label*, 2016) 14 <http://www.ditchthelabel.org/wp-content/uploads/2016/04/Annual-Bullying-Survey-2016-Digital.pdf > accessed 18 October 2016. See also, Ditch the Label, 'The annual bullying survey 2015' (*Ditch the Label*, 2015) 18 <http://ditchthelabel.org/downloads/abs2015.pdf> accessed 26 February 2016.

90 Ditch the Label, 'The annual bullying survey 2018' (*Ditch the Label*, 2018) 13 <https://www.ditchthelabel.org/wp-content/uploads/2018/06/The-Annual-Bullying-Survey-2018-2.pdf> accessed 12 December 2018.

91 Ditch the Label, 'The annual bullying survey 2020' (*Ditch the Label*, 2020) 16 <https://dtl-beta-website-assets.s3.amazonaws.com/The_Annual_Bullying_Survey_2020_2_a8a474bb3f.pdf> accessed 12 October 2023.

communications device such as a phone or laptop[92].' Indicating, that such behaviour is still present online.

Cyberbullying can have detrimental effects on those who are subjected to it:

> The psychological harm inflicted by cyberbullying, just like bullying, is reflected in low self-esteem, school failure, anger, anxiety, depression, school avoidance, school violence, and suicide.[93]

Further research conducted by John *et al* found that individuals who were bullied online were twice as likely to self-harm or attempt suicide.[94] Clearly, those subjected to such behaviour(s) online are having their right to bodily autonomy violated and in turn, their right to privacy is interfered with. A notion which was supported by the ECtHR in *Buturugă v Romania*,[95] where it was held that the states failure to investigate cyberbullying/cyber-violence in the context of domestic violence, was an infringement on the applicant's human rights under Article 8 (the right to privacy) and Article 3 (the prohibition of torture).

Despite cyberbullying being evidently present on social media sites, as noted above, bullying is not new and no matter how we change the law to prohibit such behaviour, including both in the physical and virtual world, we will never eradicate bullying or cyberbullying as it's embedded within societal norms. Consequently, the law will always fail to tackle these issues, so long as societal norms remain stagnant. As highlighted by Reed and Murray, it is only through changes in societal norms, such as better education, that legal provisions will take effect.[96] However, changing such norms in the online world is inherently difficult:

> In the physical world norms have developed (or have evolved) incrementally over centuries. In the modern world individuals normally live their lives in a mature and stable normative landscape. The norms which defined the landscape do change of course, but they change gradually and over time scales which allow most people to come to terms with their slow evolution. Cyberspace is arguably different because online communities are still in their formative phase [...] in human terms, we might see cyberspace as a young adult, taking its origins as being from the mid-1990s. Or we could say that cyberspace really began when the internet became a genuine medium of mass communication, in which case we might set its birthday as Tuesday, 26 August 2008 when Facebook announced that it had 100 million users. If

92 OFCOM, 'Online Nation' (*OFCOM*, 2022) 8 https://www.ofcom.org.uk/__data/assets/pdf_file/0023/238361/online-nation-2022-report.pdf accessed 15 November 2023.

93 Qing Li, 'Cyberbullying in High Schools: A Study of Students' Behaviours and Beliefs about This New Phenomenon' (2010) 19(4) J. Aggress Maltreat 372, 374.

94 Ann John *et al*, 'Self-Harm, Suicidal Behaviours, and Cyberbullying in Children and Young People: Systematic Review' (2018) 20(4) J Med Internet Res <https://www.jmir.org/2018/4/e129/> accessed 12 December 2018.

95 *Buturugă v Romania* (no. 56867/15, 11 June 2020).

96 Chris Reed and Andrew Murray, *Rethinking the Jurisprudence of Cyberspace* (Elgar 2018) 103.

so, cyberspace is not even a teenager and most of the largest communities therein are around a similar age.[97]

If this is the case, we have yet to set a precedent on what is or is not socially acceptable on social media sites, the law can only do so much—we need to change societal norms first, rather than attempting to restrict the right to freedom of expression or engage in arbitrary use of the criminal law.

2.4.2.2 Violence Against Women and Girls Online

The notion that online abuse is embedded in societal norms is further illustrated in the use of social media to abuse women online. In recent years, we've witnessed an increase in the use of social media sites to send abusive and often misogynistic comments to women online, with research undertaken by Amnesty International exposing that in the United Kingdom, of those women who took part in the study, 1 in 5 had experienced online abuse.[98] The abuse experienced by these women had a direct impact on their right to privacy, shown further in Figure 2.2.

Of those women who took part in the study, 55% of participants in the United Kingdom stated that they had experienced anxiety, stress, or panic attacks as a result of online abuse;[99] with a further 24% of those surveyed across Denmark, Italy, New Zealand, Poland, Spain, Sweden, the United Kingdom and United States feeling that their family's safety was at risk. But it was not just their right to bodily autonomy, as protected under Article 8, that was infringed. The findings of the study also uncovered that 76% of those who said they had been subjected to online abuse had, as a result of the conduct they were subjected to, changed how they used social media. Online abuse which results in the user changing their online habits not only infringes Article 8 but also Article 10—freedom of expression. As noted by Bernal, '[t]o exclude oneself from the internet is to put oneself at a massive disadvantage'.[100]

However, like cyberbullying discussed above, violence against women and girls is not new. In fact, it is rife in every aspect of society, including, for example, the police.[101] This is again another example of a behaviour which is inherent within society and consequently, will also be mirrored in the online world. Without changing attitudes and norms amongst citizens, no legislation will adequately deal with the 'darker' side of social media.

97 Ibid., 108.

98 Amnesty International UK, 'Online abuse of women widespread in UK' (*Amnesty International*, 2017) <https://www.amnesty.org.uk/online-abuse-women-widespread> accessed 3 October 2018.

99 Ibid.,

100 Bernal (n 4) 19.

101 His Majesty's Inspectorate of Constabulary and Fire and Rescue Services, 'An inspection of vetting, misconduct, and misogyny in the police service' (*HMICFS*, 2 November 2022) <https://hmicfrs.justiceinspectorates.gov.uk/publications/an-inspection-of-vetting-misconduct-and-misogyny-in-the-police-service/> accessed 1 February 2023.

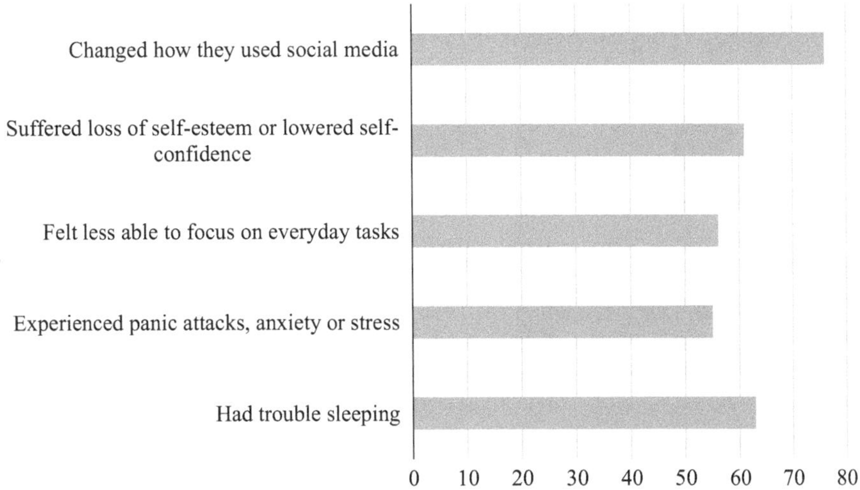

Figure 2.2 The effects of online abuse on women (in percentages) who took part in Amnesty International's research project.

Note: Amnesty International UK, 'Online abuse of women widespread in UK' (Amnesty International, 2017) <https://www.amnesty.org.uk/online-abuse-women-widespread> accessed 3 October 2018.

2.4.2.3 Summary

It cannot be disputed that online abuse does have detrimental consequences and can result in the loss of life;[102] this should never be downplayed. However, it should be that '[i]n a well-functioning liberal democratic society [...] censorship will be the exception and not the rule'.[103] With increasing pressure being placed on governments to tackle the harms associated with the online world, we need to be careful to ensure we are striking an appropriate balance between Articles 8 and 10. In which the criminal law, when underpinned by the principle of legality, should only be used in the most extreme cases to ensure freedom of expression is protected.

2.4.3 Getting the Balance Right

As new technology continues to emerge and evolve, striking the balance between privacy and freedom of expression will play a pivotal role in future regulation. It

102 For example, in 2012 Erin Gallagher (13) took her own life after being continually subjected to anonymous abusive comments on the social media site, ASK.FM. See, Greg Harkin, 'Family devastated after tragic Erin (13) takes own life after vicious online bullying' *Irish Independent* (Dublin, 29 October 2012) <https://www.independent.ie/irish-news/family-devastated-after-tragic-erin-13-takes-own-life-after-vicious-online-bullying-28824852.html> accessed 12 December 2018.

103 Newman (n 1) 680.

is generally accepted today, that there is role for the law in regulating individuals conduct online. But it should be that Articles 10 and 8 work together and:

> [g]etting that balance right is one thing for the policymakers and legislators, but it is also an issue for policing and prosecutors, to make sure that we are not criminalising the trivial, the passionate or those really strongly held views.[104]

A balance, as will be shown in the following chapters, we are currently failing to reach.

It's important that freedom of expression remains an integral concept when making decisions on how to regulate individuals on social media platforms, alongside social media companies themselves. Shutting down voices, for fear that such conduct might be objectionable, does not create change in society. Instead, it drives these voices to other areas of the internet, and this is how echo-chambers are created. We cannot challenge another person's viewpoint if we cannot hear that person's view in the first place—shutting down the conversation does not allow for public debate. As discussed in Section 2.4.1, there is a positive side to social media, which is often forgotten. We need to be able to partake in equal debate and that means, hearing opinions which maybe unfavourable for some. Freedom of expression encompasses challenging 'people by exposing them to opposing points of view'.[105]

That is not to say we should have no limits on our online communications, there is a line. But to draw that line we first need to understand that there is a difference between privacy and harm. Privacy is about protecting significant breaches of a person's personal autonomy. Harm is about reducing the impact of behaviours we may find objectionable. Indeed, the 'harms' associated with the online world means we could slip into the realms of regulating the offensive—and we have the right to be offensive. Second, any restrictions on freedom of expression need to be sufficiently clear and comply with the basic principle of legality in the criminal law, which as will be exposed in later chapters, is not always the case.

2.5 Chapter Summary

As highlighted throughout this chapter, the right to freedom of expression is considered a fundamental aspect of any democratic society. It is a right that is embedded in both national and international provisions across the globe, alongside the right to privacy. In theory, neither right is to be considered more important than the other as both are considered vital to any working and stable society. Instead, a balance must be struck between the two—something that has become extremely complex in the wake of the emergence of new technology.

104 House of Lords, *Select Committee on Communications: Freedom of Expression Online* (HL Deb 2021-04, 23) per David Tucker p.4.

105 Laidlaw (n 9) 31.

As the ways in which we can communicate become easier due to the advancement of modern technology, it has also become simpler to abuse, offend or upset another. A message sent online is a message sent to the world. In turn, as will be shown in further chapters of this monograph, there has been a growing emphasis placed on the criminal law to prohibit such speech. But much of the speech that is objected to online, such as racist and misogynistic commentary, is inherent within society and not unique to the digital world. We therefore need to be extremely careful in how we approach the criminalisation of speech to ensure privacy is maintained, whilst also ensuring we do not excessively infringe on the right to freedom of expression.

The above highlights the importance of protecting the right to freedom of expression—it can, and has been, a force for good. Yet, as noted above, it is often the 'bad' which focuses the attention of policymakers, the press and members of the public, especially when it comes to social media. Even in 'stable' states, such as that of the United Kingdom, it is vital that we do not overstep the mark and prohibit people from saying things they are perfectly within their right to say, through the coercive nature of the criminal law. Especially at a time when other rights are being restricted, such as the right to protest.[106] As will be explored in further chapters of this monograph the studies used to justify the use of the criminal law to restrict the right to freedom of expression, are significantly flawed.

Reference list

Amnesty International UK, 'Online Abuse of Women Widespread in UK' (*Amnesty International*, 2017) https://www.amnesty.org.uk/online-abuse-women-widespread

Auxier B, 'Social Media Continue to Be Important Political Outlets for Black Americans' (*Pew Research Centre*, 11 December 2020) https://www.pewresearch.org/fact-tank/2020/12/11/social-media-continue-to-be-important-political-outlets-for-black-americans/

Bates L, 'The Everyday Sexism Project' (*Everydaysexism*, 2019) https://everydaysexism.com/

Bernal P, *The Internet, Warts and All: Free Speech, Privacy and Truth* (Cambridge University Press 2018)

Boyle K, 'Thought, Expression, Association, and Assembly' in Moeckli D, Shah S & Sivakumaran S (eds), *International Human Rights Law* (Oxford University Press 2010)

Brownsword R, *Law, Technology and Society: Re-imaging the Regulatory Environment* (Routledge 2019)

Citizens Advice, 'When Can a Public Authority Interfere with Your Human Rights?' (*Citizens Advice*, 2018) https://www.citizensadvice.org.uk/law-and-courts/civil-rights/human-rights/when-can-a-public-authority-interfere-with-your-human-rights/

Communications and Digital Committee, *Free for All? Freedom of Expression in the Digital Age* (2021 – 22 HL 54).

Ditch the Label, 'The Annual Bullying Survey 2018' (*Ditch the Label*, 2018) https://www.ditchthelabel.org/wp-content/uploads/2018/06/The-Annual-Bullying-Survey-2018-2.pdf

106 See for example, The Police, Crime, Sentencing and Courts Act 2022 part 3, which has made significant changes to the right to protest in the United Kingdom.

Ditch the Label, 'The Annual Bullying Survey 2016' (*Ditch the Label*, 2016) http://www.ditchthelabel.org/wp-content/uploads/2016/04/Annual-Bullying-Survey-2016-Digital.pdf

Ditch the Label, 'The Annual Bullying Survey 2015' (*Ditch the Label*, 2015) http://ditchthelabel.org/downloads/abs2015.pdf

Foster S, 'Do We Want Free Speech or Not? Modern Challenges to Free Speech' (2019) 24(2) Cov L J 70

Gleeson J, '"(Not) Working 9–5": The Consequences of Contemporary Australian-based Online Feminist Campaigns as Digital Labour' (2016) 16(1) MIA 77 http://journals.sagepub.com/doi/pdf/10.1177/1329878X16664999

Harkin G, 'Family devastated after tragic Erin (13) takes own life after vicious online bullying' *Irish Independent* (Dublin, 29 October 2012) https://www.independent.ie/irish-news/family-devastated-after-tragic-erin-13-takes-own-life-after-vicious-online-bullying-28824852.html

His Majesty's Inspectorate of Constabulary and Fire & Rescue Services, 'An Inspection of Vetting, Misconduct, and Misogyny in the Police Service' (*HMICFS*, 2 November 2022) https://hmicfrs.justiceinspectorates.gov.uk/publications/an-inspection-of-vetting-misconduct-and-misogyny-in-the-police-service/

House of Lords, *Select Committee on Communications: Freedom of Expression Online* (HL Deb 2021-04, 23)

Human Rights Committee, 'United Nations International Covenant on Civil and Political Rights: General Comment No. 34 - Article 19: Freedoms of Opinion and Expression' (*United Nations*, 2011) https://www2.ohchr.org/english/bodies/hrc/docs/gc34.pdf

John A *et al*, 'Self-Harm, Suicidal Behaviours, and Cyberbullying in Children and Young People: Systematic Review' (2018) 20(4) J Med Internet Res https://www.jmir.org/2018/4/e129/

Laidlaw E B, *Regulating Speech in Cyberspace: Gatekeepers, Human Rights and Corporate Responsibility* (Cambridge University Press 2015)

Laurie G, *Genetic Privacy* (Cambridge University Press 2002)

Law Commission, *Abusive and Offensive Online Communications: A Scoping Report* (Law Com No 381, 2018)

Li Q, 'Cyberbullying in High Schools: A Study of Students' Behaviours and Beliefs about This New Phenomenon' (2010) 19(4) J Aggress Maltreat 372

Mchangama J, *Free Speech: A Global History from Socrates to Social Media* (Basic Books 2022)

Mendes K, Ringrose J & Keller J, *Digital Feminist Activism: Girls and Women Fight Back Against Rape Culture* (Oxford University Press 2019)

Mill J S, *On Liberty* (first published 1859, Enhanced Media 2016)

Newman S L, 'Finding the Harm in Hate Speech: An Argument against Censorship' (2017) 50(3) CJPS 679

OFCOM, 'Protecting People from Illegal Harms Online. Volume 2: The Causes and Impacts of Online Harms' (*OFCOM*, 9 November 2023) https://www.ofcom.org.uk/__data/assets/pdf_file/0019/271243/volume-2-illegal-harms-consultation-1.pdf

OFCOM, 'Online Nation' (*OFCOM*, 2022) https://www.ofcom.org.uk/__data/assets/pdf_file/0023/238361/online-nation-2022-report.pdf

Parker K, Menasce J & Anderson M, 'Amid Protests, Majorities Across Racial and Ethnic Groups Express Support for the Black Lives Matter Movement' (*Pew Research Centre*, 12 June 2020) https://www.pewresearch.org/social-trends/2020/06/12/amid-protests-majorities-across-racial-and-ethnic-groups-express-support-for-the-black-lives-matter-movement/

Reed C & Murray A, *Rethinking the Jurisprudence of Cyberspace* (Elgar 2018)

Ronson J, 'How one stupid Tweet blew up justine Sacco's life' *The New York Times Magazine* (New York, 12 February 2015) https://www.nytimes.com/2015/02/15/magazine/how-one-stupid-tweet-ruined-justine-saccos-life.html
Rowbottom J, 'To Rant, Vent and Converse: Protecting Low Level Digital Speech' (2012) 71(2) CLR 355
Sayei N, 'Alyssa Milano on the #MeToo movement: "We're not going to stand for it any more"' *The Guardian* (London, 1 December 2017) https://www.theguardian.com/culture/2017/dec/01/alyssa-milano-mee-too-sexual-harassment-abuse
Scaife L, *Handbook of Social Media and the Law* (Routledge 2015)
The Samaritans, 'Samaritans Rader' (*The Samaritans*, 10 March 2015) https://www.samaritans.org/how-we-can-help-you/supporting-someone-online/samaritans-radar
Wellman B & Haythornthwaite C (eds), *The Internet in Everyday Life* (John Wiley & Sons 2008)
Woods L, 'Social Media: It Is Not Just about Article 10' in Mangan D & Gillies L E (eds), *The Legal Challenges of Social Media* (Edward Elgar Publishing 2017)

3 Regulating Social Media
Theoretical Positioning

3.1 Introduction

Since the creation of the internet, and indeed, the expansion of new technology in general, how the internet should be regulated has often been debated in the academic literature, parliament, and society. There is no 'one size fits all' approach to the regulation of new technology, especially when it comes to social media. How we should regulate social media platforms is often debated in the mainstream press. In fact, there does not seem to be a day that goes by without a news story concerning the regulation of social media platforms—often these stories are embedded in the 'darker' side of social media, as discussed in Chapter 2. Much of the response to date has been 'more criminal law, please'.

Consequently, different theoretical positionings have emerged to try and create a universal understanding of how we *should* regulate the online world, with the first concept, cyberlibertarian theory, gaining significant coverage in the early 90s, long before the likes of social media as we know it today. Since then, further theoretical concepts have been created, including cyber-paternalism, network communitarianism, and the governance model. The first part of this chapter will discuss these theoretical concepts. It will be argued that regardless of which theory you associate with, each is underpinned by the concept of legality.

The second part of this chapter will examine the key principles of legality. Here, the chapter outlines how legality is protected on an international scale, with particular reference to the Universal Declaration of Human Rights (UDHR) and International Criminal Law (ICL), before turning to examine how legality is protected regionally by the European Convention on Human Rights (ECHR).[1] Next, Fuller's conception of legality will be outlined, drawing on the works of other legal philosophers. The final parts of this chapter will discuss what the law should strive for to be considered compatible with the principle of legality.

1 The background of the Universal Declaration of Human Rights and European Convention on Human Rights were discussed in the previous chapter.

DOI: 10.4324/9781003303688-4

3.2 Regulating Social Media

The regulation of the internet is not a new phenomenon. Since the early 90s, regulatory frameworks have emerged across the globe suggesting approaches to tackling the problems associated with the online world. Four theoretical positionings have emerged: (1) cyberlibertarianism; (2) cyber-paternalism; (3) network communitarianism; and (4) the governance model. Though each focusses on the regulation of new technology in general, each model is relevant in the wider discussions on regulating social media platforms.

3.2.1 Cyberlibertarian Theory

Early theorists, most notably John Perry Barlow through the Electronic Frontier Foundation, disputed that governments could impose legal obligations in cyberspace. Barlow, following the enactment of the Communications Decency Act 1996 in the United States, published a Declaration of Independence for Cyberspace:

> I declare the global social space we are building [the internet] to be naturally independent of the tyrannies you seek to impose on us. You have no moral right to rule us, nor do you possess any methods of enforcement we have true reason to fear.[2]

For Barlow, the internet represented a space beyond the realms of government interference, free from political gain and outside the reach of the criminal law.[3] Cyberlibertarianism

> refers to a perspective (some would say philosophy) which claims that cyberspace and the internet should be regarded as uncontrolled and unregulated electronic spaces where anyone is free to be whatever they wish and express themselves however they like.[4]

Those who endorse cyberlibertarianism suggest that the internet is a separate entity from that of the physical world, and consequently, beyond state interference. In essence, for cyberlibertarians 'earthbound' laws should not apply to cyberspace, for several reasons.[5] First, the process of law making is the exercise of power over those persons whom the state can control and for this to work, you need the consent

2 John Perry Barlow, 'A declaration of the independence of cyberspace' (*Electronic Frontier Foundation*, 8 February 1996) <https://www.eff.org/cyberspace-independence> accessed 16 November 2018. Note this declaration was originally sent *via* email.

3 For example, see the work of Esther Dyson, George Gilder, George Keyworth and Alvin Toffler, 'Cyberspace and the American Dream: A Magna Carta for the Knowledge Age' (1994) Future Insight <http://www.pff.org/issues-pubs/futureinsights/fi1.2magnacarta.html> accessed 26 September 2018.

4 David J. Bell, Brian D. Loader, Nicholas Pleace and Douglas Schuler, *Cyberculture: The Key Concepts* (Routledge 2004) 35.

5 Paul Bernal, *The Internet, Warts and All: Free Speech, Privacy and Truth* (Cambridge University Press 2018) 20.

of the people to be governed. Whereas unlike in the physical world, 'in cyberspace, physical boundaries no longer function as signposts informing individuals of the obligations assumed by entering into a new, legally significant place'.[6] Second, if the law were to be imposed on cyberspace, whose law should take precedence? You might be entering cyberspace in the United Kingdom but viewing content from the United States, with Dyson, Gilder, Keyworth, and Toffler arguing that the emergence of new technology, such as the internet, has created an 'information superhighway' which should be for the benefit of its users, and consequently beyond the reach of the law.[7] They go as far as arguing that even computer hacking is for the benefit of 'economic growth and trade leadership'.[8] But how then, should the internet be regulated?

For Post and Johnson, who endorse a cyberlibertarian approach, cyberspace should be treated as a separate geographical location, in which 'laws' or rules will evolve naturally. Like that of churches, clubs, and other social organisations, cyber institutions can create their own rules which would govern users of the internet, rather than imposing state-based law. In essence, rules or 'laws' would be created by individuals in cyberspace as opposed to a centralised legal system. Cyberlibertarianism is essentially built on the premise that the lack of borders on the internet means that regulation by states is impossible, with this theoretical perspective gaining momentum within some internet-based organisations. For instance, Wikipedia founder Jimmy Wales has been vocal about the internet being beyond the realms of state intervention.[9] Indeed, for early cyberlibertarian theorists, the internet is *akin* to the Wild West.

But this utopia of internet regulation is unattainable. As the digital world expands and creeps into every aspect of everyday life in the physical world, it is inevitable the law will intertwine with our online conduct. Most notably for Reed,[10] whilst a physical body remains in a place controlled by a government, the law will have no choice but to intervene—and nothing has demonstrated this more than social media.

3.2.2 Cyber-paternalism

Whereas those who endorse a cyberlibertarian approach to the 'regulation' of the online world believe the internet should be free from interference of traditional lawmakers, those who take a cyber-paternalistic approach, believe that the internet can, and is, regulated through its design:

6 David Johnson and David Post, 'Law and Borders: The Rise of Law in Cyberspace' (1996) 48 SLR 1367, 1389.

7 Dyson, Gilder, Keyworth and Toffler (n 3).

8 Ibid.,

9 David Golumbia, 'Cyberlibertarianism: The Extremist Foundations of "Digital Freedom"' (Clemson University, South Carolina, September 2013).

10 Chris Reed, *Internet Law: Text and Materials* (Cambridge University Press 2004) 174–175.

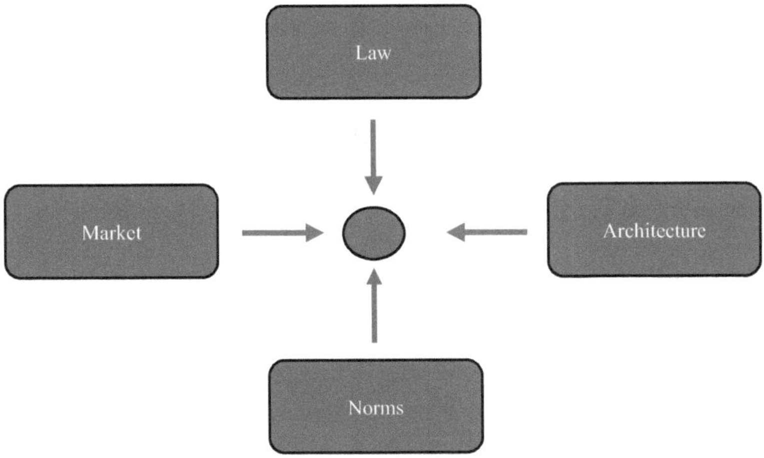

Figure 3.1 Lessig's modalities.

> Cyber-paternalists believe that the guidance of the state, or the elite, achieved through manipulation of software code or network hardware, is necessary to prevent cyberspace from becoming anarchic or simply inefficient.[11]

In essence, the 'internet is regulated by its architecture',[12] whereby we can alter and manipulate the design of an internet platform to regulate the conduct of the user, as supported in the work of Reidenberg who argues that the regulation of the online world occurs in two ways: first, through contractual agreements between internet providers; second, through its own design.[13]

For Lessig, how we regulate behaviour (whether online or offline) is embedded in the four 'Modalities of regulation'—law, markets, architecture, and norms:

> Norms constrain through the stigma that a community imposes; markets constrain through the price they exact; architectures constrain through the physical burdens they impose; and law constrains through the punishment it threatens.[14]

With each 'modalities of regulation' interacting with one another. Lessig uses what he terms a 'pathetic dot' to illustrate his argument, as shown in Figure 3.1. Here the 'pathetic dot' represents an individual, with the four 'modalities of regulation' seen as external forces which puts pressure on the 'pathetic dot' to act in a particular manner. Thereby, rather than the internet being impossible to regulate due to

11 Chris Reed and Andrew Murray, *Rethinking the Jurisprudence of Cyberspace* (Elgar 2018) 144.
12 Ibid., 149.
13 Joel Ridenberg, 'Governing Networks and Rule-Making in Cyberspace' (1996) 45 Emory Law J 911.
14 Lawrence Lessig, *Code Version 2.0* (Basic Books 2006) 124.

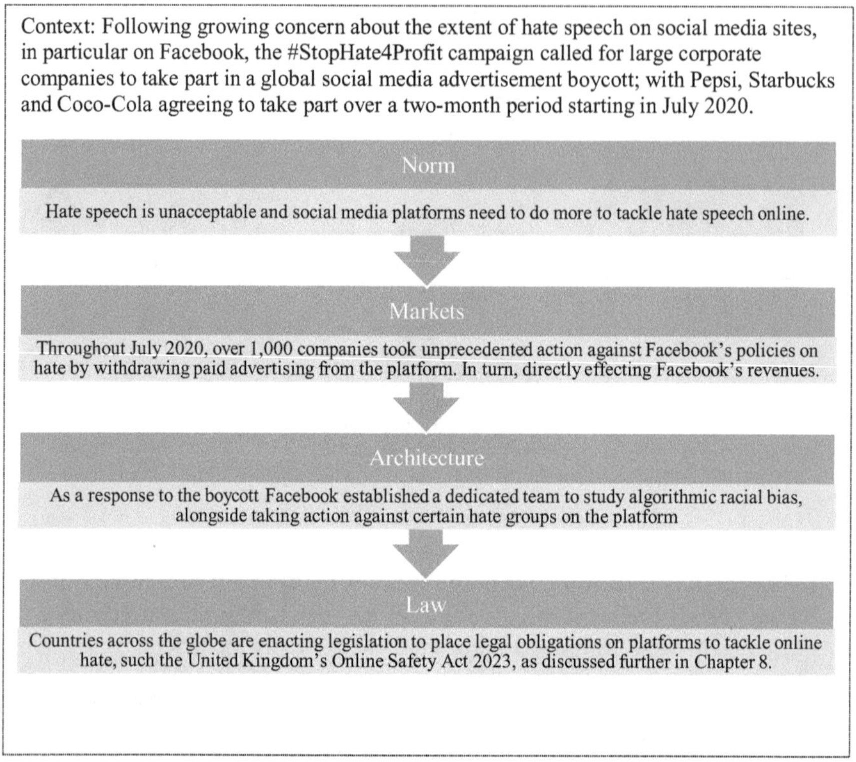

Figure 3.2 #StopHate4Profit.

its design, i.e., no borders as endorsed by cyberlibertarians, it is, in fact, regulated by its own architecture, rooted in the four 'modalities of regulation': law, markets, architecture, and norms.

An example of this can be seen in the #StopHate4Profit campaign, as illustrated in Figure 3.2.

Whereas cyberlibertarians argue that the internet is free from traditional regulation, for instance, free from the coercive nature of the law, cyber-paternalists argue that law and platform design can work together to regulate the online world. An ideal that has gained significant momentum in recent years, with governments across the globe calling for privacy-by-design. Put simply:

> In the same way that law is about control—aiming to instruct people how to behave and providing an incentive for obedience by sanctioning those who do not comply—code also controls behaviour.[15]

Though, it is important to note that for those who endorse a cyber-paternalist approach, the regulation of code should be through governments, rather than private entities.

15 Ibid., 86.

However, this approach is not without fault:

> Lessig's belief assumes three things: first, that lawmakers can always use law to exercise control over those whom it is addressed, a belief [... that is ...] distinctly over-optimistic. Second, that code can be compelled to be written in a particular way in accordance with the designs of lawmakers which [...] cannot be assumed. Third, it assumes that the law's requirements are commands, which are either complied with or flouted.[16]

For Murray, this school of thought fails to consider that 'code is not purely a design or architectural tool it also has a social or communicative function'.[17] Codes can be overwritten. Codes need constant improvement. In essence, the regulation of new technologies is significantly more complex, particularly in a social media context, than both cyberlibertarian and cyber-paternalism theory can comprehend. In fact, cyber-paternalism neglects that humans are not two-dimensional and come with their own 'baggage' which influences the regulatory world.

3.2.3 Network Communitarianism

A further theoretical approach to the regulation of the online world is network communitarianism, as coined by Murray.[18] Stemming from a mix of Actor Network Theory (ANT) (the 'theory of social transactions which accepts a role for non-human actors in any social situation')[19] and Social Systems Theory (SST) (the theory that a person's behaviour is influenced by several factors that work together as a system), Murray argues:

> when taken together they can illuminate much of our understanding of communications and social interaction in a networked environment such as the internet with a variety of actors, both human and non-human.[20]

So, whereas Lessig focussed on one 'pathetic dot', Murray argues, that by applying the theories of ANT and SST, the 'pathetic dot' cannot be considered as an individual in its own right. The individual is influenced by social transactions by non-human actors (ANT). For instance, in a social media setting, the person will be influenced by the content of their social media feed, which is based on algorithms set by the platforms themselves. Alongside being influenced by the system around them (SST), such as a person's political standpoint. In essence, the individual, or the 'pathetic dot' as termed by Lessig, must always be considered in its wider context:

16 Ibid., 151–152.
17 Andrew Murray, *Information Technology Law* (5th edn, Oxford University Press 2023) 68.
18 Andrew Murray, *The Regulation of Cyberspace: Control in the Online Environment* (Routledge 2007).
19 Murray (n 17).
20 Ibid., 69.

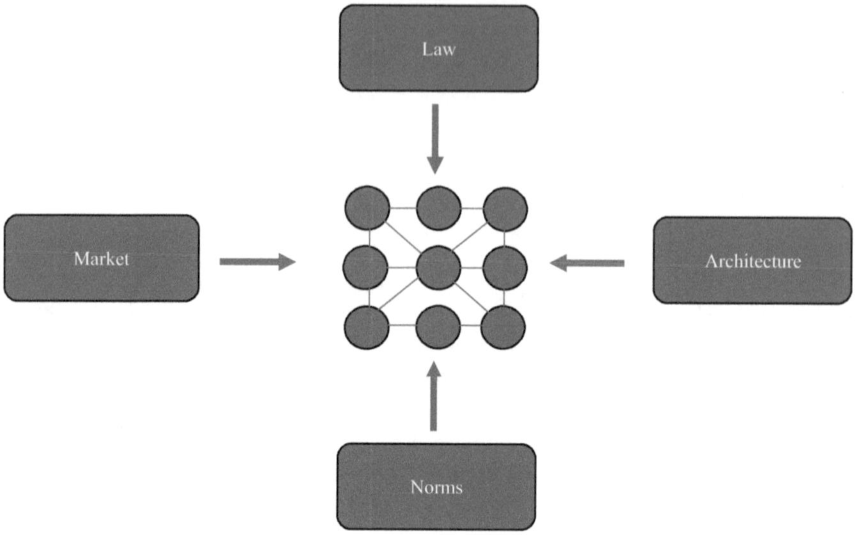

Figure 3.3 Murray's regulatory discourse.

> when one examines the modalities of regulation proposed by Lessig we find that of the four, three of them—laws, norms, and markets—are in fact a proxy for community-based control. Laws are passed by lawmakers elected by the community; markets are merely a reflection of value, demand, supply, and scarcity as reflected by the community in monetary terms; and norms are merely the codification of community values. I [Murray] recognized [*sic*] that these 'socially mediated modalities' reflected an active role for the 'dot' in the regulatory process; far from being a 'pathetic dot' which was the subject of external regulatory forces, the dot was in fact an 'active dot' taking part in the regulatory process.[21]

In essence, whereas cyber-paternalism argues that technology can be regulated and controlled through design, Murray argues that technology enables rather than constrains, in which individuals, considering their social, community, moral, political stances, etc., are active in accepting or rejecting regulation, in which we have gone from one 'pathetic dot' to a 'matrix of dots', all of which influence the four 'modalities of regulation'. See Figure 3.3. A clear example of this, as given by Murray, is the lack of prosecutions under the Obscene Publications Act 1959 and 1964 for privately viewing obscene material using an internet connection,[22] as the morality of the public has evolved to understand that the viewing of pornographic material online, is no longer morally objectionable to the point in which the law should intervene.

21 Ibid., 69–71.
22 Ibid., 62.

For Murray, this approach to online regulation includes the complexity of human nature, and the complex nature of the online world, though he does accept that criminal law is also needed—but this should only be in the most extreme cases. For both Reed and Murray we need to ensure legality in whichever form of regulatory model we subscribe to is maintained to ensure legitimacy in cyberspace.[23] The approach endorsed by both Reed and Murray as to the 'laundry list' of what legality should encompass, is discussed further in Section 3.4.

3.2.4 The Governance Model

Though access to the online world has evolved to be considered a human right in itself,[24] to enter the 'virtual world' you need to go through a gatekeeper. Two types of gatekeepers exist:

1. 'The gatekeeper that controls access to information and acts as an inhibitor by limiting access to or restricting the scope of the information';[25] and
2. A gatekeeper who acts as 'innovator, change, agent, communication channel, link, intermediary, helper, adapter, opinion leader, broker, and facilitator'.[26]

The governance model, as endorsed by Laidlaw recognises the importance of gatekeepers in the regulatory framework of the internet, in which she argues that their approach to regulation needs to be grounded in human rights and corporate responsibility, using the right to freedom of expression to illustrate why this is the case.[27]

The requirements which should be placed on gatekeepers, for Laidlaw, depend wholly on the type of gatekeeper and in particular, the impact it has on democracy, with Laidlaw suggesting that three types of gatekeepers exist:[28]

1. Micro-gatekeepers—these are sites which have some involvement in democratic discussions but are closed and restricted. An example given by Murray is the *Huffington Post*;[29]
2. Authority gatekeepers—these are high trafficking sites (i.e., a high proportion of users) but you do not need to access them to use the internet. Here, some social media platforms would be considered as an authority gatekeeper. Indeed, at the time in which Laidlaw was writing, Facebook would have been considered an

23 Reed & Murray (n 11) 200–230.

24 United Nations General Assembly (47th session) Resolution 47: The promotion, protection and enjoyment of human rights on the Internet (7 July 2021).

25 Emily B. Laidlaw, *Regulating Speech in Cyberspace: Gatekeepers, Human Rights and Corporate Responsibility* (Cambridge University Press 2015) 40.

26 Cheryl Metoyer-Duran, 'Information Gatekeepers' (1993) 28 ARIST 111, 118.

27 Laidlaw (n 25).

28 Ibid., 51–57.

29 Andrew Murray, *Information Technology Law* (4th edn, Oxford University Press 2019) 72.

authority gatekeeper. However, Facebook has grown beyond anything anyone could have expected, and today is more *akin* to a macro-gatekeeper; and

3. Macro-gatekeepers—sites, applications etc., that you must go through in order to access the online world. For example, search engines such as Google.

Using the governance model, Laidlaw accepts that the law on its own is not the most effective way to protect human rights online,[30] such as the right to freedom of expression as discussed in Chapter 2. Instead, she endorses the creation of a corporate governance model for the digital age.

The governance model is built on the acceptance that such platforms have advanced so much, that they have the power to distort public opinion and create codes for their own personal gain. Facebook, which is still the biggest social media platform to date, is a prime example of this and has been heavily criticised for distorting public views on politics, following the Cambridge Analytica scandal.[31] Consequently, the most powerful gatekeepers, such as Facebook, should have a responsibility to protect human rights with Laidlaw stating:

> Whether human rights responsibilities should be incurred and the extent of the responsibilities depends on the extent to which the gatekeeper controls deliberation and participation in the forms of meaning-making in democratic society.[32]

Maco-gatekeepers should therefore have the most responsibility in protecting human rights online. Whereas micro-gatekeepers should have the least. This approach to online regulation, as argued by Laidlaw, would be overseen by an independent digital commission (the regulator and adjudicator of the governance model), consisting of three layers of governance.

The top layer includes education, research, and policy, with emphasis placed on educating gatekeepers about the responsibilities they have in protecting human rights, alongside educating the general public. The middle layer encompasses corporate support, including policies, assessment tools, and audits. The bottom level consists of remedial mechanisms, whereby users or businesses can make complaints to the digital commission, where the complainant has been heavily impacted by the decisions or processes of the gatekeeper. The remedial mechanism would include a complaints procedure consisting of dispute resolution and in some instances, fines issued by the digital commission for a failure to comply with human rights.[33] This is the bedrock of the governance model for Laidlaw.

30 Laidlaw (n 25) 62.

31 See, for example, Nicholas Confessore, 'Cambridge Analytica and Facebook: The scandal and the fallout so far' *The New York Times* (New York, 4 April 2018) <https://www.nytimes.com/2018/04/04/us/politics/cambridge-analytica-scandal-fallout.html> accessed 23 January 2023.

32 Emily B. Laidlaw, 'A Framework for Identifying Internet Information Gatekeepers' 24 (2010) Int. Rev. Law Comput. Technol. 263, 267.

33 Laidlaw (n 25) 266.

This approach endorsed by Laidlaw is somewhat similar to Brownsword, who supports the concept of regulation over that of law, which must be grounded not just in human rights but also the principle of legality.[34] As will be discussed in Chapter 8, Laidlaw's approach seems to have been accepted by the United Kingdom's government in their quest to become one of the safest places in the world to go online—though there are some clear differences between what Laidlaw recommended and the approach undertaken in the Online Safety Act 2023 (OSA).

3.2.5 Summary

There is clearly no 'one size fits all' approach to online regulation and as the internet has advanced, governments across the globe have attempted, and failed, to keep pace. Regardless of which theoretical positioning a person assigns to, legality will underpin all these regulatory frameworks. The 'modalities of regulation' as endorsed by Lessig includes, laws and these legal provisions need to encompass key aspects of legality. Cyber-paternalism would also need to include a consideration of legality, alongside network communitarianism. Indeed, the governance model would need to be underpinned by legality to ensure the digital commission stayed within the boundaries set by Parliament.

As social media usage has increased, we have seen a rise in the use of the criminal law to tackle the 'harms' associated with the online world,[35] despite it being clear that the 'mistake of twentieth-century jurists is to assume that rules and norms are the exclusive keys to social ordering'.[36] In fact, technology can and has been used in situations, where the criminal law has failed to control the behaviours of individuals it deems as inappropriate. Brownsword uses the example of building GPS into golf buggies to stop joy ridders, as was done by Warwickshire Golf and Country Club.[37] For Brownsword, technology is the answer to where law, in particular, the criminal law, is struggling, so long as this technology conforms to legality.[38]

Any form of internet regulation, whether it be cyber-paternalism, network communitarianism or the governance model needs to be grounded in the principle of legality. Even more so, where the regulation of online behaviours is embedded in the criminal law framework.

34 Roger Brownsword, *Law, Technology and Society: Re-imaging the Regulatory Environment* (Routledge 2019).

35 See for example, The BBC, 'Teenager's life "ruined" by Live.me and Twitter "trolls"' *The BBC* (London, 24 October 2017) <http://www.bbc.co.uk/news/uk-england-41693437> accessed 5 October 2023.

36 Brownsword (n 33) 6.

37 Ibid., 7.

38 Ibid.,

3.3 The Principle of Legality

The principle of legality is considered a fundamental right, and though there is no definitive definition,[39] it is underpinned by several guiding principles: (1) society should be governed by law; (2) there should be certainty within legal provisions; and (3) the law needs to be accessible, in which members of the public should be able to govern their behaviour in accordance with clear and distinct rules. Legality is therefore a vital condition of the rule of law (the idea that we should be ruled by law, not by man).[40] It is fundamental in any democratic society and of particular importance when it comes to the criminal law where, as discussed in Section 3.5, at worst, a person's liberty could be at risk, or in some countries, a person's life. Put simply, legality prohibits the use of retrospective criminal law; it strives to ensure that laws are defined with the 'maximum clarity which the imperfect medium of language can attain';[41] it helps to protect against arbitrary punishment; and most importantly, it creates legal certainty—without which, anarchy would prevail.

For Luban legality ensures the law is action-guiding, whilst also curtailing the powers of the state.[42] Similarly, for Fuller, the law should not only be action-guiding but should also encompass a further seven principles in a bid to strive for legal excellence, all of which are discussed in Section 3.4,[43] and will be the theoretical positioning of legality underpinning later chapters. Likewise, Bingham's conception of what constitutes the rule of law encompasses key values associated with legality including the ideal that the law must be accessible and so far as possible intelligible, clear and predictable; legal questions should be answered by law and not by discretion; and the protection of human rights.[44] Legality is a principle enshrined in international and regional legal provisions.

3.3.1 Legality on an International Scale

Though there is no true consensus as to the meaning of legality, it is a principle that dates back to 440 AD. It has been embedded throughout legal provisions across the globe. For instance, the principle is enshrined in Article 11(2) of the UDHR which states that:

> [n]o one shall be held guilty of any penal offence on account of any act or omission which did not constitute a penal offence, under national or international law, at the time when it was committed. Nor shall a heavier penalty be

39 Jason N. E. Varuhas, 'The Principle of Legality' (2020) 79(3) Camb. L. J. 578.

40 Hilaire Barnett, *Constitutional & Administrative Law* (5th edn, Cavendish Publishing 2004) 69.

41 Law Commission, *Report on Conspiracy and Criminal Law Reform* (1976) Law Com No. 76 [1.8].

42 David Luban, 'Fairness to Rightness: Jurisdiction, Legality, and the Legitimacy of International Criminal Law' in Samantha Besson and John Tasioulas (eds), *The Philosophy of International Law* (Oxford University Press 2010) 34.

43 Lon L. Fuller, *The Morality of Law* (Yale University Press 1964).

44 Tom Bingham, *The Rule of Law* (Penguin 2011). See also, Lord Bingham, 'The Rule of Law' (2007) 66(1) Camb. Law. J. 67.

> imposed than the one that was applicable at the time the penal offence was committed.

A similar provision is found in Article 15(1) of the International Covenant on Civil and Political Rights (ICCPR):

> No one shall be held guilty of any criminal offence on account of any act or omission which did not constitute a criminal offence, under national or international law, at the time when it was committed. Nor shall a heavier penalty be imposed than the one that was applicable at the time when the criminal offence was committed.

In essence, the principle of legality under international human rights law prohibits the use of retrospective criminal law—which as discussed further in Section 3.3.2.1 is the ideal that a person cannot be convicted of an offence which was not a crime at the time in which the conduct took place.

The principle of legality underpins many other human rights provisions, including the right to life, the right to liberty and the right to a fair trial—rights which would be at significant risk if legality was not maintained. In fact, for the Human Rights Committee of the ICCPR even in times of national emergency, the principle of legality must be upheld.[45] However, the principle of legality encompasses more than the prohibition of retrospective criminal law. Demonstrated further when examining how legality is maintained and protected in ICL.

ICL is, at its very basic, the application of criminal law across borders,[46] consisting of three types of provisions: domestic criminal law enforced on an international scale; treaty-based criminal law (this is the creation of treaties between states which criminalises certain conduct); and pure ICL which includes crimes such as genocide, aggressive war, and crimes against humanity.[47] Legality must be obtained in the creation, publication, and application of ICL.[48]

For Corsi, who endorses a strict approach to legality in ICL, legality encompasses four key principles:[49]

1. *Nullum Crimen Sine lege*—no crime without law (discussed further in Section 3.4.1);
2. *Lex Praevia*—the prohibition of retrospective criminal law (discussed further in Section 3.3.2.1);

45 UN Human Rights Committee, 'General comment no. 29, States of emergency (article 4): International Covenant on Civil and Political Rights' (*United Nations Digital Library*, 2001) <https://digitallibrary.un.org/record/451555?ln=en> accessed 15 August 2022.

46 Douglas Guilfoyle, *International Criminal Law* (Oxford University Press 2016) 3.

47 Luban (n 42) 19.

48 Darryl Robinson, *The Principle of Legality in International Criminal Law* (Legal Studies Research Papers Series 10-08, 2010) 4.

49 Jessica Lynn Corsi, 'An Argument for Strict Legality in International Criminal Law' [2018] 49 Geo. J. Int'l L. 1321.

3. *Lex Certa*—legal certainty within criminal law provisions; and
4. *Lex Stricta*—strict interpretation of the law.

Though Corsi acknowledges that 'strict legality is not a perfect principle [...] strict legality offers the most just, most effective, most coherent, most persuasive, most legitimate and even the most moral approach to ICL'.[50]

3.3.1.1 Lex Certa

Lex certa is the principle that there should be legal certainty within legal provisions, in which a crime should be 'articulated with sufficiently precise wording so that the criminalized [*sic*] conduct is clear'.[51] The drafting of legal provisions should therefore be clear, tangible, and predictable, allowing citizens to govern their behaviour in accordance with law. In essence, a person should be able to foresee that their given actions may invoke the criminal law, as reinforced by the International Criminal Court (ICC) in *Čelebići*[52] with the ICC confirming the connection between legality and the action-guiding nature of the law: 'The purpose of this principle is to prevent the prosecution and punishment of an individual for acts which he reasonably believed to be lawful at the time of their commission.'[53]

Here the ICC upholds the idea that law must be constructed in a clear and precise manner, which without, 'no process of criminalization [*sic*] would be recognized [*sic*]'.[54] Indeed, in *Vasiljević* it was found that 'violence to life and person, prohibited under Article 3 of the GCs [Geneva Conventions], was insufficiently precise to form the basis of a criminal conviction.'[55] As noted by Bassiouni, '[t]he goals of principles of legality are to provide notice as part of general prevention, as well as notice to the accused as part of fairness'.[56]

3.3.1.2 Lex Stricta

Lex stricta is the idea that when interpreting legal provisions, judges should not over-stretch the meaning of a legal provision and instead, they should endorse a strict approach to interpretation. For Bassiouni, 'the basic rule of interpretation embodying the principle of legality is the prohibition or limitation on the use of analogy in judicial interpretation'.[57]

50 Ibid.,

51 Ibid., 1334.

52 *Čelebići Camp, Prosecutor v Delalić (Zejnil) and others,* Appeal Judgment, Case no IT-96-21-A, ICL 96 (ICTY 2001), 20 February 2001, United Nations Security Council [UNSC]; International Criminal Tribunal for the Former Yugoslavia [ICTY]; Appeals Chamber.

53 *Ibid.*, 313.

54 Robinson (n 48) 4.

55 *Prosecutor v Vasiljević*, Case no. IT-98-32-T, Trial Judgment, (Int'l Crim. Trib. for the Former Yugoslavia 29 November 2002) 193.

56 Mahmoud Cherif Bassiouni, *International Criminal Law* (3rd edn, Brill Nijhoff 2008) 98.

57 Ibid., 73.

Consequently, to adhere to the principle of legality and *lex stricta*, where there is doubt over the meaning of legal provisions, interpretation should always fall in the favour of the accused, to render it compatible with the principle of legality. For Fuller, as discussed further in Section 3.4.3, an over-expansive interpretation of criminal provisions, can result in retrospective criminal law.[58] *Lex Stricta* is therefore built on the principle that judges should not create law. However, as Hall acknowledges, the line between judicial interpretation and judicial legislating can be thin,[59] which is why the four principles endorsed by Corsi in underpinning legality in ICL must all work in tangent.[60]

3.3.1.3 Summary

To adhere to the principle of legality from an international stance, each source of law must be significantly clear to allow citizens to reasonably foresee what a given action would invoke because without clarity, free choice and autonomy are heavily affected. Further, only the law can dictate what is a criminal offence, in which retrospective criminal law is prohibited. In turn, judges need to ensure they are not going beyond reasonable interpretations of the law. But these are principles which ICL should strive to achieve whilst also keeping in mind that complete adherence may not always be achievable. There needs to be some air of flexibility in the law. Or as potently put by Corsi:

> While perfect notice—the idea that all persons know and understand all law—is a legal fiction, law makers and those with the power to punish are instructed by these two tenets to take all steps necessary to ensure maximum notice.[61]

Indeed, no valid legal system or democratic society can exist if it does not conform to legality, as reinforced in the ECHR.

3.3.2 The European Convention on Human Rights: The Fundamental Principle of Legality

The principle of legality is further enshrined in the ECHR in two fundamental ways, which as was discussed in Chapter 2, has been incorporated into UK law. First, as discussed in Section 3.3.2.1, legality has specific protection under Article 7, no punishment without law. Second, when it comes to the restriction of rights under the ECHR, at its very basic, the restriction needs to be prescribed by law, which as discussed in the previous chapter encompasses key provisions associated with the principle of legality and will not be repeated here at length.

58 Fuller (n 43) 73.
59 Jerome Hall, *General Principles of Criminal Law* (2nd edn, Lawbook Exchange Ltd 2010) 173.
60 Corsi (n 49) 1332.
61 Ibid., 1335.

3.3.2.1 Article 7: No Punishment without Law

The fundamental principle of legality has specific protection under Article 7(1), no punishment without law, in which the right is considered absolute. As a result, Article 7 cannot be infringed under any circumstances, even in times of national emergency.[62] Similar to Article 11(2) of the UDHR and Article 15(1) of the ICCPR, Article 7 of the ECHR states:

> No one shall be held guilty of any criminal offence on account of any act or omission which did not constitute a criminal offence under national or international law at the time when it was committed.

As acknowledged by the European Court of Human Rights (ECtHR) in *Kafkaris v Cyprus*:[63]

> The guarantee enshrined in Article 7, which is an essential element of the rule of law, occupies a prominent place in the Convention [ECHR] system of protection […] It should be construed and applied, as follows from its object and purpose, in such a way as to provide effective safeguards against arbitrary prosecution, conviction, and punishment.[64]

In essence, a person cannot be punished for an offence, which was not an offence at the time in which the conduct took place. With Fuller, as discussed in Section 3.4 below and Corsi as discussed above in Section 3.3.1, going as far as to argue that the law should not be overly stretched beyond its original interpretation that it *could* constitute retrospective criminal law.

Similar to ICL, Article 7 is underpinned by two overriding principles: *nullum crimen sine lege* (no crime without law) and *nulla poena sine lege* (no punishment without law).[65] As outlined by Murphy, the concept of legality within the ECHR consists of three interrelated rules: only the law can define a crime and prescribe a penalty, the prohibition of retrospective criminal law, and no harsher penalties can be given then those that are prescribed in the law at the time the offence was committed.[66] The first rule was restated and affirmed in *Kafkaris v Cyprus*,[67] where the court upheld the concept that only the law can define a crime, affirming that 'law'

62 Cian C. Murphy, 'The Principle of Legality in Criminal Law under the European Convention on Human Rights' (2010) 2 Eur. Hum. Rights Law Rev. 192, 207.

63 *Kafkaris v Cyprus* App no. 21906/04 (ECtHR, 12 February 2008).

64 Ibid., [137].

65 Susana Sanz-Caballero, 'The Principle of nulla poena sine lege Revisited: The Retrospective Application of Criminal Law in the Eyes of the European Court of Human Rights' (2017) 28(3) Eur. J. Int. Law 787, 788.

66 Murphy (n 62) 192.

67 *Kafkaris* (n 63) [137].

consists of both Acts of Parliament and the Common Law.[68] With the second and third rule, affirmed in *Welch v The United Kingdom*:

> The concept of 'penalty' in Article 7(1) is, like the notions of 'civil rights and obligations' and 'criminal charge' in Article 6(1), an autonomous Convention concept. To render the protection offered by Article 7 effective, the Court must remain free to go behind appearances and assess for itself whether a particular measure amounts in substance to a penalty within the meaning of this provision.[69]

A person therefore cannot be punished for conduct that was not considered unlawful at the time in which the behaviour took place and is further enshrined throughout the ECHR and the jurisprudence of Strasbourg under the concept of 'prescribed by law'.

3.3.2.2 Prescribed by law

As previously discussed at length in Chapter 2, the prescribed by law element preserved in the ECHR, encompasses key elements of the principle of legality. Therefore, this will not be repeated here, but as a quick reminder to the reader, legality, from a 'prescribed by law' perspective includes:

1. No punishment without law;
2. The law must be accessible;
3. The law must be action-guiding; and
4. Adequate safeguards must be in place.

If any of the above are missing, it cannot be said that the legal provision is prescribed by law and therefore it will be incompatible with the principle of legality.

3.3.2.3 Summary

The principle of legality as embedded in Article 7(1) and enshrined throughout the ECHR and the jurisprudence of the ECtHR creates legal certainty within the law and is described by Murphy as the 'hidden jewel of the Convention [ECHR]'.[70] It encompasses procedural requirements (i.e., there must first be some legal provision in place), alongside substantive or qualitative obligations, most notably accessibility and foreseeability.[71] However, as acknowledged by the ECtHR strict construction of the law is not always achievable:

68 *Sunday Times v The United Kingdom* (1979) 2 EHRR 245.
69 *Welch v The United Kingdom* App no. 17440/90 (ECtHR, 9 February 1995) 247.
70 Murphy (n 62) 206.
71 *Del Río Prada v Spain* [GC], no. 42750/09, ECHR 2013.

> As the Court has already had occasion to note, it is a logical consequence of the principle that laws must be of general application that the wording of statutes is not always precise. One of the standard techniques of regulation by rules is to use general categorisations as opposed to exhaustive lists. The need to avoid excessive rigidity and to keep pace with changing circumstances means that many laws are inevitably couched in terms which, to a greater or lesser extent, are vague.[72]

To render the legal provision compatible with the principle of legality, it must be 'sufficiently clear in the large majority of cases'.[73] Where a provision is considered overly vague, this will render it incompatible with the ECHR.

Consequently, as society changes there needs to be some form of flexibility with the law—it just cannot go beyond what would reasonably be expected, as noted by the House of Lords, in *R v Rimmington*:[74]

> no one should be punished under a law unless it is sufficiently clear and certain [...] and no one should be punished for any act which was not clearly and ascertainably punishable when the act was done [...] It is accepted that absolute certainty is unattainable, and might entail excessive rigidity since the law must be able to keep pace with changing circumstances, some degree of vagueness is inevitable. [75]

Indeed, 'those who skate on thin ice can hardly expect to find a sign which will denote the precise spot where they may fall in'.[76]

The law therefore needs to be adaptable to reflect the fast-changing nature of society; this is especially true with technology. Technology and the internet have evolved at a significant pace and in some instances, the law has struggled to keep pace.[77] The law needs to maintain some form of openness to resolve legal disputes, but this should not be at the detriment of breaching the fundamental principle of legality—particularly where the criminal law is involved.

3.4 Fuller: Legality in the Criminal Law

Legality comprises of the logic that individuals should be able to partake in any activities knowing whether their behaviour breaches the law and includes both a thin and thick approach.[78] It is closely linked to the rule of law. Scholars have

72 *Cantoni v France,* 15 November 1996, Reports of Judgments and Decisions 1996-V [31].

73 Ibid., [32].

74 *R v Rimmington, R v Goldstein* [2005] UKHL 63, [2006] 1 A.C. 459.

75 Ibid., per Lord Bingham [33].

76 *Knuller (Publishing, Printing and Promotions) Ltd. and Others Appellants v DPP Respondent* [1972] 3 W.L.R. 143, [1973] A.C. 435 per Lord Morris 463.

77 Law Commission, *Abusive and Offensive Online Communications: A Scoping Report* (Law Com No 381, 2018) [10.76].

78 Reed and Murray (n 11).

opposing views on what constitutes legality, though there are some overlaps in approaches. Reed and Murray, taking a 'thin' approach, suggest that the core 'laundry list' of ingredients that make up legality,[79] are as follows:

1. The law must be set forth in advance;
2. The law must be made public;
3. The law must be general;
4. The law must be clear;
5. The law must be stable and certain; and
6. The law must be applied to everyone according to its terms.

'Amongst the best known' approaches to legality is Fuller,[80] who argues that legality encompasses eight key principles:[81]

1. The law must be general;
2. The law must be prospective;
3. The prohibition of retrospective law;
4. The law must be clear;
5. The prohibition of contradictory law;
6. The law must not ask the impossible;
7. Consistency must be maintained in the law; and
8. The law must be fairly enforced.

For Fuller, '[a] total failure in any one of these eight directions does not simply result in a bad system of law; it results in something that is not properly called a legal system at all'.[82] Though, similar to the approach undertaken by the ECtHR, Fuller does accept that '[n]o system of law—whether it be judged made or legislatively enacted—can be so perfectly drafted as to leave no room for dispute'.[83] It will be Fuller's principle of legality which will be used as the foundations of the theoretical positioning, or as potently put by Reed and Murray the 'laundry list', for which future arguments in later chapters will be based.

3.4.1 The Law Must Be General

For the law to be general, it must be in place. A prospective that is further endorsed by the ECtHR as discussed above, and supported by Raz[84] and Reed and Murray.[85]

79 Ibid., 206.
80 Ibid., 202.
81 Fuller (n 43) 33–94.
82 Ibid., 39.
83 Ibid., 56.
84 Joseph Raz, *The Authority of Law: Essays on Law and Morality* (Clarendon Press 2012) 214.
85 Reed and Murray (n 11) 81.

In essence, for the law to be considered as general, there must be a legal provision prohibiting the behaviour under scrutiny. However, for Fuller, for the law to be general, and in turn to ensure fairness in the application of legal provisions, it is more than just being able to point to a legal provision and stating, 'there is the law'. To adhere to generality in the law, citizens must be given advanced notice of the law, in which the law should not be decided on an ad-hoc approach. In essence, for Fuller, we should not proceed on a 'case by case' account of insight into how the law should develop;[86] the law needs to be accessible for all those living within a society, underpinning the principle of fairness in the criminal law.[87]

This is an issue, which is discussed at length in Chapter 6, where it is argued that the current approach to establishing if a communication can be considered as grossly offensive or menacing, as prohibited under s.127(1) of the Communications Act 2003 (CA) is being decided on an ad-hoc basis where there is a lack of consensus amongst those working in the criminal justice system, as to what constitutes a grossly offensive or a menacing message. As a result, as highlighted throughout Chapter 6, it cannot be said that generality is being maintained in the application of current communication offences to conduct carried out on social media sites. Similar arguments are raised in Chapter 7, which explores the recent amendments made to communication law.

3.4.2 The Law Must Be Prospective

Underpinning the concept of generality in the law is the ideal that the law must also be prospective. For the law to be considered as prospective, it must be publicly accessible. For instance, you should not punish another for failure to adhere to the law if say, the law was written on the back of a post-it-note for no one to see. Though as acknowledged by Fuller, it would be impossible to educate all citizens on the meaning of every legal provision within a given legal system, he maintains that prospectivity is important because 'knowledge of the law by a few often influences indirectly the actions of many'.[88]

Luban, influenced by the work of Fuller and Raz, refers to this as 'the fair notice argument'.[89] Those who breach criminal law provisions are subject to punishment. Nonetheless, those who break legal provisions that are incapable of guiding those living in a society, should not be subject to punishment. Similar provisions are endorsed by Raz under his conception of the rule of law, who argued that 'if the law is to be obeyed it must be capable of guiding the behaviour of its subjects. It must be such that they can find out what it is and act on it'.[90] Here, a person must

86 Fuller (n 43) 46.
87 What is considered 'law' is beyond the scope of this book, but there is a wide array of literature available exploring this concept. See for instance, Raz (n 84).
88 Fuller (n 43) 51.
89 Luban (n 42) 40.
90 Raz (n 84). See also, Joseph Raz, 'The Rule of Law and its Virtue' (1977) 93 L.Q.R.195.

be seen to have been given 'constructive notice' that their actions have breached the law, or as potently put by Bingham:

> This is not because bank robbers habitually consult their solicitors before robbing a branch of the NatWest, but because many crimes are a great deal less obvious than robbery, and most of us are keen to keep on the right side of the law if we can.[91]

By adhering to this aspect of legality, it allows citizens to guide their actions in accordance with law and in turn, curtails the power of the state, whilst also allowing citizens to guide their affairs on the basis of knowing what the law is.[92] In fact, 'obedience of the law depends on humans understanding the law'.[93]

Lack of prospectivity in the law and its application to a social media context is a recurring theme throughout this monograph. For example, in Chapter 5 issues are highlighted with the terms course of conduct, harassment, and stalking, particularly where such notions are applied to the online world. It is argued that these elements which make up the offence of cyberharassment or cyberstalking are so inherently vague, it is impossible to see when a person's behaviour online will go beyond what is reasonably expected (with the Law Commission previously noting that the monitoring of a person's social media activity is generally consented to in the online world)[94] to such behaviour that warrants criminalisation.

3.4.3 The Law Cannot Be Retrospective

As a fundamental principle of not just legality, but also international law, as discussed in Section 3.3 and specifically prohibited under Article 7 of the ECHR, as examined in Section 3.3.2.1, you should not enact a law which would change the legal position of the past. For example, you cannot create a legal provision today prohibiting the sending of a communication which contains explicit language and prosecute those who sent messages yesterday which contained such language. A concept that is further supported by Fuller, who goes as far as stating that 'retrospective law is truly a monstrosity',[95] though he does maintain that there is a place

91 Bingham (n 44) 37.
92 Friedrich Hayek, *The Constitution of Liberty* (Routledge 2006) 153–154.
93 Reed and Murray (n 11) 85.
94 Law Com No 381 (n. 77) [8.171].
95 Fuller (n 43) 53.

for retrospective law in some instances, just not in the criminal law.[96] For Raz, 'one cannot be guided by retrospective law'.[97]

As discussed at length by Reed and Murray, retrospective law is not necessarily an issue when it comes to the online world,[98] however, it is not as simple as stating 'we do not create legal provisions that alter the legal position of the past'. It is also, as noted by Fuller and further supported in both the jurisprudence of the ICC and the ECtHR, important to ensure that when provisions are interpreted, they are not so generously interpreted, that they would, in theory, create retrospective criminal law:

> It is an accepted principle of interpretation that a criminal statute should be construed strictly, so that acts falling outside its normal meaning are not to be considered criminal simply because they present the same kind of danger as those described by the language of the statute.[99]

For Fuller, like the approach endorsed by the ECtHR and the ICC, the interpretation of the statute cannot go beyond what would be reasonably acceptable as this would create retrospective criminal law. Supported further by Bingham.[100] Though no matter how

> clearly drafted a legal provision may be, in any system of law, including criminal law, there is an inevitable element of judicial interpretation. There will always be a need for elucidation of doubtful points and for adaptation to changing circumstances. Indeed, in the United Kingdom, as in the other Convention States, the progressive development of the criminal law through judicial law-making is a well entrenched and necessary part of legal tradition. Article 7 of the Convention cannot be read as outlawing the gradual clarification of the rules of criminal liability through judicial interpretation from case to case, provided that the resultant development is consistent with the essence of the offence and could reasonably be foreseen.[101]

With foreseeability relating both to elements of the offence and the applicable punishment. As potently put by Bingham:

96 Fuller uses the hypothetical scenario that a legal provision is enacted declaring that from such a date no marriage will be vailed unless the marriage certificate contains a special stamp on it, provided by the state. Unfortunately, due to technical difficulties, the stamp is not available when the legal provision comes into force. Further, due to a lack of knowledge of changes to the law in this area, marriages continue to take place but without the stamp affixed to the marriage certificate. Here, it would be accepted that retrospective law could be used to alter the position of the past, to ensure the validity of these marriages. Ibid., 53–54.

97 Raz (n 84).

98 Reed and Murray (n 11) 212.

99 Fuller (n 43) 58.

100 Bingham (n 44) 45.

101 *SW v The United Kingdom, CR v The United Kingdom* App no. 20166/92 (ECtHR, 22 November 1995) [36].

> It is one thing to move the law a little further along a line on which it is already moving, or to adapt it to accord with modern views and practises; it is quite another to seek to recast the law in a radically innovative or adventurous way, because that is to make it uncertain and unpredictable.[102]

By prohibiting retrospective criminal law, due process of the law is maintained. This is an issue discussed further in Chapters 4 and 5. For instance, in Chapter 4, concerns are raised that the current use of stirring up offences under the Public Order Act 1986 (POA) in a social media context, is creating retrospective criminal law as the Act is being applied in situations which it was never intended to do, i.e., where public order was never under threat.

3.4.4 The Law Must Be Clear

Clarity is considered an 'essential ingredient'[103] of legality and is highly supported by not only legal philosophers such as Fuller but also the courts. It upholds the ideal that the law needs to be clear to ensure safeguarding 'against arbitrary punishment by governments under a cover of vague, underspecified law'.[104] This is of particular importance when it comes to legal provisions which, by utilising Bingham's argument above, is not as clear cut as robbery. Bingham suggests three essential reasons why we need clarity in the law:[105]

1. 'If you and I are liable to be prosecuted, fined or perhaps imprisoned for doing or failing to do something, we ought to be able, without undue difficulty, to find out what we must or must not do';[106]
2. So, citizens are aware of their rights and obligations; and
3. You need clarity in the law for successful trade, investment and business. No one would invest in a company or indeed, in new technology, which is based in a country that lacks clarity within its legal provisions—an important consideration when examining the OSA, as discussed in Chapter 8.

However, despite its importance in maintaining legality in the criminal law, it is a concept that is difficult to achieve. On the one hand, a strict approach would render the law rigid and difficult to adapt to the changing nature of society, with Fuller supporting in some instances, the best way to achieve clarity is to take advantage of, and incorporate into the law, common sense standards of judgment that have grown up in the ordinary life lived outside legislative halls.

102 Bingham (n 44) 46.
103 Fuller (n 43) 62.
104 Luban (n 42) 37.
105 Bingham (n 44) 37–38.
106 Ibid., 37.

On the other, an overly vague legal provision which lacks any clarity undermines the principle of fairness within the legal system and risks abuse, with Fuller clarifying that

> [i]t is a serious mistake—and a mistake made constantly—to assume that, though the busy legislative draftsman can find no way of converting his objective into clearly stated rules, he can always safely delegate this task to the courts or special administrative tribunals.[107]

Both the ICC and the ECtHR have previously rendered legal provisions incompatible with legality due to the inherent vagueness contained in the principles under scrutiny.[108] They have also come to decisions, that despite some vagueness within a provision, the legal provision under consideration does conform to legality.[109] Clarity is therefore an important principle, but it needs to be balanced. We need to strive for the 'maximum clarity which the imperfect medium of language can attain'[110] in order to render the law compatible with the principle of legality. The lack of clarity in current legal provisions used to restrict online conduct is a recurring theme throughout this monograph. For instance, in Chapter 7, issues are raised with the 'new' communication offences which became legally enforceable on the 31 January 2024. These 'new' offences were enacted following a significant review of communication law and its application to the online world. Yet as demonstrated in Chapter 7, many of the issues highlighted with previous communication offences, which are discussed in Chapter 6, remain, most notably, the inherent vagueness in terms used throughout the provisions.

3.4.5 The Law Must Be Non-contradictory

A further principle of legality as endorsed by Fuller is the prohibition of contradictory law. In essence, we cannot require a person under a legal provision to do one thing but at the same time prohibit that same conduct under another provision. Fuller provides the example of asking motorists under a statute to change their licence plates on January 1st but also prohibiting any form of labour on the same day, under another statute—this would create contradictory law.[111] Although contradictory laws have and do exist, it is important that such situations are avoided, particularly when it comes to the criminal law, to maintain the principle of legality. A contradictory law cannot be considered, at the very least, general, prospective, or clear. In fact, in 1988 and 1995, the Italian Constitutional Court ruled that ignorance of the law may constitute an excuse for the citizen when the formulation of the law is such as to lead to obscure and contradictory results.

107 Ibid.,
108 See for example, *Prosecutor v Vasiljević* (n 55); *Sunday Times* (n 68).
109 See for example, *Kafkaris* (n 63); *Čelebići Camp* (n 52).
110 Law Com No. 76 (n 41).
111 Fuller (n 43) 65–66.

But consistency is difficult to maintain in the criminal law. For instance, in Chapter 4, concerns are raised following a decision by the Crown Prosecution Service (CPS) to prosecute Liam Stacey under the POA for sending racist comments on Twitter.[112] Whereas in a similar case a decision was made to prosecute under s.127(1) of the CA,[113] with the reasoning as to why the CA was preferred over the POA being unclear and impossible to justify.

3.4.6 The Law Must not Ask the Impossible

Though Fuller accepts that his sixth principle, the law must not ask the impossible, is arguably obvious, because 'no sane lawmaker, not even the most evil dictator, would have any reason to enact such law',[114] it is not that simple. Using the example of the 'reasonable man'[115] test found in various legal instruments, including that of the criminal law, Fuller argues that by comparing a person's actions against that of the 'reasonable man', may in some circumstances ask the impossible, as his education or even personal characteristics may not meet such a requirement.[116] Yet imposing the 'reasonable man with the same characteristics as the defendant' as the boundary for criminal liability risks objectivity being lost.[117] These requirements create difficulties as it imposes some form of moral judgement upon the accused, which may be considered impossible to meet.

The use of the 'reasonable person' test in a social media context, is discussed at length in Chapter 5, in which the *mens rea* of the offence of harassment and/or stalking, is based upon knowledge. Here it must be found that the sender of a message knew or should have known that the message sent would cause alarm or distress in another, with knowledge being judged on that of the 'reasonable person'. Chapter 5 demonstrates the difficulties in applying the 'reasonable person' test to a social media context, calling instead, for the notion of the 'reasonable social media' user to be implemented. Impossibility is also discussed in Chapter 7 where it is argued that the 'new' communication offences are asking individuals to do the impossible at times.

3.4.7 The Law Must Remain Consistent

The idea that the law must remain consistent is closely related to the prohibition of retrospective law. In essence, the law should not change so regularly that citizens are unable to govern their behaviour in accordance with the rules of the land. For Murray, to adhere to legality and in turn, the rule of law, the law must have

112 *R v Liam Stacey,* Swansea Crown Court On Appeal From The Magistrates' Court, A20120033.

113 *R v Alison Chabloz,* Westminster Magistrates' Court, 11 January 2018 (unreported).

114 Fuller (n 43) 70.

115 Though Fuller users the notion of the 'reasonable man', it has come to be accepted that the appropriate term is 'reasonable person'. Reasonable person will be used throughout this monograph.

116 Fuller (n 43) 71–72.

117 Ibid.,

Table 3.1 New criminal offences created either through primary or secondary legislation between 1st June 2009 and 31st May 2014

	2010	*2011*	*2012*	*2013*	*2014*
Primary	189	31	36	66	87
Secondary	523	143	256	261	193
Total	712	174	292	327	280

Note: Ministry of Justice, 'New Criminal Offences (England and Wales)' (*Gov.uk*, 11 December 2014) <https://assets.publishing.service.gov.uk/government/uploads/system/uploads/attachment_data/file/385537/new-criminal-offences-june-2009-may-2014.pdf> accessed 22 August 2022.

some form of stability.[118] Similar provisions are endorsed by Raz.[119] For this to be achieved, the law needs to refrain from being decided on an ad-hoc approach of enlightenment, as mirrored in Fuller's first requirement of generality in the law. A key example of failure to adhere to this requirement is shown in Chapters 4, 5 and 6, following an increase in the use of the criminal law to tackle the 'harms' associated with the online world.

As society has evolved, particularly in the United Kingdom, significant weight has been placed on the law to control not only the behaviour of its citizens but to also govern public institutions, such as schools and the NHS. As noted by Sir Menzies Campbell, between 1997 and 2007 there had been '382 Acts of Parliament, including 10 health Acts, 12 education Acts, and 29 criminal justice Acts, and more than 3,000 new criminal offences' enacted.[120] Such reliance on the law, in particular, the criminal law, has continued since, peeking at over 700 new criminal offences being created in 2010 alone, demonstrated further in Table 3.1. Indeed, when it comes to the internet, there have been significant reforms in this area of the law. With research undertaken by Reed and Murray, finding that in 2017, there were 595 reported court cases on the legal database 'WestLaw' mentioning the word 'internet'; 557 containing the word 'online'.[121] As stated by Bingham '[i]t seems that legislative hyperactivity has become a permanent feature of our governance'.[122]

With a continued rhetoric, when it comes to social media, that more criminal law is needed to control what society deems as inappropriate behaviour online, consistency of the law is under significant threat. An issue highlighted further in Chapter 7 which demonstrates the continued problems in consistency and the application of communication offences, which overlap with each other. An issue which has become even more complex, given the increase in the number of communication offences available to prosecutors in the wake of the OSA.

118 Andrew D Murray, 'Mapping the rule of law for the internet' in David Mangan and Lorna E Gillies (eds), *The Legal Challenges of Social Media* (Edward Elgar Publishing 2017) 27.

119 Raz (n 84) 214–215.

120 HC Deb 11 July 2007, vol 462, col 1455.

121 Reed and Murray (n 11) 223.

122 Bingham (n 44) 40.

3.4.8 The Law Must Be Fairly Enforced

Failure in any of the previous principles of legality as discussed above all lead to one clear consequence—unfair enforcement of the law. If a legal provision lacks, for instance, generality, cannot be considered publicly accessible, contradictory and in turn lacks any form of clarity, how can actors within the criminal justice system ensure that the law is enforced fairly across all sectors of society? Unfair enforcement of the law creates criticism in the community and in turn, results in negative attitudes towards the criminal justice system. Due process cannot be maintained, as highlighted throughout various chapters of this monograph.

For Fuller:

> The most subtle element in the task of maintaining congruence between law and official actions lies, of course, in the problem of interpretation. Legality requires that judges and other officials apply statutory law, not according to their fancy or with crabbed literalness, but in accordance with principles of interpretation.[123]

The interpretation of statute law is dependent on the principle of legality being maintained in all legal provisions. Without this, there runs the real risk of the law being unfairly enforced, as discussed at length in Chapter 7 following the increased availability in communication-based offences.

However, for the law to be fairly enforced, it also needs to be accessible for those working in the criminal justice system, which has not necessarily always been the case as

> there is no comprehensive statute law database with hyperlinks which would enable an intelligent person, by using a search engine, to find out all the legislation on a particular topic. This means that the courts are in many cases unable to discover what the law is.[124]

As put by Bingham:

> [i]t must be questioned whether the current volume and style of legislation are well suited to serve the rule of law even if it is accepted, as it must be, but the subject matter of much legislation is inevitably very complex.[125]

As discussed further in various chapters of this monograph, fair enforcement of the law when it comes to social media offences, is currently under threat. What one police force or prosecutor might consider to be, as an example, harassment, another might deem it grossly offensive. A problem discussed in Chapter 5, following a decision by the CPS to prosecute Peter Nunn for sending grossly offensive communications contrary to s.127(1) of the CA as opposed to s.2 or s.4 of the Protection

123 Fuller (n 43) 82.

124 *R v Chambers* [2008] EWCA Crim 2467 per Lord Justice Toulson [68].

125 Bingham (n 44) 42.

from Harassment Act 1997.[126] A decision which the complainant, Caroline Criado-Perez, has been heavily critical of.[127]

3.4.9 Summary

Fuller's eight principles of legality are evidently mirrored in other scholars' approaches to legality and the rule of law. Each intertwine and support each other. And it will be Fuller's conception of legality which will be used to critique the application of current legal provisions to conduct carried out on social media because, as noted above, Fuller's approach to what constitutes legality is '[a]mongst the best known'.[128] However, it is important to note, like that of the approach endorsed by the ICC, the ECtHR and the domestic courts of the United Kingdom in what constitutes legality, it is accepted that it would be impossible to create legislation which is so perfectly enacted it adheres to all eight principles soundly. It is a degree of balance, and as will become apparent in later chapters, this balance is not being met when it comes to the application of the criminal law to social media conduct.

3.5 Chapter Summary

Ultimately, there is no one approach to the regulation of social media. Different theoretical positions have emerged since the creation of the internet. However, as more individuals choose to live out their lives online, significant emphasis is being placed on the criminal law to tackle the 'darker' side of human behaviour. The principle of legality is therefore fundamental in ensuring the protection of freedom of expression and the avoidance of overcriminalisation. It is a basic requirement to ensure due process—regardless of the stance taken to online regulation.

The principle of legality is therefore important in all aspects of the law, but none more so than the criminal law. The use of the criminal law, at its very basic, is a form of social control and by its very nature, it is coercive. Failure to comply can result in heavy penalties, including a loss of liberty. But despite the coercive nature of the criminal law, as discussed by Corsi,[129] societies make use of criminal law provisions for (at least) five important reasons:

1. To restrict conduct, or completely prohibit, such conduct that may cause harm to another or society in general;
2. As a form of social control;
3. To safeguard individuals from conduct that can be considered as criminal;
4. To act as a warning to others that such conduct will not be permitted; and
5. To elaborate to members of society what is considered minor or serious offences.

126 *R v Peter Nunn*, The City of London Magistrates' Court, 29 September 2014 (unreported).

127 Caroline Criado-Perez, 'A brief comment on Peter Nunn, sentenced today for Twitter abuse' (*Week Women*, 2014) <https://weekwoman.wordpress.com/2014/09/29/a-brief-comment-on-peter-nunn/> accessed 29 October 2016.

128 Reed and Murray (n 11) 202.

129 Corsi (n 49) 1340.

In essence, '[t]he wrongdoing which the criminal law seeks to punish is that which threatens the fundamental values upon which a society is founded'.[130]

Legality therefore underpins many theoretical concepts of the criminal law, including:

1. Deterrence—if a person cannot guide their behaviour in accordance with clear and distinct rules, it cannot be said that the law is action-guiding, therefore deterrence does not exist. A concept reinforced in Bingham's understanding of the rule of law, in which he argues that 'one important function of the criminal law is to discourage criminal behaviour, and we cannot be discouraged if we do not know, and cannot reasonably easily discover, what it is we should not do';[131]
1. Expression—social condemnation cannot exist without clarity in the law; and
2. Retribution—if the law is not accessible, punishment becomes random and cannot be justified.

Not only does legality underpin these theories of criminalisation, but it also acts as a shield against arbitrary interference from the state.

Although legality is an important principle when it comes to the criminal law in general, when it comes to the application of the law in cyberspace, it is vital that legality is maintained for the law to be considered as legitimate in the online world. As discussed by Reed and Murray, three factors influence the legitimacy of the law in cyberspace, all of which are underpinned by the principle of legality.[132] First, the law needs to be seen as being addressed to the online user in which the provision needs to communicate what is and is not accessible (the action-guiding nature of the law). For this to occur, the provision needs to be easily understood (clarity within the law) because

> [i]f the legitimacy of the law's claim to authority is conferred on it by the collective acceptance of those to whom it is addressed, it must follow that structuring or wording the law in a way which makes it inaccessible to those addresses will reduce the likelihood of acceptance. Achieving legitimacy requires attention to style as well as substance.[133]

Second, provisions need to be coherent with cyberspace (the consistency element); and third, provisions need to be fair and uphold natural justice (fair enforcement of the law).

Though as noted at the start of this chapter, there is not one approach to the meaning of legality, there are some clear overlaps between different theoretical

130 Carleton Kemp Allen, 'The Nature of a Crime' (1931) 13(1) J. Comp. Legislation and Int'l L.J. 1, 11.

131 Bingham (n 44) 38.

132 Reed and Murray (n 11) 206–225.

133 Ibid., 183.

approaches, including approaches endorsed by ICL and the ECHR. In particular, though not limited to:

1. Generality;
2. Clarity;
3. Prospectiveness;
4. Consistency;
5. Non-retrospective law, including, strict interpretation of legal provisions; and
6. Fair enforcement of the law.

These concepts of legality will be used in the following chapters to evaluate the current application of the criminal law's response to tackling online abuse, with particular regard to hate crime (Chapter 4), cyberharassment and cyberstalking (Chapter 5), and abusive communications (Chapter 6), before turning to evaluate the latest attempts by the former UK government to control online communications (Chapter 7). Part Three will turn to examine the regulation of social media platforms beyond the criminal law and the United Kingdom, against the principle of legality.[134]

Reference list

Allen C K, The Nature of a Crime (1931) 13(1) J Comp Legislation & Int'l L J 1
Barnett H, *Constitutional & Administrative Law* (5th edn, Cavendish Publishing 2004)
Bartlett J, *The People Vs Tech: How the Internet Is Killing Democracy (and How We Can Save It)* (Penguin 2018)
Barlow J P, 'A Declaration of the Independence of Cyberspace' (*Electronic Frontier Foundation*, 8 February 1996) https://www.eff.org/cyberspace-independence
Bassiouni M C, *International Criminal Law* (3rd edn, Brill Nijhoff 2008)
Bell D J, Loader B D, Pleace N & Schuler D, *Cyberculture: The Key Concepts* (Routledge 2004)
Bernal P, *The Internet, Warts and All: Free Speech, Privacy and Truth* (Cambridge University Press 2018)
Bingham T, *The Rule of Law* (Penguin 2011)
Brownsword R, *Law, Technology and Society: Re-imaging the Regulatory Environment* (Routledge 2019)
Corsi J L, 'An Argument for Strict Legality in International Criminal Law' (2018) 49 Geo J Int'l L 1321
Confessore N, 'Cambridge Analytica and Facebook: The Scandal and the Fallout So Far' *The New York Times* (New York, 4 April 2018) https://www.nytimes.com/2018/04/04/us/politics/cambridge-analytica-scandal-fallout.html
Criado-Perez C, 'A Brief Comment on Peter Nunn, Sentenced Today For Twitter Abuse' (*Week Women*, 2014) https://weekwoman.wordpress.com/2014/09/29/a-brief-comment-on-peter-nunn/
Duff A & Green S, *Philosophical Foundations of Criminal Law* (Oxford University Press 2013)

134 Bingham (n 44) 84.

Dyson E, Gilder G, Keyworth G & Toffler A, 'Cyberspace and the American Dream: A Magna Carta for the Knowledge Age' (1994) Future Insight http://www.pff.org/issues-pubs/futureinsights/fi1.2magnacarta.html
Fuller L L, *The Morality of Law* (Yale University Press 1964)
Golumbia D, *Cyberlibertarianism: The Extremist Foundations of "Digital Freedom"* (Clemson University, South Carolina, September 2013)
Guilfoyle D, *International Criminal Law* (Oxford University Press 2016)
Hall J, *General Principles of Criminal Law* (2nd edn, Lawbook Exchange Ltd 2010)
Hayek F, *The Constitution of Liberty* (Routledge 2006)
HC Deb 11 July 2007, vol 462, col 1455
Johnson D & Post D, 'Law and Borders: The Rise of Law in Cyberspace' (1996) 48 SLR 1367
Laidlaw E B, *Regulating Speech in Cyberspace: Gatekeepers, Human Rights and Corporate Responsibility* (Cambridge 2015)
Laidlaw E B, 'A Framework for Identifying Internet Information Gatekeepers' 24 (2010) Int Rev Law Comput Technol 263
Law Commission, *Abusive and Offensive Online Communications: A Scoping Report* (Law Com No 381, 2018)
Law Commission, *Report on Conspiracy and Criminal Law Reform* (Law Com No. 76, 1976)
Lessig L, *Code Version 2.0* (Basic Books 2006)
Lord Bingham, 'The Rule of Law' (2007) 66(1) Cam Law J 67
Luban D, 'Fairness to Rightness: Jurisdiction, Legality, and the Legitimacy of International Criminal Law' in Besson S & Tasioulas J (eds), *The Philosophy of International Law* (Oxford University Press 2010)
Metoyer-Duran C, 'Information Gatekeepers' (1993) 28 ARIST 111
Ministry of Justice, 'New Criminal Offences (England and Wales)' (*Gov.uk*, 11 December 2014) https://assets.publishing.service.gov.uk/government/uploads/system/uploads/attachment_data/file/385537/new-criminal-offences-june-2009-may-2014.pdf
Murphy C C, The Principle of Legality in Criminal Law Under the European Convention on Human Rights (2010) 2 Eur Hum Rights Law Rev 192
Murray A, *Information Technology Law* (5th edn, Oxford University Press 2023)
Murray A, *Information Technology Law* (4th edn, Oxford University Press 2019)
Murray A, *The Regulation of Cyberspace: Control in the Online Environment* (Routledge 2007)
Murray A D, 'Mapping the rule of law for the internet' in Mangan D & Gillies L E (eds), *The Legal Challenges of Social Media* (Edward Elgar Publishing 2017)
Raz J, *The Authority of Law: Essays on Law and Morality* (Clarendon Press 2012)
Raz J, The Rule of Law and its Virtue (1977) 93 LQR 195
Reed C, *Internet Law: Text and Materials* (Cambridge University Press 2004)
Reed C & Murray A, *Rethinking the Jurisprudence of Cyberspace* (Elgar 2018)
Ridenberg J, 'Governing Networks and Rule-Making in Cyberspace' (1996) 45 Emory Law J 911
Robinson D, *The Principle of Legality in International Criminal Law* (Legal Studies Research Papers Series 10-08, 2010)
Sanz-Caballero S, 'The Principle of nulla poena sine lege Revisited: The Retrospective Application of Criminal Law in the Eyes of the European Court of Human Rights' (2017) 28(3) Eur J Int Law 787
The BBC, 'Teenager's life "ruined" by Live.me and Twitter "trolls"' *The BBC* (London, 24 October 2017) http://www.bbc.co.uk/news/uk-england-41693437
UN Human Rights Committee, 'General Comment no. 29, States of Emergency (Article 4): International Covenant on Civil and Political Rights' (*United Nations Digital Library*, 2001) <https://digitallibrary.un.org/record/451555?ln=en>
Varuhas JNE, 'The Principle of Legality' (2020) 79(3) Camb LJ 578

Part II

The Application of the Criminal Law

The second part of this monograph examines three key growing concerns surrounding social media and online abuse: racist commentary online (Chapter 4), cyberharassment/cyberstalking (Chapter 5) and unlawful communications (Chapters 6 and 7). Each of these areas are explored in separate chapters in which the use of the criminal law to tackle these online 'harms' is critically examined. To do this each chapter takes a very similar format, looking at the rise in these behaviours, the criminal law framework used to prohibit such behaviour(s), followed by its application to a social media context. The final parts of each chapter highlight how the application of each of these provisions, to conduct carried out on social media platforms, creates significant issues with legality and in turn, it cannot be said that freedom of expression is being protected.

DOI: 10.4324/9781003303688-5

4 Online Racism and the Criminal Law

4.1 Introduction

Though there is no universal definition of hate speech, under international law, any advocacy of national, religious, or racial hatred that constitutes incitement to discriminate must be prohibited by law. Hate speech falls within the broad definition of a hate crime, which like that of hate speech, no universal definition exists. A hate crime can however be considered a criminal act aimed at another due to a protected characteristic associated with that person, whether real or presumed. Characteristics can include race, gender, sexuality, religion, or political affiliation, to name but a few—and what characteristics constitute a hate crime differ between states and indeed, between social media platforms.

In England and Wales, a hate crime is defined as a criminal act aimed at another which is 'motivated by hostility or demonstrates hostility towards the victim's disability, race, religion, sexual orientation, or transgender identity'.[1] Hate speech, is the use of words or expressions motivated by hostility or demonstrates hostility towards another because of a protected characteristic (i.e., disability, race, religion, sexual orientation, or transgender identity). Although hate speech constitutes more than racial hatred,[2] this chapter will focus on the use of the criminal law to target the spread of racial hatred on social media sites—something which has become a growing concern for governments across the globe.[3]

The use of the criminal law in prohibiting racist speech is uncontroversial, as restricting this form of speech is designed to protect the vulnerable.[4] In fact, 'bigoted views can be restricted as they bring about harm and add little to public

1 The Crown Prosecution Service, 'Hate crime' (*CPS.gov*, 2018) <https://www.cps.gov.uk/hate-crime> accessed 9 December 2018.

2 Significant criticism has been raised against the current approach to hate crime in England and Wales, where it has been argued that the current approach has created a hierarchy of offences. For further discussion on this see, Catherine Heard 'Hate crime law: more incrementalism or time for reform?' (2014) (7) Arch. Rev 4.

3 See Chapter 9 for an in-depth discussion on different government approaches to combatting hate online.

4 Erick Bleich, 'The Rise of Hate Speech and Hate Crime Laws in Liberal Democracies' (2011) 37(6) J Ethn Migr Stud 917, 918.

DOI: 10.4324/9781003303688-6

discussion'.[5] However, like that of other online expressions, which will be discussed in later chapters, we need to distinguish between racist speech, which is prohibited in many countries across the globe, including England and Wales and offensive speech, which as outlined in Chapter 2, is a protected form of speech. Or as potently put by Murray 'sometimes there is a fine line between political speech and hate speech'[6] and as noted by Newman, there is no bright light to distinguish between the two.[7]

The following chapter will start by exploring the rise in hate crime in England and Wales, before turning to examine hate speech on social media sites which has caught the attention of the public in recent years and, in turn, placed increasing pressure on the government and the criminal justice system to do more. The hate characteristic which will be of focus, is racially motivated incidences. This chapter will then proceed to discuss how the criminal justice system is attempting to tackle racist speech. To do this, the criminal law will be outlined and examined, exploring how legal provisions are currently being utilised to govern racist speech on social media sites, with reference to s.66 of the Sentencing Act 2020 (SA), s.31 and 32 of the Crime and Disorder Act 1988 (CDA) and s.18 and 19 of the Public Order Act 1986 (POA).[8] Although other criminal law provisions exist to prohibit racist speech, focus is placed on these provisions as their use to prosecute comments made online, as will be examined in the final parts of this chapter, poses significant issues with the principle of legality and freedom of expression. Further, the stirring up offences contained in s.18 and 19 of the POA are priority illegal content under the Online Safety Act 2023 (OSA)—discussed further in Chapter 8.

4.2 The Rise in Hate

The number of hate crimes recorded by the police in England and Wales has risen year-on-year, as illustrated further in Figure 4.1, with 124,091 hate crimes recorded in the year ending March 2021, rising to 155,841 by the end of March 2022.[9] Though arguably the rise in the number of recorded hate crime offences could be associated with a drive towards a better understanding of the offence across police forces, certain political events (the EU referendum in 2016; terrorist attacks in 2017; and the COVID-19 pandemic in 2020) have created spikes within the data.[10]

5 Melina Bell, 'John Stuart Mill's Harm Principle and Free Speech: Expanding the Notion of Harm' (2021) 33(2) Utilitas 162, 173.

6 Andrew Murray, *Information Technology Law: The Law and Society* (5th edn, Oxford University Press 2023) 126.

7 Stephen L. Newman, 'Finding the Harm in Hate Speech: An Argument against Censorship' (2017) 50(3) CJPS 679, 681.

8 Though similar provisions are contained in the Racial and Religious Hatred Act 2006, there is little evidence in the literature of its use in cases related to social media.

9 Home Office, 'Hate crime, England and Wales, 2021 to 2022' (*Gov.uk*, 6 October 2022) <https://www.gov.uk/government/statistics/hate-crime-england-and-wales-2021-to-2022/hate-crime-england-and-wales-2021-to-2022> accessed 1 February 2023.

10 Ibid.,

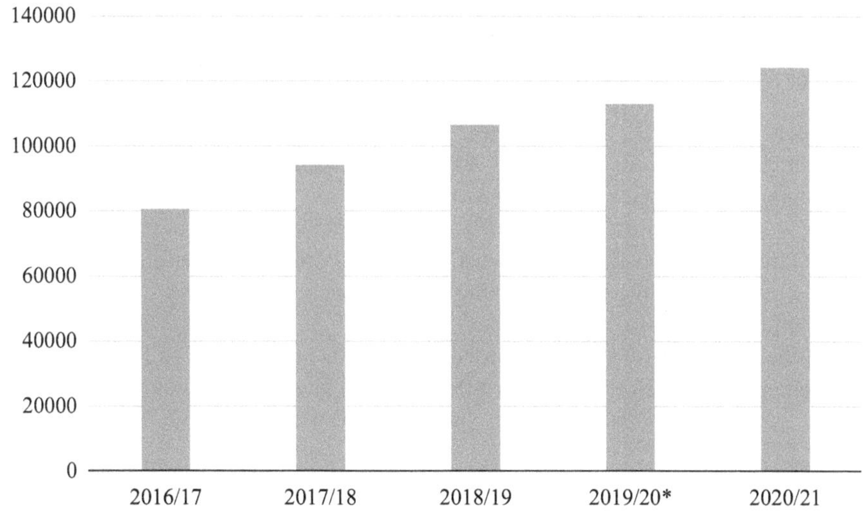

Figure 4.1 Number of hate crimes recorded by the police in England and Wales from 2016 to 2022.

Note: Home Office, 'Hate Crime, England and Wales, 2021 to 2022' (*Gov.uk*, 6 October 2022) <https://www.gov.uk/government/statistics/hate-crime-england-and-wales-2021-to-2022/hate-crime-england-and-wales-2021-to-2022> accessed 1 February 2023. *Does not include figures from Greater Manchester due to technology issues.

Similar trends have also been mirrored in the online world, where the volume of online hate is enormous,[11] with Stephens arguing that:

> [t]here seems to be something about the internet that encourages toxicity in public discourse amongst a small but vocal minority, who seem to have nothing better to do than resort to hateful, racist, misogynistic comments and personal attacks, something that would never be tolerated in the offline world.[12]

Indeed, during the global COVID-19 pandemic, online hate rose by 20%.[13] Reflected further in the high number of posts removed from social media platforms, for breaching hate speech policies. For Minto:

11 Dr Fernand de Varennes, 'Recommendations made by the Forum on Minority Issues at its thirteenth session on the theme "Hate speech, social media and minorities"' (*United Nations*, 26 January 2021) <https://documents-dds-ny.un.org/doc/UNDOC/GEN/G21/017/29/PDF/G2101729.pdf?OpenElement> accessed 19 August 2022.

12 Hugh Stephens, 'Facebook: Who bears responsibility for them?' (*The International Forum for Responsible Media Blog*, 25 November 2021) <https://inforrm.org/2021/11/25/facebook-who-bears-responsibility-for-them-hugh-stephens/> accessed 28 June 2022.

13 Michael Baggs, 'Online hate speech rose 20% during pandemic: "We've normalised it"' *The BBC* (London, 15 November 2021) <https://www.bbc.co.uk/news/newsbeat-59292509> accessed 28 June 2022.

> Gone are the days when far-right information and propaganda were confined to the back rooms of seedy pubs and clubs. This material can now be accessed 24/7 with a few swipes of your phone.[14]

Between January 2020 and December 2021, Facebook removed a total of 192.6 million pieces of content from its site for breaching hate speech policies.[15] In the same period, Twitter removed around 7 million pieces of hateful content;[16] YouTube removed just over 334,000 videos it deemed hateful or abusive;[17] and TikTok removed over 7 million pieces of content considered hateful.[18] Of all hate crimes recorded by the police and removed by social media sites, racially motivated offences are the most common.[19]

Of the 155,841 hate crimes recorded by the police in England and Wales in the year ending March 2022, nearly two-thirds (70%) were racially motivated, with a

14 Rachel Minto found in Sally Weale, 'Children coming across far-right material when researching, teachers say' *The Guardian* (London, 18 April 2022) <https://www.theguardian.com/politics/2022/apr/18/children-coming-across-far-right-material-when-researching-teachers-say> accessed 28 June 2022.

15 Statista, 'Global number of hate speech-containing content removed by Facebook from 4th quarter 2017 to 1st quarter 2022' (*Statista*, 1 June 2022) <https://www.statista.com/statistics/1013804/facebook-hate-speech-content-deletion-quarter/> accessed 19 August 2022. Note, Facebook defines hate speech: 'as a direct attack against people – rather than concepts or institutions—on the basis of what we call protected characteristics: race, ethnicity, national origin, disability, religious affiliation, caste, sexual orientation, sex, gender identity, and serious disease'. See, Meta 'Hate speech' (*Meta*, 2022) <https://transparency.fb.com/en-gb/policies/community-standards/hate-speech/> accessed 1 September 2022.

16 Twitter, 'Rules enforcement' (*Twitter*, 2022) <https://transparency.twitter.com/en/reports/rules-enforcement.html#2020-jan-jun> accessed 1 September 2022. Note: Twitter defines hateful content as 'violence against or directly attack or threaten other people on the basis of race, ethnicity, national origin, caste, sexual orientation, gender, gender identity, religious affiliation, age, disability, or serious disease'. See, Twitter, 'Hateful conduct policy' (*Twitter*, 2022) <https://help.twitter.com/en/rules-and-policies/hateful-conduct-policy> accessed 1 September 2022.

17 Google, 'YouTube community guidelines enforcement' (*Google*, 2022) <https://transparencyreport.google.com/youtube-policy/removals?hl=en&total_channels_removed=period:2021Q1&lu=channels_by_reason&channels_by_reason=period:2021Q4> accessed 1 September 2022. Note: YouTube defines hateful or abusive content as 'violence or hatred against individuals or groups based on any of the following attributes: age, caste, disability, ethnicity, gender identity and expression, nationality, race, immigration status, religion, sex/gender, sexual orientation, victims of a major violent event and their kin [and] Veteran status'. See, YouTube, 'Hate speech policy' (*YouTube*, 2022) <https://support.google.com/youtube/answer/2801939?hl=en-GB> accessed 1 September 2022.

18 TikTok, 'Community guidelines enforcement report' (*TikTok*, 2022) <https://www.tiktok.com/transparency/en/community-guidelines-enforcement-2021-4/> accessed 1 September 2022. Note, TikTok defines hateful behaviour as 'content that attacks, threatens, incites violence against, or otherwise dehumanizes [*sic*] an individual or a group on the basis of the following protected attributes: race, ethnicity, national origin, religion, caste, sexual orientation, sex, gender, gender identity, serious disease, disability [and] immigration status'. See, TikTok, 'Community guidelines: Hateful behaviour' (*TikTok*, 2022) <https://www.tiktok.com/community-guidelines?lang=en#38> accessed 1 September 2022.

19 Home Office (n 9).

19% rise in racially motivated offences compared to the previous year.[20] Although it is impossible to quantify how many of these reported racially motivated offences could be labelled as 'hate speech', studies have shown that racially motivated hate speech is also on the increase, particularly on social media sites.[21] For instance, '[b] etween 2019 and mid-2021, on average there was a new post [online] about race or ethnicity-based hate speech every 1.7 seconds' with a total of 50.1 million racially motivated posts being published during the same data period.[22]

The use of social media platforms to send racist commentary is not a new phenomenon. Indeed, comments made by the then Director of Public Prosecutions, Alison Saunders, in 2017 supported the use of the criminal law to tackle this form of abuse:

> When an ever-greater amount of our time is spent online, it is only right that we [The Crown Prosecution Service] do everything possible to ensure that people are protected from abuse that can now follow them everywhere via the screen of their smartphone or tablet. Whether shouted in their face on the street, daubed on a wall or tweeted into their living room, hateful abuse can have a devastating impact on victims.[23]

Similar arguments were raised again in 2021, following the use of social media sites, such as Twitter and Instagram, to send racist abuse to several footballers following England's defeat in the 2020 Euros.[24] Twitter in the immediate aftermath of the racist abuse removed over 1,000 racist tweets.[25] With many calling for stronger criminal provisions to make such conduct illegal.

4.3 Racism and the Criminal Law

Until the mid-1960s, the UK government was against the specific regulation of racist behaviour, with the Race Relations Act 1965 being the first attempt to regulate such conduct. Since then, numerous provisions have been put in place to combat

20 Ibid.,

21 Baggs (n 13).

22 Bandwatch Academy, 'Uncovered: Online hate speech in the Covid era' (*Bandwatch Academy*, 2022) <https://www.brandwatch.com/reports/online-hate-speech/view/> accessed 1 September 2022.

23 Alison Saunders, 'Hate is hate. Online abusers must be dealt with harshly' *The Guardian* (London, 21 August 2017) <https://www.theguardian.com/commentisfree/2017/aug/20/hate-crimes-online-abusers-prosecutors-serious-crackdown-internet-face-to-face> accessed 26 September 2018.

24 Alastair Jamieson, 'Bukayo Saka, Jadon Sancho and Marcus Rashford racially abused online after England Euro 2020 final defeat' *The Independent* (London, 12 July 2021) <https://www.independent.co.uk/news/uk/home-news/bukayo-saka-racist-england-sancho-rashford-b1882303.html> accessed 28 June 2022.

25 Simon Evans, 'English football faces up to global nature of online hate' (*Reuters*, 26 May 2022) <https://www.reuters.com/lifestyle/sports/english-football-faces-up-global-nature-online-hate-2021-07-13/> accessed 29 June 2022.

racism within society with three approaches being endorsed in the criminal law framework: aggravated offences, enhanced offences, and stirring up racial hatred.

4.3.1 Aggravated Offences: Sentencing Act 2020

S.66 of the SA, previously s.145–146 of the Criminal Justice Act 2003, allows the court to 'uplift' a perpetrators sentence where it is found that the criminal act (and it can be any criminal act, including the sending of unlawful communications online) was aggravated by hostility towards the complainant because of the complainants, whether real or presumed, protected characteristic, including race. In essence, the provision places an obligation on all sentencing judges or magistrates to take into consideration any hostility towards the complainant which was based on the complainant's (whether real or presumed) protected characteristic. If proven, a judge or magistrate is under an obligation to 'treat the fact that the offence is aggravated by hostility of any of those types [protected characteristics] as an aggravating factor and must state in open court that the offence is so aggravated'.[26] Put simply, where it is found that hostility was also present during the commission of a criminal offence, the court has the power to increase a person's punishment.

S.66 of the SA applies to any criminal offence, including, as noted above, communication offences. So, for instance, where a person sends a grossly offensive communication online contrary to s.127(1) of the Communications Act 2003 (CA), as discussed further in Chapter 6, the maximum custodial sentence which can be given is six months. However, where it is deemed an aggravated offence, the maximum custodial sentence is two years, as illustrated in *R v John Nimmo*[27] where the defendant's sentence was increased by 50% for every communication he sent contrary to s.127(1) of the CA, which was found to be motivated by racial hostility.[28] The meaning of hostility and racial group are discussed further in the following section.

4.3.2 Enhanced Offences: Crime and Disorder Act 1988

Unlike s.66 of the SA which allows for sentences to be uplifted for all criminal offences which are motivated or demonstrate hostility towards a person because of their real or presumed protected characteristic, s.31 and 32 of the CDA is only applicable in very defined circumstances.

26 Sentencing Act 2020 s.66(2).

27 *R v John Nimmo,* Newcastle Crown Court, 10 February 2017 (unreported). See also, Sandra Laville, 'Internet troll who sent labour MP antisemitic abused is jailed' *The Guardian* (London, 10 February 2017) <https://www.theguardian.com/uknews/2017/feb/10/internet-troll-who-sent-labour-mp-anti-semitic-messages-is-jailed> accessed 13 January 2020.

28 Note, this case was prior to the enactment of the Sentencing Act 2020 and therefore s.145–146 of the Criminal Justice Act 2003 was applied.

First, it only applies to certain specific criminal offences, including,[29]

1. Public Order Offences (s.31 CDA); and/or
2. Harassment or Stalking (s.32 CDA).

Second, the aggravating factor of the offence is limited to race or religion only.[30] For the purpose of this chapter, only race will be examined. For other types of aggravated offences, or other protected characteristics beyond race and religion, the SA, discussed above, would apply.

In essence, the base offence must be proven first (for instance, the *actus reus* and *mens rea* contained in s.2 of the Protection from Harassment Act 1997 (PfHA)—discussed further in the following chapter) followed by the elements which make up the enhanced offence (motivation or demonstration of hostility—as discussed further below) which will ultimately be for the jury or magistrate to decide. In a social media context, both s.31 and 32 of the CDA can, and have, been used to control online communications. For example, in 2012 Liam Stacey was successfully prosecuted for an offence contrary to s.19 and 4A of the POA,[31] with his sentence being enhanced under s.31 of the CDA. *R v Stacey* is discussed further in Section 4.4.

Under s.31(1) of the CDA:

> a person is guilty of an offence under this section if he commits (a) an offence under section 4 of the Public Order Act 1986 (fear or provocation of violence); (b) an offence under section 4A of that Act (intentional harassment, alarm or distress); or (c) an offence under section 5 of that Act (harassment, alarm or distress), which is racially or religiously aggravated for the purposes of this section.

Likewise, under s.32(1) of the CDA, a person is guilty if they commit an offence under either s.2, 2A, 4, or 4A of the PfHA which was racially or religiously hostile. In essence, for an enhanced sentence under either of these provisions (where the aggravated factor is race) it must be found that the action of the defendant amounts to a 'demonstration' or 'motivation' of 'hostility' to a 'racial group':

> An offence is racially or religiously aggravated [...] if (a) at the time of committing the offence, or immediately before or after doing so, the offender demonstrates towards the victim of the offence hostility based on the victim's membership (or presumed membership) of a racial or religious group; or (b)

29 Note, the Crime and Disorder Act 1988 also applies to a further two offences: assault (s.28) and criminal damage (s.29). However, these have not been included as they are not applicable to social media.

30 These provisions were updated in 2006 to include religion following the implementation of Racial and Religious Hatred Act 2006. And further updated in 2008 to include sexual ordination following the implementation Criminal Justice and Immigration Act 2008.

31 *R v Liam Stacey,* Swansea Crown Court On Appeal From The Magistrates' Court, A20120033.

> the offence is motivated (wholly or partly) by hostility towards members of a racial or religious group based on their membership of that group.[32]

Hostility is the base of the offence, though there is no definition of hostility either within the CDA or other relevant statutes. Instead, it has come to be accepted that hostility takes its everyday meaning and can include 'ill-will, spite, contempt, prejudice, unfriendliness, antagonism, resentment, and dislike'.[33] However, the hostility must be experienced at the time in which the offence took place or immediately prior to or after the event by establishing a connection between the hostility and the offence in question. Where hostility is not proven, it will be concluded that the actions of the defendant were not aggravated.[34] Similarly, a wide definition has been given to the term 'racial group' which is defined in the CDA as a 'group of persons defined by reference to race, colour, nationality (including citizenship), or ethnic or national origins'.[35] This includes, in certain circumstances, Gypsies and travellers, refugees, or asylum seekers, in which the courts have endorsed a flexible approach to the term.[36]

Next, the conduct either needs to demonstrate hostility or be motivated by hostility. Consequently, two limbs of the offence have been created: (1) motivation or (2) demonstration. Due to the difficulties in proving motivation, most cases are brought before the courts under the demonstration limb. To demonstrate hostility, this will usually involve words, though songs may also constitute a demonstration of hostility. This is a purely objective test, for the jury or magistrate to decide. The complainant's perceptions are not considered. Indeed, as is explored in Section 4.5.2 the demonstration of hostility has been heavily criticised as being overly expansive, rendering issues with legality.[37]

Whereas the motivation limb 'requires that the offence be motivated, in whole or in part, by hostility'[38] and requires a subjective analysis of the defendant's motivation. In essence, to be motivated by hostility it must be found that the person harboured some form of ill-will etc., towards the person, based on evidence gathered by the police, which can include social media postings or being part of a particular group. In addition, when it comes to demonstrating hostility, there must be an identifiable complainant—an element not needed under the motivation limb.

32 Crime and Disorder Act 1988 s28(1).
33 The Crown Prosecution Service, 'Hate crime' (*CPS.gov*, 2022) <https://www.cps.gov.uk/crime-info/hate-crime> accessed 24 October 2022.
34 Note, an individual could still be prosecuted for the base offence.
35 Crime and Disorder Act 1998 s28(4).
36 *Rogers* [2007] UKHL 8, [2007] 2 A.C. 62 [12].
37 Law Commission, *Hate Crime Laws: Final Report* (Law Com No 401, 2021) [9.38].
38 Ibid., [2.26].

4.3.3 Public Order Act 1986

The POA, in particular provisions related to 'stirring up' hate,[39] is considered one of the most controversial provisions in combatting hate crime, as it significantly impedes the right to freedom of expression—though considered by many as a legitimate interference. Its use in the criminal law framework is rare, although it has and continues to be used in a social media context, despite, as will be explored in Section 4.5, issues with the principle of legality, particularly given that the stirring up offences discussed in this section, do not explicitly cover internet communications.

The purpose behind the creation of the POA was:

> to abolish the common law offences of riot, rout, unlawful assembly and affray and certain statutory offences relating to public order; to create new offences relating to public order; to control public processions and assemblies; to control the stirring up of racial hatred.[40]

The POA was considered an opportunity to strengthen the law and 'aimed at protecting those in [...] communities who [were] most vulnerable to loutish and abusive behaviour'.[41] It introduced six types of conduct under the provision of 'stirring up racial hatred'.[42] However, the two which will be of focus for the purposes of this chapter are:

1. Using threatening, abusive, or insulting words or behaviour or displaying written material which is threatening, abusive, or insulting—s.18; and
2. Publishing or distributing written material which is threatening, abusive, or insulting—s.19.

Each of these sections will be taken in turn.

4.3.3.1 Section 18

S.18(1) makes it a criminal offence to use or display:

> written material or behaviour which is threatening, abusive or insulting [...] which intends thereby to stir up racial hatred, or having regard to all circumstances racial hatred is likely to be stirred up.

39 The provisions themselves were originally only aimed at stirring up racial hatred but have since been expanded to include religion and most recently, sexual ordination. See, Public Order Act 1986 Part 3A.

40 Public Order Act 1986.

41 HC Deb 13 January 1986, vol 89, col 793.

42 Similar provisions exist for religion and sexual orientation.

Although this does not include words or behaviours used or displayed as part of a programme. For example, those used in a TV drama.[43]

For a prosecution to be brought under this provision, consent of the Attorney General is needed, alongside establishing the *actus reus* and *mens rea* of the offence and disproving any defences raised, though the Crown Prosecution Service (CPS) also maintain:

> In deciding upon the public interest of charging these offences it is essential that prosecutors keep in mind that in a free, democratic, and tolerant society people are able to robustly exchange views, even when these may cause offence.[44]

The following discussion will outline the *actus reus* and *mens rea* of the offence under s.18 of the POA, before turning to examine the defences contained within the provision.

4.3.3.1.1 ACTUS REUS

The *actus reus* of s.18(1) comprises of several elements:

1. To use or display written material or behaviour. Neither display nor behaviour are defined in the POA. Instead, they are given their ordinary English meaning. I.e., their dictionary meaning. Whereas written material is defined as, 'signs or other visual representation',[45] and includes, as noted by the court in *R v Sheppard,*[46] articles sent in electronic form. However, it is important to note that live video broadcasts, which are common on most social media platforms, are not considered written material for the purposes of this section;[47]
2. The material or behaviour needs to be threatening, abusive, or insulting. These words take their 'ordinary English meaning' and are a question based on fact.[48] Though it's important to note, that the conduct does not have to fall under all three categories, one is sufficient. In essence, no specific definition is given in law as to the meaning of threatening, abusive, or insulting—though as discussed in Chapter 2, all could be considered as protected under the right to freedom of expression. Indeed, how terms such as threatening and abusive work in relation to freedom of expression is still not fully understood within

43 Public Order Act 1986 s18(6).

44 The Crown Prosecution Service, 'Racist and religious hate crime – prosecution guidance' (*CPS.gov*, 3 March 2022) <https://www.cps.gov.uk/legal-guidance/racist-and-religious-hate-crime-prosecution-guidance> accessed 24 October 2022.

45 Public Order Act 1986 s.29.

46 *R v Sheppard* [2010] EWCA Crim 65.

47 Law Com No 401 (n 37) [10.387].

48 The Crown Prosecution Service, 'Racist and religious hate crime – prosecution guidance' (*CPS.gov*, 2017) <http://www.cps.gov.uk/legal/p_to_r/racist_and_religious_crime/#a07> accessed 22 October 2017.

the criminal law framework.[49] The issues this renders with legality are discussed in Section 4.5.1.1;

3. The material must or will likely stir up racial hatred. As with the CDA discussed above, racial hatred is considered 'hatred against a group of persons [...] defined by reference to colour, race, nationality (including citizenship) or ethnic or national origins',[50] though the hatred needs to be directed at a group, 'not merely an individual'.[51] In which a literal approach should not be used when it comes to 'race'. Words such as 'African, foreigners, and immigrants' can fall within the definition of race.[52] In *R v Martin Hartshorn*[53] it was held that the use of the word 'Paki' amounted to racial hatred under the provision. The term likely is defined as 'more than merely possible or likely'[54] but does not include 'liable to';[55] and
4. The material or behaviour may be displayed in a public or private setting, however, 'no offence is committed where the words or behaviour are used, or the written material is displayed, by a person inside a dwelling and is not heard or seen except by other persons in that or another dwelling'.[56]

If all the factors above are established the *actus reus* for a s.18(1) offence under the POA can be proven. A failure to establish one or more of the elements discussed above would result in no criminal offence contrary to this provision.

4.3.3.1.2 MENS REA

In addition to establishing the factors discussed above, it must also be found that the accused also had the specific *mens rea* to commit an offence under s.18(1) of the POA. Two *mens rea* elements exist under this provision: (1) intent or (2) awareness. Intent under this provision can be summarised as to 'intentionally denigrate or demean those against whom the words are directed'.[57] Here intent is judged wholly objectively. Whereas to prove the *mens rea* of 'awareness' it must be found that the defendant 'intended his conduct to be, or was aware that it might be, "threatening,

49 Law Com No 401 (n 37) [10.96].
50 Public Order Act 1986 Part 3 s.17.
51 Law Com No 401 (n 37) [2.134].
52 The Crown Prosecution Service, 'Violent extremism and related criminal offences' (*CPS.gov*, 2017) <https://www.cps.gov.uk/publications/prosecution/cases_of_inciting_racial_and_religious_hatred_and_hatred_based_upon_sexual_orientation.html> accessed 22 October 2017.
53 *R v Martin Hartshorn,* Grimsby Crown Court, 4 November 2011 (unreported). See also, Dave Higgens, 'Man jailed for riot race-hate posts' *The Independent* (London, 4 November 2011) <http://www.independent.co.uk/news/uk/crime/man-jailed-for-riot-race-hate-posts-6257282.html> accessed 22 October 2017.
54 Violent Extremism and Related Criminal Offences (n 52).
55 *Parkin v Norman* [1983] QB 92, 100.
56 Public Order Act 1986 s.18(4).
57 David Ormerod and Karl Laird, *Smith, Hogan and Ormerod's Criminal Law* (16 edn, Oxford University Press 2021) [31.10.1.2].

abusive, or insulting" and that having regard to all the circumstances, racial hatred was likely to be stirred up thereby'.[58]

4.3.3.1.3 DEFENCES

Two specific defences are contained in s.18. First, it is a

> defence for the accused to prove that he was inside a dwelling and had no reason to believe that the words or behaviour used, or the written material displayed, would be heard or seen by a person outside that or any other dwelling.[59]

Second,

> A person who is not shown to have intended to stir up racial hatred is not guilty of an offence under this section if he did not intend his words or behaviour, or the written material, to be, and was not aware that it might be, threatening, abusive or insulting.[60]

If either of these defences can be proven, no offence has taken place contrary to s.18(1) of the POA. However, the defence itself does not apply to material published on the internet, even if such a communication was sent from a private dwelling, i.e., a person sending a message on X in the comfort of their own home.[61] In addition, the law has extended to note that this defence does not apply to all private conversations, where in *BC v Chief Constable of Scotland* it was held that the sending of messages on WhatsApp which were deemed as sexist, racist, antisemitic, homophobic, and disablist between a group of police officers, was not inherently private.[62]

4.3.3.2 Section 19

Whereas s.18(1) of the POA makes it an offence to display certain material, s.19(1) makes it a criminal offence to publish:

> written material which is threatening, abusive, or insulting [...] which intends thereby to stir up racial hatred, or having regard to all circumstances racial hatred is likely to be stirred up.

Like previous sections, the following discussion will outline the *actus reus* and *mens rea* of the offence under s.19(1) of the POA. Before turning to outline the defence contained within this provision.

58 Ibid., [31.10.1.3].
59 Public Order Act 1986 s.18(4).
60 Ibid., s.18(5).
61 *Sheppard* (n 46).
62 Law Com No 401 (n 37) [10.293].

4.3.3.2.1 ACTUS REUS

The *actus reus* under s.19(1) comprises of three elements:

1. The distribution or publication of written material. Here, it must be found that written material, which includes 'signs or other visual representation' is published or distributed to the 'public or a section of the public'.[63] As noted above, this would include material published on the internet, though would not necessarily cover emojis, memes, or gifs;[64]
2. The material published needs to be threatening, abusive, or insulting. As previously discussed, these words will take their 'ordinary English meaning' and are a question based on fact;[65] and
3. The material must or will likely stir up racial hatred. Again, the same meaning is given to this element of the offence as with s.18 of the POA discussed in Section 4.3.3.1.1.

If all three elements are established the *actus reus* of a s.19(1) public order offence can be proven. A failure to establish one or more of the elements discussed above would result in no criminal offence contrary to s.19(1) of the POA.

4.3.3.2.2 MENS REA

The *mens rea* under s.19(1) of the POA is the same as s.18(1)—done with intent or an awareness that a person's actions were likely to stir up racial hatred.

4.3.3.2.3 DEFENCE

Whereas s.18 of the POA has several defences, s.19 contains one very specific defence:

> In proceedings for an offence under this section it is a defence for an accused who is not shown to have intended to stir up racial hatred to prove that he was not aware of the content of the material and did not suspect, and had no reason to suspect, that it was threatening, abusive or insulting.[66]

If such can be proven, no offence will be deemed to have taken place contrary to s.19(1) of the POA.

63 Public Order Act 1986 s.29.
64 Law Com No 401 (n 37) [10.323].
65 Racist and Religious Hate Crime – Prosecution Guidance (n 48).
66 Public Order Act 1986 s.19(2).

4.3.4 Summary

It is clear that the provisions discussed above all have one recurring theme: they are all embedded in vague terminology. Concerns can be raised with the lack of understanding as to what is meant by the terms threatening, abusive, or insulting. Indeed, as outlined in Section 4.5.1.1, each of these types of communications could fall within the right to freedom of expression. For the Law Commission, during their consultation into hate crime laws:

> [w]here the law affects a right like freedom of expression, uncertainty, and misunderstandings of the law can create a 'chilling effect' on the exercise of that right, even if that interference itself is permissible, proportionate, and justified.[67]

Further, as illustrated in Section 4.5.2, issues can also be raised with proving the 'demonstration of hostility'. Issues, which undermine the principle of legality, especially when examining the use of the POA in a social media context.

4.4 Social Media, Racism, and the Criminal Law

As social media usage has increased, and in turn, we have seen a rise in the use of such platforms to send racist communications, we have witnessed the use of all the provisions discussed above, being used to arrest, prosecute, and/or convict individuals for their online conduct. In particular, the use of s.18 and 19 of the POA. For instance, these provisions were used to arrest several individuals in the wake of England's defeat in the 2020 Euros.

Following the defeat, some individuals took to social media platforms to vent their frustrations, and in some cases, grounding these frustrations in racist commentary. For example, in the aftermath of the defeat, three individuals were arrested on suspicion of committing an offence contrary to either s.18 or 19 of the POA—though none were actually convicted of a POA offence.[68]

In *R v Paul Bradbury*,[69] the defendant was originally arrested on suspicion of displaying written material which was threatening, abusive, or insulting which was likely to stir up racial hatred contrary to s.18(1) of the POA after posting the following message: 'Whites only in future'. Similarly, in *R v Hugh Laidlaw* the defendant was originally arrested under s.18 of the POA for allegedly posting images and comments containing racist material on Facebook. Whereas in *R v*

67 Law Com No 401 (n 37) [10.10].

68 National Police Chiefs' Council, 'Hate crime investigation following Euro 2020 final leads to 11 arrests' (*NPCC*, 5 August 2021) <https://news.npcc.police.uk/releases/hate-crime-investigation-following-euro-2020-final-leads-to-11-arrests> accessed 24 October 2022.

69 *R v Paul Bradbury,* Telford Magistrates' Court, 4 March 2022 (unreported). See also, The Crown Prosecution Service, 'Former councillor pleads guilty to posting racist comments about England football players' (*CPS.gov*, 4 March 2022) <https://www.cps.gov.uk/west-midlands/news/former-councillor-pleads-guilty-posting-racist-comments-about-england> accessed 24 October 2022.

Scott McCluskey,[70] the defendant was arrested under suspicion of displaying threatening, abusive, or insulting written material that was likely to stir up racial hatred contrary to s.19(1) of the POA, for using Facebook to make several racist remarks.

However, in all the cases outlined above, each defendant was later charged and pleaded guilty to sending a grossly offensive message contrary to s.127(1) of the CA as opposed to a public order offence. Similarly, in *DPP v Bussetti*,[71] the defendant was originally arrested contrary to an offence under the POA. He was later charged and convicted of committing an offence contrary to s.127(1) of the CA, after sharing a video on WhatsApp of a bonfire party, mocking the deaths of those who died in a tower block fire in London, in 2017.

There have however been some successful prosecutions in a social media context under the 'stirring up' offences contained in the POA.[72] For example, s.19(1) of the POA was successfully utilised in *R v Liam Stacey*, who in 2012, was convicted and sentenced to 56 days in jail for sending racist and obscene tweets, following the collapse of Bolton Wanderers star Fabrice Muamba during a football match.[73] Fabrice Muamba went into cardiac arrest live on television. For six minutes medical staff helped to resuscitate him on the pitch.[74] Following the incident, Stacey took to his Twitter page to make several racist comments about the footballer including: 'LOL [Laughing out Loud]. Fuck Muamba he's dead [*sic*]!!! #Haha' and 'Go suck a nigger [*sic*] dick you fucking aids-ridden cunt'. Stacey made numerous tweets of this nature not only aimed at Fabrice Muamba but also other individuals who attempted to defend the footballer.[75]

Despite some successful prosecutions, it is not always clear why the POA is used in some cases, but not in others, as highlighted in Section 4.5.3 below. This creates inconsistencies within the law. The issues this presents with the principle of legality, alongside the vagueness inherent in the provisions outlined thus far, are discussed in the final parts of this chapter.

70 *R v Scott McCluskey,* Warrington Magistrates' Court, 8 September 2021 (unreported). See also, The Crown Prosecution Service, 'Cheshire man sentenced for racist abuse of England players' (*CPS .gov*, 8 September 2021) <https://www.cps.gov.uk/mersey-cheshire/news/cheshire-man-sentenced-racist-abuse-england-players> accessed 24 October 2022.

71 *DPP v Kingsley Anthony Smith* [2017] EWHC 359 (Admin).

72 Another 'successful' case includes *R v Stuart Sutton,* Bolton Crown Court, 21 December 2022 (unreported). See, Greater Manchester Police, 'Wigan man jailed for posting racist and anti-Semitic material' (*GMP*, 22 December 2022) <https://www.gmp.police.uk/news/greater-manchester/news/news/2022/december/wigan-man-jailed-for-posting-racist-and-anti-semitic-material/> accessed 1 Feb 2023.

73 *Stacey* (n 31).

74 BBC Sport, 'Bolton's Fabrice Muamba collapses during Spurs-Bolton match' *The BBC* (London, 17 March 2012) <http://www.bbc.co.uk/sport/football/17417973> accessed 10 January 2018.

75 Steven Morris, 'Student jailed for racist Fabrice Muamba tweets' *The Guardian* (London, 27 March 2012) <https://www.theguardian.com/uk/2012/mar/27/student-jailed-fabrice-muamba-tweets> accessed 10 January 2018.

4.5 Hate Crime, Legality, and the Criminal Law

It is important that the law, in particular the criminal law, is not used to send social signals as to what is or is not socially acceptable; or to impose unjust moral standpoints on freedom of expression. Though the effect of hate speech, in particular, racist speech online should not, and cannot, be downplayed, the use of such words is a systematic societal issue—not an issue that is unique to the digital world. And no matter how much criminal law is imposed on the digital world, legal norms will not take effect until societal norms have evolved.

Yet there continues to be a strong emphasis on using the criminal law to tackle such comments being made on social media sites despite, as will be outlined below, issues with the principle of legality in the legal provisions discussed above and their application to social media. Indeed, as will be shown, some of these provisions are being stretched well beyond their original purpose and in turn, threatening freedom of expression. The discussion below argues that such use of the criminal law, in the context of social media, is not prospective or clear. As a result, inconsistent application of the law, across the criminal justice system, is leading to unfair enforcement of the law—all of which undermines the principle of legality.

4.5.1 The Issue of Language

As was touched upon in Section 4.3, all the provisions discussed in this chapter are grounded in vague terminology. In particular, issues can be raised with the use of the terms insulting, abusive, and hostile—all of which are deemed to take their ordinary English meaning, rendering issues with the principle of legality and the right to freedom of expression. Each of these terms will be taken in turn and examined against the principle of legality in the following discussion.

In addition, the use of the term threatening, as found in both s.18(1) and 19(1) of the POA, renders issues with clarity in the criminal law. However, the issues the term threatening posses with legality and the right to freedom of expression will not be discussed here. Instead, the issue of quantifying what a threatening communication is, will be discussed in Chapters 6 and 7, where a similar term, threat, was used in s.1 of the Malicious Communications Act 1988 and now contained in s.181 of the OSA, alongside 'menacing' in s.127(1) of the CA.

4.5.1.1 The Meaning of Abusive and Insulting

Both s.18(1) and 19(1) of the POA prohibit either the use or display (s.18) or the publishing or distributing (s.19) of material which can be considered either abusive or insulting and has the possibility of stirring up racial hatred. Yet neither of these terms, abusive or insulting, are defined within the POA, nor are defined elsewhere in the criminal law framework. Consequently, both abusive and insulting are deemed to take their ordinary English meaning. I.e., insulting can be defined, with reference to the Cambridge English Dictionary, as 'rude or offensive';[76] abusive can be defined as 'using rude and offensive words'.[77]

The lack of a definition as to the meaning of abusive and insulting has surprisingly been supported by those working in the judicial system. For instance, in *Brutus v Cozens*,[78] the House of Lords, who in the context of s.5 of the Public Order Act 1936[79] supported their inherent vagueness:

> We were referred to a number of dictionary meanings of 'insult' such as treating with insolence or contempt or indignity or derision or dishonour or offensive disrespect. Many things otherwise unobjectionable may be said or done in an insulting way. There can be no definition. But an ordinary sensible man knows an insult when he sees or hears it [...] vigorous and it may be distasteful or unmannerly speech or behaviour is permitted so long as it does not go beyond any one of three limits. It must not be threatening. It must not be abusive. It must not be insulting. I see no reason why any of these should be construed as having a specially wide or a specially narrow meaning. They are all limits easily recognisable by the ordinary man. [80]

Indeed, the House of Lords even went as far as suggesting that prohibiting such behaviour did not restrict the right to freedom of speech.[81]

However, it is important to note that the Public Order Act 1936, to which the House of Lords was referring to in *Brutus* predates the European Convention on Human Rights (ECHR) and therefore such terminology may not be complicit with human rights today. This is especially true given that the European Court of Human Rights (ECtHR), as discussed in detail in Chapter 2, has acknowledged that free speech includes the right to be offensive.[82] And most recently, the UK Divisional Court suggested that free speech encompasses the right to be abusive.[83] This has been further supported by the Law Commission which noted, in their recent review on hate crimes that prohibiting insulting material, is not necessarily compatible with Article 10 of the ECHR[84] and indeed, raised concerns about the lack of understanding as to when certain conduct will invoke s.18 or 19 of the POA:

76 Cambridge Dictionary, 'Insult' (*Cambridge Dictionary*, 2023) <https://dictionary.cambridge.org/dictionary/english/insult> accessed 8 February 2023.

77 Cambridge Dictionary, 'Abusive' (*Cambridge Dictionary*, 2023) <https://dictionary.cambridge.org/dictionary/english/abusive> accessed 8 February 2023.

78 *Brutus v Cozens* [1973] A.C. 854.

79 Which prohibited an individual in a public place or at any public meeting from using 'threatening, abusive or insulting words or behaviour with intent to provoke a breach of the peace'.

80 *Brutus* (n 78) per Lord Reid at 862.

81 Ibid.,

82 *Handyside v The United Kingdom* (1976)1 EHRR 737 [49].

83 *Katherine Elizabeth Scottow v Crown Prosecution Service* [2020] EWHC 3421 (Admin) per Justice Warby [43].

84 Law Com No 401 (n 37) [10.305].

> We believe that insofar as the law itself may be giving rise to ill-founded complaints that interfere with the right to freedom of expression, the correct way to address this is by clearly defining the limits of the stirring up offences […] It is possible that in making clear that the law does not criminalise—for instance—discussion of controversial topics, our recommendations will enable law enforcement officials to deal robustly in dismissing complaints that do not amount to a criminal offence.[85]

Yet, as it currently stands, s.18 and 19 of the POA continues to be used to prohibit online communications.

With such a reliance on the 'ordinary meaning' of the terms insulting and abusive it is difficult to see how consistency can be maintained in the law, particularly when it comes to social media posts, where in the online world, it is expected that there should be a little bit of give-and-take. What one police force or Crown Prosecutor might consider to be insulting or abusive another may not, and the boundary between protected speech and non-protected speech is not easily identifiable. To set the bar so low, in this instance, abusive or insulting, when it comes to speech, is to risk the overcriminalisation of free speech. As potently put by the Law Commission:

> We are wary of attempts to use the criminal law merely to send a social signal, whether that is towards making the expression of certain views unacceptable, or, conversely, to encourage social acceptance of a broader range of views.[86]

The issues of consistency and clarity are further reflected in the notion of what constitutes hostility.

4.5.1.2 The Meaning of Hostility

Hostility is 'central to […] hate crime provisions in England and Wales'.[87] Yet, there is no definition in law as to the meaning of hostility, 'nor is there any standard direction on what types of conduct it might encompass'.[88] This is despite hostility being central to all the provisions discussed above. With a lack of a statutory definition, alongside a failure of the courts to establish the boundaries of what constitutes hostility, the term has been given its ordinary dictionary meaning:

85 Ibid., [10.38].

86 Ibid., [10.34].

87 Mark Austin Walters, 'Conceptualizing Hostility for Hate Crime Law: Minding the Minutiae When Interpreting s.28(1)(a) of the Crime and Disorder Act 1998' (2014) 34 (1) Oxf. J. Leg. Stud 47, 48.

88 Ibid.,

> Hostility is not defined in the legislation. Consideration should be given to ordinary dictionary definitions, which include ill-will, ill-feeling, spite, prejudice, unfriendliness, antagonism, resentment, and dislike.[89]

This is a particularly wide net to set, with Walters stating: '[i]t is likely that many fact finders will simply turn to colloquial understandings of what they think hostility means when determining an offender's guilt'.[90] In turn, it cannot be said that consistency is being maintained in the criminal law. What one 'fact finder' may consider 'ill-will, ill-feeling, spite, prejudice, unfriendliness, antagonism, resentment, and dislike' and therefore hostile, another may not—a reoccurring theme throughout this monograph. This becomes more complex when the provisions discussed above are applied to communications sent *via* social media. Here messages can be exchanged in an instant, but the context behind such messages can be lost, and not fully understood to those on the outside looking in.

Indeed, the ordinary dictionary meaning given to hostility includes many natural human emotions or behaviours, which all fall within the boundaries of freedom of expression, such as 'unfriendliness' and 'dislike'. Such reliance on a term, without a clear understanding of its meaning, can result in broad interpretations and at its very basic, for the law to be compatible with the principle of legality, provisions need to be clear and prospective. How can digital citizens guide their online behaviour in accordance with the law, if the law itself casts such a wide net over what might be considered hostile?

What might be considered as 'dislike' or 'contempt' to one, and therefore hostile, to another might be considered political debate to others. As noted at the start of this chapter, for Murray, 'sometimes there is a fine line between political speech and hate speech'.[91] Therefore to centralise criminal provisions around hostility, without clearly defining the boundaries of the term, is to risk unfair enforcement of the law.

4.5.1.3 Summary

It is evidently clear, from discussions above, that the provisions used to tackle racist comments online are embedded in significantly vague terminology, creating inconsistencies in their application to complaints of social media (mis)usage. This, in turn, renders issues with the principle of legality. For legality to be maintained in the criminal law, criminal provisions should not be so overly vague. Yet, the use of the terms insulting, abusive or hostile have no true definition in law. Consequently, the law, as it currently stands, does not allow individuals to reasonably understand when their behaviour may fall foul of the law. Again, concerns which have not gone unnoticed with the Law Commission previously noting:

89 Hate Crime (n 33).
90 Walters (n 87).
91 Murray (n 6).

> [we] are deeply concerned that misrepresentation as to the sort of conduct which are caught by the stirring up offences may lead people to avoid saying things that they are perfectly entitled to express.[92]

This is even more concerning given, as illustrated above, the dictionary definitions of these terms, insulting, abusive, or hostile could all fall within the right to freedom of expression. Democracy depends on people being free to express, debate, and criticise opposing viewpoints and social media is an important aspect of this, where political debate or contentious issues are often discussed. And it is only through having these discussions that we can attempt to seek 'truth'.[93] Indeed, the ECtHR in *Jersild v Denmark* maintains speech that is intended to inform rather than offend attracts greater protection, even if it could be construed as racist.[94]

In turn, when complaints are made to police forces, Crown Prosecutors, etc., without a clear understanding as to when a communication crosses the threshold to be considered so insulting, abusive, or hostile to invoke the criminal law, results in consistency being lost. Consistency is needed, to ensure public faith in the criminal justice system is maintained. Similar arguments are raised in Chapters 6 and 7. These concerns are further amplified when considering what is meant by demonstrating hostility under the CDA and the SA.

4.5.2 Demonstrating Hostility

The demonstration of hostility is central to both s.66 of the SA and s.31 and 32 of the CDA. Yet it has been heavily criticised in the literature for being overly broad and in turn, overly expansive:

> What constitutes the 'hostility' needed to prove 'racial aggravation', for example, is a source of contention for legal practitioners. In some local magistrates' courts, for example, there is: 'a current of opinion, most often expressed by stipendiary magistrates and defence solicitors, that the law came down rather hard on people who, in the course of "normal working class mayhem" as one person put it, uttered words which were part of their natural vocabulary'.[95]

Consequently, there is growing evidence that 'any reference to the victim's ethnicity in connection with an offence' proves 'the necessary element of racial

92 Law Com No 401 (n 37) [10.12].

93 John Stuart Mill, *On Liberty* (first published 1859, Enhanced Media 2016) 40.

94 *Jersild v Denmark* (1994) (Application no. 15890/89).

95 David Gadd, 'Aggravating Racism and Elusive Motivation' (2009) 49 Brit. J. Criminol 755, 758. Quote found in Elizabeth Burney and Gerry Rose, *Racist Offences: How Is the Law Working?; the Implementation of the Legislation on Racially Aggravated Offences in the Crime and Disorder Act 1998* (Home Office Research Study, 2002) 20.

aggravation' due to the purely objective test used to determine what constitutes the demonstration of hostility.[96]

By assessing the demonstration of hostility objectively, it has resulted in conflicting opinions:

> those who adopt a free-speech outlook are likely to view a (potentially) hateful act as non-hostile whereas those who adopt a critical race theory perspective may view it as definitely hostile and hateful. These differing positions are often irreconcilable.[97]

This is even more glaringly apparent in a social media situation where, as noted previously and will be echoed throughout this monograph, context is often lost and opposing viewpoints are easy to come by—whether a person is actively looking for it or not. Indeed, many social media platforms use algorithms to bring content to a user's attention, without the user actively searching for such content. The role social media platforms play in spreading such content is discussed further in Chapter 8.

Consequently, the broad understanding as to what is considered the demonstration of hostility gives actors working in the criminal justice system significant discretion as to what constitutes not only the meaning of 'hostility', but also the demonstration requirement. In turn, this poses significant issues with the principle of legality. As discussed at length in Chapter 3: 'legality requires that judges and other officials apply statutory law, not according to their fancy or with crabbed literalness, but in accordance with principles of interpretation'.[98] Yet, growing research suggests that 'crabbed literalness' is being applied to such offences.[99]

For The Alan Turing Institute, if the bar is not set high enough in determining what is considered as demonstrating hostility, there is the real possibility of the overcriminalisation of speech.[100] When it comes to social media, something said in the heat of the moment, to hostile and hateful behaviour, is not easily identifiable and embedding legal provisions in broad concepts, such as 'demonstrating hostility' undermines the action-guiding nature of the law. In turn, due process is at risk, alongside the coercive nature of the law being abused. Anecdotally it has been suggested that

> defendants have tended to contest this inference, often proving willing to admit to the basic (non-racially aggravated) charges, but unwilling to plead guilty to their racially aggravated equivalents.[101]

96 Ibid.,
97 Law Com No 401 (n 37) [9.51].
98 Lon L Fuller, *The morality of law* (Yale University Press 1964) 82.
99 Walters (n 87).
100 Law Com No 401 (n 37) [9.51].
101 Gadd (n 95).

4.5.3 Going Beyond Parliament's Intent: The Purpose of the POA

The purpose and rationale behind the creation of the POA was, and still preserves to be, to maintain public order, with Moses J in *Dehal v CPS* when referring to the POA stating:

> the criminal law should not be invoked unless and until it is established that the conduct which is the subject of the charge amounts to such a threat to public order as to require the invocation of the criminal as opposed to the civil law.[102]

Indeed, for Dorfman, the POA should not be used in the context of social media, especially when there is no element of violence present. Using *R v Stacey*, discussed above, Dorfman argues that:

> [U]sing the Public Order Act for a few nasty Twitter comments which never advocated violence or hatred against specific people is a downright abuse of the law; at no time was 'public order' ever threatened in any reasonable conception of the term.[103]

For Dorfman, the conduct of Stacey lacked the relevant criteria needed to prosecute an individual under the POA, despite Stacey's successful conviction. Though his behaviour was abusive and had a racial element, his actions did not stir up racial hatred nor could it be said that it was likely to cause racial hatred and consequently, criminal proceedings under the POA was an abuse of the law. Similar arguments were raised by Walters:

> Though this author is in agreement with the outpouring of social disapproval and the prosecution of offenders who commit public order offences aggravated by racism, it is felt that the sentence in this case [*Stacey*] was grossly disproportionate to the seriousness of the offence [...] it is worth highlighting a couple of points about the punishment of hate crime offenders. First, it cannot be denied that without effective punishments the criminalization [*sic*] of hate would become almost meaningless, and as such those who transgress hate crime laws must face some form of punishment. Still, some judicial restraint is required if hate crime laws are not to become simply another means through which we add to an ever growing prison population.[104]

Indeed, the CPS guidelines on social media related offences from 2016 stated:

102 *Dehal v Crown Prosecution Service* [2005] EWHC 2154 per Moses J [5].

103 Rosalee Dorfman, 'Can you say "social media prosecutions" with a straight face? The Crown Prosecution Service can' (2013) LSLCJR <http://criminology.leeds.ac.uk/2013/09/05/social-media-prosecutions/> accessed 20 October 2016.

104 Walters (n 87) 57.

> [P]articular care should be taken in dealing with social media cases in this way because public order legislation is primarily concerned with words spoken or actions carried out in the presence or hearing of the person being targeted (i.e. where there is physical proximity between the speaker and the listener).[105]

Yet in December 2022, Stuart Sutton was convicted under s.19(1) of the POA for sending racist and antisemitic material on social media, despite no complaint being made and his posts only coming to the attention of the police following the gathering of intelligence in another matter. He received a 16 month custodial sentence.[106]

The CPS maintain that the POA should only be used in limited circumstances, but it is not clear from the guidelines what constitutes these conditions. The guidelines mention the difference between 'irritating, contentious, unwelcome, and provocative' conduct and behaviour that provokes violence, though no clarity is given beyond this.[107] Although the Law Commission notes:

> [t]he offences [stirring up hate] do not criminalise conduct expressing or inciting hostility or hatred towards specific individuals. Rather, they address conduct intended or likely to cause others to hate entire national or ethnic groups.[108]

Further, the POA was used on several occasions in 2021 to arrest several individuals for sending racist abuse online aimed at three English footballer's, as mentioned in Section 4.4 above. The National Police Chief's Council confirmed in the aftermath of England's defeat in the 2020 Euros, 11 arrests, of which three individuals were originally arrested under provisions contained in the POA, yet later convicted under s.127(1) of the CA, as shown in Table 4.1. The use of the POA in some circumstances, and not in others, evidently leaves open, once again, questions on consistency within this area of the criminal law. For instance, the POA was applied in *Stacey* but not in the matter of *R v Alison Chabloz*.[109]

Chabloz, who defines herself as a holocaust revolutionist, as discussed further in Chapter 6, uploaded several YouTube videos online mocking Holocaust victims and survivors, including Anne Frank. In these videos, she also made 'expressions of anti-Semitic hatred',[110] in which it was put before the court that her actions were

105 The Crown Prosecution Service, 'Guidelines on prosecuting cases involving communications sent via social media' (*CPS.gov*, 2016) <https://www.cps.gov.uk/legal-guidance/guidelines-prosecuting-cases-involving-communications-sent-social-media> accessed 10 January 2018. Note, this is no longer included in the most updated version of the guidelines.

106 *Sutton* (n 72).

107 Racist and Religious Hate Crime – Prosecution Guidance (n 48).

108 Law Com No 401 (n 37) [2.126].

109 *R v Alison Chabloz,* Westminster Magistrates' Court, 11 January 2018 (unreported).

110 ITV News, 'Blogger "mocked Anne Frank and Holocaust survivors" court told' *ITV News* (London, 11 January 2018) <http://www.itv.com/news/2018-01-11/blogger-mocked-anne-frank-and-holocaust-survivors-court-told/> accessed 11 January 2018.

Table 4.1 Arrests and outcomes following the abuse of three England footballer's in the aftermath of the Euros 2020

Details	*Arrest*	*Outcome*
Scott McCluskey arrested for posting racist and insulting comments on Facebook.	Arrested on suspicion of displaying threatening, abusive, or insulting written material that is likely to stir up racial hatred contrary to s.19(1) of the POA.	Convicted of a s.127(1) offence, in which s.66 of the SA was applied to 'uplift the punishment'.
Paul Bradbury arrested for posting 'whites only in the future'.	Arrested on suspicion of committing an offence contrary to s.18(1) of the POA.	Convicted of a s.127(1) offence.
Hugh Laidlaw arrested for posting comments and images online of a racist nature.	Arrested on suspicion of committing an offence contrary to s.18(1) of the POA.	Convicted of a s.127(1) offence.

'designed to provoke maximum upset and discomfort'.[111] Despite this, she was convicted for sending grossly offensive material *via* a communications network contrary to s.127(1) of the CA, even though her actions, by applying the case of *Stacey,* could be *akin* to an offence under the POA.

Both Stacey's and Chabloz's conduct can be considered abusive, at the very least, and arguably demonstrated racial hatred. Stacey used the phrase, 'go suck a nigger [*sic*] dick you fucking aids-ridden cunt'. Chabloz used several antisemitic terms, including comparing Auschwitz concentration camp to 'a theme park just for fools'. The main issue falls on whether the behaviour of both defendants was likely to stir up racial hatred. In *Stacey,* the CPS concluded that this was the case, though this is not fully accepted in the academic literature.[112] Whereas in *Chabloz* it is unclear if this was considered by the CPS as charges were brought under a different Act of Parliament. If the rationale of *Stacey* is applied, it is hard to distinguish why Chabloz was not prosecuted for a breach of s.19(1) of the POA. This is even more apparent as during the hearing held at Westminsters' Magistrates' Court, one of Chabloz's videos was played to the court, which was met with a round-of-applause by supporters in the public gallery.[113] Arguably, the POA fails to uphold the key principles of legality, as it is uncertain when this Act will be used in a social media context, as there is a lack of consistency as to when the CPS will press charges under the POA for social media related offences.

It could be said that the law has adapted to the changing nature of society, by using the POA to criminalise conduct carried out online. The adaptation of the law to fit the changing circumstances of a given society conforms to the principle of

111 Ibid.,
112 Dorfman (n 103).
113 ITV News (n 110).

legality as affirmed in *SW and CR v The United Kingdom*,[114] so long as the change is foreseeable. However, under the POA the prohibited conduct must incite public disorder as public order is considered a 'fundamental social good'.[115] Whereas public disorder affects the social wellbeing of a society, this was reflected in the tough stance taken by Parliament during the implementation of the POA:

> The 1986 Act like its predecessor is very much a pragmatic reaction to recent events. It was passed against the background of inner-city disturbances, soccer hooliganism, racist marches, and racist attacks on members of the ethnic minorities, mass industrial picketing and the resurgence of major public demonstrations. The Act is not founded on well articulated premises addressing the scope and ambit of the criminal law in the area of public order.[116]

The purpose of the POA was to regulate public disorder in particular, riots, football hooliganism, and protests which were very much apparent during the early 1980s, an element missing in the matter of *Stacey*. Yet the Act has been used to cover social media related offences, despite the POA being enacted at least 20 years before the likes of X, Facebook, Snapchat, TikTok, etc. For Haralambous and Geach this is a clear breach of the rule of law, and consequently undermines the principle of legality in the criminal law,[117] as the law, at the very least should 'be accessible and so far, as possible intelligible, clear, and predictable',[118] arguably areas missing regarding the use of the POA in a social media context, in turn, creating retrospective criminal law.

4.6 Chapter Summary

As eloquently put by the Law Commission,

> [w]hile there is a role for legislation in underpinning this and criminalising the most serious forms of hate speech, we consider that the most important driver of this change must be social change.[119]

Yet, there continues to be significant emphasis being placed on the criminal law to tackle racist commentary online. This could be partly because the underlying message behind hate crime legislation is:

> [t]o deal with this uniquely expressive form of offending, society needs to send an equally forceful and unambiguous message in reply and, for

114 *SW v The United Kingdom, CR v The United Kingdom* App no. 20166/92 (ECtHR, 22 November 1995).

115 HC Deb 13 January 1986, vol 89, cols 793.

116 Jim Driscoll, 'Protest and Public Order: The Public Order Act 1986' (1987) Sep J.S.W.L. 280.

117 Nicola Haralambous and Neal Geach, 'Online Harassment and Public *Dis*-order' (2010) 174 CL&J 409, 411.

118 Lord Bingham, 'The Rule of Law' (2007) 66(1) CLJ 67, 69-70.

119 Law Com No 401 (n 37) [10.40].

advocates of hate crime legislation, the criminal law is an important, if not the only, suitable means of communication.[120]

But the criminal law cannot always be the answer. We need to ensure that due process is being maintained.

This is a key element missing in the current use of applying the provisions discussed in this chapter to conduct carried out online. This chapter has demonstrated key issues with the principle of legality. In particular,

- The vagueness inherent in the language used in all three provisions. Of note, is the lack of understanding as to what is deemed insulting, abusive, or hostile. The definitions that have been given to these terms by the jurisprudence of the courts, through the application of their ordinary English meaning, can all fall within the right to freedom of expression;
- The demonstration of hostility as contained in s.66 of the SA and s.31 and 32 of the CDA. This poses significant issues with the principle of legality as there is a lack of understanding as to what constitutes the demonstration limb of the offence. What one police force might consider the demonstration of hostility, another may not, leading to inconsistent application of the law; and
- The application of stirring up offences going beyond Parliament's original intent, creating retrospective criminal law.

Many of the issues highlighted in this chapter are also applicable to other areas of the criminal law and its application to social media abuse.

Reference List

Baggs M, 'Online hate speech rose 20% during pandemic: "We've normalised it"' *The BBC* (London, 15 November 2021) https://www.bbc.co.uk/news/newsbeat-59292509

Bandwatch Academy, 'Uncovered: Online Hate Speech in the Covid Era' (*Bandwatch Academy*, 2022) https://www.brandwatch.com/reports/online-hate-speech/view/

BBC Sport, 'Bolton's Fabrice Muamba collapses during Spurs-Bolton match' *The BBC* (London, 17 March 2012) http://www.bbc.co.uk/sport/football/17417973

Bell, M, 'John Stuart Mill's Harm Principle and Free Speech: Expanding the Notion of Harm' (2021) 33(2) Utilitas 162

Bingham B, *The Rule of Law* (Penguin 2011)

Bleich E, 'The Rise of Hate Speech and Hate Crime Laws in Liberal Democracies' (2011) 37(6) J Ethn Migr Stud 917

Burney E & Rose G, *Racist Offences: How Is the Law Working?; the Implementation of the Legislation on Racially Aggravated Offences in the Crime and Disorder Act 1998* (Home Office Research Study, 2002)

Cambridge Dictionary, 'Abusive' (*Cambridge Dictionary*, 2023) https://dictionary.cambridge.org/dictionary/english/abusive

120 Bill Dixon and David Gadd, 'Getting the Message? "New" Labour and the Criminalization of "Hate"' (2006) 6(3) CCJ 309, 310.

Cambridge Dictionary, 'Insult' (*Cambridge Dictionary*, 2023) https://dictionary.cambridge.org/dictionary/english/insult

de Varennes F, 'Recommendations Made by the Forum on Minority Issues at Its thirteenth Session on the Theme "Hate Speech, Social Media and Minorities"' (*United Nations*, 26 January 2021) https://documents-dds-ny.un.org/doc/UNDOC/GEN/G21/017/29/PDF/G2101729.pdf?OpenElement

Discourse and Barring Service, 'About Us' (*Gov.uk,* 2022) https://www.gov.uk/government/organisations/disclosure-and-barring-service/about

Dixon B & Gadd D, 'Getting the Message? "New" Labour and the criminalization of "Hate"' (2006) 6(3) CCJ 309

Dorfman R, 'Can You Say "Social Media Prosecutions" with a Straight Face? The Crown Prosecution Service Can' (2013) LSLCJR http://criminology.leeds.ac.uk/2013/09/05/social-media-prosecutions/

Driscoll J, 'Protest and Public Order: The Public Order Act 1986' (1987) Sep JSWL 280

Evans S, 'English Football Faces up to Global Nature of Online Hate' (*Reuters*, 26 May 2022) https://www.reuters.com/lifestyle/sports/english-football-faces-up-global-nature-online-hate-2021-07-13/

Fuller L L, *The Morality of Law* (Yale University Press 1964)

Gadd D, 'Aggravating Racism and Elusive Motivation' (2009) 49 Brit J Criminol 755

Google, 'YouTube Community Guidelines Enforcement' (*Google*, 2022) https://transparencyreport.google.com/youtube-policy/removals?hl=en&total_channels_removed=period:2021Q1&lu=channels_by_reason&channels_by_reason=period:2021Q4

Greater Manchester Police, 'Wigan Man Jailed for Posting Racist and Anti-Semitic Material' (*GMP*, 22 December 2022) https://www.gmp.police.uk/news/greater-manchester/news/news/2022/december/wigan-man-jailed-for-posting-racist-and-anti-semitic-material/

Haralambous N, & Geach N, 'Online Harassment and Public *Dis*-order' (2010) 174 CL&JW 409

HC Deb 13 January 1986, vol 89, col 793

Heard C, 'Hate Crime Law: More Incrementalism or Time for Reform?' (2014) (7) Arch Rev 4

Higgens D, 'Man jailed for riot race-hate posts' *The Independent* (London, 4 November 2011) http://www.independent.co.uk/news/uk/crime/man-jailed-for-riot-race-hate-posts-6257282.html

Home Office, 'Hate Crime, England and Wales, 2021 to 2022' (*Gov.uk*, 6 October 2022) <https://www.gov.uk/government/statistics/hate-crime-england-and-wales-2021-to-2022/hate-crime-england-and-wales-2021-to-2022>

ITV News, 'Blogger "mocked Anne Frank and Holocaust survivors" court told' *ITV News* (London, 11 January 2018) http://www.itv.com/news/2018-01-11/blogger-mocked-anne-frank-and-holocaust-survivors-court-told/

Jamieson A, 'Bukayo Saka, Jadon Sancho and Marcus Rashford racially abused online after England Euro 2020 final defeat' *The Independent* (London, 12 July 2021) https://www.independent.co.uk/news/uk/home-news/bukayo-saka-racist-england-sancho-rashford-b1882303.html

Laville S, 'Internet troll who sent labour MP antisemitic abused is jailed' *The Guardian* (London, 10 February 2017) https://www.theguardian.com/uknews/2017/feb/10/internet-troll-who-sent-labour-mp-antisemitic-messages-is-jailed

Law Commission, *Hate Crime Laws: Final Report* (Law Com No 401, 2021)

Lord Bingham, 'The Rule of Law' (2007) 66(1) CLR 67

Meta, 'Hate Speech' (*Meta*, 2022) https://transparency.fb.com/en-gb/policies/community-standards/hate-speech/

Mill J S, *On Liberty* (first published 1859, Enhanced Media 2016)

Morris S, 'Student jailed for racist Fabrice Muamba tweets' *The Guardian* (London, 27 March 2012) https://www.theguardian.com/uk/2012/mar/27/student-jailed-fabrice-muamba-tweets

Murray A, *Information Technology Law: The Law and Society* (5th edn, Oxford University Press 2023)

National Police Chiefs' Council, 'Hate Crime Investigation Following Euro 2020 Final Leads to 11 Arrests' (*NPCC*, 5 August 2021) https://news.npcc.police.uk/releases/hate-crime-investigation-following-euro-2020-final-leads-to-11-arrests

Newman S L, 'Finding the Harm in Hate Speech: An Argument against Censorship' (2017) 50(3) CJPS 679

Ormerod D & Laird K, *Smith, Hogan and Ormerod's Criminal Law* (16 edn, Oxford University Press 2021)

Saunders A, 'Hate is hate. Online abusers must be dealt with harshly' *The Guardian* (London, 21 August 2017) https://www.theguardian.com/commentisfree/2017/aug/20/hate-crimes-online-abusers-prosecutors-serious-crackdown-internet-face-to-face

Statista, 'Global number of hate speech-containing content removed by Facebook from 4th quarter 2017 to 1st quarter 2022' (*Statista*, 1 June 2022) https://www.statista.com/statistics/1013804/facebook-hate-speech-content-deletion-quarter/

Stephens H, 'Facebook: Who Bears Responsibility for Them?' (*The International Forum for Responsible Media Blog*, 25 November 2021) https://inforrm.org/2021/11/25/facebook-who-bears-responsibility-for-them-hugh-stephens/

The Crown Prosecution Service: 'Former Councillor Pleads Guilty to Posting Racist Comments about England Football Players' (*CPS.gov,* 4 March 2022) https://www.cps.gov.uk/west-midlands/news/former-councillor-pleads-guilty-posting-racist-comments-about-england

The Crown Prosecution Service, 'Racist and Religious Hate Crime - Prosecution Guidance' (*CPS.gov*, 3 March 2022) https://www.cps.gov.uk/legal-guidance/racist-and-religious-hate-crime-prosecution-guidance

The Crown Prosecution Service, 'Hate Crime' (*CPS.gov*, 2022) https://www.cps.gov.uk/crime-info/hate-crime

The Crown Prosecution Service: 'Cheshire Man Sentenced for Racist Abuse of England Players' (*CPS.gov*, 8 September 2021) https://www.cps.gov.uk/mersey-cheshire/news/cheshire-man-sentenced-racist-abuse-england-players

The Crown Prosecution Service, 'Hate Crime' (*CPS.gov*, 2018) https://www.cps.gov.uk/hate-crime

The Crown Prosecution Service: 'Racist and Religious Hate Crime - Prosecution Guidance' (*CPS.gov*, 2017) http://www.cps.gov.uk/legal/p_to_r/racist_and_religious_crime/#a07

The Crown Prosecution Service: 'Violent Extremism and Related Criminal Offences' (*CPS.gov*, 2017) https://www.cps.gov.uk/publications/prosecution/cases_of_inciting_racial_and_religious_hatred_and_hatred_based_upon_sexual_orientation.html

The Crown Prosecution Service: 'Guidelines on Prosecuting Cases Involving Communications Sent Via Social Media' (*CPS.gov*, 2016) https://www.cps.gov.uk/legal-guidance/guidelines-prosecuting-cases-involving-communications-sent-social-media

TikTok, 'Community Guidelines Enforcement Report' (*TikTok*, 2022) https://www.tiktok.com/transparency/en/community-guidelines-enforcement-2021-4/

TikTok, 'Community Guidelines: Hateful Behaviour' (*TikTok*, 2022) https://www.tiktok.com/community-guidelines?lang=en#38

Twitter, 'Hateful Conduct Policy' (*Twitter*, 2022) https://help.twitter.com/en/rules-and-policies/hateful-conduct-policy

Twitter, 'Rules Enforcement' (*Twitter*, 2022) https://transparency.twitter.com/en/reports/rules-enforcement.html#2020-jan-jun

Walters M A, 'Conceptualizing Hostility for Hate Crime Law: Minding the Minutiae When Interpreting Section 28(1)(a) of the Crime and Disorder Act 1998' (2014) 34 (1) Oxf J Leg Stud 47

Weale S, 'Children coming across far-right material when researching, teachers say' *The Guardian* (London, 18 April 2022) https://www.theguardian.com/politics/2022/apr/18/children-coming-across-far-right-material-when-researching-teachers-say

YouTube, 'Hate Speech Policy' (*YouTube*, 2022) https://support.google.com/youtube/answer/2801939?hl=en-GB

5 Harassment and Stalking Online

5.1 Introduction

As the world becomes more digital, behaviours which were once traditionally linked to the physical world have now emerged online. Nothing has demonstrated this more than harassment and stalking, where social media has become a relatively new way to harass or stalk another.[1] Both these behaviours have become interchangeable in recent years, and as will be discussed in later parts of this chapter, confusion between the two has caused problems across the criminal justice system, particularly where these behaviours are conducted online. Indeed, for the Law Commission, harassment and stalking are difficult to define and in turn, create significant issues in determining when someone's behaviour constitutes harassment or stalking.[2] Issues which become even more glaringly apparent in the context of social media.

Though not universally defined, harassment has come to be considered as 'repeated attempts to impose unwanted communications and contact upon a victim in a manner that could be expected to cause distress or fear in any reasonable person'.[3] Whereas stalking can be considered as 'a constellation of behaviours in which one individual inflicts on another repeated unwanted intrusions and communications'.[4] Behaviours can include, to name but a few, 'following a person [...] monitoring the use by a person of the internet, email or any other form of electronic communication [... and] watching or spying on a person'.[5] With the aid of social media, both these behaviours can now occur around the clock. Stalking and harassment are therefore not only mirrored online but arguably are more intense.[6] In the online context, these behaviours have been coined as cyberharassment and

1 Azy Barak, 'Sexual Harassment on the Internet' (2005) 23(1) Soc. Sci. Comput. Rev 77.

2 Law Commission, *Abusive and Offensive Online Communications: A Scoping Report* (Law Com No 381, 2018) [8.9].

3 The Crown Prosecution Service, 'Stalking and harassment' (*CPS.gov*, 23 May 2018) <https://www.cps.gov.uk/legal-guidance/stalking-and-harassment> accessed 2 December 2022.

4 Michele Pathé and Paul Mullen, 'The Impact of Stalkers on their Victims' (1997) 170(1) BJPsych 12.

5 Protection from Harassment Act 1997 s.2A(3).

6 Subhajit Basu and Richard Jones, 'Regulating Cyberstalking' 2007 7(2) JILT <https://warwick.ac.uk/fac/soc/law/elj/jilt/2007_2/basu_jones/> accessed 2 December 2022.

DOI: 10.4324/9781003303688-7

cyberstalking—though like that of harassment and stalking, both terms are inherently difficult to define.

How the law has attempted to regulate cyberharassment and cyberstalking will be explored in the following chapter. First, the rise in these types of behaviours will be outlined. Next, the Protection from Harassment Act 1997 (PfHA) which criminalises the conduct of harassment and stalking, both online and offline, will be discussed. Taking each of these behaviours in turn, with specific reference to s.2, 2A, 4, and 4A of the Act, the law will be outlined with case examples, before then turning to highlight the issues of applying the PfHA to the online world and the principle of legality. Like that of previous chapter, it will be argued that imposing such provisions, without adhering to the principle of legality, poses significant issues with freedom of expression.

5.2 The Rise in Cyberharassment and Cyberstalking

Though difficult to define, research suggests that the prevalence of cyberharassment and cyberstalking is high, with stalking by an unknown person being more common online than in the offline world.[7] Both cyberharassment and cyberstalking encompass a range of behaviours online, including, though not limited to: repeated unwanted communications, false profiles, the publication of false information, unauthorised access to personal information, and covert surveillance.[8] The difficulty in defining the two behaviours means that much of the research into cyberharassment and cyberstalking is grouped together. Indeed, as discussed in later parts of this chapter, law enforcement often use these terms interchangeably. Even before the likes of Facebook and X, which are still to-date, two of the biggest social media platforms in the world, both cyberstalking and cyberharassment were causing growing concerns at the turn of the millennium.[9]

In 2003, Bocij conducted a survey examining the extent of cyberstalking online.[10] Bocij, using an email snowballing sample,[11] surveyed 169 participants aged between 16 and 84 years old.[12] Bocji asked several questions related to certain types of conduct associated with stalking online, to determine if a person had been cyberstalked. For example, the receiving of threatening or abusive emails and the spread of false information. Using this approach Bocji was able to conclude that

7 Law Com No 381 (n 2).
8 Ibid., [8.13].
9 For a brief discussion on the history and development of the internet see Chapter 1.
10 Paul Bocij, 'Victims of Cyberstalking: An Exploratory Study of Harassment Perpetrated via the Internet' (2003) 8(10) First Monday <http://firstmonday.org/ojs/index.php/fm/article/view/1086> accessed 30 October 2016.
11 In a snowballing sample the researcher recruits a few participants, who in turn recommends others to take part in the study. See, Earl Babbie, *The Basics of Social Research* (Cengage Learning 2007) 205.
12 Bocij (n 10).

21.9% of his participants had been subjected to what he deemed cyberstalking.[13] Interestingly, although social media existed at the time in which Bocji conducted his research, its use and prevalence were nothing compared to today.

The expansion of both social media platforms and users, has revolutionised how cyberstalking and cyberharassment can be conducted. Research undertaken on behalf of Bedfordshire University by Brown, Maple and Short in 2011, found that of the 353 participants who took part in their survey, 92% had been subjected to some form of cyberstalking or cyberharassment, with social media playing a pivotal role.[14] As social media usage has increased, evidence suggests so has cyberharassment and cyberstalking, with research conducted by the Pew Research Centre in 2022 concluding that 41% of Americans had experienced cyberharassment or cyberstalking. Although the Pew Research Centre accepted that this figure was the same as 2017, their study suggests that such behaviours have intensified.[15] Of those who took part in the study conducted by the Pew Research Centre, 75% had experienced such behaviours on some form of social media.[16]

Even in the United Kingdom, there is growing evidence that cyberharassment and cyberstalking are prevalent on social media platforms, with the Criminal Justice Inspectorates and HM Crown Prosecution Service Inspectorate (HMCPSI) finding that of the 112 cases they reviewed as part of a larger investigation into the criminal justice systems response to harassment and stalking, 82 cases (73%) had a social media element.[17] The HMCPSI concluded that harassment and stalking are increasingly likely to involve some form of digital media.[18] Indeed, during the COVID-19 pandemic, social media was one of the main methods utilised to stalk and harass others.[19] As shown in Figure 5.1, cyberharassment is considered the second most common form of online abuse experienced by complainants in the United Kingdom, following research conducted by the Victims Commissioner.

Despite a growing perception that cyberharassment and cyberstalking are on the increase, which in turn has placed growing pressure on the criminal justice system

13 Ibid.,

14 Antony Brown, Carsten Maple and Emma Short, 'Cyberstalking in the United Kingdom: An Analysis of the ECHO Pilot Survey' (*University of Bedfordshire National Centre for Cyberstalking Research*, 2011) 9 <https://www.beds.ac.uk/__data/assets/pdf_file/0003/83109/ECHO_Pilot_Final.pdf> accessed 25 October 2016.

15 Emily A. Vogels, 'The State of Online Harassment' (*Pew Research Centre*, 13 January 2021) <https://www.pewresearch.org/internet/2021/01/13/the-state-of-online-harassment/> accessed 2 December 2022. To conduct this research, the centre surveyed 10,093 adults in the United States, over a period of three days. Participants were representative of the demographic of the United States.

16 Ibid.,

17 Criminal Justice Inspectorates and HM Crown Prosecution Service Inspectorate, 'Living in fear – the police and CPS response to harassment and stalking' (*justiceinspectorates.gov*, July 2017) 27 <http://www.justiceinspectorates.gov.uk/hmicfrs/wp-content/uploads/living-in-fear-the-police-and-cps-response-to-harassment-and-stalking.pdf> accessed 29 November 2017.

18 Ibid., 28

19 Suzy Lamplugh Trust, 'Unmasking Stalking: A Changing Landscape' (*Suzy Lamplugh Trust*, April 2021) <https://www.suzylamplugh.org/Handlers/Download.ashx?IDMF=fcfb781a-f614-48c8-adcf-4cfa830c16a7> accessed 1 December 2023.

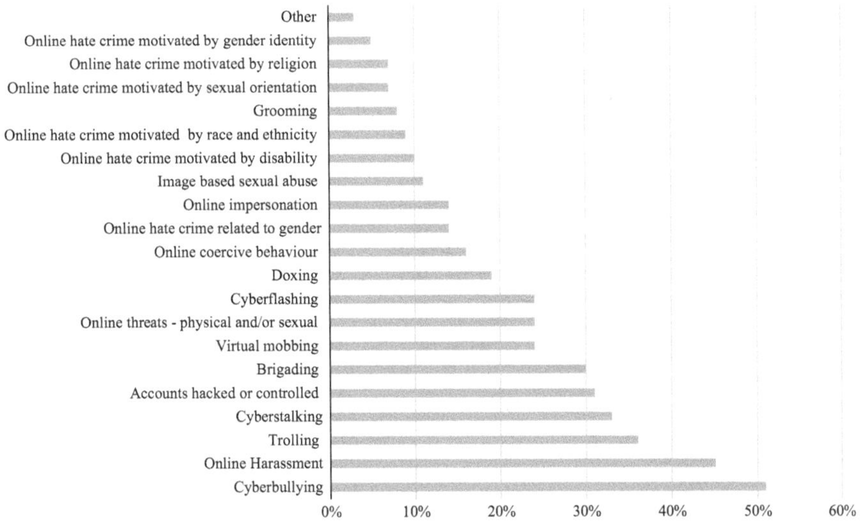

Figure 5.1 Types of online abuse experienced by victims in the United Kingdom as of February 2022.

Note: Madeline Storry & Sarah Poppleton, 'The Impact of Online Abuse: Hearing the Victims Voice' (*Victims Commissioner*, 1 June 2022) <https://s3-eu-west-2.amazonaws.com/jotwpublic-prod-storage-1cxo1dnrmkg14/uploads/sites/6/2022/05/Hearing-the-Victims-Voice.pdf> accessed 2 December 2022. This data was collected from 534 surveys.

and the government to do more to tackle this 'harm' associated with the online world, studies report varying levels. As noted by Davidson *et al*, these levels can fluctuate from 9% to 46.7%, dependent entirely on the definition given to these behaviours.[20] For instance, Brown *et al*, as discussed above, defined cyberharassment as 'the intimidation, repeated or otherwise, of one individual by another or a group perpetrated through or utilising electronic means'.[21] Whilst defining cyberstalking as 'a course of action that involves more than one incident perpetrated through or utilising electronic means, that causes distress, fear, or alarm'.[22] Whereas the Pew Research Centre did not specifically define cyberharassment, and instead encompassed a list of six behaviours they associated with cyberharassment: (1) offensive name-calling; (2) purposeful embarrassment; (3) stalking; (4) physical threats; (5) harassment over a sustained period of time and; (6) sexual harassment.

20 Julia Davidson, Sonia Livingtone, Sam Jenkins, Anna Gekoki, Clare Choak, Tarela Ike and Kirsty Phillips, 'Adult Online Hate, Harassment & Abuse: A Rapid Evidence Assessment' (*Gov.uk*, June 2019) 4 <https://assets.publishing.service.gov.uk/government/uploads/system/uploads/attachment_data/file/811450/Adult_Online_Harms_Report_2019.pdf> accessed 2 December 2022.

21 Brown, Maple and Short (n 14) 4.

22 Ibid.,

Each of these definitions used in the research examining the extent and effect of cyberharassment and cyberstalking has issues. As discussed in Chapter 2, being offensive falls within a person's right to freedom of expression and therefore to define cyberharassment as 'offensive name calling' does not truly reflect the reality of cyberharassment. In turn, a perception is created that cyberharassment and/or cyberstalking are significant issues, and places increasing pressure on the criminal justice system to do more to tackle this 'growing' problem within society, which might not necessarily be true, posing significant issues with freedom of expression. Yet, the issue with how we define such behaviours does not relate solely to academic research on cyberharassment and/or cyberstalking. The PfHA, which criminalises harassment and stalking in England and Wales, whether online or offline, also fails to grasp the complexity of these behaviours.

5.3 The Protection from Harassment Act 1997

Before the creation and implementation of the PfHA the behaviours associated with stalking and harassment were considered by some as already prohibited under the criminal law of England and Wales. For instance, during a parliamentary debate, MPs argued that the conduct of harassment and stalking was criminalised contrary to, the Malicious Communications Act 1988, the Telecommunications Act 1984 (both discussed further in the following chapter), and the Public Order Act 1986 (as discussed in the previous chapter), to name but a few.[23] In addition, the House of Lords expanded the scope of the law to prohibit harassment and/or stalking under the Offences against the Person Act 1861 (OAPA),[24] in *R v Ireland*.[25]

Despite some consensus prior to the implementation of the PfHA that the conduct of stalking and harassment was already contrary to law,[26] during a Home Office consultation committee, it was found that the criminal law was, in fact, inadequate in protecting individuals from this form of abuse:

> Though the offences under the Public Order Act may provide a sanction against stalkers in some instances, offences under the provisions of sections 4 and 4A would be committed only if the stalker intended his behaviour to cause the victim to believe that immediate violence would be used (section 4) or if harassment, alarm or distress is caused (section 4A). There are problems also in applying other aspects of the criminal law against stalkers. The Malicious Communications Act 1988 requires that the article sent must be indecent or grossly offensive. It must also be proved that the sender's purpose was to cause distress or anxiety. In the situations where stalkers continually

23 Home Office, *Stalking A Consultation Paper* (11 July 1996).

24 Mary Baber and Helena Jeffs, 'Stalking, Harassment and Intimidation and the Protection from Harassment Bill' (Research Paper 96/115, 13 December 1996) 5–6.

25 *R v Ireland* [1998] A.C. 147.

26 Home Office (n 23).

send greetings cards, flowers or other unsolicited gifts, such intent cannot be proven.[27]

Further, the *ad-hoc* nature of attempting to apply criminal provisions to stalking and harassment resulted in failures by the criminal justice system to adequately protect individuals from this type of conduct—issues the PfHA sought to amend.

5.3.1 Creation and Purpose

The anti-stalking Bill, as it was first known, was introduced into the House of Commons on 5 December 1996, a year before the first form of social media, as we know it today, was made available to the public—sixdegrees.com.[28] The Bill itself, which was to become the PfHA, was rushed through Parliament by the then conservative government, in a bid to prove to voters that the government was being tough on crime.[29] The PfHA received Royal Assent in March 1997 with an overall objective of protecting 'victims of harassment [... including from ...] so-called stalking behaviour, racial harassment, or anti-social behaviour by neighbours'.[30]

The Act itself consists of several provisions and imposes both criminal and civil law repercussions, with the remainder of this chapter focussing on the criminal provisions contained within the Act.[31] It is also important to note that the Act, in its original format, was not without fault. In particular, the Act treated stalking and harassment as the same offence—despite concerns being raised at the time that both conducts were inherently different. This was amended in 2012 under the Protection of Freedoms Act s.111(1) (PFA). However, as will be explored below, the difference between the two, harassment and stalking, by law, is not easily identifiable as they are heavily interlinked. In fact, as the law currently stands, to prosecute an individual for stalking, harassment first needs to be proven. Consequently, this has led to confusion across the criminal justice system—and even more confusion has been created, where such behaviours are conducted on the likes of social media.[32]

5.3.2 Harassment

S.1(1) of the PfHA prohibits 'a course of conduct which amounts to harassment of another, and which he knows or ought to know amounts to harassment of the other'. The provision covers both harassment from a single person and harassment

27 Ibid., [3.4–3.6].

28 Laura Scaife, *Handbook of Social Media and the Law* (Routledge 2015) 4.

29 Judith Gowland, 'Protection from Harassment Act 1997: The "new" Stalking Offences' (2013) 77(5) J. Crim. Law 387.

30 HL Deb 24 January 1997, vol 1, col 917.

31 Where such as offence demonstrates hostility because of a persons, whether real or presumed characteristic, a person may be charged under s.32 of the Crime and Disorder Act 1988 (as discussed in Chapter 4).

32 Criminal Justice Inspectorates and HM Crown Prosecution Service Inspectorate (n 17).

by a group,[33] which has come to be known in the online world as 'dog-piling'. S.2(1) of the Act makes it a specific criminal offence to subject another to this behaviour, with a maximum custodial sentence of six months. However, if the behaviour can be regarded as a hate crime, as discussed in the previous chapter, a higher sentence can be imposed.[34]

The following discussion, taking a similar format to Chapter 4, will outline the *actus reus* and *mens rea* of the offence of harassment, before turning to examine the defences contained within the provision. Later sections of this chapter will expose significant issues with the language contained in the PfHA.

5.3.2.1 Actus Reus

The *actus reus* contained in s.1(1) of the PfHA consists of two elements: a course of conduct and harassment. A course of conduct is defined in s.7(3) of the Act: 'a "course of conduct" must involve—(a) in the case of conduct in relation to a single person [...] conduct on at least two occasions in relation to that person'.[35] The Crown Prosecution Service (CPS) have further confirmed that a course of conduct can include speech.[36] In essence, 'each separate complained about behaviour must directly link back to the accused'.[37]

A minimum requirement has therefore been set by Parliament as to when certain behaviours may invoke the PfHA, though in reality, for a course of conduct to be found in relation to a single individual, it is likely that more than two incidents would need to be proven:

> I fully accept that the incidents which need to be proved in relation to harassment need not exceed two incidents, but, as it seems to me, the fewer the occasions and the wider they are spread the less likely it would be that a finding of harassment can reasonably be made.[38]

In most circumstances, for the courts to be satisfied that a course of conduct is present, contact must occur on more than two occasions, as acknowledged further by Agate and Ledward.[39]

33 Which was substituted into the Act by the Serious Crime and Police Act 2005.
34 This would be under either s.66 of the Sentencing Act 2020 or s.32 of the Crime and Disorder Act 1988. See Chapter 4.
35 Protection from Harassment Act 1999 s.7(3). This was substituted into the Act by the Serious Organised Crime and Police Act 2005.
36 Stalking and Harassment (n 3).
37 Laura Bliss, 'The Protection from Harassment Act 1997: Failures by the Criminal Justice System in a Social Media Age' 83(3) J. Crim. Law 217.
38 *Lau v DPP* [2000] 1 F.L.R 799 (DC) per Mr Justice Schiemann [15].
39 Jennifer Agate and Jocelyn Ledward, 'Social Media: How the Net is Closing in on Cyber Bullies' (2013) 24(8) ELR 263.

In *Pratt v DPP*,[40] it was found that two incidents which occurred almost three months apart were 'close to the line' but did not cross the threshold to establish a course of conduct. Whereas in *Kelly v DPP*,[41] three phone calls in the space of five minutes, constituted a course of conduct. This leaves many questions unanswered when it comes to prosecuting harassment conducted online. For example, could it be said that contact once a month over the course of a year can amount to harassment, or would it need to be more contact over a shorter period of time? What if the person being targeted online did not know about the messages sent until months after the event? Indeed, in *Wass v DPP* the Divisional Court considered two incidents that occurred within minutes of each other as a course of conduct.[42]

The second part of the *actus reus* is harassment. Under the PfHA harassment is defined as 'alarming' a person or causing them 'distress'.[43] No further information is given within the Act itself as to the meaning of these terms and will simply be based on the evidence, in which the behaviour needs to be oppressive or unreasonable:

> Harassment is a persistent and deliberate course of unreasonable and oppressive conduct, targeted at another person, which is calculated to and does cause that person alarm, fear or distress.[44]

Simply put, there must be evidence that the course of conduct was calculated to cause alarm or distress, which was oppressive or unreasonable—here it must be found that the reasonable person in the same position as the complainant would find that harassment had taken place. Further, there must be proof of distress. If from the evidence provided it can be established that the complained about behaviour amounts to alarming or distressing a person, the condition of harassment will be satisfied. Though

> [w]here the quality of the conduct said to constitute harassment is being examined, courts will have in mind that irritations, annoyances, even a measure of upset, arise at times in everybody's day-to-day dealings with other people. Courts are well able to recognise the boundary between conduct which is unattractive, even unreasonable, and conduct which is oppressive and unacceptable. To cross the boundary from the regrettable to the unacceptable the gravity of the misconduct must be of an order which would sustain criminal liability.[45]

40 *Pratt v DPP* [2001] EWHC 483.
41 *Kelly v DPP* [2002] EWHC 1428.
42 *Wass v DPP* (QBD, 11 May 2000).
43 Protection from Harassment Act 1997 s.7(2).
44 *Hayes v Willoughby* [2013] UKSC 17.
45 *Majrowski v Guy's and St Thomas's NHS Trust* [2006] UKHL 34, [2007] 1 A.C. 224 per Lord Nicolls of Birkenhead [30].

This has created issues when it comes to cyberharassment. In many cases of online harassment, the conduct is 'disturbing, unpleasant, and may transgress the norms of socially acceptable' behaviour but *should not* cross the harassment threshold to warrant criminal law intervention. But as discussed in Section 5.2 above, it is often found that those who experience 'disturbing' or 'unpleasant' conduct online, consider this to be harassment.[46] And it is studies like these which are often used as evidence to support stronger criminal legal provisions to tackle the 'harms' associated with the online world.

5.3.2.2 Mens Rea

The *mens rea* of harassment is based on the construction of knowledge. Under the PfHA in order to bring a successful prosecution for harassment it must be proven that the defendant, 'knows or ought to know that the[ir] behaviour would amount to [the] harassment of another'.[47] The concept of 'knowledge' is based on the reasonable person test.[48] Essentially, it must be established that the average sober person in the same position as the defendant would come to the knowledge *or should have* come to the knowledge, that their conduct would amount to the harassment of another individual.

This is a purely objective test, where the feelings or perceptions of the accused are not taken into consideration. As noted by Herring:

> [i]t means that it is no defence for a defendant to claim, "I thought I was expressing my love for the victim; I did not realize [*sic*] she would find this distressing" if it would have been obvious to a reasonable person that the conduct was harassing.[49]

The Court of Appeal confirmed in *R v Colohan*,[50] that the characteristics of the accused, in this case, their mental health, could not be taken into consideration.

If both the *actus reus* and *mens rea* of the offence can be found, then the defendant is liable for a breach of s.1(1) of the PfHA, which as noted above is made a criminal offence under s.2(1) of the Act: 'a person who pursues a course of conduct in breach of [s.1(1) or (1A)] is guilty of an offence'—provided the defences discussed below do not apply. However, if the behaviour can be considered as putting someone in fear of violence, this may constitute a more serious offence under s.4(1) of the PfHA, as discussed further in Section 5.3.4.1.

46 For example, in the research conducted by the Pew Research Centre as discussed in Section 5.2, the definition given to harassment included offensive name calling.

47 Protection from Harassment Act 1997 s.1(1)(b).

48 For a discussion on the reasonable person see, Reid Griffith Fontaine, *The Mind of the Criminal: The Role of Developmental Social Cognition in Criminal Defense Law* (Cambridge University Press 2012) 13.

49 Jonathan Herring, *Criminal Law: Text, Cases and Materials* (10th edn, Oxford University Press 2022) 348.

50 *R v Colohan* [2001] EWCA Crim 1251.

5.3.2.3 Defences

It is important to note that s.1(3) of the PfHA specifically outlines three situations in which it will be found that there was no course of conduct present, and therefore no offence:

1. Where the course of conduct is done for the purpose of preventing or detecting a crime. For example, police officers investigating a criminal act would be able to rely upon this defence. Although it could, in some circumstances, be raised by journalists or neighbourhood watch bodies as a defence, although it would need to be proven that a person acted rationally in the circumstances;[51] or
2. Where the course of conduct is 'pursued under any enactment or rule of law or to comply with any condition or requirement imposed by any person under any enactment'.[52] For example, a company acting within their legal boundaries; or
3. Where the course of conduct can be considered reasonable—with reasonability being assessed by the courts.

It is under the reasonability defence element, where consideration should be given by actors within the criminal justice system to Articles 10 and 8 of the European Convention on Human Rights (ECHR), as confirmed by the European Court of Human Rights (ECtHR) in *Pal v The United Kingdom*[53]—a case discussed at length in Section 5.4. Indeed, journalistic speech will fall within the defence of reasonableness, provided such behaviour is not so unreasonable.[54] If any of the defences above are found, it will be concluded that a course of conduct cannot be established, and consequently, no offence has taken place under this Act.

5.3.3 Stalking

As previously noted, before 2012 stalking and harassment were treated as the same offence, and both governed under s.1(1) of the PfHA and criminalised under s.2(1). Although there is evidence that stalking offences were successfully prosecuted under s.2(1), many of those subjected to such behaviour felt that they were not taken seriously by those working in the criminal justice system.[55] In 2012, the conduct of stalking was implemented into the Act under s.111(1) of the PFA, following concerns that the PfHA did not cover stalking sufficiently.[56]

The provision came into force in November 2012, with a maximum custodial sentence of six months and like that of harassment, a higher sentence can

51 *Hayes* (n 44).
52 Protection from Harassment Act 1997 s.1(3)b.
53 *Pal v The United Kingdom* (44261/19) [2021] 11 WLUK 603 Times, 9 February 2022 (ECHR).
54 *Carina Trimingham v Associated Newspapers Limited* [2012] EWHC 1296 (QB).
55 Stalking and Harassment (n 3).
56 Neil MacEwan, 'The New Stalking Offences in English law: Will They Provide Effective Protection from Cyberstalking?' (2012) 10 Crim.L.R. 767.

be imposed if it falls within the definition of a hate crime.[57] Although there were calls for cyberstalking to be included as a separate offence under the PfHA, this was rejected at the time of amendment, with MacEwan noting that it was 'naïve, to view it simply as "old behaviour using new technologies"'.[58] Below, the *actus reus* and *mens rea* of the offence is set out, before turning to outline the defences available under this provision. As will be shown in later sections, the interchangeable nature of harassment and stalking poses significant issues with the principle of legality.

5.3.3.1 Actus Reus

S.2A(1) of the PfHA takes a very similar format to that of s.1(1). However, the *actus reus* consists of three elements in total:

1. A course of conduct—what is considered a course of conduct under this section of the Act has the same meaning as s.1(1) discussed above—where one single person is the perpetrator, there must be at least two incidents which are close in time;
2. The course of conduct must amount to harassment—again, the same meaning is given to harassment as with s.1(1), as discussed in Section 5.3.2.1; and
3. The behaviour must amount to stalking.

There is no definition of stalking contained in the Act, instead, a non-exhaustive list of behaviours is included:

> (a) following a person; (b) contacting, or attempting to contact, a person by any means; (c) publishing any statement or other material—(i) relating or purporting to relate to a person, or (ii) purporting to originate from a person; (d) monitoring the use by a person of the internet, email, or any other form of electronic communication; (e) loitering in any place (whether public or private); (f) interfering with any property in the possession of a person; [and] (g) watching or spying on a person.[59]

Consequently, 'as "stalking" lacks an overarching definition, it is not always clear when harassing conduct has crossed the threshold into stalking'.[60] This confusion is an inevitable consequence as upon the drafting s.2A of the PfHA, as noted above, a person cannot be stalked if harassment is not present. Indeed, it becomes further complicated in a social media context, where the 'monitoring of social media is

57 Where such as offence demonstrates hostility because of a persons, whether real or presumed characteristic, a person may be charged under s.32 of the Crime and Disorder Act 1988 (as discussed in Chapter 4).

58 MacEwan (n 56) 768.

59 Protection from Harassment Act 1997 s.2A(3).

60 Law Com No 381 (n 2) [8.171].

typically expected and consented to within social media environments'.[61] The issue with what constitutes stalking by law, is discussed further in Section 5.5.1.3.

5.3.3.2 Mens Rea

Like that of harassment, as discussed above, the *mens rea* for the offence of stalking is based on the construction of knowledge. Here, it must be found that the perpetrator 'knows or ought to know that the[ir] behaviour would amount to [the] harassment of another'.[62] Like that of harassment, if the behaviour can be considered as putting someone in serious fear of violence or serious alarm or distress, this may constitute a more serious offence under s.4A of the PfHA, as discussed further in Section 5.3.4.2.

5.3.3.3 Defences

As with harassment, it is a defence to stalk another where the course of conduct is done for the purpose of preventing or detecting a crime; or is 'pursued under any enactment or rule of law or to comply with any condition or requirement imposed by any person under any enactment';[63] or where it can be considered reasonable in the circumstances to pursue a course of conduct.

5.3.4 Putting Someone in Fear of Violence and/or Serious Alarm or Distress

A more serious offence of both harassment and stalking are contained in the PfHA: putting someone in fear of serious violence as outlined in s.4 (harassment) or putting someone in fear of serious violence or alarm or distress as prohibited under s.4A (stalking) of the Act. Both provisions carry a custodial sentence of up to ten years.[64]

5.3.4.1 Section 4: Putting Someone in Fear of Violence (Harassment)

For an offence to be established under s.4(1) of the PfHA—the harassment of another which puts someone in fear of violence—first the *actus reus* for harassment, as discussed above in Section 5.3.2.1, must be present (a course of conduct that amounts to harassment). In addition to the above, it must also be found that the defendant:

61 Ibid., [8.174].

62 Protection from Harassment Act 1997 s.2A 2(c).

63 Ibid., s.1(3)b.

64 Until 2017, the maximum custodial sentence under s.4 or s.4A of the Protection from Harassment Act 1997 was five years. This was changed under s.175(1)a & s.175(1)b of the Policing and Crime Act 2017.

> ought to know that his course of conduct will cause another to fear that violence will be used against them if a reasonable person in possession of the same information would think that the course of conduct would cause the other so to fear on that occasion.

Put simply, 'it has to be shown that the accused has undertaken a course of conduct which on at least two occasions led the victim to fear that violence would be used against them'.[65] With the court confirming in *Henley* that the conduct needs to be more than merely frightening, it must put the complainant in fear of violence.[66] Further, it must be directed at an individual. So, for example, in a social media context, an offence would not be committed under this provision, if the message was not directly aimed at another.

To determine if a course of conduct, which amounts to harassment can be considered as putting a person in fear of violence, the reasonable person test is used. Here, it must be found that the reasonable sober person in the same position as the complainant would have feared that violence might be used against them. Like that of the reasonable person test contained in s.1(1) and 2A(1), the perpetrator's perceptions are not taken into consideration, nor their personal characteristics, such as their mental health.

5.3.4.2 Section 4A: Putting Someone in Fear or Serious Alarm or Distress (Stalking)

Like that of s.4(1), for an offence to be established under s.4A(1), it must first be found that an offence contrary to s.2A(1) of the PfHA (a course of conduct, that amounts to harassment and can be considered as stalking) is present. However, whereas s.4(1) is only applicable to situations which cause a person fear of violence, s.4A(1) also covers the situation whereby a person fears 'serious alarm or distress'. Put simply, an offence under s.4A(1) can be conducted in one of two ways (provided an offence under s.2A(1) is also present):

1. It causes another to fear, on at least two occasions, that violence will be used against them; or
2. It causes another serious alarm or distress which has a substantial adverse effect on his or her usual day-to-day activities.

The first way of committing the offence, similar to s.4(1), is based on the reasonable person in the same position as the complainant, though personal characteristics are not taken into account. The second is more complex, and 'recognises the overall

65 Herring (n 49) 350.
66 *Henley* [2000] Crim LR 582.

emotional and psychological harm that stalking may cause to victims, even where an explicit fear of violence is not created by each incident of stalking behaviour'.[67]

To cause serious alarm or distress it must be found that the conduct had a 'substantial adverse effect on his or her usual day-to-day activities' with no further information contained within the Act. Indeed, neither 'alarm' nor 'distress' are appropriately defined. However, guidelines issued by the Home Office, have attempted to bring some clarity to the phrase by suggesting that such behaviour may include:

> (a) the victim changing their routes to work, work patterns, or employment; (b) the victim arranging for friends or family to pick up children from school (to avoid contact with the stalker); (c) the victim putting in place additional security measures in their home; (d) the victim moving home; (e) physical or mental ill-health; (f) the deterioration in the victim's performance at work due to stress; (g) the victim stopping/or changing the way they socialise.[68]

It is important to note this list is non-exhaustive and could perhaps include removing oneself from a social media platform. Though it must be proven that the accused knows or ought to know that the course of conduct will cause another alarm or distress and again, is based on the reasonable person, with the court clarifying in *Pendlebury v DPP:*

> it is not enough to establish that conditions existed which might reasonably have engendered a fear of violence: that will be so in many stalking cases. The matter is one of fact. The Bench must be sure that such a fear was actually engendered.[69]

5.3.4.3 Section 4 and 4A: Defences

Both provisions also carry a defence element:

> It is a defence for a person charged with an offence under this section (4 and 4A) to show that—(a) his course of conduct was pursued for the purpose of preventing or detecting crime, (b) his course of conduct was pursued under any enactment or rule of law or to comply with any condition or requirement imposed by any person under any enactment, or (c) the pursuit of his course of conduct was reasonable for the protection of himself or another or for the protection of his or another's property.[70]

67 Stalking and Harassment (n 3).

68 Home Office, 'Circular: a change to the Protection from Harassment Act 1997' (*Gov.uk*, 16 October 2012) <https://www.gov.uk/government/publications/a-change-to-the-protection-from-harassment-act-1997-introduction-of-two-new-specific-offences-of-stalking> accessed 4 December 2023.

69 *Pendlebury v DPP* [2018] EWHC 3567 (Admin) per Lord Justice Irwin [29].

70 Protection from Harassment Act 1997 s.4(3) & 4A(4).

If any of the defences above can be established, no offence has taken place.

5.3.5 Summary

Though s.2 and 2A of the PfHA are intended to cover different types of behaviours, it is clear that both overlap, and the distinction between harassment and stalking is not easily identifiable. Both provisions, however, have been used in a social media context, though a recent case before the ECtHR has confirmed that even when it comes to harassment or stalking, freedom of expression may well constitute reasonability to pursue a course of conduct against another, particularly when it comes to comments made on social media against individuals in the public domain.[71] Discussed further in the following section.

5.4 The Protection from Harassment Act 1997 and Social Media

There is growing evidence of all the provisions discussed above being used in the context of social media. For instance, a successful prosecution for an offence contrary to s.2 of the PfHA, involving social media and new technologies, can be seen in *R v Danevska*,[72] where it was held that exceptional as a restraining order was (including prohibiting the defendant from publishing any information about the complainant and other interested persons) it was justified.

What is believed to be the first successful prosecution of an individual under s.2 of PfHA for cyberharassment occurred in 2009, where Kelly Houghton was jailed for three months for sending death threats *via* Facebook.[73] Following the implementation of stalking into the PfHA in 2012, the first successful conviction for cyberstalking under s.4A occurred in 2014, following an individual using computer software to spy on women in their own homes.[74]

Since then, there have been numerous successful convictions under s.4A of the PfHA, including *R v Gomes*,[75] where a conviction contrary to s.4A was upheld by the court following the defendant bombarding the complainant with emails, telephone calls, and other online communications. Similarly, in *R v Patterson*[76] the court upheld the conviction of the defendant under s.4A of the PfHA, where the defendant made unwanted contact on social media with the complainant over several years, expressing his love for her:

71 *Pal* (n 53).

72 *R v Danevska* [2017] EWCA Crim 1084.

73 Helen Carter, 'Teenage girl is first to be jailed for bullying on Facebook' *The Guardian* (London, 21 August 2009) <https://www.theguardian.com/uk/2009/aug/21/facebook-bullying-sentence-teenage-girl> accessed 7 February 2023.

74 *R v Andrew Meldrum*, Woolwich Crown Court, 30 May 2014 (unreported). See also, Nicola Fifield, 'Cyber stalker bugged women's computers to spy on them in their bedrooms' *The Telegraph* (London, 30 May 2014) <http://www.telegraph.co.uk/news/uknews/crime/10866262/Cyber-stalker-bugged-womens-computers-to-spy-on-them-in-their-bedrooms.html> accessed 25 October 2016.

75 *R v Gomes* [2022] EWCA Crim 1395.

76 *R v Patterson* [2021] EWCA Crim 1475.

> She [the complainant] has frequently had panic attacks and had to undergo intensive group therapy. The repeated contacts with the victim and family members—even those living in other countries—has caused embarrassment and shame. The appellant's conduct has caused Miss Neilson [the complainant] to become isolated from friends and has harmed romantic relationships. Her work as an artist requires a public presence and reliance on social media as well as galleries. The appellant's behaviour has interfered with her ability to use Facebook and Instagram to promote her work online—an essential tool.[77]

Arguably, these cases warranted criminalisation under the PfHA. However, issues arise in the context of messages sent on social media sites, where a course of conduct maybe apparent, but that course of conduct is reasonable. For example, sending multiple messages which might not be offensive or even abusive but might cause another distress—an issue the ECtHR had to consider in *Pal v The United Kingdom*.[78]

Pal concerned a dispute between a journalist and a barrister (who was also considered a part-time journalist), with several comments being exchanged between the two on Twitter around transgender rights. Several of these disputes had been republished in the national press. However, at no point could it be said that the comments made between the two were either abusive or relentless. Despite this, the initial complainant, known only as AB, reported the tweets and an article written by Pal on the dispute to the Metropolitan Police (MET). Pal was later arrested and charged under s.2 of the PfHA. The charge was however eventually dropped by the CPS after they concluded, upon review of the case, that the article published was 'an informative piece' and the tweets complained about, could not amount to harassment. Despite no further action being taken by the CPS, Pal and their legal team issued legal proceedings before the Courts against the MET for various procedural breaches, including a breach of Article 10 of the ECHR. The grounds for review were rejected by both the County Court and the High Court, with an appeal to the Court of Appeal and the Supreme Court being disallowed. Pal and their legal team were later granted permission to have their case heard before ECtHR.

The ECtHR concluded that there had been a breach of Article 10, in which the domestic courts of the United Kingdom, when coming to their decision on the previous appeals, had not taken Article 10 into account: '[n]o consideration appears to have been given to the subject matter of the applicant's article or tweets, and whether they could be said to have contributed to a debate of general interest'.[79] The ECtHR distinguished between the public and private nature of the complained about conduct. When a person can be considered as a public person, such as a

77 Ibid., per Justice Holgate [6].
78 *Pal* (n 53).
79 Ibid., [59].

journalist, 'they could be subject to wider limits of acceptable criticism than the ordinary person'.[80]

In light of the finding of the ECtHR in *Pal* and on further reflection of *Magyar Helsinki Bizottsag v Hungary*,[81] it could be argued that in certain circumstances, posts on social media platforms, or tweets/Xs, could be considered as journalistic material. However, it is important to note that in disputes of a private nature involving words or actions 'the courts are unlikely to side with free speech once they are satisfied that there is a strong *prima facie* claim for harassment'.[82] But

> [w]hat the judgment in *Pal* has done, however, is to concentrate the minds of the police and the courts, at first instance and on appeal and review, on the importance of human rights in the interpretation and application of domestic law, both generally and in the area of free speech, in our case specifically the law of harassment.[83]

In essence, the PfHA should only be used, particularly in cases related to public communications, where the rights of the complainant have been disproportionally affected and consideration must always be given to tolerance and open debate. Failure to do so may render a decision incompatible with human rights.

5.5 The Protection from Harassment Act, Social Media, and Legality

Like that of hate speech, the effects of cyberharassment and/or cyberstalking cannot be downplayed, with its effects well documented within the literature.[84] However, the PfHA was never created with the digital world in mind, and many of the justifications for the use of vague terminology, surrounds the stalking or harassment of another in the physical world. Although it is accepted that some form of vagueness is inherent within the criminal law when it comes to offences such as harassment and/or stalking, particularly in the context of social media, which is not as clear cut, to use Bingham's example, as offences such as robbery,[85] it is vitally important that we strive for the 'maximum clarity which the imperfect medium of language can attain'.[86]

The discussion below will outline two key issues with the PfHA and its application in a social media context, against the principle of legality:

1) Terminology issues; and
2) The reasonable person test.

80 Ibid.,

81 *Magyar Helsinki Bizottsag v Hungary* (2020)71 E.H.R.R. 2.

82 Steve Foster, 'The Right to Criticise and the Determination of Permissible Expression in Harassment Cases: The Decision in Pal v United Kingdom' (2022) 6 E.H.L.R 571, 580.

83 Ibid., 586.

84 See for example, Law Com No 381 (n 2) Chapter 8.

85 Tom Bingham, *The Rule of Law* (Penguin 2011) 37.

86 Law Commission, *Report on Conspiracy and Criminal Law Reform* (1976), Law Com No. 76 [1.8].

Though arguably when viewed on their own, these factors might not be considered significant enough to breach the principle of legality, when looked at in the round, with social media in mind, legality is not being maintained.

5.5.1 Terminology Issues

As noted in earlier parts of this chapter, the PfHA contains several ambiguous terms, in particular, a course of conduct, harassment, and stalking. This is significant, considering the purpose behind the PfHA is to protect 'victims of harassment [... including from ...] so-called stalking behaviour'.[87] Neither behaviour can be present if there is no course of conduct. Yet the meaning of these terms, course of conduct, harassment and stalking are not clearly defined, nor is it clear what the difference is between harassment and stalking—an issue that becomes relevant in a social media context, where, as noted earlier, 'the general monitoring of a person's social media page, is expected, and somewhat consented to'.[88]

For the law to adhere to the principle of legality, and in turn, protect freedom of expression online, citizens need to be able to guide their behaviour in accordance with clear and understandable legal provisions—this is a fundamental aspect of legality—and something the PfHA lacks.

5.5.1.1 A Course of Conduct

As outlined above, to establish a course of conduct, it must be found that there is a connection between at least two incidents, which causes another harassment. Although, in reality, it is likely that there needs to be more than two incidents.[89] In addition, these incidents need to be close in time—though each event does not need to amount to a crime in itself.[90] And like that of harassment, discussed in Section 5.5.1.2 below, what constitutes a course of conduct, has been decided in an ad-hoc manner, with no clear meaning as to what is meant by a course of conduct, despite it being accepted that we should not rely on a 'case by case' account of insight into how the law should develop.[91]

In *Hills*,[92] the court concluded that two violent incidents between a couple over the course of six months, could not constitute a course of conduct, as the couple had separated and later reconciled. Similarly, in *Lau v DPP*,[93] a slap, followed by a threat four months later, was not considered a course of conduct. Interestingly in *Lau*, it was not just that the two incidents were not close in time, but consideration

87 HL Deb (n 30).
88 Law Com No 381 (n 2) [8.174].
89 *Lau* (n 38) [15].
90 *Jones v DPP* [2010] EWHC 523 (Admin), [2011] 1 W.L.R. 833 at [27].
91 Lon L Fuller, *The Morality of Law* (Yale University Press 1964) 46.
92 *Hills* [2001] Crim LR 318.
93 *Lau* (n 38).

was also given to the different nature of the events in question by the court.[94] However, the court did maintain that an incident once a year could, in certain circumstances, constitute a course of conduct.[95] For example, a threat once a year on a person's birthday.

Much of the case law relating to a course of conduct stems from contact that has occurred in the physical world, as opposed to the virtual. There are some cases, where 'modern' communications are at play, such as in *Kelly v DPP*,[96] where it was held that three voicemail messages which were left in the space of five minutes and listened to consecutively, did amount to a course of conduct. However, the finding of the court in *Kelly,* leaves open several questions in the context of communications aided by social media. For example:

1. Would, hypothetically, five messages sent over the course of a few months, which are all read by a person on the same day, constitute a course of conduct, if it caused another harassment? In the social media world, it is common for individuals to have hundreds, if not millions of followers online, and it might be that some communications are not always seen at the point in which they were sent;
2. In the context of X, where communications are limited to a maximum of 240 characters,[97] would several Xs linked together, commonly referred to as a 'thread' constitute a course of conduct? Or could they be perceived as one incident?
3. In the context of Instagram, where video content has a specified time limit, would two videos continuing a conversation be enough to constitute a course of conduct?

To prove a course of conduct in the physical world seems significantly easier, compared to that of the online world. In the online world consideration must be given to the innerworkings of the platform, the limits placed on users by platform creators, such as character limitations or time constraints, and the need to ensure consistency across the criminal justice system. Here, as discussed at length in Chapter 3, we need to ensure that terms such as course of conduct are not so generously overly interpreted that they would, in theory, create retrospective criminal law.

As noted by Murray and Reed 'obedience of the law depends on humans understanding the law'.[98] As echoed earlier, on its own, the lack of understanding as to what is meant by a course of conduct might not be so vague to render the PfHA incompatible with legality, but when examined in light of the other vague terminology used in the Act, issues with legality become obviously apparent.

94 Ibid., [16].
95 Ibid., [15].
96 *Kelly* (n 41).
97 Note, there are some exceptions to the 240 character rule.
98 Chris Reed and Andrew Murray, *Rethinking the Jurisprudence of Cyberspace* (Elgar 2018) 85.

5.5.1.2 The Meaning of Harassment

Like that of a course of conduct, harassment has no specific meaning in law, despite its importance within the legal provision, because without it, no offence will be committed under any of the provisions discussed in earlier parts of this chapter. Instead, the Act considers harassment to be conduct that causes another alarm or distress. There has been some attempt to clarify the meaning of alarm and distress, with the court in *Majrowski v Guy's and St Thomas's NHS Trust* suggesting that anxiety is considered something just short of a recognised psychiatric illness.[99] Whilst defining distress as 'oppressive and unreasonable behaviour'.[100]

There is, therefore, a degree of subjectivity in the law when it comes to harassment—which is not necessarily a breach of the principle of legality because, as noted by Lord Morris 'those who skate on thin ice can hardly expect to find a sign which will denote the precise spot where he [...] will fall in'.[101] Horder interprets the 'thin ice' principle to mean, that where a person's behaviour is on the borders of criminal liability, they risk criminalisation because

> [u]nless the criminal law occasionally resorts to such open-ended terms [...], it would have to rely on immensely detailed and lengthy definitions which might be so complicated as to restrict the intelligibility of the law.[102]

Indeed, for Foster, the vagueness of the term harassment does not warrant issues with legality due to extensive case law.[103]

For the Court of Appeal in *Thomas v News Group Newspapers*

> "Harassment" is [...] a word which has a meaning which is generally understood. It describes conduct targeted at an individual which is calculated to produce the consequences [...] which is oppressive and unreasonable.[104]

But even so, difficulties have emerged—which in turn, creates issues with legality. For Fuller, for the law to adhere to the principle of legality, the criminal justice system must refrain from deciding cases in an ad-hoc manner. As argued above, we should not decide legal disputes case-by-case—there needs to be some consistency in law and in turn this will ensure that fair enforcement of the law is being upheld—maintaining due process. For this to happen, and to avoid a case-by-case approach of enlightenment, we need clarity. However, the difficulty in establishing the meaning of harassment, particularly in the context of social media, means that

99 *Majrowski* (n 45) per Auld LJ [45].

100 Ibid., per Auld LJ May LJ [82].

101 *Knuller (Publishing, Printing and Promotions) Ltd. and Others v DPP* [1972] 3 W.L.R. 143, [1973] A.C. 435 per Lord Morris 463.

102 Jeremy Horder, *Ashworth's Principles of Criminal Law* (8th edn, Oxford University Press 2016) 27.

103 Foster (n 82) 573.

104 *Thomas v News Group Newspapers Ltd* [2001] EWCA Civ 1233 [30]. This definition was further endorsed by the Supreme Court in *Hayes* (n 44).

not only are cases being decided in an *ad-hoc* manner, but also legal provisions are being applied inconsistently—a similar argument raised in the previous chapter. Demonstrated further in *R v Peter Nunn*.[105]

In 2013, Nunn was arrested and convicted under s.127(1) of the Communications Act 2003 (CA)—an Act discussed further in Chapter 6—for sending grossly offensive messages, including threats of rape, to Caroline Criado-Perez, an active feminist campaigner. Caroline Criado-Perez had launched a public campaign to get the author, Jane Austin, printed on banknotes in England and Wales. Following this public campaign, she was subjected to a crusade of abuse online, in particular, from Peter Nunn:

> He dug up my work history. He dug up my relationship and family history. He dug up my family's work history—including publishing home addresses. He wrote reams of blogs about me and my every public move. He made numerous videos about me. He set up numerous [T]witter accounts all of which spoke almost exclusively about me. In these same [T]witter accounts he detailed the best way to rape and drown a witch, alongside repeatedly naming me as the head of the 'witches' coven'. He also boasted on [T]witter in the same account about having bought a gun, and wondered 'how much death' this gun could buy him.[106]

Despite a clear course of conduct being present, which arguably amounted to harassment and caused Caroline Criado-Perez distress, Nunn was prosecuted and found guilty of sending grossly offensive communications contrary to s.127(1) of the CA, receiving a six week custodial sentence.[107]

It's difficult, by applying the language of the PfHA, to determine why Nunn was pursued under a communication offence, as opposed to the PfHA, a decision Caroline Criado-Perez has been heavily critical of:

> This man [Peter Nunn] made me fear for my life as no-one ever has before. I felt he was a clear and present threat to me. He made me scared to go outside, to appear in public. He stopped me being able to sleep; being able to work. He seemed obsessed enough to carry out his threats. I am glad he has been found guilty. I am glad he cannot contact me again. I hope he has learnt from this. But I think the CPS got the charge wrong. I don't feel they understood what happened to me.[108]

As noted in Chapter 3, unfair enforcement of the law creates criticism in the community, which is evidently clear here. In turn, consistency is lost. All issues which

105 *R v Peter Nunn*, The City of London Magistrates' Court, 29 September 2014 (unreported).

106 Caroline Criado-Perez, 'A brief comment on Peter Nunn, sentenced today For Twitter abuse' (*Week Women*, 2014) <https://weekwoman.wordpress.com/2014/09/29/a-brief-comment-on-peter-nunn/> accessed 29 October 2016.

107 *Peter Nunn* (n 105).

108 Criado-Perez (n 106).

undermine the principle of legality. The lack of understanding as to what constitutes harassment is even more apparent when attempting to differentiate between stalking and harassment.

5.5.1.3 The Meaning of Stalking

Like that of harassment, there is no specific meaning given to stalking. Instead, as outlined in Section 5.3.3.1, the PfHA gives a non-exhaustive list of behaviours which may constitute stalking, many of which are applicable to cyberstalking. However, to prove a person's behaviour crosses the line into stalking, due to the construct of s.2A of the PfHA, first harassment needs to be proven. Consequently, harassment and stalking are inherently interlinked, as legally you cannot be stalked, if harassment is not present. As a result, the difference between harassment and stalking is not easily identifiable.

The lack of understanding between these two behaviours was exposed in a HMCPSI report, as previously mentioned, into the PfHA in July 2017, where it was found that there was a serious lack of understanding between police forces as to what constitutes harassment or stalking, specifically the difference between the two behaviours. From a study of 112 police reports, across six forces, it was found that the police had failed to deal with any of these cases correctly and in many cases, misunderstood the severity of the behaviour being reported:

> [L]egislation does not exhaustively define stalking or the particular circumstances that make stalking different from those of harassment. Therefore, without any additional clarification, what differentiates harassment and stalking can be open to interpretation and result in confusion.[109]

Consequently, a recommendation was put forward in the report for a more definitive definition of stalking to try and combat this issue. However, the law is still yet to change, meaning that both behaviours are often misunderstood, and when it comes to social media, the difference between the two is often important to ensure faith in the criminal justice system is maintained.

As discussed in Chapter 3:

> The most subtle element in the task of maintaining congruence between law and official actions lies, of course, in the problem of interpretation. Legality requires that judges and other officials apply statutory law, not according to their fancy or with crabbed literalness, but in accordance with principles of interpretation.[110]

However, where two elements of an offence, such as harassment and stalking are so inherently interlinked, a clear understanding of the boundaries between the two

109 Criminal Justice Inspectorates and HM Crown Prosecution Service Inspectorate (n 17) 24.
110 Fuller (n 91) 82.

becomes significantly difficult to articulate. In turn, the action-guiding nature of the law, which is an important aspect of legality, is heavily affected. At its very basic, citizens should be able to guide their behaviour in accordance with clear and distinct rules. Without further clarity on the difference between stalking and harassment, particularly considering the ease with which both behaviours can now be conducted, renders significant issues with the principle of legality and in turn, poses a threat to freedom of expression.

5.5.1.4 Summary

As echoed earlier, taking each element of the offence discussed above on its own, might not necessarily render the PfHA incompatible with legality, but when considered together, the ambiguity present does not maintain the action-guiding nature of the law:

> [i]f the legitimacy of the law's claim to authority is conferred on it by the collective acceptance of those to whom it is addressed, it must follow that structuring or wording the law in a way which makes it inaccessible to those addresses will reduce the likelihood of acceptance. Achieving legitimacy requires attention to style as well as substance.[111]

By not being able to clearly differentiate between harassment and stalking, this creates issues for those working in the criminal justice system. This is heighten further given the lack of understanding as to what constitutes harassment, stalking, or indeed, a course of conduct, risking the overcriminalisation of speech. We need legality to ensure due process. We need legality to ensure the protection of freedom of expression, particularly where an offence is grounded in the reasonable person.

5.5.2 The Reasonable Person

As noted in Chapter 3, for Fuller, comparing a person's actions against that of the reasonable person, may in some circumstances be asking the impossible, as his education or even personal characteristics may not meet such a requirement. Yet imposing the 'reasonable man with the same characteristics as the defendant' as the boundary for criminal liability risks objectivity being lost.[112] There is clearly a balance that needs to be struck.

In cases of harassment and/or stalking, whether online or offline, the reasonable person test is used to determine the *mens rea* element of the offence: 'knows or should have known'. This is a purely objective test, in which the characteristics of the accused are not taken into consideration, such as a persons mental health. This objective approach to determining the *mens rea* of harassment and/or stalking has been supported and justified by the courts:

111 Reed and Murray (n 98) 183.
112 Fuller (n 91) 71–72.

> The conduct at which the Act [PfHA] is aimed, and from which it seeks to provide protection, is particularly likely to be conduct pursued by those of obsessive or otherwise unusual psychological make-up and very frequently by those suffering from an identifiable mental illness. Schizophrenia is only one such condition which is obviously very likely to give rise to conduct of this sort. We are satisfied that to give the Act the construction for which Mr. Butterfield contends [a subjective test] would be to remove from its protection a very large number of victims and indeed to run the risk of significantly thwarting the purpose of the Act.[113]

However, in the context of social media, the reasonable person test may well be flawed, particularly where the elements which make up a certain offence, are inherently vague, as with the PfHA. Indeed, to ensure legality is maintained, not only should we perhaps re-examine what constitutes harassment, a course of conduct and stalking, in a social media context, but we should also rethink what we consider to be the reasonable person. Here, the offence of cyberharassment or cyberstalking should not be judged solely on the reasonable person, but instead should be based on the *reasonable social media user*—to ensure the law is not asking the impossible. This was an approach endorsed by the Supreme Court in *Stoker v Stoker*[114] when determining if comments made on Facebook, amounted to defamation.

Mrs Stoker, in a series of posts on Facebook about her ex-husband's new partner, used the phrase 'he tried to strangle me', which Mr Stoker took to mean, he tried to kill her. He issued proceedings for defamation. The issue before the court turned on the meaning of strangle, with the case eventually making its way to the Supreme Court. The Justices of the Supreme Court, in determining the meaning of strangle, confirmed that the proper approach to determining its meaning, was not with reference to a dictionary, as was done by the judge at first instance, but to consider how it would be understood by the ordinary reasonable reader, taking into account context as 'a factor of considerable importance'.[115] And for the Supreme Court Justices, this meant taking into account social media where communications are considered a casual medium, 'in the nature of conversation rather than carefully chosen expression'.[116] Indeed, a person's reaction to a communication online is 'impressionistic and fleeting'.[117] Consequently, the reasonable person, in the context of *Stoker* was considered by the court to be the reasonable social media user:

> The touchstone remains what would the ordinary reasonable reader consider the words to mean [...] All of this, of course, emphasises that the primary role of the court is to focus on how the ordinary reasonable reader would construe the words. And this highlights the court's duty to step aside from a

113 *Colohan* (n 50) per Justice Hughes [18–19].
114 *Stoker v Stoker* [2020] A.C. 593.
115 Ibid., [40].
116 Ibid., [43].
117 Ibid., [44].

> lawyerly analysis and to inhabit the world of the typical reader of a Facebook post. To fulfil that obligation, the court should be particularly conscious of the context in which the statement was made.[118]

When it comes to complaints of harassment involving written words, 'there are two relevant persons: the writer and the subject of what is written'.[119] When it comes to complaints of cyberharassment and/or cyberstalking, this is even more complex as the context of social media involves 'tactics employed by [...] parties [which] can be belligerent and robust'.[120]

Embedding the concept of the reasonable person, into the reasonable social media user, will allow the court to be more considerate of the nature of social media and how it's different from the physical world. In turn, it will allow the court to also ensure that the definition of harassment is not 'given an interpretation which restricts the right of freedom of expression, save in so far as this is necessary in order to achieve a legitimate aim'.[121] Particularly given that:

> there may be some concern that the permissible limits of free speech—including the right to shock and offend, particularly in the context of public interest speech—might be given less credence in harassment cases.[122]

5.6 Chapter Summary

There is no doubt that the increase in new technology has revolutionised the ways in which harassment and stalking can occur. Its effects on those subjected to such behaviour are well documented. Although, as noted in earlier parts of this chapter, there are legitimate concerns over the definitions used in the research into this area of online abuse—something we need to consider given that it is often studies such as those discussed in Section 5.2 above, which are used to justify the increased usage of the criminal law to tackle cyberharassment and cyberstalking.

The effects of those subjected to intense cyberharassment or cyberstalking cannot, and should not be downplayed, but as highlighted throughout this chapter, and echoed throughout this monograph, we need to be careful in applying the coercive nature of the criminal law to such conduct. Broadly speaking, the discussions above have highlighted two wider points of concern:

1. The use of vague terminology throughout the PfHA. Of particular concern, is the use of the terms course of conduct, harassment, and stalking. What constitutes a course of conduct has been debated at length in cases before the judicial system, but as discussed in Section 5.5.1.1, there is little discussion

118 Ibid., [38].
119 *Carina Trimingham* (n 54) per Justice Tugendhat [57].
120 Foster (n 82) 572.
121 *Thomas* (n 104).
122 Foster (n 82) 572.

of how a course of conduct is established in the virtual world. For example, as noted above, '[i]n the context of X, where communications are limited to a maximum of 240 characters, would several Xs linked together, commonly referred to as a "thread" constitute a course of conduct? Or could they be perceived as one incident?'. Further issues were also highlighted with the term's harassment and stalking. Indeed, as illustrated above stalking and harassment are two different types of behaviours, but the law is struggling to articulate the difference between the two. And it's important that those working in the criminal justice system can articulate the differences between these two behaviours, to ensure trust is maintained in a given society and to uphold the principle of legality; and

2. The reasonable person test. Though a longstanding provision in the criminal law, the use of the reasonable person test, as discussed in Section 5.5.2, when applied in a social media context presents significant issues. This is particularly acute in the digital world, where context is everything but is often lost. To better protect the right to freedom of expression and adhere to the principle of legality, it has been suggested that we should, when considering online behaviours, ground the reasonable person test into that of the reasonable social media user—to better protect not only the right to freedom of expression but also to help maintain the principle of legality.

The concerns raised above, mirror many of the previous concerns highlighted in Chapter 4 and are further reflected in Chapters 6 and 7 in the application of communication offences to the online world.

Reference list

Agate J & Ledward J, 'Social Media: How the Net Is Closing in on Cyber Bullies' (2013) 24(8) ELR 263

Babbie E, *The Basics of Social Research* (Cengage Learning 2007)

Baber M & Jeffs H, Stalking, *harassment and Intimidation and the Protection from Harassment Bill* (Research Paper 96/115, 13 December 1996)

Barak A, 'Sexual Harassment on the Internet' (2005) 23(1) Soc Sci Comput Rev 77

Basu S & Jones R, 'Regulating Cyberstalking' (2007) 7(2) JILT https://warwick.ac.uk/fac/soc/law/elj/jilt/2007_2/basu_jones/

Bingham T, *The Rule of Law* (Penguin 2011)

Bliss L, 'The Protection from Harassment Act 1997: Failures by the Criminal Justice System in a Social Media Age' (2019) 83(3) J Crim Law 217

Bocij P, 'Victims of Cyberstalking: An Exploratory Study of Harassment Perpetrated via the Internet' (2003) 8(10) First Monday http://firstmonday.org/ojs/index.php/fm/article/view/1086

Brown A, Maple C & Short E, 'Cyberstalking in the United Kingdom: An Analysis of the ECHO Pilot Survey' (*University of Bedfordshire National Centre for Cyberstalking Research*, 2011) https://www.beds.ac.uk/__data/assets/pdf_file/0003/83109/ECHO_Pilot_Final.pdf

Carter H, 'Teenage girl is first to be jailed for bullying on Facebook' *The Guardian* (London, 21 August 2009) https://www.theguardian.com/uk/2009/aug/21/facebook-bullying-sentence-teenage-girl

Criado-Perez C, 'A Brief Comment on Peter Nunn, Sentenced Today for Twitter Abuse' (*Week Women*, 2014) https://weekwoman.wordpress.com/2014/09/29/a-brief-comment-on-peter-nunn/
Criminal Justice Inspectorates & HM Crown Prosecution Service Inspectorate, 'Living in Fear – the Police and CPS Response to Harassment and Stalking' (*justiceinspectorates.gov*, July 2017) http://www.justiceinspectorates.gov.uk/hmicfrs/wp-content/uploads/living-in-fear-the-police-and-cps-response-to-harassment-and-stalking.pdf
Davidson J, Livingtone S, Jenkins S, Gekoki A, Choak C, Ike T & Phillips K, 'Adult Online Hate, Harassment & Abuse: A Rapid Evidence Assessment' (June 2019) https://assets.publishing.service.gov.uk/government/uploads/system/uploads/attachment_data/file/811450/Adult_Online_Harms_Report_2019.pdf
Fontaine R G, *The Mind of the Criminal: The Role of Developmental Social Cognition in Criminal Defense Law* (Cambridge University Press 2012)
Foster S, 'The Right to Criticise and the Determination of Permissible Expression in Harassment Cases: The Decision in Pal v United Kingdom' (2022) 6 EHLR 571
Fuller L L, *The Morality of Law* (Yale University Press 1964)
Gowland J, 'Protection from Harassment Act 1997: The "New" Stalking Offences' (2013) 77(5) J Crim Law 387
Herring J, *Criminal Law: Text, Cases and Materials* (10th edn, Oxford University Press 2022)
HL Deb 24 January 1997, vol 1, col 917
Home Office, 'Circular: A Change to the Protection from Harassment Act 1997' (*Gov.uk*, 16 October 2012) https://www.gov.uk/government/publications/a-change-to-the-protection-from-harassment-act-1997-introduction-of-two-new-specific-offences-of-stalking
Home Office. *Stalking A Consultation Paper* (11 July 1996)
Horder J, *Ashworth's Principles of Criminal Law* (8th edn, Oxford University Press 2016)
Law Commission, *Abusive and Offensive Online Communications: A Scoping Report* (Law Com No 381, 2018)
Law Commission, *Report on Conspiracy and Criminal Law Reform* (Law Com No. 76, 1976)
MacEwan N, 'The New Stalking Offences in English Law: Will They Provide Effective Protection from Cyberstalking?' (2012) 10 Crim L R 767
Pathé M & Mullen P, 'The Impact of Stalkers on Their Victims' (1997) 170(1) B J Psych 12
Reed C & Murray A, *Rethinking the Jurisprudence of Cyberspace* (Elgar 2018)
Scaife L, *Handbook of Social Media and the Law* (Routledge 2015)
Storry M & Poppleton S, 'The Impact of Online Abuse: Hearing the Victims Voice' (*Victims Commissioner,* 1 June 2022) https://s3-eu-west-2.amazonaws.com/jotwpublic-prod-storage-1cxo1dnrmkg14/uploads/sites/6/2022/05/Hearing-the-Victims-Voice.pdf
Suzy Lamplugh Trust, 'Unmasking Stalking: A Changing Landscape' (*Suzy Lamplugh Trust*, April 2021) https://www.suzylamplugh.org/Handlers/Download.ashx?IDMF=fcfb781a-f614-48c8-adcf-4cfa830c16a7
The Crown Prosecution Service, 'Stalking and Harassment' (*CPS.gov*, 23 May 2018) https://www.cps.gov.uk/legal-guidance/stalking-and-harassment
Vogels E A, 'The State of Online Harassment' (*Pew Research Centre*, 13 January 2021) https://www.pewresearch.org/internet/2021/01/13/the-state-of-online-harassment/

6 Abusive Communications Online

The Development of Communication Law

6.1 Introduction

The regulation of communications between private individuals is not a new phenomenon. As outlined in Chapter 2, since as early as 1884 the criminal law has been used to regulate private communications following the enactment of the Post Office Protection Act, which criminalised the sending of grossly offensive material *via* the postal system.[1] Since then, following a rise in new technologies, the law has developed to expand beyond just the postal system, to fax machines, to telephone calls and in 2001, expanded to cover all forms of electronic communications. In England and Wales, until 2024, only two communication offences existed: the Malicious Communications Act 1988 (MCA) and s.127(1) and (2) of the Communications Act 2003 (CA). As will be outlined in later parts of this chapter, both provisions are very similar, though there are some subtle, but important, differences between the two. However, through the creation and implementation of the Online Safety Act 2023 (OSA), this has since expanded. Some of these 'new' communication offences, will be outlined in Chapter 7. In addition, on 31 January 2024, several changes were made to the MCA and s.127(2) of the CA, as will be outlined in the discussions below.

As discussed in Chapter 1, the rise and dominance of the internet has placed unprecedented pressure on governments, law enforcement, and society to do more to tackle the growing concerns related to the use of the internet, in particular, social media sites to send abusive communications to another. Since, we have witnessed a significant rise in the use of communication offences to prosecute and criminalise comments made online, with, as noted by Gibbon, these provisions often being associated with offensive language that generally only becomes a problem when it is imposed on a person who finds it objectionable.[2] In recent years there have been several calls for the repeal and replacement of both the MCA and s.127 of the CA.[3]

1 Post Office Protection Act 1884 s.4(1)c.

2 Thomas Gibbon, 'Case Comment: Grossly Offensive Communications' (2006) 11(4) Comms L 136, 137 (note).

3 The Law Commission in 2021 recommended their repeal and replaced with a new 'harms' offence. This is discussed in length in Chapter 7.

DOI: 10.4324/9781003303688-8

The discussion below will provide a brief history of communication offences, before turning to outline the continued usage of social media platforms to send inappropriate communications to another, often referred to as online abuse. Following on from this, the offences which were originally prohibited under the MCA and s.127(1) and (2) of the CA, until very recently, will be discussed, with particular reference to the language contained within these two provisions—highlighting issues with legality. Therefore allowing, in Chapter 7, direct comparisons between the 'new' communication offences and the 'old' offences.

6.2 A Brief History of Communication Law

As noted above, the regulation of communications is not new with discussions predating many forms of technology we know today. For example, as previously discussed in Chapter 2, in 1857 an individual was given a 21 month custodial sentence for writing offensive words relating to Christianity on a gate, under the Common Law.[4] Since, criminal law provisions have continued to be created and utilised in an attempt to regulate communications, including speech, articles sent through the postal system and the prohibition of certain conduct or the display of certain material(s), including such material(s) online.

What is believed to be the first parliamentary attempt to use the criminal law to limit communications can be traced back to 1884 with the enactment of the Post Office Protection Act which made it an offence to send 'any words, marks, or designs of an indecent, obscene, or grossly offensive character'.[5] Yet, even in 1884, concerns were raised that prohibiting the sending of material which could be defined as 'indecent, obscene, or grossly offensive' was so significantly vague, that it would render issues with consistency across the criminal justice system:

> 'grossly offensive' might be taken to mean something very different to indecent or obscene [...] it might happen that one man would use words—for instance, he might write 'swindler' or 'liar' upon the outside of a letter—which were not really indecent or obscene, only what they would call vulgar [...] the words 'grossly offensive' might be taken to mean something that was extremely offensive to the person who received it, although it did nothing more than lacerate the feelings of the person receiving it.[6]

Concerns, which later became a reality.[7]

Yet the use of these terms, indecent, obscene or grossly offensive, have continued to be utilised to control communications—both verbal and written. Indeed,

4 John Stuart Mill, *On Liberty* (first published 1859, Enhanced Media 2016) 25.

5 Post Office Protection Act 1884 s.4(1)c.

6 HC Deb, 09 August 1884, vol 292, col 371.

7 Graham Smith, 'From telegram to tweet: Section 127 and all that' (*Cyberlegal*, 22 February 2015) <https://www.cyberleagle.com/2015/02/from-telegram-to-tweet-section-127-and.html> accessed 5 July 2022.

not only did the law expand to prohibit the types of materials which could be sent, but it also extended beyond just the postal system to include new forms of technology. For instance, the Post Office Amendment Act 1935 made it an offence for any person to send:

> (1) any message by telephone which is grossly offensive or of an indecent, obscene, or menacing character; or (2) sends any message by telephone, or any telegram, which he knows to be false, for the purpose of causing annoyance, inconvenience, or needless anxiety to any other person; or (3) persistently makes telephone calls without reasonable cause and for any such purpose as aforesaid.[8]

Similar provisions were contained in the Telecommunications Act 1984 (TCA) which prohibited a person from sending:

> (a) by means of a public telecommunication system, a message or other matter that is grossly offensive or of an indecent, obscene, or menacing character; or (b) sends by those means, for the purpose of causing annoyance, inconvenience, or needless anxiety to another, a message that he knows to be false or persistently makes use for that purpose of a public telecommunication system.[9]

The only element that has altered since 1884, is the mode of communication, with the content of such communications which are prohibited by law remaining, until recently, relatively consistent—the prohibition of sending material which can be defined as:

- Grossly offensive;
- Indecent;
- Obscene;
- Threatening;
- Menacing; or
- False.

Until 31 January 2024, such content continued to be prohibited under both the MCA and s.127(1) and (2) of the CA—two provisions which became prominent in attempting to tackle the 'harms' associated with the online world.

6.3 The Rise in Online Abuse

There does not seem to be a day in the United Kingdom where a story does not emerge concerning the rise in the use of social media platforms to send abusive

8 Post Office Amendment Act 1935 s.10(2).
9 Telecommunication Act 1984 s.34(1)a & s.34(1)b.

communications to another. Put simply, the rise in online abuse is well documented with ex-police Chief Constable for Essex, Stephen Kavanagh, suggesting that the current research into online abuse is only 'the tip of the iceberg', with police forces unable to cope with the scale of abuse that is currently taking place online.[10]

In October 2017 the BBC released a Freedom of Information request exposing the number of police reports made between 2015 and 2016, where malicious communications were the main element of the crime being complained about.[11] 38 out of 42 police forces responded to the request. The BBC uncovered that between 2015 and 2016, there had been an increase of 36,462 police reports involving malicious communications. Similarly, in 2017 Amnesty International, a human rights organisation, conducted an IPSOS[12] poll examining the extent to which women are abused *via* social media.[13] The poll was conducted across eight states looking at the online experiences of women aged between 18 and 55. 23% of those surveyed stated that they had, on at least one occasion, experienced online abuse. A significant proportion of this abuse was sent *via* Facebook. Findings which were further mirrored in research by Amnesty International following the 2017 General Election in the United Kingdom.[14]

The growing trend in the use of the internet to send abusive communications to others is further highlighted throughout research studies exploring online abuse. For instance, in 2018 OFCOM conducted a representative sample of 1,882 UK adults on their experiences of seeing abusive content online, finding that 53% of internet users had viewed hateful content, a rise of 6% on the previous year.[15] Further research conducted on behalf of OFCOM in 2020 revealed that 63% of adults had experienced at least one harmful incident in the previous 12 months.[16] Outside of the United Kingdom, studies further support the concept that online abuse is prevalent. For example, in 2018 Pacheco and Melhuish conducted a representative sample of 1,001 adults in New Zealand (weighted to ensure representation of minorities) on their experience of seeing or encountering online abuse.

10 The BBC, 'Teenager's life "ruined" by Live.me and Twitter "trolls"' *The BBC* (London, 24 October 2017) <http://www.bbc.co.uk/news/uk-england-41693437> accessed 30 January 2018.

11 Ibid.,

12 IPSOS is a global market research firm.

13 Amnesty International, 'Amnesty reveals alarming impact of online abuse against women' (*Amnesty International*, 20 November 2017) <https://www.amnesty.org/en/latest/news/2017/11/amnesty-reveals-alarming-impact-of-online-abuse-against-women/> accessed 9 December 2018.

14 Amnesty International UK, 'Black and Asian women MPs abused more online' (*Amnesty International*, 2017) <https://www.amnesty.org.uk/online-violence-women-mps> accessed 9 December 2018. See also, Laura Bliss, 'Abuse of women MPs is not just a scandal – it's a threat to democracy' *The Conversation* (London, 17 July 2017) <https://theconversation.com/abuse-of-women-mps-is-not-just-a-scandal-its-a-threat-to-democracy-80781> accessed 9 December 2018.

15 OFCOM, '"Adults"' media use and attitudes report' (*OFCOM*, 25 April 2018) <https://www.ofcom.org.uk/__data/assets/pdf_file/0011/113222/Adults-Media-Use-and-Attitudes-Report-2018.pdf> accessed 1 May 2023.

16 OFCOM, '"Adults"' media use and attitudes report 2020/21' (*OFCOM*, 28 April 2021) <https://www.ofcom.org.uk/__data/assets/pdf_file/0025/217834/adults-media-use-and-attitudes-report-2020> accessed 1 May 2023.

28% of participants in Pacheco and Melhuish's study had seen or encountered online abuse.[17] Similarly, in Australia, in 2022, 40% of adults, rising to 50% for the LGBTQIA+ and the CALD (culturally and linguistically diverse) community, had experienced an online harm in the prior 12 months.[18]

It is figures such as these, that are often used to justify stronger criminal provisions in tackling the use of the internet, in particular social media, to send abusive communications. However,

> [p]art of the challenge is that, at present, the data, tools, processes and systems needed to effectively and accurately monitor online abuse are not fully available and the field is beset with terminological, methodological, legal and theoretical challenges.[19]

Consequently, the evidence that is often used to demonstrate the rise in the use of the internet to send abusive communications is 'incomplete and inadequate'.[20] The Alan Turing Institute estimates that the prevalence of abusive communications on online platforms, which is serious enough to warrant intervention, is extremely low, giving an estimate of 0.001% to 1%.[21] Given that the required proof of wrongdoing by platforms to justify the removal of such content compared to that of the criminal law is significantly lower, this is concerning.

Reviewing several academic and thinktanks studies into online abuse, including some of those discussed above, The Alan Turing Institute were able to conclude that 0.001% to 1% of content on mainstream platforms, such as Facebook, Twitter, etc., could be labelled as abusive. Though do acknowledge that 'some users and events generate far more abuse, such as prominent figures (e.g., MPs) and terror attacks'.[22] Consequently, as noted in the report:

> [t]he available evidence is fragmented, incomplete and inadequate for understanding the prevalence of online abuse. Appropriate statistics are difficult

17 Pacheco Edgar and Neil Melhuish, 'NetSafe online hate speech: A survey on personal experiences and exposure among adult New Zealanders' (*Netsafe*, 12 December 2018) <https://netsafe.org.nz/wp-content/uploads/2016/12/Netsafe-insights-into-New-Zealands-experience-of-online-hate-speech.pdf> accessed 1 May 2023.

18 Michelle Rowland, 'Release of the 2022 National Online Safety Survey' (*Ministers for Infrastructure, Transport, Regional Development, Communications and the Arts*, 14 November 2022) <https://minister.infrastructure.gov.au/rowland/media-release/release-2022-national-online-safety-survey> accessed 11 May 2023.

19 Bertie Vidgen, Helen Margetts and Alex Harris, 'How Much Online Abuse Is There? A Systematic Review of Evidence for the UK. Policy Briefing – Full Report' (*The Alan Turing Institute*, 2020) 3 <https://www.turing.ac.uk/sites/default/files/2019-11/online_abuse_prevalence_full_24.11.2019_-_formatted_0.pdf> accessed 3 May 2023.

20 Ibid., 4.

21 Ibid.,

22 Ibid., 28.

> to find and, in many cases, are not provided with the necessary contextual information to fully interpret them.[23]

Put simply, 'in the absence of a mature and rigorous evidence base, it is hard to see how [...] the broad ambitions to regulate [are justified]'.[24] Nothing demonstrated this more than the MCA and s.127(1) and (2) of the CA. Both provisions contained, and still do, despite recent changes, overly broad language, creating significant issues with the principle of legality and in turn, threatens the right to freedom of speech.

6.4 The Malicious Communications Act 1988

One of the most significant changes to communication law came in 1988 with the enactment of the MCA following growing concerns relating to the rise in the use of the postal system to send poison pen letters or articles to another.[25] Many of these communications, though distressing upon the receiver, fell outside the realms of criminal law intervention. For instance, the case of a coffin being sent through the postal system, though distressing upon the recipient, was not unlawful.[26] This small but significant void in the law paved the way for the creation and enactment of the MCA.

The MCA is a prime example of Parliament's attempt to keep pace with how individuals communicate with each other. Though was never intended, in its original form, to cover electronic communications, as during the Law Commission's consultation into poison-pen letters, an informed decision was made not to include electronic communications within the scope of the proposed offence.[27] This was amended under s.43 of the Criminal Justice and Police Act 2001, in which 'electronic communications' was substituted into the Act.

As technology has evolved and how we communicate with others has changed, the MCA has undergone several alterations in an attempt to keep pace with changing technology. Before 2015, the MCA was a summary only offence, carrying a maximum custodial sentence and a limitation period of six months.[28] This was,

23 Ibid., 4.

24 Victoria Nash, 'Revise and Resubmit? Reviewing the 2019 Online Harms White Paper' 2019 11(1) J. Media Law 18, 24.

25 Poison pen letters and articles are considered 'a communication, written or otherwise, which is grossly offensive, or of an indecent, shocking, or menacing character, [which is sent] for the purpose of causing needless anxiety or distress [to the recipient] or any other person'. See, Law Commission, *Report on Poison-Pen Letters* (Law Com No 147 1985) [2.1].

26 HC Deb 12 February 1988, vol 127, col 620.

27 During the Law Commission's consultation into poison-pen letters, a decision was made not to include electronic communications within the scope of their proposed bill. See, Law Com No 147 (n 25) [2.11]. This was amended under s.43 of the Criminal Justice and Police Act 2001.

28 Limitation periods are 'the time limit within which the state may prosecute a particular crime'. See, George P. Fletcher, *Basic Concepts of Criminal Law* (Oxford University Press 1998) 10.

however, later raised as an issue before Parliament by former MP Angie Bray,[29] with the MCA now an either-way offence as per s.15 of the Criminal Justice and Courts Act 2015. Increasing the maximum custodial sentence under the MCA from six months to two years.

Until 31 January 2024, the MCA made it an offence:

> to send to another (a) a letter, electronic communication or article of any description which conveys—(i) a message which is indecent or grossly offensive; (ii) a threat; or (iii) information which is false and known or believed to be false by the sender; or (b) any article of electronic communication which is, in whole or part, of an indecent or grossly offensive nature.[30]

Today, by way of the OSA, the MCA only covers indecent or grossly offensive communications. Threatening and false communications are now prohibited under s.181(1) and 179(1) of the OSA—discussed further in Chapter 7. However, as highlighted above, for the purposes of this chapter, the MCA in its original format will be examined, to highlight why it was thought necessary to amend and allow for further discussions to take place in the following chapter directly comparing the 'old' offences to the 'new' offences.

6.4.1 Actus Reus

The *actus reus* of the MCA consists of two elements. First, the message must be sent in the form of a letter, electronic communication, or an article. What is a letter is self-explanatory, whereas what constitutes an 'electronic communication' or indeed an 'article', is a little more complex. S.2A of the MCA defines an electronic communication as: '(a) any oral or other communication by means of an electronic communications network; and (b) any communication (however sent) that is in electronic form'. Therefore, messages sent on social media platforms would constitute an electronic communication under the MCA. Whereas 'article' takes a very broad meaning and is defined under the Obscene Publications Act 1959 and 1964 (OPA) as anything 'containing or embodying matter to be read or looked at or both, any sound record, and any film or other record of a picture or pictures'.[31]

29 Ministry of Justice, 'Impact assessment: Making the offence in s.1 of the Malicious Communications Act 1988 a triable either-way offence' (*Gov.uk*, 30 May 2014) <https://assets.publishing.service.gov.uk/government/uploads/system/uploads/attachment_data/file/321285/malicious-communicationss-impact-assessment.pdf> accessed 11 July 2021. In her constituency, the police had attempted to prosecute an individual under s.15 of the Sexual Offences Act 2003, following an adult male sending sexually explicit text messages to a young female. Yet the prosecution failed because a meeting had not taken place, and therefore no offence had occurred under the Sexual Offences Act 2003. No further action could be taken against the individual as the six month limitation period under the Malicious Communications Act 1988 had surpassed.

30 Malicious Communications Act 1988 s.1(1).

31 Obscene Publications Act 1959 and 1964 s.1(2).

It could therefore include, and has included, the sending of an image showing an aborted foetus.[32]

Further, to fulfil the first part of the *actus reus*, the message *must be sent to another*, meaning in theory a public message sent on X, would not be covered under this Act of Parliament. In essence, the MCA *should* only cover communications sent privately. However, there are examples in which the MCA has been applied incorrectly, such as the matter of an anonymous blogger, who in July 2011 received a caution under the MCA for sending false information *via* a communications network.[33] The anonymous blogger had made allegations that a contestant on the television show 'Britain's Got Talent' had been groomed for the show by key organisers. The allegations were later proven to be false. Nevertheless, on a literal interpretation of the MCA, the matter was indeed pursued under the wrong Act of Parliament—the communication was never sent to another.[34] A similar case was brought before the courts in 2018, where Michael John Paley had his conviction quashed for sending a false communication under the MCA after it was concluded that the *actus reus* of the MCA had not been met, as the message had not been sent to another.[35]

Second, for the *actus reus* of the MCA to be fulfilled, it must have been found, in its original format, that the message sent was either indecent, grossly offensive, threatening, or false. As noted above, today the message would need to be considered either indecent or grossly offensive to fulfil this part of the *actus reus*.

As highlighted in the Act itself—'whole or in part'—it is sufficient if only part of the message falls into one of these categories. For example, it does not need to be found that the whole communication was grossly offensive to breach the MCA. How these terms are defined will be critically examined in later parts of this chapter. However, it is important to note that the actual criminal offence is in the physical sending of a letter, electronic communication, or an article which conveys a prohibited message. Consequently, there is no need for the intended message to be received by the recipient. This allows the law to intervene even if the communication is intercepted by a third party or the recipient does not receive the intended communication.[36]

6.4.2 Mens Rea

For the *mens rea* to be established it must be concluded that the *purpose* on behalf of the sender was to cause anxiety or distress upon the intended recipient,

32 *Connolly v DPP* [2007] EWHC 237 (Admin), [2008] 1 W.L.R. 276 (DC). Discussed further in Section 6.4.2.

33 Press Association, 'Britain's Got Talent blogger cautioned by police' *The Guardian* (London, 3 July 2011) <https://www.theguardian.com/tv-and-radio/2011/jul/03/britains-got-talent-blogger-cautioned> accessed 1 May 2018.

34 Laura Scaife, *Handbook of Social Media and the Law* (Routledge 2015) 165–166.

35 *R v Michael John Paley,* Preston Crown Court, 26 April 2018 (unreported).

36 Law Com No 147 (n 25) [4.4].

the recipient does not need to demonstrate they experienced anxiety or distress.[37] Though the MCA contains no definition of anxiety or distress, previous cases of harassment have shed some light on the meaning given to these terms by the courts. As affirmed in *Majrowski v Guy's and St Thomas's NHS Trust*,[38] anxiety is considered something just short of a recognised psychiatric illness. Whereas distress can be defined as 'oppressive and unreasonable behaviour'.[39] If a person sends a communication for another purpose, for instance in the form of a lawful protest, no offence would have occurred under the MCA; an issue the High Court considered in *Connolly v DPP*.[40]

Connolly, a Roman Catholic pro-life campaigner, sent several pictures to pharmacies of aborted foetuses clearly depicting the limbs and faces of the foetus, as a form of protest against the issuing of the morning after-pill by the pharmacies. One image sent by Connolly included a photograph of a deceased 21-week-old foetus with the face and limbs visible to the person viewing the image. The court concluded that the images were indecent and grossly offensive and therefore the *actus reus* of the MCA was present. However, the matter turned on whether the images were sent for the purpose of causing anxiety or distress, and therefore unlawful under the MCA. Or, were in fact, sent as a form of lawful protest:

> A person who sends an indecent or grossly offensive communication for a political or educational purpose will not be guilty of the offence unless it is proved that his purpose was also to cause distress or anxiety.[41]

For the High Court, a balance had to be struck between Connolly's right to freedom of expression and freedom of thought, conscience and religion as protected under Articles 10 and 9 of the European Convention on Human Rights (ECHR) respectively, alongside the 'rights of others', which would allow the court to legitimately infringe upon these Article rights.[42] On balance, the High Court upheld the decision of Coventry Crown Court and came to the judgment that the images sent were sent for the purpose of causing anxiety or distress contrary to the MCA:

> We have seen the photographs. They are close-up colour photographs of dead 21-week-old foetuses. The face and limbs are clearly visible. One of them is a close-up showing an abortion taking place. They are shocking and disturbing. That is why Mrs Connolly sent them to the pharmacists.[43]

37 Malicious Communications Act 1988 s.1(1).

38 *Majrowski v Guy's and St Thomas's NHS Trust* [2005] EWCA Civ 251, [2005] Q.B 848 per Auld LJ [45].

39 Ibid., per May LJ [82].

40 *Connolly* (n 32).

41 Ibid., per Dyson LJ [9].

42 See Chapter 2 for an in-depth discussion of when a person's right to freedom of expression can be restricted.

43 *Connolly* (n 32) [11].

Put simply, if the sending of such material is not sent for the purpose of causing anxiety or distress, no offence has taken place contrary to the MCA.

6.4.3 Summary

The scope of the MCA has shifted dramatically since its conception by the Law Commission in the early 1980s. We have moved away from covering communications sent *via* the postal system only, to electronic communications, including communications sent *via* social media. However, as will be discussed in later parts of this chapter, significant issues with legality were, and continue to be raised against the MCA, particularly the prohibited conduct under the Act. Issues which are also directly applicable to s.127(1) and (2) of the CA.

6.5 The Communications Act 2003: S.127 (1) and (2)

Following the turn of the new millennium, the use of technology, the internet and electronic communications increased dramatically across the globe; a trend that continues to occur even to this day. Like the reasoning behind the creation of the MCA, concerns were raised that the law was struggling to keep pace with the advancement of new technology, in particular, how individuals communicated:

> We are living at a time of revolution in the ways in which we communicate. The worlds of telephone, broadcasting, mobile communications, and the internet are changing and converging with astonishing speed. Meanwhile, our current regulatory framework was designed for a different age. We need to update the framework of regulation, and put in place a system that recognises the current fast-changing picture and can cope with the inevitability of change in years to come.[44]

In 2003 Parliament enacted s.127(1) and (2) of the CA to address the issues raised above.[45]

The CA had several aims. The Act created The Office of Communications commonly referred to as OFCOM, paved the way for the switch from analogue to digital television broadcasting and ensured universal internet access. In addition, it repealed and replaced s.43 of the TCA, discussed in Section 6.2, with s.127 (1) and (2) of the CA.

Under s.127(1) of the CA, it is a criminal offence to send:

> (a) by means of a public electronic communications network a message or other matter that is grossly offensive or of an indecent, obscene or menacing character; or (b) causes any such message or matter to be so sent.

44 HC Deb 12 December 2000, vol 359, col 481.

45 HC Deb 12 December 2000, vol 359, col 483.

Whereas under s.127(2), initially it was an offence to send, 'for the purpose of causing annoyance, inconvenience, or needless anxiety to another [...] by means of a public electronic communications network' a knowingly false message; or caused 'such a message to be sent; or persistently made use of a public electronic communications network'. However, s.127(2)a and 127(2)b have since been repealed.[46] False communications are now criminalised under s.179(1) of the OSA, discussed in Chapter 7. Below, the *actus reus* and *mens rea* of s.127(1) of the CA will be examined, alongside s.127(2) of the CA (as originally enacted).

6.5.1 Actus Reus

The *actus reus* of the offences prohibited under s.127 of the CA constitutes two elements. First, a message must be sent *via* a public electronic communications network. Like that of the MCA the offence is in the sending of the message (provided all other elements of the offence are present). There is no need for the message to be received. The term 'message' is given a wide definition in law, and includes text, images, or videos. However, the message *must be* sent *via* a public electronic communications network, defined under s.32 of the CA as:

> (a) a transmission system for the conveyance, by the use of electrical, magnetic or electro-magnetic energy, of signals of any description; and (b) such of the following as are used, by the person providing the system and in association with it, for the conveyance of the signals: (i) apparatus comprised in the system; (ii) apparatus used for the switching or routing of the signals; (iii) software and stored data.

A public electronic communications network therefore includes the internet and social media. However, a public electronic communications network does not include the sending of material *via* Bluetooth or Apple Airdrop.

Secondly, to prove the *actus reus* of an offence under s.127(1) of the CA, the content of such messages sent must be either grossly offensive or of an indecent, obscene or menacing character. Under s.127(2), as originally enacted, the message must have been false—terminology which will be defined and examined further in Section 6.6.2.

6.5.2 Mens Rea

Under s.127(1) of the CA as enacted by Parliament, no *mens rea* was present. Traditionally speaking, criminal acts which lack a definitive *mens rea* are considered strict liability offences.[47] However, using the principles laid out in *Sweet v*

46 S.127(2)c of the Communications Act 2003 remains but has been amended to cover hoax phone calls.

47 Strict liability offences are not dependent on actual negligence or intent to harm.

Parsley,[48] the House of Lords substituted a *mens rea* element into s.127(1) of the CA, arguing that it would be illogical for there to be no *mens rea* present:

> Parliament cannot have intended to criminalise the conduct of a person using language which is, for reasons unknown to him, grossly offensive to those to whom it relates, or which may even be thought, however wrongly, to represent a polite or acceptable usage. On the other hand, a culpable state of mind will ordinarily be found where a message is couched in terms showing an intention to insult those to whom the message relates or giving rise to the inference that a risk of doing so must have been recognised by the sender.[49]

Therefore, for an individual to be held liable for a breach of s.127(1) of the CA, it must be found that a grossly offensive, menacing, indecent, obscene message was sent to another recklessly or with intent.

The *mens rea* element of s.127(1) of the CA has recently been reaffirmed by the Divisional Court in *DPP v Kingsley Anthony Smith*[50] and *DPP v Bussetti*.[51] Although both *Kingsley* and *Bussetti* concerned the sending of grossly offensive communications, it has been confirmed by the Crown Prosecution Service (CPS), in their 2023 social media prosecution guidelines (and discussed further in later parts of this chapter) that messages which can be defined as 'indecent, obscene, or of a menacing character' contain the same *mens rea* element—intention or recklessness.[52]

Whereas for an offence to have occurred under s.127(2) of the CA, prior to the passing of the OSA, the defendant's purpose must have been to cause annoyance, inconvenience, or needless anxiety to another. As with the MCA, anxiety was considered as something just short of a recognised psychiatric illness, with the court noting in *Scottow v CPS*:

> what Parliament intended was to proscribe a course of persistent conduct the sole purpose of which is to cause annoyance, anxiety, or inconvenience by virtue of its persistence, rather than its informational content.[53]

Indicating that annoyance or inconvenience was met when a false message was persistently sent.

48 *Sweet v Parsley* [1969] 2 W.L.R. 470, [1970] A.C. 132. The principles created by the House of Lords in this case allow the court to substitute a *mens rea* into an offence where the criminal act is considered a true crime as opposed to a regulatory offence.

49 *DPP v Collins* [2006] UKHL 40, [2006] 1 W.L.R. 2223 per Lord Bingham of Cornhill [11].

50 *DPP v Kingsley Anthony Smith* [2017] EWHC 359 (Admin).

51 *DPP v Busetti* [2021] EWHC 2140 (Admin).

52 The Crown Prosecution Service, 'Social media and other electronic communications (*CPS.gov*, 9 January 2023) <https://www.cps.gov.uk/legal-guidance/social-media-and-other-electronic-communications> accessed 1 May 2023.

53 *Scottow v CPS* [2020] EWHC 3421 per Mr Justice Warby [32].

6.5.3 Summary

Both the MCA and s.127(1) and (2) of the CA, in their original format, covered very similar behaviours. Put simply, under the MCA it was an offence to send to another a message which could be labelled as:

- Threatening;
- False;
- Indecent; or
- Grossly offensive.

The following were prohibited under s.127(1) and (2) of the CA:

- Grossly offensive;
- Indecent;
- Obscene;
- Menacing; or
- False communications.

The behaviours prohibited under each provision above, caused difficulties for those working in the criminal justice system, with Scaife going as far as arguing that both provisions had become 'interchangeable in recent years'.[54] And even with the modest changes made to both provisions under the OSA, issues remain, particularly with the prohibited behaviours under both provisions and the principle of legality.

6.6 Communication Offences, Social Media, and Legality

Under the principle of legality, as discussed in Chapter 3, legal provisions need to be constructed in a clear and precise manner, for an individual to be liable for a criminal offence. In *Kafkaris v Cyprus*[55] the European Court of Human Rights (ECtHR) noted that

> [a]n individual must know from the wording of the relevant provision and, if need be, with the assistance of the courts' interpretation of it, what acts and omissions will make him criminally liable and what penalty will be imposed.[56]

As will be demonstrated below, the above is not being adhered to in the application of communication offences to complaints of social media abuse. In summary, issues will be exposed with the principle of legality, in particular, the interchangeable nature of the MCA and s.127 of the CA and the meaning of the prohibited

54 Scaife (n 34) 166.
55 *Kafkaris v Cyprus* App no. 21906/04 (ECtHR, 12 February 2008).
56 Ibid., [140].

conduct under each provision, despite the creation and implementation of social media prosecution guidelines in an attempt to bring some form of clarity to this area of the law. Indeed, as will be highlighted further in Chapter 7, many of these issues remain despite changes made by the OSA.

6.6.1 Malicious Communications Act v Communications Act

There are clearly some parallels between both the MCA and s.127 of the CA. Which, as alluded to in earlier parts of this chapter, has caused issues for actors working in the criminal justice system. In essence, both provisions have become interchangeable.[57] And even with the removal of false and threatening communications, overlaps remain. Yet, it is vital that a distinction is made between the two, as they both carry different sentencing tariffs.[58]

One justification for the overlap between the two offences is that the MCA was never intended to cover electronic communications, with little attention being given to other offences at the time of amendment (mainly, s.43 of the TCA) which already prohibited such conduct. However, there are some subtle, but significant, differences between the two. The biggest being the type of communications covered under each provision. The MCA covers all forms of communication, including the postal system. Whereas s.127(1) and (2) of the CA only covers public electronic communications. In essence, a message sent through the postal system which, as an example, is grossly offensive, is only prohibited under the MCA. Further, as noted previously, the MCA should only be used to cover private communications, yet this has clearly not always been the case, with the Law Commission warning that interpretive challenges are likely to remain without amendments to these provisions.[59] And this becomes even more significant given that the passing of the OSA has complicated this further, as discussed at length in Chapter 7.

In an attempt to bring some clarity to this area of the law the CPS created and implemented social media prosecution guidelines in 2013, following cases such as *Chambers v DPP*[60] and *R v Woods*[61]—cases discussed further below. The guidelines were later updated in 2016, 2018, January of 2023 and most recently, in early 2024 following the enactment of the OSA.[62] The purpose of the guidelines was to

57 Scaife (n 34).

58 The maximum custodial sentence which can be imposed on a person under either s.127(1) or (2) of the Communications Act 2003 is six months. Under s.1 of the Malicious Communications Act 1988, the maximum custodial sentence is two years.

59 Law Commission, *Abusive and Offensive Online Communications: A Scoping Report* (Law Com No 381, 2018) [4.30].

60 *Chambers v DPP* [2012] EWHC 2157 (Admin), [2013] 1 W.L.R. 183.

61 *R v Matthew Woods*, Chorley Magistrates' Court, 8 October 2012 (unreported).

62 The Crown Prosecution Service, 'Communication offences' (*CPS.gov,* 2024) <https://www.cps.gov.uk/legal-guidance/communications-offences#:~:text=the%20summary%20offences%20under%20section,of%20a%20public%20communications%20network.> accessed 11 March 2024.

bring consistency to this area of the law but as will be demonstrated below, this has not necessarily been the case.

In their updated social media guidelines in 2018, the CPS attempted to highlight the key differences between both the MCA and s.127 of the CA, in particular the difference between the mental elements of the crime.[63] Yet, the Law Commission in their scoping report, found examples of the two provisions continuing to be applied incorrectly, concluding that there is considerable overlap between the 'two offences in the online context and the purpose of this has not been convincingly demonstrated or explained'.[64] This renders issues with the principle of legality, as consistency and clarity are lost because

> if you and I are liable to be prosecuted, fined or perhaps imprisoned for doing or failing to do something, we ought to be able, without undue difficulty, to find out what we must or must not do.[65]

If actors in the criminal justice system are struggling to comprehend the differences between both the MCA and s.127 of the CA, and now also the OSA, it is difficult to see how communication offences under any of these provisions can be considered action-guiding — as noted further in Chapter 7. This is even more important because, as noted above, there are differences in the sentencing tariffs contained in each provision. Indeed, for Bakalis, both the MCA and s.127 of the CA have become fragmented and outdated.[66]

Significant weight is therefore given to prosecutorial discretion in determining the best course of action when it comes to complaints regarding behaviours carried out on social media platforms—all of which undermine the principle of legality. The confusion between the MCA and s.127 of the CA, and their place within the wider criminal law framework, is further amplified by the ambiguous terminology contained in each provision—in both their original enactment and their current format.

6.6.2 Terminology

Under the MCA, it was an offence to send a message which could be categorised as indecent or grossly offensive, a threat, or a message which could be deemed to be false. S.127(1) and (2) of the CA prohibited the sending of a message which was false, grossly offensive, indecent, obscene, or of a menacing character. For the Law Commission, prior to the OSA, we had 'a set of offences that [were]

63 The Crown Prosecution Service, 'Guidelines on prosecuting cases involving communications sent via social media' (*CPS.gov,* 21 August 2018) <https://www.cps.gov.uk/legal-guidance/social-media-guidelines-prosecuting-cases-involving-communications-sent-social-media> accessed 11 October 2018.

64 Law Com No 381 (n 59) [4.148].

65 Tom Bingham, *The Rule of Law* (Penguin 2011) 37.

66 Chara Bakalis, 'Rethinking Cyberhate Laws' (2017) 27(1) Inf. Commun. Technol. Law 86.

remarkably broad' where such speech was criminalised online, but not necessarily in the 'offline' world.[67] For instance, a person who writes a grossly offensive communication which they save to 'the cloud' could be liable for a s.127(1) offence. However, if the same communication was typed and then printed, being stored in a draw, no offence would have been committed. In addition, the vague terminology contained in both provisions, as outlined below, placed significant weight on prosecutorial discretion and court cases to clarify the meaning of these terms—all of which undermined the principle of legality. Indeed,

> The breadth of the offence[s] and its potential for infringing freedom of expression, can also be seen in the fact that what is outlawed is communications that are 'grossly offensive', 'indecent', or 'obscene'. It is difficult to see how proscribing such words would come within the Art 10(2) exceptions, as it is unlikely to be seen as 'necessary in a democratic society' to outlaw communications simply based on their gross offensiveness, indecency, or obscenity.[68]

Issues that remain, despite the modest changes made by the passing of the OSA as the behaviours in the quote above, are still prohibited between both the MCA and s.127(1) of the CA.

6.6.2.1 Indecent

The conduct of sending an indecent communication is criminalised under both the MCA and s.127(1) of the CA. Neither Act defines the term indecent, and its meaning is unclear. Though it could include conversations between individuals online or the sharing of materials. Indecent is also considered as being less serious than obscene but more serious than grossly offensive.[69] In *Collins*, indecency was understood to be 'in the eye of the beholder; but the law has generally treated them as a matter of objective fact to be determined by contemporary standards of decency'.[70] The CPS prosecuting guidelines on social media offences also continues to fail to fully articulate the meaning of this term, with the courts arguing that the term indecent takes its ordinary meaning in a given society and will be subjective depending on the context of a case.[71]

Under the principle of legality, as discussed in Chapter 3, legal provisions must be clear and certain so citizens can abide by the law. The idea that the term indecent will be subjective depending on the case before the courts creates uncertainty within the law. Indeed, for the Law Commission, the failure of legislation to define the term indecent and subsequently the failure of the courts to bring clarity to this

67 Law Com No 381 (n 59) [4.63].
68 Bakalis (n 66) 107.
69 The meaning of 'grossly offensive' is discussed further in Section 6.6.2.6.
70 *Collins* (n 49) per Sedley LJ [10].
71 Ibid.,

area of the law, creates interpretative challenges for all.[72] In particular, the difference between obscene, indecent and grossly offensive—there is a threshold between these behaviours, but that threshold is difficult to articulate. Further, by simply stating that indecent takes its ordinary dictionary meaning, here under the Oxford English Dictionary the term would be defined as 'morally offensive', little clarity is brought to this area of the law. Indeed, should something that is morally offensive, be criminal? For instance, Gillespie argues that indecency could include the sharing of naked images of adults, even where both adults consent to the sending of these images, commonly referred to as sexting:

> The meaning of indecent must mean that virtually any sexualised conversation could be captured by this offence, even that between consenting adults—something that would seem on the face of it extraordinary.[73]

This is especially true in matters concerning social media, as what one person may find indecent another may not. Arguably, the sending of an image online of a female without a burqa on could be considered as indecent by some.[74]

6.6.2.2 Obscene

Like that of indecent, there is no definition of obscene contained in s.127(1) of the CA, which prohibits the sending of such material. However, obscene is defined elsewhere within the criminal law framework of England and Wales, in particular under s.1(1) of the OPA which defines obscene as

> an article [which …] tend[s] to deprave and corrupt persons who are likely, having regard to all relevant circumstances, to read, see or hear the matter contained or embodied in it.[75]

Indeed, in most cases where a communication can be labelled as obscene, it is likely to be pursued under s.1(1) of the OPA as opposed to s.127(1) of the CA, as illustrated in *R v Smith (Gavin)*.[76]

Put simply, a matter will be obscene if the content goes beyond what is acceptable in a multicultural society and has the possibility of corrupting public morality.[77] In essence, it is a context-dependant crime, in which the matter is left to the judge, magistrates or jury to decide if such material is obscene. Though the term, obscene lacks a clear definition, the concept that obscene material is based on acceptability in a multicultural society, as opposed to the context of the case, like that of the term

72 Law Com No 381 (n 59) [6.84].
73 Alasdair A. Gillespie, 'Obscene Conversations, the Internet and the Criminal Law' (2014) 5 Crim. L.R. 350, 359.
74 See, for example, *LCG v OVD* [2023] EWHC 2058 (KB).
75 Obscene Publications Act 1959 and 1964 s.1(1).
76 *R v Smith (Gavin)* [2012] EWCA Crim 398, [2012] 1 W.L.R. 3368.
77 *Handyside v The United Kingdom* (1976)1 EHRR 737.

indecent, allows for the law to adapt to the changing nature of society. As Lord Bingham argues:

> it is accepted that absolute certainty is unattainable, and might entail excessive rigidity since the law must be able to keep pace with changing circumstances, some degree of vagueness is inevitable and development of the law is a recognised feature of common law courts.[78]

Indeed, what may have been classified as obscene 30 years ago, may not be today. But this creates issues when it comes to the publication of obscene material online, whereby anyone could read or see the material. This could mean, in the case of a communication sent on a social media site, obscene could be judged as corrupting the morality of those aged 13 or above.[79] In turn, this poses a significant issue with the right to freedom of expression.

6.6.2.3 False

Both the MCA and s.127(2) of the CA prohibited the sending of a message which could be categorised as false, in which it would have been for the judge, magistrate, or jury to determine if such a message could be considered as false. Such as, Stephen Dure, who in 2018, received a 15 week custodial sentence for making false claims on a social media platform that another person was engaging in paedophilic activities contrary to s.127(2) of the CA.[80]

The use of social media platforms to send false communications has created, and still creates significant debate in the United Kingdom. In particular, concerns have been raised around the use of social media sites to distribute mis/disinformation, commonly referred to as fake news and the use of fake social media accounts to send abusive communications.[81] Resulting in calls, prior to the OSA, for the meaning of false communications to also cover fake social media accounts and/or mis/disinformation under both the MCA and s.127(2) of the CA.

In 2016 following the CPS updating their prosecuting guidelines on social media offences, emphasis was placed on tackling the growing trend of anonymous online profiles:

78 *R v Rimmington, R v Goldstein* [2005] UKHL 63, [2006] 1 A.C. 459 per Lord Bingham [35].

79 This is because in most cases the minimum age for a person to be on a social media platform is 13.

80 *R v Stephen Dure,* Southampton Magistrates' Court, 3 September 2018 (unreported). See also, *The BBC*, 'Southampton "paedophile hunter" jailed for false claim' *The BBC* (London, 3 September 2018) <https://www.bbc.co.uk/news/uk-england-hampshire-45395607> accessed 16 May 2023.

81 James Titcomb, 'Facebook admits up to 270m users are fake and duplicate accounts' *The Telegraph* (London, 2 November 2017) <https://www.telegraph.co.uk/technology/2017/11/02/facebook-admits-270m-users-fake-duplicate-accounts/> accessed 30 April 2018. In May 2017 Facebook detailed the scale of abuse on its site. Their study found that in one three month period, they had removed 583 million fake accounts. See, Dave Lee, 'Facebook details scale of abuse on its site' *The BBC* (London, 15 May 2018) <http://www.bbc.co.uk/news/technology-44122967> accessed 29 May 2018.

> Online communication is developing at such a fast pace, new ways of targeting and abusing individuals online are constantly emerging [...]. Offenders can mistakenly think that by using false online profiles and creating websites under a false name their offences are untraceable.[82]

This statement given by the CPS during the release of their updated guidelines in 2016, proceeded to give examples of behaviours which were criminalised under both the MCA and s.127(2) of the CA, including the creation of a fake social media profile containing false information.

In August 2018 the guidelines were further updated regarding what could constitute a false communication online:

> The act of setting up a false social networking account or website, or the creation of a false or offensive profile or alias could amount to a criminal offence, depending on the circumstance.[83]

The CPS supported the idea that false communications could encompass fake online profiles 'depending on the circumstance[s]'.[84] But was this a step too far in interpreting the meaning of false under both the MCA and s.127(2) of the CA? Arguably, such a situation could be considered as an arbitrary use of the law.

Though the author does not justify the sending of highly abusive communications on social media platforms through fake profiles, there is a place for anonymity online. For instance, during times of conflict, individuals may be prosecuted for speaking out against their State, here anonymity protects these individuals. Even in more 'stable' societies, anonymity is needed. For example, a person may have to hide their true identity, such as gender or sexuality, from their loved ones for religious or cultural reasons. Indeed, for the Law Commission, prior to the OSA, they noted:

> The offence under section 127(2)(a) and (b) of the CA 2003 is particularly broad, as it may be committed simply where the intention is to cause another person 'inconvenience'. This means that many hoaxes and 'April fools' jokes posted online could technically fall within the offence if the intention is to cause 'inconvenience' to another.[85]

Consequently, the lack of understanding as to what constituted a false message allowed those working in the criminal justice system to stretch its meaning beyond

82 David Barrett, 'Faking social media accounts could lead to criminal charges' *The Telegraph* (London, 3 March 2016) <https://www.telegraph.co.uk/news/uknews/crime/12180782/Faking-social-media-accounts-could-lead-to-criminal-charges.html> accessed 30 April 2018.

83 Guidelines on Prosecuting Cases Involving Communications Sent via Social Media (n 63) [8].

84 Ibid.,

85 Law Com No 381 (n 59) [11.136].

Parliament's original intent, with the ECtHR noting that false communications are actually protected under the ECHR.[86]

Under a strict interpretation of the term false, petty behaviours post break up, could have constituted a criminal offence under the MCA or s.127(2) of the CA. For example, in *R v Alexandra Gregory*[87] the defendant received a four month custodial sentence, suspended for 18 months, after sending hundreds of false messages online to her ex-partner claiming to be pregnant. Although such behaviour cannot be condoned, it is legitimate to ask if the criminal law should have been invoked in this matter. That is not to say the CPS would prosecute in all cases, but significant weight was/is given to prosecutorial discretion.

As previously noted, the sending of false material online is now criminalised under s.179(1) of the OSA in a bid to bring more certainty to this area of the law, and to tackle more directly mis/disinformation. Chapter 7 will highlight that the issues raised above remain even with the changes made to this area of the criminal law.

6.6.2.4 Threatening

The MCA made it a criminal offence to convey a message which was threatening. Like many of the prohibited behaviours previously discussed, there was no legal definition of what constituted a threat. A problem not just unique to communication offences, but also other offences in the criminal law framework of England and Wales.[88]

From case law, and the 2023 CPS prosecuting guidelines on social media offences, it was difficult to articulate what was meant by a threatening message contrary to the MCA. The key issue was whether the communication could have been considered as a credible threat. An issue which is incredibly difficult to determine in an online context where tone is important, but impossible to understand:

> Unlike a face to face communication, there is no accompanying tone, volume or gesture, and no visible reactions from others (such as cowering or hiding). It can be difficult, then, to infer from an online communication whether, for example, a threat is sincere, or whether a person intends to be menacing or grossly offensive, rather than humorous.[89]

Easy access to the online world has also created new forms of threatening behaviour(s). For example, doxing (sharing a person's home address); dogpiling

86 *Salov v Ukraine* (2007) 45 EHRR 51.

87 *R v Alexandra Gregory*, Worcester Magistrates' Court, 2 November 2023 (unreported). See also, Maisie Olah and Alice Cullinane, 'Redditch nurse faked having baby during lockdown' *The BBC* (London, 2 November 2023) <https://www.bbc.co.uk/news/uk-england-hereford-worcester-67264932> accessed 4 December 2023.

88 Peter Alldridge, 'Threats Offences – A Case for Reform' (1994) 8 Crim. LR 176.

89 Law Com No 381 (n 59) [7.83].

(encouraging others to target a specific person online); and sharing of personal intimate images (commonly referred to as revenge porn). Issues like that of false messages which have not gone unnoticed, resulting in the removal of this prohibited behaviour from the MCA and a 'new' offence of sending a threatening message being created under s.181(1) of the OSA which, as discussed in Chapter 7, has attempted to clarify what is meant by the term threat.

6.6.2.5 Menacing

Both the MCA and s.127(1) of the CA prohibit the sending of a communication that can be defined as menacing. Like that of the other prohibited behaviours above, neither provision defines the term. Instead, it has been for the courts to set the boundaries as to what is a menacing message, to ensure that a distinction can be made between a menacing message and a message which can be labelled as a joke or satire humour, even if it is ill-thought-out on behalf of the sender. The leading case on what constitutes a menacing message is *Chambers v DPP*.[90]

Paul Chambers, in January of 2010, took to Twitter to vent his frustration following the closure of Doncaster Robin Hood Airport due to bad weather, tweeting: 'Crap! Robin Hood Airport is closed. You've got week [*sic*] and a bit to get your shit together otherwise I'm blowing the airport sky high'. Quite quickly the tweet, after being brought to the attention of airport officials, was dismissed as a credible threatening communication i.e., it was not a threat of terrorism. Had airport officials considered it a credible threat, the tweet would have been reported to the Ministry of Defence.[91] Foster gives three arguments as to why the threat was deemed non-credible by authorities.[92] First, the communication was posted on Twitter for widespread reading, which would be considered as unusual in the context of threatening to blow an 'airport sky high'. Second, when examining the discourse of the tweet, the language and grammar were inconsistent with the intention of terrorism. Third, it would have been unusual for a person to threaten terrorism in such a way which makes the sender so easily identifiable. Instead, the tweet was reported to the police, with Chambers later being arrested and convicted of sending a menacing message contrary to s.127(1) of the CA.[93]

Chambers and his legal team appealed his conviction in the Crown Court[94] arguing that his message was intended as a joke, and therefore he did not have the relevant *actus reus* needed to commit the offence. This was dismissed by the Crown

90 *Chambers* (n 60).

91 Alisdair A. Gillespie, 'Twitter, Jokes and the Law' (2012) 76(5) J. Crim. Law. 364 (note).

92 Steve Foster, 'Freedom of Expression: Is There a Human Right to Make a Joke?' (2012) 17(2) Cov. L. J. 97, 99.

93 *R v Paul Chambers*, Doncaster Magistrates' Court, 10 May 2010 (unreported).

94 *Chambers v DPP*, Doncaster Crown Court, 3 March 2011 (unreported). For a discussion of the case history see, Gervase de Wilde, 'News: "Twitter joke" case goes to the high court' (*The International Forum for Responsible Media Blog*, 8 February 2012) <https://inforrm.org/2012/02/08/news-twitter-joke-case-goes-to-the-high-court-gervase-de-wilde/> accessed 2 May 2018.

Court who concluded that the tweet sent by Chambers was 'menacing in its content and obviously so. It could not be more clear. Any ordinary person reading this would see it in that way and be alarmed'.[95] He was later permitted to appeal before the High Court.[96]

The original case heard before the High Court was subjected to a second appeal after an agreement was unable to be reached in the first case.[97] Unlike the finding of the Crown Court, the High Court came to the judgment that Chambers' comments, though 'ill-thought-out', were intended as a joke and as a result quashed his conviction in July 2012, concluding that:

> Satirical, or iconoclastic, or rude comment, the expression of unpopular or unfashionable opinion about serious or trivial matters, banter or humour, even if distasteful to some or painful to those subjected to it should and no doubt will continue at their customary level, quite undiminished by this legislation [s.127(1) of the CA].[98]

Put simply, a menacing message means to create a 'sense of apprehension or fear' in another.[99]

The judgment of the High Court has been praised with Foster going as far as arguing that *Chambers* was a 'victory for common sense',[100] after many commentators were heavily critical of the criminal justice system's approach to the matter.[101] Indeed, Gillespie states: 'Paul Chambers should have been told to be careful about his choice of tweets, but prosecution was unnecessary, especially given that nobody took it seriously'.[102] It is estimated that the total cost of this case to the taxpayer was around £18,000.[103]

The vagueness inherent in the term menacing gives actors in the criminal justice system significant discretion, which in turn breaches the principle of legality and threatens freedom of expression. This has however been recognised, to a degree, as an issue, particularly after the finding of the High Court in *Chambers*, resulting in the creation of the CPS social media prosecution guidelines. Yet, there is still a lack of understanding as to the meaning of many of the provisions contained in both

95 Per Judge Jacqueline Davis found in Scaife (n 34) 135.
96 *Chambers* (n 60) per Lord Judge [28].
97 Scaife (n 34) 135.
98 *Chambers* (n 60) per Lord Judge [28].
99 Social Media and other Electronic Communications (n 52).
100 Foster (n 92) 101.
101 Lilian Edwards, 'Section 127 of the Communications Act 2003: Threat or Menace?' (2012) 23(4) Comp. & L. 22.
102 Gillespie (n 91) 368.
103 Nick Cohen, '"Twitter joke" case only went ahead at insistence of DPP' *The Guardian* (London, 28 July 2012) <https://www.theguardian.com/law/2012/jul/29/paul-chambers-twitter-joke-airport> accessed 2 May 2018.

the MCA and s.127(1) of the CA, including that of menacing—as demonstrated in *DPP v Kingsley Anthony Smith*.[104]

In September 2016, Kingsley was acquitted by Luton Magistrates' Court on four counts of sending a menacing message after he had posted several extremist comments on the social media site, YouTube. Including a comment threatening to place an IED on the doorstep of the then Prime Minister, David Cameron. For Luton Magistrates' Court, the comments made were grounded in 'anger and bigotry' but ultimately 'represented empty bombast or ridiculous banter'—which, by applying the outcome of *Chambers*, was not an offence under s.127(1) of the CA. An appeal was sought and granted before the High Court.

The grounds for appeal centred around the assessment in establishing if a communication is menacing, with the High Court concluding that Luton Magistrates' Court had made a mistake:

> the District Judge had first to consider whether the sole element of the *actus reus* that was in dispute on each charge, namely whether the relevant message was of the proscribed character, was proved. To answer that he had to ask himself whether, as a question of fact, taking account of the context and all relevant circumstances, and applying the standards of a reasonable person in an open and just multi-racial (and, I would add, multi-faith) society, it was proved that a particular message [...] was of a menacing character—i.e. would have created a sense of apprehension or fear in a person of reasonable fortitude who received or read it. However, it is clear that the District Judge failed to ask himself that question at all.[105]

Under the principle of legality, the law should be action-guiding, in which citizens should be able to make decisions on their everyday behaviour in accordance with clear and distinct legal provisions. Yet, there are ample examples of the courts struggling to come to a definitive decision on what constitutes a menacing message. Indeed,

> [i]f judicially trained minds can come to diametrically opposite conclusions on the same set of facts it is obvious that expressions such as [...] 'menacing' are so vague that there is no manageable standard by which a person can be said to have committed an offence.[106]

These issues are even more glaringly apparent when trying to qualify the meaning of grossly offensive.

104 *Kingsley Anthony Smith* (n 50).
105 Ibid., per Mr Justice Sweeney [36–37].
106 *Shreya Singhal v Union of India* (2013) 12 S.C.C. 73 [82].

6.6.2.6 Grossly Offensive

The use of the term grossly offensive to label and prohibit certain types of communications, is a significantly controversial term, with concerns being raised that it infringes on the right to freedom of expression: 'Some of us might draw the boundary in one place, whilst others who are particularly concerned about the development of electronic communications might draw it in another'.[107] Though it is important to note, that 'grossly offensive messages do not contribute much to improving our knowledge or participation as citizens in a democracy. Rather, their effect is to distort communications'.[108]

Yet as outlined in Section 6.2 above, its use can be traced back to as early as 1884, where even then, concerns were raised.[109] Yet the term continues to be used in modern communication offences, in which grossly offensive is considered to take its ordinary English meaning.[110] Today, under both the MCA and s.127 (1) of the CA, it is an offence to send a message *via* a (public) electronic communications network[111] that can be labelled as grossly offensive.

For those working in the criminal justice system, it is denoting what is meant by the term grossly offensive, which has caused opposing opinions. For example, in *Collins*, the Magistrates' Court[112] and the High Court[113] would have acquitted Collins for leaving racially aggravated messages on an answering machine, after concluding that the messages did not constitute a grossly offensive communication. Whereas the House of Lords concluded that the messages were of a grossly offensive nature.[114] For Sedley LJ in the High Court, a message could be labelled as grossly offensive when it breached the 'standards of an open and just multiracial society'.[115] Whereas for Lord Bingham in the House of Lords, grossly offensive comments were defined as 'highly abusive, insulting, pejorative, [and of an] offensive character'.[116]

Though *Collins* concerned telephone conversations, the issues relating to what is deemed a grossly offensive message, have become even more apparent following the increased usage of social media. Further complicated by the fact that it can be easy to see why a decision was made in one case, such as that of *R v Woods*,[117] but

107 Law Com No 147 (n 25) [3.6].
108 Gibbon (n 2) 137.
109 HC Deb 09 August 1884, vol 292, col 371.
110 *Connolly* (n 32) per Lord Justice Dyson [10].
111 The Malicious Communications Act 1988 covers electronic communications. S.127 of the Communications Act 2003 covers public electronic communications. Therefore, the Malicious Communications Act 1988 would cover Bluetooth or Apple Airdrop, but s.127 of the Communications Act 2003 would not.
112 *R v Collins*, Leicestershire Magistrates' Court, 4 October 2004 (unreported).
113 *DPP v Collins* [2005] EWHC 1308 (Admin).
114 Ibid., per Sedly LJ [11].
115 *Collins* (n 49) per Lord Bingham of Cornhill [11].
116 Ibid., per Lord Bingham of Cornhill [13].
117 *Matthew Woods* (n 61).

less so in others, such as that of Thomas. Supported further by the Law Commission who found that there is 'controversy over charging and prosecution decisions'.[118] This mirrors similar arguments made in Chapter 4.

On 1 October 2012 a five-year-old child, April Jones, was reported missing by her parents.[119] The case quickly caught the attention of the national press along with her picture being actively shared across social media sites.[120] During the evening of 1 October, Matthew Woods made several remarks on Facebook about the missing schoolchild, before going on to make comments about Madeleine McCann, a child who went missing in Portugal in 2007: 'I woke up this morning in the back of a transit van with two beautiful little girls, I found April in a hopeless place'. 'Could have just started the greatest Facebook argument EVER [*sic*]. April fools, who wants Maddie? I love April Jones'.[121] The next day, 50 people descended on Woods' home, after his address was leaked online, resulting in the police having to arrest him for his safety. He was later rearrested and convicted of sending grossly offensive messages contrary to s.127(1) of the CA.

Whereas a different approach was undertaken in the matter of Daniel Thomas.[122] Thomas took to Twitter following the divers Tom Daley and Peter Waterfield coming fourth during the 2012 London Olympics: 'if there is any consolation for finishing fourth at least [*sic*] daley and waterfield [*sic*] can go bum each other #teamHIV'. Although, unlike in the case of *Woods*, Thomas' message was labelled as 'offensive' as opposed to grossly offensive:

> There is no doubt that the message posted by Mr Thomas was offensive and would be regarded as such by reasonable members of society. But the question for the CPS is not whether it was offensive, but whether it was so grossly offensive that criminal charges should be brought. The distinction is an important one and not easily made.[123]

There is therefore a line between offensive and grossly offensive commentary. However, it is hard to distinguish when a comment crosses the appropriate threshold to warrant criminalisation.[124] For Lilienthal and Ahmad, the distinction simply falls on whether the reasonable person would find the communication grossly

118 Law Com No 381 (n 59) [5.73].

119 Telegraph Reporters, 'What happened to murdered April Jones and who is Mark Bridger?' *The Telegraph* (London, 20 June 2017) <https://www.telegraph.co.uk/news/0/happened-murdered-april-jones-mark-bridger/> accessed 29 April 2018.

120 The following day Mark Bridger was arrested, and later convicted of April's murder.

121 Steven Morris and Dan Sabbagh, 'April Jones: Matthew Woods jailed over explicit Facebook comments' *The Guardian* (London, 8 October 2012) <https://www.theguardian.com/uk/2012/oct/08/april-jones-matthew-woods-jailed> accessed 29 April 2018.

122 The Crown Prosecution News Brief, 'DPP statement on Tom Daley Case and social media prosecutions' (*CPS.gov*, 2012) <http://blog.cps.gov.uk/2012/09/dpp-statement-on-tom-daley-case-and-socialmedia-prosecutions.html> accessed 29 April 2018.

123 Ibid.,

124 Laura Bliss, 'The Crown Prosecution Guidelines and Grossly Offensive Comments: An Analysis' (2017) 9(2) J. Media Law 173, 177.

offensive.[125] However, as argued by Gillespie certain sectors of society will always deem a message more grossly offensive than others, bringing issues in prosecuting abusive comments under these legal provisions.[126] Indeed, for Rowbottom, the current use of grossly offensive in governing online behaviour can be 'overly expansive and catch statements that might not warrant such serious treatment',[127] which in turn has a chilling effect on free speech.[128] There is no distinction as to when a message goes from one being of an offensive nature to one so grossly offensive the criminal law should intervene.

Under the principle of legality:

> no one should be punished under a law unless it is sufficiently clear and certain [...] and no one should be punished for any act which was not clearly and ascertainably punishable when the act was done.[129]

The matters of *Woods* and Thomas highlight the lack of clarity contained in both the MCA and s.127(1) of the CA in terms of what is considered grossly offensive behaviour. Again, issues which have already been recognised, and were a further rationale for the creation and implementation of prosecuting guidelines. Yet, issues remain, even with the guidelines in place. For instance, the case of *R v Alison Chabloz*,[130] a case discussed previously in Chapter 4, illustrates the continued misunderstanding of the term grossly offensive.

Chabloz who defines herself as a Holocaust revolutionist published several videos on the social media site YouTube. These videos contained footage of Chabloz performing songs set to the beat of traditional Jewish folk music, which contained antisemitic hate. 'Campaign Against Antisemitism', a not-for-profit organisation, had made numerous complaints to the police about Chabloz's behaviour online, yet no further action was taken by authorities. In fact, at one point Chabloz herself approached her local police force claiming she was being harassed online by the Jewish community, leading to the police believing she was a victim of online abuse.[131] As a result, 'Campaign Against Antisemitism' chose to pursue a private prosecution against Chabloz, before the CPS eventually took over the handling of the case. Had 'Campaign Against Antisemitism' not pursued the private prosecution, the limitation period under s.127(1) of the CA would have lapsed.[132]

125 Gary Lilienthal and Nehaluddin Ahmad, 'Hate Crime and Social Media in the UK' (2016) 22(7) CTLR 188, 191.
126 Alisdair A. Gillespie, 'Offensive Communications and the Law' (2006) (17)8 ELR 236, 237.
127 Jacob Rowbottom, 'To Rant, Vent and Converse: Protecting Low Level Digital Speech' (2012) 71(2) Cambridge L.J. 355, 375.
128 Ibid., 378.
129 *Rimmington* (n 78) per Lord Bingham [33].
130 *R v Alison Chabloz*, Westminster Magistrates' Court, 11 January 2018 (unreported).
131 Laura Bliss, 'Social Media: "A Theme Park just for Fools"' (2018) 82(4) J. Crim. Law. 301, 303 (note).
132 Under s.127(1) and (2) of the Communications Act 2003 when an incident is reported to the police, the police have six months to act. Failure to act during this time will result in no further criminal proceedings.

Before Westminster Magistrates' Court, Chabloz was found guilty of sending grossly offensive content contrary to s.127(1) of the CA, a decision later upheld by the Crown Court,[133] with the prosecution successfully arguing that:

> [t]he songs, specifically the language used within them, have been carefully considered and composed with the language chosen deliberately [...] They are anti-Semitic, they are targeting the Jewish people as a whole and use both their content and their tone to ensure maximum offence.[134]

Further clarity as to the meaning of grossly offensive was sought in *DPP v Busetti*,[135] where it was concluded that:

> the sender must intend or be aware that the message is not simply offensive but grossly offensive. The fact that the message is in bad taste, even shockingly bad taste, is not enough.[136]

Yet as potently put by Allen:

> in today's new, challenging digital environment, the existing body of legislative instruments, including the Communications Act 2003, do not provide for the degree of harmonisations, clarity nor necessary efficiency to meet the demands which cases such as *Woods* are placing on them.[137]

6.7 Chapter Summary

From the discussion above, a spectrum has been created within the criminal justice system when it comes to inappropriate communications. In essence, there is a hierarchy of unlawful communications as demonstrated in Figure 6.1, but the boundary between them is impossible to quantify. For instance, offensive comments are beyond the scope of the law, but grossly offensive commentary is not. Yet there is no clear distinction in law as to when offensive conduct will be deemed grossly offensive:

> Or, to put the matter more pertinently, if the two cases of *Woods* [the person who sent grossly offensive messages relating to the missing school girl, April Jones] and Thomas [the footballer who sent a homophobic tweet about the divers Tom Daley and Peter Waterfield] are on either side of a line between

133 *R v Alison Chabloz,* Southwark Crown Court, 13 February 2019 (unreported).

134 Bliss (n 131) 302.

135 *Busetti* (n 51).

136 Ibid., per Lord Justice Bean [35].

137 David Allen Green, 'The "Twitter joke trial" returns to the high court' (*NewStatesman*, 22 June 2012) <https://www.newstatesman.com/blogs/david-allen-green/2012/06/twitter-joke-trial-david-allen-green> accessed 30 April 2018.

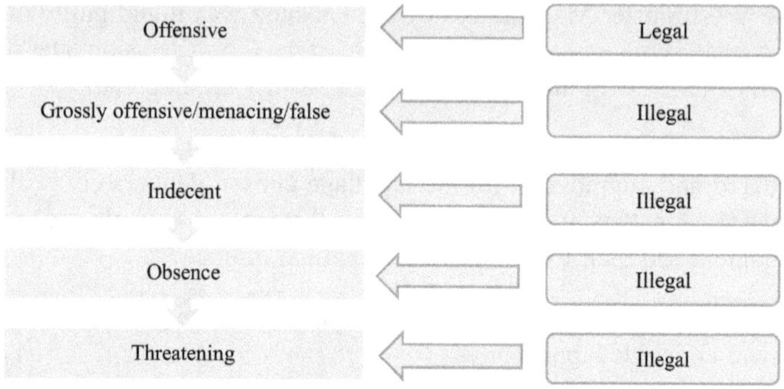

Figure 6.1 The hierarchy of lawful and unlawful communications.

> comments which are merely offensive and those which are grossly so, the question arises as to where that line lies.[138]

This, in turn, has a direct effect on the principle of legality.

As discussed at length in Chapter 3, the law must be certain, clear, and accessible to uphold the principle of legality in the criminal law. Both the MCA and s.127(1) and (2) of the CA in their original format, lacked clarity in terms of what constituted either a threatening, false, indecent, grossly offensive, obscene, or menacing communication, which in turn left both citizens and those working in the criminal justice system at a disadvantage. Indeed,

> [t]he confusion in the case law would seem to demonstrate that emphasising the importance of context in the CPS guidelines may still not be sufficient to guide decisions to charge and prosecute.[139]

For Rowbottom, and noted in Chapter 1,

> [t]he legal responses can, however, seem heavy-handed for what might have been a statement made with little thought while the speaker was sat at a desk at home. Words typed in seconds followed by hitting the enter key can lead to a criminal record or costly civil litigation.[140]

Clearly both the continued use of the MCA and s.127(1) of the CA, even with the minor changes under the OSA, is contentious, for the reasons highlighted throughout this chapter. Significant concern relates mainly to the continued vague language contained in both provisions, concerns which have not gone unnoticed, but have not been addressed in the 'new' communication offences, to which Chapter 7 will turn to examine.

138 Bliss (n 124) 177.
139 Law Com No 381 (n 59) [5.76].
140 Rowbottom (n 127) 356.

Reference list

Alldridge P, 'Threats Offences – A Case for Reform' (1994) 8 Crim LR 176

Amnesty International, 'Amnesty Reveals Alarming Impact of Online Abuse against Women' (*Amnesty International*, 20 November 2017) https://www.amnesty.org/en/latest/news/2017/11/amnesty-reveals-alarming-impact-of-online-abuse-against-women/

Amnesty International UK, 'Black and Asian Women MPs Abused More Online' (*Amnesty International*, 2017) https://www.amnesty.org.uk/online-violence-women-mps

Bakalis C, 'Rethinking Cyberhate Laws' (2017) 27(1) Inf Commun Technol Law 86.

Barrett D, 'Faking Social Media Accounts Could Lead to Criminal Charges' *The Telegraph* (London, 3 March 2016) https://www.telegraph.co.uk/news/uknews/crime/12180782/Faking-social-media-accounts-could-lead-to-criminal-charges.html

Bingham T, *The Rule of Law* (Penguin 2011)

Bliss L, 'Social Media: "A Theme Park Just for Fools"' (2018) 82(4) J Crim Law 301

Bliss L, 'Abuse of women MPs is not just a scandal – it's a threat to democracy' *The Conversation* (London, 17 July 2017) https://theconversation.com/abuse-of-women-mps-is-not-just-a-scandal-its-a-threat-to-democracy-80781

Bliss L, 'The Crown Prosecution Guidelines and Grossly Offensive Comments: An Analysis' (2017) 9(2) J Media Law 173

Cohen N, '"Twitter joke" case only went ahead at insistence of DPP' *The Guardian* (London, 28 July 2012) https://www.theguardian.com/law/2012/jul/29/paul-chambers-twitter-joke-airport

de Wilde G, 'News: "Twitter Joke" Case Goes to the High Court' (*The International Forum for Responsible Media Blog,* 8 February 2012) https://inforrm.org/2012/02/08/news-twitter-joke-case-goes-to-the-high-court-gervase-de-wilde/

Edgar P & Melhuish N, 'NetSafe Online Hate Speech: A Survey on Personal Experiences and Exposure Among Adult New Zealanders' (*Netsafe*, 12 December 2018) https://netsafe.org.nz/wp-content/uploads/2016/12/Netsafe-insights-into-New-Zealands-experience-of-online-hate-speech.pdf

Edwards L, 'Section 127 of the Communications Act 2003: Threat or Menace?' (2012) 23(4) Comp L 22

Fletcher G P, *Basic Concepts of Criminal Law* (Oxford University Press 1998)

Foster S, 'Freedom of Expression: Is There a Human Right to Make a Joke?' (2012) 17(2) Cov L J 97

Geach N & Haralambous N, 'Regulating Harassment: Is the Law Fit for the Social Networking Age?' (2009) 73(3) J Crim Law 241

Gibbon T, 'Case Comment: Grossly Offensive Communications' (2006) 11(4) Comms L 136

Gillespie A A, 'Obscene Conversations, the Internet and the Criminal Law' (2014) 5 Crim LR 350

Gillespie A A, 'Twitter, Jokes and the Law' (2012) 76(5) J Crim Law 364 (note)

Gillespie A A, 'Offensive Communications and the Law' (2006) 17(8) ELR 236, 237

Green D A, 'The "Twitter Joke Trial" Returns to the High Court' (*NewStatesman*, 22 June 2012) https://www.newstatesman.com/blogs/david-allen-green/2012/06/twitter-joke-trial-david-allen-green

HC Deb 09 August 1884, vol 292, col 371

HC Deb 12 December 2000, vol 359, col 481 - 483

HC Deb 12 February 1988, vol 127, col 620

Law Commission, *Abusive and Offensive Online Communications: A Scoping Report* (Law Com No 381, 2018)

Law Commission, *Report on Poison-Pen Letters* (Law Com No 147, 1985)

Lee D, 'Facebook details scale of abuse on its site' *The BBC* (London, 15 May 2018) http://www.bbc.co.uk/news/technology-44122967

Lilienthal G & Ahmad N, 'Hate Crime and Social Media in the UK' (2016) 22(7) CTLR 188

Mill J S, *On Liberty* (first published 1859, Enhanced Media 2016)

Ministry of Justice, 'Impact Assessment: Making the Offence in Section 1 of the Malicious Communications Act 1988 a Triable Either-Way Offence' (*Gov.uk*, 30 May 2014) https://assets.publishing.service.gov.uk/government/uploads/system/uploads/attachment_data/file/321285/malicious-communicationss-impact-assessment.pdf

Morris S & Sabbagh D, 'April Jones: Matthew Woods jailed over explicit Facebook comments' *The Guardian* (London, 8 October 2012) https://www.theguardian.com/uk/2012/oct/08/april-jones-matthew-woods-jailed

Nash V, 'Revise and Resubmit? Reviewing the 2019 Online Harms White Paper' 2019 11(1) J Media Law 18

OFCOM, *"Adults"' Media Use and Attitudes Report 2020/21* (*OFCOM*, 28 April 2021) https://www.ofcom.org.uk/__data/assets/pdf_file/0025/217834/adults-media-use-and-attitudes-report-2020-21.pdf

OFCOM, *"Adults" Media Use and Attitudes Report* (*OFCOM*, 25 April 2018) https://www.ofcom.org.uk/__data/assets/pdf_file/0011/113222/Adults-Media-Use-and-Attitudes-Report-2018.pdf

Olah M & Cullinane A, 'Redditch nurse faked having baby during lockdown' *The BBC* (London, 2 November 2023) https://www.bbc.co.uk/news/uk-england-hereford-worcester-67264932

Press Association, 'Britain's Got Talent blogger cautioned by police' *The Guardian* (London, 3 July 2011) https://www.theguardian.com/tv-and-radio/2011/jul/03/britains-got-talent-blogger-cautioned

Rowbottom J, 'To Rant, Vent and Converse: Protecting Low Level Digital Speech' (2012) 71(2) Cambridge LJ 355

Rowland M, 'Release of the 2022 National Online Safety Survey' (*Ministers for Infrastructure, Transport, Regional Development, Communications and the Arts*, 14 November 2022) https://minister.infrastructure.gov.au/rowland/media-release/release-2022-national-online-safety-survey

Scaife L, *Handbook of Social Media and the Law* (Routledge 2015)

Smith G, 'From Telegram to Tweet: Section 127 and All That' (*Cyberlegal*, 22 February 2015) https://www.cyberleagle.com/2015/02/from-telegram-to-tweet-section-127-and.html

Telegraph Reporters, 'What happened to murdered April Jones and who is Mark Bridger?' *The Telegraph* (London, 20 June 2017) https://www.telegraph.co.uk/news/0/happened-murdered-april-jones-mark-bridger/

The BBC, 'Southampton "paedophile hunter" jailed for false claim' *The BBC* (London, 3 September 2018) https://www.bbc.co.uk/news/uk-england-hampshire-45395607

The BBC, 'Teenager's life "ruined" by Live.me and Twitter "trolls"' *The BBC* (London, 24 October 2017) http://www.bbc.co.uk/news/uk-england-41693437

The Crown Prosecution News Brief, 'DPP Statement on Tom Daley Case and Social Media Prosecutions' (*CPS.gov,* 2012) http://blog.cps.gov.uk/2012/09/dpp-statement-on-tom-daley-case-and-socialmedia-prosecutions.html

The Crown Prosecution Service, 'Communication Offences' (*CPS.gov,* 2024) https://www.cps.gov.uk/legal-guidance/communications-offences#:~:text=the%20summary%20offences%20under%20section,of%20a%20public%20communications%20network

The Crown Prosecution Service, 'Social Media and other Electronic Communications' (*CPS.gov,* 9 January 2023) https://www.cps.gov.uk/legal-guidance/social-media-and-other-electronic-communications

The Crown Prosecution Service, 'Guidelines on Prosecuting Cases Involving Communications Sent via Social Media' (*CPS.gov,* 21 August 2018) https://www

.cps.gov.uk/legal-guidance/social-media-guidelines-prosecuting-cases-involving-communications-sent-social-media

Titcomb J, 'Facebook admits up to 270m users are fake and duplicate accounts' *The Telegraph* (London, 2 November 2017) https://www.telegraph.co.uk/technology/2017/11/02/facebook-admits-270m-users-fake-duplicate-accounts/

Vidgen B, Margetts H & Harris A, 'How Much Online Abuse Is There? A Systematic Review of Evidence for the UK. Policy Briefing – Full Report' (*The Alan Turing Institute*, 2020) https://www.turing.ac.uk/sites/default/files/2019-11/online_abuse_prevalence_full_24.11.2019_-_formatted_0.pdf

Walden I, *Computer Crime and Digital Investigations* (2nd edn, Oxford University Press 2016)

7 The Future of Communication Law

7.1 Introduction

The previous chapter has highlighted significant flaws with the application of communication offences to the digital world. Flaws and concerns which have not gone unnoticed. Following an examination by the Law Commission, starting in 2018 and ending in 2021, into the application of the criminal law to conduct carried out online, in particular social media, the Commission noted several areas of reform. These reforms included the repeal of both the Malicious Communications Act 1988 (MCA) and some aspects of s.127 of the Communications Act 2003 (CA).[1] In addition, several 'new' criminal offences were recommended in the Law Commission's final report into criminalising online conduct. Some of which, were originally included in the first draft of the Online Safety Bill 2022 (OSB) but were eventually removed. Others were included either in the first draft of the Bill or added at a later date, following a lengthy passage through Parliament.[2] However, some of these 'new' offences, are not necessarily new, and as will be shown, in many cases, such offences already existed in the criminal law framework of England and Wales before the passing of the OSB—now the Online Safety Act 2023 (OSA).

Of note, and will be the focus of further discussions below are:

- The original recommendation to create a new harms-based offence to replace both the MCA and s.127(1) of the CA, which has since been rejected;[3]
- Specifically criminalising the sending of threatening communications;[4] and
- Prohibiting the use of a communications device to distribute false information.[5]

1 Note, the Law Commission recommended keeping aspects of s.127(2)c of the Communications Act 2003 to cover hoax communications to the emergency services. See, Law Commission, *Modernising Communications Offences: A final report* (Law Com. No.399, 2021) [3.94].

2 For instance, an offence of encouraging or assisting another to self-harm was added late to the Bill following a House of Lords Amendment.

3 Originally included in the Online Safety Bill 2022 c.150.

4 Online Safety Act 2023 s.179.

5 Ibid., s.181.

DOI: 10.4324/9781003303688-9

Other offences have also been included in the OSA, but will not be examined here, as they are beyond the scope of this monograph.

The majority of the OSA is aimed at search services or user-to-user services, which is discussed at length in the following chapter, as opposed to being aimed at online users. However, the OSA also includes several changes to the criminal law framework, including the offences mentioned above. The following discussion will take each of these offences in turn, and like that of previous chapters, the *actus reus* and *mens rea* of each offence will be outlined. It will expose, as with previous discussions, growing concerns around the use of the criminal law to limit freedom of expression, particularly given the vague terminology used in many of the offences discussed below. In essence, this chapter will demonstrate how the concerns previously raised around communication offences, in both this monograph and by other scholars,[6] including the Law Commission themselves,[7] have not necessarily been addressed by these 'new' criminal offences.

7.2 The Harms Offence

One of the most significant recommendations made by the Law Commission in their final report into communication offences was the recommendation to replace both the MCA and s.127(1) of the CA with a new harms-based offence. As noted in the previous chapter, the Law Commission had found several issues with both the MCA and s.127(1) of the CA,[8] many of which have previously been highlighted. To summarise, the main concerns were:

- The interchangeable nature of the law—the Law Commission noted that both the MCA and s.127 of the CA were often used interchangeably and at times, one was used instead of the other by mistake;
- The vague language in the offences—both offences use vague terminology, which has created confusion for actors in the criminal justice system; and
- Inconsistent prosecutions—the lack of understanding as to the meaning of many of the terms contained in each offence, has led to enforcement by law officials in some cases, but not in others. As a result, discretion is being heavily relied upon and not everyone can be said to be equal before the law.

These concerns, amongst others, led the Law Commission to recommending the repeal of current communication offences in England and Wales, suggesting a new harms-based offence to replace them.

6 See, for example, Laura Higson-Bliss, 'Prosecuting Communication Offences: Interpreting the Harms Model' (2022) 4 CLR 271; Graham Smith, 'Licence to chill' (*Cyberleagle*, 22 November 2021) <https://www.cyberleagle.com/2021/11/licence-to-chill.html> accessed 7 August 2023.

7 Law Commission, *Abusive and Offensive Online Communications: A Scoping Report* (Law Com No 381, 2018) Chapters 4 and 5.

8 Ibid.,

This recommendation was originally accepted by the government at the time and was included in the first draft of the OSB. However, as will be discussed in later sections, significant concerns were raised by politicians, civil liberty organisations, and scholars that the issues previously raised by the Law Commission against current communication offences had not been addressed. In turn, legality was not being maintained, due to vague terminology and as a result freedom of expression was under significant threat. In January 2023, the former conservative government removed this provision, stating:

> The harmful communications offence was one of three new communications offences included in the Bill, alongside the false and threatening communications offences. Since then, stakeholders and members of the public have expressed concerns that the harmful communications offence has the potential to produce unintended consequences on freedom of expression and potentially criminalise legal and legitimate speech on the basis that it has caused someone offence [...] The harmful communications offence has been removed from the Bill.[9]

The MCA and s.127(1) of the CA therefore remain good law, though slightly amended following the passing of the OSA, despite concerns which have been raised against their use to restrict what individuals do or say online. However, it is important to still discuss the recommended offence and its ultimate demise as it illustrates the need to ensure the creation of offences which are grounded in the principle of legality and do not undermine the right to freedom of expression. Further, many of the issues highlighted with this offence, apply to the other 'new' offences contained in the OSA.

The Law Commission recommended, and was accepted initially, the replacement of both the MCA and s.127(1) of the CA with a provision which would have made it a criminal offence to do the following:

> A person commits an offence if—(a) the person sends a message [...] (b) at the time of sending the message—(i) there was a real and substantial risk that it would cause harm to a likely audience, and (ii) the person intended to cause harm to a likely audience, and (c) the person has no reasonable excuse for sending the message.[10]

Below the *actus reus* and *mens rea* of the offence prohibited under the harms offence is outlined. Later parts of this chapter will expose significant issues with the language contained within this recommended criminal law provision, the

9 Department of Digital, Culture, Media and Sport, 'Overview of expected impact of changes to the Online Safety Bill' (*Gov.uk,* 18 Jan 2022) <https://www.gov.uk/government/publications/online-safety-bill-supporting-documents/overview-of-expected-impact-of-changes-to-the-online-safety-bill#:~:text=The%20adult%20safety%20duties%20(%27legal,and%20freedom%20of%20expression%20dutie> accessed 7 August 2023.

10 Online Safety Bill 2022 c.150(1).

principle of legality and the right to freedom of expression, resulting in its removal from the OSB.

7.2.1 Actus Reus

The *actus reus* of the proposed offence consisted of several elements:

1. The sending of a communication;
2. The communication sent must have been likely to cause harm;
3. The likely harm caused must be to a likely audience; and
4. The communication was sent without reasonable excuse.

Under the proposed offence the sending of a message was considered the situation whereby a person:

> (a) causes a communication (including an oral communication) to be sent, transmitted, or published by electronic means, or (b) causes a letter or a thing of any other description to be sent.[11]

Therefore merging the types of communications prohibited between both the MCA and s.127(1) of the CA. Second, for the communication to be considered as likely to cause harm, it must have been found that the communication sent was likely to cause 'psychological harm amounting to at least serious distress'.[12] Therefore, physical, or financial harm, was specifically excluded from the offence. However, as will be highlighted in Section 7.5.1.1, no further information was given as to what would be considered as 'psychological harm amounting to at least serious distress' and instead, serious distress was seen to take its ordinary meaning within the boundaries of the criminal law. This rendered significant issues with the right to freedom of expression and legality.

Third, the likely (undefined) harm had to have the possibility of harming a likely audience, which was defined under the OSB as someone likely to 'encounter' the message.'[13] Which in the context of some social media platforms, could have been anyone in the world. As was noted in the proposed offence itself, the communication sent did not actually need to cause harm, there merely needed to be the possibility that harm might have been inflicted.[14] Last, to fulfil the *actus reus* of the proposed offence, the communication must have been sent without any lawful excuse.

Though initially there was some confusion with the requirement of proving lawful excuse (for some stakeholders, they interpreted this element as a defence), the Law Commission later re-emphasised this was not a defence. Instead, lawful excuse

11 Ibid., c.153(3).
12 Ibid., c.150(4).
13 Ibid., c.150(2).
14 Ibid., c.150(1)b(i).

formed part of the *actus reus* of the offence.[15] Put simply, it would have been for the *prosecution* to prove that the communication sent *was not sent* with a lawful excuse as opposed to the *defence* proving that the communication *was sent* with a lawful excuse. Like that of the other provisions contained in the offence, there was little information given as to what would constitute a reasonable excuse.

7.2.2 Mens Rea

Under first drafts of the proposed offence one of two *mens rea* elements needed to be proven: intent or awareness. Like that of other offences discussed in previous chapters, awareness was considered the situation whereby the person sending the message 'knew' or 'should have known' that harm might have been inflicted upon the likely audience—though as noted above, harm did not need to be proven. However, in an attempt to reduce the scope and reach of the offence, in the final proposal, and was repeated in the OSB, the *mens rea* was reduced to one of 'intent' only,[16] with intent being given its ordinary meaning. I.e,. the sending of a harmful communication in which the aim or purpose was to cause, or likely to cause, harm amounting to at least serious distress.

7.2.3 Summary

Though it is accepted that the repeal of both the MCA and s.127(1) of the CA was a significant step forward in protecting freedom of expression, the proposed replacement, the harms offence, was flawed from its conception. The offence contained significantly vague terminology which would have led to a repeat of how the MCA and s.127(1) of the CA were, and still are, being applied to offences carried out *via* social media. It could not be said that the proposed offence adhered to some of the basic principles of legality, as discussed in Section 7.5 below. Consequently, as has been noted above, the offence was removed from the OSB in January 2023 in a bid to protect the right to freedom of expression. However, as will be shown further, many of the provisions contained in this harms offence, are repeated in other offences contained in the OSA, despite obvious flaws.

7.3 False Communications

As highlighted in the previous chapter, both the MCA and s.127(2) of the CA prohibited the sending of a false communication, with some successful prosecutions under both provisions for doing such an act. However, issues with the use of social media platforms to send mis/disinformation; false advertisements leading to financial loss; and fear-mongering communications, to name but a few, led the Law Commission to note:

15 Law Com. No.399 (n 1) [3.67].
16 Online Safety Bill 2022 c.150(1)b(ii).

> The online environment provides a fertile ground for the spread of false communications, and creates particular challenges for the criminal law in seeking to curtail the most serious conduct.[17]

Many of the challenges in attempting to tackle these issues, were discussed in Chapter 6.

Despite false communications being prohibited under both the MCA and s.127(2) of the CA, the Law Commission, following their 2018 scoping report, recommended the creation of a 'new' false communication offence, which was heavily supported by stakeholders during the consultation period.[18] Since then a 'new' offence of sending a false communication has been created through the enactment of s.179 of the OSA replacing similar provisions in the MCA and s.127(2) of the CA. Under s.179(1) of the OSA it is an offence to intentionally send a knowingly false message, which has the possibility of causing 'non-trivial psychological or physical harm' to a likely audience and is sent without reasonable excuse—mirroring many of the requirements of the proposed harms offence, discussed above. The following section will outline the *actus reus* and *mens rea* of the offence contained in s.179(1) of the OSA.

7.3.1 Actus Reus

The *actus reus* of the offence prohibited under s.179(1) of the OSA consists of the following:

1. A person sends a message.[19] A message under this provision can be sent in several ways:[20]
   (a) Through the sending, transmission, or publishing of an electronic communication; or
   (b) Sending or giving a letter; or
   (c) Causing a letter or electronic communication to be sent. However, it is important to note, that an internet service provider is exempt and is therefore not criminally liable for the sending of a false message under this provision.[21]
2. The message sent conveys false information.[22] No definition is contained in the provision as to the meaning of false, though it is likely that the term false will take its ordinary English meaning. I.e., something that is untrue.

17 Law Com No 381 (n 7) [11.2].
18 Law Com. No 399 (n 1) [3.69].
19 Online Safety Act 2023 s.179(1)a.
20 Ibid., s.182(3).
21 Ibid., s.182(4).
22 Ibid., s.179(1)b.

3. The message sent caused 'non-trivial psychological or physical harm'.[23] It is not clear from the wording of the Act if such harm has to be inflicted or the mere possibility, rendering issues for future interpretation. In addition, both 'non-trivial psychological' and 'physical' harm are left undefined. Once again, this renders issues with interpretation in the future. This is discussed further in Section 7.5.1.2 below.
4. The 'non-trivial psychological or physical harm' inflicted must be to a 'likely audience'.[24] Here, a likely audience is defined as anyone who would encounter the message, including the likelihood that they would encounter the message through another person sharing or forwarding the communication.[25] This creates further issues with the principle of legality, discussed in Section 7.5.2.
5. There must be no 'reasonable excuse' for sending such a message.[26] Like that of the proposed harms offence discussed above, this will be for the prosecution to prove. Similarly, there is no further definition given as to what might be considered a reasonable excuse, giving wide discretionary powers to actors in the criminal justice system.

The OSA also attempts to set some parameters of the offence, including confirming that such an offence includes the sharing of hyperlinks.[27] The issue as to whether the sharing of a hyperlink can constitute an electronic communication has previously caused some judicial discussion, following Alison Chabloz (discussed in Chapter 6) appealing her conviction under s.127(1) of the CA. The premise of her appeal centred on the argument that a hyperlink did not constitute the sending of an electronic communication, even if the content was grossly offensive. This was rejected by the High Court.[28]

Further parameters of the offence include: (1) a statute of limitation period in which prosecutors must bring a case within six months of being notified of such conduct.[29] However, such proceedings cannot be brought more than three years after the commission of the offence.[30](2) 'Recognised news publishers' are exempt from the offence, alongside those holding a licence under the Broadcasting Act 1990 or 1996, those who hold a multiplex licence, a provider of an on-demand service, or 'a film made for cinema to members of the public'.[31]

If all the above elements can be proven and no exemptions contained in the Act are applicable (i.e., falling outside the scope of the limitation period, or the content

23 Ibid., s.179(1)c.
24 Ibid.,
25 Ibid., s.182(5).
26 Ibid., s.179(1)d.
27 Ibid., s.182(7).
28 *Chabloz v Crown Prosecution Service* [2019] EWHC 3094 (Admin).
29 Online Safety Act 2023 s.179(7).
30 Ibid., s.179(8).
31 Ibid., s.180.

is distributed by a 'recognised news publisher', etc,.) the *actus reus* of the offence is present.

7.3.2 Mens Rea

The *mens rea* of the offence contained in s.179(1) of the OSA consists of two elements. First, the message sent must be knowingly false. Under the criminal law framework of England and Wales, knowledge as part of the *mens rea* element is common. For instance, knowledge needs to be proven for an offence of harassment, as discussed in Chapter 5. Therefore, it must be proven that the defendant knows or should have known that the message sent was indeed false. I.e., they knew or should have known that a communication was untrue. Like that of harassment, this will be based on the reasonable person test.

Essentially, it must be established that the average person in the same position as the defendant would have come to the knowledge *or should have* come to the knowledge, that their conduct would result in the sending of a message which was false. At the time of writing, it is unclear if an objective or subjective test will be applied. If the test is subjective, it would result in the characteristics of the accused being taken into consideration when determining if they should have known that the message sent was false. For instance, taking into consideration the religious beliefs of the sender. Whereas an objective test would not, mirroring the law of harassment.

Second, the message sent must be sent with intent, to ensure those who mistakenly share false messages online, do not fall foul of the law. This is an important safeguard given that recent research has uncovered a lack of knowledge by online users in knowing when something online is false:

> OFCOM's study reveals that 30% of UK adults who go online (14.5 million) are unsure about, or don't even consider, the truthfulness of online information. A further 6%—around one in every twenty internet users—believe everything they see online.[32]

Consequently, for the *mens rea* to be present under this offence, the knowingly false message must be sent with the aim or purpose of causing 'non-trivial psychological or physical harm'. Those who share such content recklessly, would not have committed an offence under this provision, therefore providing some safeguards to ensure the overcriminalisation of speech does not occur.

32 OFCOM, 'The genuine article? One in three internet users fail to question misinformation' (*OFCOM*, 22 March 2022) <https://www.ofcom.org.uk/news-centre/2022/one-in-three-internet-users-fail-to-question-misinformation> accessed 7 August 2023.

7.3.3 Summary

It is difficult to see how this 'new' offence differs from previous provisions contained in the MCA and s.127(2) of the CA. A concern that is discussed further in Section 7.5, especially given that the meaning of false still renders issues with legality. An issue which is also relevant when examining the 'new' offence of sending a threatening communication.

7.4 Threatening Communications

Like that of false communications, concerns have previously been raised that the criminal law, in particular the MCA and s.127(1) of the CA, were struggling to keep pace with the use of the internet to send threatening communications to another. Issues often arose between what could be considered a 'credible threat', what is merely menacing or a communication that does not fall within the scope of the criminal law at all, such as that of *Chambers v DPP*.[33] Acknowledged further by the Association of Police and Crime Commissioners in response to the Law Commission's consultation on communication offences:

> Whilst we recognise that existing communications offences explicitly cover threats […] As commissioners of victims services, we recognise the impact that a communication that contains threatening or menacing language may have on victims, and would argue that in order for the offence to adequately capture and describe the nature of wrongdoing, there is sufficient justification for a specific offence relating to threats to be created.[34]

Similar support was received by other stakeholders who argued that the use of a communications device to send threatening messages was well documented,[35] though as highlighted in the previous chapter, figures varied depending on the study. Indeed, human right's organisations specialising in the right to freedom of expression were in support of such an offence provided it was used to target only *genuine threats* and was not *already prohibited* under the criminal law framework of England and Wales.[36]

A new offence therefore of sending a threatening message has been incorporated into s.181 of the OSA, replacing the sending of a threatening message under the MCA, though as noted below, menacing messages, as prohibited under s.127(1) of the CA, remains. Under s.181(1) of the OSA, it is an offence to send a message which conveys a threat of death or serious harm.

33 *Chambers v DPP* [2012] EWHC 2157 (Admin); [2013] 1 W.L.R. 1833.
34 Law Com. No. 399 (n 1) [3.102].
35 Ibid., [3.103–3.106].
36 Ibid., [3.114].

7.4.1 Actus Reus

Two elements need to be established to prove the *actus reus* of the offence of sending a threatening communication contray to s.181(1) of the OSA. First, the person must send a message.[37] The meaning of a 'message' is given the same definition as discussed above and includes the sending of a hyperlink.[38] Second, the message must convey a threat of death or serious harm.[39] Death needs no further explanation. However, serious harm is defined under s.181(2) of the OSA, as either:

- Serious injury amounting to grievous bodily harm within the meaning of the Offences against the Person Act 1861 (OAPA);[40]
- Rape;[41]
- Assault by penetration within the meaning of section 2 of the Sexual Offences Act 2003 (SOA);[42] or
- Serious financial loss.[43]

The inclusion of 'rape' or 'assault by penetration within the meaning of section 2 of the' SOA as amounting to serious harm under this offence, is arguably in direct response to the growing use of social media platforms to target women online.[44] Many studies highlight how women are subjected to a unique form of abuse online, often grounded in threats of sexual violence.[45] Similarly, serious financial loss[46] has arguably been included due to growing concerns around sextortion.[47] These forms of serious harm will not be discussed any further.

However, defining serious harm as a 'serious injury amounting to [...] GBH' is significantly more complex. GBH is defined in *DPP v Smith*[48] as anything amounting to 'really serious harm' and would include actions such as threatening to break

37 Online Safety Act 2023 s.181(1)a.
38 Ibid., s.182(7).
39 Ibid., s.181(1)b.
40 Ibid., s.181(2)a.
41 Ibid., s.181(2)b.
42 Ibid., s.181(2)c.
43 Ibid., s.181(2)d.
44 Law Com. No. 399 (n 1) [3.106].
45 Amnesty International UK, 'Black and Asian women MPs abused more online' (*Amnesty International*, 2017) <https://www.amnesty.org.uk/online-violence-women-mps> accessed 9 December 2018.
46 Note, there is a defence available under s.181(3) of the Online Safety Act 2023 for this particular 'harm': 'it is a defence for the person to show that the threat was used to reinforce a reasonable demand, and the person reasonably believed that the use of the threat was a proper means of reinforcing the demand'. This does not apply to any other harm contained in the section.
47 Kevin Ashford, '"Sextortion"' cases on the increase, police say, leaving victims embarrassed and out of pocket' *ITV News* (London, 7 September 2022) <https://www.itv.com/news/2022-09-07/sextortion-cases-on-the-increase-leaving-victims-embarrassed-and-out-of-pocket> accessed 16 August 2023.
48 *DPP v Smith* [1961] A.C. 290.

another's wrist or ankle, etc. However, the meaning of GBH was extended in 1997 following the case of *R v Ireland*[49] to include psychiatric harm. The meaning of psychiatric harm and the issues it renders with legality is discussed in Section 7.5.1.3.

7.4.2 Mens Rea

The *mens rea* of the offence can be committed in one of two ways:

1. At the time of sending the message the defendant intended that the individual encountering the message would fear that the threat would be carried out;[50] or
2. The defendant '[...] was reckless as to whether an individual encountering the message would fear that the threat would be carried out.'[51]

Put simply, the threatening message must be sent with intent or be sent recklessly. Intent, like that of previous offences discussed, would amount to a defendant's aim or purpose being to cause fear to another that a threat of death or serious harm would be carried out. Whereas recklessness is considered as taking an unjustified risk.

7.4.3 Summary

The inclusion of an offence to specifically criminalise the sending of a threatening message has been welcomed by many, particularly organisations campaigning for better protection for women online.[52] Although the offence under s.181 of the OSA contains more clarity than other provisions discussed above, and in previous chapters, in which Parliament has attempted to set the parameters of the offence during its construction, there are some small, but significant issues as will be demonstrated further in the following sections.

7.5 The 'New' Communication Offences, Freedom of Expression, and Legality

As noted in the introduction, the proposed harms offence and the 'new' communication offences have all been created following an extensive review of communication offences by the Law Commission between 2018 and 2021. However, many of the issues highlighted in the Law Commission's review were not adequately tackled in the proposed harms offence, resulting in its eventual removal, to protect freedom of expression. Of significant worry, though not the only worry, was the use of the terms harm, likely audience and reasonable excuse. Yet, such terminology

49 *R v Ireland* [1998] A.C. 147.
50 Online Safety Act 2023 s.181(1)c(i).
51 Ibid.,
52 Law Com. No.399 (n 1) [3.106].

has continued to be included in the 'new' communication offences which became legally enforceable on the 31 January 2024. These 'new' offences therefore do not tackle many of the issues previously highlighted with communication law and the principle of legality. Of concern, and will be the focus of the remainder of this chapter are:

- The lack of understanding as to the meaning of the types of 'harms' prohibited under all provisions discussed above, in which it cannot be said that the law is sufficiently clear;
- The wide definition given to likely audience and how this is asking the impossible;
- The use of the term reasonable excuse and the wide discretionary powers given to actors in the criminal justice system to establish what constitutes a reasonable excuse; and
- The continued interchangeable nature of the law.

7.5.1 The Types of Harms Prohibited and Lack of Clarity

Harm reduction in the criminal law is not a new phenomenon and arguably underpins the criminal law framework of states across the globe as 'the criminal law system is primary instrumentality for preventing people from intentionally or recklessly harming one another'.[53] But the use of such a principle in determining what should or should not be a criminal offence is contentious as the boundary between what is harmful and what is not is not easily identifiable:

> At first sight, the harm principle seems a straightforward and unproblematic guide to legislative decision-making. Indeed, insofar as our models for understanding harm are broken bones and stolen purses, we can preserve that illusion and think of harm as a simple, determinate, even purely empirical notion [... however ...] harm is a very complex concept with hidden normative dimensions, and that partly because of this, the harm principle cannot be applied in a plausible way in large ranges of circumstance without supplementary criteria (or 'mediating maxims'), some of which are provided by independent moral principles.[54]

Put simply, what might be considered as harmful to one person, might not be to another. Mirroring many of the concerns previously raised against the use of the terms offensive and grossly offensive in the criminal law, as was discussed in the

53 Joel Feinberg, *Harm to Others: The Moral Limits of the Criminal Law* (Oxford University Press 1984) 31. For an alternative approach to harm-based reduction, see, Andy Phippen and Louisa Street, 'Addressing online harms in policy - failing to learn from the war on drugs' (2023) 34(2) Ent. L.R. 36.

54 Ibid., 214.

previous chapter. Arguably, the boundaries of what constitutes harmful behaviour become even more complex, when attempting to apply this notion to speech.

Under each provision discussed above, the purpose is, or was, to prohibit some form of harm to another, with each provision being given a different definition of harm. Regardless of how this harm was or is defined under each provision, as will be shown below, each renders significant issues with legality in which it cannot be said that clarity is being maintained in the criminal law.

7.5.1.1 The Meaning of Harm under the Harms Offence

As outlined in Section 7.2.1, harm, under the proposed harms offence, was defined as 'psychological harm amounting to at least serious distress'. Though arguably this was an improvement on the original definition given by the Law Commission—'psychological harm amounting to at least serious *emotional* distress'[55]—issues remained, particularly given that no further explanation was given to the term 'serious distress':

> the meaning of the level of harm in the Law Commission's recommended reform—'serious distress'—is never specifically defined by the Commission. Instead, there just needs to be a 'real or substantial risk of harm'. Consequently, without a clear understanding as to what is actually meant by 'serious distress', particularly in the context of speech, how does the criminal justice system ensure that trivial harms are not criminalised and, in turn, ensure the right to freedom of expression is not infringed?[56]

As a result, it could not be said that the proposed harms offence to replace both the MCA and s.127(1) of the CA brought any form of clarity to this area of the law and instead, merely repeated issues which had previously been raised. I.e., the use of vague terminology with no measurable standard to restrict the right to freedom of expression.

Though the use of the term 'serious distress' is often used in criminal law provisions in England and Wales and therefore could be considered a quantifiable standard by which criminal liability could be measured, the Law Commission made it clear in their proposal that current definitions of the term should not be written into the offence: '[we are] not recommending that all existing definitions of "serious distress" be copied immutably [... into ...] the offence—it is a different offence, and importantly one based on *likely* harm'.[57] The lack of a clear understanding of the term harm became even more concerning given that under the proposed offence, harm did not need to be proven, there just needed to be the possibility

55 Law Commission, *Harmful Online Communications: The Criminal Offences. A Consultation Paper* (Law Com. No.248, 2020) [5.51].

56 Higson-Bliss (n 6) 281.

57 Law Com. No.399 (n 1) [2.52].

that someone from the likely audience, discussed further below, *could* have been harmed by the communication:

> The problem of an inadequate definition is particularly acute because on the drafting there is no need to prove that any harm to any person was caused to that level. It is enough that this level of harm was likely to be caused to anyone likely to see the message. With some forms of communication on the open internet that would mean that it would be enough that the most vulnerable person who might have seen it would be seriously distressed, so long as the message was sent with intent. That would appear to be a threshold that is very easily met.[58]

In essence, there just needed to be the mere possibility that some form of undefined harm could have been inflicted upon another by the sending of the communication.

It is for this main reason that the harms offence was removed from the OSB in January 2023. The lack of a clear understanding as to what would constitute a harmful message meant that there was a real risk that the right to freedom of expression would be infringed:

> In fact, under the Commission's final recommendation those who knowingly express[ed] provocative views online, for instance, individuals who express[ed] controversial views on transgender rights, on both sides of the debate, could well [have found] themselves falling foul of the law.[59]

Yet vague terminology has continued to be used in the 'new' communication offences.

7.5.1.2 The Meaning of Non-trivial Psychological or Physical Harm when Sending a False Communication

It is an offence under s.179(1) of the OSA to send a false communication which has the intention to cause another 'non-trivial psychological or physical harm'. No definition is contained in the Act as to what constitutes this type of harm. Though in defence of lawmakers, the provision directly reflects the recommendation made by the Law Commission, in which no definition was given to this aspect of the *actus reus*. Consequently, a false message sent with the intent to cause 'non-trivial psychological or physical harm' is unlawful; false communications sent with the intent to cause trivial 'psychological or physical harm' is not. However, the line between the two is not easily identifiable. Even non-trivial physical harm, which on the face of it might be considered easy to prove, renders some issues with the interpretation of the offence.

58 Higson-Bliss (n 6) 281.
59 Ibid., 283.

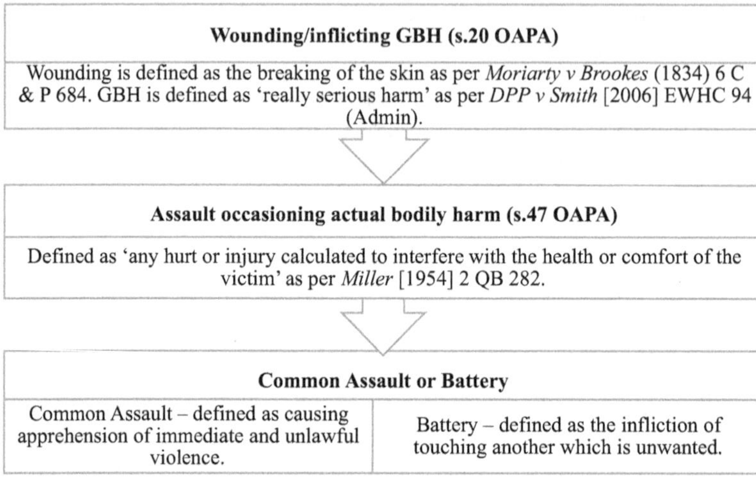

Figure 7.1 Defining non-fatal offences and physical harm.

The use of the criminal law to prohibit the physical harm of another is far from a new phenomenon and exists in many criminal law provisions, particularly in the context of non-fatal offences against the person in England and Wales. This has led to a hierarchy of different types of non-fatal physical harm being created through the enactment of legislation and extensive case law, as shown further in Figure 7.1. At the 'bottom' of this hierarchy is battery, which is defined as 'any intentional touching of another person without the consent of that person and without lawful excuse. It need not necessarily be hostile, rude, or aggressive'.[60] Essentially, the touching of a person on the shoulder, could well constitute battery. In terms of s.179(2)a of the OSA, could it be said that 'battery' would constitute a non-trivial physical harm? Or should the line be drawn at GBH which is defined as 'really serious harm',[61] such as broken or fractured bones? Though, it might be difficult to envision a false communication being sent which causes another some form of physical harm. An example would be the sending of a message which knowingly falsely accuses another of being a paedophile, which is common in the online world,[62] and results in the accused receiving some form of physical injury.

Currently, the difference between a 'non-trivial physical harm' and a 'physical harm' is not clear, creating issues in ensuring the law is action-guiding. Recent events have showcased how the digital world can often merge into the physical

60 *Faulkner v Talbot* [1981] 3 All ER 468 per Lord Lane CJ 471.

61 *Smith* (n 48).

62 Rowena Mason, 'Social media users warned against falsely calling politicians paedophiles' *The Guardian* (London, 8 July 2014) <https://www.theguardian.com/media/2014/jul/08/social-media-users-warned-against-making-false-allegations-mps-paedophiles> accessed 9 August 2023.

world.[63] However, this concern could be easily resolved by clearly stipulating where in the hierarchy of non-fatal offences, a 'non-trivial physical harm' would fall. This will likely be made clearer as cases come before the criminal justice system. Though this does to some degree undermine the principle of legality, in which Fuller notes that we should refrain from case-by-case enlightenment as to the boundaries of the law,[64] this breach is not as significant when compared to the inclusion of the term 'non-trivial psychological' harm.

Though, like that of physical harm, the prohibition of causing someone psychological harm is not new to the criminal law, in which many criminal law provisions prohibit conduct that causes another psychological harm, how this offence is phrased is new. For the Law Commission, during the drafting of the provision on which s.179 of the OSA is based, 'non-trivial psychological harm' was likely 'to include, for example, distress and anxiety, but not annoyance or inconvenience'.[65] It is not quite clear why then, distress or anxiety were not used, as opposed to 'non-trivial psychological harm'. Had distress or anxiety been used, this would have mirrored other criminal law provisions, which in turn would have brought some clarity to this area of the law—whilst at the same time, upholding the principle of legality. For instance, in relation to the Protection from Harassment Act 1997, as discussed in Chapter 5, distress is considered 'oppressive and unreasonable behaviour'.[66] Though it should be noted, when it comes to speech, this could be setting the bar far too low. Nonetheless, there is a definition. Whereas 'non-trivial psychological harm' is not defined.

Like that of 'non-trivial physical harm', it is now for actors in the criminal justice system to set the precedent as to the meaning of this term—which is not easy when there is little case law or previous precedent. Indeed, in the most updated version of the Crown Prosecution Service (CPS) guidelines on communication offences, which were updated following the enactment of the OSA, it simply states, in relation to 'non-trivial psychological harm': 'Prosecutors should be clear when making a charging decision about what the evidence is concerning the suspect's intention and how what was intended was not "trivial", and why not'.[67] Concerns as to where the criminal law should draw the boundaries with psychological harm are further illustrated in how 'serious harm' is defined in s.181(2) of the OSA—the prohibition of sending threatening communications.

63 See, for example, The BBC, 'Rishi Sunak criticises social media craze that led to disorder' *The BBC* (London, 15 August 2023) <https://www.bbc.co.uk/news/uk-england-hampshire-66508164> accessed 15 August 2023.

64 Lon L. Fuller, *The Morality of Law* (Yale University Press 1964) 46.

65 Law Com. No.399 (n 1) [6.46].

66 *Majrowski v Guy's and St Thomas's NHS Trust* [2005] EWCA Civ 251, [2005] Q.B 848 per Auld LJ [82].

67 The Crown Prosecution Service, 'Communication offences' (*CPS.gov,* 2024) <https://www.cps.gov.uk/legal-guidance/communications-offences> accessed 12 March 2024.

7.5.1.3 The Meaning of 'Serious Harm' and Threatening Communications

Whereas the proposed harms offence and the offence of sending a false communication contains or contained no clear definition of the prohibited harm, the provision prohibiting threatening communications does:

> Serious harm means—(a) serious injury amounting to grievous bodily harm (GBH) within the meaning of the Offences against the Person Act 1981, (b) rape, (c) assault by penetration within the meaning of section 2 of the Sexual Offences Act 2003, or (d) serious financial loss.[68]

As noted above, s.181(2)b, c, and d is relatively clear, s.181(2)a—GBH—is less so. GBH is defined in *DPP v Smith*[69] as 'really serious harm'. For instance, a fractured skull or a broken bone would constitute GBH. Though it is likely, in keeping with the extensive case law on the meaning and limits of GBH, the threatening harm does not need to be 'life-threatening, dangerous, or permanent, nor [is it a] requirement that the victim [would] require treatment or that the harm [would] have lasting consequences' to constitute GBH.[70] On a literal interpretation of s.181(2) of the OSA, this may not be of concern. For instance, if someone sends a message online threatening to break another's leg or threatening to 'kick someone's head in' this would amount to serious harm under this provision.

However, as previously noted, in 1997 the meaning of GBH was extended by the House of Lords in *R v Ireland*[71] to include psychiatric injury, which must amount to a recognised psychiatric illness or condition.[72] Though it is important to note that '[h]ysterical and nervous conditions and mere emotions such as fear, distress, or panic do not constitute psychiatric injury'.[73] In the context of communications online, the psychological harm that can be inflicted on another has been well documented.[74] However, communications sent threatening to inflict psychiatric injury less so, though it is not impossible to envision. For example, could it be said that a communication threatening to 'drive someone crazy' or to put another in a 'mental institution' could amount to the psychological harm of another, and if it could, is this foreseeable? What about in the context of a communication which puts a person in fear of their own life but does not directly go far enough to threaten 'really serious harm' or the other prohibited threats under this provision?

68 Online Safety Act 2023 s.181(2).
69 *Smith* (n 48).
70 *Bollom* [2003] EWCA Crim 2846 per Justice Fulford [53].
71 *Ireland* (n 49).
72 *R v D* [2006] EWCA Crim 1139.
73 Michael Allen and Ian Edwards, *Criminal Law* (16th edn, Oxford University Press 2021) 436.
74 Amnesty International, 'Amnesty reveals alarming impact of online abuse against women' (*Amnesty International*, 20 November 2017) <https://www.amnesty.org/en/latest/news/2017/11/amnesty-reveals-alarming-impact-of-online-abuse-against-women/> accessed 9 December 2018.

Like that of the other harms explored above, it will likely be for actors in the criminal justice system to set the parameters of the offence in the wake of vague terminology. This creates significant issues with legality and the criminal law.

7.5.1.4 Summary

Clarity, as noted by Fuller, is an 'essential ingredient' in the criminal law and without it, the law fails in its mission to be 'action-guiding'.[75] As discussed in Chapter 3, the law needs to be action-guiding so that individuals can govern their behaviour in accordance with clear and distinct rules and ultimately not fall foul of the law simply because the legal provision was insufficiently clear. This is of further importance when it comes to criminal provisions which are not necessarily obvious, as is common with offences related to communications, where a balancing act must be made. Or as potently put by Lord Bingham and highlighted at various points in this monograph:

> This is not because bank robbers habitually consult their solicitors before robbing a branch of the NatWest, but because many crimes are a great deal less obvious than robbery, and most of us are keen to keep on the right side of the law if we can.[76]

In essence, we need to strive for the 'maximum clarity which the imperfect medium of language can attain'[77]—something the proposed harms offence failed to do resulting in its quick demise. Yet vague terminology continues to plague the 'new' communication offences.

The use of the terms 'non-trivial psychological or physical harm' and 'serious harm' fails to bring any form of consistency or clarity to this area of the law. In turn, concerns raised previously with the prohibition of false and threatening communications under both the MCA and s.127(1) and (2) of the CA remain. As a result, it cannot be said that legality is being maintained, which like that of other offences examined throughout this monograph, poses a threat to the right to freedom of expression as actors within the criminal justice system attempt to interpret the different meanings given to harm under the 'new' offences.

Failure to create a clear and distinct understanding of any legal provision, particularly when it comes to the coercive nature of the criminal law, further results in personal autonomy being heavily impeded, alongside that of the right to freedom of expression. For Husak,[78] to ensure the protection of personal autonomy against that of the criminal law, 'the law should only intervene where the harm under question

75 Fuller (n 64) 62.

76 Tom Bingham, *The Rule of Law* (Penguin 2011) 37.

77 Law Commission, *Report on Conspiracy and Criminal Law Reform* (1976) Law Com No. 76 [1.8].

78 Douglas Husak, *Overcriminalization: The Limits of the Criminal Law* (Oxford University Press 2008) 82.

prohibits a behaviour which can be defined as a non-trivial harm'.[79] Yet, a lack of understanding as to what would have constituted 'serious distress' in the context of the harms offence and a continued lack of understanding as to what constitutes 'serious harm' or 'non-trivial psychological or physical harm' does not ensure that only non-trivial harms would or could have been criminalised. The lack of clarity as to the meaning of harm under the harms offence and the prohibition of sending false or threatening communications is amplified further when trying to understand the meaning of the likely audience and reasonable excuse.

7.5.2 The Meaning of Likely Audience: Asking the Impossible

Under both the proposed harms offence and the offence prohibiting the sending of a false communication the harm imposed needs to be directed at a likely audience. A likely audience under s.179(2) of the OSA is defined as:

> if, at the time the message is sent, it is reasonably foreseeable that the individuals—(a) would encounter the message, or (b) in the online context, would encounter a subsequent message forwarding or sharing the content of the message.

Similarly, under the proposed harms offence a likely audience was defined as anyone likely to 'encounter' the message.[80] And as briefly discussed in Section 7.2.1 above, in the context of social media, this could have been anyone in the world. Rendering this requirement of the offence significantly broad:

> While this may seem reasonable [the use of the term likely audience], it creates major difficulties on such platforms as Twitter which are open to all. The likely audience is everyone in the world. Given the ease of mass communication, it is a simple matter for a lobby group, such as those asserting the importance of gender identity theory, to organise as part of any audience—making them a 'likely' part of that audience. If they so desire, they can assert the right of transwomen to be potential partners for lesbians and claim that those who disagree are harming them. This is not a fictitious example: it occurs on a regular basis.[81]

The likely audience could therefore be, or could have been under the harms offence, individuals who were more prone to offence; users who are as young as 13 years of age;[82] or political alliances who firmly disagree(d) with the speaker's position. For example, a message sent stating, 'a man cannot become a woman' maybe deemed as false by some. However, for others, this might be considered a true statement but

79 Higson-Bliss (n 6) 280.
80 Online Safety Bill 2022 c.150(4).
81 Law Com. No.399 (n 1) [2.149] (LGB Alliance).
82 In most cases, social media users need to be at least 13 years old.

could be interpreted as causing 'non-trivial psychological harm'.[83] A message sent online is a message sent to the world. Indeed, a likely audience would also include those who are forwarded such content but who might have not necessarily come across it themselves.

The lack of an understanding as to who the likely audience would have been under the harms offence, coupled with the lack of a definitive definition of what constituted harm, was a further reason for its removal from the OSB. However, such a requirement remains in relation to false communications. Not only does its inclusion in s.179(1) of the OSA seriously risk the overcriminalisation of speech, but it also undermines the principle of legality by asking an online user to do the impossible—send a false communication that does not cause 'non-trivial psychological or physical harm' to another—this 'other' being anyone in the world.

Put simply, the false communications offence is asking individuals not to send a message which can be deemed as false, which might cause 'non-trivial psychological or physical harm' to another. In the online world, where context is everything, but is often lost, jokes, banter or distasteful humour could fall within the scope of this offence, given that the likely audience pool is so large. For example, messages sent on April Fool's Day are often false but justified as being a 'joke' or 'banter'. Some of these 'jokes' or 'banter' have the real possibility of causing 'non-trivial psychological or physical harm', such as a 'fake' pregnancy announcement. Where the likely audience could well be anyone in the world, it would not take long to find someone who would be highly upset by such a message, which may go beyond 'trivial psychological harm', especially for those experiencing pregnancy loss or infertility.

Arguably, the example given above of an April Fool's Day joke would be protected under the further requirement placed on the prosecution to prove that the message was sent without reasonable excuse but as explored in Section 7.5.3, issues are also present with this element of the offence. Consequently, it cannot be said that the use of the term likely audience ensures 'the chances of the provision being used arbitrarily is minimal or, indeed, the criminal law will only be invoked as a last resort'.[84]

7.5.3 The Meaning of Reasonable Excuse: Discretion in the Criminal Law

The requirement that a harmful or false message was or is sent without reasonable excuse was contained in both the proposed harms offence and in the offence of sending a false message under s.179(1) of the OSA. During the Law Commission's consultation into these 'new' communication offences, the inclusion of reasonable excuse was hailed by many as safeguarding the right to freedom of expression.[85]

83 This could fall under 'reasonable excuse' but there are issues with this element of the offence also. See Section 7.5.3..

84 Higson-Bliss (n 6) 285.

85 Law Com. No. 399 (n 1) [3.62].

The requirement that the burden of proof is placed on the prosecution to demonstrate, beyond all reasonable doubt, that a message is sent without reasonable excuse was considered as a further aid in protecting the right to freedom of expression: 'In essence, the sender will be entitled to be acquitted unless the prosecution proves that the sender did not have a legitimate or reasonable reason for sending a communication'.[86] Yet, even with this safeguard, the harms offence was still considered a threat to freedom of expression, resulting in its quick demise. And when examining how reasonable excuse has been applied in other areas of the criminal law, it is easy to see why reasonable excuse may not be an adequate safeguard for freedom of expression where the *actus reus* of an offence is so significantly broad.

Reasonable excuse is not new to the criminal law. Similar provisions are found in other areas of the criminal law framework. For instance, reasonable excuse is found in s.1 of the Prevention of Crime Act 1953 whereby a person can rely on reasonable excuse if caught in a public place with a weapon—though this forms part of the defence, rather than the *actus reus*. However, the most well-known connotation of reasonable excuse is its use during the COVID-19 pandemic in England. Under provisions contained in the Health Protection (Coronavirus, Restrictions) (England) Regulations (Health Protection Regulations) 2020, individuals could only leave their homes where they had a 'reasonable excuse to do so'. Although this might have seemed a legitimate requirement—i.e., only to leave one's home when there was no other option—given the global threat to public health, what resulted was an inconsistent application of the criminal law. In some instances, the term was interpreted broadly; in others, a very narrow definition was given.[87] One particular case which caught the national headlines was the issuing of fines to two individuals after they drove five miles to go for a walk by Derbyshire Police who subsequently issued an apology.[88]

Following growing concerns around the inconsistent application of the Health Protection Regulations, the CPS undertook a review of all fines issued under this provision. Concluding that 279 charges (18%) issued under the Health Protection Regulations between March 2020 and March 2021 were unlawful.[89] If such mistakes can take place where a person's basic civil liberties are at risk, how do we ensure similar mistakes are not made with s.179 of the OSA where freedom of expression is at risk?

Put simply, what one police force might see as a reasonable excuse to send a knowingly false message with the intent to cause 'non-trivial psychological or

86 Higson-Bliss (n 6) 287.

87 For an in-depth discussion see, Adam Wagner, *Emergency State: How We Lost Our Freedoms in the Pandemic and Why It Matters* (Penguin 2022).

88 The BBC, 'Covid: Women fined for going for a walk receive police apology' *The BBC* (London, 11 January 2021) <https://www.bbc.co.uk/news/uk-england-derbyshire-55625062> accessed 9 August 2023.

89 Lizzie Dearden, 'All 270 charges brought under Coronavirus Act wrongful, official review finds' *The Independent* (London, 14 May 2021) <https://www.independent.co.uk/news/uk/home-news/coronavirus-act-prosecutions-wrongful-cps-review-b1847194.html> accessed 9 August 2023.

physical harm to another', another may not. This leads to a real possibility that a lack of consistency will be applied in the future interpretation of this offence. Once again, this mirrors concerns with how communication law has previously been applied. For example, what one police force considered grossly offensive another did not.[90] Significant discretion is therefore given to law enforcement, judges, prosecutors, and juries in determining what constitutes a reasonable excuse. Although prosecution guidelines could be created to bring some clarity and consistency to this area of the law, previous attempts to do this with the MCA and s.127(1) and (2) of the CA proves this is easier said than done.[91]

For the Law Commission:

> When deciding whether the defendant had a reasonable excuse for sending or posting the communication, the court must have regard to whether the communication was, or was meant as, a contribution to a matter of public interest.[92]

This was reaffirmed during debates following the inclusion of both the harms offence and the false communications offence in the first draft of the OSB:

> In addition, the court cannot find someone guilty of the harm-based offence or false communications offence if they have a reasonable excuse. A reasonable excuse would include if the communication was or was intended as a contribution to the public interest.[93]

However, what is in the public interest is already a requirement prosecutors need to take into consideration with any criminal offence, including communication offences.[94] Yet, as shown in the previous chapter, there are some cases where it is not always clear why it was in the public interest to pursue a prosecution, *Chambers v DPP* being a prime example of this.[95] Therefore, reasonable excuse does not add any further requirements than already found in the criminal law framework of England and Wales. Further, as noted in Chapter 2, any case that comes before the police, prosecutors or judges, needs to be decided taking into consideration human rights, including the right to freedom of expression, as discussed in Chapter 2. It's difficult to see how a reasonable excuse will provide a better safeguard than

90 As discussed at length in Chapter 6.

91 Ibid,.

92 Law Com. No.248 (n 55) [4.59].

93 Chris Philp, 'Update on the law commission's review of modernising communications offences' (*Gov.uk*, 4 February 2022) <https://questions-statements.parliament.uk/written-statements/detail/2022-02-04/hcws590#:~:text=In%20addition%2C%20the%20court%20cannot,contribution%20to%20the%20public%20interest> accessed 5 August 2023.

94 The Crown Prosecution Service, 'Guidelines on prosecuting cases involving communications sent via social media' (*CPS.gov*, 21 August 2018) [8] <https://www.cps.gov.uk/legal-guidance/social-media-guidelines-prosecuting-cases-involving-communications-sent-social-media> accessed 11 October 2018.

95 *Chambers* (n 33).

human rights law. Therefore, in reality, this element adds very little to ensure the overcriminalisation of speech does not occur and instead gives significant discretion to those working in the criminal justice system to determine what is a reasonable excuse for sending a false communication. For instance, utilising a previous example above, would April Fool's Day be considered a reasonable excuse for sending a false message that amounts to 'non-trivial psychological or physical harm' to another?

As a result, overreliance is therefore still being placed on the discretion of actors in the criminal justice system (the police, prosecutors, and judges) to determine the parameters of this offence. In essence, not tackling the issues raised against previous communication offences where grossly offensive was/is being interpreted narrowly for some but broadly for others.[96]

7.5.4 The Online Safety Act 2023 and Its Place in the Criminal Law Framework of England and Wales: The Interchangeable Nature of the Law

The MCA and s.127(1) and (2) of the CA, as discussed in Chapter 6, have been heavily criticised for their interchangeable nature, where the difference between the two provisions was/is difficult to articulate. This has been made even more complex following the passing of the OSA. For ease the following are criminalised across the MCA, s.127(1) of the CA and s.179 and 181 of the OSA:

- False communications are specifically criminalised under:
  - S.179(1) of the OSA—which prohibits the sending of a knowingly false communication to a likely audience to cause 'non-trivial psychological or physical harm' to another which is sent without reasonable excuse by any form of communication.
- Threatening/menacing communications are prohibited under:
  - S.127(1) of the CA—which prohibits the sending of a menacing message *via* a public electronic communications network with intent or was reckless in the sending of such a message.
  - S.181(1) of the OSA—which prohibits the sending of a threatening message that amounts to a threat of death or serious harm, which is sent recklessly or with intent by any form of communication.
- Indecent communications are prohibited under:
  - S.1(1)(i) of the MCA—which prohibits the sending of an indecent message to another for the purpose of causing 'anxiety or distress' by any form of communication.
  - S.127(1) of the CA—which prohibits the sending of an indecent message *via* a public electronic communications network with intent or was reckless in the sending of such a message.
- Grossly offensive communications are prohibited under:

96 See the previous chapter.

    - S.1(1)(i) of the MCA—which prohibits the sending of a grossly offensive message to another for the purpose of causing 'anxiety or distress' by any form of communication.
    - S.127(1) of the CA—which prohibits the sending of a grossly offensive message *via* a public electronic communications network with intent or was reckless in the sending of such a message.
- Obscene communications are prohibited under:[97]
    - S.127(1) of the CA—which prohibits the sending of an obscene message *via* a public electronic communications network with intent or was reckless in the sending of such a message.

The differences between all these provisions are not obviously clear. Each prohibits very similar behaviour(s) which in turn will lead to continued confusion across the criminal justice system. Mirroring previous concerns around the interchangeable nature of the MCA and s.127 of the CA.[98] For instance, as noted by the CPS, in their current communication offences prosecuting guidelines, following the enactment of the OSA:

> There is a clear overlap with the offence of sending a menacing communication under section 127(1)(a) Communications Act 2003 [and sending threatening communications]. Focus should be on whether the summary section 127 offence reflects the seriousness of the offence and provides adequate powers of sentence. The section 127 offence may also be available where the legal elements of the section 181 offence are not established.[99]

Arguably, the interchangeable nature of the criminal law has been complicated further by the 'new' communication offences.

As it currently stands there is a significant overlap between three communication offences, which is no doubt going to cause issues as cases proceed through the criminal justice system. Put simply, it cannot be said that consistency will be maintained. One police force might pursue a threatening communication under s.181 of the OSA, another might use s.127(1) of the CA, labelling it a menacing communication. This becomes of concern given the difference in punishment; s.181 of the OSA is an either-way offence; s.127(1) is a summary-only offence. It is vital, therefore, that the two are not used interchangeably and if there is a difference between the two offences, this is highlighted clearly in prosecution guidance. Though as was highlighted in the previous chapter, even with prosecution guidance, where legislation is enacted which is embedded in vague terminology mistakes will continue to take place, even with such guidance.

97 Note, obscene material is also prohibited under the Obscene Publications Act 1959 and 1964, though beyond the scope of this monograph.

98 See the previous chapter.

99 Communication Offences (n 67).

This could be easily resolved by amending s.127(1) of the CA to remove reference to menacing communications and removing other overlaps across all three statutes. However, arguably, as has been hinted above, these 'new' communication offences, add very little to our understanding as to when a communication, particularly one sent online where context is everything but is often lost, invokes the criminal law. In essence, many of the issues that have been raised with how the criminal law applies to online communications, even following the passing of the OSA, remain.

7.6 Chapter Summary

The acknowledgement by the Law Commission that previous communication offences were inadequate in dealing with the complexity of online communications, especially those that take place on social media, was a welcomed conclusion. However, it is evident from the discussions above that many of the issues concerning the right to freedom of expression and the principle of legality which have been raised in previous chapters have not been adequately addressed in the passing of these 'new' offences. Instead, many of the issues highlighted throughout this monograph, are also applicable to the OSA.

As discussed above, the 'new' provisions contained in the OSA render significant issues with clarity in the criminal law, particularly around the notion of what constitutes a harmful communication. Even grounding offences around the notion of likely audience or reasonable excuse does not adequately protect the right to freedom of expression. Put simply, it cannot be said that legality is protected under any of these 'new' provisions.

If anything, the law is now more complex than ever. These 'new' offences continue to be grounded in vague terminology, which in turn will result in uncertainty, significant weight being given to the discretion of actors in the criminal justice system to establish the boundaries of the offences and in turn, will lead to inconsistency in the enforcement of the law.

Reference list

Allen M & Edwards I, *Criminal Law* (16th edn, Oxford University Press 2021)
Amnesty International, 'Amnesty Reveals Alarming Impact of Online Abuse against Women' (*Amnesty International*, 20 November 2017) https://www.amnesty.org/en/latest/news/2017/11/amnesty-reveals-alarming-impact-of-online-abuse-against-women/
Amnesty International, 'Black and Asian women MPs abused More Online' (*Amnesty International*, 2017) https://www.amnesty.org.uk/online-violence-women-mps
Ashford K, '"Sextortion" cases on the increase, police say, leaving victims embarrassed and out of pocket' *ITV News* (London, 7 September 2022) https://www.itv.com/news/2022-09-07/sextortion-cases-on-the-increase-leaving-victims-embarrassed-and-out-of-pocket
Bingham T, *The Rule of Law* (Penguin 2011)
Dearden L, 'All 270 charges brought under Coronavirus Act wrongful, official review finds' *The Independent* (London, 14 May 2021) https://www.independent.co.uk/news/uk/home-news/coronavirus-act-prosecutions-wrongful-cps-review-b1847194.html

Department of Digital, Culture, Media & Sport, 'Overview of Expected Impact of Changes to the Online Safety Bill' (*Gov.uk*, 18 January 2021) https://www.gov.uk/government/publications/online-safety-bill-supporting-documents/overview-of-expected-impact-of-changes-to-the-online-safety-bill#:~:text=The%20adult%20safety%20duties%20(%27legal,and%20freedom%20of%20expression%20duties

Feinberg J, *Harm to Others: The Moral Limits of the Criminal Law* (Oxford University Press 1984)

Fuller L L, *The Morality of Law* (Yale University Press 1964)

Higson-Bliss L, 'Prosecuting Communication Offences: Interpreting the Harms Model' (2022) 4 CLR 271

HM Government, 'Internet Safety Strategy – Green Paper' (*HM Government,* 2017) https://assets.publishing.service.gov.uk/government/uploads/system/uploads/attachment_data/file/650949/Internet_Safety_Strategy_green_paper.pdf

Husak D, *Overcriminalization: The Limits of the Criminal Law* (Oxford University Press 2008)

Law Commission, *Modernising Communications Offences: A Final Report* (Law Com. No.399, 2021)

Law Commission, *Harmful Online Communications: The Criminal Offences. A Consultation Paper* (Law Com. No.248, 2020)

Law Commission, *Abusive and Offensive Online Communications: A Scoping Report* (Law Com No 381, 2018)

Law Commission, *Report on Conspiracy and Criminal Law Reform* (Law Com No. 76, 1976)

Mason R, 'Social media users warned against falsely calling politicians paedophiles' *The Guardian* (London, 8 July 2014) https://www.theguardian.com/media/2014/jul/08/social-media-users-warned-against-making-false-allegations-mps-paedophiles

OFCOM, 'The Genuine Article? One in Three Internet Users Fail to Question Misinformation' (*OFCOM*, 22 March 2022) https://www.ofcom.org.uk/news-centre/2022/one-in-three-internet-users-fail-to-question-misinformation

Phippen A & Street L, 'Addressing online harms in policy - failing to learn from the war on drugs' (2023) 34(2) Ent. L.R. 36

Smith G, 'Licence to Chill' (*Cyberleagle*, 22 November 2021) https://www.cyberleagle.com/2021/11/licence-to-chill.html

The BBC, 'Rishi Sunak criticises social media craze that led to disorder' *The BBC* (London, 15 August 2023) https://www.bbc.co.uk/news/uk-england-hampshire-66508164

The BBC, 'Covid: Women fined for going for a walk receive police apology' *The BBC* (London, 11 January 2021) https://www.bbc.co.uk/news/uk-england-derbyshire-55625062

The Crown Prosecution Service, 'Communication Offences' (*CPS.gov,* 2024) https://www.cps.gov.uk/legal-guidance/communications-offences

The Crown Prosecution Service, 'Guidelines on Prosecuting Cases Involving Communications Sent via Social Media' (*CPS.gov,* 21 August 2018) https://www.cps.gov.uk/legal-guidance/social-media-guidelines-prosecuting-cases-involving-communications-sent-social-media

Wagner A, *Emergency State: How We Lost Our Freedoms in the Pandemic and Why It Matters* (Penguin 2022)

Part III

The Criminal Law and Beyond

Whereas Part Two of this monograph focussed primarily on three significant 'harms' associated with the online world, hate speech (Chapter 4), cyberharassment/cyberstalking (Chapter 5) and unlawful communications (Chapters 6 and 7) from a criminal law perspective, Part Three turns to examine the regulation of social media platforms themselves as opposed to individuals. Chapter 8 discusses the latest attempts by the former United Kingdom's government to control unlawful or in some cases, legal but harmful, content online through the creation and implementation of the Online Safety Act 2023. Whereas Chapter 9 explores other countries' approaches to regulating social media. Like that of Part Two, issues with legality and freedom of expression will be critically highlighted. The final chapter, Chapter 10, provides a conclusive summary of the arguments made in the monograph, suggesting recommendations for tackling the 'harms' associated with the online world.

DOI: 10.4324/9781003303688-10

8 Regulating Online Behaviour in the United Kingdom

Beyond the Criminal Law

8.1 Introduction

Whereas Part Two of this monograph focussed primarily on three significant 'harms' associated with the online world, hate speech (Chapter 4), cyberharassment and cyberstalking (Chapter 5), and abusive messages (Chapters 6 and 7) from a criminal law perspective, Part Three turns to examine the regulation of social media platforms themselves as opposed to individuals. This chapter will therefore outline and critically discuss parts of the Online Safety Act 2023 (OSA) and the obligations it places on social media companies to do more to tackle the 'harms' associated with the online world, including those discussed in previous chapters.[1] It is also important to note that, unlike the legal provisions discussed previously, the extent of the OSA extends to Scotland and Northern Ireland, as opposed to just England and Wales.

As will be discussed below, the OSA was born out of the former conservative government's bold plans to become one of the safest places in the world to go online.[2] First envisioned in 2017, the premise of the Act grew significantly; eventually receiving Royal Assent in October 2023. Like many of the legal provisions discussed throughout this monograph, its incorporation into the legal framework of the United Kingdom has been contentious. Many charities and government officials have hailed it as a benchmark for other countries in tackling online 'harms';[3]

1 As noted below, it does regulate more than social media platforms. Arguably however, it has been created with social media in mind.

2 HM Government, 'Internet Safety Strategy – Green Paper' (*Gov.uk*, October 2017) 1 <https://assets.publishing.service.gov.uk/government/uploads/system/uploads/attachment_data/file/650949/Internet_Safety_Strategy_green_paper.pdf> accessed 22 September 2023.

3 See, for example, National Society for the Prevention of Cruelty to Children, 'The Online Safety Bill has been passed in "a momentous day for children"' (*NSPCC,* 19 September 2023) <https://www.nspcc.org.uk/about-us/news-opinion/2023/2023-09-19-the-online-safety-bill-has-been-passed-in-a-momentous-day-for-children/> accessed 22 September 2023; Department for Science, Innovation and Technology and The Rt Hon Michelle Donelan MP, 'Britain makes internet safer, as Online Safety Bill finished and ready to become law' (*Gov.uk*, 19 September 2023) <https://www.gov.uk/government/news/britain-makes-internet-safer-as-online-safety-bill-finished-and-ready-to-become-law> accessed 22 September 2023.

DOI: 10.4324/9781003303688-11

others have considered it a significant threat to the right to freedom of expression.[4] Further, the Act has also been criticised for being unworkable,[5] which in turn breaches the principle of legality as the OSA is, to put it simply, asking the impossible.

Due to the extensive requirements set out under the Act, not all aspects of the OSA will be explored in this chapter—to examine the full requirements under the Act is a book within itself. And even then, the Act is only a very small proportion of the overall reach of this complex piece of legislation. Many of the requirements detailed in the OSA are awaiting further guidance from OFCOM, the United Kingdom's communications regulator, who are essentially the guardians of the OSA, alongside the Secretary of State. It will likely be years before we know the true extent of the OSA and if it has achieved its ultimate aim of ensuring the United Kingdom is one of the safest places in the world to go online.

The discussion below will start by briefly examining the development and history of the OSA. Next, the parameters of the OSA will be outlined, focussing on who falls within its scope. It's important to note, though was often missed in conversations during its passing through Parliament, that the obligations under the OSA extend beyond just social media platforms:

> Almost every online platform that allows user-to-user engagement or search will be caught by the OSB [Online Safety Bill]. From Wikipedia, to Mumsnet, to Minecraft, to Signal, to Tinder, to your local community forum, every online platform or communication channel around the globe which 'targets the United Kingdom' will have to comply with an increasingly onerous array of obligations.[6]

Following on from this, the requirements placed on those who fall within a Category 1 or 2B provider will be discussed—known as the duties. The final parts of this chapter will expose how the Act creates a significant threat to the right to freedom of expression and issues with the principle of legality.

8.2 The Development of the Online Safety Act 2023

Attempting to make the online world safer had been on the former conservative government's agenda since 2017, following a general election in which promises were made to voters:

4 James Baker, 'ORG warns of threat to privacy and free speech as Online Safety Bill is passed' (*Open Rights Group*, 19 September 2023) <https://www.openrightsgroup.org/press-releases/org-warns-of-threat-to-privacy-and-free-speech-as-online-safety-bill-is-passed/> accessed 22 September 2023.

5 Nadia Tymkiw, 'The Online Safety Bill and the risk of unintended consequences' (*The International Forum for Responsible Media Blog*, 11 April 2023) <https://inforrm.org/2023/04/11/the-online-safety-bill-and-the-risk-of-unintended-consequences-nadia-tymkiw/> accessed 6 October 2023.

6 Ibid.,

> to educate today's young people in the harms of the internet and how best to combat them; introduce an industry-wide levy which could in the future be underpinned with legislation; and protect the vulnerable and give people confidence to use the internet without fear through initiatives like a code of practice for social media companies.[7]

Further promising to 'make Britain the best place in the world to setup and run a digital business, while simultaneously ensuring that Britain is the safest place in the world to be online'.[8] The recommendations put forward in the Green Paper focussed on five key areas:

1. A strategic response which would require cooperation between stakeholders, including industry bodies, what was the UK Council for Child Internet Safety (now the UK Council for Internet Safety), government departments and public sector organisations;
2. Working with industry to make the online world safer by encouraging the use of technology to make platforms better, including the creation of a 'social media levy';
3. Embedding the use of safety by design for new and future online platforms;
4. Supporting children, parents, and carers by empathising the critical role education can play in protecting individuals from the 'harms' associated with the online world; and
5. Putting mechanisms in place to respond to online 'harms'.

An extremely modest plan in comparison to what was eventually produced.

In April 2019 the Online Harms White Paper was released, with its recommendations being labelled as 'ambitious'.[9] The key recommendation made in the paper focussed on holding social media companies directly accountable for content on their sites, as opposed to stakeholders working together which was emphasised in the 2017 Green Paper. The paper supported a 'new regulatory framework' to ensure the United Kingdom became 'the safest place in the world to go online'.[10]

What was first envisioned as a modest 20 page piece of legislation imposing a duty of care on some online platforms to protect its users from the 'harms' associated with the online world during its early conception, has changed dramatically. A draft version of the Online Safety Bill (OSB) was made publicly available in 2021 consisting of 145 pages. Since then, the magnitude and reach of the proposed framework has significantly increased, with a final version of the draft Bill being put before Parliament in March 2022. The OSB eventually passed through all

7 The Conservative and Unionist Party Manifesto, 'Forward together: Our plan for a stronger Britain and a prosperous future' (*The Conservative and Unionist Party*, 2017) 79 <https://ucrel.lancs.ac.uk/wmatrix/ukmanifestos2017/localpdf/Conservatives.pdf> accessed 22 September 2023.

8 HM Government (n 2) 3.

9 HM Government, 'Online Harms White Paper' (CP 57, 2019) 1.

10 Ibid.,

stages of the parliamentary process on 19 September 2023, receiving Royal Assent on 26 October 2023. It is now just over 280 pages in length with Paul Scully ex-MP noting at the time of its passing:

> The Bill is world-leading, and the legislative framework established by it will lead to the creation of a profoundly safer online environment in this country. It will kickstart change where that is sorely needed, and ensure that our children are better protected against pornography and other content that is harmful to them.[11]

8.3 The Parameters of the Online Safety Act 2023

The OSA imposes duties on providers of services regulated by the OSA to identify, mitigate, and manage the risks of 'harms' associated with the online world. The OSA applies to two types of online providers: user-to-user services and search services. Search services 'means an internet service that is, or includes, a search engine'.[12] For the purposes of this monograph, search services will not be discussed any further.[13]

Whereas, a user-to-user service is defined as:

> an internet service by means of which content that is generated directly on the service by a user of the service, or uploaded to or shared on the service by a user of the service, may be encountered by another user, or other users, of the service.[14]

With s.2(2) of the Act clarifying:

> (a) For the purposes of subsection (1) it does not matter if content is actually shared with another user or users as long as a service has a functionality that allows such sharing; (b) it does not matter what proportion of content on a service is content described in that subsection.

The scope of user-to-user services includes companies based both inside and outside of the United Kingdom, though the user-to-user service must have links to the United Kingdom. I.e., a significant number of users are based in the United Kingdom[15] or the United Kingdom is a target market.[16] During the parliamentary process, a further element was added to extend the scope of bodies falling within the definition of a user-to-user service to include services

11 HC Deb 12 September 2023, vol 737, col 799.
12 Online Safety Act 2023 s.2(4).
13 Under the Online Safety Act 2023, search services have several different requirements to adhere to as per Chapter 3 of the Act.
14 Online Safety Act 2023 s.2(1).
15 Ibid., s.4(5)a.
16 Ibid., s.4(5)b.

(a) capable of being used in the United Kingdom by individuals, and (b) there are reasonable grounds to believe that there is a material risk of significant harm to individuals.[17]

Essentially allowing smaller companies who pose significant 'harm' to fall within the scope of the OSA, therefore extending the reach of the obligations under the Act from just the biggest platforms to include, smaller platforms that carry high risks.

However, it is important to note the requirements under the OSA do not extend to services which provide email only, text messaging only or one-to-one telephone calls only.[18] Furthermore, the following are also exempt from its requirements: review websites, internal business communications, public bodies, Parliament, those with foreign sovereign power, or journalistic material from a recognised news publisher.[19] However, it does apply to services which provide pornographic material.[20] As a result of the wide definition given to user-to-user services under the Act, many of the biggest social media platforms we know and use today, will fall within its scope. However, the duties applied to each platform will depend on how the platform is categorised, with the OSA setting out two categories for a user-to-user service:

- Category 1 platform—a user-to-user service which will have the most extensive duties to comply with.
- Category 2B platform—a user-to-user service which will only have to adhere to the minimum requirements set out under the Act.

Arguably, a third category exists, a platform that does not fall within the definition of a user-to-user service under the Act (though might be considered a social media platform) and therefore does not need to adhere to the requirements under the OSA.

As noted above, those who fall within the scope of a Category 1 service will have the most significant duties imposed upon them, as outlined in Section 8.4.2 below. Those who fall within a Category 2B service, will still have duties to adhere to, but these will not be as extensive. The idea is to ensure the protection of competition, whilst also ensuring the state does not overburden small user-to-user services or start-up companies. At the time of writing, who falls within each of these categories is still to be decided by the Secretary of State.[21] In addition, platforms which can be accessed by children (regardless of whether children SHOULD be on the platform) will also have additional duties, as set out in Section 8.4.3 below. And to complicate

17 Ibid., s.4(6).
18 Ibid., Schedule 1.
19 Newspaper publisher is defined under s.56(1) of the Online Safety Act 2023.
20 Online Safety Act 2023 Part 5.
21 Ibid., schedule 11 sets out the parameters that must be followed in determining what is a Category 1 and Category 2B platform.

things further, under provisional guidance created by OFCOM (currently under consultation at the time of writing) these platforms will be split into a further four subgroups (following the outcome of their risk assessment, discussed further in Section 8.4.1.1) to determine how they must adhere to their duties under the Act.[22]

8.4 The Requirements under the Online Safety Act 2023

The OSA is underpinned by three key objectives:[23]

1. Tackling illegal content and activity across user-to-user services;
2. Protecting children from 'harm' (this has become the mantra behind much of the discussions around the OSA during its conception and final passage through Parliament); and
3. Empowering users to make informed decisions about the content they wish to see by giving users greater control, whilst also protecting the right to freedom of expression and the right to privacy.

Failure to comply with the requirements set out under the OSA can have significant consequences for those who fall within its scope. OFCOM can impose a fine of up to £18 million or 10% of annual turnover (whichever is higher) on services for failure to comply with their duties (discussed further below) and in certain situations, company officials can be given a custodial sentence, for instance, for providing false information around child exploitation material online.[24]

The requirements placed on those who fall within the definition of a user-to-user service are extensive with the previous government, in their press release following the passing of the OSA noting:

> Social media platforms will be expected to: remove illegal content quickly or prevent it from appearing in the first place, including content promoting self-harm; prevent children from accessing harmful and age-inappropriate content; enforce age limits and age-checking measures; ensure the risks and dangers posed to children on the largest social media platforms are more transparent, including by publishing risk assessments, [and] provide parents and children with clear and accessible ways to report problems online when they do arise.[25]

22 OFCOM, 'Consultation at a glance: our proposals and who they apply to' (*OFCOM,* 9 November 2023) [1.3–1.4] <https://www.ofcom.org.uk/__data/assets/pdf_file/0023/271175/Consultation-at-a-glance-our-proposals-and-who-they-apply-to.pdf> accessed 1 January 2024.

23 Naomi Kilcoyne, 'The Online Safety Bill: Everything in Moderation?, Parts I and II, Content and Context' (*The International Forum for Responsible Media Blog*, 8 December 2022) <https://inforrm.org/2022/12/08/the-online-safety-bill-everything-in-moderation-parts-i-and-ii-content-and-context-naomi-kilcoyne/> accessed 5 October 2023.

24 Online Safety Act 2023 s.69.

25 Department for Science, Innovation and Technology (n 3).

The extent of these requirements, known as duties, is wholly dependent on how the user-to-user service is categorised.

8.4.1 The Minimum Duties for Category 1 and Category 2B Platforms

S.7(2) of the OSA sets out the minimum duties all user-to-user services must adhere to, regardless of whether they are a Category 1 or 2B platform. As a minimum, all providers must:

1. Complete an illegal content risk assessment;[26]
2. Act upon illegal content on their site, following the risk assessment;[27]
3. Create a content reporting procedure;[28]
4. Create a complaints procedure;[29]
5. Protect, maintain, and balance the right to freedom of expression and privacy;[30] and
6. Make and keep written records of all risk assessments.[31]

8.4.1.1 Complete an Illegal Content Risk Assessment

With guidance and a timeframe set out by OFCOM, all user-to-user services must complete an illegal content risk assessment. They must keep the risk assessment updated (reflecting any changes made by OFCOM as to what is considered illegal content) and before making any significant changes to a platform, the user-to-user service must carry out 'a sufficient illegal content risk assessment relating to the impacts of that proposed change'.[32] This is one of the broadest requirements under the OSA with s.9(5) setting out what should be included in the risk assessment, including, though not limited to, the level of risk to the user in being presented with illegal content, the different ways the service is used, how this can impact online 'harms' and the nature and severity of an online 'harm'.

For this to work, the OSA splits illegal content, for adults,[33] into two subcategories:

26 Online Safety Act 2023 s.9.
27 Ibid., s.10.
28 Ibid., s.20.
29 Ibid., s.21.
30 Ibid., s.22.
31 Ibid., s.23.
32 Ibid., s.9(4).
33 Content is categorised differently for children as discussed in Section 8.4.3.1.

1. Priority illegal content—content that is terrorist related,[34] offences related to child sexual exploitation,[35] or content related to any of the following:[36] suicide assistance; threats to kill; public order offences, including harassment and stalking; drugs or psychoactive substances; firearms or other weapons; assisting illegal immigration; human trafficking; sexual exploitation; sexual images, including extreme pornography and image-based abuse; proceeds of crime; fraud and some matters related to financial services; foreign interference; and inchoate offences. In essence, as highlighted in Section 8.4.1.2, priority content should not be seen by users.
2. Other illegal content—any other material which does not fall within the definition of priority illegal content but is a criminal offence under any of the criminal law frameworks of Scotland, Northern Ireland, England and Wales. In essence, as highlighted in Section 8.4.1.2, users should be limited from seeing other illegal content. The issues this poses with legality, in particular, not asking the impossible, is discussed in Section 8.5.1.2.

Draft guidance, created and issued by OFCOM, is currently under consultation at the time of writing to support user-to-user services in complying with this duty, in the form of a rubrics (labelled as risk profiles by OFCOM). In essence, using these risk profiles, companies will use evidence to determine if their platform poses a low, medium, or high risk of exposure to such content with OFCOM providing a four-step process.[37] Following the risk assessment, services must then act upon reducing illegal content on their sites.

8.4.1.2 Act upon Illegal Content on Their Site, Following the Risk Assessment

All user-to-user services must

> take or use proportionate measures relating to the design and operation of the service to (a) prevent individuals from encountering priority illegal content by means of the service, (b) effectively mitigate and manage the risk of the service being used for the commission or facilitation of a priority offence, as identified in the most recent illegal content risk assessment of the service, and (c) effectively mitigate and manage the risks of harm to individuals, as identified in the most recent illegal content risk assessment of the service.[38]

34 Online Safety Act 2023 Schedule 5.
35 Ibid., Schedule 6.
36 Ibid., Schedule 7.
37 OFCOM, 'Protecting people from illegal harms online. Annex 5: Service Risk Assessment Guidance' (*OFCOM*, 9 November 2023) 10–11 <https://www.ofcom.org.uk/__data/assets/pdf_file/0020/271163/annex-5-illegal-harms-consultation.pdf> accessed 1 January 2024.
38 Online Safety Act 2023 s.10(2).

Put simply, all user-to-user services must act to ensure users do not encounter illegal content and such content is 'swiftly' taken down. Though no definition is given to the term 'swift' under the OSA, or under draft guidance issued by OFCOM. Yet, when it comes to priority illegal content, services must act to 'minimise the length of time' in which such content is present on their site.[39] The problems this causes with legality are discussed further in Section 8.5.2.1. To assist with this duty, codes of practice will be issued by OFCOM, based on whether a platform is considered low, medium, or high risk of exposure to the 'harm' being assessed.

8.4.1.3 Create a Content Reporting Procedure

All services have '[a] duty to operate a service using systems and processes that allow users and affected persons to easily report' illegal content.[40] The procedure must be available to both users of the service and non-users provided the non-user is based in the United Kingdom and is either:

> (a) the subject of the content, (b) a member of a class or group of people with certain characteristic targeted by the content, (c) a parent of, or other adult with responsibility for, a child who is a user of the service or is the subject of the content, or (d) an adult providing assistance in using the service to another adult who requires such assistance, where the other adult is a user of the service or is the subject of the content.[41]

These processes must be easy to use, including for children.

8.4.1.4 Create a Complaints Procedure

A complaints procedure must be created by all user-to-user services. The system created must allow for a complaint to be made in the following situations:[42]

1. Content on a platform that is considered as illegal content;
2. The failure of the platform in not complying with its duties to remove illegal content (s.9), its duties in relation to content reporting (s.16) or not complying with its duties to protect freedom of expression or privacy (s.18(2) and (3));
3. Content that is removed from the platform that has been considered illegal;
4. Complaints with regards to action taken against a user, such as a suspension; and
5. The use of proactive technology in removing or restricting content or the technology has been used in a way that is not compatible with the terms of service.

39 Ibid., s.10(3)a.
40 Ibid., s.20(2).
41 Ibid., s.20(5).
42 Ibid., s.21(4).

The procedure must be 'easy to access, easy to use (including by children), and transparent.'[43]

8.4.1.5 Protect, Maintain, and Balance the Right to Freedom of Expression and Privacy

A duty which has been imposed on all platforms, which was used by the previous government as a counterargument to many of the critics of the OSA, is the duty to protect, maintain and balance the right to freedom of expression and privacy. In essence, every user-to-user service, '[w]hen deciding on, and implementing, safety measures and policies' have a duty to take into account 'the importance of protecting users' right to freedom of expression within the law' whilst also taking into account users' privacy.[44] Put simply, decisions made by user-to-user services must be underpinned by the right to freedom of expression and the right to privacy, including if reported content should be removed from a site, essentially allowing for the privatisation of speech regulation. The issues this creates are discussed further in Section 8.5.1.2.

8.4.1.6 Make and Keep Written Records of All Risk Assessments

The final duty imposed upon all user-to-user services is to keep the following written records:

1. The risk assessment carried out in relation to illegal content (as discussed in Section 8.4.1.1 above) and the children's risk assessment (as discussed in Section 8.4.3.1 below) including details about how the assessment was carried out and its findings;[45] and
2. A written record of how the service has complied with its duties against the recommendations made by OFCOM through codes of practice or if they choose a different route of complying with the duty other than the one recommended by OFCOM, what measures they put in place.[46]

These records must be accurate and made available to OFCOM, with OFCOM having powers under the OSA to enforce user-to-user services to hand these documents over.[47]

43 Ibid., s.21(2)c.
44 Ibid., s.22.
45 Ibid., s.23(2).
46 Ibid., s.23(3).
47 Ibid., Chapter 6.

8.4.2 Category 1

In addition to the duties above, those user-to-user services which are defined as a Category 1 platform, must also adhere to additional duties. As noted earlier, at the time of writing, it is not yet known who will fall within the scope of a Category 1 platform, though it is expected that the biggest platforms, such as Facebook, X, TikTok, etc., will likely be a Category 1 service for the purpose of the OSA. In essence, it will be for the Secretary of State, with guidance from OFCOM to determine the boundaries of a Category 1 platform, considering the number of users, functionalities of the services, how easy content is distributed on the platform and any other factors OFCOM feel are relevant.[48] A Category 1 services must, in addition to the above:

1. Adhere to additional requirements under the illegal content risk assessment duty;[49]
2. Empower adult users;[50]
3. Protect content of democratic importance, protect content generated by a news publisher, and protect journalistic content;[51]
4. Adhere to additional duties around content reporting;[52]
5. Adhere to additional requirements under the duty to protect freedom of expression and privacy;[53] and
6. Adhere to additional requirements under the duty to keep records.[54]

8.4.2.1 Adhere to Additional Requirements under the Illegal Content Risk Assessment Duty

In addition to the duties around illegal content risk assessments, as discussed above in Section 8.4.1.1, Category 1 platforms also have

> [a] duty to summarise in terms of service the findings of the most recent illegal content risk assessment of a service (including as to levels of risk and as to nature, and severity, of potential harm to individuals).[55]

8.4.2.2 Empower Adult Users

The overall purpose of the user empowerment duty is to allow users of a Category 1 service, greater control of the content they see on a platform. In essence, the

48 Ibid., Schedule 11.
49 Ibid., s.10(9).
50 Ibid., s.14.
51 Ibid., s.14; s.18 and s.19.
52 Ibid., s.21(6).
53 Ibid., s.22(4), (6) and (7).
54 Ibid., s.23(9) and (10).
55 Ibid., s.10(9).

attention of all users will be actively drawn to mechanisms on a Category 1 site to reduce the likelihood of seeing, or a user given a warning before viewing, content that may contain any of the following:

- the encouragement, promotion or instructions for suicide, self-harm, or eating disorder material;[56] or
- content that is abusive and targets, or incites hatred against another because of any of the following characteristics: race, religion, sex, sexual orientation, disability, or gender reassignment.[57]

The mechanisms used by a Category 1 service will be proportionate, taking into consideration the size and capacity of the service and the findings of a user empowerment risk assessment.[58] These mechanisms must be clearly stated in the platform's terms of service.[59]

In addition, all users will also be given the option to filter out non-verified users from both interacting with the user and encountering content that has been uploaded by a non-verified user.[60] Consequently, all Category 1 services will now need to give users an option to verify their identity. It is important to note, that this duty is not about putting blame on users for not taking control of their own settings, but instead ensuring platforms allow users to make informed decisions on what they would like to see online.

8.4.2.3 Protect Content of Democratic Importance, Protect Content Generated by a News Publisher, and Protect Journalistic Content

All Category 1 services are also under a duty to protect content of democratic importance. To do this, the platform must have proportionate systems in place (depending on the size and capacity of said platform) to ensure freedom of expression is considered when making decisions on the removal of democratically important content or where action is taken against a user regarding content of democratic importance.[61] S.17(7) defines content of democratic importance as content produced either by a news publisher or a user and 'the content is or appears to be specifically intended to contribute to democratic political debate in the United Kingdom or part or area of the United Kingdom'. This duty should be applied in a 'diverse'[62] manner to a wide array of political opinions and how a service is adhering to this duty must be clearly displayed in its terms of service.[63]

56 Ibid., s.16(3).
57 Ibid., s.16(4) and (5).
58 Ibid., s.16(1).
59 Ibid., s.15(7).
60 Ibid., s.64.
61 Ibid., s.17.
62 Ibid., s.17(3).
63 Ibid., s.17(4).

Similarly, Category 1 services are also under a duty to protect journalistic content to ensure freedom of expression is considered when making decisions on the removal of journalistic content or where action is taken against a user regarding journalistic material.[64] This includes a dedicated complaints procedure for the removal of journalistic material and a complaints procedure for a user who has action taken against them, for content they consider to fall within the scope of journalistic content.[65] For the purposes of this duty, journalistic content is considered material produced by a recognised news publisher or a user which is produced 'for the purpose of journalism; and the content is UK-linked'.[66] For instance, the United Kingdom is a target market or 'content is or is likely to be of interest to a significant number of United Kingdom users'.[67] How a company determines if content falls within the meaning of journalistic content must be clearly noted in its terms of service.[68]

In addition, in a further bid to protect the right to freedom of expression, Category 1 services are also obligated to protect content generated by a recognised news publisher. A recognised news publisher is defined as:

> any of the following entities—(a) the British broadcasting cooperation, (b) Sianel Pedwar Cymru, (c) the holder of a licence under the Broadcasting Act 1990 or 1996 who publishes news-related material in connection with the broadcasting activities authorised under the licence [... and are not excluded under further provisions contained in s.56(2) and (3) of the Act].[69]

In essence, any content that is produced by an organisation that falls within the definition above, is entitled to notice where a platform is considering taking action against them (i.e., removing such content from its site or restricting access). S.18 of the OSA sets out the procedure which must be taken to 'give notice'.[70]

8.4.2.4 Adhere to Additional Duties around Content Reporting

In addition to the requirements set out in Section 8.4.1.4 above, Category 1 services must also have a dedicated complaints procedure for grievances related to their

- User empowerment duty;
- The protection of material of democratic importance, news publisher content or journalistic material; and

64 Ibid., s.19.
65 Ibid., s.19(3).
66 Ibid., s.19(10).
67 Ibid., s.19(11).
68 Ibid., s.19(8).
69 Ibid., s.55(1).
70 There are some situations in which notice does not need to be given, for example, '[i]f a provider of a service reasonably considered that the provider would incur criminal or civil liability'. Ibid., s.18(4).

- Protecting the right to freedom of expression and privacy.

8.4.2.5 Adhere to Additional Requirements under the Duty to Protect Freedom of Expression and Privacy

All Category 1 services, in addition to the duties discussed in Section 8.4.1.5 above, must carry out an assessment on the impact any safety measures or policies will have on the right to freedom of expression and privacy, with particular regard to content produced by news publishers or journalistic content.[71] This assessment must be kept up-to-date and be publicly available,[72] including a statement on the positive steps the platform has taken to protect the right to both privacy and freedom of expression.[73]

8.4.2.6 Adhere to Additional Requirements under the Duty to Keep Records

S.23(9) sets out

> a duty to make and keep written record, in an easily understandable form, of all aspects of every assessment under section 14 (assessments related to adult user empowerment duty set out in section 15(2)), including details about how the assessment was carried out and its findings.

A copy of this written record must also be supplied to OFCOM.[74]

8.4.3 Accessed by Children

All user-to-user services are under a legal obligation to protect children on their platforms with a higher level of protection for children than adults online being considered a priority under the OSA.[75] Therefore, all services will need to carry out a children's access assessment to determine if children can access the platform or part of the platform. In which

> [a] provider is only entitled to conclude that it is not possible for children to access a service, or part of it, if age verification or age estimation is used on the service with the result that children are not normally able to access the service or that part of it.[76]

In essence, all services that fall within the scope of the OSA will need to embed some form of age verification software into their platform. The issues this presents

71 Ibid., s.22(4).
72 Ibid., s.22(6).
73 Ibid., s.22(7).
74 Ibid., s.23(10).
75 Ibid., s.11(1).
76 Ibid., s.13(6).

with legality are discussed further in Section 8.5.1.1. Regardless, many of the social media platforms we use today will have to adhere to additional duties to protect children online as most allow users to access their sites from the age of 13.

OFCOM, in their draft guidance on protecting children from online harms has proposed a two-stage test to adhere to this requirement. First, user-to-user services must determine if a child can access the service; Second, the child user service condition must be met.[77] I.e. a significant number of children can access the site and/or the service is likely to attract children.[78] In which OFCOM endorse a context specific approach in determining if the child user service condition is met.[79]

A user-to-user service will, therefore, be considered as 'likely to be accessed by children' in one of the following three situations:

1. The children's access assessment concludes that it is possible that a significant number of children in the United Kingdom can access all or part of the service or the service 'is a kind likely to attract a significant number of children' and there is no mechanism to stop children accessing the platform;[80]
2. Where a service fails to carry out a children's access requirement;[81] or
3. Following a failure to carry out a children's access requirement OFCOM concludes, after an investigation, that the service is likely to be accessed by children.[82]

The children's access assessment must be done at least once a year,[83] or before any major changes are made to the platform[84] or where concerns have been raised that the age verification software being used, is not working.[85] As noted above, draft guidance is currently under consultation at the time of writing with final guidance due to be published by April 2025, to help platforms adhere to this duty.

Following a children's access assessment, if it is found that a user-to-user service is available to children or is likely to be accessed by children, the platform must then adhere to a further two duties:

1. Carry out a children's risk assessment; and
2. Protect children from the 'harms' associated with the online world.

77 OFCOM, 'Children's Access Assessments: Draft guidance for consultation' (*OFCOM*, 8 May 2024) 11-24 <https://www.ofcom.org.uk/siteassets/resources/documents/consultations/category-1-10-weeks/284469-consultation-protecting-children-from-harms-online/associated-documents/a5-draft-childrens-access-assessments-guidance.pdf?v=336055?> accessed 10 July 2024.

78 Online Safety Act 2023 36(3).

79 OFCOM (n 77) [4.8].

80 Online Safety Act 2023 s.37(2).

81 Ibid., s.37(3).

82 Ibid., s.37(4).

83 Ibid., s.36(3).

84 Ibid., s.36(4)a.

85 Ibid., s.36(4)b.

Failure to adhere to these duties, as noted earlier in this chapter, can result in heavy fines, or in some cases, the imprisonment of company officials.

8.4.3.1 Children's Risk Assessment

A user-to-user service which is likely to be accessed by children is under an obligation to carry out a risk assessment, specifically looking at the risks to children on their site, even if children would not ordinarily be on the site—there needs to just be the mere possibility following the children's access assessment. This risk assessment needs to be kept updated and OFCOM notified of any significant changes.[86] Like that of an illegal content risk assessment, where a user-to-user service wishes to make a change to its platform, it must carry out a fresh children's risk assessment, before any changes are done.[87]

S.11(6) of the OSA outlines what should be considered during a children's risk assessment. This includes taking into account the risks posed by different age groups (because what might be harmful to a 13 year old, might be very different to someone aged 17). In addition, '[t]he level of risk of harm to children presented by content that is harmful to children which particularly affects individuals with a certain characteristic or members of a certain group' must also be considered when carrying out a risk assessment.[88] And like that of an illegal content risk assessment, further guidance and support will be issued by OFCOM to help support user-to-user services in carrying out this duty.

Whereas illegal content in relation to adults, as discussed in Section 8.4.1.1 above is split into two types (priority and illegal harms), for children, three categories of 'harm' are created under the OSA:

1. Primary priority content—as set out in s.61 this includes: pornographic material,[89] and content that encourages, promotes or provides instructions for suicide or an act of deliberate self-injury or an eating disorder. This material should never be seen by children.
2. Priority content which children should be protected from seeing - this includes:[90]
    a. Content that targets a person or incites hatred against another because of their race, religion, sex, sexual orientation, disability, or gender reassignment;
    b. Bullying content which either conveys a serious threat, can be labelled as 'humiliating or degrading' or forms 'part of a campaign of mistreatment';

86 Ibid., s.11(3).
87 Ibid., s.11(4).
88 Ibid., s.11(6)d.
89 Note there are some exceptions as set out in s.61(6) Ibid.,
90 Ibid., s.62.

c. 'Content which promotes or provides instructions for an act of serious violence against people' or depicts such content;
d. Content that promotes violence against animals, including providing instructions;
e. 'Content which encourages, promotes or provides instructions for a challenge or stunt highly likely to result in serious injury to the person who does it or to someone else'; or
f. 'Content which encourages a person to ingest, inject, inhale, or in any other way self administer—(a) physical harmful substance; (b) a substance in such a quantity as to be physically harmful.'

3. Non-designated content that is harmful to children—content that 'presents a material risk of significant harm to an appreciable number of children in the United Kingdom'[91]—come to be known as the lawful but awful clause. Put simply, this is content that is perfectly legal but could be considered as harmful to children, in which platforms will need to notify OFCOM if they come across such material. Examples given by OFCOM include: 'lived experience posts, speaking about anxiety, depression and self-harm' and 'posts portraying feelings of loneliness or inadequacy as helpless and/or deserved'.[92] The issues this presents with the right to freedom of expression and legality are discussed in Section 8.5.2.2.

These 'harms' must be considered during the children's risk assessment and all services likely to be accessed by children must also have a separate content reporting system for the above content.[93] What services must do when they come across these types of 'harms', is discussed in the following section.

8.4.3.2 Duty to Protect Children

Following the children's risk assessment, all platforms are under an obligation to protect children from the harms highlighted during their risk assessment. To do this, they must

> take or use proportionate measures relating to the design or operation of the service to effectively—(a) mitigate and manage the risks of harm to children in different age groups […] and mitigate the impact of harm to children in different age groups.[94]

91 Ibid., s.60(2)c.

92 OFCOM, 'Protecting children from harms online: Volume 3: The causes and impacts of online harms to children' (*OFCOM*, 8 May 2024) 7.9.64 <https://www.ofcom.org.uk/siteassets/resources/documents/consultations/category-1-10-weeks/284469-consultation-protecting-children-from-harms-online/associated-documents/vol3-causes-impacts-of-harms-to-children.pdf?v=336052> accessed 10 July 2024.

93 Online Safety Act 2003s.20(4).

94 Ibid., s.12(2)a.

In addition, all platforms must

> operate a service using proportionate systems and processes designed to—(a) prevent children of any age from encountering, by means of the service, primary priority content that is harmful to children; (b) protect children in age groups judged to be at risk from other content that is harmful to children.[95]

What will be considered as proportionate will be determined by the outcome of the risk assessment and the size and capacity of the user-to-user service, with support from OFCOM. To aid in these duties, services must use age-verification.[96] Indeed, in draft guidance issued by OFCOM, emphasis is placed on making it harder for children to see 'harmful' content on user-to-user services, even if it can be considered age-appropriate.[97]

All services will also be under a duty to ensure they include in their terms of service details:

- Outlining 'how children of any age are to be prevented from encountering primary priority content';[98]
- 'How children of any age groups judged to be at risk of harm from priority content [...] are protected from encountering it';[99] and
- 'How children in age groups judged to be at risk of harm from non-designated content [...] are prevented from encountering it'.[100]

Furthermore, Category 1 services have '[a] duty to summarise in the terms of service the findings of the most recent children's risk assessment of a service (including as the levels of risk and as to nature, and severity, of potential harm to children)'.[101] Failure to adhere to these requirements can have significant consequences, including the imprisonment of company officials in certain circumstances.

8.4.4 Summary

The passing of the OSA has been hailed by government officials as being the first country in the world to legislate against the 'harms' associated with the online world,[102] though this is not necessarily true, as shown in Chapter 9. Yet, it will not

95 Ibid., s.12(3).
96 Ibid., s.12(4).
97 OFCOM, 'Webinar 3: An Introduction to the Draft Children's Safety Codes of Practice' (*YouTube*, 6 June 2024) <https://www.youtube.com/watch?v=A2mRx5Myj0I&list=PLq5bH2AD0WPmlNP-reUIKf04KkaZ2ZpOA&index=12 accessed 10 July 2024.
98 Online Safety Act 2023 s.12(9)a.
99 Ibid., s.12(9)b.
100 Ibid., s.12(9)c.
101 Ibid., s.10(9).
102 Department for Digital, Culture, Media and Sport and The Rt Hon Nadine Dorries MP, 'World-first online safety laws introduced in Parliament' (*Gov.uk*, 17 March 2022) <https://www.gov.uk/government/news/world-first-online-safety-laws-introduced-in-parliament> accessed 5 January 2023.

be known for several years as to whether the United Kingdom has achieved its aim of becoming one of the safest places in the world to go online, as the OSA is only a small aspect. Full guidance giving life to the OSA is expected in due course from OFCOM, with all guidance not expected to be completed and fully enforced by OFCOM until at least early 2026.

Nonetheless, from the discussions above, it is clear the requirements under the Act are extensive, overly complex and are embedded in vague terminology—despite the duties to protect the right to freedom of expression and the right to privacy. And as will be demonstrated below, we are slowly moving towards the privatisation of speech, in which services that fall within the scope of the OSA, will become the judge and jury of what is, or is not, legal online.

8.5 The Online Safety Act 2023, Freedom of Expression, and Legality

The implementation of the OSA is contentious, to say the least. It was considered by the acting conservative government at its time of passing as a significant step forward in protecting individuals, particularly children, from the 'harms' associated with the online world. Yet 'despite good intentions, the legislation in its current form leaves many open questions and has the potential to lead to unintended and damaging outcomes'.[103] The Act itself is merely a skeleton, with further guidance to come from both the Secretary of State and OFCOM to add meat to its bones.

As will be shown below, the OSA is so overly complex, legality is at risk, and in turn, there is no doubt that freedom of expression is under threat. To demonstrate this, two areas will be of focus (though there is much more that renders issues with legality, such as the discretionary powers given to the Secretary of State and OFCOM—but is beyond the scope of this chapter):

1. Asking the impossible—of focus will be the requirement to use proactive technology to ensure children cannot access a site, or certain areas of a site, and asking social media companies to become the judge and jury in deciding if a post could reasonably be considered as a breach of the law (in either Scotland, Northern Ireland, or England and Wales); and
2. Lack of clarity in the law—of focus will be the requirement to remove illegal content 'swiftly' and 'non-designated content that is harmful to children'.

8.5.1 Asking the Impossible: Proactive Technology and Interpreting the Criminal Law

Throughout various points in the OSA, the requirements being placed on those who fall within its scope is impossible to adhere to. Two significant impossibilities

103 Tymkiw (n 5).

relate to the reliance on proactive technology to verify a user's age and the requirement that platforms, when reviewing flagged content, interpret the criminal law framework of the whole of the United Kingdom—to determine if such content flagged, can reasonably be considered as 'unlawful'. A task which has proven difficult for the courts themselves, as highlighted throughout earlier chapters of this monograph.

8.5.1.1 Proactive Technology and Verification of Age

As noted above, one of the underlying principles of the OSA is to protect young people from the 'harms' associated with the online world. All platforms will therefore be asked to carry out a children's access assessment to see if children can access their platform, even if they are not meant to be on the service. To do this, as required under the OSA, all user-to-user services will need to embed proactive technology, such as facial recognition, into their sites—failure to do so, would mean that a site will automatically be considered as 'likely to be accessed by children'. Essentially, legally enforcing minimum age requirements upon platforms, following their failure to ensure only those who are age appropriate can use their service. I.e., those over the age of 13.[104] To adhere to this requirement, platforms could use their own proactive technology, provided it is adequate, or technology endorsed by OFCOM. Using the technology endorsed by OFCOM would automatically render the platform compatible with this duty.

OFCOM, in their draft guidance which is currently under consultation at the time of writing, emphasis that to meet this duty, the age-assurance/verification software needs to be 'highly effective'.[105] To be considered 'highly effective' such software needs to be technically accurate, robust, reliable, and fair (i.e., ensuring the software is not biased).[106] Yet, such technology, although does exist to some extent, is not 'yet foolproof, or indeed accurate enough to determine someone's exact age'.[107] Indeed, technology endorsed age-verification systems to solve the issues of the digital world are not new. For instance, in 2017 under the Digital Economy Act, Parliament attempted to enforce age-verification on pornographic sites (in a bid to stop young people from accessing such sites).[108] However, this section of the Act was eventually repealed due to the 'complexities of implementation'.[109]

104 OFCOM, 'Children and parents: Media use and attitudes report 2022' (*OFCOM*, 30 March 2022) <https://www.ofcom.org.uk/__data/assets/pdf_file/0024/234609/childrens-media-use-and-attitudes-report-2022.pdf> accessed 31 January 2024.

105 OFCOM (n 77) [3.2].

106 Ibid., [3.4].

107 Andy Phippen, 'Online safety bill: why making the UK the "safest place to go online" is not as easy as the government claims' *The Conversation* (London, 26 September 2023) <https://theconversation.com/online-safety-bill-why-making-the-uk-the-safest-place-to-go-online-is-not-as-easy-as-the-government-claims-214290> accessed 1 January 2024.

108 Digital Economy Act 2017 s.16.

109 Phippen (n 107).

Subsequently, platforms are being required to do something that is currently unfeasible—verify the age of all their users to ensure children cannot access material which they should not be seeing—whilst also upholding the right to freedom of expression, which includes the exchanging of ideas. This could result in some individuals being prevented from accessing perfectly lawful content they are entitled to see, especially given that sites will be forced to prevent those under 18 from seeing lawful but awful material (an issue discussed further in Section 8.5.2.2). Making this aspect of the OSA currently unworkable and even if such technology did exist in the future, '[t]hese measures, if effectively implemented, will make it more difficult for young people to stumble across content meant for adults, but they will not prevent the determined teenager'.[110] Therefore not addressing the issues it set out to do i.e., stopping young people from seeing 'harmful' material online.

8.5.1.2 Interpreting the Criminal Law

Tackling illegal content online, like protecting children online, is another key principle behind the creation and implementation of the OSA. All services, regardless of category, are under a legal obligation to remove illegal content, upon notice of its existence, from its platform. As noted in Section 8.4.1.1 above, illegal content, in relation to adults, is split into two categories: priority illegal content and 'other' unlawful content—with significant weight being given to the ideal that priority illegal content should not be seen by users, but both being subject to the same takedown requests under the OSA. Similarly, for those platforms that are likely to be accessed by children, illegal content—primary priority content, priority content and 'non-designated content that is harmful to children' (discussed further in Section 8.5.2.2)—also needs to be removed/restricted from platforms. Current draft guidance from OFCOM attempts to clarify how this duty will work for content likely to be accessed by children. However, the focus of this section will be on the removal of illegal content for adults.[111]

Under current draft guidance, to adhere to this duty in relation to adults, platforms will be required to essentially interpret priority illegal content in line with the criminal law framework of the whole of the United Kingdom, to determine if a complained about post, or a post flagged by AI technology, could be considered as reasonably unlawful. To do this platforms must have 'reasonable grounds to infer that content is illegal content' based on 'reasonably available information'.[112] Reasonable grounds, as set out in s.192 of the OSA, might exist where a platform:

110 Ibid.,

111 OFCOM, 'Protecting people from illegal harms online. Volume 5: How to judge whether content is illegal or not? (Illegal Content Judgements Guidance)' (*OFCOM*, 9 November 2023) <https://www.ofcom.org.uk/__data/assets/pdf_file/0023/271148/volume-5-illegal-harms-consultation.pdf> accessed 31 January 2024.

112 Online Safety Act 2023 s.192.

> a) has reasonable grounds to infer that the conduct or behaviour element (the act which constitutes the offence) is present and satisfied; b) has reasonable grounds to infer that the mental element (the state of mind when committing the offence) is present and satisfied, and; c) does not have reasonable grounds to infer that a defence to the offence may be successfully relied upon.[113]

Though,

> [i]t is important to note that 'reasonable grounds to infer' is a new legal threshold and is different from the 'beyond reasonable doubt' threshold used by the criminal courts. The 'beyond reasonable doubt' threshold is a finding that only UK courts can reach. When the 'beyond reasonable doubt' threshold is found in UK courts, the person(s) responsible for the relevant illegal activity will face criminal conviction. However, when services have established 'reasonable ground to infer' that content is illegal according to the Act, this does not mean that the user will necessarily face any criminal liability for the content and nor is it necessary that any user has been prosecuted or convicted of a criminal offence in respect of such content. When services make an illegal content judgement in relation to particular content and have reasonable grounds to infer that the content is illegal, the content must however be taken down.[114]

'Reasonably available information' could include content information, complaints information, user profile information, user profile activity, and published information.[115]

Essentially, a burden of proof less than the balance of probabilities has been created. To aid with this process, OFCOM proposes, in relation to restricting content for adults, providing in-depth guidance on the interpretation of the priority offences set out in Schedules 5 to 7 of the OSA only, with some minor exceptions.[116] Whereas when it comes to 'other' illegal content, this only needs to be removed by services when a reasoned decision is put to them by law enforcement, or the material is found unlawful by a court of law.[117]

Put simply, under this duty, platforms, with some support from OFCOM (depending on whether the complained about conduct is a priority offence or another offence), in legal terms, will need to interpret the *actus reus, mens rea*, and defences of a specific criminal act, against a piece of content that has come to their attention. A complex task for those working in the criminal justice system, let alone

113 OFCOM (n 111) [26.12].

114 Ibid., [26.14].

115 Ibid., [26.26].

116 OFCOM anticipate creating guidelines for the following non-priority offences: false communications; epilepsy trolling; assisting another to self-harm; and cyberflashing. Ibid., [26.72].

117 Ibid., [26.281].

private entities, with OFCOM noting that: 'some areas the law is uncertain and in others it is highly technical'.[118]

This includes interpreting all the provisions which have been discussed in previous chapters of this monograph. With hate related speech, cyberharassment and cyberstalking being defined under the Act as priority illegal content, with guidance for these offences being issued from OFCOM. Communication offences as set out in Chapter 6, including the 'new' offences discussed in Chapter 7, fall within the definition of 'other' illegal content and therefore no guidance will be available, with the exception of false communications. Instead, platforms will be required to remove 'other' illegal content when a reasoned decision is put to them by law enforcement, or the material is found unlawful by a Court of Law.[119] Despite, as an example, as highlighted in Chapter 6, the continued difficulties the criminal justice system has experienced in determining what is a 'grossly offensive' communication.

As highlighted throughout this monograph, often those working in the criminal justice system, have found it difficult to interpret these provisions accurately due to the vague terminology used in each provision. For instance, as discussed in Chapter 5, the European Court of Human Rights found the United Kingdom in breach of Article 10 of the European Convention on Human Rights, for its failure to consider the right to freedom of expression in the pursuement of another for harassment.[120] If judicially trained minds can come to diametrically opposite opinions on a set of facts, where the burden of proof is much higher (beyond all reasonable doubt) how are social media platforms supposed to adhere to these requirements? Yet, repeated failure to remove such content, will result in further action being taken. Leading essentially to the privatisation of speech as platforms are asked to do the impossible—police speech based on provisions grounded in vague terminology.

This will no doubt lead to unintended consequences. As eloquently put by Tymkiw:

> Would algorithms remove all tweets promoting an environmental protest, on the basis that they are procuring a Public Order offence? Would the UK population be banned from discussing the unofficial '420 Day', where cannabis producers, consumers, and advocates around the world celebrate marijuana use? Would discussion online about the pros and cons of buying shares be removed for potential violation of financial services legislation?[121]

Concerns which have been echoed in other jurisdictions that have used similar approaches to regulating the online world, such as Germany, as discussed further in the following chapter. McMillan highlights, with regards to Germany, that without

118 Ibid., [26.61].
119 Ibid., [26.281].
120 *Pal v The United Kingdom* (44261/19) [2021] 11 WLUK 603 Times, 9 February 2022 (ECHR).
121 Tymkiw (n 5).

legal expertise, private companies have 'suddenly become the investigator, judge, and executioner' of what is or is not lawful to say online.[122]

8.5.1.3 Summary

The requirements that user-to-user services use proactive technology and the further requirement to interpret the criminal law framework of the United Kingdom are, put simply, asking platforms to do the impossible. And as discussed in Chapter 3, the law should not, and cannot, in accordance with the principle of legality, ask the impossible. But the OSA does. It emphasises the importance of proactive technology in protecting young people from the 'harms' associated with the online world. Yet such technology is not yet 'foolproof' and failure to adhere to these requirements can invoke significant consequences, including prison sentences for company officials in certain circumstances.

In addition to this, private companies are being asked to do the work of the courts, in determining if certain content breaches the criminal law framework of the United Kingdom, whilst also balancing the right to freedom of expression. A duty, like ensuring children cannot access material they should not, that needs to be adhered to, otherwise, a platform could be in breach of the OSA. However, this monograph has proven the difficulties for judicially trained minds in determining if a message is unlawful or indeed, protected under the right to freedom of expression. A clear example of this is *Chambers v DPP*,[123] discussed in Chapter 6. Both the Magistrates' Court and the Crown Court concluded that Chambers had sent a menacing message contrary to s.127(1) of the Communications Act 2003. The High Court quashed his conviction.

Now, under the OSA, companies such as Meta, X, and Snapchat will have to make these decisions. And although the takedown of material user-to-user services deem unlawful will not necessarily meet the threshold of a criminal offence (as it's based on 'reasonableness' as opposed to 'beyond all reasonable doubt') private entities are now the gatekeepers of the right to freedom of expression. Though it could be suggested that since the creation of social media, this has always been the case, as platforms have always been able to determine what is and is not acceptable on their services in accordance with their own terms of service, there are ample examples of platforms getting this wrong. For example, Facebook has previously come under heavy criticism for removing the American Declaration of Independence, after AI technology flagged it as hate speech.[124] If platforms have struggled to interpret their own terms of service against the right to freedom of

122 Robert Gorwa, 'Elections, institutions, and the regulatory politics of platform governance: The case of the German NetzDG' (2021) 45(6) Telecommun. Policy 1, 7.

123 *Chambers v DPP* [2012] EWHC 2157 (Admin), [2013] 1 W.L.R. 183.

124 The BBC, 'Facebook finds Independence document "racist"' *The BBC* (London, 5 July 2018) <https://www.bbc.co.uk/news/technology-44722728> accessed 1 January 2024.

expression, how are they now expected to interpret the criminal law framework of the whole of the United Kingdom? And to complete this task 'swiftly'?

8.5.2 The Repetition of Vague Terminology

Similar to other provisions discussed in this monograph, the OSA is grounded in vague terminology, with this being a deliberate and considered choice by lawmakers, to ensure the OSA is as future proofed as possible.[125] Though it is accepted that the OSA needs to be able to adapt to future changes in technology, especially as we move away from Web 2.0 into Web 3.0 and the Metaverse, some elements are so inherently vague, it renders significant issues with legality. Of concern, and will be the focus of this section, are the meanings, or in some cases, lack of meaning, given to the terms 'swiftly' and 'harmful material to children'.

8.5.2.1 The Meaning of Swiftly

As noted in Section 8.4.1.2 above, platforms will be under a duty to remove unlawful content, which has come to the company's attention, either by AI technology or being flagged by users, 'swiftly'. The OSA does not define the term 'swiftly' and provisional guidance from OFCOM indicates that they too, will not define what it considers as 'swift'. Justifying this decision by stating:

> we are conscious of the risk of perverse outcomes if the regulator were to suggest a one size fits all approach to deadlines for content moderation processes, including those for complaints. It could lead to resources being diverted from types of illegal content which due to their virality cause harm to very many people, resources being diverted from very serious harms which are not illegal (for example, harms to children), or to decisions being made incorrectly due to time pressures.[126]

Instead, it will be for platforms themselves, to determine in their own risk assessments (as part of additional duties under the OSA) what they consider to be 'swift' in the takedown of illegal content:

> To comply with these duties, OFCOM recommends that all large and multi-risk services set and record internal content policies, measure their performance against recorded targets, adequately resource their moderation

125 HC Deb 12 September 2023, vol 737, col 802.

126 OFCOM, 'Protecting people from illegal harms online. Volume 4: How to mitigate the risk of illegal harms – the illegal content Codes of Practice' (*OFCOM*, 9 November 2023) [16.100] <https://www.ofcom.org.uk/__data/assets/pdf_file/0022/271147/volume-4-illegal-harms-consultation.pdf> accessed 1 January 2024.

functions and ensure employees involved in content moderation receive appropriate training.[127]

An interesting approach, given in the past, Facebook has been heavily criticised for its failure to act 'swiftly' in the takedown of a video livestreaming a terrorist attack in Christchurch, New Zealand in 2019.[128] The attack was livestreamed on Facebook, later being removed by the company around an hour after the event occurred. However, other Facebook users continued to reupload the video to the platform, along with the video being uploaded on other sites. In the first, 24 hours after the Christchurch attack, Facebook removed 1.5 million copies of the video from its site.[129] Though not considered by many, as quick enough.[130]

Even in other jurisdictions outside of the United Kingdom, there are inconsistencies in what is considered a reasonable time limit to remove unlawful content online. For example, in France, legislation was struck down by the French Constitutional Court for its requirement that social media platforms remove 'manifestly unlawful content' within 24 hours.[131] Yet in Germany, similar legislation requires platforms to remove such material within 24 hours of being notified of its existence.[132] Both approaches are discussed further in Chapter 9.

The ambiguous nature of the term 'swiftly'—which is defined by the Cambridge English Dictionary as 'quickly or immediately'[133]—does not ensure the law is action-guiding. What one platform might consider as 'swift' another may not. What OFCOM might consider 'swift' in one incident, might not be the same in another. Yet, consistent failure to remove unlawful content in a 'swift' manner can result in heavy fines under the OSA. This becomes of even more concern when platforms will be required to remove 'harmful content' for children, 'swiftly', given that what constitutes harmful in this instance, also lacks any form of clarity in the law.

127 Jessica Kingsbury, 'Ofcom's "Roadmap to Regulation" underway with its consultation on illegal harms duties under the Online Safety Act' (*The International Forum for Responsible Media Blog*, 17 January 2024) <https://inforrm.org/2024/01/17/ofcoms-roadmap-to-regulation-underway-with-its-consultation-on-illegal-harms-duties-under-the-online-safety-act-jessica-kingsbury/> accessed 17 January 2024.

128 Paul Karp, 'Australia passes social media law penalising platforms for violent content' *The Guardian* (London, 4 April 2019) <https://www.theguardian.com/media/2019/apr/04/australia-passes-social-media-law-penalising-platforms-for-violent-content> accessed 30 April 2019.

129 Jim Waterson, 'Facebook removed 1.5m videos of New Zealand terror attack in first 24 hours' *The Guardian* (London, 17 March 2019) <https://www.theguardian.com/world/2019/mar/17/facebook-removed-15m-videos-new-zealand-terror-attack> accessed 26 March 2019.

130 Karp (n 128).

131 Online Hate Bill (2020) Article 1. See, Chapter 9 for an in-depth discussion on the French approach to regulating the online world.

132 Act to Improve Enforcement of the Law in Social Networks (2017) s.3(2). See, Chapter 9 for an in-depth discussion on the German approach to regulating the online world.

133 Cambridge Dictionary, 'Swiftly' (*Cambridge Dictionary*, 2024) <https://dictionary.cambridge.org/dictionary/english/swiftly> accessed 29 January 2024.

8.5.2.2 Harmful Material to Children

For those platforms which can be accessed by those under the age of 18, they will be under a further duty to remove material that presents 'a material risk of significant harm to an appreciable number of children in the United Kingdom'.[134] In essence, material which is essentially legal but could be considered harmful, may, in some instances, be requested to be removed from user-to-user services. Some aspects of this provision are defined under the OSA, others, at the time of writing, are under consultation.[135]

The OSA defines harm under this provision as 'physical or psychological harm' taking into consideration '(a) the nature of the content; (b) the fact of its dissemination; (c) the manner of its dissemination', under s.234(2) and (3) of the Act. A significantly broad definition, in which platforms will also have to interpret 'non-designated content that is harmful to children' considering different characteristics of child users, such as age, race, disabilities, etc., with company officials facing imprisonment for repeated failures to protect children from the online world.[136] Current draft guidance issued by OFCOM has also attempted to bring some further clarity to this provision, by defining the following:[137]

- 'significant harm to an appreciable number of children in the United Kingdom' is to be interpreted as causing 'lasting or severe impacts on children's wellbeing, attitudes or behaviour, or attitudes or behaviours likely to cause harm to others'; and
- Material risk means the likelihood of it occurring.

In essence, it will be for platforms, with some support from OFCOM to determine what might be completely lawful, but harmful for children, rendering issues with the over-censorship of speech, using a 4 stage test:[138]

1 What harms are arising from children viewing online content?
2 Is the harm identified significant?
3 Is there a material risk of that harm occurring?
4 Are an 'appreciable' number of children in the United Kingdom affected?

Even with this further draft guidance from OFCOM, issues remain. For example, the following could all fall within the 'lawful but awful' clause for children:

- Material promoting gambling;
- Material discussing current events, for instance, ongoing wars, terrorist attacks, or criminal behaviour; or
- The promotion of unhealthy food.

134 Online Safety Act 2023 s.60(2)c.
135 OFCOM (n 92).
136 Online Safety Act 2023 s.11(6)d.
137 OFCOM (n 92) [7.9.13 - 7.9.14].
138 Ibid., [7.9.16].

This is just a small list of examples which could fall within the scope of the wide discretion given to platforms to consider what is lawful, but harmful to children as

> [p]latforms will be expected to make rules around content they deem might be harmful to certain users, and censor it before it can be published. As a result, this might also prevent children from accessing information related to gender and sexuality that could be caught up in the filtering and monitoring systems platforms will put in place. Without a clear definition of what harmful content is, it will be down to platforms to guess—and with moving goalposts, depending on the government of the day.[139]

As noted above, examples given by OFCOM which can amount to 'lawful but harmful' content to children could include: 'lived experience posts, speaking about anxiety, depression and self-harm' and 'posts portraying feelings of loneliness or inadequacy as helpless and/or deserved'. I.e., depressive content.[140] Of greater concern, is the recent discussions around the banning of under 16s from all social media platforms, which at the time of writing, was being mooted by some MPs[141] and being discussed in the media.[142] This has since been dismissed by the Labour Government, following a change in leadership.[143] However, the wide-reaching nature of the OSA and its grounding in vague terminology leaves the real possibility of it being open to abuse in the future.

8.5.2.3 Summary

Vague terminology is an issue that has been raised throughout this monograph and although the OSA was purposely created to be as broad as possible, to future-proof the Act, legality is clearly undermined. Platforms will be required to Act 'swiftly' to remove unlawful content, but no parameters have been set, either in statute or guidance as to what constitutes 'swiftly'. If user-to-user services are to adhere to the extensive duties to remove illegal content from their sites, they need to understand the full scope of their requirements, including, how long platforms have to remove such unlawful content. This is particularly acute where a law is complex and right and wrong are not easily identifiable and at the heart of decisions, is the right to freedom of expression. Indeed,

139 Phippen (n 107).

140 OFCOM (n 97).

141 The BBC, 'MP urges social media ban for under 16s' *The BBC* (London, 25 January 2024) <https://www.bbc.co.uk/news/av/uk-politics-68085378> accessed 31 January 2024.

142 Aletha Adu and Dan Milmo, 'Rishi Sunak considers curbing social media use for under-16s' *The Guardian* (London, 14 December 2023) <https://www.theguardian.com/media/2023/dec/14/rishi-sunak-considers-curbing-social-media-use-under-16s> accessed 31 January 2024.

143 Nathan Standley, 'Starmer against "simply banning" children's phones' *The BBC* (London, 12 July 2024) <https://www.bbc.co.uk/news/articles/cd1rklm6433o#:~:text=Prime%20Minister%20Sir%20Keir%20Starmer,dangers%22%20posed%20to%20children%20online> accessed 12 July 2024.

> Free speech as we understand it today relies on specific legal processes of protection that have developed over centuries. It's not an abstract idea, but one grounded in very particular social and legal practices.[144]

To understand these 'social and legal practices', time is needed. Time which platforms might not necessarily have, depending on how 'swiftly' is interpreted in future complaints that come to the attention of OFCOM. Although OFCOM will not be reviewing individual complaints, they will be making decisions on whether a platform has adhered to its duties under the Act, including the requirement to remove illegal content 'swiftly' even if we do not know what constitutes 'swift'. With the notion of 'swiftly' becoming even more complex in the requirement to remove legal, but harmful content from services.

Under initial drafts of the OSA, the 'lawful but awful' clause, as it became known, applied to both adults and children, with a decision later being made to remove this requirement in relation to adults, due to concerns over the right to freedom of expression.[145] Yet, it remains in place for children and what constitutes 'harmful' under this provision is anything that can amount to 'physical or psychological harm' with some, though inherently vague, parameters being set. For instance, considering the nature of the content of the communication. The justification for such vagueness is to allow for action to be taken quickly to protect young people from the 'harms' associated with the online world. But like that of adults, children are entitled to the right to freedom of expression, regardless of frontiers and to disseminate information.

8.6 Chapter Summary

Only time will tell if the OSA will ultimately achieve its aim of the United Kingdom becoming one of the safest places in the world to go online. But what this chapter has demonstrated is the complex nature of this piece of legislation, with only a small glimpse of just some aspects of the Act. As noted at the start of this chapter, to examine every aspect of the Act would be a book in itself. And yet, there is still so much unknown around the Act, as we await further concrete guidance from OFCOM to bring the OSA to life.

However, from what we do know about the Act thus far, it's overly complex, asking the impossible in some instances and grounded in vague terminology—risking the real possibility of the over-censorship of speech as we allow private entities

144 Phillip Seargeant, 'How AI threatens free speech – and what must be done about it' *The Conversation* (London, 18 January 2024) <https://theconversation.com/how-ai-threatens-free-speech-and-what-must-be-done-about-it-221330> accessed 31 January 2024.

145 Department of Digital, Culture, Media and Sport, 'Overview of expected impact of changes to the Online Safety Bill' (*Gov.uk,* 18 Jan 2022) <https://www.gov.uk/government/publications/online-safety-bill-supporting-documents/overview-of-expected-impact-of-changes-to-the-online-safety-bill#:~:text=The%20adult%20safety%20duties%20(%27legal,and%20freedom%20of%20expression%20duties> accessed 7 August 2023.

to do the work of public organisations. Of particular concern, as highlighted above, is the requirement for platforms to use proactive technology that does not yet fully exist, therefore asking platforms to do the impossible—embed proactive technology in their platforms which stops young people from accessing information or content they are not entitled to see. As noted by Seargeant, in relation to AI technology in general: 'The consequences of this are that AI-driven filters can often lean towards censoring content which is neither illegal or offensive'.[146] In which 'it's entirely possible that many young adults would be falsely identified as being minors—which might prevent them from accessing legal content'.[147]

Further concerns have been raised with the requirement for user-to-user services to remove unlawful content from their sites in line with the criminal law framework of Scotland, Northern Ireland, and England and Wales, swiftly. And when it comes to children, removing harmful, but legal content from their view. All of which threatens the right to freedom of expression.

But despite the former government preaching that the United Kingdom is the first country in the world to legislate against the 'harms' associated with the online world, this is not necessarily true, as shown in the following chapter. Many of the issues highlighted in this chapter, also apply to other countries' approaches to regulating the online world, in particular, social media platforms.

Reference list

Adu A & Milmo D, 'Rishi Sunak considers curbing social media use for under-16s' *The Guardian* (London, 14 December 2023) https://www.theguardian.com/media/2023/dec/14/rishi-sunak-considers-curbing-social-media-use-under-16s

Baker J, 'ORG Warns of Threat to Privacy and Free Speech as Online Safety Bill Is Passed' (*Open Rights Group*, 19 September 2023) https://www.openrightsgroup.org/press-releases/org-warns-of-threat-to-privacy-and-free-speech-as-online-safety-bill-is-passed/

Cambridge Dictionary, 'Swiftly' (*Cambridge Dictionary*, 2024) https://dictionary.cambridge.org/dictionary/english/swiftly

Department for Science, Innovation and Technology & The Rt Hon Michelle Donelan MP, 'Britain makes internet safer, as Online Safety Bill finished and ready to become law' (*Gov.uk*, 19 September 2023) https://www.gov.uk/government/news/britain-makes-internet-safer-as-online-safety-bill-finished-and-ready-to-become-law

Department for Digital, Culture, Media & Sport & The Rt Hon Nadine Dorries MP, 'World-first Online Safety Laws Introduced in Parliament' (*Gov.uk*, 17 March 2022) https://www.gov.uk/government/news/world-first-online-safety-laws-introduced-in-parliament

Gorwa R, 'Elections, institutions, and the regulatory politics of platform governance: The case of the German NetzDG' (2021) 45(6) Telecommun Policy 102

HC Deb 12 September 2023, vol 737, col 799 - 802

HM Government, *Online Harms White Paper* (CP 57, 2019)

HM Government, 'Internet Safety Strategy – Green Paper' (*Gov.uk*, October 2017) https://assets.publishing.service.gov.uk/government/uploads/system/uploads/attachment_data/file/650949/Internet_Safety_Strategy_green_paper.pdf

146 Seargeant (n 144).
147 Phippen (n 107).

Karp P, 'Australia passes social media law penalising platforms for violent content' *The Guardian* (London, 4 April 2019) https://www.theguardian.com/media/2019/apr/04/australia-passes-social-media-law-penalising-platforms-for-violent-content

Kilcoyne N, 'The Online Safety Bill: Everything in Moderation? Parts I and II, Content and Context' (*The International Forum for Responsible Media Blog*, 8 December 2022) https://inforrm.org/2022/12/08/the-online-safety-bill-everything-in-moderation-parts-i-and-ii-content-and-context-naomi-kilcoyne/

Kingsbury J, 'Ofcom's "Roadmap to Regulation" Underway with Its Consultation on Illegal Harms Duties Under the Online Safety Act' (*The International Forum for Responsible Media Blog*, 17 January 2024) https://inforrm.org/2024/01/17/ofcoms-roadmap-to-regulation-underway-with-its-consultation-on-illegal-harms-duties-under-the-online-safety-act-jessica-kingsbury/

National Society for the Prevention of Cruelty to Children, 'The Online Safety Bill Has Been Passed in "a Momentous Day for Children"' (*NSPCC,* 19 September 2023) https://www.nspcc.org.uk/about-us/news-opinion/2023/2023-09-19-the-online-safety-bill-has-been-passed-in-a-momentous-day-for-children/

OFCOM 'Children's Access Assessments: Draft guidance for consultation' (*OFCOM*, 8 May 2024) https://www.ofcom.org.uk/siteassets/resources/documents/consultations/category-1-10-weeks/284469-consultation-protecting-children-from-harms-online/associated-documents/a5-draft-childrens-access-assessments-guidance.pdf?v=336055

OFCOM, 'Protecting children from harms online: Volume 3: The causes and impacts of online harms to children' (*OFCOM*, 8 May 2024) https://www.ofcom.org.uk/siteassets/resources/documents/consultations/category-1-10-weeks/284469-consultation-protecting-children-from-harms-online/associated-documents/vol3-causes-impacts-of-harms-to-children.pdf?v=336052

OFCOM, 'Consultation at a Glance: Our Proposals and Who they Apply to' (*OFCOM,* 9 November 2023) https://www.ofcom.org.uk/__data/assets/pdf_file/0023/271175/Consultation-at-a-glance-our-proposals-and-who-they-apply-to.pdf

OFCOM, 'Protecting People from Illegal Harms Online. Volume 4: How to Mitigate the Risk of Illegal Harms – the Illegal Content Codes of Practice' (*OFCOM*, 9 November 2023) https://www.ofcom.org.uk/__data/assets/pdf_file/0022/271147/volume-4-illegal-harms-consultation.pdf

OFCOM, 'Protecting People from Illegal Harms Online. Volume 5: How to Judge Whether Content Is Illegal or Not? (Illegal Content Judgements Guidance)' (*OFCOM*, 9 November 2023) https://www.ofcom.org.uk/__data/assets/pdf_file/0023/271148/volume-5-illegal-harms-consultation.pdf

OFCOM, 'Protecting People from Illegal Harms Online. Annex 5: Service Risk Assessment Guidance' (*OFCOM*, 9 November 2023) https://www.ofcom.org.uk/__data/assets/pdf_file/0020/271163/annex-5-illegal-harms-consultation.pdf

OFCOM, 'Children and Parents: Media Use and Attitudes Report 2022' (*OFCOM*, 30 March 2022) https://www.ofcom.org.uk/__data/assets/pdf_file/0024/234609/childrens-media-use-and-attitudes-report-2022.pdf

Phippen A, 'Online safety bill: why making the UK the 'safest place to go online' is not as easy as the government claims' *The Conversation* (London, 26 September 2023) https://theconversation.com/online-safety-bill-why-making-the-uk-the-safest-place-to-go-online-is-not-as-easy-as-the-government-claims-214290

Seargeant P, 'How AI threatens free speech – and what must be done about it' *The Conversation* (London, 18 January 2024) https://theconversation.com/how-ai-threatens-free-speech-and-what-must-be-done-about-it-221330

Standley N, 'Starmer against "simply banning" children's phones' *The BBC* (London, 12 July 2024) https://www.bbc.co.uk/news/articles/cd1rklm6433o#:~:text=Prime%20Minister%20Sir%20Keir%20Starmer,dangers%22%20posed%20to%20children%20online

The BBC, 'MP urges social media ban for under 16s' *The BBC* (London, 25 January 2024) https://www.bbc.co.uk/news/av/uk-politics-68085378

The BBC, 'Facebook finds Independence document "racist"' *The BBC* (London, 5 July 2018) https://www.bbc.co.uk/news/technology-44722728

The Conservative and Unionist Party Manifesto, 'Forward Together: Our Plan for a Stronger Britain and a Prosperous Future' (*The Conservative and Unionist Party*, 2017)

Tymkiw N, 'The Online Safety Bill and the Risk of Unintended Consequences' (*The International Forum for Responsible Media Blog*, 11 April 2023) https://inforrm.org/2023/04/11/the-online-safety-bill-and-the-risk-of-unintended-consequences-nadia-tymkiw/

Waterson J, 'Facebook removed 1.5m videos of New Zealand terror attack in first 24 hours' *The Guardian* (London, 17 March 2019) https://www.theguardian.com/world/2019/mar/17/facebook-removed-15m-videos-new-zealand-terror-attack

9 International Perspectives on Social Media and the Law

9.1 Introduction

The issues discussed throughout this monograph thus far on the problems the digital world has created are not unique to the United Kingdom as 'free speech is not restricted by geographical boundaries any longer: with both positive and negative developments having a spill over effect worldwide'.[1] At the time of writing, governments across the globe are attempting to tackle the 'harms' associated with the digital age, through both the criminal law and social media regulation. This chapter explores the regulation of social media beyond the United Kingdom, focussing on France, Germany, and India. Each of these states, including the United Kingdom, have taken similar approaches to regulating (mis)behaviours online. However, the outcome of each approach has been significantly different—the rationale as to why these examples have been chosen.

In France, many provisions contained in the Online Hate Bill 2020 (OHB) were struck down by the French Constitutional Court for breaching the right to freedom of expression as guaranteed under both Article 10 of the European Convention on Human Rights (ECHR) and Article 11 of the Declaration of the Rights of Man and of the Citizen of 1789 (DRMC). Whereas in Germany the Act to Improve Enforcement of the Law in Social Networks 2017 (NetzDG), which inspired France's OHB, proved highly controversial when it was first enacted. Yet, NetzDG remains good law, despite continued concerns that the provision breaches both the right to freedom of expression and legality. Finally, India's Information Technology Act 2008 (IT Act) and the most recent enactment of the Information Technology (Intermediary Guidelines and Digital Media Ethics Code) Rules 2021 (the Rules), will be examined, providing a clear example as to why we should be careful when it comes to regulating the online world; whether through the criminal law or the regulation of platforms.

1 Raghav Mendiratta, 'India's IT Rules, 2021: Incompatible with the ICCPR and a fatal blow to democratic discourse' (*The International Forum for Responsible Media Blog*, 19 May 2021) <https://inforrm.org/2021/05/19/indias-it-rules-2021-incompatible-with-the-iccpr-and-a-fatal-blow-to-democratic-discourse-raghav-mendiratta/> accessed 31 July 2023.

DOI: 10.4324/9781003303688-12

9.2 France: Regulating Hate Speech

Like that of the United Kingdom, French officials have attempted to be proactive in trying to combat online hate, particularly in the wake of several terrorist attacks in and around France. France, again like the United Kingdom, is party to the ECHR and therefore is lawfully obliged to protect the right to freedom of expression, as guaranteed under Article 10. In addition, the right to freedom of expression and communication, as noted in the introduction, is further protected under Article 11 of the DRMC which states that:

> The free communication of ideas and of opinions is one of the most precious rights of man. Any citizen may therefore speak, write and publish freely, except what is tantamount to the abuse of this liberty in the cases determined by Law.

This right extends to internet communications.[2]

However the right to freedom of expression and communications is not absolute under Article 11 of the DRMC, and it can, and has been limited, but it must be necessary, appropriate, and proportionate to the objective being pursued.[3] For the Constitutional Council of France the right to freedom of expression constitutes 'a fundamental freedom, all the more precious as its exercise is one of the essential guarantees of respect for other rights and freedoms and for national sovereignty'.[4]

A balance must therefore be struck between a person's right to freedom of expression and communication and conduct that undermines public order and the rights of third parties.[5] A balance which was not struck during the creation and implementation of the OHB by the French officials resulting in its quick demise following the Constitutional Court of France concluding that many of the provisions set out in the Bill were unconstitutional. Indeed, for Senate Bruno Retailleau , the decision of the French Constitutional Court represented a total 'decapitation of the law' after six Articles were struck down and five were amended by the Court.[6]

9.2.1 The Online Hate Bill: Creation and Purpose

The OHB stemmed from a 2018 report into tackling racism and Antisemitism online.[7] The report recommended the creation of stronger laws to impose stricter requirements on social media platforms to report online hate, without any delay,

2 Decision no. 2020-801 DC of 18 June 2020 [4].
3 French Constitution Article 34.
4 Decision no. 84-181 DC of 11 October 1984.
5 French Constitution Article 34.
6 Decision no. 2020-801 (n 2)
7 Karim Amellal, Laetitia Avia and Gil Taïeb, 'Strengthen the fight against racism and antisemitism on the internet' (*gouvernement.fr*, 20 September 2018) <https://www.gouvernement.fr/sites/default/files/contenu/piece-jointe/2018/09/rapport_visant_a_renforcer_la_lutte_contre_le_racisme_et_lantisemitisme_sur_internet_-_20.09.18.pdf> accessed 6 June 2023 (translated 6 June 2023).

following growing concerns around the radicalisation of young people and terrorist, racist, antisemitic, and hateful content on the internet:[8]

> In this mechanism of hatred, the internet plays a key role, both by offering an unlimited number of spaces for expression, by ensuring the longevity of hate speech, but also as an instrument of propagation, 'virality', and amplification of hate speech.[9]

Such content as listed above, was seen not only as harmful to those who were subjected to such abuse but also posed a significant threat to democracy.

The purpose behind the OHB, following the 'Strengthen the fight against racism and antisemitism on the internet' report was to impose specific rules upon social media platforms to do more to tackle content that 'incited hatred and discrimination based on race, religion, gender, sexual orientation, or disability'.[10] In essence, pushing 'technology companies to regulate hate content more forcefully and to repress certain types of behaviours more quickly'.[11] Many of the proposed elements of the Bill would have made reforms to Article 6 of the 2004 Digital Economy Act in France. The Bill passed on 13 May 2020, with 255 officials in favour, 155 opposing, and 47 abstaining its passing.[12] Five days after its passing, 60 Senators appealed its implementation before the French Constitutional Court.[13]

The Bill itself imposed several obligations on social media platforms,[14] including though not limited to, the creation of more formalised notification methods which must be easily accessible and useable for all social media users in France; and increased transparency over content removal decisions and blocking content on mirror sites.[15] However, the most controversial obligation imposed under the OHB, under Article 1, was the removal of manifestly unlawful content from platforms within 24 hours of notification that such content existed, or in the case of child sexual abuse or terrorist content, one hour. Failure to comply could have

8 Ibid.,

9 Ibid.,

10 Giovanni Ziccardi, *Online Political Hate Speech in Europe: The Rise of New Extremisms* (Edward Elgar Publishing 2020) 153.

11 Ibid.,

12 Assemblée nationale, 'PPL visant à lutter contre les contenus haineux sur internet: adoption en lecture définitive' (*Assemblée nationale*, 13 May 2020) <https://www.assemblee-nationale.fr/dyn/actualites-accueil-hub/ppl-visant-a-lutter-contre-les-contenus-haineux-sur-internet-adoption-en-lecture-definitive> accessed 6 June 2023 (translated 6 June 2023).

13 Jacob Schulz, 'What's going on with France's online hate speech law?' (*LawFare*, June 23 2020) <https://www.lawfareblog.com/whats-going-frances-online-hate-speech-law> accessed 6 June 2023.

14 At the time in which the Bill was passed, and subsequently stuck down, it was not exactly known which internet platforms would fall within its scope. The law stipulated that only sites that surpassed a certain metric of monthly traffic would have been subjected to its provisions, but the figures were never determined. However, it is very likely it would have covered many of the social media platforms we use today. See, Assemblée nationale (n 12).

15 Ibid.,

resulted in the imprisonment of company officials or fines of up to 250,000 euros for each breach.[16] It is Article 1, which will be the focus of the remaining conversation on France's OHB.

9.2.2 The Online Hate Bill and Its Demise

As noted above, Article 1 of the OHB imposed significant obligations on platforms, including the removal of manifestly unlawful content within 24 hours. In essence, content that was considered manifestly unlawful, either by users or AI technology, *had to be removed*. Content that did not reach the threshold to be considered as manifestly unlawful could legally remain on the site (though platforms could have still removed such content, particularly if it breached a platforms community guidelines). Manifestly unlawful content included:

> Hate related content, which targeted an individual or a group on their origin, ethnicity, race, religion, gender, or gender identity, sexual orientation, or disability; or content that glorified criminal offences; or material that either denied or trivialised genocide, war crimes, slavery, or crimes against humanity; or sexual harassment, including the removal of pornography which could be accessed by minors.[17]

Like that of the United Kingdom's Online Safety Act (OSA), a hierarchy of illegal content was created. In the scenarios listed above, such content, under the OHB, had to be removed—there were no exceptions. So even in the hypothetical situation that content was not removed in the short timeframe due to technical issues, this would have still rendered a breach of Article 1 of the OHB. In addition, child pornography and terrorist content had to be removed, once reported, within one hour (under previous provisions, platforms were already under a legal duty to remove such content within 24 hours).[18] As with 'manifestly unlawful' content, there were no exceptions. Following the passing of the Bill, the French Constitutional Court was asked to consider Articles 1, 4, 5, 7, and 8.

The premise of the argument before the Constitutional Court, primarily focussed on the right to freedom of expression and communication, as protected under Article 11 of the DRMC. Considerable concern was raised against Article 1 of the OHB, with arguments being put forward that the lack of safeguards, such as there being no extension to the 24 hour rule, posed a significant threat to freedom of expression. In addition, arguments were put forward that the Bill created legal provisions which were, simply put, asking the impossible, which as explored in further detail in Section 9.2.4 below, is a fundamental breach of the principle of legality. These were concerns, which were echoed in the local media following the passing of the Bill, with ARTICLE 19 (a human rights organisation) being highly critical:

16 Schulz (n 13).
17 Assemblée nationale (n 12).
18 Decision no. 2015-125 DC of 5 February 2015.

> ARTICLE 19 is concerned that the French proposal of a Bill on Countering Online Hatred [...] fails to comply with international free expression standards. We warn that the Bill entrenches private censorship of a wide range of illegal content at the expense of the courts, the 24 hour-time limit on content removals is too short and the sanctions meted out to eradicate online 'hatred' are disproportionate.[19]

Other arguments put to the Court included a lack of understanding in some of the language used in the provision; some of the provisions contained in the Bill, already existing in the legal framework of France, creating contradictory law; and the possibility of excessive removal of online content—all of which threatened the right to freedom of expression.

Though the French Constitutional Court accepted that the OHB was enacted with good intention, they found that many of the prohibited types of content under the Bill undermined the right to freedom of expression and communication.[20] The Court concluded that in many of the Articles under scrutiny: 'the legislator ha[d] made an attack on [...] freedom of expression and communication which [was] not appropriate, necessary, and proportionate to the aim pursued'.[21] Many of the Articles contained in the provision, including that of manifestly unlawful content, were struck down as being unconstitutional with the Court noting:

> However, on the one hand, the determination of the unlawful nature of the content in question is not based on its manifest nature. It is subject to the sole discretion of the administration. On the other hand, the initiation of an appeal against the withdrawal request is not suspensive and the one-hour period given to the publisher or the host to withdraw or make inaccessible the targeted content does not allow him to obtain a decision from the judge before being forced to withdraw it. Finally, the host or publisher who does not comply with this request within this period may be sentenced to one year's imprisonment and a fine of 250,000 euros.[22]

Consequently, the Bill itself only made some very modest reforms after the judgment of the French Constitutional Court.[23]

9.2.4 The French Approach, Legality, and Freedom of Expression

The French approach to attempting to tackle the 'harms' associated with the online world posed significant issues with the right to freedom of expression. The Bill itself was highly controversial both inside and outside of France, with the European

19 ARTICLE 19, 'France: Analysis of draft hate speech bill' (*article19.org*, 3 July 2019) <https://www.article19.org/resources/france-analysis-of-draft-hate-speech-bill/> accessed 4 July 2023.

20 Decision no. 2020-801 (n 2) [13].

21 Ibid., [8].

22 Ibid., [7].

23 Schulz (n 13).

Commission warning that implementation of the Bill was 'likely [a] violation of the e-commerce directive and the Charter of Fundamental Rights of the European Union'.[24] Although the case before the Constitutional Court primarily focussed upon the right to freedom of expression and communication, as protected under Article 11 of the DRMC and Article 10 of the ECHR, there were also issues with the principle of legality—an argument supported further by ARTICLE 19.[25]

For example, the OHB was, at times, asking the impossible: 'these provisions would impose constraints on all publishers and hosts that are impossible to satisfy and would in doing so disregard the principle of equality before public charges'.[26] Under the principle of legality for a legal provision to adhere to this principle, it cannot ask the impossible, with Fuller giving the example of legally obliging someone to do something in one Act of Parliament but prohibiting them from doing the same in another.[27] In this case, it was put to the Court, that the requirements imposed under the Bill were impossible for some organisations to meet, and in turn, not only did this undermine the principle of legality but also equity before the law. An argument, that was accepted by the French Constitutional Court. Yet, as noted in Chapter 8, impossibility remains present in the United Kingdom's OSA.

Further, the requirement that content which could have been considered as 'manifestly unlawful' having to be removed within 24 hours, was also asking the impossible. During these 24 hours, moderators on platforms, or perhaps in some cases AI technology, would have had to determine, by reference to the criminal code of France, if the content flagged did constitute something that can be considered as manifestly unlawful. An impossible task to be completed in one 24 hour period. This therefore raises further legitimate concerns with how the United Kingdom have attempted to regulate the online world, through the creation and implementation of the OSA, as discussed in the previous chapter. Put simply, where a platform is given a time limit and failure to adhere to that time limit can result in heavy fines, there is a real and significant threat that platforms will remove such content anyway, to avoid punishment:

> given the difficulties in assessing the manifestly illegal nature of the content reported within the time limit, the penalty incurred from the first breach and the absence of any specific cause for exoneration from liability, the contested provisions can only encourage online platform operators to withdraw the content reported to them, whether or not it is manifestly illegal.[28]

24 European Union Agency for Fundamental Rights, 'French Bill on combating hate content on the internet' (*FRA*, 2019) <https://fra.europa.eu/en/content/french-bill-combating-hate-content-internet> accessed 4 July 2023.

25 ARTICLE 19 (n 19).

26 Decision no. 2020-801 (n 2) [3].

27 Lon L. Fuller, *The Morality of Law* (Yale University Press 1964) 70.

28 Decision no. 2020-801 (n 2) [19].

In essence, we run the real risk of lawful content being taken down when companies are asked to do the work of the Courts, without the expertise and coercive punishment is implemented. Indeed, as noted by Ziccardi in relation to the OHB:

> The approach of entrusting exclusively to platforms the duty to discipline and deciding on content had long been criticised, and raises problems of censorship with the risk that platforms under threat of sanction will be led to delete as much content as possible.[29]

The OHB also failed to adequately set the boundaries of the law, with it being suggested, using the provision of hateful and sexual material as an example (content which under the Bill was labelled as manifestly unlawful), before the Court that the offence was not 'defined in sufficiently clear and precise terms [... and ...] it could result from simple negligence'.[30] Similarly, in other parts of the provision, the word 'promptly' was included. For instance, under the Bill, platforms were under an obligation to promptly inform 'competent authorities about illegal content notified to them'.[31] Failure to adhere to this could have resulted in fines or in some cases imprisonment. However, no meaning was given to this term, with little clarity found elsewhere in the legal framework of France, as to the meaning of 'promptly'. For example, in a copyright case in 2012, five days were seen to fall outside the scope of 'promptly'.[32]

Despite the OHB no longer being legally enforceable, if it ever was, this Bill provides a leading example of why we need to be careful when it comes to regulating the online world, whether regulating platforms or individuals. Many of the concerns raised against this provision in France, resulting in its ultimate demise, applies to the OSA in the United Kingdom. As potently put by the Constitutional Court:

> The attack on the freedom of expression and communication therefore did not result from the fact that the unlawful remarks had to be withdrawn by the operators of the online platform, but from the fact that the system adopted by the legislator could only lead to also entail the withdrawal of lawful remarks for the sole reason that they have been the subject of a report.[33]

Concerns which have been echoed in Germany, following the creation and implementation of NetzDG, the inspiration behind France's attempt, and failure to regulate the online world through the OHB.

29 Ziccardi (n 10) 152.
30 Decision no. 2020-801 (n 2) [11].
31 ARTICLE 19 (n 19).
32 Schulz (n 13).
33 Decision no. 2020-801 (n 2) [8].

9.3 The German Approach: NetzDG

The French approach to online regulation, as noted above, was highly influenced by legislation passed in Germany in 2017: Gesetz zur Verbesserung der Rechtsdurchsetzung in sozialen Netzwerken (The Act to Improve Enforcement of the Law in Social Networks), commonly referred to as NetzDG. The rationale behind the implementation of the Act was to attempt to force companies to be more proactive, rather than reactive to the 'harms' associated with the online world, in particular hate speech, whilst also finding the balance between regulation and the right to freedom of expression.

Like France and the United Kingdom, Germany is party to the ECHR and therefore must comply with the protection of both the right to freedom of expression and the right to privacy, as outlined in Chapter 2. In addition, freedom of expression is further protected under Article 5(1) of the German Constitution:

> Every person shall have the right freely to express and disseminate his opinions in speech, writing and pictures and to inform himself without hindrance from generally accessible sources.

This right is not absolute and can be limited, as noted in Article 5(2): 'These rights shall find their limits in the provisions of general laws, in provisions for the protection of young persons and in the right to personal honour'. A clear example of this is NetzDG, which imposes several obligations on social media companies to combat hate speech, amongst other illegal behaviours, on their platforms. Most notably, NetzDG forces platforms to remove or block 'manifestly unlawful content' from its site in Germany within 24 hours of being 'flagged'[34]—a requirement which will be the focus of this section. Although this approach is similar to the OHB in France, as discussed in Section 9.2 above, unlike in France, NetzDG has stood the test of time and remains good law to this day—despite concerns being raised at the time that the right to freedom of expression and legality would be impinged. NetzDG is an important and significant legal provision as it is 'probably the first regulation in the world to directly proscribe how large user-generated content platforms moderated harmful content'.[35]

9.3.1 NetzDG: Creation and Purpose

In 2017 the Federal Government of Germany passed NetzDG which became legally binding on 1 January 2018,[36] following voluntary codes of practise failing to tackle

34 'Flagged' is a term used to describe when a communication online has been reported to the network host.

35 Robert Gorwa, 'Elections, institutions, and the regulatory politics of platform governance: The case of the German NetzDG' (2021) 45(6) Telecommun. Policy 1.

36 The Act was implemented in October 2017 however social media sites had three months from the implementation of the Act to comply with its conditions, before it became legally enforceable on 1 January 2018.

the ever-growing presence of online hate in Germany.[37] There were several aims behind the creation of the Act, most notably, forcing social media companies to be more proactive in removing illegal content from their sites, in particular, hate speech; impose punishments, including fines and in some cases imprisonment upon platforms who were slow to remove such content; and attempt to tackle the growing issue of 'fake' online news—which was being used as justification to target minority groups in and around Germany.[38] Leading the Federal Government of Germany to decide to legislate against fake news and online hate,[39] after finding that:

> Online discussions are often aggressive, abusive and hateful [...] Hate crime may seriously threaten the peace in a liberal, open, and democratic society if it's not suppressed and prosecuted effectively.[40]

As a direct result, the German government created and passed NetzDG through Bundestag in less than five months.[41]

NetzDG specifically applies:

> to telemedia service providers which, for profit-making purposes, operate internet platforms which are designed to enable users to share any content with other users or to make such content available to the public.[42]

This includes social media platforms, with over 2 million users, though like that of the OSA in the United Kingdom, journalistic content is exempt. Systematic breaches of the NetzDG by a platform can result in fines of up to 50 million euros.[43]

9.3.2 NetzDG: Initial Concerns

The Act itself imposes several obligations on those who fall within its scope including, though not limited to, transparent reporting systems for both users and non-users; a designated domestic point of contact to receive information requests from

37 Gorwa, (n 35) 4.

38 Fake news can be defined 'as information distributed via a medium – often for the benefit of specific social actors – that then proves unverifiable or materially incorrect'. See, Simeon Yates, '"Fake news" – why people believe it and what can be done to counter it' *The Conversation* (London, 13 December 2016) <https://theconversation.com/fake-news-why-people-believe-it-and-what-can-be-done-to-counter-it-70013> accessed 31 July 2018.

39 Lizzie Dearden, 'Germany to fine social networks up to €50m for not taking down illegal "fake news" posts' *The Independent* (London, 5 April 2017) <https://www.independent.co.uk/news/world/europe/germany-fake-news-social-networks-fine-facebook-50-million-euros-illegal-content-hate-speech-angela-a7668731.html> accessed 31 July 2018.

40 Federal Ministry of Justice and Consumer Protection, 'Questions and answers: Act to Improve Enforcement of the Law in Social Networks' (*German Government*, 2017) <https://www.bmjv.de/SharedDocs/FAQ/EN/NetzDG/NetzDG.html> accessed 31 July 2018.

41 This was arguably rushed through Bundestag due to an upcoming election.

42 Act to Improve Enforcement of the Law in Social Networks (2017).

43 If an honest mistake is made or a platform overlooks an item by error, there is no liability.

German law enforcement (with responses expected within 48 hours); six monthly transparency reports; and most notably the removal of unlawful types of content in a specified time period—the focus of this section.

For content to be considered unlawful under NetzDG, the conduct itself must breach the German Criminal Code. Under the German Criminal Code, multiple offences are prohibited which apply to the online world. Where a behaviour can constitute a criminal offence under the Penal Code, platforms must remove or block such content within a specified time limit. Put simply, platforms when making decisions on what content can be viewable in Germany, must make these decisions taking into account the German Criminal Code.

The specified time limit for social media companies to remove illegal content from their sites in Germany was, and still is, dependent upon the classification of the unlawful content. The Act places different time constraints on the removal of 'manifestly unlawful content' and content which is simply 'unlawful'. 'Manifestly unlawful content' must be removed or blocked from sites within 24 hours of being 'flagged'. Whereas content which is considered as 'unlawful' contrary to the German Criminal Code, must be removed within seven days.[44] Though unlike France's OHB, this seven day period can be extended.[45] Yet the Act did not, and still does not, clarify the difference between 'manifestly unlawful content' and 'unlawful content'—which raised significant concerns for many human rights organisations following the passing of the Act:

> The distinction between 'manifestly unlawful' and 'unlawful' is not clear, and difficult for a private enterprise to accurately predict: the Act provides no guidance to social networks on how they should differentiate 'manifestly unlawful content' from 'unlawful content', and contains no duty on the part of social networks to consider user's rights to freedom of expression when making these (even though law enforcement authorities are required to consider this when acting to restrict freedom of expression pursuant to the GCC [German Criminal Code]).[46]

Unlike that of the United Kingdom's OSA, there are no provisions contained in NetzDG protecting the right to freedom of expression. Indeed, there is no punishment for platforms that block or delete content which is *not* unlawful, only for systematically *failing* to block or delete content that is contrary to the German Criminal Code.

44 Act to Improve Enforcement of the Law in Social Networks (2017) s.3(3).

45 William Echikson and Olivia Knodt, 'Germany's NetzDG: A key test for combatting online hate' (*CEPS*, November 2018) 3 <https://papers.ssrn.com/sol3/papers.cfm?abstract_id=3300636> accessed 4 July 2023.

46 ARTICLE 19, 'Germany: The Act to Improve Enforcement of the Law in Social Networks' (*article19.org*, August 2017) 19 <https://www.article19.org/wpcontent/ uploads/2017/09/170901-Legal-Analysis-German-NetzDG-Act.pdf> accessed 13 July 2018.

For the campaign group ARTICLE 19, the passing of NetzDG was a clear violation of the right to freedom of expression:

> ARTICLE 19 finds the Act, taken overall, to be dangerous to the protection of freedom of expression in Germany, and we are particularly concerned that countered with much weaker institutional and legal safeguards for the protection of human rights are looking at this Act as a model for increasing intermediary liability.[47]

In essence, for ARTICLE 19, the requirements placed on platforms under NetzDG would mean that social media sites would have to act *too* quickly in the removal of content from their servers as there was (and remains) no guidance in place to protect freedom of speech. Consequently, arguments were raised suggesting social media sites would be more likely to be over-cautious creating an 'environment wherein lawful content is routinely blocked or removed as a precaution'.[48] This perspective was supported further by David Kaye, the former United Nations Special Rapporteur on Freedom of Expression:

> With these 24 hour [to] seven day deadlines—if you are a company you are going to want [to] avoid fines and bad public branding of your platform. If there is a complaint about a post you are just going to take it down. What is in it for you to leave it up? I think the result is likely to be greater censorship.[49]

Put bluntly, concerns were raised that a 'deletion orgy' would be an inevitable consequence of such vague terminology.[50] A concern which was echoed by the French Constitutional Court when striking down similar provisions in the French OHB.

These concerns surrounded the lack of clarity contained in the Act, in particular, the difference between unlawful content (which needs to be removed within seven days) and manifestly unlawful content (which needs to be removed within 24 hours). Neither were defined, therefore undermining the principle of legality. Though it is accepted that some form of vagueness is inherent in the law, as it would be impossible to legislate for all matters, citizens, or in the case of NetzDG social media platforms, must be able to govern their behaviour in accordance with law. In turn, the law must not ask the impossible. Here, arguably the requirements under NetzDG could be seen as breaching key components of the principle of legality, in particular, clarity, the action-guiding nature of the law and not asking the impossible. Indeed, following the implementation of NetzDG, several individuals in Germany had their social media accounts blocked by social media companies,

47 Ibid., 24.

48 Ibid., 1.

49 David Kaye, the UN's Special Rapporteur on Freedom of Expression. See, Patrick Evans, 'Will Germany's new law kill free speech online?' *The BBC* (London,18 September 2017) <https://www.bbc.co.uk/news/blogs-trending-41042266> accessed 31 July 2018.

50 Gorwa (n 35) 7.

who had perhaps been overcautious in their approach to interpreting the German Criminal Code, due to the vague requirements contained in NetzDG.[51]

9.3.3 NetzDG: Initial Concerns versus Reality

The passing of NetzDG has been described by Gorwa, as 'the best of bad options [*sic*]'[52], though since its enactment, 26 countries have passed similar legislation.[53] Yet, roughly six years since the implementation of NetzDG responses have been mixed:

> Supporters see the legislation as a necessary and efficient response to the threat of online hatred and extremism. Critics view it as an attempt to privatise a new 'draconian' censorship regime, forcing social media platforms to respond to this new painful liability with unnecessary takedowns.[54]

Some studies have shown that mass removal of content, which was the biggest fear following the enactment of NetzDG has not been proven.[55] Whilst on the other hand, arguments have been put forward that:

> Technology companies have no doubt acted cautiously under [N]etzDG by removing infringing content to avoid possible sanctions, which has led to satire material being incorrectly removed.[56]

At the point of writing, no consensus has been reached as to whether NetzDG has restricted the right to freedom of expression, or indeed, if the legislation has ultimately achieved its aim of reducing hate speech online. The German government has promised a study exploring the impact of this piece of legislation, but this is still forthcoming. To date, only two fines have been issued under the provision, both against Facebook and neither related to manifestly illegal or unlawful content.[57]

51 Philip Oltermann, 'Tough new German law puts tech firms and free speech in spotlight' *The Guardian* (London, 5 January 2018) <https://www.theguardian.com/world/2018/jan/05/tough-new-german-law-puts-tech-firms-and-free-speech-in-spotlight> accessed 6 August 2018; Joseph Nasr, 'Beatrix von Storch: German police accuse AfD politician of hate incitement over anti-Muslim tweet' *The Independent* (London, 2 January 2018) <https://www.independent.co.uk/news/world/europe/beatrix-von-storch-germany-afd-anti-muslim-twitter-north-rhine-westphalia-new-years-eve-a8138086.html> accessed 31 July 2018.

52 Gorwa (n 35) 10.

53 Rachel Griffin, 'New school speech regulation as a regulatory strategy against hate speech on social media: The case of Germany's NetzDG' (2022) 46(9) Telecommun. Policy 1, 6.

54 Echikson and Knodt (n 45) i.

55 Ibid.,

56 Written evidence from techUK (FEO0062). See, House of Lords, *Select Committee on Communications and Digital: Freedom of Expression Online* (HL Deb 2021-04, 23).

57 One fine related to misleading users into reporting content under Facebook's community guidelines as opposed to NetzDG (arguably to mislead officials on figures) and the other fine related to transparency reports.

Since its initial implementation, NetzDG has undergone multiple amendments, including additional transparency requirements; expanding the reach of the Act to include video sharing websites; and most controversially, requiring companies to send the personal information of users to the Federal German Police, who have had content removed for hate speech violations. This controversial amendment to NetzDG was reviewed by the German Courts, following Google, the owner of YouTube, taking legal action.[58] The German Administrative Court granted interim relief.

For Bayer, NetzDG has failed in its mission to stop extremist online behaviour, arguing that 'simply deleting or blocking this [extremist] content did not help to stop the tide of increasing extremist expressions in Germany'.[59] With Facebook, in the six months following the implementation of NetzDG, removing 21.2% of material reported for breaching the German Criminal Code; YouTube 27.1%; and Twitter 10.8%.[60] Put simply, NetzDG has not necessarily resulted in the mass take-down of content. To put this into context:

> Facebook's NetzDG transparency report shows that 74 pieces of content were blocked in Germany for inciting hatred. Within the same period, it removed a total of 6 million pieces of content globally for violating its community rules on hate speech.[61]

However, despite changes made to the law, NetzDG still offers no definitive definition of 'manifestly illegal content' and content that is simply 'unlawful'. With the list of criminal offences covered under this provision still being significantly broad, and continues to grow, as new criminal provisions make their way onto the statute book.[62]

9.3.4 NetzDG, Legality, and Freedom of Expression

NetzDG has created mixed debates since its implementation into the legal system of Germany. The law tackles the hosts of unlawful content, rather than the publisher, putting the onus on companies to remove illegal content from their sites rather than pursuing the individual who posted such commentary. A notion which Balkin defines as 'new school speech regulation'.[63] Put simply, new school speech

58 Reuters, 'Google takes legal action over Germany's expanded hate-speech law' (*Reuters*, 27 July 2021) <https://www.reuters.com/technology/google-takes-legal-action-over-germanys-expanded-hate-speech-law-2021-07-27/> accessed 1 January 2024.

59 Judit Bayer, 'Germany: New law against right-wing extremism and hate crime' (*The International Forum for Responsible Media Blog*, 24 April 2021) <https://inforrm.org/2021/04/24/germany-new-law-against-right-wing-extremism-and-hate-crime-judit-bayer//> accessed 7 July 2023.

60 Echikson and Knodt (n 45) 7.

61 Ibid., 8.

62 For instance, the criminal code has been updated to include explicitly criminalising threats of sexual assault.

63 Jack M. Balkin, 'Old-School/New-School Speech Regulation' (2014) 127(8) Harv. Law Rev. 2296.

regulation describes 'state regulation which coerces intermediaries into restricting speakers instead of coercing speakers directly'.[64] In essence, we have moved away from the regulation of individuals (old-school regulation), as discussed in Part Two of this monograph, to the regulation of platforms themselves (new school speech regulation). Yet this approach to regulation is contentious, with NetzDG providing a leading example with its compliance with the right to freedom of expression, and in turn, legality, being debatable.

For McMillian, although NetzDG is compatible with Article 5 of the German Constitution, due to the previous jurisprudence of the German courts, the provision itself remains incompatible with Article 10 of ECHR for two significant reasons. First, the vague language contained in the provision does risk the probability of over-blocking or the removal of lawful speech, without due process.[65] The vagueness inherent in the provision means that platforms are unable to guide their actions in accordance with clear and definitive laws, which in turn, risks them being overly cautious. McMillian further argues that the requirements imposed on platforms under NetzDG are not proportionate and in turn, are not taking the least intrusive method possible, to render it compatible with the right to freedom of expression:

> [NetzDG] suffers from the issues all laws passed with short notice and out of fear suffer from: it is vague, overbroad, and pins the moral blame for very real issues on the wrong individuals [...] The ECtHR should find that this law violates Article 10 for all criminal provisions other than those which implicate imminent violence or threats to government agencies.[66]

Second, the privatisation of the administration of justice is taking place: 'Now, without the guidance that years of judicial experience would provide, tech companies are sent out to sea to determine what content is manifestly unlawful'.[67] As potently put by McMillian

> whether something is insulting or defamatory to an individual is hard enough to determine [by the courts]. Whether it is insulting or defamatory enough to fine an intermediary is another question altogether.[68]

Without legal expertise, private companies have 'suddenly become the investigator, judge, and executioner' of what is or is not illegal to say or do online.[69] In turn, risking freedom of expression.

But as noted above, these concerns could arguably be misleading, with arguments raised by Griffin that privatisation concerns, such as those raised by

64 Griffin (n 53) 4.

65 Imara McMillan, 'Enforcement Through the Network: The Network Enforcement Act and Article 10 of the European Convention on Human Rights' 2019 20(1) Chic. J. Int. Law 252, 289.

66 Ibid.,

67 Ibid., 265.

68 Ibid., 284.

69 Gorwa (n 35) 7.

McMillian and Gorwa, are overstated.[70] Further, the lack of evidence available on the outcome of the implementation of NetzDG, means that conclusions are unable to be drawn. Although broadly speaking 'complaints are typically resolved within 24 hours and around a quarter result in removal; this has been interpreted as evidence both for overblocking and against it'.[71] In essence, a firm consensus cannot be drawn, though Griffin goes as far as suggesting that NetzDG, put simply, does not work.[72]

9.4 India: When Things Go Really Wrong

Like that of other jurisdictions examined throughout this monograph, the Indian government has also followed suit in attempting to regulate the online world. The Indian government has taken a two-stage approach to regulating unacceptable behaviour online in particular, unacceptable behaviour on social media platforms. First through the implementation of criminal law provisions to regulate individual conduct. Of note and will be the focus in the first part of this section, is the IT Act, which was amended in 2008 to include an offence of sending grossly offensive, menacing or false messages.[73] Second, through the regulation of intermediaries, including social media platforms, following the creation and implementation of the Rules, which will be the focus towards the end of this section. Though India is a very different jurisdiction compared to the European countries which have previously been examined, India has been chosen as a case study to illustrate why we need to be careful when it comes to regulating the online world, particularly when the rest of the globe are looking for inspiration in their own fight against the 'harms' associated with the internet.

In a well-functioning democracy such as France, Germany, and the United Kingdom (as discussed above and in the previous chapters) the use of vague terminology in legal provisions, might not necessarily be a cause for serious concern (though as highlighted throughout this monograph the author disputes this). However, when less functioning democracies take inspiration from these more 'stable' jurisdictions, the issues presented throughout this monograph become even more glaringly apparent, with India being a clear example of this, where there has been a growing trend in silencing voices across the state, particularly those online.[74]

Whereas Germany, France, and the United Kingdom, are all party to the ECHR, India is not. However, India is a signatory to the United Nations Universal

70 Griffin (n 53) 1.
71 Ibid., 10.
72 Ibid., 11.
73 Information Technology Act 2000 s.66A.
74 Toshaar Trivedi and Achyutam Bhatnagar, 'India: The dubious dilemma, muzzling fake news or gagging freedom of speech and expression' (*The International Forum for Responsible Media Blog*, 12 May 2021) <https://inforrm.org/2021/05/12/india-the-dubious-dilemma-muzzling-fake-news-or-gagging-freedom-of-speech-and-expression-toshaar-trivedi-and-achyutam-bhatnagar/> accessed 7 July 2023.

Declaration of Human Rights (UDHR) and acceded to the International Covenant on Civil and Political Rights (ICCPR) in 1948 and 1979 respectively. These provisions were discussed in Chapter 2. Both the UDHR and the ICCPR have had significant influence on the Indian Constitution—the supreme law of India—which amongst other protected rights, includes the protection of the right to freedom of expression: 'All citizens shall have the right (a) to freedom of speech and expression'.[75] There are however limits:

> Nothing in sub clause (a) of clause (1) shall affect the operation of any existing law, or prevent the state from making any law, in so far as such law imposes reasonable restrictions on the exercise of the right conferred by the said sub clause in the interests of the sovereignty and integrity of India, the security of the state, friendly relations with foreign states, public order, decency, or morality or in relation to contempt of court, defamation, or incitement to an offence.[76]

Of note, is the limitation on expressions which have the possibility to create public disorder or can be considered to impede on the decency or morality of Indian citizens, with the government of India using these limitations to justify the wide-reaching nature of the law under s.66A of the IT Act and most recently, during the enactment of the Rules in 2021.

9.4.1 The Information Technology Act 2000/2008

The construction of the IT Act in India is loosely based on the United Nations Model Law on Electronic Commerce 1996, with India at the time of passing the IT Act, being the 12th country in the world to pass legislation aimed at combatting the 'harms' associated with the online world. The Act itself was significantly updated in 2008 and contains more than just criminal law provisions—somewhat similar to the OSA in the United Kingdom. It covers, though not limited to, updating the law to recognise the importance of digital signatures; computer misuse; data offences; and specific criminal offences *akin* to communication offences which were found in the criminal law framework of England and Wales, as critically examined in Chapter 6. Of all the provisions contained in the IT Act, the most controversial was s.66A, which was written into the Act in 2008, after arguments were made that the law in India did not adequately protect individuals from online abuse.

S.66A of the IT Act stated:

> Any person who sends, by means of a computer resource or a communication device (a) any information that is grossly offensive or has [a] menacing character; or (b) any information which he knows to be false, but for the purpose of causing annoyance, inconvenience, danger, obstruction, insult, injury,

75 Indian Constitution Article 19(1).
76 Ibid., 19(2).

> criminal intimidation, enmity, hatred, or ill will, persistently by making use of such computer resource or a communication device [will be prosecuted].

A provision *akin* to s.127 of the Communications Act 2003 (CA) and the Malicious Communications Act 1988 (MCA) in their original format. However, there was a small, but significant difference between this provision and the provisions which were found in the law of England and Wales; the sending of false communications goes beyond annoyance or inconvenience, to include, amongst other situations, insult, hatred, or ill-will in India. In essence, the purpose of the provision was to criminalise the conduct of sending grossly offensive, menacing, or false messages online, to help reduce the use of the internet being used to abuse others.

The implementation of this provision into the IT Act was contentious. The lack of understanding as to what constituted a grossly offensive, menacing, or false message left some commentators concerned it would be abused.[77] A substantial concern given that the criminal justice system in India is overly slow, and open to abuse.[78] Concerns which quickly became a reality as s.66A of the IT Act proved to be 'totally subjective and depend[ed] upon individual judges'.[79] Consistency was therefore lost, and discretion, resulting in the abuse of s.66A, was rampant, which in turn, infringed dramatically on the right to freedom of expression across India. In many cases, individuals were often charged for an offence under s.66A of the IT Act, with other offences being tagged on, such as 'offending religious feelings'[80] or 'defiling sacred objects'.[81]

An example of the abuse of s.66A of the IT Act can be seen in the prosecution of Shaheen Dhada who, in 2012, posted the following, after the death of a controversial political figure in India:

> Every day thousand[s] of people die. But still the world moves on [...] Just due to one politician dead. A natural death. Everyone goes crazy [...] Respect is earned not given out, definitely not forced. Today Mumbai shuts down due to fear not due to respect.[82]

Soon after she posted the message on Facebook Dhada received phone calls from friends telling her to remove the message for fear of retaliation by authorities. Later the same day, she was arrested for her own safety after supporters of the political

77 Nikhil Moro, 'Social Media and Criminal Libel in India: A Case for Reform' (2020) 25(4) Comms L 209.

78 Madan B. Lokur, 'Justice Lokur: Our fundamental rights to free speech and protest are being eroded and mauled' *The Wire* (India, 12 October 2020) <https://thewire.in/rights/fundamental-rights-free-speech-protest> accessed 30 July 2023.

79 Moro (n 77) 218.

80 Indian Penal Code of 1860 s.295.

81 Ibid., s.295A.

82 Rajini Vaidyanathan, 'India Facebook arrests: Shaheen and Renu speak out' *The BBC* (London, 26 November 2012) <https://www.bbc.co.uk/news/world-asia-india-20490823> accessed 8 August 2018.

figure, took offence to the Facebook post. The next day after being held in a police cell overnight, Dhada was rearrested for violating s.295a of the Indian Penal Code, later being charged with a subsequent offence under s.66A of the IT Act. In addition, a friend of Dhada, Renu Srinivasan, was also charged with the same offence under the IT Act for liking, sharing, and commenting on the original post.[83] This is one of many examples in which s.66A, since its implementation into the IT Act in 2008, was used to limit the right to freedom of expression in India with the provision often being considered to be on 'shaky ground[s] constitutionally' due to the ambiguous terminology used in the provision.[84]

In 2013, the Supreme Court of India was asked to consider the compatibility of s.66A of the IT Act against the Indian Constitution, with significant weight being placed on its incompatibility with the right to freedom of expression in *Shreya Singhal v Union of India*.[85] In addition, the Court also examined if the provision complied with the principle of legality, in that could it be said that the provision was clear and predictable:

> a penal law is void for vagueness if it fails to define the criminal offence with sufficient definiteness. Ordinary people should be able to understand what conduct is prohibited and what is permitted. Also, those who administer the law must know what offence has been committed so that arbitrary and discriminatory enforcement of the law does not take place.[86]

The Supreme Court of India therefore looked at the language contained in s.66A including the meaning of the terms 'grossly offensive' and 'menacing', which as discussed in Chapter 6, has been of significant judicial debate in England and Wales.

The lack of a clear understanding as to what was meant by these terms for the Supreme Court of India raised legitimate concerns that the terms were too ambiguous, and consequently did not comply with the principle of legality and in turn, were incompatible with Article 19 of the Indian Constitution:

> Quite apart from this, as has been pointed out above, every expression used is nebulous in meaning. What may be offensive to one may not be offensive to another. What may cause annoyance or inconvenience to one may not cause annoyance or inconvenience to another. Even the expression 'persistently' is completely imprecise—suppose a message is sent twice, can it be said that it was sent 'persistently'? Does a message have to be sent (say) at least eight times, before it can be said that such message is 'persistently' sent? There is

83 Ibid.,

84 Moro (n 77) 220.

85 *Shreya Singhal v Union of India* (2013) 12 S.C.C. 73.

86 Ibid., [52].

> no demarcating line conveyed by any of these expressions—and that is what renders the Section unconstitutionally vague.[87]

Interestingly, the Supreme Court of India was concerned about the term 'persistently' and its lack of a definitive definition, a concern which has been echoed by ARTICLE 19 concerning Germany's NetzDG, as noted earlier in this chapter. To further examine the principle of legality and the prohibited behaviours under s.66A of the IT Act, the Indian Supreme Court examined in detail two cases stemming from English Law: *DPP v Collins*[88] and *Chambers v DPP*[89]—both of which were discussed in Chapter 6.

For the Supreme Court of India, the cases of *Collins* and *Chambers* provided undisputable evidence of the ambiguity of the terms grossly offensive and menacing and their lack of compatibility with the basic fundamental principles of legality, and in turn, threatened the right to freedom of expression:

> These two cases illustrate how judicially trained minds would find a person guilty or not guilty depending upon the Judge's notion of what is 'grossly offensive' or 'menacing'. In *Collins'* case, both the Leicestershire Justices and two Judges of the Queen's Bench would have acquitted Collins whereas the House of Lords convicted him. Similarly, in the *Chambers* case, the Crown Court would have convicted Chambers whereas the Queen's Bench acquitted him. If judicially trained minds can come to diametrically opposite conclusions on the same set of facts it is obvious that expressions such as 'grossly offensive' or 'menacing' are so vague that there is no manageable standard by which a person can be said to have committed an offence or not to have committed an offence.[90]

The clear lack of a definition as to what constituted a menacing or grossly offensive communication, led the Supreme Court of India to conclude that s.66A of the IT Act did not comply with the notion of legality in the criminal law and ran the real risk of breaching the right to freedom of expression. In essence, alongside the lack of clarity contained in the provision, s.66A provided no safeguards to protect the right to freedom of expression, therefore conflicting with Article 19 of the Indian Constitution. Consequently, this section of the IT Act was considered void by the Indian Supreme Court and struck down in 2015.[91] This was a most welcomed decision: '[t]he ruling is a huge victory for free speech in India.'[92] However, though

87 Ibid., [76].

88 *DPP v Collins* [2006] UKHL 40.

89 *Chambers v DPP* [2012] EWHC 2157 (Admin), [2013] 1 W.L.R. 183.

90 *Shreya Singhal* (n 85) [82].

91 Ibid., [119]. Other provisions were also struck down including the Kerala Police Act 2011. In addition, the Indian Supreme Court read down the meaning of s.79 of the Information Technology Act 2008.

92 Stephen Mathias, 'Section 66A Judgment' (2015) 31(4) C.L.S. Rev 573.

'widely celebrated as a landmark judgment on free speech',[93] in reality, this was not necessarily the case.

Research undertaken by the People's Union for Civil Liberties uncovered that five years after the Supreme Court struck down the provision, 800 cases related to s.66A, were pending before the Indian Courts.[94] This was despite s.66A technically no longer being on the statute book: 'right from the police station, to trial courts, and all the way to High Courts, one finds that s.66A is still in use despite it being denied a place on the statute book'.[95]

The research conducted by the People's Union for Civil Liberties provided concrete evidence that the change in law was not being implemented on the ground, particularly by law enforcement, with numerous examples being given in the report of the provision still being enforced even by the courts:

> We conducted advanced searches over two online databases; a popular public access platform (IndianKanoon) and another requiring subscription (Supreme Court Cases Online), to survey the extent to which Section 66-A continues to haunt the legal system. The data range for the period between January 2018 to September 2018 (till date) from IndianKanoon lists a total of 45 cases, while that for Supreme Court Cases Online from March 2015 till September 2018 is a total of 21 cases.[96]

As a consequence of these findings, the People's Union for Civil Liberties approached the Indian Supreme Court in 2021 with a writ asking for further directions as to the constitutional compatibility of the continued use of s.66A of the IT Act, with the Supreme Court ordering in October 2022 all pending cases, investigations and prosecutions under s.66A to be closed.[97]

It is not fully known why s.66A of the IT Act continued to be used across the criminal justice system in India, despite it being struck down. To explore this further, would be beyond the scope of this monograph. However, what is demonstrated is the importance of restricting the implementation of ambiguous legislation, which renders the possibility of abuse, especially when less democratic countries are taking inspiration from other jurisdictions. Illustrated further in the Indian government's attempt to directly regulate platforms, in particular social media companies through the creation of the Rules.

93 Sanjana Srikumar and Tanmay Singh, 'Third time lucky?: Section 66A and the afterlife of strategic litigation' (*Global Freedom of Expression*, 6 January 2023) <https://globalfreedomofexpression.columbia.edu/updates/2023/01/third-time-lucky-section-66a-and-the-afterlife-of-strategic-litigation/> accessed 1 July 2023.

94 Ibid.,

95 Abhinav Sekhri and Apar Gupta, 'Section 66A and other legal zombies' (*Global Freedom of Expression*, 6 January 2023) 3 <https://globalfreedomofexpression.columbia.edu/wp-content/uploads/2015/06/Sec.-66A-and-Other-Legal-Zombies.pdf> accessed 2 July 2023.

96 Ibid.,

97 Srikumar and Singh (n 93).

9.4.2 The Misguided War: Protecting Freedom of Expression

In recent years, the citizens of India have witnessed a decline and breakdown in civil liberties and the right to freedom of expression across India with a 'particularly disturbing trend [in] the silencing of voices through targeted prosecutions of activists, students, academics, protesters, and journalists'.[98] Most recently though, through the creation and implementation of the Rules in 2021, which, unlike India's previous approaches to regulating the online world, as discussed above, targets platforms as opposed to individuals, with criminal sanctions imposed on platforms for a failure to adhere to these new obligations. Although arguably this approach is *akin* to NetzDG and the OSA, the Rules do not contain any safeguards to ensure the protection of the right to freedom of expression. In turn, the provision has been drafted so significantly broad in places, that it renders issues with the principle of legality, posing a further threat to democracy across India.

The Rules, which became legally binding on 25 May 2021, were created by the Central Government of India through secondary legislation. In essence, the Rules were created by the government as opposed to Parliament, using the provision laid out in s.87(1) of the IT Act which states that: 'The Central Government may, by notification in the Official Gazette and in the Electronic Gazette, make rules to carry out the provisions of this [IT] Act'. Put simply, the Rules did not have any parliamentary oversight or debate before their implementation and although perfectly legal, this approach has been heavily criticised, particularly given the number of bodies who find themselves within the scope of the Rules. Alongside the vague nature of the provision which leaves the real possibility of the provision being abused by the state, as legality is not being maintained.

Like the OSA and NetzDG, the Rules impose extensive obligations on platforms within its scope to remove certain types of content within a specified time limit. Such content includes, though not limited to: discrimination; depiction of illegal or harmful substances; imitable behaviour; nudity; language; and sex.[99] Though, unlike the OSA and NetzDG, the Rules extend beyond intermediaries such as social media platforms and search services, to include traditional forms of media, such as news sites and on demand platforms, for instance, Netflix and Amazon.[100] Consequently, the platforms which fall within the scope of the Rules have been heavily criticised for their failure to protect and maintain the right to freedom of expression.[101]

98 Mendiratta (n 1).

99 Information Technology (Intermediary Guidelines and Digital Media Ethics Code) Rules 2021 rule 2H.

100 Ibid., rule 3. Although the use of the Rules to regulate the press and on demand services is of serious concern, this will not be explored any further here as it is beyond the scope of this monograph. Though it is important to note, some of the provisions which are of most concern when it comes to the regulation of social media platforms, are also applicable to these other types of intermediaries.

101 For example, Trivedi and Bhatnagar (n 74).

The Rules were created to specifically deal with the 'harms' associated with the online world, in particular, the 'harms' carried out on social media platforms. The Rules split social media companies into two categories: (1) a 'social media intermediary' and (2) a 'significant social media intermediary'. A 'social media intermediary' is defined as:

> an intermediary which primarily or solely enables online interaction between two or more users and allows them to create, upload, share, disseminate, modify, or access information using its services.[102]

A 'significant social media intermediary' is defined as a: 'social media intermediary having number of registered users in India above such threshold as notified by the Central Government'.[103] With the Internet Freedom Foundation noting that '[t]hese categories bring a high level of government discretion in determining which platforms need to comply with what regulations'.[104] Since the Central Government of India has set this bar at 5 million users. A number which has been set, with no parliamentary oversight. Of even more concern, the Central Government of India, under Rule 6(1) can declare a 'social media intermediary' as a 'significant social media intermediary' even if they do not reach the 5 million user threshold. In essence, the Indian government is free to declare a platform a 'significant social media intermediary' giving the Central Government of India wide discretionary powers, which undermines the principle of legality:

> The most subtle element in the task of maintaining congruence between law and official actions lies, of course, in the problem of interpretation. Legality requires that judges and other officials apply statutory law, not according to their fancy or with crabbed literalness, but in accordance with the principle of interpretation.[105]

Without clear and distinct rules as to when a body is considered a 'significant social media intermediary' platforms are under a continued arbitrary threat and accordingly, it cannot be said, that the law is action-guiding.

This is of particular concern because, like that of the OSA, those platforms which can be defined as a 'significant social media intermediary' have considerably more obligations to adhere to. As a minimum all social media intermediaries must, though not limited to:

- Publish all policies on their website;

102 Information Technology (Intermediary Guidelines and Digital Media Ethics Code) Rules 2021 rule 2W.

103 Ibid., rule 2V.

104 Krishnesh Bapat, Anushka Jain, Apar Gupta and Tanmay Singh, 'Deep dive: How the intermediaries rules are anti-democratic and unconstitutional' (*Internet Freedom Foundation*, 27 February 2021) <https://internetfreedom.in/intermediaries-rules-2021/> accessed 31 July 2023.

105 Fuller (n 27) 82.

- Have rules that prohibit 'the display, upload, modify, publish, transmit, store, update, or share any information that':[106]
  - Belongs to another;
  - 'is obscene, pornographic, paedophilic, invasive of another's privacy including bodily privacy, insulting, or harassing on the basis of gender, racially or ethnically objectionable, relating or encouraging money laundering or gambling';[107]
  - Is harmful to children;
  - Can be labelled as mis/disinformation; and
  - 'Threatens the unity, integrity, defence, security, or sovereignty of India, friendly relations with foreign states, or public order, or causes incitement to the commission of any cognisable offence, or prevents investigation of any offence, or is insulting to other nation'.[108]
- Produce yearly reports;
- Keep on file any material removed for 180 days for use by law enforcement; and
- Assist law enforcement within 72 hours.

In addition, where such material is reported, the platform must remove such material within 24 hours. Failure to adhere can result in heavy fines and like that of the OHB in France, there are no exceptions. Running the real risk of the over-removal of lawful speech, for fear of repercussions for leaving unlawful content viewable. This approach was seen as incompatible with the right to freedom of expression in France and is of great concern in India, due to continued evidence that democracy is slowly being eroded in this state.

Further additional requirements for a 'significant social media intermediary' include, though not limited to:[109]

- Appointing a chief complaints officer in India;
- Nominating a contact person in India who can help support the police and must be available 24 hours a day, seven days a week. They must also be different from the chief complaints officer and be based in India; and
- Publishing monthly compliance reports.

The requirements, therefore, placed on social media platforms are extensive, whether a 'social media intermediary' or a 'significant social media intermediary'. No further information is given in the Rules as to the meaning of the prohibited content. For example, it is not known what is harmful to children, what

106 Information Technology (Intermediary Guidelines and Digital Media Ethics Code) Rules 2021 rule 3(1)b.
107 Ibid., rule 3(1)b(ii).
108 Ibid., rule 3(1)b(vii).
109 Ibid., rule 4.

would constitute nudity as opposed to art or what would be considered fake news. These are left to the interpretation of the Indian Authorities, mainly the Central Government of India, The Ministry of Electronics and Information Technology, and the Ministry of Information and Broadcasting, with no parliamentary oversight. Once again, giving significant discretionary powers to government bodies, which in turn risks the law being unfairly enforced, especially where clear and distinct legal rules are lacking. All of which do not comply with the principle of legality.

The lack of understanding as to the extent and boundaries of many of the provisions contained in the Rules has led to significant criticism.[110] Many commentators are heavily critical of the Rules, for fear, like that of s.66A of the IT Act, the provision will lead to an abuse of power, particularly given the lack of legality contained throughout.[111] Concerns which are slowly starting to become a reality, following the Ministry of Information and Broadcasting asking social media platforms to remove adverts for a BBC documentary which was critical of the Indian Prime Minister, Narendra Modi, under the proviso that it threatened the 'unity, integrity, defence, security, or sovereignty of India' contrary to Rule 3(1)b(ii). In response

> [b]oth Twitter and YouTube quickly complied with the government's censorship requests. Posts on about 50 Twitter accounts were removed, with activists, politicians and even Hollywood actors among those affected, as well as an unspecified number of YouTube channels.[112]

By January 2023, action had been taken 'against 94 YouTube channels, 19 social media accounts and 747 URLs on the government's request' for failure to adhere to the Rules.[113] Indeed, at the time of writing more than 18 petitions challenging the constitutionality of the Rules before the Indian High Courts have been submitted,[114] with the Foundation for Independent Journalism wishing to get a Supreme Court decision declaring the Rules void due to their incompatibility with Article 19 of the Indian Constitution. Due to the inherently slow nature of the justice system in India, only time will tell if these Rules will remain legally binding.

110 See for instance, Bapat, Jain, Gupta and Singh, (n 104); Devdutta Mukhopadhyay, 'Now streaming: The chilling effect of the new IT rules' *The Indian Express* (India, 11 March 2021) <https://indianexpress.com/article/opinion/columns/new-it-rules-ott-platforms-content-youtube-netflix-amazon-7222873/> accessed 31 July 2023; Trivedi and Bhatnagar (n 74).

111 Mendiratta (n 1).

112 Hannah Ellis-Petersen, 'Indian ban on BBC Modi film puts Musk's Twitter "free speech" to the test' *The Guardian* (London, 25 January 2023) <https://www.theguardian.com/world/2023/jan/25/india-ban-on-bbc-modi-film-elon-musk-twitter-free-speech-emergency-laws#:~:text=The%20response%20by%20the,be%20banned%20on%20social%20media> accessed 31 July 2023.

113 Ibid.,

114 Apar Gupt, 'The IT Rules fail the test of constitutionality' *The Indian Express* (India, 1 October 2021) <https://indianexpress.com/article/opinion/columns/the-it-rules-fail-the-the-test-of-constitutionality-7542621/> accessed 31 July 2023.

9.4.3 The Indian Approach, Legality, and Freedom of Expression: Why We Should Care

The inclusion of India as a case study in this monograph as opposed to another European state may seem out of the ordinary, but it provides a prime example as to why states across the globe need to be careful when it comes to the regulation of the online world. Those countries which are more 'democratic' or 'stable' need to be careful they do not set a precedent, by using vague terminology in their own provisions, which could be abused by other states, looking for 'inspiration' in attempting to tackle the 'harms' associated with the digital world. The approach by officials in India reflects many approaches undertaken across Europe, with s.66A of the IT Act directly mirroring, in their original format, both the MCA and s.127(1) and (2) of the CA in England and Wales. Yet, in India this provision was heavily abused in restricting the right to freedom of expression, resulting in s.66A eventually being declared unconstitutional by the Indian Supreme Court. However, it remained on the statute book until very recently.[115]

A significant justification for s.66A being struck down by the Indian Supreme Court was the inherent vagueness contained in the provision:

> The court found that the provision was vague and overbroad—a significant finding given that the provision was overwhelmingly used to crush political dissent, and silence social and political commentary on social media and the internet.[116]

At no point could it be said that the provision complied with any of the requirements of the principle of legality. The vagueness in the terms contained in the provision lead to an abuse of power and in India, '[a]ny law which does not strike a proper balance between freedoms guaranteed and social control permitted by the clauses in Article 19 is an unreasonable restriction'.[117] Yet, such laws remain in England and Wales. But most concerningly the Indian government continues to create provisions similar to European approaches, which are so inherently vague they render incompatible with the basic principles of legality. Nothing demonstrates this more than the recent enactment of the Rules by the Central Government of India.

9.5 Chapter Summary

Regulating the online world, in particular, social media has proven a headache for governments across the globe, with there being no general consensus on how the digital world *should* be regulated. The countries discussed in this chapter were all chosen because of the similarities in approaches each has taken in regulating social media platforms. Yet the outcomes have differed dramatically. In addition,

115 Srikumar and Singh (n 93).

116 Ibid.,

117 *Chintaman Rao v State of Madhya Pradesh*, 1950 SCC 118.

the issues highlighted above, have also been raised against the United Kingdom's OSA—as discussed at length in Chapter 8.

In France, the French Constitutional Court struck down the OHB, following its failure to adequately set limitations to ensure the law was not abused and in turn, protected the right to freedom of expression. Significant weight was also given to the lack of legality within the provision, which was open to abuse. Whereas in Germany, similar provisions, despite initial concerns around freedom of expression and legality, remains legally enforceable. Indeed, research into NetzDG since its enactment has been mixed—neither confirming nor denying if the provision has hindered the right to freedom of expression. In India, similar legislation to both France, Germany, and the United Kingdom has been implemented into the legal framework of India, but such use of these provisions, many of which mirror European approaches to regulating the online world, have been heavily abused.

Without a consensus as to how the online world should be regulated, the issues highlighted above are going to remain and perhaps this is not an issue one country can tackle on its own. However, what is clear though, from the discussions above, is the need to ensure less 'stable' societies do not follow suit in enacting legislation grounded in vagueness and discretion, as has been the case in the United Kingdom, France, and Germany. India is a prime example of why we need to be careful when enacting legislation which does not comply with the principle of legality when the rest of the world is looking for inspiration in their own fight against tackling the 'harms' associated with the online world.

Reference list

Amellal K, Avia L & Taïeb G, 'Strengthen the Fight against Racism and Antisemitism on the Internet' (*gouvernement.fr*, 20 September 2018) https://www.gouvernement.fr/sites/default/files/contenu/piece-jointe/2018/09/rapport_visant_a_renforcer_la_lutte_contre_le_racisme_et_lantisemitisme_sur_internet_-_20.09.18.pdf

ARTICLE 19, 'France: Analysis of Draft Hate Speech Bill' (*article19.org*, 3 July 2019) https://www.article19.org/resources/france-analysis-of-draft-hate-speech-bill/

ARTICLE 19, 'Germany: The Act to Improve Enforcement of the Law in Social Networks' (*article19.org*, August 2017) https://www.article19.org/wpcontent/ uploads/2017/09/170901-Legal-Analysis-German-NetzDG-Act.pdf

Assemblée nationale, 'PPL visant à lutter contre les contenus haineux sur internet: adoption en lecture définitive' (*Assemblée nationale*, 13 May 2020) https://www.assemblee-nationale.fr/dyn/actualites-accueil-hub/ppl-visant-a-lutter-contre-les-contenus-haineux-sur-internet-adoption-en-lecture-definitive

Balkin J M, 'Old-School/New-School Speech Regulation' (2014) 127(8) Harv Law Rev 2296

Bapat K, Jain A, Gupta A & Singh T, 'Deep Dive: How the Intermediaries Rules Are Anti-Democratic and Unconstitutional' (*Internet Freedom Foundation*, 27 February 2021) https://internetfreedom.in/intermediaries-rules-2021/

Bayer J, 'Germany: New Law against Right-Wing Extremism and Hate Crime' (*The International Forum for Responsible Media Blog*, 24 April 2021) https://inforrm.org/2021/04/24/germany-new-law-against-right-wing-extremism-and-hate-crime-judit-bayer//

Dearden L, 'Germany to fine social networks up to €50m for not taking down illegal "fake news" posts' *The Independent* (London, 5 April 2017) https://www.independent.co.uk

/news/world/europe/germany-fake-news-social-networks-fine-facebook-50-million-euros-illegal-content-hate-speech-angela-a7668731.html

Echikson W & Knodt O, 'Germany's NetzDG: A Key Test for Combatting Online Hate' (*CEPS*, November 2018) https://papers.ssrn.com/sol3/papers.cfm?abstract_id=3300636

Ellis-Petersen H, 'Indian ban on BBC Modi film puts Musk's Twitter "free speech" to the test' *The Guardian* (London, 25 January 2023) https://www.theguardian.com/world/2023/jan/25/india-ban-on-bbc-modi-film-elon-musk-twitter-free-speech-emergency-laws#:~:text=The%20response%20by%20the,be%20banned%20on%20social%20media

European Union Agency for Fundamental Rights, 'French Bill on Combating Hate Content on the Internet' (*FRA*, 2019) https://fra.europa.eu/en/content/french-bill-combating-hate-content-internet

Evans P, 'Will Germany's new law kill free speech online?' *The BBC* (London, 18 September 2017) https://www.bbc.co.uk/news/blogs-trending-41042266

Federal Ministry of Justice and Consumer Protection, 'Questions and answers: Act to Improve Enforcement of the Law in Social Networks' (*German Government*, 2017) https://www.bmjv.de/SharedDocs/FAQ/EN/NetzDG/NetzDG.html

Fuller L L, *The Morality of Law* (Yale University Press 1964)

Gorwa R, 'Elections, Institutions, and the Regulatory Politics of Platform Governance: The Case of the German NetzDG' (2021) 45(6) Telecommun Policy 1

Griffin R, 'New School Speech Regulation as a Regulatory Strategy against Hate Speech on Social Media: The Case of Germany's NetzDG' (2022) 46(9) Telecommun Policy 1

Gupt A, 'The IT Rules fail the test of constitutionality' *The Indian Express* (India, 1 October 2021) https://indianexpress.com/article/opinion/columns/the-it-rules-fail-the-the-test-of-constitutionality-7542621/

House of Lords, *Select Committee on Communications: Freedom of Expression Online* (HL Deb 2021-04, 23)

Lokur M B, 'Justice Lokur: Our fundamental rights to free speech and protest are being eroded and mauled' *The Wire* (India, 12 October 2020) https://thewire.in/rights/fundamental-rights-free-speech-protest

Mathias S, 'Section 66A Judgment' (2015) 31(4) CLS Rev 573

McMillan I, 'Enforcement Through the Network: The Network Enforcement Act and Article 10 of the European Convention on Human Rights' 2019 20(1) Chic J Int Law 252

Mendiratta R, 'India's IT Rules, 2021: Incompatible with the ICCPR and a Fatal Blow to Democratic Discourse' (*The International Forum for Responsible Media Blog*, 19 May 2021) https://inforrm.org/2021/05/19/indias-it-rules-2021-incompatible-with-the-iccpr-and-a-fatal-blow-to-democratic-discourse-raghav-mendiratta/

Moro N, 'Social Media and Criminal Libel in India: A Case for Reform' (2020) 25(4) Comms L 209

Mukhopadhyay D, 'Now streaming: The chilling effect of the new IT rules' *The Indian Express* (India, 11 March 2021) https://indianexpress.com/article/opinion/columns/new-it-rules-ott-platforms-content-youtube-netflix-amazon-7222873/

Narayan V, 'Man booked under IT Act for "defaming" CM Devendra Fadnavis' *The Times of India* (Mumbai, 10 July 2015) https://timesofindia.indiatimes.com/india/Manbooked-under-IT-Act-for-defaming-CM-Devendra-Fadnavis-intweet/articleshow/48011122.cms

Nasr J, 'Beatrix von Storch: German police accuse AfD politician of hate incitement over anti-Muslim tweet' *The Independent* (London, 2 January 2018) https://www.independent.co.uk/news/world/europe/beatrix-von-storch-germany-afd-anti-muslim-twitter-north-rhine-westphalia-new-years-eve-a8138086.html

Oltermann P, 'Tough new German law puts tech firms and free speech in spotlight' *The Guardian* (London, 5 January 2018) https://www.theguardian.com/world/2018/jan/05/tough-new-german-law-puts-tech-firms-and-free-speech-in-spotlight

Reuters, 'Google Takes Legal Action over Germany's Expanded Hate-Speech Law' (*Reuters*, 27 July 2021) https://www.reuters.com/technology/google-takes-legal-action-over-germanys-expanded-hate-speech-law-2021-07-27/

Schulz J, 'What's Going on with France's Online Hate Speech Law?' (*LawFare*, 23 June 2020) https://www.lawfareblog.com/whats-going-frances-online-hate-speech-law

Sekhri A & Gupta A, 'Section 66A and Other Legal Zombies' (*Global Freedom of Expression*, 6 January 2023) https://globalfreedomofexpression.columbia.edu/wp-content/uploads/2015/06/Sec.-66A-and-Other-Legal-Zombies.pdf

Srikumar S & Singh T, 'Third Time Lucky?: Section 66A and the Afterlife of Strategic Litigation' (*Global Freedom of Expression*, 6 January 2023) https://globalfreedomofexpression.columbia.edu/updates/2023/01/third-time-lucky-section-66a-and-the-afterlife-of-strategic-litigation/

The Local, 'This is what Facebook moderators in Germany have to deal with' *The Local* (Stockholm, 16 December 2016) https://www.thelocal.de/20161216/this-is-what-facebook-moderators-in-berlin-have-to-deal-with

Trivedi T & Bhatnagar A, 'India: The Dubious Dilemma, Muzzling Fake News or Gagging Freedom of Speech and Expression' (*The International Forum for Responsible Media Blog*, 12 May 2021) https://inforrm.org/2021/05/12/india-the-dubious-dilemma-muzzling-fake-news-or-gagging-freedom-of-speech-and-expression-toshaar-trivedi-and-achyutam-bhatnagar/

Vaidyanathan R, 'India Facebook arrests: Shaheen and Renu speak out' *The BBC* (London, 26 November 2012) https://www.bbc.co.uk/news/world-asia-india-20490823

Yates S, '"Fake news" – why people believe it and what can be done to counter it' *The Conversation* (London, 13 December 2016) https://theconversation.com/fake-news-why-people-believe-it-and-what-can-be-done-to-counter-it-70013

Ziccardi G, *Online Political Hate Speech in Europe: The Rise of New Extremisms* (Edward Elgar Publishing 2020)

10 Conclusion

10.1 Introduction

There is no doubt that there is a 'darker' side to the online world, that is not under dispute in this monograph. But what is under dispute, is the continuous rhetoric that the only way to tackle these issues is by using the criminal law. The criminal law is a coercive tool and when embedded in vague terminology, is open to abuse. What is often lost in these conversations, as was highlighted in Chapter 2, is the 'good' that can come from engaging with the online world. The continued echo that there 'should be a law against that' is never going to tackle societal norms. Indeed, as noted by Reed and Murray, there seems to be a clash between legal norms and the norms of cyberspace:

> From a lawmaker's perspective, legal norms are by definition supremely authoritative in the geographical territory of their law systems. But does this weight of authority carry over into cyberspace? How well does the law compete with other norms we have described? The answer seems to be: not terribly well.[1]

This is complicated further when legal norms are being applied to online abuse, an ambiguous term, as discussed in Chapter 6, that often invokes strong opinions on both sides of the debate.

Throughout various chapters of this monograph, it has been shown that many of the provisions currently used to tackle behaviours such as racist commentary on social media sites (Chapter 4), the use of such sites to subject another to cyber-harassment and/or cyberstalking (Chapter 5), or just to abuse another (Chapters 6 and 7), are currently not adhering to the principle of legality. And as highlighted in Chapter 3, and reflected throughout this monograph, legality needs to be maintained because 'legality offers the most just, most effective, most coherent, most persuasive, most legitimate and even the most moral approach'[2] to the regulation of individual behaviour. Something that is of utmost importance when it comes to

1 Chris Reed and Andrew Murray, *Rethinking the Jurisprudence of Cyberspace* (Elgar 2018) 120.
2 Jessica Lynn Corsi, 'An Argument for Strict Legality in International Criminal Law' [2018] 49 Geo. J. Int'l L. 1321.

DOI: 10.4324/9781003303688-13

restricting a fundamental democratic right, such as the right to freedom of expression. Indeed, the current approach to regulating the online world clearly highlights that 'this inconsistent over-action, under-action, lack of action, shows that we need a principled framework, centred in human rights',[3] including maintaining the principle of legality.

The following discussion highlights the importance of tackling societal norms to combat the 'harms' associated with the online world. It is only when societal norms are on par with what is expected of human behaviour, that legal norms will take effect. To do this, Chapter 10 provides a keyword analysis of the National Education Curriculum in England for both primary and secondary school students. It highlights the lack of emphasis placed on education to improve digital etiquette within society and how, instead, emphasis is being placed on the criminal law to tackle these issues. Final discussions outline recommendations for the issues highlighted throughout this monograph on how we can appropriately regulate online behaviour, whilst adhering to the principle of legality and protecting freedom of expression.

10.2 Applying the Criminal Law to the Online World

As noted in Chapter 1, there does not seem to be a day that goes by where there is not a news story highlighting the 'darker' side of the online world, followed by the cry 'there should be a law against that'. In turn, as demonstrated in previous chapters, the criminal law framework has been adapted to fit the digital age—and in many cases, resulting in the principle of legality being undermined and freedom of expression being threatened. Yet, there seems to be something about the online world, where societal and legal norms come into direct conflict with one another.

Using child sexual abuse material as an example, Reed and Murray demonstrate the difficulties in applying legal norms to cyberspace. The conduct, the sharing of child sexual abuse material, is a behaviour where there is a general consensus, internationally, that 'the creation, communication and (in most countries) possession of such images should be a serious criminal offence'.[4] Indeed, as noted by Reed and Murray, it is a type of online harm that is often reported in the mainstream press, especially in the United Kingdom. As a result, 'it seems likely that most UK residents know that child sexual abuse images contravene the law'.[5] For instance, as noted in Chapter 8 one of the key objectives of the United Kingdom's Online Safety Act 2023 (OSA), is to protect children from the 'harms' associated with the online world, including child sexual abuse material. Yet, the sharing of such material online is, and continues to be, common. Leading Reed and Murray to conclude:

3 Ellie Judson, 'The Online Safety Bill: Demos Position Paper' (*Demos*, April 2022) 7 <https://demos.co.uk/wp-content/uploads/2023/02/OSB-position-paper.pdf> accessed 14 March 2024.

4 Reed and Murray (n 1) 121.

5 Ibid.,

> It is clear that legal norms are losing out to technically mediated cyberspace norms even in this field, which is normally seen as the most heinous of criminal offences. [...] it seems clear that technology is driving a whole new set of online social norms. Whereas there is a strong legal (and offline social) norm against the production and distribution of child sexual abuse images, the ready availability of high quality cameras and the prevalent cyber-norm of producing and sharing sexualised images digitally has led to cyber-normalisation of sharing of indecent images irrespective of whether these are images of adults or minors.[6]

Similar arguments can be raised about copyright material, where there seems to be a genuine consensus amongst online users, that the illegal distribution, downloading and uploading of copyright material, is acceptable online behaviour but stealing property in the physical world is not, and in many cases, would constitute theft.

If legal norms are struggling to take effect in the online world against behaviour(s) where there is a genuine offline norm that such conduct is unacceptable, such as the sharing of child sexual abuse imagery, why would this be any different in relation to behaviour(s) where there is no genuine consensus that such behaviour is unacceptable, either offline or online? A prime example of this would be the sending of grossly offensive communications, which are prohibited, as outlined in Chapter 6, under both the Malicious Communications Act 1988 (MCA) and s.127(1) of the Communications Act 2003 (CA). Though it is illegal to send such a communication that can be defined as grossly offensive, there is no genuine societal norm that such behaviour is unacceptable both in the offline world and the online world. Whereas, as discussed above, in the offline world, when it comes to child sexual abuse imagery, there is a consensus that such conduct is wholly unacceptable—yet such behaviour continues to take place online.

What is therefore missing from many of the discussions around tackling the 'harms' associated with the online world, is the importance of changing societal norms. Instead, especially in England and Wales, and arguably societies across the globe we have seen a call for stronger legal provisions because after all, 'there should be a law against that'. Arguably because the law can be considered a 'quick fix' as it is perceived that something is being done as opposed to tackling the root cause of the problem.[7] This is despite, as was discussed and highlighted in Chapter 2 the importance of tackling social norms first: 'day to day we are more likely to alter our behaviour to comply with social norms in preference to legal norms or legal rules'.[8]

This is evidently clear when it comes to the use of the criminal law to tackle racist conduct, both online and offline, as highlighted in Chapter 4. The use of the internet, in particular, social media to send racist abuse to another is a prevalent

6 Ibid., 122–123.

7 Stephen L. Newman, 'Finding the Harm in Hate Speech: An Argument against Censorship' (2017) 50(3) CJPS 679, 687.

8 Reed and Murray (n 1) 103.

issue for governments across the globe. For instance, as outlined in Chapter 4, year-on-year, police forces in England and Wales have witnessed an increase in crimes where there is a hate crime element, with race being the most recorded hate crime characteristic.[9] Similar trends have also been mirrored on social media platforms.[10] Yet, such conduct is not unique to the digital world, racist behaviour is inherent within society, social media has just magnified the problem for the world to see. No longer, is it easy to ignore that such traits exist amongst humankind. Yet, no matter how much criminal law is enacted to tackle the problem of racist speech online, without changing societal norms, such behaviour will continue to flourish. Indeed, '[o]ne cannot blame Facebook for the dark side of human nature'.[11]

But how then do we tackle societal norms, to ensure we are getting the best out of legal provisions? The answer is simple, but not an overnight fix, nor something that can necessarily be pointed to by policymakers to state: 'look what we have done'—which we have seen with the previous UK government's stance following the passing of the OSA. The answer is simply, better, and stronger digital education:

> One way of shifting the social norm in the direction of law is via education. If cyberspace actors can be convinced that their conduct is morally wrong, rather than merely unlawful, enough of them may change their behaviour in the direction of the legal norm to start change in the social norm.[12]

Reed and Murray give the example of education programmes being run by the music industry to equate 'file sharing with crimes against physical property, such as theft which are widely accepted as being morally wrong'.[13] The belief that education plays an important role in tackling social and legal norms is supported further by Newman. Newman, using hate speech as an example, suggests that if the censorship of such conduct is about supporting 'public good', which is the notion often used to support the use of legal provisions to prohibit such conduct, then surely, education is better than the coercive nature of the criminal law? As echoed in Chapter 2:

> Words can be hurtful. They can antagonize [*sic*], they can wound. But words also inform and educate. They can dispel prejudice and reveal unanticipated

9 Home Office, 'Hate crime, England and Wales, 2021 to 2022' (*Gov.uk*, 6 October 2022) <https://www.gov.uk/government/statistics/hate-crime-england-and-wales-2021-to-2022/hate-crime-england-and-wales-2021-to-2022> accessed 1 February 2023.

10 Matthew L. Williams, Pete Burnap, Amir Javed, Han Liu and Sefa Ozalp, 'Hate in the Machine: Anti-Black and Anti-Muslim Social Media Posts as Predictors of Offline Racially and Religiously Aggravated Crime' 60(1) BJC 93.

11 Hugh Stephens, 'Those problematic comments on Facebook – Who bears responsibility for them?' (*Hugh Stephens Blog*, 22 November 2021) <https://hughstephensblog.net/2021/11/22/those-problematic-comments-on-facebook-who-bears-responsibility-for-them%EF%BF%BC/> accessed 14 March 2024.

12 Reed and Murray (n 1) 129.

13 Ibid., 129.

Table 10.1 Keyword and phrase search National Primary Curriculum for England

Keyword/phrase	*Mentions in the National Primary Curriculum for England*	*Why it was mentioned*
Social Media	0	NA
Internet	4	Privacy and safety – 1 How the internet works – 1 Using the internet – 1 Miscellaneous reference – 1
Computer	7	Computer science – 2 Writing computer programmes – 1 Computer networks – 1 Computer design – 1 Computer mapping – 1 Miscellaneous reference – 1
Online	5	Online technology and privacy – 1 Miscellaneous reference – 4
Technology	27	Design and technology – 14 Using technology – 7 Communication and Technology – 2 Respect and technology – 2 Miscellaneous reference – 2

*Note:*Department for Education, 'The National Curriculum in England: Key stages 1 and 2 framework document' (*Gov.uk,* September 2013) <https://assets.publishing.service.gov.uk/media/5a81a9abe5274a2e8ab55319/PRIMARY_national_curriculum.pdf> accessed 14 March 2024.

> affinities. We need strong laws to protect us from violence and discrimination, but laws that protect us from hurtful words tend to silence us as well as our antagonists, because they close down rather than open conversations and because they teach good citizens to call the police rather than do the hard work of combating hateful ideas.[14]

Yet, education is given very little consideration in debates surrounding the 'harms' associated with the online world. For instance, the National Educational Curriculum for primary school children (aged 4–11) in England does not contain the phrase 'social media' in any of its 201 pages, as shown further in Table 10.1. Similarly, the National Educational Curriculum for England for secondary school students (aged 11–16) does not mention the phrase 'social media' anywhere in its 105 pages, as demonstrated in Table 10.2.

Indeed, as shown in Tables 10.1 and 10.2, little consideration is given to digital etiquette compared to technology usage in general. For instance, in the primary school National Curriculum, reference is made only once that students should be

14 Newman (n 7) 693.

Table 10.2 Keyword and phrase search National Secondary Curriculum for England

Keyword/phrase	*Mentions in the National Secondary Curriculum for England*	*Why it was mentioned*
Social Media	0	NA
Internet	0	NA
Computer	10	Computer science – 4 Writing computer programmes – 1 Understanding computer systems – 3 Creativity and computers – 1 Miscellaneous reference – 1
Online	3	Online technology and privacy – 1 Miscellaneous reference – 2
Technology	22	Design and technology – 13 Using technology – 2 Communication and Technology – 2 Safety and technology – 2 Creativity and technology – 1 Development of technology – 1 Miscellaneous reference – 1

Note: Department for Education, 'The National Curriculum in England: Key stages 3 and 4 framework document' (*Gov.uk*, December 2014) <https://assets.publishing.service.gov.uk/media/5da7291840f0b6598f806433/Secondary_national_curriculum_corrected_PDF.pdf> accessed 14 March 2024.

able to 'use technology safely, respectfully and responsibly; recognise acceptable/unacceptable behaviour; identify a range of ways to report concerns about content and contact'.[15] For secondary school children, they should

> understand a range of ways to use technology safely, respectfully, responsibly, and securely, including protecting their online identity and privacy; recognise inappropriate content, contact and conduct and know how to report concerns.[16]

This is mentioned twice in the document. Whereas, how to use new technology more broadly is referenced fourteen times for primary school children; and eleven times for secondary school children. Given the extent of new technology and its impact on every aspect of everyday life, as highlighted in Chapter 1, it is regrettable

15 Department for Education, 'The National Curriculum in England: Key stages 1 and 2 framework document' (*Gov.uk*, September 2013) accessed 14 March 2024.

16 Department for Education, 'The National Curriculum in England: Key stages 3 and 4 framework document' (*Gov.uk*, December 2014) 86 <https://assets.publishing.service.gov.uk/media/5da7291840f0b6598f806433/Secondary_national_curriculum_corrected_PDF.pdf> accessed 14 March 2024.

that more emphasis is not placed on the importance of digital education, beyond just how to use new technology, to encompass more discussions on how individuals present themselves and interact with others online. Whereas in Australia, for example, significant emphasis has been placed on education to help tackle online 'harms', through the creation and implementation of the eSafety Commissioner.

In Australia, educational initiatives to combat inappropriate conduct online is regarded as 'one of the most important elements of crime prevention'.[17] Programmes have been created by the federal government such as 'Thinkuknow', targeting both school-aged children and the wider public in digital etiquette.[18] An approach endorsed further in the work of the e-Safety Commissioner, who has the power to investigate complaints related to conduct carried out online, and report the most serious incidents to the police, whilst tackling the less serious complaints by working with online platforms. The e-Safety Commissioner also has significant responsibility for the education of Australian Citizens to 'have safer, more positive online experiences'.[19] This includes, though not limited to:[20]

- Free presentations to pre-service teachers in their final year of tertiary study;
- Free webinars to upskill university support staff about the latest online safety risks to young adults and strategies to deal with them;
- Creating best practice frameworks;
- Classroom resources; and
- Specific training tools for schools and universities.

Whereas education has been placed at the center of combatting online 'harms' in Australia, in the United Kingdom, significant weight has been placed on the criminal law to tackle such issues, as noted in various chapters of this monograph, and clearly demonstrated during the passing of the OSA where weight was placed on regulating social media companies through coercive enforcement,[21] as opposed to tackling the societal issues and education. This has been reflected further in other countries approaches to tackling online abuse, as shown in Chapter 9.

At the time of writing, the Department of Education for England is consulting on draft guidance for those working in the education system on 'Relationships

17 Parliament of Australia, *Cyber Safety – Joint Select Committee High-wire act: Cyber-safety and the young Interim report* (June 2011) [11.18].

18 Thinkuknow, 'What we see, say, do online' (*Thinkuknow.org.au*, 2018) <https://www.thinkuknow.org.au/what-we-see-say-do-online> accessed 27 July 2018.

19 eSafety Commissioner, 'eSafety Education' (*Australian Government*, 2024) <https://www.esafety.gov.au/educators#:~:text=eSafety%27s%20education%20and%20training%20team,strategies%20to%20deal%20with%20them.> accessed 14 March 2024.

20 Ibid.,

21 Department for Science, Innovation and Technology and The Rt Hon Michelle Donelan MP, 'Overwhelming support for Online Safety Act as rules making UK the safest place in the world to be online become law' (*Gov.uk*, 27 October 2023) <https://www.gov.uk/government/news/overwhelming-support-for-online-safety-act-as-rules-making-uk-the-safest-place-in-the-world-to-be-online-become-law> accessed 14 March 2024.

Education, Relationships and Sex Education and Health education.'[22] The guidance covers an array of topics, including some aspects of online 'harms'. However, much of the focus in the guidance is placed on teaching young people the legal provisions prohibiting 'harms' such as the sharing of sexual imagery, cyberharassment and violent material as opposed to digital citizenship, with strict age-limits being imposed on teachers as to when certain content can be discussed.[23] A similar approach is endorsed in the Department of Education's 'Teaching online safety it schools' guidance, where significant emphasis is placed on the 'bad' rather than the 'good' when it comes to interacting with the online world. In essence, what digital education we have is extremely limited.[24]

Education is vital in the discussion of online behaviours, a discussion that is often missing in wider discourse around our online interactions. That is not to say there is no place for the criminal law when it comes to conduct carried out online. There will always be certain behaviours that warrant criminal law intervention. However, where the criminal law is used, it needs to be grounded in legal provisions which comply with the principle of legality—something we are currently failing to do.

10.3 Protecting Legality in the Criminal Law on Social Media

The previous chapters have examined several 'harms' associated with the online world, and the criminal law's application to prohibit such conduct, from the principle of legality. This monograph started by setting the theoretical background to the thesis of this book—the criminal law framework, as it currently stands, when applied to certain conduct carried out on social media does not uphold the very basics of legality. In turn, the right to freedom of expression, as protected under both national and international law, as outlined in Chapter 2, is under threat. As discussed in Chapter 3, legality comprises of several underlying principles, though there is no true consensus as to what these underlying principles are, there are significant overlaps. As noted in Chapter 3, '[a]mongst the best known'[25] approaches to legality, is Lon Fuller.[26] And it was Fuller's understanding of legality which was

22 Department for Education, 'Draft Relationships Education, Relationships and Sex Education (RSE) and Health education: Statutory guidance for governing bodies, proprietors, head teachers, principals, senior leadership teams, teachers' (*Gov.uk*, 16 May 2024) <https://consult.education.gov.uk/rshe-team/review-of-the-rshe-statutory-guidance/supporting_documents/Draft%20RSE%20and%20Health%20Education%20statutory%20guidance.pdf> accessed 10 July 2024. Note, this guidance was produced under the previous government and therefore might not be implemented by the new Labour government.

23 For example, 'the concepts and laws relating to harmful sexual behaviour' cannot be taught before the age of 11. Ibid., 27.

24 Department for Education, 'Teaching online safety in schools' (*Gov.uk*, 12 January 2023) <https://www.gov.uk/government/publications/teaching-online-safety-in-schools/teaching-online-safety-in-schools#curriculum-context> accessed 10 July 2024.

25 Reed and Murray (n 1) 202.

26 Lon L. Fuller, *The Morality of Law* (Yale University Press 1964) 202.

used to critique the current use of law to govern individual conduct online, in Part Two and Three of this monograph. As a reminder to the reader, Fuller argued that legality encompassed eight key principles:[27]

1. The law must be general;
2. The law must be prospective;
3. The prohibition of retrospective law;
4. The law must be clear;
5. The prohibition of contradictory law;
6. The law must not ask the impossible;
7. Consistency must be maintained in the law; and
8. The law must be fairly enforced.

Each chapter that followed, exposed, with examples, the current problems in applying the criminal law to conduct carried out on social media platforms.

Chapter 4 discussed the issues of attempting to tackle racist behaviour on social media platforms, through the application of s.66 of the Sentencing Act 2020 (SA), s.31 and 32 of the Crime and Disorder Act 1988 (CDA) and s.18 and 19 of the Public Order Act 1986 (POA) to a social media context. Though, as noted in Chapter 4, the arguments made in this chapter, are not unique to conduct carried out on social media, social media has made the issue more complex, where context is everything but is often lost in the virtual world. Indeed, the Law Commission, during their extensive review of hate crimes more generally, stated:

> [we] are deeply concerned that misrepresentation as to the sort of conduct which are caught by the stirring up offences may lead people to avoid saying things that they are perfectly entitled to express.[28]

An issue that gained significant conversation in April 2024, following a new criminal provision being enacted in Scotland, the Hate Crime and Public Order (Scotland) Act 2021, which 'introduces the offence of "stirring up hatred" – either in person or online – related to age, disability, religion, sexual orientation, transgender identity or being intersex.'[29]

Three significant concerns with legality were raised in Chapter 4: (1) the inherent vagueness of key aspects of s.66 of the SA, s.31 and 32 of the CDA and s.18 and 19 of the POA; (2) consistency being lost; and (3) retrospective law, following the application of stirring up offences to the online world, going beyond Parliament's original intent. In recent years, calls have been made to make significant changes

27 Ibid., 33–94.

28 Law Commission, *Hate Crime Laws: Final Report* (Law Com No 401, 2021) [10.12].

29 Laura Higson-Bliss, 'Scotland's hate crime law: the problem with using public order laws to govern online speech' *The Conversation* (London, 23 April 2024) <https://theconversation.com/scotlands-hate-crime-law-the-problem-with-using-public-order-laws-to-govern-online-speech-227951> accessed 10 July 2024.

to this area of the law, to ensure a better balance between freedom of expression and the criminal law.[30] Most notably, to remove reference to insulting behaviour, following a general consensus that such conduct falls within the realms of freedom of expression.[31] But if we want to strive for the 'maximum clarity which the imperfect medium of language can attain',[32] we also need to consider the continued use of the terms hostile and abusive. Both terms are inherently vague and impossible to quantify in the online world where, as echoed earlier, context is everything but is often lost. Indeed, in today's liberal and democratic society, where free speech needs to be protected, it is legitimate to ask if we should even criminalise hostile and abusive speech.

The lack of consensus as to the meaning of key terms in the provisions discussed in Chapter 4, has led to inconsistency in the criminal law. As highlighted in Chapter 4, it is not quite clear why in one case, as an example, the POA was used, as seen in *R v Liam Stacey.*[33] Whereas in another matter, such as *R v Alison Chabloz,*[34] s.127(1) of the CA was applied. Both cases could be considered abusive, at the very least, and arguably demonstrated racial hatred—yet different legal provisions were applied and the reasoning as to why, is unclear. In fact, it is of concern that provisions such as the POA are being used in a social media context, as arguably this is creating retrospective criminal law. The purpose of the POA is to regulate public disorder, in particular, riots, football hooliganism, protests, etc. Yet the Act has been used to cover social media related offences, where public disorder is not under threat.

Similar issues were raised in Chapter 5 over the use of vague terminology, most notably, a course of conduct, harassment and stalking in the Protection from Harassment Act (PfHA). And like that of Chapter 4, these arguments are not unique to the digital world but have become more complex in the wake of new technology. Following significant amendments to the PfHA to include stalking as a separate offence under the Act in 2012, arguments were made at the time that cyberstalking should be included as a separate offence, with MacEwan noting that it was 'naïve, to view it simply as "old behaviour using new technologies"'.[35] As new technology has advanced it has become easier to harass and stalk another. Further, given the added complex notion that 'the general monitoring of a person's social media page, is expected, and somewhat consented to',[36] we are at an opportune moment to consider the specific criminalisation of cyberharassment and cyberstalking, to address the issues highlighted in Chapter 5. Ensuring that any provisions created,

30 Ibid.,

31 Law Com No 401 (n 28) [10.305].

32 Law Commission, *Report on Conspiracy and Criminal Law Reform* (1976) Law Com No. 76 [1.8].

33 *R v Liam Stacey,* Swansea Crown Court On Appeal From The Magistrates' Court, A20120033.

34 *R v Alison Chabloz,* Westminster Magistrates' Court, 11 January 2018 (unreported).

35 Neil MacEwan, 'The New Stalking Offences in English Law: Will They Provide Effective Protection from Cyberstalking?' (2012) 10 Crim.L.R. 767, 768.

36 Law Commission, *Abusive and Offensive Online Communications: A Scoping Report* (Law Com No 381, 2018) [8.174].

are not grounded in vague terminology, and are targeting the most extreme cases. This is not creating new law; it is simply bringing clarity to this area of the law.

In turn, as highlighted in Chapter 5, to protect legality further, we need to reconsider what constitutes the reasonable person in a social media context and ensure the law is not asking the impossible. Though a longstanding provision in the criminal law, the use of the reasonable person test, as discussed at length in Chapter 5, when applied in a social media context significant issues can be raised. To better protect the right to freedom of expression and adhere to the principle of legality, it is suggested that we should, when considering online behaviours, ground the reasonable person test into that of the reasonable social media user.

Of all the provisions used to govern conduct carried out on social media, the most controversial is the continued usage of the MCA and s.127(1) of the CA, as it cannot be said that generality in the law is being adhered to. Significant concern was raised in Chapter 6 with the terms threatening, indecent, obscene, false, menacing, and grossly offensive. Concerns that have not gone unnoticed with the Law Commission in 2021, calling for the repeal of both provisions.[37] Both provisions are grounded in terminology that is difficult to define, and even with the modest changes made by the passing of the OSA, the issues raised in Chapter 6 have not been addressed. As a result, heavy reliance is being placed on the discretion of law enforcement, Crown Prosecutors, judges, and juries on what constitutes, as an example, a menacing, or indeed, grossly offensive communication—resulting in inconsistent application of the law. Even prosecuting guidelines have not brought generality to this area of the law. For instance, as discussed in Chapter 6, even with prosecuting guidelines being in place, mistakes continue to be made by those working in the criminal justice system, such as in the case of *Alison Chabloz*.[38]

The Law Commission's recommendation to repeal both the MCA and s.127(1) of the CA was welcomed by many,[39] after they concluded that both provisions were not suited for the digital world.[40] This is something that needs to be revisited, and although their recommended replacement, the harms model, did not necessarily address previous concerns with communication offences—it is clear that both the MCA and s.127(1) of the CA pose such significant issues with legality, they need to be replaced to ensure fair enforcement of the law in the future. Currently, what one police force, Crown Prosecutor, judge, or jury may consider, as an example, grossly offensive, another may not—it cannot be said, that the MCA and s.127(1) of the CA are action-guiding.

If history has taught us anything, it is that creating communication offences, grounded in vagueness only undermines freedom of expression and does not protect those who are subjected to the most horrific forms of abuse. And even

37 Law Commission, *Modernising Communications Offences: A Final Report* (Law Com. No.399, 2021) [2.1].

38 *Chabloz* (n 34).

39 Law Commission, *Harmful Online Communications: The Criminal Offences. A Consultation Paper* (Law Com. No.248, 2020) Chapter 5.

40 Law Com No 399 (n 37).

with prosecuting guidelines in place, overreliance on discretion results in unfair enforcement of the law and in turn prospectivity being threatened. Yet, the 'new' communication offences, as outlined in Chapter 7, still pose significant issues with legality. All the provisions discussed in Chapter 7, are embedded in vague terminology, most notably what constitutes 'harm' under each of the offences discussed in Chapter 7. 'Harm' under the harms-offence was defined as 'psychological harm amounting to at least serious distress'[41]—a significantly wide definition and one of the reasonings for its removal from the Online Safety Bill 2022 before its passing through Parliament.[42] Nonetheless, both s.179 and 181 of the OSA, lack clarity in relation to the 'harms' they prohibited. Under s.179 'harm' is defined as 'non-trivial psychological or physical harm'.[43] Whereas s.181(2) defines harm as:

> (a) serious injury amounting to grievous bodily harm (GBH) within the meaning of the Offences against the Person Act 1981, (b) rape, (c) assault by penetration within the meaning of section 2 of the Sexual Offences Act 2003, or (d) serious financial loss.[44]

All of these render issues with legality and do not sufficiently overcome the arguments made against previous communication offences, most notably the MCA and s.127(1) of the CA. Clarity, as noted by Fuller, is an 'essential ingredient'[45] in legality and yet, it's a significant issue highlighted throughout each chapter of this monograph, and should have been at the forefront of lawmakers' minds, when creating and implementing these 'new' communication offences.

The lack of understanding as to the meaning of 'harm' is also reflected in the notion of 'likely audience'. A term used in the recommended harms-offence and repeated in s.179 of the OSA—despite concerns over how inherently broad the term was, leading to individuals being asked to do the impossible—send a communication that does not cause 'psychological harm amounting to at least serious distress' to a likely audience, with this likely audience being anyone in the world.[46] In turn, many of the issues highlighted in Chapter 6 remain, and in fact, are complicated further, given that we now have a range of communication offences which overlap, as opposed to just two offences. The interchangeable nature of communication-based law therefore remains, despite the passing of the OSA.

41 Online Safety Bill 2022 c.150(4).

42 Department of Digital, Culture, Media and Sport, 'Overview of expected impact of changes to the Online Safety Bill' (*Gov.uk,* 18 Jan 2022) <https://www.gov.uk/government/publications/online-safety-bill-supporting-documents/overview-of-expected-impact-of-changes-to-the-online-safety-bill#:~:text=The%20adult%20safety%20duties%20(%27legal,and%20freedom%20of%20expression%20dutie> accessed 7 August 2023.

43 Online Safety Act 2023 s.179(1)c.

44 Ibid., s.181(2).

45 Fuller (n 26) 62.

46 Laura Higson-Bliss, 'Prosecuting Communication Offences: Interpreting the Harms Model' (2022) 4 CLR 27, 284.

These 'new' provisions need to be applied carefully going forwards and where issues remain, these are highlighted to lawmakers, to ensure future amendments can address the concerns raised in Chapter 7. The 'harms' prohibited under the remaining offences need to be clarified further, to ensure discretion in the criminal justice system is not being heavily relied upon, and the notions of 'reasonable excuse' and 'likely audience' need to be replaced with concepts which create consistency within the law and are not open to abuse. As shown during the COVID-19 pandemic, 'reasonable excuse' creates more issues than it seeks to resolve, resulting in negative attitudes towards the law and those enforcing it.

The concerns raised in Part Two of this monograph, are not just unique to the United Kingdom, or indeed, the criminal law. Legality is also a significant issue in the recent attempt to regulate user-to-user services, in particular, social media platforms through the creation and implementation of the OSA—the previous UK government's attempt to become one of the 'safest places in the world to go online',[47] as discussed in Chapter 8. The OSA is an extensive piece of legislation which will have far-reaching consequences, that are still not yet fully understood. The Act is grounded in vagueness and at times, asking the impossible. Of significant concern, is the requirement that user-to-user services do the work of law enforcement, judges, and juries in interpreting the criminal law framework of Scotland, Northern Ireland, England and Wales in determining if a complained about behaviour constitutes an offence in any of these jurisdictions. This includes all the provisions discussed in previous chapters of this monograph—despite, as highlighted in Chapter 8, those working in the criminal justice system, who have significant expertise that private platforms do not, struggling to apply the law correctly at times. In turn, we are seeing the privatisation of speech. The issues this presents with legality are further amplified by the vagueness and lack of clarity contained in many of the provisions in the OSA.

But despite the previous UK government suggesting we are world-leading in the fight against the 'harms' associated with the online world, this is not necessarily true and many of the issues highlighted in Chapter 8 have also been raised against other countries' approaches to regulating the internet, in particular, social media platforms, as shown in Chapter 9. Chapter 9 examined three other countries' attempts to regulate social media platforms: France, Germany, and India. And as highlighted in Chapter 9, each was chosen due to their similarities, not only with each other but also with the United Kingdom's OSA and each posed, or still poses, significant issues with legality.

In France, the French Constitutional Court struck down the Online Hate Bill, following its failure to adequately set limitations to ensure the law was not abused and in turn, protected the right to freedom of expression.[48] Significant weight was

47 HM Government, 'Internet Safety Strategy – Green paper' (*Gov.uk*, October 2017) 1 <https://assets.publishing.service.gov.uk/government/uploads/system/uploads/attachment_data/file/650949/Internet_Safety_Strategy_green_paper.pdf> accessed 22 September 2023.

48 Decision no. 2020-801 DC of June 18, 2020.

also given to the lack of legality within the provision. In particular, like that of the United Kingdom's OSA, asking platforms to do the impossible by requiring them to interpret the criminal code of France against reported content to determine if such content should be removed—giving platforms 24 hours to make such a determination. Whereas in Germany, a similar provision, NetzDG, despite initial concerns around freedom of expression and legality, remains legally enforceable. Indeed, research into NetzDG since its enactment has been mixed—neither confirming nor denying if the provision has hindered the right to freedom of expression,[49] though it is accepted that legality is still a significant issue. In India, similar legislation to both France, Germany, and the United Kingdom has been implemented into the legal framework of India, but these provisions, many of which mirror European approaches to regulating the online world, have been heavily abused due to its lack of legality.

Of significant concern, as was discussed at length in Chapter 9, was the abuse of s.66A of the Information Technology Act 2000/2008 (IT Act) which mirrored both the MCA and s.127(1) and (2) of the CA (in their original enactment) by Indian authorities. The lack of understanding as to what constituted, as an example, a grossly offensive communication under the IT Act, led to significant abuse by authorities across India, leading the Supreme Court of India to strike down the provision for being incompatible with both the right to freedom of expression and legality.[50] Yet, legal provisions continue to be created in India, mirroring many European approaches to regulating the online world, which are being heavily abused, such as the recent creation and implementation of the Information Technology (Intermediary Guidelines and Digital Media Ethics Code) Rules 2021—which is very similar to the OSA and NetzDG. India is a prime example of why we need to be careful when enacting legislation which does not comply with the principle of legality when the rest of the world is looking for inspiration in their own fight against the 'harms' associated with the online world.

The principle of legality is fundamental in any legal framework but is of particular importance in criminal law provisions. It is vital, that when criminal law is produced, it is created with as much legal certainty as possible, especially when it's being used to restrict a fundamental right such as the right to freedom of expression. A balance we are struggling to make when it comes to conduct carried out on social media. Many of the discussions, as highlighted throughout this monograph, when it comes to social media, focus on the 'darker' side, and it is this 'darker' side that is used to justify the coercive nature of the criminal law. But if we really do want to become one of the 'safest places in the world to go online' whilst also protecting democracy and the right to freedom of expression, we need to ensure that the criminal law is only used in the most extreme circumstances, and where it is used, it is grounded in the principle of legality. Whilst also acknowledging the criminal

49 Rachel Griffin, 'New school speech regulation as a regulatory strategy against hate speech on social media: The case of Germany's NetzDG' (2022) 46(9) Telecommun. Policy 1, 10.

50 *Shreya Singhal v Union of India* (2013) 12 S.C.C. 73.

law cannot be used to send social signals about what is or is not acceptable online, education needs to play a part—an element that is heavily missing across wider conversations on regulating the online world, in particular, social media platforms.

Reference list

Corsi J L, 'An Argument for Strict Legality in International Criminal Law' (2018) 49 Geo J Int'l L 1321

Department for Education, 'Draft Relationships Education, Relationships and Sex Education (RSE) and Health education: Statutory guidance for governing bodies, proprietors, head teachers, principals, senior leadership teams, teachers' (*Gov.uk*, 16 May 2024) https://consult.education.gov.uk/rshe-team/review-of-the-rshe-statutory-guidance/supporting_documents/Draft%20RSE%20and%20Health%20Education%20statutory%20guidance.pdf

Department for Education, 'Teaching online safety in schools' (*Gov.uk*, 12 January 2023) https://www.gov.uk/government/publications/teaching-online-safety-in-schools/teaching-online-safety-in-schools#curriculum-context

Department for Education, 'The National Curriculum in England: Key Stages 3 and 4 Framework Document' (*Gov.uk,* December 2014) https://assets.publishing.service.gov.uk/media/5da7291840f0b6598f806433/Secondary_national_curriculum_corrected_PDF.pdf

Department for Education, 'The National Curriculum in England: Key Stages 1 and 2 Framework Document' (*Gov.uk,* September 2013) https://assets.publishing.service.gov.uk/media/5a81a9abe5274a2e8ab55319/PRIMARY_national_curriculum.pdf

Department for Science, Innovation and Technology and The Rt Hon Donelan M MP, 'Overwhelming support for Online Safety Act as rules making UK the safest place in the world to be online become law' (*Gov.uk*, 27 October 2023) https://www.gov.uk/government/news/overwhelming-support-for-online-safety-act-as-rules-making-uk-the-safest-place-in-the-world-to-be-online-become-law

Department of Digital, Culture, Media & Sport, 'Overview of expected impact of changes to the Online Safety Bill' (*Gov.uk,* 18 Jan 2022) https://www.gov.uk/government/publications/online-safety-bill-supporting-documents/overview-of-expected-impact-of-changes-to-the-online-safety-bill#:~:text=The%20adult%20safety%20duties%20(%27legal,and%20freedom%20of%20expression%20duties

eSafety Commissioner, 'eSafety Education' (*Australian Government*, 2024) https://www.esafety.gov.au/educators#:~:text=eSafety%27s%20education%20and%20training%20team,strategies%20to%20deal%20with%20them

Fuller L L, *The Morality of Law* (Yale University Press 1964)

Griffin R, 'New School Speech Regulation as a Regulatory Strategy against Hate Speech on Social Media: The Case of Germany's NetzDG' (2022) 46(9) Telecommun Policy 1

Higson-Bliss, 'Scotland's hate crime law: the problem with using public order laws to govern online speech' *The Conversation* (London, 23 April 2024) https://theconversation.com/scotlands-hate-crime-law-the-problem-with-using-public-order-laws-to-govern-online-speech-227951

Higson-Bliss L, 'Prosecuting Communication Offences: Interpreting the Harms Model' (2022) 4 CLR 271

HM Government, 'Internet Safety Strategy – Green Paper' (*Gov.uk*, October 2017) https://assets.publishing.service.gov.uk/government/uploads/system/uploads/attachment_data/file/650949/Internet_Safety_Strategy_green_paper.pdf

Home Office, 'Hate Crime, England and Wales, 2021 to 2022' (*Gov.uk*, 6 October 2022) https://www.gov.uk/government/statistics/hate-crime-england-and-wales-2021-to-2022/hate-crime-england-and-wales-2021-to-2022

Judson E, 'The Online Safety Bill: Demos Position Paper' (*Demos*, April 2022) https://demos.co.uk/wp-content/uploads/2023/02/OSB-position-paper.pdf
Law Commission, *Hate Crime Laws: Final Report* (Law Com No 401, 2021)
Law Commission, *Modernising Communications Offences: A Final Report* (Law Com. No.399, 2021)
Law Commission, *Harmful Online Communications: The Criminal Offences. A Consultation Paper* (Law Com. No.248, 2020)
Law Commission, *Abusive and Offensive Online Communications: A Scoping Report (Law Com No 381, 2018)*
Law Commission, *Report on Conspiracy and Criminal Law Reform* (Law Com No. 76. 1976)
MacEwan N, 'The New Stalking Offences in English Law: Will They Provide Effective Protection from Cyberstalking?' (2012) 10 Crim.L.R. 767
Newman S L, 'Finding the Harm in Hate Speech: An Argument against Censorship' (2017) 50(3) CJPS 679
Parliament of Australia, *Cyber Safety - Joint Select Committee High-wire act: Cyber-Safety and the Young Interim Report* (June 2011)
Reed C & Murray A, *Rethinking the Jurisprudence of Cyberspace* (Elgar 2018)
Stephens H, 'Those Problematic Comments on Facebook—Who Bears Responsibility for Them?' (*Hugh Stephens Blog*, 22 November 2021) https://hughstephensblog.net/2021/11/22/those-problematic-comments-on-facebook-who-bears-responsibility-for-them%EF%BF%BC/
Thinkuknow, 'What We See, Say, Do Online' (*Thinkuknow.org.au*, 2018) https://www.thinkuknow.org.au/what-we-see-say-do-online
Williams M L, Burnap P, Javed A, Liu H & Ozalp S, 'Hate in the Machine: Anti-Black and Anti-Muslim Social Media Posts as Predictors of Offline Racially and Religiously Aggravated Crime' 60(1) BJC 93

Index

FSC
www.fsc.org
MIX
Papier | Fördert gute Waldnutzung
FSC® C083411

Kolibri 360 GmbH
Ferdinand-Jühlke-Straße 7
99095 Erfurt, Deutschland
produktsicherheit@kolibri360.de